The scheme for the "System of Ophthalmology" is as follows.

Vol. I. THE EYE IN EVOLUTION

Vol. II. THE ANATOMY OF THE VISUAL SYSTEM

Vol. III. NORMAL AND ABNORMAL DEVELOPMENT
- Pt. 1. Embryology
- Pt. 2. Congenital Deformities

Vol. IV. THE PHYSIOLOGY OF THE EYE AND OF VISION

Vol. V. OPHTHALMIC OPTICS AND REFRACTION

Vol. VI. OCULAR MOTILITY AND STRABISMUS

Vol. VII. THE FOUNDATIONS OF OPHTHALMOLOGY
Heredity, Pathology, Methods of Diagnosis, General Therapeutics

Vol. VIII. DISEASES OF THE OUTER EYE
- Pt. 1. Conjunctiva
- Pt. 2. Cornea and Sclera

Vol. IX. DISEASES OF THE UVEAL TRACT

Vol. X. DISEASES OF THE RETINA

Vol. XI. DISEASES OF THE LENS AND VITREOUS; GLAUCOMA AND HYPOTONY

Vol. XII. NEURO-OPHTHALMOLOGY

Vol. XIII. THE OCULAR ADNEXA
- Pt. 1. Diseases of the Eyelids
- Pt. 2. Lacrimal, Orbital and Para-orbital Diseases

Vol. XIV. INJURIES
- Pt. 1. Mechanical Injuries
- Pt. 2. Non-mechanical Injuries

Vol. XV. OUTLINE OF SYSTEMIC OPHTHALMOLOGY AND GENERAL INDEX

SYSTEM OF OPHTHALMOLOGY

EDITED BY

SIR STEWART DUKE-ELDER

VOL. XIII

THE OCULAR ADNEXA

By

Sir Stewart Duke-Elder

G.C.V.O., F.R.S.

and

Peter A. MacFaul

M.B., B.S., F.R.C.S.

Ophthalmic Surgeon, Middlesex Hospital, and the Royal National Orthopædic Hospital, London; Honorary Research Assistant, Institute of Ophthalmology, University of London

WITH 1,300 ILLUSTRATIONS AND 6 COLOURED PLATES

PART I

DISEASES OF THE EYELIDS

ST. LOUIS

THE C. V. MOSBY COMPANY

1974

First published 1974

ISBN 0 85313 787 0

ISSN 0082 1195

MADE AND PRINTED IN GREAT BRITAIN

PREFACE

A STUDY of diseases of the lids, lacrimal apparatus, orbit and para-orbital areas comprises a vast medley of unrelated conditions, some of them merely of local interest, others essentially dermatological, and many of them having widespread systemic implications. In the last category the greatest stress has been directed to their ophthalmological aspects, but sufficient attention has been given to their general nature to give a relatively complete clinical picture as well as to serve as a guide to the appropriate treatment. It is true that in many cases the ætiology is obscure; when this is so our ignorance is pointed out and the various hypotheses that have been suggested from time to time have been indicated and their virtues and deficiencies analysed in the light of modern knowledge.

Since our study of the clinical methods of diagnosis in Volume VII of this *System* was limited to the examination of the eye itself, a considerable amount of attention has been given to this part of the subject, particularly when specific techniques are employed in the elucidation of diseases of the lacrimal apparatus and the orbit. As in the other volumes of this *System* this book is essentially clinical and does not pretend to give the surgical procedures in detail; for this, textbooks on ophthalmic surgery, of which many excellent treatises exist, or in some cases on plastic surgery should be consulted. An indication is given only of the outlines of the many various procedures which have been or are advocated with the appropriate references to the literature and general indications of their applicability to any particular circumstance. In order to maintain the completeness aimed at throughout this *System*, a note on the history of many subjects and bibliographies of them all have been included.

It has once again been found advisable for ease of manipulation to expand one volume of the *Text-Book of Ophthalmology* into two, the first dealing with diseases of the lids and the second with the remainder of our subject. The two, however, form one composite whole and when the same condition is discussed in different localities, cross-references are frequent to avoid reduplication of the text or of the bibliographies.

STEWART DUKE-ELDER.

INSTITUTE OF OPHTHALMOLOGY
UNIVERSITY OF LONDON
1974

ACKNOWLEDGMENTS

THE WRITER of a book of this type must pile up a mountain of debt to hosts of his colleagues who have contributed much to the advancement of our specialty. This I have freely done but in each case the acknowledgement has been made in the bibliographies. My debt for illustrations is more technical, but again a very great many of my colleagues have been most liberal in supplying me with illustrations for the multitude of conditions described in these pages; again, they are all acknowledged in the legends to the figures. So far as clinical pictures are concerned, my greatest debt is to Dr. Peter Hansell and his staff at the Institute of Ophthalmology for giving me permission to take what I wanted from their ever-growing collection, as well as to Dr. Peter Borrie of St. Bartholomew's Hospital, and to Professor Arthur J. Rook of Addenbrooke's Hospital, Cambridge, for allowing me to make free use of illustrations from his *Textbook of Dermatology*. For the illustrations of tumours I have been liberally supplied by Dr. M. Lederman of London, Professor A. Mortada of Cairo, and Dr. D. Silva of Mexico. For pathological illustrations I have as usual received many from Norman Ashton, Professor of Pathology at the Institute of Ophthalmology, and Lorenz E. Zimmerman of the Armed Forces Institute of Pathology in Washington. Dr. Glyn Lloyd of Moorfields Eye Hospital, and Dr. Stephen Trokel of New York have supplied me with many excellent radiographs. Finally, Dr. F. N. L. Poynter, lately Director of the Wellcome Institute of the History of Medicine, has filled many gaps in my sequence of historical personalities to whom we owe much.

For the somewhat arduous job of reading proofs I have to thank Sir Allen Goldsmith and Miss Mina H. T. Yuille for their welcome help. As with other volumes in this *System* my indebtedness is as great as ever to my secretary, Miss Rosamund Soley, for preparing the manuscript and the figures, reading the proofs, undertaking the immense task of verifying the bibliographies and compiling the index.

Finally, Mr. Ronald Deed of my publishers, Henry Kimpton, has as usual condoned all my vagaries and granted all my requests with a smile.

CONTENTS OF PART 1

SECTION I

DISEASES OF THE EYELIDS

CHAPTER I

CIRCULATORY AND SECRETORY DISORDERS

CHAPTER II

INFLAMMATIONS OF THE LIDS

PAGE

Chapter III

Dermatoses of Unknown or Varied Ætiology

Chapter IV

The Lids in Systemic Disease

Chapter V

Atrophies, Hypertrophies, Degenerations, Pigmentations

CHAPTER VI

DISORDERS OF THE EYEBROWS AND LASHES

CHAPTER VII

CYSTS AND TUMOURS

Chapter VIII

Motor Disorders and Deformations of the Eyelids

CONTENTS OF PART 2

SECTION II

DISEASES OF THE LACRIMAL APPARATUS

Chapter IX

Diseases of the Lacrimal Gland

Chapter X

Diseases of the Lacrimal Passages

SECTION III

DISEASES OF THE ORBIT

Chapter XI

General Considerations

Chapter XII

Disturbances of the Circulation

PAGE

CHAPTER XIII

INFLAMMATIONS OF THE ORBIT

PAGE

CHAPTER XVI

CYSTS AND TUMOURS

SECTION IV

DISEASES OF THE PARA-ORBITAL REGIONS

APPENDIX

DISPLACEMENTS OF THE GLOBE

SECTION I

DISEASES OF THE LIDS

Fig. 1.—Antonio Scarpa [1747–1832].

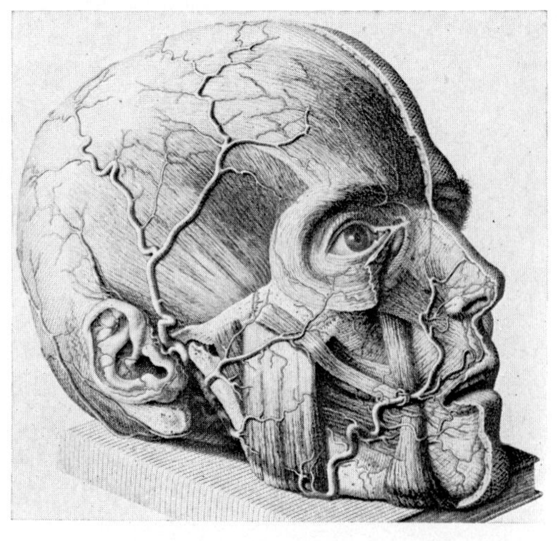

Fig. 2.—One of the delightful copper engravings illustrating the anatomy of the face and lids prepared by Faustino Anderloni under Scarpa's direction for his *Saggio di osservazioni e d'esperienze sulle principali malattie degli occhi*, Pavia (1801).

CHAPTER I

CIRCULATORY AND SECRETORY DISORDERS

THE unusually talented Venetian, ANTONIO SCARPA [1747–1832] (Fig. 1), a brilliant anatomist and surgeon and equally famous as a skilled ophthalmologist, is a suitable introduction to this Volume. His classical work, *Saggio di osservazioni e d'esperienze sulle principali malattie degli occhi*, which ran through five Italian editions (1801–1836), four French (1802–1839), two German (1803–1823), two English (1806–1818) and one Dutch (1812), represented the highest and last exposition of the Galenic tradition in ophthalmology with its humoral philosophy. The book was written with remarkable clarity and is essentially practical, dealing with those conditions to which he himself had " sedulously and repeatedly attended " ; it exercised a unique influence over the whole civilized world for more than a generation. Its first six chapters were devoted to diseases of the lids and lacrimal apparatus and, as were all Scarpa's writings, it was illustrated with his clear and delightful drawings irreproachably accurate in detail, for the reproduction of which he himself trained Faustino Anderloni to execute the copper engravings (Fig. 2). It is interesting that he advocated couching in the surgical treatment of cataract despite Daviel's exposition of extraction published half a century previously, and because of his reputation his advocacy delayed the adoption of the new operation for 50 years. His interests, of course, were not confined to ophthalmology; his greatest work was the magnificent *Tabulæ nevrologicæ* in which the nerves of the heart were delineated for the first time. He discovered the membranous labyrinth, the naso-palatine nerve and originated the procedure of iridodialysis, and after him are named the triangle of the thigh and staphyloma posticum verum.

Scarpa had an interesting life. At 18 years of age he received his medical degree at Padua from his teacher, Morgagni. At the age of 20 he was appointed professor of anatomy and theoretical surgery at the University of Modena and in 1783 he was elected to the Chair of Anatomy of his parent University of Padua. Here anatomy flourished more than in any other centre, for Scarpa was given permission from the Emperor of Austria to obtain for dissection the bodies of all who died in the state hospital; his teaching was also revolutionary, its main theme being the relation of structure to function as an introduction to surgery. When Napoleon invaded Northern Italy and sacked Padua, his soldiers threw away many of the anatomical specimens which Scarpa had carefully preserved in spirits of wine—and drank the wine. Adapting himself to the new ruler, however, in 1793 he was made Rector of the University which he ruled with an authority so dictatorial that he died in his palatial villa outside the city walls unloved and alone—and blind. After his burial his body was disinterred and dissected in the anatomical amphitheatre he himself had built.

This Section is confined to diseases of the lids: for the anatomy of the lids, see Vol. II, p. 503 ; for their development, Vol. III, p. 231 ; and for their congenital deformities, Vol. III, p. 827.

GENERAL CONSIDERATIONS

The eyelids are the meeting place of so many different structures, each with a characteristic pathology, that a description of their diseases is apt to form a somewhat disjointed and untidy section. Here skin diseases of only incidental ophthalmological interest become related to conjunctival diseases with all their important sequelæ. In the structures of the lids are included numerous specialized organs readily liable to pathological processes of

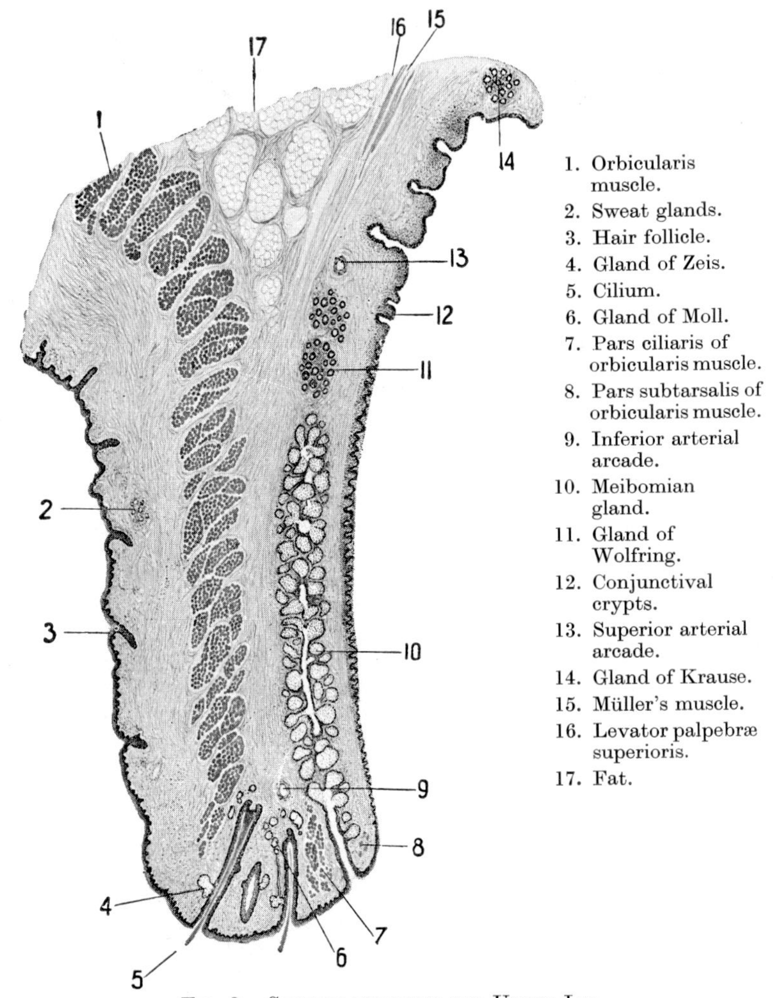

Fig. 3.—Section through the Upper Lid.

many types; and defects in the neuromuscular apparatus may have complications more important than in most parts of the body. Moreover, the obviousness of diseases of the lids to the patient and the disastrous consequences which pathological events occurring in them may have upon the eye make their early diagnosis and adequate treatment unusually urgent.

From the pathological point of view the lids may be considered in three layers (Fig. 3): (1) a superficial cutaneous layer comprising skin and its associated appendages, cellular and muscular tissue; (2) an intermediate fibrous layer comprising the tarsus and deep cellular tissue; and (3) the inner conjunctival layer which has already been considered in an earlier Volume.[1] In many respects the marginal area must be considered separately; it forms the meeting place of the three, its pathology is complicated by the presence of specialized structures, and it is important in its close relations to the cornea. To a large extent affections of these different regions retain their local individuality. In the superficial layer the skin differs from that of the rest of the body in the thinness and fineness of its texture, in its loose attachment to the underlying tissue and in the absence of fat in the subcutaneous tissue—a significant feature precluding an accumulation of adipose tissue in the obese which would, at the least, make the function of the lids mechanically impossible. Although fat is absent, the lids are rich in loose cellular tissue rendering them easily distended by œdema or blood and aiding the spread of inflammatory processes. The intermediate layer, on the other hand, is of dense fibrous consistency, and is separated from the superficial by a line of cleavage between the muscles and the tarsal plate so definite that pathologically and surgically they can be treated as two different entities. While the superficial layer is largely related to the integument and its morbid processes, the deeper layer shares in the diseases of the conjunctiva; and the lid-margins, a densely packed area where the hairs are specialized as cilia, the apocrine glands as glands of Moll and the sebaceous glands as Zeis's and meibomian glands, share in both the superficial and the conjunctival diseases or may initiate and maintain pathological processes of their own.

DISORDERS OF THE CIRCULATION

The vascular system of the skin consists of a rich anastomosis in the deeper parts of the corium from which single vessels run upwards to form a second network in the upper layers of the corium; thence terminal arterioles ascend into the papillæ to end in capillary loops. From these, collecting venules form venous networks parallel to the surface and finally discharge into large veins in the subcutaneous tissues. There are abundant arterio-venous connections and sphincters in the arterioles can divert the arterial blood directly into the venous circulation, by-passing the capillaries. When the capillary circulation is closed in this way the skin is pale; if the blood-flow is rapid and the capillary circulation is open, the skin is pink and hot; if the flow is sluggish, the blood is de-oxygenated and the skin is blue.

ACTIVE HYPERÆMIA (ERYTHEMA) OF THE LIDS is due to overfilling of the arterial and capillary circulation, producing a bright red colour of the skin, but is of little clinical consequence. It may be symptomatic of a general disturbance, as fever or intoxication (*toxic erythema*), the result of a local

[1] Vol. VIII.

irritation, such as heat, actinic rays or an irritant (*physical* or *chemical erythema*), or may be due to a regional *inflammatory* condition such as a dermatitis or a deep inflammation of the lids themselves, of the conjunctiva (as a gonococcal or diphtheritic conjunctivitis), of the eye (as a panophthalmitis), of the orbit (as a cellulitis or periostitis), of the lacrimal sac, or of neighbouring sinuses (as an empyema of the frontal). Further, an *allergic erythema* may occur, and an emotional or *reflex erythema*, particularly in subjects of an unstable vasomotor disposition. Any subjective symptoms such as itching or burning are usually slight although some peeling may result. The treatment depends essentially on the elimination of the cause, but local applications such as calamine lotion may be soothing.

PASSIVE HYPERÆMIA of the lids, due to overfilling of the venous circulation, producing a dark cyanotic colour of the skin usually with some œdema, although of little importance in itself may be a valuable sign of deeper mischief. It may also be of systemic origin, part of a general cyanosis of cardiac, pulmonary or cachectic origin; to this class belongs the intense hyperæmia of the lids which forms an early sign of typhus (Himly, 1843) and cholera (von Graefe, 1866). When of local origin it denotes a deeply seated obstruction of the circulation such as thrombosis of the orbital veins or the cavernous sinus, an arterio-venous aneurysm, a tumour or abscess of the orbit. Rarely the venous obstruction is further afield, as at the jugular vein in cervical adenitis (Kalt, 1892), or may be due to traumatic compression of the chest.[1] It may also be a sequel to prolonged bandaging or long-continued blepharospasm.

HYPERÆMIA OF THE LID-MARGIN

HYPERÆMIA OF THE LID-MARGIN or VASOMOTOR BLEPHARITIS is a relatively common affection, typified in the appearance of red-rimmed eyes wherein the margins of the lids look unpleasant, slightly swollen and red. Even in the chronic cases a discharge does not exist, but the blood vessels become chronically engorged and dense arborizations of neo-capillaries may appear, making the hyperæmic condition permanent. The symptoms are few and limited to a mild discomfort or heaviness and a feeling of dry itching; the most distressing feature, indeed, is the appearance of bleariness which may be intensified by a similar condition of the conjunctiva.

It is caused by congestion of the superficial blood vessels and occurs readily in persons of a blond complexion with a delicate skin. In such predisposed individuals the hyperæmia may appear in early youth and last a lifetime; in others it is produced on the most trivial insult, such as tiring the eyes, weeping, a vitiated atmosphere or a sleepless night. These exciting causes may be conveniently grouped into four categories:

[1] Vol. XIV, p. 728.

(*a*) *direct irritation*, such as exposure to wind, sunlight, dust, fumes or a stuffy, smoke-laden atmosphere, or a *local infection* in which case excessive infestation by the *Demodex folliculorum* should be remembered[1];

(*b*) *reflex irritation*, as from eye-strain due to refractive errors or muscular imbalance;

(*c*) *metabolic and constitutional causes*, particularly chronic indigestion, constipation or alcoholic excess;

(*d*) *toxic and allergic influences*, either from infections particularly in the neighbouring regions as in the nose, throat and teeth, or from foreign proteins to which a susceptibility exists.

Treatment should be directed towards the cause of the irritation, whether it is direct, reflex, constitutional, toxic or allergic. Ideally the offending agent should be eliminated or, if this is not easy or practicable, its effects

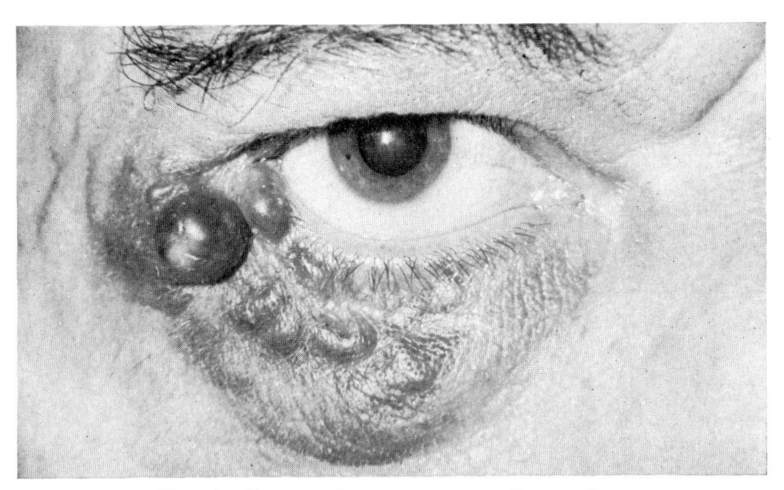

Fig. 4.—Cirsoid Aneurysm of the Lower Lid (P. MacFaul).

should be counteracted as much as possible. Irritants in the environment should be avoided, refractive errors corrected, digestive troubles put right, foci of infection dealt with and desensitization to allergic factors attempted. As a general rule, local collyria—which should always be bland and never too astringent—are of little and transient value.

A peculiar case of VARICOSE VEINS in the lids was described by Halasa and Matta (1964). On bending the head down or on looking to the right, marked swellings appeared in the upper and lower lids; these venous dilatations were seen by angiography and dissection to be due to a congenital absence of any direct communication between the venous plexuses of both lids and the superficial facial system of veins, with the exception of a small branch to the pterygoid plexus.

A very rare anomaly is the presence of a RACEMOSE (CIRSOID) ANEURYSM associated with a similar malformation in the orbit; it may give rise to recurrent

[1] p. 226.

hæmorrhages (Fig. 4). An aneurysm associated with the palpebral branch of the ophthalmic artery in a luetic female was described by Lewis (1943). Arterio-venous aneurysms in the anterior part of the orbit may also invade the lids.[1]

TELANGIECTASES

A telangiectasia is a localized dilatation of the cutaneous blood vessels, either arterioles, capillaries or venules, tracing fine lines on the surface of the skin. The condition may be generalized (*essential telangiectasia*), affecting large areas of the body particularly the limbs, a condition of unknown ætiology. *Secondary telangiectases* occur in many other diseases such as rosacea, dermatomyositis, lupus erythematosus, scleroderma, which are noted elsewhere, and following therapeutic irradiation of neoplasms of the lids.[2]

Three syndromes, however, which affect the lids should be noted.

BLOOM'S SYNDROME (1966) is comprised essentially of *congenital telangiectasia* with erythema and stunted growth. It is inherited, particularly in Jewish males, as an autosomal recessive condition showing a high frequency of chromosomal aberrations (German, 1969). The syndrome appears during infancy or early childhood; telangiectatic erythematous patches appear in a butterfly area over the cheeks and nose and involving the lower lids and sometimes the forehead and arms, associated with slight scaling, all of which is exacerbated by sunlight; the appearance somewhat resembles lupus erythematosus. The birth weight is low and in early life there is a moderate proportionate dwarfism, but growth resumes in later life. It is interesting that the mortality from neoplasic diseases such as leucæmia is significantly increased.

HEREDITARY HÆMORRHAGIC TELANGIECTASIA (the *Rendu-Osler-Weber disease*), a condition which is transmitted as an autosomal dominant trait,[3] may affect the lids (Ravault and Bonamour, 1959; Petrozzi and Carbone, 1968; Wolper and Laibson, 1969) where a veritable telangiectomatous tumour may be formed (Reeh and Swan, 1950). The lids may also be affected in ATAXIA-TELANGIECTASIA wherein the telangiectasia, mainly conjunctival in distribution but later affecting a butterfly area on the face, is associated with a progressive ataxia of the cerebellar type[4] (Boder and Sedgwick, 1958). The condition has been associated with a deficiency of immunoglobulin E (Ammann *et al.*, 1969).

ANGIOMA SERPIGINOSUM, considered by some to be a type of telangiectasia, is discussed at a later stage.[5]

HÆMORRHAGE INTO THE LID

SMALL PETECHIAL HÆMORRHAGES into the skin of the lids have no particular ophthalmological significance whether they occur spontaneously as in other parts of the skin or the conjunctiva or appear in severe systemic infections (septicæmia) or in the hæmorrhagic diatheses (purpura, scurvy, Barlow's disease, hæmophilia) (Stephenson, 1915; Blake, 1921; Leffertstra, 1962; Palmer, 1963; and others) (Figs. 5 and 6); multiple hæmorrhages

[1] p. 848.
[2] Vol. XIV, p. 956.
[3] Vol. III, 910; Vol. VIII, p. 30.
[4] Vol. VIII, p. 32.
[5] p. 500.

into the lids may be an early sign of systemic amyloidosis (Ramsdell and Winder, 1970) and are characteristic of poisoning by benzene.[1] MASSIVE SUFFUSIONS into the loose subcutaneous tissues are frequently of more significance. In the majority of cases they are of traumatic origin, either direct as from a blow over the eye, or indirect from a fracture of the orbit, the nose, the base of the skull or the nasal sinuses (Fig. 7). More rarely spontaneous bleeding is induced by a sudden increase in blood pressure

FIGS. 5 and 6.—HÆMORRHAGES IN SCURVY
(C. A. L. Palmer).

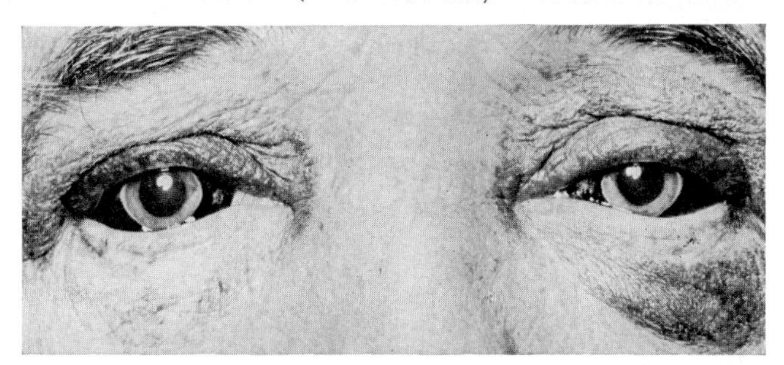

FIG. 5.—The initial appearance showing bilateral hæmorrhages in the subconjunctival and orbital tissues, in a man aged 69.

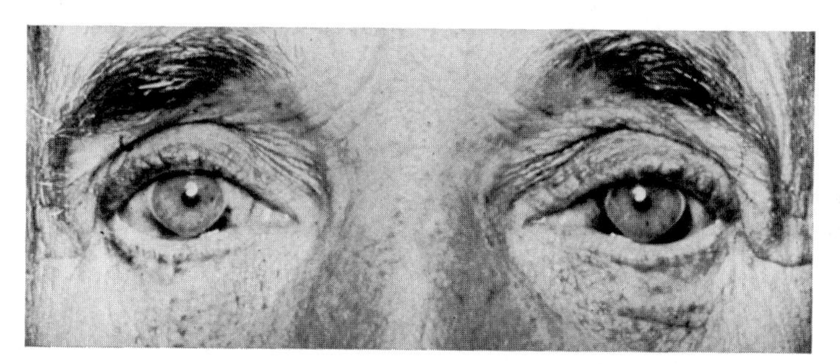

FIG. 6.—The appearance of the patient one month later.

associated with venous compression due to violent coughing, vomiting or mechanical compression of the chest (compression cyanosis)[2] (Wagenmann, 1900–24); such an accident has happened in the aged and arteriosclerotic after unwise physical effort at work, at stool or in bed. In many cases, of course, the source of the bleeding is orbital, whence, after some delay in traversing the tarso-orbital fascia, the blood eventually becomes evident as an ecchymosis of the lids; thus it is frequently the presenting sign in patients with metastatic neuroblastomata[3] (Albert *et al.*, 1967).

[1] Vol. XIV, p. 1266.
[2] Vol. XIV, p. 728.
[3] p. 1154, Fig. 1211.

When the origin of the hæmorrhage is local, the ecchymosis may be confined to a relatively small region of the lid, but if bleeding is profuse the blood tends to diffuse widely through the loose subcutaneous cellular tissue until it is held up by the anchoring of the fascia at the eyebrows preventing spread on to the forehead, and at the naso-jugal and malar folds preventing extension to the cheek and upper lip. Within this territory the lids may become so swollen as to close the eye and a tense hæmatoma may be formed; from it the easiest egress is over the bridge of the nose into the tissue of the lids of the other eye (Fig. 8). As the skin over the nose is thick the blood is not seen under it, so that the continuity of the seepage from one eye to the other may not be apparent and a second and independent hæmorrhage on

FIGS. 7 and 8.—PALPEBRAL HÆMORRHAGES.

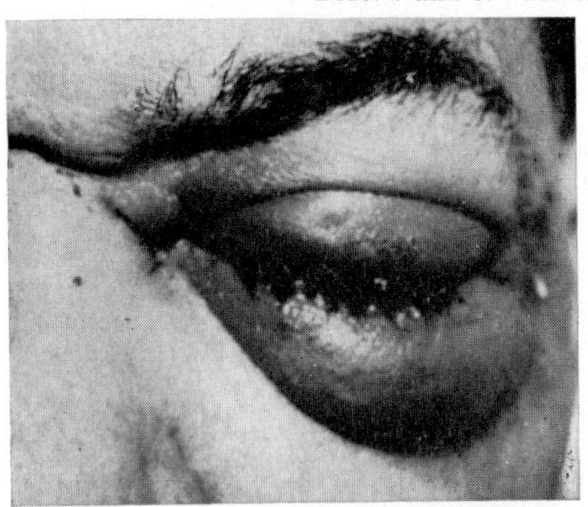
FIG. 7.—Hæmorrhage into the lids after a blow, showing gross swelling.

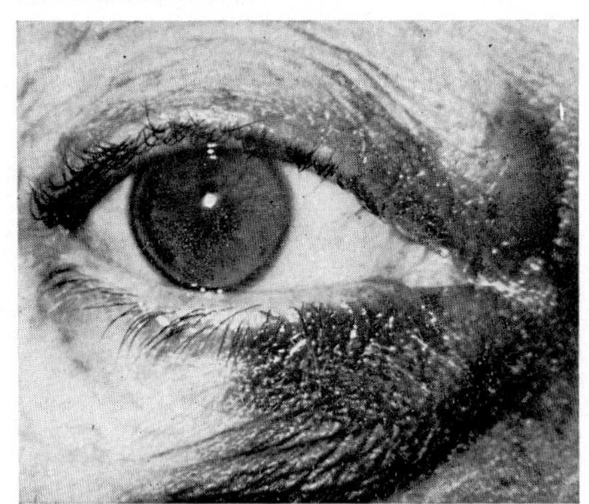
FIG. 8.—The seepage of blood across the nose from an injury of the other eye by a blow from a fist, causing hæmorrhage in the lids of the undamaged eye.

this side may be suggested. The common occurrence of such a spread in conditions in which the trauma is known to be limited to one side, as after the enucleation of an eye, disproves the necessity for this suspicion.

From the diagnostic point of view the most important type of long-distance bleeding is that occurring some little time after a *fracture of the base of the skull*,[1] Here, sometimes a few days after the injury, extravasated blood seeps along the floor of the orbit and appears usually towards the nasal side in the tissues of the subconjunctiva and lower lid. In 100 basal fractures Liebrecht (1906) found that 34% had bleeding into the lids (22 unilateral and 12 bilateral), 10% had subconjunctival hæmorrhage in addition, and 4% subconjunctival ecchymoses alone. He also found that the mortality of cases with hæmorrhage into the lid was greater than of those without (35 to 22), so that some prognostic significance may be attached to such bleeding. A fracture of the orbital wall also tends to show a certain localization in the bleeding (Morison, 1894; Kehl, 1921–23); from the orbital roof blood tends to infiltrate along the levator muscle into the upper lid, from the apex along the lateral rectus, and from the floor into the lower lid. When the bleeding is derived from the

[1] Vol. XIV, p. 287.

soft tissues of the lids, the hæmatoma is pretarsal, when it is derived from a fracture of the orbital wall or the base of the skull, blood is found both in front of and behind the orbital septum; since a retroseptal hæmatoma in the upper lid may be confined to tissue above the tarsus, a differential diagnosis can sometimes be made only by doubly everting the lid (Honegger, 1969).

The prognosis of bleeding into the lids is good unless secondary infection is admitted either through a solution in the continuity of the skin or the mucous membrane of a sinus; in this event cellulitis and an abscess of the lid may occur. For this reason any attempt at evacuation of the blood by incision or puncture should be discouraged. Apart from this, the suffusion is self-limiting and absorbs in a week or so. Owing to the obviously disfiguring appearance of a " black eye ", however, as it changes from black to green and yellow, treatment is frequently demanded. Cold water compresses are as good treatment as any (or raw beef steak!), and are probably as effective as more impressive old-fashioned applications such as lotions of lead and opium, hamamelis, or ammonium chloride and alcohol. Painting the skin with guaiacol may help to obscure the discoloration by its blanching effect, as also may the liberal use of cosmetic applications, aided by wearing dark glasses.

Albert, Rubenstein and Scheie. *Amer. J. Ophthal.*, **63**, 727 (1967).
Ammann, Cain, Ishizaka *et al.* *New Engl. J. Med.*, **281**, 469 (1969).
Blake. *Amer. J. Ophthal.*, **4**, 736 (1921).
Bloom. *J. Pediat.*, **68**, 103 (1966).
Boder and Sedgwick. *Pediatrics*, **21**, 526 (1958).
German. *Amer. J. hum. Genet.*, **21**, 196 (1969).
von Graefe. *v. Graefes Arch. Ophthal.*, **12** (2), 198 (1866).
Halasa and Matta. *Arch. Ophthal.*, **71**, 176 (1964).
Himly. *Die Krankheiten u. Missbildungen d. mensch. Auges*, Berlin, 210 (1843).
Honegger. *Klin. Mbl. Augenheilk.*, **155**, 68 (1969).
Med. Mschr., **23**, 441 (1969).
Kalt. *Ann. Oculist.* (Paris), **107**, 41 (1892).
Kehl. *Beitr. klin. Chir.*, **123**, 203 (1921).
Virchows Arch. path. Anat., **246**, 194 (1923).
Leffertstra. *Ophthalmologica*, **143**, 131 (1962).
Lewis. *J. med. Ass. Georgia*, **32**, 185 (1943).
Liebrecht. *Arch. Augenheilk.*, **55**, 36 (1906).
Morison. *Lancet*, **1**, 16 (1894).
Palmer. *Brit. J. Ophthal.*, **47**, 692 (1963).
Petrozzi and Carbone. *Arch. Oftal. B. Aires*, **43**, 371 (1968).
Ramsdell and Winder. *Sth. med. J.*, **63**, 822 (1970).
Ravault and Bonamour. *J. Méd. Lyon*, **40**, 693 (1959).
Reeh and Swan. *Trans. Amer. Acad. Ophthal.*, **54**, 312 (1950).
Stephenson. *Ophthalmoscope*, **13**, 132 (1915).
Wagenmann. *v. Graefes Arch. Ophthal.*, **51**, 550 (1900).
Graefe-Saemisch Hb. d. ges. Augenheilk., 3rd ed., *Verletzungen d. Auges*, Berlin, **3**, 2113 (1924).
Wolper and Laibson. *Arch. Ophthal.*, **81**, 272 (1969).

ŒDEMA

VASCULAR ŒDEMA

The laxity and distensibility of the subcutaneous connective tissue of the lids make them a ready site for the accumulation of œdematous fluid. This, like extravasated blood, is prevented by the fascial attachments from spreading to the forehead or cheek beyond the brows and the naso-jugal and malar folds, but within this region the swelling may be so acute—sometimes from the most trivial causes—as to close the eye mechanically and spread to the conjunctiva as a chemosis, producing a clinical picture much more frightening both to the patient and the practitioner than its importance warrants (Fig. 9). In itself, œdema of the lids is not a disease-entity but forms a symptom of many pathological conditions generally characterized by increased capillary permeability and these it may be convenient to summarize here. Two main clinical types, however, exist: inflammatory œdema, when the skin is tense, hot and shiny; and non-inflammatory (pale, white or cold œdema), which is passive or static in origin (" puffy eyelids "). When either is on the increase the skin is smooth and tense, but on the commencement of subsidence a fine wrinkling indicates that resolution is on the way—an important sign that the height of the

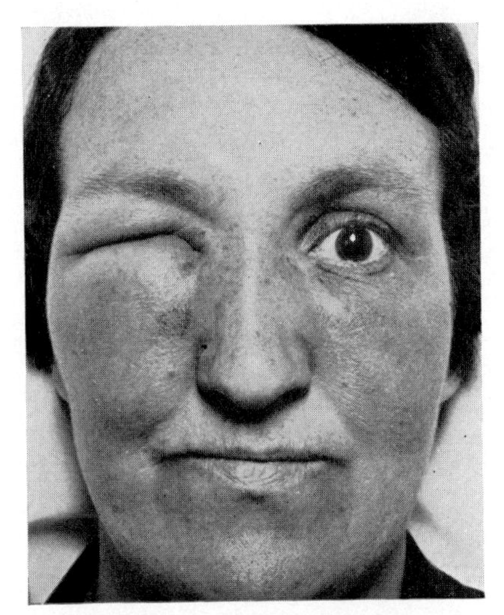

Fig. 9.—Palpebral Œdema.
A recurrent œdema of self-inflicted traumatic origin (W. H. Brown).

crisis is past. A chronically recurrent œdema may have the effect of producing laxity and atrophy of the palpebral tissues in the condition of blepharo-chalasis.[1]

(1) Œdema due to local inflammation may arise from a trauma or any inflammatory process in the vicinity of the lids—either in the lids themselves (sometimes from a very small focus such as a hardly detectable stye or a small chalazion), in the lacrimal fossa (dacryocystitis), in the eye (particularly corneal ulcers, glaucoma, panophthalmitis), in the orbit (particularly cellulitis, tenonitis, myositis, periostitis), in the cranial cavity (cavernous sinus thrombosis or, as a temporary phenomenon, meningitis from middle ear infection) (Deutsch, 1922; Richter, 1927), or in the neighbouring nasal sinuses. Finally, an œdema of the upper lid is one of the

[1] p. 350.

regular symptoms of a septic infection of the scalp, while a seborrhœic disposition with dandruff of the hair and eczema are not uncommon ætiological factors. In some infections of the lids œdema may form the first symptom (zoster, Löwenstein, 1933).

Most of these œdemas come on acutely and disappear rapidly when the exciting inflammatory process dies down, leaving no sequelæ.

The differential diagnosis in these cases sometimes presents problems. The first step should be to examine the eye, sometimes a matter of considerable difficulty necessitating the use of retractors. Inflammatory conditions of the conjunctiva or

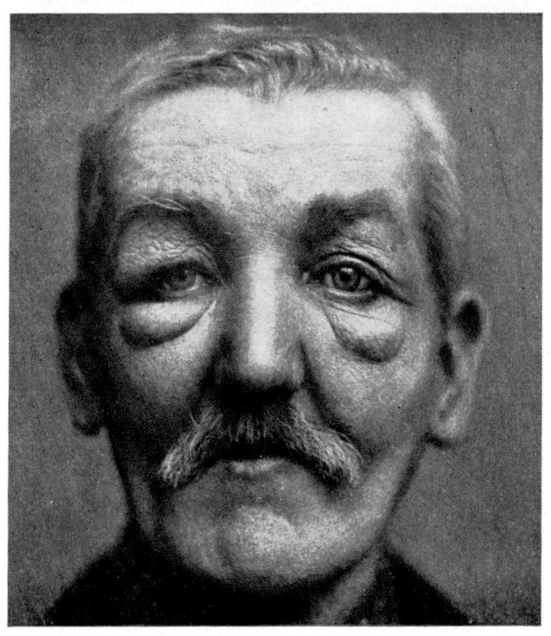

Fig. 10.—Renal Œdema (Maitland Ramsay).

the eyeball can thus be detected, while the presence of chemosis, proptosis and immobility of the globe with a normal anterior segment provides a clue to orbital mischief. When the inflammation is in the lid the accompanying massive œdema may obscure its origin; the minute point of entry of an insect's sting, for example, may be difficult to detect, or an early stye or meibomian inflammation may be revealed only by a localized point of tenderness to palpation when the whole lid is carefully explored. A hard, sensitive area at the inner angle below the medial ligament suggests a dacryocystitis, above it a sinusitis, while induration and pain on palpation over the orbital margin suggest a periostitis. In erysipelas the swelling and redness are uniform and the skin feels thick and dense; while chronic eczema is frequently overlooked, since it may be betrayed only by an insignificant roughness on the skin's surface.

(2) A TOXIC ŒDEMA may be due to a multitude of causes, among which may be noted (*a*) *toxins*, either metabolic such as occur in renal disease

wherein œdema of the lids is a prominent feature (Fig. 10), or exogenous poisons such as thallium (Beach, 1933) or iodine (Hewkley, 1888).

(*b*) *Organismal toxins* are also important, for œdema of the lids accompanies many acute infections—diphtheria, rheumatic fever, scarlet fever, influenza (Denti, 1890; S. Phillips, 1895; Spriggs, 1908; Elgood, 1909; and others), inflammations of the gall-bladder (Berardinelli, 1951) or ulcerative colitis (Ellis and Gentry, 1964). It may be the first symptom in typhoid fever (Beach, 1933), a sequel of relapsing fever (Elliot, 1920), and is a characteristic sign in infectious mononucleosis (Hoagland's sign) (McCarthy and Hoagland, 1964).

(*c*) *Parasitic toxins* have been found to cause periodic attacks of œdema of the lids, lasting from a few days to some weeks, particularly in children (Pajtás, 1959; Conrads and Heinmüller, 1959).

These are typified in trypanosomiasis[1] or in nemathelminthic infestations. This sign is important in trichiniasis, wherein the swollen lids and typical slit-like palpebral aperture form one of the earliest symptoms and should always be considered as suggestive in regions where the infestation is rife (Drake *et al.*, 1935; Cogan, 1969); in sporadic cases in non-endemic areas the diagnosis is frequently difficult (Chhung Hin, 1945). Œdema of the lids is also very prominent in filariasis; in infestation by the *Loa loa* the œdematous CALABAR SWELLINGS and evanescent AMBULANT ŒDEMAS are characteristic[2] and in *Wuchereria bancrofti* infestation and dracontiasis the swelling of the lids may be enormous and intense. Giardiasis may also be the cause of recurrent œdema of the lids in children (Toselli *et al.*, 1965), as also may ascariasis, in which case the condition is frequently unilateral (Soulas, 1954; Gogina, 1964; Castellazzo, 1967). In malaria also, particularly in cachectic states, palpebral œdema may be prominent (Bunton, 1909; Blank, 1920).

(3) A STATIC ŒDEMA of the lids may be due to any local or general cause which impedes the venous return. Among *local causes* must be mentioned long-continued blepharospasm, wherein the palpebral veins are compressed by the orbicularis muscle: thus œdema is frequently seen acutely in children with phlyctenular keratitis, and may persist for years in cases wherein constant blepharospasm accompanies trachoma and may not disappear until relief of the symptoms (de'Cori, 1935). It may also appear as a result of a scar at the temporal margin of the lids, as of a Krönlein operation. A tumour in the depths of the orbit may have the same effect, and it may be seen to an alarming degree in endocrine exophthalmos (Usdan *et al.*, 1972) (Plate V); operating at a further distance, compression of the jugular veins (Kalt, 1892), cicatrization arising from suppurating cervical lymph nodes (Debaisieux, 1893), traumatic compression of the chest or a mediastinal tumour may similarly cause a palpebral œdema.

General stasis, as in heart failure, may cause an œdema of the lids, sometimes of a livid type; although it usually commences in the dependent parts such as the ankles, it may be first evident, particularly in children, on the face.

[1] p. 186. [2] p. 188.

(4) Œdema of the lids due to BLOOD DISEASES is seen in the anæmias and chlorosis; here the œdema is not usually extreme and may be characterized by a pale bluish appearance.

(5) ALLERGIC AND VASOMOTOR ŒDEMAS can be divided into two types: urticaria affecting mainly the dermis and the subcutaneous tissues, and angio-œdema in which the subcutaneous tissues rather than the dermis are preferentially involved; the two types may be combined in the same patient.

Urticaria (*nettlerash, hives*) is an acute œdema characterized by the sudden appearance of pink or white weals which are due to the massive transudation of fluid from the capillaries into the dermis and subcutaneous tissues in response to the local release of histamine-like substances from the

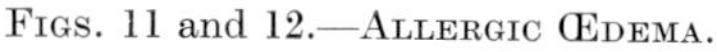
FIGS. 11 and 12.—ALLERGIC ŒDEMA.

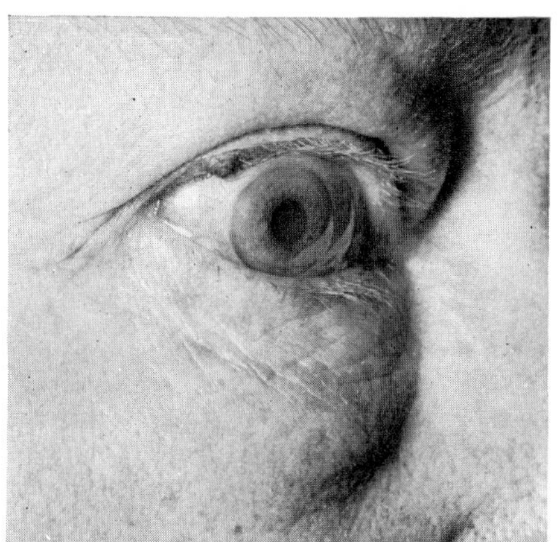

FIG. 11.—Simple allergic œdema in a woman aged 63 (Inst. Ophthal.).

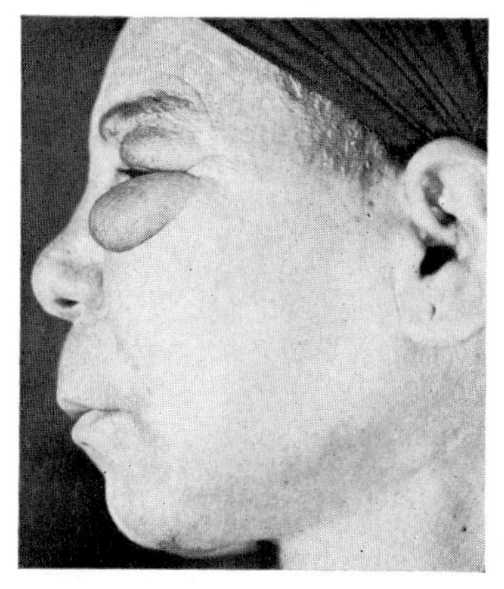

FIG. 12.—Giant urticarial œdema of undetermined ætiology (R. P. Wilson).

mast cells of the skin. Similar lesions can be produced in the normal skin by the intradermal injection of histamine or histamine-liberators and the reaction is frequently inhibited by sufficient doses of antihistamine drugs. The weals appear rapidly, often with an intense irritation and a variable halo of erythema; they vary considerably in size and tend to coalesce, and often after a few hours disappear spontaneously without trace, frequently, however, to recur; on the other hand, the condition may be more chronic and the attack may persist for weeks or months, new lesions appearing at irregular intervals. Elsewhere in the body the lesions usually have a sharply defined edge but in the lids the œdema tends to be more generalized (Fig. 11).

The phenomenon is usually a hypersensitivity-reaction of the anaphylactic (atopic) type and is seen typically in serum sickness but, in addition to the ingestion of foreign protein, it may be caused by a host of allergens of which the most common are foods, drugs, pollens, dusts, and other plant and animal products (Derbes and Engelhardt, 1944; and others). Almost any food can be the cause, but fish, shellfish, nuts, pork, eggs, mushrooms and fruit are common offenders (*e.g.*, mushrooms, Montresor, 1950; apricot jam, Strebel, 1952). These substances all contain proteins, and although many drugs are not antigenic in themselves, they may become so after conjugation with a protein in the body. Typical examples are the reactions which may follow the administration of aspirin (Warin, 1960), the sulphonamides or penicillin (Gordon, 1946; and many others). Non-allergic factors figure less frequently in the ætiology and are usually vague and indeterminate. Infections are said to be a frequent cause (30%, Fink and Gay, 1934) and parasitic infestations are important in localities where these are prevalent. An endocrine influence is suggested by its occurrence in menstruation or at the menopause; while thyroid insufficiency may enter into the ætiology as well as stress and emotional factors. When these various suggestions are considered it is not surprising that in many cases no firm ætiology can be established (90%, Hopkins and Kesten, 1934; 70% of chronic cases, Green *et al.*, 1965).

Angio-œdema (*giant urticaria*), a condition which is frequently called *angioneurotic œdema* (an unfortunate term with its suggestion of a neurosis), is a more chronic condition of vasomotor origin affecting the subcutaneous tissue; the lesions are tense, rounded and non-pitting, lacking the sharply demarcated borders of urticaria (Fig. 12); the overlying skin is normal and itching is usually absent. It may occasionally be due to local reflex action, as, for example, a reflex from nasal disease (Dunn, 1892); thus it has followed the lodgement of a foreign body (a peanut) in the nose for a period of 17 years, the removal of which cured the condition (Pruzanski, 1951). In the great majority of cases, however, it is an expression of generalized instability of the vasomotor system due to causes which are essentially unknown, but which have been variously ascribed to allergic, endocrine or toxic disturbances. An atopic hypersensitivity-reaction is undoubtedly a common cause, the œdema corresponding to the type of allergy seen in asthma or hay fever. The protein acting as an antigen may be inhaled (dust, pollen, etc.), ingested as a food, or injected (such as antitoxins); when it is known or suspected, the ætiology can be demonstrated by scratching it through the epidermis into the dermis whereafter a weal will appear. It is, however, rare that an ætiology can be so definitely established.

In the view of some writers most cases were at one time considered to be of toxic origin, particularly from the teeth, tonsils or bowel, and the beneficial effects of removal of the teeth or tonsils and elaborate methods to eliminate intestinal intoxication occasionally—but only very occasionally—justified their faith. Others consider any œdema not bacterial or mechanical to be allergic in origin. Among the endocrine glands the thyroid has excited most interest, and both hypothyroidism, particularly marked after thyroidectomy (Weiss and King, 1932), and hyperthyroidism (Michail,

1931) have been arraigned. Moreover, the attacks are apt to be seen in those of a nervous temperament and may be brought about by stress, emotional or hysterical crises (Potvin, 1936). Physical agents such as light, heat and cold act as ætiological factors in certain cases. In all these views there is probably some truth; for these idiopathic œdemas are the local reaction of an unstable capillary endothelium or an unbalanced vasomotor mechanism to various agents ranging from circulating toxins or proteins for which an allergy has been acquired, to traumatic, thermal or even emotional stimuli. In many of the cases, however, the closest search will reveal no cause whatever. It is noteworthy that many cases occur in patients with a personal or family history of allergy; a rare hereditary form has been observed with a strong familial tendency and, although the genetic mechanism may be complex, a dominant transmission is usually evident (J. M. Phillips and Barrows, 1922; Dunlop and Lemon, 1929). It has been ascribed to a deficiency of a globulin which is an inhibitor of C1 esterase, one of the many components of complement (Austin and Sheffer, 1965).

Angio-œdema may assume many clinical aspects, coming and going suddenly or lasting for days or weeks. The attacks may occur irregularly and apparently without cause, or be cyclic, appearing at the menstrual periods or in the early mornings (Clegg, 1936). If they persist or recur over a long period, sometimes measured in years, permanent hypertrophic changes tend to occur in the tissues, providing a histological picture resembling chronic exudative and proliferative inflammation sometimes with hyperpigmentation (Michail, 1931). In the recurrent type, after years of repeated attacks, the redundant skin and subcutaneous tissue so formed may hang in folds which can be gathered up by the fingers in the quiet periods, suddenly to be filled out by œdema when an attack recurs (Lawson and Sutherland, 1906).

The great majority of patients exhibiting the somewhat distressing symptom of "puffy eyelids" are women, usually highly strung and nervous, many of them mentally strained, emotionally unhappy or overworked, a number of them giving a history of an acute or chronic infection such as ulcerative colitis. The only symptom may be a swelling or puffiness of the eyelids with periodic acute exacerbations and occasionally eczema of the skin, a condition which persists indefinitely without obvious cause. In such cases, of course, allergy to an external irritant, an article of diet or a drug such as the sulphonamides must be excluded; but although a contact dermatitis[1] may account for many cases presenting this syndrome, a seborrhœic background is not uncommon and a possible clinical association is a long period of sensitization to an organism such as the streptococcus. Recovery, however, is rarely easy or rapid.

Two particular varieties of angio-œdema deserve passing mention.

QUINCKE'S ŒDEMA FUGAX (1882), first noted by Goltz (1880) and Bannister (1880), is a paroxysmal fleeting condition resulting in circumscribed swellings of the skin and subcutaneous tissues, and sometimes of mucous membranes. It shares the ætiological basis and clinical features of angioneurotic œdema generally, but is often hereditary (Bruun and Dragsted, 1951), is more common in males than females and generally affects the young, the attacks sometimes persisting intermittently for years, usually recurring at irregular intervals. An allergic background is the rule, a neurotic temperament is common, and frequent activating stimuli are cold, trauma and emotion.

[1] p. 58.

The swelling comes on suddenly and may appear red and hyperæmic, standing out prominently from the surrounding skin, affecting the lids sometimes of one eye, sometimes of both, and it may spread to the conjunctiva producing a chemosis (Weekers and Barac, 1936), or to the orbit with a resulting proptosis and occasionally disturbances of ocular motility (Benedict, 1958; Dimshits and Khelle, 1961). Other areas of skin besides the lids are affected—the lips, cheeks, backs of the hands, penis, scrotum, or the whole of a limb, while mucous membranes other than the conjunctiva may also be implicated—the tongue and glottis (which has proved fatal) and the mucosa of the bronchi, the alimentary or the urinary tract—which again has proved fatal (Stähelin, 1903). Ocular complications are few, but a papillœdema (Handwerck, 1907) and narrowing or widening of the pupil (Sacquépée, 1909) have been recorded. The search for a causal factor is often unrewarding but in all cases investigation should be made for any allergic factor or for the presence of infections or parasites.

VASOMOTOR MIGRAINE forms another rare but interesting clinical entity. We have already seen that the presumptive ætiology of migraine—vasomotor instability in the brain with vascular spasm followed by dilatation—is closely akin to that of angio-œdema.[1] In some cases of migraine this cerebral reaction is associated with a similar but less marked vasomotor disturbance on the affected side of the face characterized by an initial pallor followed by flushing. In vasomotor migraine this peripheral reaction is intensified and the usual symptoms of migraine, headaches, scotomata and teichopsia are accompanied by a marked and transient œdema of the lids (Doyne, 1887; de Schweinitz, 1888; Robinson, 1888).

The puffiness of the lids, to some extent due to œdema, met with in *thyrotoxicosis* (Enroth's sign) will be discussed subsequently.[2] In *myasthenia gravis* œdema of the lids may also be an early symptom (Klar, 1930).

Sclerœdema, a brawny œdema wherein the fluid consists of depolymerized hyaluronic acid, will be considered at a later stage in the section on connective-tissue diseases.[3]

Any *treatment* for œdema of the lids is to be directed towards the cause if this can be found. In the inflammatory type the source of disturbance should be attacked while the local condition of the lids usually reacts best to the application of moist heat. In the non-inflammatory types an endeavour should be made to trace the underlying ætiology. Any focus of infection should be eliminated, desensitization attempted with great care for it is to be noted that such patients are usually abnormally reactive; intoxications should be treated, sources of allergy either as foods, drugs or exogenous irritants traced and avoided or the patient desensitized, and other exciting causes eliminated as effectively as possible. Even with the best will in the world, however, such a search will often be unavailing, and if tentative conclusions are arrived at it may be found impossible to deal effectively with the objects of suspicion. In this event every effort should be made to tone up the general system by a complete temporary reorientation of life by rest, a change of interest and environment, and the elimination so far as is possible of physical or emotional strain. Such treatment, of course, is by no means always easy or possible. So far as medicaments are concerned, local

[1] Vol. XII, p. 550. [2] p. 943. [3] p. 320.

applications are of little value, although carbolized calamine lotion or saturated sodium bicarbonate may relieve the itching of an urticarial condition. In the more acute types symptomatic relief may be obtained by the subcutaneous injection of adrenaline (0·5 ml. of 1:1,000 soln.) which may be repeated in 15 or 30 minutes. Kafka (1937) recommended ephedrine sulphate by mouth, Wise (1936) pilocarpine or atropine. In more chronic cases antihistamine preparations taken orally[1] are sometimes of considerable value in allergic œdema or, if these are not effective, corticosteroids. In the chronic stages when hypertrophic changes have developed, surgical extirpation of the redundant subcutaneous tissue may be advisable, a procedure which frequently results in considerable functional and cosmetic improvement (Montresor, 1950; New and Kirch, 1961; Hallett and Mitchell, 1968; and others).

LYMPHŒDEMA

Lymphœdema is due to a defect in lymphatic drainage which in its recurrent or chronic forms leads to the development of a firm, non-pitting swelling associated with epidermal thickening and a rich subepidermal cellular accumulation particularly of lymphocytes and plasma cells (*solid œdema*) (Figs. 13 to 15). Eventually a condition of permanent hypertrophy or elephantiasis[2] may result. This sequence is due to the stagnation of an extracellular fluid rich in proteins which leads to the cellular infiltration and the stimulation of fibroblasts.

A lymphœdema of a *primary* nature may appear at birth or in later life due to defective development of the lymphatic vessels in embryonic life. The occurrence of this type of "essential" elephantiasis as a congenital condition was established by Nonne (1891) and Milroy (1892–1928), while the delayed type was annotated by Meige (1898–1933), and it is variously known as MILROY'S DISEASE or the NONNE-MILROY-MEIGE DISEASE. The anomaly has a strong hereditary tendency, being transmitted usually as an autosomal dominant but sometimes as an X-chromosomal recessive character (Cockayne, 1933). The condition is usually seen in the lower extremities and genitalia but may affect the face and lids or the conjunctiva (Tabbara and Baghdassarian, 1972). Various anomalies have been reported with the condition—distichiasis, blepharoptosis, ectropion of the lower lid, pterygium colli, spinal anomalies and others.[3]

A *secondary* lymphœdema is more common and may be due to several factors, among the most common of which are malignant disease, surgical obstruction as by a Krönlein operation, irradiation, skin diseases and

[1] Chlorpheniramine maleate (Piriton) 4 mg. three or four times a day, triprolidine hydrochloride (Actidil) 2·5 mg. three or four times a day, promethazine hydrochloride (Phenergan) 25 to 50 mg. at night.

[2] p. 356.

[3] Bloom (1941), Campbell (1945), Kinmonth *et al.* (1957), Falls and Kertesz (1964), Robinow *et al.* (1970), Hoover and Kelley (1971).

FIGS. 13 and 14.—LYMPHŒDEMA (Inst. Ophthal.).

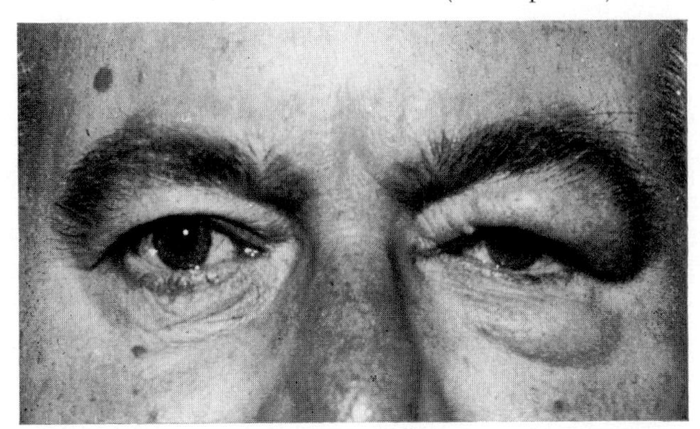

FIG. 13.—A unilateral case.

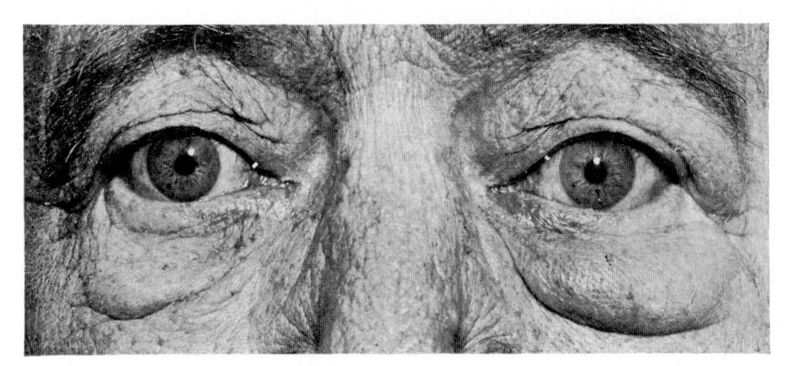

FIG. 14.—A bilateral case.

FIG. 15.—SOLID ŒDEMA.

In a patient with recurrent erysipelas. Note the swellings of both lids of the right eye and the entire cheek down to the naso-labial fold.

various types of inflammation (Borrie and Taylor, 1962). Among the last, the solid œdema associated with recurrent erysipelas is important,[1] but the ætiology of many cases is very difficult to trace. It is frequently difficult to decide whether the lymphœdema is the primary event which predisposes to bacterial infection or the latter determines the œdema; the œdema may precede the infection, but it is undoubtedly the case that infection is a common sequel of the œdema, and if recurrences of cellulitis ensue the lymphatic obstruction and œdema are augmented. It is probable, however, that the elephantoid state does not arise from lymphœdema alone; for this reason the condition and its treatment will be more fully discussed under the heading of elephantiasis nostras of the lid.[2]

Elephantiasis due to filarial infestation is well known in tropical countries usually attacking the lower extremities (*elephantiasis lymphangioides* or *filariosa*), but it rarely affects the lids; Becker (1895) described such a case which showed marked hypertrophic changes wherein the lymphatics were enormously enlarged, fibrous tissue had proliferated in bands and networks and the epidermis had hypertrophied. A similar hypertrophic solid œdema with the formation of much new connective tissue has also been noted with leishmaniasis of the nose[3] (*elephantiasis leishmaniana*).

THE MELKERSSON-ROSENTHAL SYNDROME

This is a rare combination, originally described by Rossolino (1901), comprising recurrent attacks of facial palsy associated with marked œdema of the face, particularly of the lids (Melkersson, 1928) and furrowing of the tongue (Rosenthal, 1930). Initially the facial palsy which involves a certain degree of lagophthalmos may be preceded or accompanied by an acute attack of œdema, non-pitting, non-inflammatory and painless resembling angio-œdema. The first attack which often appears in childhood tends to clear up, but the affection is habitually recurrent and after several episodes the œdema becomes chronic, the subcutaneous tissues showing cellular infiltration and fibrosis, and the œdema, at first soft, assumes a rubbery consistency. The lingua plicata is congenital and frequently hereditarily transmitted (Carr, 1965); its association with the other two manifestations of the triad is obscure.

A considerable number of reports has appeared in the literature,[4] and ocular complications include keratitis sicca (Koch, 1951), corneal opacities (Gassler and Berthold, 1961), papillœdema (Kolle, 1963), retrobulbar neuritis, disturbances of the cranial nerves (Steinvorth, 1958) and sheathing of the peripheral retinal vessels (Fötzsch, 1967). The ætiology is unknown, and the several theories advanced have little or no foundation. Treatment can only be symptomatic, but a good response has been reported to triamcinolone (Cerimele and Serri, 1965). If the swelling becomes chronic and permanent the removal of excess tissue of the lips (New and Kirch, 1961) or lids (Hallett and Mitchell, 1968) may be advisable.

Austin and Sheffer. *New Engl. J. Med.*, **272**, 649 (1965).
Bannister. *Chic. med. Rev.*, **1**, 281 (1880).
Beach. *Amer. J. Ophthal.*, **16**, 119 (1933).
Becker. *v. Graefes Arch. Ophthal.*, **41** (3), 169 (1895).
Benedict. *Amer. J. Ophthal.*, **45** (2), 43 (1958).
Berardinelli. *Lyon chir.*, **46**, 349 (1951).
Blank. *Dtsch. Arch. klin. Med.*, **132**, 179 (1920).
Bloom. *N.Y. J. Med.*, **41**, 856 (1941).

[1] p. 93. [2] p. 356. [3] p. 184.
[4] Luescher (1949), Curtin (1956), Steinvorth (1958), Marinković and Stevanović (1959), Hofmann (1963), Kolle (1963), Paton (1965), Kunstadter (1965), Fötzsch (1967), Hallett and Mitchell (1968).

Borrie and Taylor. *Brit. J. Derm.*, **74**, 453 (1962).
Bruun and Dragsted. *Acta allerg.* (Kbh.), 281 (1951).
Bunton. *Brit. med. J.*, **1**, 308 (1909).
Campbell. *Univ. Mich. med. Bull.*, **11**, 69 (1945).
Carr. *Ohio St. med. J.*, **61**, 709 (1965).
Castellazzo. *Minerva oftal.*, **9**, 146 (1967).
Cerimele and Serri. *Arch. Derm.*, **92**, 695 (1965).
Chhung Hin. *Brit. med. J.*, **2**, 219 (1945).
Clegg. *Trans. ophthal. Soc. U.K.*, **56**, 259 (1936).
Cockayne. *Inherited Anomalies of the Skin and its Appendages*, London (1933).
Cogan. *Neurology of the Ocular Muscles*, 2nd ed., Springfield, 49 (1969).
Conrads and Heinmüller. *Klin. Mbl. Augenheilk.*, **135**, 496 (1959).
de'Cori. *Boll. Oculist.*, **14**, 1395 (1935).
Curtin. *Irish J. med. Sci.*, No. 365, 235 (1956).
Debaisieux. *Arch. Ophtal.*, **13**, 392 (1893).
Denti. *Ann. Ottal.*, **19**, 77 (1890).
Derbes and Engelhardt. *Sth. med. J.*, **37**, 729 (1944).
Deutsch. *Mschr. Ohrenheilk.*, **56**, 686 (1922).
Dimshits and Khelle. *Vestn. Oftal.*, No. 3, 22 (1961).
Doyne. *Brit. med. J.*, **1**, 1106 (1887).
Drake, Hawkes and Warren. *J. Amer. med. Ass.*, **105**, 1340 (1935).
Dunlop and Lemon. *Amer. J. med. Sci.*, **177**, 259 (1929).
Dunn. *Amer. J. Ophthal.*, **9**, 134 (1892).
Elgood. *Brit. med. J.*, **1**, 88 (1909).
Elliott. *Tropical Ophthalmology*, London, 505 (1920).
Ellis and Gentry. *Amer. J. Ophthal.*, **58**, 779 (1964).
Falls and Kertesz. *Trans. Amer. ophthal. Soc.*, **62**, 248 (1964).
Fink and Gay. *Bull. Johns Hopk. Hosp.*, **55**, 280 (1934).
Fötzsch. *Arch. Psychiat. Nervenkr.*, **210**, 169 (1967).
Gassler and Berthold. *Klin. Mbl. Augenheilk.*, **139**, 44 (1961).
Gogina. *Oftal. Zh.*, No. 5, 341 (1964).
Goltz. *Dtsch. med. Wschr.*, **6**, 225 (1880).
Gordon. *J. Amer. med. Ass.*, **131**, 727 (1946).
Green, Koelsche and Kierland. *Ann. Allergy*, **23**, 30 (1965).
Hallett and Mitchell. *Amer. J. Ophthal.*, **65**, 542 (1968).
Handwerck. *Münch. med. Wschr.*, **54**, 2332 (1907).
Hewkley. *Brit. med. J.*, **1**, 1160 (1888).
Hofmann. *Klin. Mbl. Augenheilk.*, **142**, 1039 (1963).
Hoover and Kelley. *Trans. Amer. opththal. Soc.*, **69**, 293 (1971).
Hopkins and Kesten. *Arch. Derm. Syph.* (Chic.), **29**, 358 (1934).
Kafka. *Med. Rec.*, **146**, 441 (1937).
Kalt. *Ann. Oculist.* (Paris), **107**, 41 (1892).
Kinmonth, Taylor, Tracy and Marsh. *Brit. J. Surg.*, **45**, 1 (1957).
Klar. *Klin. Mbl. Augenheilk.*, **85**, 224 (1930).
Koch. *Ber. dtsch. ophthal. Ges.*, **57**, 269 (1951).
Kolle. *Hautarzt*, **14**, 318 (1963).
Kunstadter. *Amer. J. Dis. Child.*, **110**, 559 (1965).
Lawson and Sutherland. *Trans. ophthal. Soc. U.K.*, **26**, 12 (1906).
Löwenstein. *Med. Klin.*, **29**, 9 (1933).
Luescher. *Schweiz. med. Wschr.*, **79**, 1 (1949).
McCarthy and Hoagland. *J. Amer. med. Ass.*, **187**, 153 (1964).
Marinković and Stevanović. *Ann. Derm. Syph.* (Paris), **86**, 281 (1959).
Meige. *Presse méd.*, **2**, 341 (1898).
Rev. neurol. (Paris), **1**, 70 (1933).
Melkersson. *Hygiea*, **90**, 737 (1928).
Michail. *Z. Augenheilk.*, **73**, 337 (1931).
Milroy. *N.Y. med. J.*, **56**, 505 (1892).
J. Amer. med. Ass., **91**, 1172 (1928).
Montresor. *Atti Soc. oftal. ital.*, **12**, 84 (1950).
New and Kirch. *J. Amer. med. Ass.*, **100**, 1230 (1961).
Nonne. *Virchows Arch. path. Anat.*, **125**, 189 (1891).
Pajtás. *Ophthalmologica*, **137**, 35 (1959).
Paton. *Amer. J. Ophthal.*, **59**, 705 (1965).
Phillips, J. M. and Barrows. *Genetics*, **7**, 573 (1922).
Phillips, S. *Brit. med. J.*, **1**, 194 (1895).
Potvin. *Bull. Soc. belge Ophtal.*, No. 72, 68 (1936).
Pruzanski. *J. Amer. med. Ass.*, **147**, 1234 (1951).
Quincke. *Mh. prakt. Derm.*, **1**, 129 (1882).
Richter. *Münch. med. Wschr.*, **74**, 1755 (1927).
Robinow, Johnson and Verhagen. *Amer. J. Dis. Child.*, **119**, 343 (1970).
Robinson. *Brit. med. J.*, **1**, 1006 (1888).
Rosenthal. *Z. ges. Neurol. Psychiat.*, **131**, 475 (1930).
Rossolino. *Neurol. Zbl.*, **20**, 744 (1901).
Sacquépée. *Bull. Soc. méd. Hôp. Paris*, **28**, 639 (1909).
de Schweinitz. *Amer. J. Ophthal.*, **5**, 170 (1888).
Soulas. *Bull. Soc. hellén. Ophtal.*, **22**, 75 (1954).
Spriggs. *Brit. med. J.*, **2**, 1744 (1908).
Stähelin. *Z. klin. Med.*, **49**, 461 (1903).
Steinvorth. *Klin. Mbl. Augenheilk.*, **133**, 105 (1958).
Strebel. *Int. Arch. Allergy*, Suppl. 64, 67 (1952).
Tabbara and Baghdassarian. *Amer. J. Ophthal.*, **73**, 531 (1972).
Toselli, Bertoni and Volpi. *Ann. Ottal.*, **91**, 774 (1965).
Usdan, Bowen, Fischer and Morgan. *Arch. Ophthal.*, **87**, 596 (1972).

Warin. *Brit. J. Derm.*, **72**, 350 (1960).
Weekers and Barac. *Bull. Soc. belge Ophtal.*, No. 73, 29 (1936).
Weiss and King. *Ohio St. med. J.*, **28**, 341 (1932).
Wise. *Year Book of Dermatology*, Chicago, 6 (1936).

DISORDERS OF SECRETION

The Sweat Glands

The sweat glands, interpreted in the widest sense, are of two types—eccrine and apocrine glands.

The ECCRINE SWEAT GLANDS are distributed all over the skin; their function is to secrete a watery fluid which cools the body by evaporation, an activity controlled by the thermo-regulating centre in the hypothalamus through the sympathetic nervous system. Each gland is composed of a basal coil which forms the secretory portion lying in the deeper parts of the dermis or between it and the subcutaneous fat, and a long duct running through the dermis and epidermis to open at a pore on the surface (Fig. 24). The secretory cells form a single layer composed of two types of cell, small dark elements containing mucopolysaccharides and larger clear cells containing glycogen (Montagna *et al.*, 1953). On secreting, both these substances are discharged into the lumen and are propelled to the surface by an outer layer of contractile myo-epithelial cells (Dobson *et al.*, 1958; Hurley and Witkowski, 1962).

The APOCRINE GLANDS, on the other hand, are limited to certain parts of the skin, mainly the axillæ, the groin, the ano-genital region and more sparsely on the face; modifications constitute the mammary glands, the ceruminous glands in the external auditory meatus and the glands of Moll in the eyelids (Fig. 3). The glands represent the odoriferous glands of the lower animals; they commence functional activity at puberty and cease at the climacteric. The secretory portion of the gland is in the form of a coil in the dermis and consists of two layers of cells, an inner of secretory and an outer of myo-epithelial cells. The wide duct runs not to the surface of the skin but into a pilosebaceous follicle. The secretion is milky and viscous and while the process is continuous, discharge is intermittent and is due to contraction of the duct, usually as a result of emotional stimuli (Hurley and Shelley, 1954).

OLIGHIDROSIS AND ANHIDROSIS

A *deficiency* (OLIGHIDROSIS, ὀλίγος, little; ἱδρώς, sweat) or *absence* (ANHIDROSIS) *of the secretion of the sweat glands* of the lids is occasionally of importance since it tends to produce dryness of the skin and disturbance of the vitality of the lashes; as a general disorder it may involve the impairment of heat-regulation and the development of hyperpyrexia. It may occur in a great variety of conditions (Shelley *et al.*, 1950). In some, no morphological abnormalities of the glands are demonstrable: hypothyroidism, alcoholic polyneuritis, sympathetic paralysis, after the administration of drugs blocking acetylcholine such as atropine, and in states of dehydration and heat stroke. More dramatic are those conditions wherein pathological changes in the skin are evident. In certain congenital anomalies such as hereditary anhidrotic ectodermal dysplasia the eccrine sweat glands may be hypoplasic or absent and sometimes the apocrine glands also,[1] while in

[1] Sunderman (1941), Upshaw and Montgomery (1949), Korting and Salfeld (1961), Greither and Tritsch (1963).

congenital ichthyosis the sweat glands and the sebaceous glands are atrophied (Tendlau, 1902). A diminution or lack of sweating may also occur in certain chronic skin diseases, particularly those involving scarring, such as some forms of the ichthyosiform dermatoses, scleroderma, lichen sclerosus et atrophicus, angiokeratoma corporis diffusum of Fabry, progressive facial hemiatrophy, senile atrophy of the skin and in radio-dermatitis.

Plugging of the ducts of the eccrine glands and a consequent diminution of sweating may occur in conditions of hyperkeratosis, as in psoriasis or atopic eczema. In miliaria such a condition is marked when the ducts become blocked by keratotic plugs. When the block occurs at the orifices of the ducts (*miliaria crystallina*) crops of minute vesicles appear containing sweat usually during an acute febrile illness; the condition is asymptomatic and tends to disappear after some days. When the blockage occurs within the epidermis crops of discrete erythematous papulo-vesicles appear which may become pustular (*miliaria rubra: prickly heat*). The condition occurs in people from temperate climates while staying in the tropics and resolves in a cool environment. Anhidrosis is present in the areas involved during an attack and sometimes persists.

Local amelioration in conditions of olighidrosis may be gained by attempting to replace the natural secretions by suitable applications such as equal parts of glycerin and lanolin with 2% salicylic acid, but cure is unknown.

HYPERHIDROSIS

An *over-activity of the sweat glands* of the lids is usually of nervous origin. Ætiologically two types can be differentiated. The more obvious follows an *irritative nervous lesion* somewhere in the sympathetic supply, either centrally

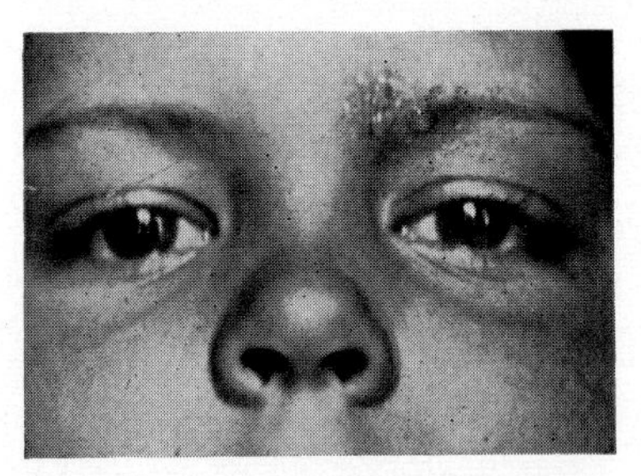

FIG. 16.—LOCALIZED HYPERHIDROSIS.

In a patch on the medial part of the left eyebrow; the cause was undetermined (R. L. and R. L. Sutton).

in the brain-stem or cord, or peripherally in the cervical sympathetic or in the facial or trigeminal nerves which carry the secretory fibres. Hypersecretion may thus be associated with hemiplegia, epilepsy, meningitis, tabes, syringomyelia, familial dysautonomia,[1] cerebral traumatism or tumour, affections of the thorax or neck, or peripheral lesions of the Vth or

[1] Vol. III, p. 914.

VIIth nerves. This type of hyperhidrosis typically affects one side of the face. Less clearly understood, but possibly frequently due to sympathetic disturbance, is the *idiopathic* variety which usually attacks young people and may be familial and congenital. It frequently has a localized patchy distribution affecting, for example, a relatively small area underneath the eyebrows, whereon drops of sweat constantly appear on very slight physical or emotional stimuli (von Graefe, 1858; Sutton, 1912; Agnello, 1947) (Fig. 16). The disease is of more interest than importance, but if the secretion is profuse the lid may become macerated or eczematous.

In these cases, *treatment* by frequent washing followed by astringents such as 1% formalin or alum may keep the tendency to hypersecretion in check, an injection of atropine gives temporary amelioration, while radiation with X-rays may inhibit it, at any rate for a considerable time (Pirie, 1936; and others) but the depth at which the sweat glands lie make their destruction difficult without injuring the overlying skin. In the most extreme cases resort may be had to sympathetic ganglionectomy to relieve the condition: for hyperhidrosis of the head and neck alcoholic injection or removal of the stellate ganglion is the most satisfactory procedure (see Pearl and Shapiro, 1935; Adson, 1936; Haxton, 1948).

Vörner (1907) described an interesting case wherein a congenital hyperhidrosis of the lids became converted into an anhidrosis during an attack of pneumonia, to revert to the original condition on convalescence.

HÆMATIDROSIS

In the very rare condition of *bloody sweat*, blood is mixed with the normal secretion of the sweat-glands. This has been said to have appeared on the lids as red drops in fevers, emotional states or at the menstrual periods in nervously unstable subjects (Hebra and Kaposi, 1872), in epileptics (Messedaglia and Lombroso, 1869), or most commonly in purpura.

CHROMIDROSIS

CHROMIDROSIS (χρῶμα, colour; ἱδρώς, sweat) (CHROMOCRINIA, Le Roy de Méricourt, 1861; MELANIDROSIS; BLEPHARO-MELÆNA), *the secretion of blue or black sweat,* is another rarity characterized by the appearance of blue-black spots in the axillæ or on the face including the lids (Fig. 17); occasionally the colouring is uniform, producing a lacquer-like surface. The lower lids are preferentially affected, and here the intensity of colouring is always greater if all four lids are involved; only exceptionally is the upper lid affected first (Blanchard and Maillard, 1907), and only rarely does the condition spread farther afield to the nose and face. The coloration appears usually with profuse perspiration at intervals over a period varying from some months to some years, but occasionally remains permanently over this period. The colour can be removed by wiping with oily applications, but it returns, sometimes after some minutes, sometimes after some hours, and occasionally only after several days.

The condition was first described by James Yonge (1709) and Le Roy de Méricourt and Robin (1863) contributed an elaborate study analyzing the 22 cases reported up to that time. There are some curious features in its occurrence. It practically always affects women, and dwellers near the sea have provided the majority of cases, especially in particular localities of England, Ireland and France, such as Plymouth, Dublin and Brest[1]; it has also been noted in America (Mitchell, 1898; Heidingsfeld, 1902) and Japan (Matsuhashi, 1920), but the references in the literature, plentiful in the middle of

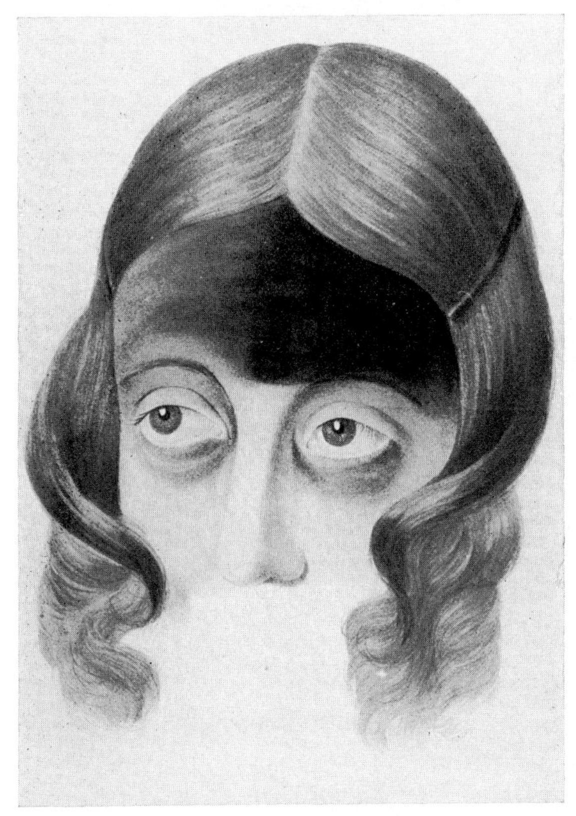

FIG. 17.—CHROMIDROSIS.

The classical illustration of Teevan of a girl aged 15. The pigmentation was shown not to be artificial; it appeared under observation 5 hours after the forehead had been wiped clean.

last century, have largely disappeared. Bonnet (1952) recorded its occurrence after zoster and following an injury to the ophthalmic nerve. Finally, it is quite clear that many of the cases are properly *pseudo-chromidrotic*, the colouring being due to arte-facts, particularly cosmetics such as carbon and indigo, which on occasion have been applied with the greatest persistence for years in order to simulate the appearance of disease (von Graefe, 1864; Rothmund, 1866; Wilhelmi, 1880) or simply to dramatize the appearance. Certain cases, however, have been so thoroughly and independently investigated under controlled conditions that they must be accepted as genuine

[1] Banks (1858), Lyons (1858), Godefroy (1863), Moerloose (1864), Warlomont (1864), Le Roy de Méricourt (1861–64), and others.

(Teevan, 1845). A useful dodge was suggested by Spring of Liège (1861) to differentiate the true from the self-induced chromidrosis: he painted the lids with collodion and accepted as genuine those cases wherein the spots of colour appeared below this layer.

A considerable amount of speculation has been devoted to the nature of the pigmentation in the genuine cases; but Shelley and Hurley (1954), in a case occurring on the cheek, showed that the secretion was derived from the apocrine glands which contained granules of lipofuscin. The secretion thus occurs only between puberty and the climacteric and appears as a result of emotional stimuli, epinephrine or mechanical irritation of the myo-epithelium of the gland. The coloration is due to the different states of oxidation of the

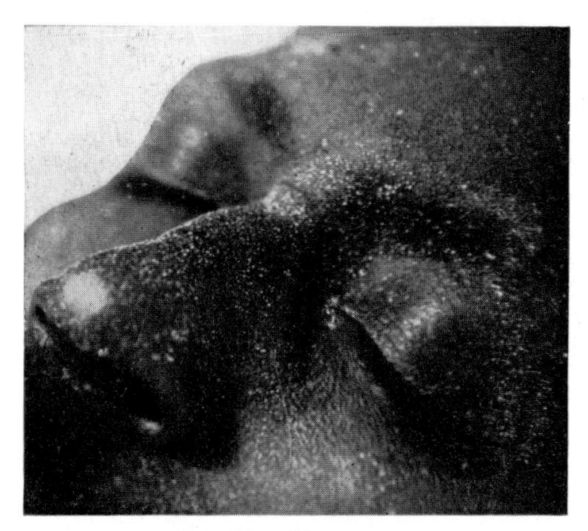

Fig. 18.—Uridrosis.
In a case of uræmia (R. L. and R. L. Sutton).

lipofuscin, the more this occurs the darker the colour. Treatment consists merely in wiping away the spots with oil and treating any underlying cause such as anæmia or hysteria.

Other cases occur wherein the sweat becomes coloured, usually red or yellow, due to contamination with pigment formed by various chromogenic bacteria, especially corynebacteria, or fungi (Trommsdorff, 1904).

URIDROSIS is an extremely rare condition sometimes affecting the eyelids, in which urea is excreted in excessive quantity by the sweat glands in such conditions as uræmia or cholera; when the sweat evaporates a hoar-frost-like deposit of crystals or powdery masses of sodium chloride and urea are left on the skin (Büttner and Robbers, 1935) (Fig. 18).

RETENTION CYSTS of the sweat glands will be noted subsequently.[1]

TUMOURS and CYSTS of the sweat glands and their homologues (Moll's glands) will also be considered subsequently.[2]

[1] p. 391. [2] pp. 394, 458.

Adson. *J. Amer. med. Ass.*, **106,** 360 (1936).
Agnello. *Riv. oto-neuro-oftal.*, **22,** 222 (1947).
Banks. *Dubl. quart. J. med. Sci.*, **25,** 257 (1858).
Blanchard and Maillard. *Bull. Acad. Méd.* (Paris), **58,** 527 (1907).
Bonnet. *Bull. Soc. Ophtal. Fr.*, 109 (1952).
Büttner and Robbers. *Klin. Wschr.*, **14,** 372 (1935).
Dobson, Formisano, Lobitz and Brophy. *J. invest. Derm.*, **31,** 147 (1958).
Godefroy. *Ann. Oculist.* (Paris), **50,** 172 (1863).
von Graefe. *v. Graefes Arch. Ophthal.*, **4** (2), 254 (1858).
Klin. Mbl. Augenheilk., **2,** 386 (1864).
Greither and Tritsch. *Arch. klin. exp. Derm.*, **216,** 50 (1963).
Haxton. *Brit. med. J.*, **1,** 636 (1948).
Hebra and Kaposi. *Virchows Hb. d. spec. Path. Therap.*, 2nd ed., Erlangen, **3** (1), 78 (1872).
Heidingsfeld. *J. Amer. med. Ass.*, **39,** 1519 (1902).
Hurley and Shelley. *J. invest. Derm.*, **22,** 143 (1954).
Hurley and Witkowski. *J. invest. Derm.*, **39,** 329 (1962).
Korting and Salfeld. *Derm. Wschr.*, **144,** 1141 (1961).
Le Roy de Méricourt. *Bull. Acad. Méd.* (Paris), **26,** 773 (1861).
Mémoir sur la chromidrose, Paris (1864).
Le Roy de Méricourt and Robin. *Ann. Oculist.* (Paris), **50,** 5, 110, 267 (1863).
Lyons. *Dubl. Hosp. Gaz.*, **5,** 147 (1858).
Matsuhashi. *Jap. J. Derm. Urol.*, **20,** 296 (1920).
Messedaglia and Lombroso. *Riv. clin. di Bologna:* Ref. in *Virchow-Hirschs Jber.*, **2,** 551 (1869).
Mitchell. *Phila. med. J.*, **1,** 117 (1898).
Moerloose. *Ann. Oculist.* (Paris), **52,** 205 (1864).
Montagna, Chase and Lobitz. *J. invest. Derm.*, **20,** 415 (1953).
Pearl and Shapiro. *Ann. Surg.*, **102,** 16 (1935).
Pirie. *Canad. med. Ass. J.*, **34,** 301 (1936).
Rothmund. *Klin. Mbl. Augenheilk.*, **4,** 103 (1866).
Shelley and Hurley. *Arch. Derm. Syph.* (Chic.), **69,** 449 (1954).
Shelley, Horvath and Pillsbury. *Medicine* (Balt.), **29,** 195 (1950).
Spring (1861). *See* Le Roy de Méricourt and Robin. *Ann. Oculist.* (Paris), **50,** 22 (1863).
Sunderman. *Arch. intern. Med.*, **67,** 846 (1941).
Sutton. *J. Amer. med. Ass.*, **59,** 1193 (1912).
Teevan. *Med.-chir. Trans.*, **10,** 611 (1845).
Tendlau. *Virchows Arch. path. Anat.*, **167,** 465 (1902).
Trommsdorff. *Münch. med. Wschr.*, **51,** 1285 (1904).
Upshaw and Montgomery. *Arch. Derm. Syph.* (Chic.), **60,** 1170 (1949).
Vörner. *Dtsch. med. Wschr.*, **33,** 2090 (1907).
Warlomont. *Ann. Oculist.* (Paris), **52,** 97 (1864).
Wilhelmi. *Klin. Mbl. Augenheilk.*, **18,** 252 (1880).
Yonge. *Phil. Trans.*, **26,** 424 (1709).

The Sebaceous Glands

SEBACEOUS GLANDS are alveolar holocrine glands without a lumen, their secretion being formed by the decomposition of their cells. They are present in the skin of the lateral parts of the lids and are also of ophthalmological importance in their meibomian modifications in the tarsus. In the medial halves of the lids, however, the glands are replaced by single cells secreting sebum lying in the deepest layer of the epithelium (Wolff, 1951). A typical sebaceous gland consists of several lobules, each without a lumen and each evacuating its secretion through a sebaceous duct into a pilosebaceous follicle (Fig. 24). The alveolus is surrounded by a fibrous capsule, a membrana propria and an epithelial layer, each continuous with the corresponding layers of a hair follicle, while the centre of the alveolus is filled with indistinctly staining polyhedral cells containing a mixture of lipids, which are continuously renewed by the proliferating basal elements. There is no autonomic nerve supply and the function of the glands depends on endocrine stimuli, but autonomic fibres supply the hair follicles below the entry of the sebaceous glands.

The sebum contains esterified cholesterol but no free cholesterol. Small amounts of phospholipids are present and also triglycerides and waxes and squalene, an acyclic triterpene. The free fatty acids found in the surface lipids are probably formed either by the action of lipases or by rancidification (Montagna, 1962).

Excellent and detailed studies of the anatomy and physiology of these glands have been written by Way (1931) and Suskind (1951).

ASTEATOSIS (*xerosis*), a condition characterized by a deficiency of sebum, may be primary, associated with atrophic conditions of the skin of the lids, or secondary and symptomatic (senile, post-inflammatory, myxœdematous, etc.); the skin is dry and hard and may become scaly and fissured. It is of little ophthalmological importance and is treated symptomatically by the application of lubricant creams (yellow paraffin and lanolin, almond oil, etc.).

SEBORRHŒA (*sebum*, grease; ῥεῖν, to flow), *a functional overactivity of the sebaceous glands together with the eccrine and apocrine glands resulting in an abnormally greasy skin*, is of unusual interest in ophthalmology for the condition predisposes to practically every other skin disease, and blepharitis is a frequent accompaniment. It has a hereditary predisposition multifactorial in type and, in so far as the sebaceous glands are controlled primarily by the sex hormones, particularly the androgens and œstrogens, the cycle of sexual evolution would seem to be of importance in its incidence, the affection appearing just before puberty and subsiding with age; the greaseless skin of the infant and the withered skin of the aged contrast forcibly with the oily skin of the adolescent and the soft turgid skin of the adult. The nose, scalp, chest and back are preferentially affected, whereon the excessive production of oil causes the skin to assume a shiny, greasy appearance with coarse, widely dilated follicular orifices many of which contain comedones. Sometimes, in addition to the oiliness, there is a habitual accumulation of scales and crusts, particularly in the scalp, eyebrows and lashes (*dandruff*) and, as so often happens, the delicate skin of the lids frequently shows an exaggerated reaction characterized sometimes by a fine, bran-like scaliness and often by a puffy, œdematous swelling.

The affection is functional and unaccompanied by inflammatory changes; the normal sebaceous secretion is increased but no abnormalities in the composition of the sebum have been conclusively demonstrated (Moschella, 1971). The consequence is that organisms flourish which are normally saprophytic on the skin, particularly the pityrosporon, the acne bacillus and the staphylococcus, while the liability to the invasion of parasites is increased, rendering the individual prone not only to typical seborrhœic dermatitis but to most forms of dermatoses.

The first essential of *treatment* should theoretically be the control of the underlying metabolic condition and, since we are unaware of its nature, it is understandable that permanent cure is a rarity; therapeusis must depend on persistent and frequently disappointing measures of alleviation. It must be remembered that in established cases if the outlook for cure is bad, that which follows neglect is worse. Among these measures, regulation of the diet used to be advocated, with the curtailment of carbohydrates, fats and alcohol and an increase in proteins with an abundance of salads and fresh fruit (Barber, 1929–34; Sabouraud, 1936); there is, however, no substantial evidence to support this view. Although the condition is associated with endocrine function and sexual maturity, in spite of much research and experimentation no clearly cut relationship has been established and no pragmatic therapeutic results have emerged; all attempts at endocrine treatment have on the whole given results which are unsatisfactory, irregular and unpredictable (Cohen, 1941; Ingram, 1957–59).

Among the local applications none is specific for the seborrhœic state. Sulphur[1] is the most useful, either as a powder, a lotion or a soap; this not only diminishes the sebaceous secretion but also inhibits the growth of micro-organisms therein. The addition of a penetrant wetting agent may enhance the activity of such preparations (MacKee *et al.*, 1945), and after sulphur has been applied for some 10 days and the skin becomes tight and drawn, a glycerin and salicylic acid ointment is most effective. Sometimes alcoholic solutions of resorcinol (1 to 10%) are more efficient than sulphur, and selenium sulphide (2·5%) has also been used. Shampoos containing resorcinol, sulphur, salicylic acid or hexachlorophene are indicated in the treatment of the head.

ACNE VULGARIS

If the seborrhœic state has become well developed, a pilosebaceous unit may become blocked by hyperkeratotic plugging of the orifice and inspissated secretion to form the well-known *comedo* (Lat., a glutton) or " black-head ", crops of which constitute the condition of ACNE VULGARIS (Fig. 19). In this condition the production of sebum is usually increased

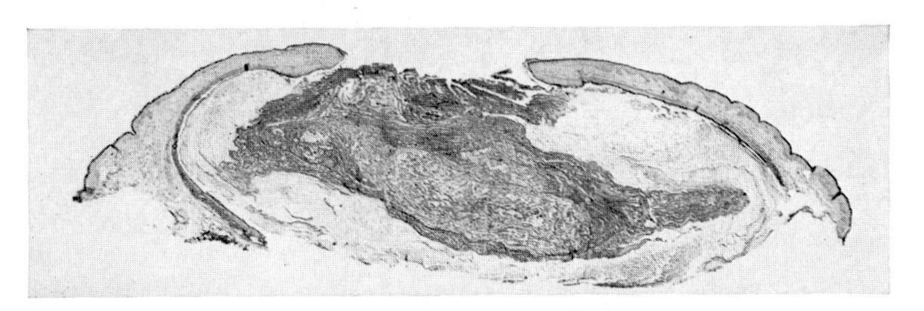

FIG. 19.—COMEDO OF THE EYELID.

Owing to the occlusion of the follicular orifice, keratotic debris and inspissated secretion is retained (L. E. Zimmerman).

but, although the composition of the fatty acids remains essentially unaltered (Pochi and Strauss, 1964), there is evidence that the squalene and wax and steroid esters are increased (Cotterill *et al.*, 1972); its retention is aided by intrafollicular hyperkeratosis (Van Scott and McCardle, 1956). An indurated inflammation may occur round the comedones (ACNE INDURATA) composed of a predominantly lymphocytic perifollicular infiltrate of the nature of a foreign-body tissue-reaction to the lipids of the comedo.

[1] Sulphur and salicylic ointment, 2%; as a lotion, sulphurated potash 4·6, zinc sulphate 4·6, camphor water to 100; as a paste, sulphur 6, resorcin 6, zinc oxide 12, hydrous wool fat 12, soft paraffin to 100.

Sometimes suppuration may lead to a superficial cutaneous pustule or abscess (ACNE PUSTULOSA); this occurs after the follicular wall has ruptured (Strauss and Kligman, 1958–60). The infiltration now consists of neutrophils, lymphocytes, plasma cells and histiocytes with giant cells formed as a reaction to particles of keratin (Fig. 20), and the lesion leaves a permanent pitted scar behind.

The sebaceous material is usually packed with the *Corynebacterium acnes* (Kirschbaum and Kligman, 1963; Marples and Izumi, 1970) associated with the

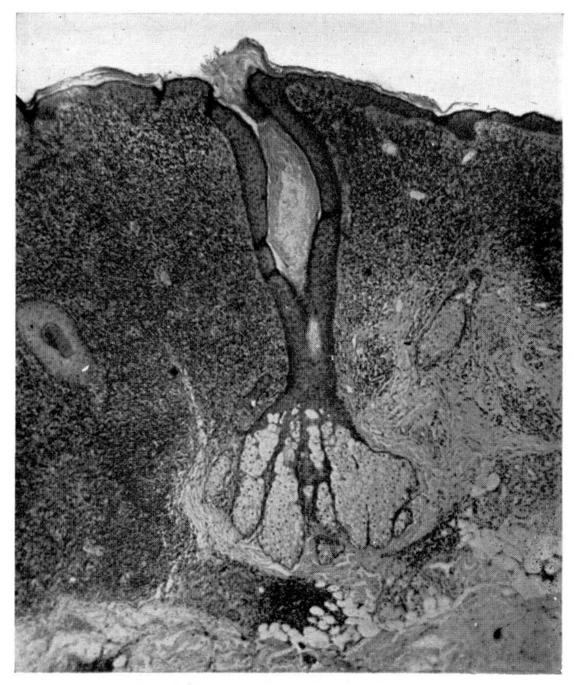

FIG. 20.—PUSTULAR ACNE.

The ostium of the hair follicle is plugged with a mass of horny cells and inspissated sebum. The surrounding dermis shows an intense acute inflammatory reaction (H. and E. × 30) (G. H. Percival).

Staphylococcus epidermidis albus (Strauss and Kligman, 1960), an organism which becomes particularly evident if suppuration occurs, while an acarian parasite, the *Demodex folliculorum*,[1] may inhabit the mouth of the follicle. Sabouraud (1936) considered that the *C. acnes* caused the excessive secretion of sebum, but more probably it flourishes where the excessive secretion exists, although it may contribute to its retention by exciting the laminated epithelial proliferation at the mouth of the gland.

The malady begins typically at or after puberty under the stimulus of androgens secreted by the testes in males and the adrenal cortex in females.

[1] p. 226.

Comedones show no individualistic features on the lids. In young people they do not occur on the palpebral parts of the lids, but appear there in the aged, their formation being due not so much to oversecretion as to retention of sebum owing to atrophy of the muscles of the hair-follicles (Csillag, 1929). In adolescents they may form large colonies on the side of the nose and on the inner half of the upper lid near the brows, usually distributed symmetrically on both sides. The condition is obstinate, not readily amenable to treatment, but fortunately tends to disappear spontaneously with the passage of time, usually about the age of 25; it does occur, although rarely, after 30 years of age (Cohen, 1945; Forbes, 1946); a rare juvenile form may occur in young children. There is evidence of a hereditary predisposition (Hellier, 1939), particularly in the juvenile type (Schleicher, 1966), and there is no sex discrimination.

Treatment should consist of the expression of the comedones to get rid of the depots of lipids before they excite a reaction. This can be done by an extractor (such as a narrow glass tube with thick edges pressed steadily round the comedo) when a cocoon-like sebaceous plug is squeezed out, cream in colour except for the portion filling the mouth of the follicle which is black owing to the presence of a dark pigment, an oxidation product of keratin. In the event of suppuration, incision should be resorted to only when necessary as the artificial scar is apt to be more obvious than the natural. A useful preparation is sulphur as a lotion, an ointment or a paste,[1] but good results have been claimed with vitamin A-acid (Plewig and Braun-Falco, 1971) or topical antibiotics[2] (Cunliffe, 1973). Small doses of X-rays may prevent recurrences by producing atrophy of the sebaceous follicles but is erratic in its effects (Way, 1960); ultraviolet radiation is said to be effective. Systemic tetracyclines which decrease the concentration of free fatty acids may be of value (Freinkel *et al.*, 1965). Œstrogens lower the formation of sebum when given in sufficient amounts and have been used cyclically in females but are contra-indicated in males; in the dosage required they are usually inadvisable. On the whole, however, treatment is difficult and unrewarding (Mills and Kligman, 1972; Hall-Smith and Marks, 1973).

ROSACEA, a condition associated with a paralytic distension of the superficial vessels of the skin with a secondary hypertrophy of the sebaceous glands, affects the lid-margins rather than the lids themselves; it will therefore be discussed later.[3]

The enormous hypertrophy of the sebaceous glands associated with rosacea and hypertrophy of the connective tissue which may occur on the nose, known as RHINOPHYMA, may spread to the lower lid (Fig. 21). The condition is symptomless and treatment, if indicated for cosmetic reasons, is by plastic surgery.

In SENILE SEBACEOUS HYPERPLASIA (SENILE SEBACEOUS NÆVUS) one sebaceous gland is involved; it consists of greatly hypertrophied lobules grouped around one or several ducts (Gilman, 1937); the condition is a hyperplasia and not an adenoma (Ramos e Silva and Portugal, 1953; Braun-Falco and Thianprasit, 1965). It forms

[1] p. 30. [2] Neomycin, chloramphenicol, etc. [3] p. 221.

a soft, elevated, yellowish nodule, slightly umbilicated at the exit of the duct, occurring on the face, especially the forehead.

The eczema-like inflammatory condition of SEBORRHŒIC DERMATITIS,[1] as well as MILIA[2] and SEBACEOUS CYSTS[3] in the lids will be noted later.

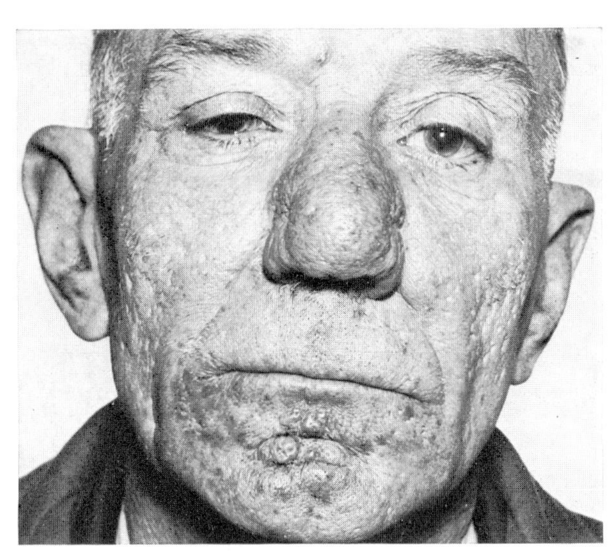

FIG. 21.—ROSACEA WITH RHINOPHYMA.
In a man aged 56 (Inst. Ophthal.).

A HYPERSECRETION OR SEBORRHŒA OF THE MEIBOMIAN (TARSAL) GLANDS is a not uncommon condition, producing as its main symptom the deposition along the lid-margins of white, frothy meibomian secretion which tends to collect at the canthi particularly in the morning (" sleep "). The tendency may be associated with the seborrhœic state, sharing in its general ætiology, in which case it is usually most obvious at puberty and sometimes at the climacteric (Krückmann, 1922), or it may be an isolated and local seborrhœa found particularly in persons of middle or advanced age (Gifford, 1921; Cowper, 1922; Linksz, 1942); in the latter case the phenomenon may be one of atonic passive retention rather than hypersecretion. The secretion, instead of being semi-solid and of cream-cheese consistency, becomes grumous, yellow and oily, somewhat resembling pus, and may be so profuse as to cause a blurring of vision and polyopia (Hollenhorst and Dyer, 1962). In severe cases the glands become hyperplasic and the tarsal plate swollen, while a seborrhœic squamous blepharitis and a chronic *conjunctivitis meibomiana*[4] may result (Elschnig, 1908; Wirth, 1924); in most instances, however, the condition is probably aggravated by chronic infection (*chronic meibomitis,*

[1] p. 267. [2] p. 395.
[3] p. 395. [4] Vol. VIII, p. 80.

Gifford, 1921).[1] Treatment of the seborrhœa, apart from measures directed against the general constitutional condition, should consist of alkaline lotions (sod. bicarbonate) and frequent expression of the glands by massage; this is done either by pressing on the skin surface of both everted lids approximated together with the fingers, or massaging the conjunctival surface of the lids with a glass rod, milking the tarsal glands against a finger laid on the skin surface (Fig. 22). It is astonishing how much secretion of an oily or fatty, pus-like nature can be expelled in this way and how much relief its removal brings to the patient. In the most severe cases the tarsus may be split into two leaves and scraped with a spoon (Filatov, 1922). Ointments, particularly with a base of soft paraffin, are contra-indicated since their imperviousness dams back the secretion; a bismuth subnitrate cream[2] is usually well tolerated, or a sulphur preparation[3] may be indicated.

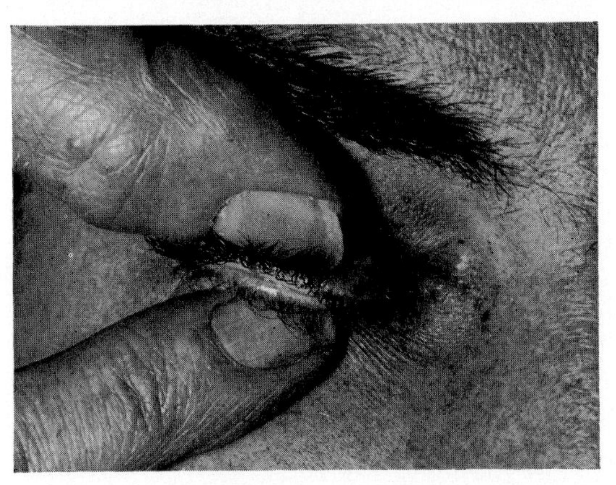

FIG. 22.—MASSAGE OF THE LIDS TO EXPRESS THE SECRETION OF THE MEIBOMIAN GLANDS.

Retention of this meibomian secretion may produce a condition of ACNE OF THE MEIBOMIAN GLANDS or COMEDO OF A GLAND. Further retention may lead to inspissation of the secretion as elongated yellow deposits shining through the palpebral conjunctiva (*meibomian "infarcts"*) seen particularly in old people in whom no other signs of seborrhœa occur. As a rule no symptoms arise therefrom, but if they are transformed by the deposition of lime salts into hard masses (*meibomian lithiasis*; λίθος, a stone), they may break through the conjunctiva and, abrading the cornea, cause considerable irritation and even ulceration. Such masses should be removed with a curette after incision.

These meibomian infarcts are not to be confused with the smaller and more superficial concretions (*lithiasis conjunctivæ*) which develop in Henle's glands and glands of new formation in the conjunctiva[4].

[1] p. 249. [2] Zinc oxide 1, bismuth subnitrate 1, ung. aquæ rosæ 10.
[3] Sulph. præcip. 1; adipis benzoin 10. [4] Vol. VIII, p. 585.

A rare type of secretory anomaly of the meibomian glands was described by Leibiger (1961) occurring in a middle-aged man. Cyst-like lumps appeared in the lids from which a greenish comedo-thread of sebaceous secretion almost 3 cm. long could be expressed. The remaining glands also yielded similar although shorter filaments; two such episodes with an interval of 6 years were observed. Bacterial examination was negative and anatomical examination eliminated a chalazion.

Barber. *Trans. med. Soc. Lond.*, **52,** 165 (1929).

Trans. ophthal. Soc. U.K., **54,** 426 (1934).

Braun-Falco and Thianprasit. *Arch. klin. exp. Derm.*, **221,** 207 (1965).

Cohen. *Brit. J. Derm.*, **53,** 231, 269 (1941); **57,** 10 (1945).

Cotterill, Cunliffe, Williamson and Bulusu. *Brit. med. J.*, **3,** 444 (1972).

Cowper. *Amer. J. Ophthal.*, **5,** 25 (1922).

Csillag. *Derm. Wschr.*, **88,** 609 (1929).

Cunliffe. *Brit. med. J.*, **4,** 667 (1973).

Elschnig. *Dtsch. med. Wschr.*, **34,** 1133 (1908).

Filatov. *Klin. Mbl. Augenheilk.*, **69,** 657 (1922)

Forbes. *Brit. J. Derm.*, **58,** 298 (1946).

Freinkel, Strauss, Yip and Pochi. *New Engl. J. Med.*, **273,** 850 (1965).

Gifford. *Amer. J. Ophthal.*, **4,** 489, 566 (1921).

Gilman. *Arch. Derm. Syph.* (Chic.), **35,** 633 (1937).

Hall-Smith and Marks. *Brit. J. Hosp. Med.*, **10,** 410 (1973).

Hellier. *Brit. J. Derm.*, **51,** 109 (1939).

Hollenhorst and Dyer. *Trans. Amer. Acad. Ophthal.*, **66,** 501 (1962).

Ingram. *Arch. Derm.*, **76,** 157 (1957).

Brit. med. J., **2,** 1167 (1959).

Kirschbaum and Kligman. *Arch. Derm.*, **88,** 832 (1963).

Krückmann. *Arch. Augenheilk.*, **91,** 167 (1922).

Leibiger. *Klin. Mbl. Augenheilk.*, **138,** 876 (1961).

Linksz. *Arch. Ophthal.*, **28,** 959 (1942).

MacKee, Wachtel, Karp and Herrmann. *J. invest. Derm.*, **6,** 309 (1945).

Marples and Izumi. *J. invest. Derm.*, **54,** 252 (1970).

Mills and Kligman. *Brit. J. Derm.*, **86,** 620 (1972).

Montagna. *The Structure and Function of the Skin*, 2nd ed., N.Y. (1962).

Moschella. *Dermatology in General Medicine* (ed. Fitzpatrick *et al.*), N.Y., 717 (1971).

Plewig and Braun-Falco. *Hautarzt*, **22,** 341 (1971).

Pochi and Strauss. *J. invest. Derm.*, **43,** 383 (1964).

Ramos e Silva and Portugal. *Ann. Derm. Syph.* (Paris), **80,** 121 (1953).

Sabouraud. *Nouvelle pratique dermatologique* (ed. Darier *et al.*), Paris, **7,** 1 (1936).

Schleicher. *Derm. Mschr.*, **155,** 909 (1966).

Strauss and Kligman. *J. invest. Derm.*, **30,** 51 (1958).

Arch. Derm., **82,** 779 (1960).

Suskind. *J. invest. Derm.*, **17,** 37 (1951).

Van Scott and McCardle. *J. invest. Derm.*, **27,** 405 (1956).

Way. *Arch. Derm. Syph.* (Chic.), **24,** 353 (1931).

Arch. Derm., **81,** 141 (1960).

Wirth. *Arch. Augenheilk.*, **94,** 73 (1924).

Wolff. *Brit. J. Derm.*, **63,** 296 (1951).

FIG. 23.—FERDINAND HEBRA
[1816–1880].
From a lithograph by Edward Kaiser (1856); Wellcome Institute of the History of Medicine.

CHAPTER II

INFLAMMATIONS OF THE LIDS

THIS Chapter is suitably introduced by the likeness of one of the brilliant figures who made the Medical School of Vienna the centre of medical progress in the middle of the nineteenth century. Much of our study of the diseases of the lids concerns affections of the skin. The science of modern dermatology may be said to date from the work of the Yorkshireman, Robert Willan [1757–1812] (Fig. 235), who classified cutaneous diseases according to their clinical appearances—papular, squamous, exanthematous, bullous, and so on, a classification which is still to some extent alive. Thereafter, the French school, led by Jean-Louis Alibert [1768–1837], was dominated by the concept of the humoral diatheses. Subsequently in Vienna FERDINAND VON HEBRA [1816–1880] (Fig. 23) founded the histological school of dermatology, whereby the classification of diseases of the skin was based on their pathological anatomy.[1] During his life his influence was immense, and his lectures, full of genial humour, made his clinic one of the most popular in Vienna; and after his death his reputation as a dermatologist has been justly augmented and maintained. His work was extended and to some extent completed by his son, Hans von Hebra [1847–1902] in Vienna who wrote a classical text-book on the systemic aspects of cutaneous diseases.

Owing to the differences in their clinical significance it is convenient to discuss inflammatory conditions in the lids separately as they affect three different regions—the superficial structures of the lids, the lid-margins and the tarsus.

General Pathology

SUPERFICIAL INFLAMMATIONS

In general terms an inflammation of the superficial tissues of the lids may start in the epidermis or in the dermis and subcutaneous tissues; the first type is usually conditioned by external irritants—physical, mechanical, chemical or bacterial; the second is due to the introduction of irritants, toxic or bacterial, into the deeper tissues either locally through the epidermis by a fissure or puncture, or alternatively, by way of the lymph or blood-stream. The pathological changes which constitute the inflammatory reaction are essentially similar in the skin to those occurring in any other organ, but the peculiar structure of this tissue involves certain modifications which must be studied here.

A note on the micro-anatomy of the skin may not be out of place here as a knowledge of it is essential to understand pathological processes (Figs. 24–5). It is composed of two layers—an ectodermal investment of epidermis overlying a mesodermal corium or dermis—the interface between the two being not regular but marked by innumerable papillæ, wherein finger-like projections of the corium fit into corresponding depressions on the deep surface of the epidermis. The *epidermis* in the lids consists of four

[1] Hebra. *Z. kön. Ges. Aerzte Wien*, **1**, 34, 142, 211 (1845).

layers, three of them derived from the deep basal layer of palisaded cells (stratum germinativum), and represent various phases of transformation into horny cells—the stratum spinosum (or rete malpighii) of soft polyhedral cells bound together by fine fibrils (prickle cells), the stratum granulosum of granular lozenge-shaped cells, and finally the protective stratum corneum with its keratinized cells without nuclei or protoplasm. The *dermis* or *corium* is composed essentially of white fibrous tissue running in the main

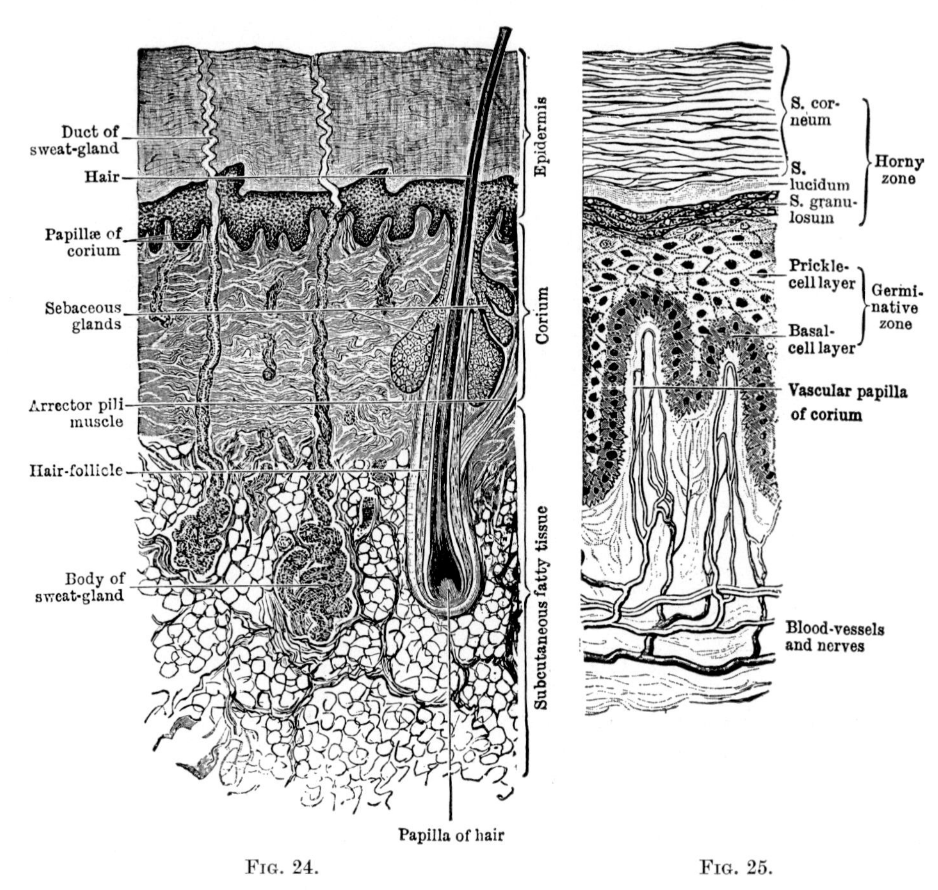

Fig. 24. Fig. 25.

Fig. 24.—*General Structure of the Skin,* showing particularly the relations of the sweat glands and hair follicles.

Fig. 25.—*Intimate Structure of the Skin,* showing the 5 layers of the typical epidermis and the arrangement of the papillæ of the corium. In the eyelids the stratum lucidum does not occur (Cunningham's *Textbook of Anatomy,* Oxford Univ. Press).

parallel to the surface bound together by abundant elastic fibres. In this are found the vessels forming the superficial subcutaneous plexus; from it vessels as well as nerves run up into the papillæ; dipping down into it from the epidermis are found the epithelial appendages—hair follicles, sweat and sebaceous glands. The anatomy of the integument at the lid-margin and the conjunctiva has already been described in a previous Volume.[1]

[1] Vol. II, pp. 523, 541.

The essential inflammatory response to any irritant occurs in the corium, where the two components—the vasodilatation with its associated serous or cellular exudate and the proliferation of the fixed cells—are identical to the reaction in other organs. In the epidermis the changes are secondary to events in the deeper tissues, but they differ as they affect the inert horny cells of the stratum corneum or the metabolically active prickle cells.

In the stratum corneum four types of change may occur:

(1) *Hyperkeratosis* or excessive cornification, in which, owing to stimulation of the basal (germinal) layer, an excessive number of cells is formed and their cohesion increased, so that the horny layer becomes hard, thickened and consolidated.

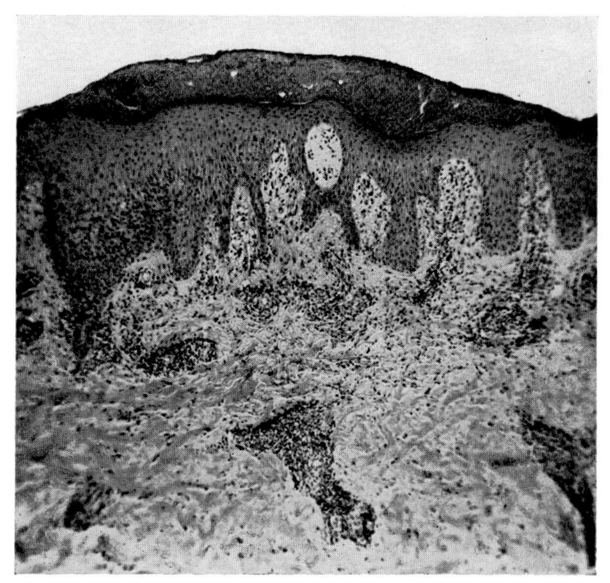

FIG. 26.—ACANTHOSIS IN A CASE OF NUMMULAR INFECTIVE ECZEMA.

On the surface of the epithelium, which shows considerable acanthosis, there is a mass of dried exudate. The dermis shows perivascular lymphocytic infiltration (H. and E. × 50) (G. H. Percival, *Atlas of Histopathology of the Skin*, Livingstone).

(2) *Parakeratosis*, characterized by an excessive formation of cells in the horny layer which show imperfect cornification, a retention of their nuclei, and become loosely united with an absence of the granular layer. This is usually caused by intracellular œdema in the underlying prickle cells, and results in an acceleration of the normal process of desquamation, so that the superficial cells form friable scales which are readily exfoliated in masses ranging in size from minute furfuraceous fragments to large sheets of horny cells. The process is essentially a superficial catarrh, well exemplified in such conditions as squamous eczema, psoriasis, or seborrhœic dermatitis.

(3) *Dyskeratosis*, a rarer change, wherein anomalies occur in the process of keratinization of individual epidermal cells, as is seen, for example, in the transformation of horny cells into molluscum bodies.

(4) *Atrophy*, characterized by diminished cornification and a diminished formation of cells.

In the deeper layers of prickle cells the essential changes are three:

(1) *Hypertrophy* (or *acanthosis*, ἄκανθα, a prickle), wherein the whole thickness of the cellular epidermis becomes increased (Fig. 26). It is due to irritation of the basal germinal layer, insufficient to cause a degeneration on the one hand or a neoplasic proliferation on the other. As the basal cells proliferate, the daughter cells do not become transformed into horny cells but, like the basal cells, keep on proliferating, both outwards towards the surface, and inwards so that the papillary formation is accentuated. This is seen in its simple form in chronic conditions such as warty growths, tuberculosis, syphilis or elephantiasis, associated with hyperkeratosis in acanthosis nigricans, or with dyskeratosis in molluscum contagiosum.

(2) *Atrophy*, wherein the prickle cells shrivel and the subcutaneous fat is lost, as is seen in senile and neuropathic conditions or as a result of malnutrition or pressure.

(3) *Dystrophy*, occurring in conditions of intracellular œdema, when the cells become swollen, spongy and devitalized.

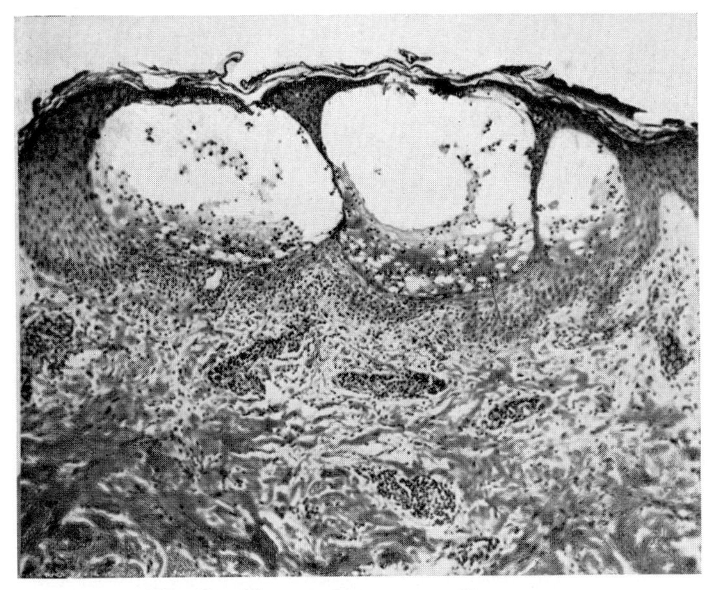

Fig. 27.—Vesiculation of the Epidermis.

In a case of contact eczema. Most of the epidermis is destroyed and there is considerable reaction in the subepidermal tissue (× 70) (G. H. Percival).

Superficial irritants produce a reaction typified in the catarrhal type of dermatitis. In its milder forms it is characterized by a dilatation of the papillary plexus which, if generalized, produces an *erythema* or, occurring particularly in foci round the follicles which become hyperæmic and swollen, gives rise to the formation of flattened red spots (*macules*) or, if there is serous exudate and cellular hyperplasia, elevated *papules* or *nodules*. A mild irritation may pass away and resolution may be complete, even although the exudate and cellular proliferation in the corium have produced considerable swelling; but if the intracellular œdema has interfered sufficiently with the epidermal cells, parakeratosis and scale-formation may result, while pigmentary changes may occur owing to the formation of granules of melanin in the cells of the basal layer. In more intense irritations the exudate may be sufficient to rupture the intercellular bridges so that fluid appears in the epidermis in mass and the red, solid papule becomes surrounded with a vesicle (*papulo-vesicular inflammation*). If the irritant is more intense and widespread, the red angry blush of the erythema is followed by parakeratosis which may then lead to extensive scaling in sheets, and œdema in this event produces crops of vesicles which may coalesce into large bullæ (Figs. 27–8); moreover, in such an exudative inflammation the œdematous fluid may

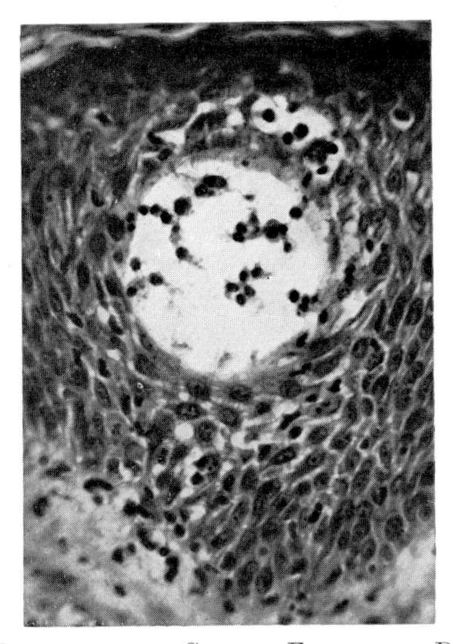

Fig. 28.—Vesiculation in Contact Eczematous Dermatitis.

High-power view of the condition of the epidermis in the neighbourhood of the vesicle. The appearance suggests liquefaction of the cells resulting in the formation of a large cavity (× 277) (G. H. Percival).

exude from the surface to constitute a *weeping eruption* or, becoming secondarily infected, form *crusts* and *scabs*. On the other hand, if the irritant is mild and prolonged in its action, a *granulomatous inflammation* occurs, characterized by a dense and persistent cellular deposit in the corium composed essentially of small round cells and plasma cells. Finally, when the irritant is sufficiently intense to produce a purulent infiltration and massive death of the tissues, the local destruction leads to the formation of an *ulcer*.

These varying types of superficial reaction can be seen in streptococcal infections: the dry crêpe-like response with scaling in superficial streptococcal dermatitis; the development of superficial vesicles with the subsequent formation of greenish-yellow crusts in impetigo contagiosa; the more violent reaction wherein local necrosis leads to ulceration surrounded by a deep inflammatory zone, in ecthyma; while if the infection is introduced into the deeper layers, the serous and exudative response of erysipelas ensues.

DEEP INFLAMMATIONS

Deep inflammations of the skin start in the dermis and affect the epidermis secondarily. The causative irritant may be introduced locally from without or may arrive through the circulation. In the first case it may be toxic, as in the stings of insects, but is more usually organismal—as the streptococcus in erysipelas, the treponema in chancre, Ducrey's bacillus in chancroid, the tubercle bacillus in lupus vulgaris, and so on. Blood infection may be either toxic or organismal. The first is exemplified in the eruptions due to drugs or the exanthemata: in these the reaction is usually acute. The organismal reaction may be of any type and is usually of more serious import. In either case the vascular reaction consists of vasodilatation, œdema whereby the fibres are separated, the cutaneous tissue becomes tense and the fluid appears in the epidermal layer, hæmorrhages by diapedesis or rupture, and finally, occlusion of the vessels resulting in necrosis and gangrene. The infiltrative changes are at first perivascular and usually perifollicular, and eventually may form a massive cellular infiltrate without reference to vessels and glands, acute processes being characterized primarily by polymorphonuclear cells, the more chronic by a rising proportion of plasma cells, lymphocytes, eosinophils, mast and giant cells with a sprinkling of endothelioid cells of reticulo-endothelial origin.

The toxic dermatoses in general assume one of four broad types—erythematous, urticarial, purpuric or exfoliative. The *erythematous* type is the simplest, in which case the evidence of inflammation may be limited to the appearance of redness due to dilatation of the vessels of the dermis, as in the erythematous rash of scarlet fever; in the more severe infiltrative types, a leucocytic infiltration and œdema lead to the accumulation of exudate beneath the epidermis and under its superficial layer, as is seen in erythema multiforme. The *urticarial* types are characterized by localized areas of transient œdema appearing as weals, as in the urticarias. In the *purpuras* the erythematous stage is followed by hæmorrhages into the dermis; while the association of erythema with profuse *scaling* is typified in the dermatitis exfoliativa. The organismal dermatoses produced by the deposition of living organisms in the tissues of the dermis take on many forms which require separate pathological study, varying from a transient erythema, as in the rose spots of typhus, to the acute necrotizing inflammations characterizing anthrax or glanders, or the chronic proliferative granulomata seen in tuberculosis, syphilis, leprosy, or frambœsia.

Subcutaneous inflammations of the lids follow the lines of general pathology, so that only the points of topographical interest in the more acute forms need be noted here.

CELLULITIS AND ABSCESS OF THE LIDS

A purulent inflammation of the lids may be derived from three sources:

(1) The introduction of infection from without, as through an erosion or puncture. It will be remembered that an abscess develops readily after a trauma which has resulted in the development of a hæmatoma in the substance of the lid (Figs. 29–30).

(2) Metastatic deposition through the blood-stream, as in pyæmia.

(3) Local spread from neighbouring inflammations, either

(*a*) from the skin of the lid or its associated glands, as in eczema, impetigo, erysipelas, furuncle or hordeolum or from the neighbouring skin (Fig. 31);

(*b*) from the conjunctiva, as in gonococcal infections;

(*c*) from the orbit, the orbital periosteum, the lacrimal passages or accessory nasal sinuses.

Figs. 29 and 30.—Superficial Abscesses of the Lids.

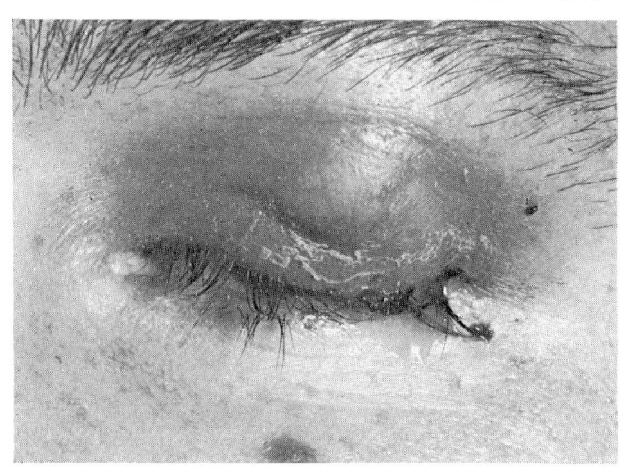

Fig. 29.

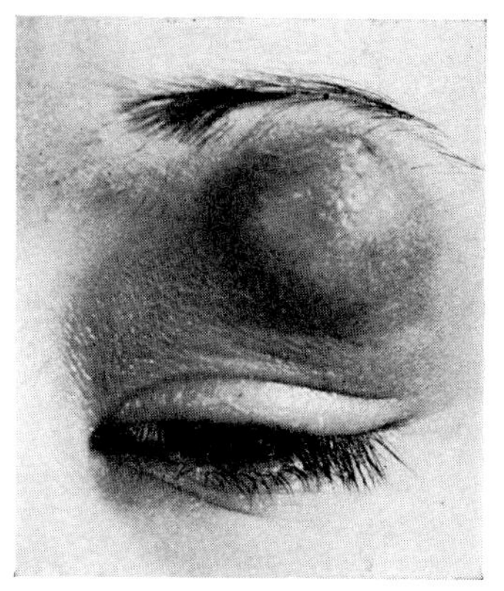

Fig. 30.

Fig. 29.—Three small pieces of stone were retained in the upper lid after a fall from a bicycle. The abscess formed 3 weeks after the healing of the laceration.

Fig. 30.—Abscess of the upper lid and brow in a girl resulting from plucking her eyebrows with tweezers which were shown to be bacteriologically dirty (P. A. MacFaul).

In the early stages of cellulitis the area of brawny infiltration characteristic of this type of infection in all parts of the body may in the lids be associated with a massive degree of swelling, the intense inflammatory œdema sometimes making localization of the trouble difficult. This is accompanied by swelling and tenderness of the regional lymph nodes in the pre-auricular and mandibular areas and general systemic disturbances of fever and prostration. In the later stages of abscess-formation the purulent infiltration tends to remain circumscribed and usually permeates and breaks through the skin, a process which may be associated with much sloughing; sometimes, however, it may show rapid and extensive spread, since the fascia and muscular aponeuroses may become necrotic. Pus may then track superficially up the forehead, along the temple or down the face, while it may spread deeply into the orbit, in which event orbital and intracranial complications are to be feared. In these cases, of course, general symptoms of illness and pyrexia are prominent.

It is interesting how extensive may be the suppuration from relatively small beginnings. Thus in chickenpox a small pock infected with a hæmolytic streptococcus has resulted in an abscess which spread over the temple and required drainage above the ear (de Schweinitz and Fewell, 1930). The opening of a hordeolum has produced a cellulitis which spread to the orbit (Mylius, 1925; Green, 1926), while infection of sores on the face, tracking through the orbit, has resulted in thrombosis of the cavernous sinus and death from meningitis (Rockliffe, 1907). A similar fatal meningitis has resulted from a gonococcal lid-abscess following ophthalmia neonatorum (Schall, 1922).

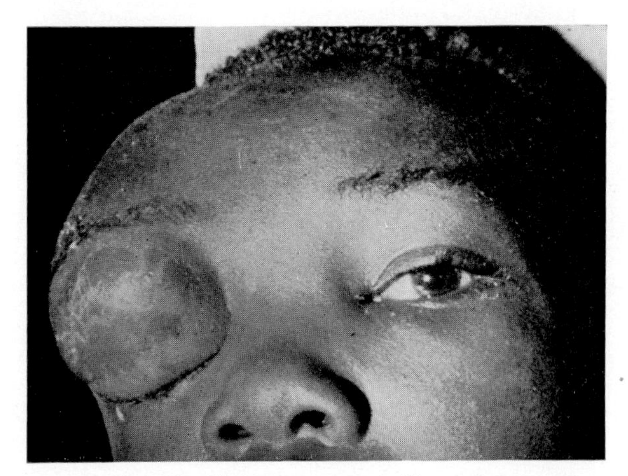

FIG. 31.—CELLULITIS OF THE UPPER LID.

The cellulitis was secondary to an abscess in the scalp in the right frontal area (G. G. Bisley).

Treatment should be on general surgical lines. In the first place, particularly if the infection is known, antibiotic therapy should be vigorously exploited. In most cases the infection can thus be controlled, but local measures may also be necessary—the application of moist heat until the presence of pus can be ascertained, to be followed by incision and drainage, together with appropriate general measures. Incision should be made only if pus has accumulated, but the immense swelling of the loose tissues of the lids may make localization difficult in the early stages. It is to be remembered that any incision should be parallel to the lid-margin in order not damage the fibres of the orbicularis muscle more than is necessary. When the skin of the lids has already sloughed, every care should be taken to protect the eyeball both from immediate damage and from subsequent exposure in the cicatricial stage. The most effective measure to ensure this is by tarsorrhaphy with, if necessary, a skin-graft over the raw area on the surface of the lid as soon as granulation commences; even if this graft does not " take " or becomes eventually unsatisfactory, it can readily be repeated at a later date, having served its essential purpose of minimizing cicatricial shrinkage and the dangers of lagophthalmos.

NECROSIS AND GANGRENE OF THE LIDS

Gangrene, that is, death of a considerable mass of tissue, may be either dry or moist. DRY GANGRENE (or NECROSIS), wherein death is followed by mummification, is rare in the lids. It occurs in the absence of infection when the tissues are drained of fluids prior to their death, as is seen in an interruption of the circulation by trauma or complete thrombosis, in crushing injuries, in burning, freezing or electrical injuries.[1] Two cases recorded by El-Naggar (1971) followed a neglected orbital cellulitis which caused thrombosis of the supra-orbital artery. Necrosis has followed the prolonged application of ice in an infant with ophthalmia (Plaut, 1900; Koelle, 1902; Würdemann, 1932). In these cases the affected area of the lid becomes browny-red and the skin mummifies, becoming dry and wrinkled in texture and of a black colour. As a rule the process is circumscribed.

MOIST GANGRENE, on the other hand, is the more common manifestation. It is a septic or putrefying process associated with a foul, offensive odour wherein, after a stage of fulminating swelling and œdema, the skin becomes discoloured black, green or yellow, the regional lymph nodes become involved, and generalization of the infection is usually shown by the development of the symptoms of asthenic fever (Fig. 75). The middle region of the lid is usually involved for, owing to its good blood-supply, the marginal strip is frequently spared; between this and the orbital rim, however, the skin, subcutaneous tissue and orbicularis muscle may slough entirely away, the gangrenous area having usually a well-defined outline (Figs. 32–5). The process of sloughing may take several weeks and it may extend so deeply that bone is eventually laid bare and even invaded, but when the slough separates there remains a raw granulating surface. The final result may give the impression that the lids had been cut cleanly away at the orbital margins. In the meantime the eye itself is by no means immune, for an infective keratitis frequently occurs, occasionally of such severity as to lead to irreparable intraocular and visual damage.

Sometimes an infection of this type is fatal (Fig. 36); at other times, after the sloughs have been cast away, the raw granulating area slowly heals, it may be after some months, leaving surprisingly little scarring with usually some degree of lagophthalmos owing to destruction of the orbicularis muscle or ectropion due to cicatrization. Thence, of course, further ocular complications may arise. The end-result varies with the extent and virulence of the process, but a gangrene so extensive as to involve the lids, a large area of the face, the bone of the lower orbital margin, the inner structures of

[1] Vol. XIV, pp. 776, 818.

FIGS. 32 to 35.—GANGRENE OF THE LID.

Due to secondary infection of a surgical wound in a woman aged 45 (A. Mortada).

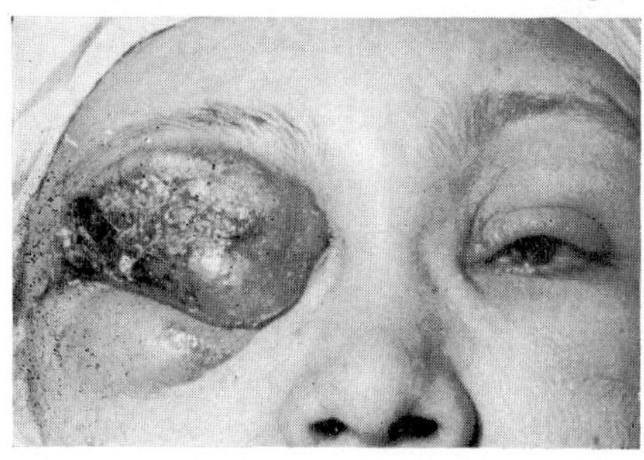

FIG. 32.—The appearance two weeks after incising an abscess of the lid.

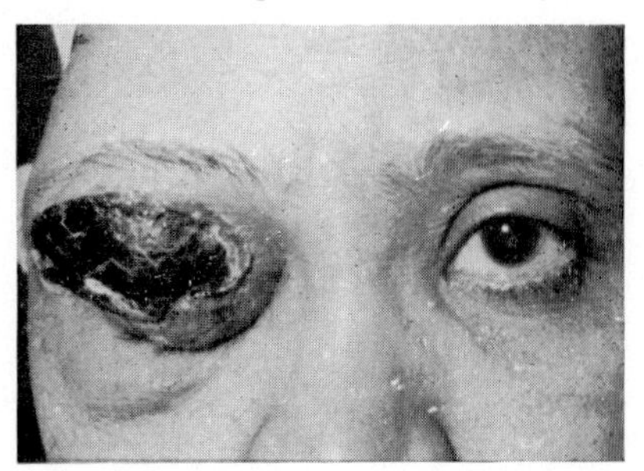

FIG. 33.—One week after treatment there is a line of demarcation between the living and the gangrenous skin.

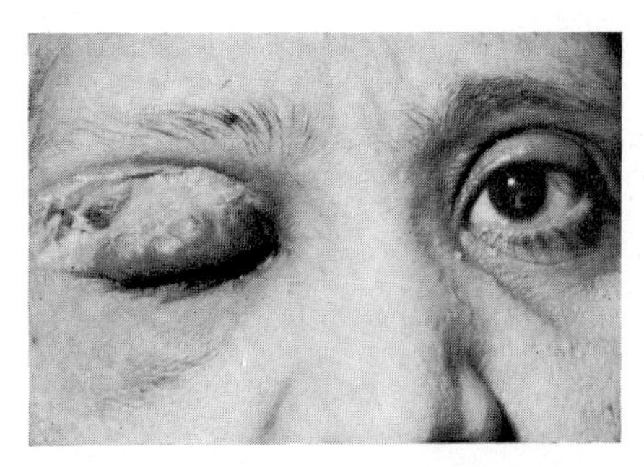

FIG. 34.—One week later a large ulcer appeared after separation of the gangrenous area.

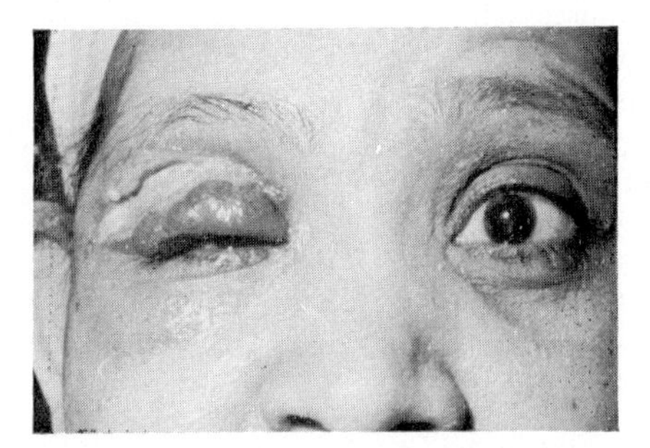

FIG. 35.—Two weeks later the ulcer showed epithelialization at its edges and fresh granulation tissue at the base.

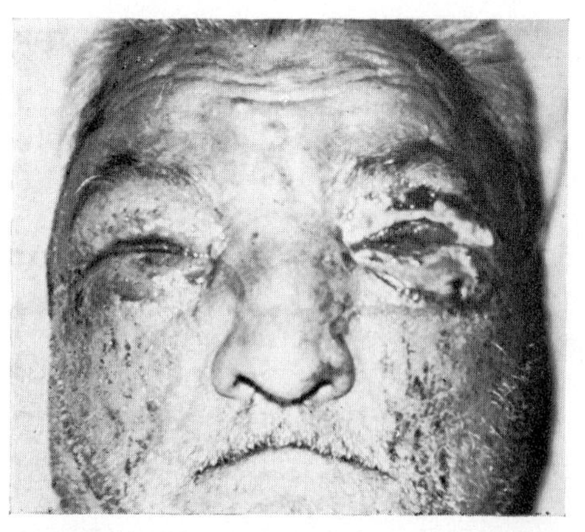

FIG. 36.—BILATERAL STREPTOCOCCAL GANGRENE OF THE EYELIDS.

In a man aged 78 who was found unconscious on the floor of his apartment. After treatment with penicillin and methicillin the condition had somewhat subsided but 5 days later he died (J. Ross and P. A. Kohlhepp).

the nose and the soft and hard palate may be survived (Ljubimov, 1935); alternatively, the end-result may be an adherent scar continuous with the skin of the forehead and the cheek, covering the partially destroyed contents of the orbit and the remnants of a destroyed eye.

The *ætiology* of gangrene of the lids varies, but today the condition is much more rare than it used to be; usually it was caused by an infection of considerable virulence in debilitated persons—marasmics, luetics, alcoholics, diabetics, or those, particularly children, suffering from infectious diseases such as measles or scarlet fever. In such patients it has followed erysipelas, impetigo or eczema of the lids (Hilbert, 1883; v. Michel, 1901; Possek, 1907). It has been a complication of smallpox (Landesberg, 1874; Adler, 1874), or chickenpox (Wintersteiner, 1909), or has occurred as a metastatic infection in septicæmia (Valude, 1890; Mitvalský, 1893). It has also followed severe conjunctival infections such as diphtheritic or gonococcal conjunctivitis, in which, indeed, sloughing of the four lids has been recorded (Elschnig, 1893; Steffens, 1900; Pes, 1904; Axenfeld, 1907; Narog, 1948). It has also complicated a traumatic hæmatoma in an arteriosclerotic which became infected (Schott, 1966), or followed the incision of an abscess (Mortada, 1964), or an infected wound (Marshall, 1960). The most common infecting organisms are the streptococcus (Valude, 1890; Morax, 1902; Eppenstein, 1914; Ross and Kohlhepp, 1973), the staphylococcus (Possek, 1907; Plocher, 1916), or a mixture of these (Givner, 1939); occasionally the pneumococcus has been found (Narog, 1948), the diphtheria bacillus (Lichtenstein, 1919; Weinberg, 1921), and as a rarity *Proteus* (Mohamed, 1934; Kamel, 1953; Parunović, 1973); but in many instances no definite ætiology has been determined and bacteriological investigation has proved surprisingly negative. Much more rarely gangrene may follow an unusual non-organismal trauma, such as an insect bite (Pes, 1904), and it has been reported following cauterization by a crystal of copper sulphate left carelessly in the fornix after the treatment of trachoma (Sédan, 1935) or as a result of dyeing the lashes[1] (Forbes and Blake, 1934).

"EGYPTIAN GANGRENE." A characteristic form of gangrene has been reported from Northern Africa from time to time, which seems to be peculiar to this region; nine such cases were collected by El-Seesy (1937) in Egypt, and the condition is seen westwards as far as Morocco (Pagès and Viennot-Bourgin, 1948). The affection appears to develop spontaneously and without apparent cause at any age in either sex. One eye is affected, the onset is acute, sloughing and discharge are profuse and necrosis occurs in a few days, usually, however, sparing the lid-margins; but the tissues of the globe are unharmed, and the end-result is remarkably good, both as regards cosmetic appearance and function (Fig. 37). Organisms of the *Proteus* group have been found in some of those cases which are sensitive to streptomycin (Barrada and Mohamed, 1935; Kamel, 1953), but it may be that the disease is of viral origin and is related to the necrotizing desert sore.

NOMA is a term frequently used to designate a particularly virulent type of gangrene of indefinite ætiology occurring in ill-nourished infants, especially those recovering from infectious diseases, such as measles[2]; it occurs most commonly today in marasmic children in Africa, Asia and tropical America. The organisms found are varied, but inanition and leucopenia seem to be the necessary forerunners and are probably the determining

[1] p. 68.

[2] Jessop (1894), Marlow (1901), Francke (1908), Stewart (1931), Pereira and Conti (1954), and others.

factors. While usually met with in children, it is still encountered in adults in tropical countries, sometimes as a complication of kala-azar, and was seen in the Second World War among the starved inmates of prison camps; Chaddah and Khanna (1968) reported a case in a child with acute leucæmia. The gangrenous process affects the mouth (*cancrum oris*) and the genitalia more commonly than the lids. Occasionally the infant may remain well, and, after some months during which the lids have sloughed away and much of the orbital contents, including the eye, may have been destroyed, the end-result may be a smooth scar over the entrance to the orbit to which the stump of the eye may be adherent (Jessop, 1894). More usually, however, the child is flabby and colourless, lethargic and afebrile, showing no resistance or immunological reaction. Deep sloughing ulcers appear on the lids, sometimes symmetrically on both sides, the skin and orbicularis muscle together with

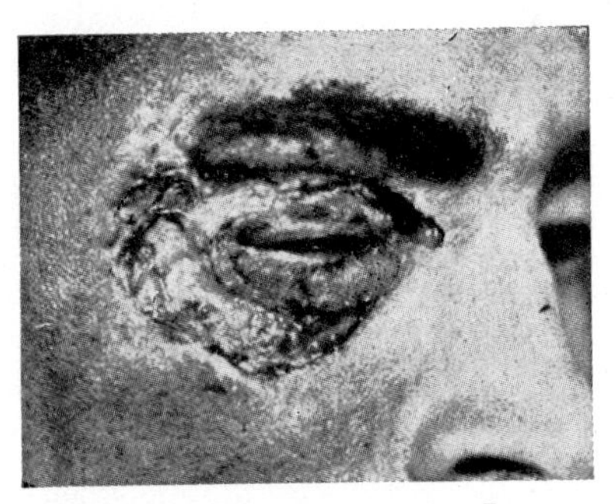

FIG. 37.—GANGRENE OF THE LIDS.

In an Egyptian male aged 30 years. An œdematous swelling of the lids gave place to discoloration and sloughing of the superficial tissues of both lids, but sparing the marginal area. The sloughing eventually exposed the malar bone, but after 3 months a granulating surface was Thiersch-grafted (R. P. Wilson).

the region of the lacrimal sac may slough entirely away as a foul ashy-grey pultaceous mass with an offensive discharge showing only a moderate reaction at the line of demarcation, extensive sinuses may develop down to the bone, while other areas in the body, particularly the nose, ears and mouth, begin to break down, and too often the infant, despite all attempts at stimulation, showing all the symptoms of collapse with a subnormal temperature, slowly and listlessly dies.

The *treatment* of gangrene of the lids varies considerably with the type of infection and the reaction of the patient. Antibiotic drugs or one of the sulphonamides should, if the organismal cause is susceptible to their influence, be exploited to the full, both by local application and general administration; such treatment, where it is applicable and employed during the early stages of the disease, may bring about immediate demarcation and rapid separation of the sloughs with marked improvement in the general condition (Vaizey, 1946; Mackay, 1949; and others). With energetic antibiotic treatment coupled with a full high-protein diet the mortality of noma, previously a highly lethal disease, has been considerably reduced, although it sometimes happens that even although the toxæmia decreases and the lesion clears up, the patient dies from the inanition which caused the disease.

If the condition of the patient is good, local treatment to the gangrenous area may be conservative. If, however, the serious aspect of the case lies less in the virulence of the local infection than in the debility and lack of response in the patient—usually an infant—and if antibiotic therapy is not effective, more prompt and energetic surgical action must be taken if the child's life is to be saved. Under an anæsthetic the sloughs as well as any diseased bone should be freely removed by cutting or scraping until healthy bleeding tissue is reached, and the area frequently dusted with antibiotic powders (penicillin and sulphathiazole, etc.) or washed with antiseptics such as peroxide of hydrogen (1 in 10), permanganate of potash or sublimate (1 in 3,000). Every endeavour should be made to preserve the cornea from damage, while the general resistance should be maintained by stimulants and abundant fluid nourishment. If recovery occurs, subsequent plastic operations or grafting, sometimes on a considerable scale, may be required to reconstitute the lids.

FOREIGN-BODY GRANULOMATA

GRANULOMATA are small, rapidly growing, inflammatory pseudo-tumours, usually attaining the size of a pea, somewhat soft in consistency and frequently associated with considerable swelling of the lid. The skin over the tumour is normal and no tenderness is present; sometimes multiple tumours may occur (Rifat, 1932). Histologically they are made up of dense

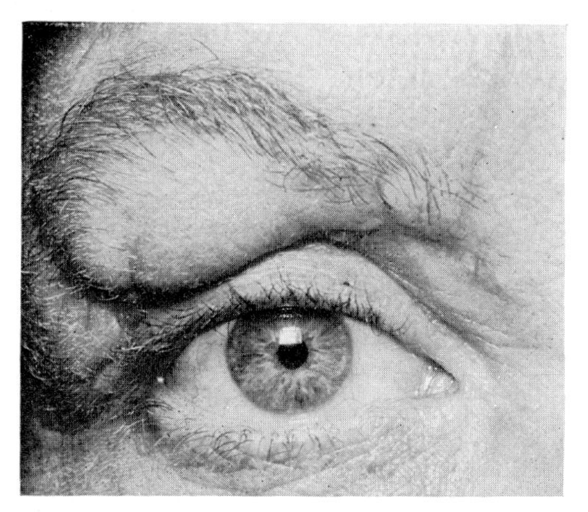

FIG. 38.—FOREIGN-BODY GRANULOMA.

After a wound in an aeroplane accident, sutured as a first-aid measure. The mass was subsequently found to contain dirt, rock and bits of metal (M. J. Reeh, *Treatment of Lid and Epibulbar Tumors*, C. Thomas, Springfield).

connective tissue infiltrated especially in the perivascular areas with lymphocytes and occasional plasma cells. Frequently giant cells indicate that the tumour is in reality a tissue-defence against a foreign body (Fig. 38)—aniline (Mylius, 1926), drops of oil (Mita, 1931), talc (Sysi, 1950), or paraffin (Uhthoff, 1905; Adler, 1905; Müller, 1905; Sallmann, 1928); Reese (1936) observed its occurrence repeatedly after the local injection of old solutions of procaine hydrochloride, and Vrabec (1967) with sulphonamide powder. A *lipogranuloma* has been reported following the retention of a piece of straw in the lid (Vancea and Balan, 1961).

An allergic granulomatous reaction may occur in response to certain substances, such as some of the dyes used in tattoos, zirconium or beryllium. In these cases the reaction is of the tuberculoid type, consisting of epithelioid cells with or without giant cells and sometimes with caseation; phagocytosis of the foreign substance is usually slight or absent.

Histological examination is necessary for a differential diagnosis from tumours; treatment is by excision.

Specific types of granuloma are dealt with subsequently—granuloma pyogenicum,[1] granuloma annulare.[2]

Adler. *Vjschr. Derm. Syph.*, Sept.-Abdruck., Wien (1874).
Zbl. prakt. Augenheilk., **29**, 104 (1905).
Axenfeld. *Die Bakteriologie in d. Augenheilkunde*, Jena (1907).
Barrada and Mohamed. *Bull. ophthal. Soc. Egypt*, **28**, 46 (1935).
Chaddah and Khanna. *Orient. Arch. Ophthal.*, **6**, 179 (1968).
Elschnig. *Klin. Mbl. Augenheilk.*, **31**, 191 (1893).
El-Naggar. *Bull. ophthal. Soc. Egypt*, **64**, 513 (1971).
El-Seesy. *Bull. ophthal. Soc. Egypt*, **30**, 120 (1937).
Eppenstein. *Z. Augenheilk.*, **32**, 16 (1914).
Forbes and Blake. *J. Amer. med. Ass.*, **103**, 1441 (1934).
Francke. *Klin. Mbl. Augenheilk.*, **46** (2), 432 (1908).
Givner. *Arch. Ophthal.*, **21**, 715 (1939).
Green. *Amer. J. Ophthal.*, **9**, 34 (1926).
Hilbert. *Zbl. prakt. Augenheilk.*, **7**, 293 (1883).
Jessop. *Trans. ophthal. Soc. U.K.*, **14**, 22 (1894).
Kamel. *Bull. ophthal. Soc. Egypt*, **44**, 39 (1953).
Koelle. *Ein Fall v. Lidgangrän nach Scharlach mit conjunctivitis diphtherica* (Diss.), Giessen (1902).
Landesberg. *Beitr. z. variolösen Ophthalmie*, Elberfeld, 20 (1874).
Lichtenstein. *Klin. Mbl. Augenheilk.*, **63**, 684 (1919).
Ljubimov. *Vestn. Oftal.*, **6**, 126 (1935).
Mackay. *Brit. med. J.*, **1**, 223 (1949).
Marlow. *Ophthal. Rec.*, **10**, 636 (1901).
Marshall. *Diseases of the Skin*, Edinb., 187 (1960).
von Michel. *Arch. Augenheilk.*, **42**, 4 (1901).
Mita. *Klin. Mbl. Augenheilk.*, **86**, 59 (1931).
Mitvalský. *Klin. Mbl. Augenheilk.*, **31**, 18 (1893).
Mohamed. *Rep. mem. ophthal. Lab. Giza*, **9**, 142 (1934).
Morax. *Ann. Oculist.* (Paris), **127**, 43 (1902).
Mortada. *Brit. J. Ophthal.*, **48**, 114 (1964).
Müller. *Zbl. prakt. Augenheilk.*, **29**, 106 (1905).
Mylius. *Klin. Mbl. Augenheilk.*, **74**, 781 (1925).
Z. Augenheilk., **56**, 302 (1925); **59**, 64 (1926).
Narog. *Klin. oczna*, **18**, 511 (1948).
Pagès and Viennot-Bourgin. *Maroc méd.*, **27**, 478 (1948).
Parunović. *Amer. J. Ophthal.*, **77**, 543 (1973).
Pereira and Conti. *Arch. Oftal. B. Aires*, **29**, 503 (1954).
Pes. *Z. Augenheilk.*, **12**, 438 (1904).
Plaut. *Klin. Mbl. Augenheilk.*, **38**, 35 (1900).
Plocher. *Klin. Mbl. Augenheilk.*, **57**, 51 (1916).
Possek. *Klin. Mbl. Augenheilk.*, **45** (1), 211 (1907).
Reese. *J. Amer. med. Ass.*, **107**, 937 (1936).
Rifat. *Ann. Oculist.* (Paris), **169**, 198 (1932).
Rockliffe. *Trans. ophthal. Soc. U.K.*, **27**, 189 (1907).
Ross and Kohlhepp. *Ann. Ophthal.* (Chic.), **5**, 84 (1973).

[1] p. 501. [2] p. 292.

Sallmann. *Z. Augenheilk.*, **65**, 298 (1928).
Schall. *Klin. Mbl. Augenheilk.*, **69**, 597 (1922).
Schott. *Industr. Med. Surg.*, **35**, 27 (1966).
de Schweinitz and Fewell. *Arch. Ophthal.*, **3**, 383 (1930).
Sédan. *Rev. int. Trachome*, **12**, 133 (1935).
Steffens. *Klin. Mbl. Augenheilk.*, **38**, 339 (1900).
Stewart. *Trans. ophthal. Soc. U.K.*, **51**, 599 (1931).
Sysi. *Acta ophthal.* (Kbh.), **28**, 257 (1950).
Uhthoff. *Berl. klin. Wschr.*, **42**, 1461 (1905).
Vaizey. *Brit. med. J.*, **2**, 14 (1946).
Valude. *Ann. Oculist.* (Paris), **103**, 204 (1890).
Vancea and Balan. *Arch. Ophtal.*, **21**, 767 (1961).
Vrabec. *Cs. Oftal.*, **23**, 253 (1967).
Weinberg. *v. Graefes Arch. Ophthal.*, **104**, 345 (1921).
Wintersteiner. *Z. Augenheilk.*, **21**, 268 (1909).
Würdemann. *Injuries of the Eye*, 2nd ed., London, 17 (1932).

DERMATITIS DUE TO IRRITATIVE FACTORS

ECZEMATOUS DERMATITIS

The term ECZEMA (ἐκζέω, to boil over) has been used to denote in a wide sense *any wet or scaly inflammation of the skin*, the nature and cause of which are unknown; it became the habit to label a similar condition *dermatitis* when the cause was known, but the two terms are now generally used as synonyms. Somewhat analogous to the term " rheumatism ", " eczema " may therefore mean almost anything or nothing, and yet may be said to embrace some 30% of all skin diseases. It forms a dermatological scrap-heap into which any superficial dermatitis of unknown cause may be legitimately cast, and out of which, as knowledge progresses, certain conditions, the ætiology and nature of which have become known, are extracted. The largest group to be picked out is that of contact or trade dermatitis, that is, eczematous dermatitis due to contact with some irritant encountered in work or applied to the skin, and as time goes on the residuum which defies satisfactory analysis will undoubtedly grow less. In this section we shall therefore consider the general nature of all eczematous dermatoses as they affect the lids, taking up the ætiology of the known types subsequently.

ECZEMATOUS DERMATITIS constitutes an inflammatory reaction characterized by the multiform clinical picture we have already outlined, exhibiting erythema, papules, vesicles, and scaly, weeping or crusted patches associated with a varying degree of itching or burning. These manifestations form different phases of the eruption, all of which need not be present and none of which commonly occurs by itself; the typical lesions, however, are at first erythematous and later papular, vesicular, scaly, and pustular if secondarily infected, going through the evolution illustrated in Fig. 39. They may be acute, subacute or chronic in type, but in habit

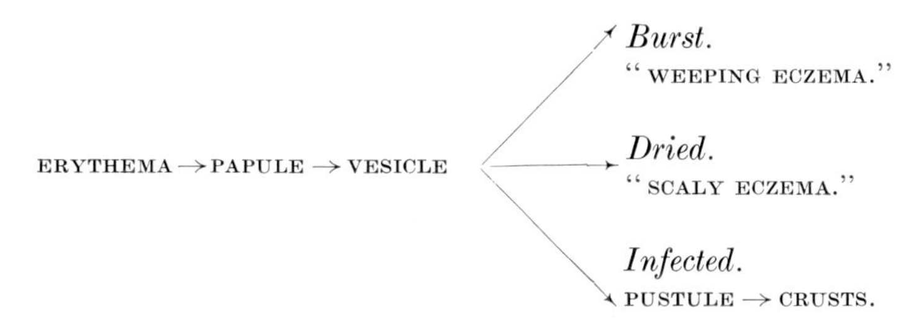

FIG. 39.—THE EVOLUTION OF ECZEMATOUS DERMATITIS.

they are usually noted for their chronicity and their liability to recurrences. The less severe forms are characterized by dry red erythematous patches which burn and itch in the lids, frequently associated with a considerable amount of œdema. The eruption may be of brief duration and resolve without trace, but the more usual tendency is chronicity with intermissions and relapses, the disease eventually evolving into a squamous scaly type with a varying degree of thickening and infiltration, or the surface may remain red and weeping.

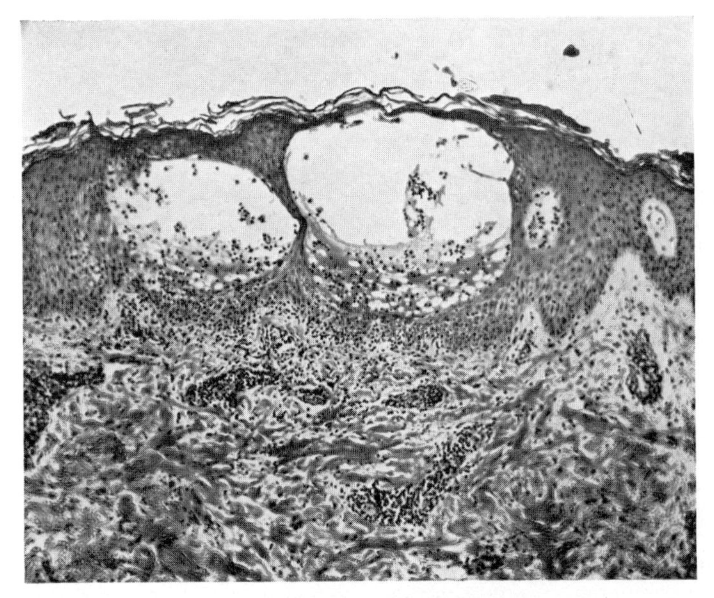

Fig. 40.—Eczematous Dermatitis.

There is fully-developed vesiculation which involves almost the entire thickness of the epidermis. The vessels in the dermis are dilated and there is a marked perivascular infiltration of leucocytes (× 70) (G. H. Percival).

In the more acute forms the usual sequence is the appearance of a bright red area with a diffuse border, in which minute nodules may appear or pin-head vesicles or bullæ are formed by intercellular followed by intracellular œdema in the epidermis (Fig. 41). If the number of vesicles is great they are eventually separated from each other only by thin septa formed by walls of resistant epidermal cells, thus constituting a multilocular bulla. The vesicles contain œdematous fluid, disintegrated epithelial cells and a few lymphocytes, eosinophils and neutrophils (Fig. 40). At the same time the cells of the stratum corneum may become keratotic and the upper dermis shows vascular dilatation with a perivascular infiltration of lymphocytes, eosinophils and neutrophils. Eventually the vesicles burst, liberating initially the clear fluid of a " weeping eczema ", but eventually, in the more acute cases, a turbid yellow exudate which forms crusts or, becoming infected, enters a purulent stage (Fig. 42). As this dries up, a squamous or scaly stage is reached, and then suddenly when almost healed the whole breaks down again violently, the cycle of vesicles and crusts repeating itself persistently for months (Figs. 43–4). Any of these phases, it is true, may resolve, in some cases rapidly and without trace, but more usually slowly after many relapses, leaving a chronic infiltration. In the subacute phase the vesicles are smaller but appear in the subepidermal tissues. In the chronic stages vesicles are absent, but acanthosis, hyperkeratosis and parakeratosis are prominent. A dense capillary bed appears and the perivascular infiltrate is predominantly composed of lymphocytes but eosinophils, histiocytes and fibroblasts are

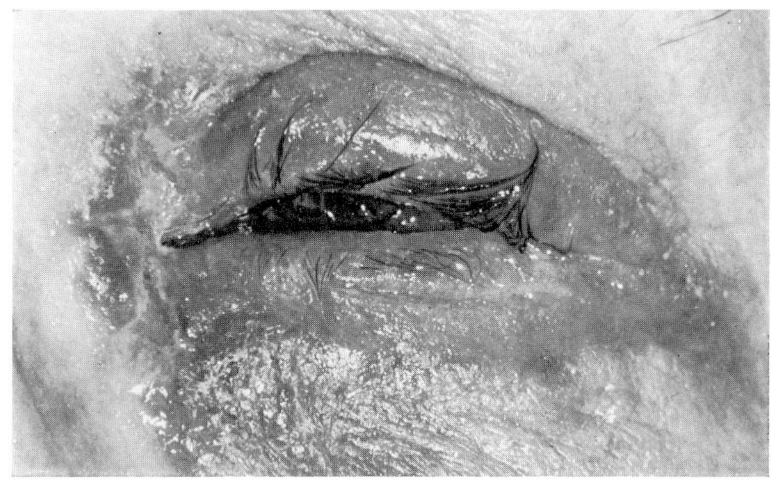

FIG. 41.—ALLERGIC ECZEMA.
In the acute œdematous stage (Inst. Ophthal.).

present; neutrophils are absent.[1] A new horny epidermal layer may be formed and cast off several times, appearing as a scaly erythematous patch before the normal appearance and texture return; thickening and infiltration may persist, or the skin may become dry and inelastic, developing deep fissures. Finally, in very old chronic cases the skin becomes thick with a dull, scaly, quadrillated surface (LICHENIFICATION) or shows a diffuse hypertrophy (ELEPHANTIASIS).

The lids are a common site for such a dermatitis, sometimes occurring as a local or solitary lesion, sometimes forming part of a facial eczema, or participating in a generalized disorder. Their predilection is doubtless due to the delicacy of the skin in this region—certainly the most delicate of the exposed areas; moreover, their involvement is readily remarked and the

[1] Winer and Lipschultz (1952), Flegel and Plötz (1963), Neumann and Winter (1965), Lever (1967).

FIGS. 42 to 44.—ALLERGIC ECZEMA (Department of Dermatology, Vanderbilt Clinic, N.Y.).

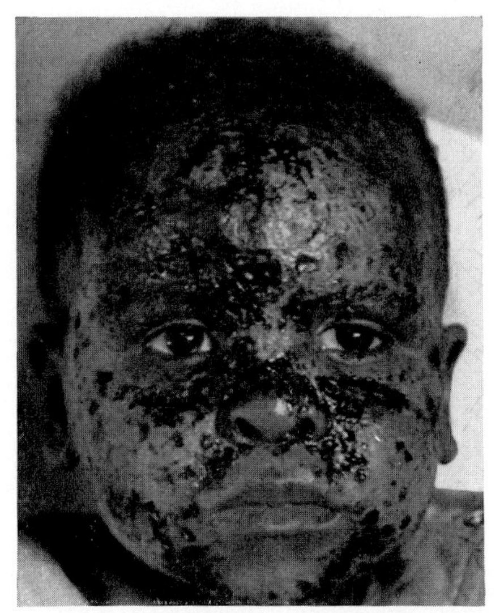

FIG. 42.—Allergic eczema in the acute pustular stage.

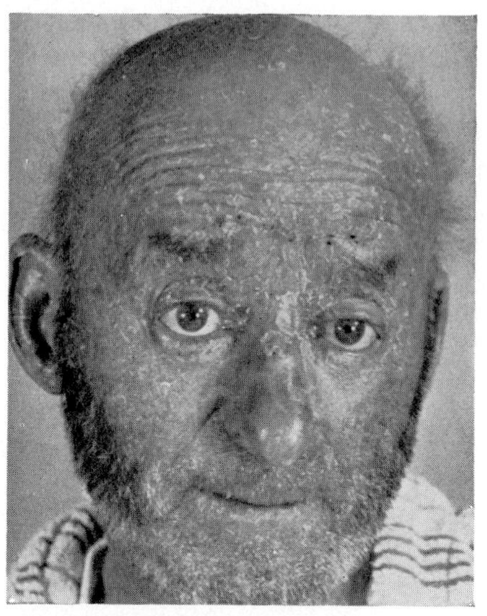

FIG. 43.—Allergic eczema in the chronic exfoliative stage.

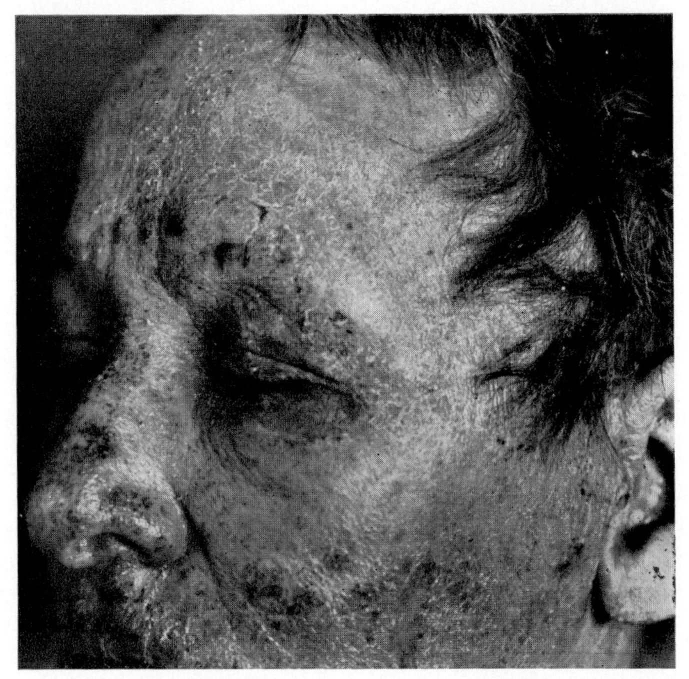

FIG. 44.—Allergic eczema in the subacute stage. Note the loss of the outer half of the eyebrows from rubbing.

tendency for the development of œdema is great. When they are affected, an associated blepharitis and conjunctivitis are frequent, and a keratitis, sometimes with corneal ulceration, a not uncommon accompaniment. Occasionally iritis occurs.

Ætiology. Two important factors, each mediated by a different mechanism, enter into the ætiology of eczematous dermatitis: primary irritation and allergy.

A *primary irritant dermatitis* is caused by the direct irritation of the epidermal cells; it is of the nature of a trauma and, although the effect varies considerably in different individuals, it can occur in anybody if the irritant is allowed to act in sufficient concentration for a sufficiently long time. Moreover, it tends to be confined to the area irritated. This region, however, is prone to react readily to further irritation, and minor traumata such as scratching increase and prolong the inflammation, in which case the skin tends to acquire an allergic sensitivity.

An *allergic dermatitis* may be due to the acquirement of hypersensitivity by the skin after repeated exposures to an allergen. Such sensitizers are of many types, sometimes a single element such as nickel or highly complex chemicals such as drugs and plastics, or even bacterial proteins. To excite this reaction these substances must be absorbed into the epidermis where they become conjugated with epidermal protein and therefore capable of provoking the formation of antibodies which enter the general circulation, travelling in association with the lymphocytes. When it reaches the skin the antibody becomes fixed to the epidermal cells and, on combining with fresh antigen, produces the eczematous reaction. When such a hypersensitivity is derived from actual contact the condition is termed *contact dermatitis*; on the other hand, a reaction resembling the anaphylactic type may develop after the absorption of chemical substances of a protein nature or capable of conjugating with the proteins of the body, as is seen typically in the drug eruptions (*dermatitis medicamentosa*). A chronic bacterial source of infection in a proximal or distant focus may be an ætiological factor, the importance of which is variously estimated by different writers (*infective dermatitis*); but although organismal allergy, toxic conditions, and even emotional factors may require consideration, it seems probable that in comparison with exogenous antigens the circulating or endogenous causes are few.

A second type of allergic reaction is seen in *atopic eczema*. Atopy is a genetically determined disorder with an increased liability to form reagin antibodies involving an increased susceptibility to such allergies as hay fever, asthma and urticaria; it has a marked familial incidence and is particularly severe when both parents have been affected and it has a basis of an inherited genetic defect. Such atopic patients usually suffer from several allergies, frequently with eosinophilia, and the skin becomes red and rough with small cracks oozing serum, often with innumerable papules (Plate I,

Figs. 3 and 4). It may occur at any age: *infantile eczema* usually starts about the third month of life and although it affects the cheeks and the forehead, the eyelids are spared. In childhood the eruption involves primarily the flexures, but in adults (*Besnier's prurigo*) the face and trunk may be affected. The general tendency, however, in all types of atopic eczema is towards recovery. Ocular complications are frequent, such as cataract, usually in the form of a dense subcapsular plaque,[1] an atopic kerato-conjunctivitis,[2] conical cornea, uveitis,[3] and a retinal detachment.

The *treatment* of eczematous dermatitis can only be effective in curing the condition and preventing recurrences if the underlying ætiological factors can be discovered and eliminated—a condition which is by no means always realizable. Any intercurrent infection, such as seborrhœic dermatitis of the scalp, must be treated before the lesion in the lids will respond, and in generalized cases in the absence of ætiological clues, rest, care of the diet, the avoidance of excessive alcohol and freedom from nervous strain are important. It can be understood that in resistant and recurrent cases these therapeutic desiderata are best obtained in a hospital or special clinic.

Local treatment can follow no hard and fast rules, and innumerable antidotes have been recommended. In general terms, weak applications should be tried first lest a more severe irritation be superimposed on the original; careful watch should be kept on the development of the case so that the therapeusis can be altered if necessary; soap and water are generally to be avoided and cold cream should preferably be used for cleansing purposes.

In the early stages when erythema, papulation or vesiculation is present, calamine lotion is the most useful[4] or, if this proves to be too drying, an oily calamine lotion.[5] In the weeping phase lotions of lead subacetate (2%), aluminium acetate (5%), silver nitrate (0·5%) or potassium permanganate (0·1%) are frequently effective. In the later scaly and crusted phases a zinc cream[6] or zinc paste[7] may be applied. In the chronic stages when the skin is thickened and lichenified and when itching is intense, coal tar (3 to 12%) may be added to the pastes or creams.

Dramatic results have sometimes been obtained with topical cortico-steroids which, by blocking the inflammatory allergic reaction, may bring about complete relief, so long as their administration is continued—but only so long; a permanent cure, however, cannot be expected unless the exciting agent is removed at the same time. Hydrocortisone may be used, or one of its more powerful derivatives incorporating fluorine atoms, such as betamethasone valerate (Betnovate). In severe or spreading cases systemically administered corticosteroids are often helpful given in large

[1] Vol. XI, p. 196. [2] Vol. VIII, p. 446. [3] Vol. IX, p. 499.

[4] Calamine 15, zinc oxide 5, bentonite 3, sodium citrate 0·5, liq. phenol 0·5, glycerin 5, water to 100.

[5] Calamine 5, wool fat 5, zinc stearate 2, oleic acid 1, light liq. paraffin 45, lime water to 100.

[6] Zinc oxide 32, oleic acid 0·5, arachis oil 32, wool fat 8, lime water to 100.

[7] Zinc oxide 25, starch 25, white soft paraffin to 100.

FIGS. 45 and 46.—PRIMARY HERPES SIMPLEX IN THE WISKOTT-ALDRICH SYNDROME.

In a Negro boy aged 3 with a history of eczema, thrombocytopenia with episodes of epistaxis and melæna and recurrent infections (S. M. Podos *et al.*).

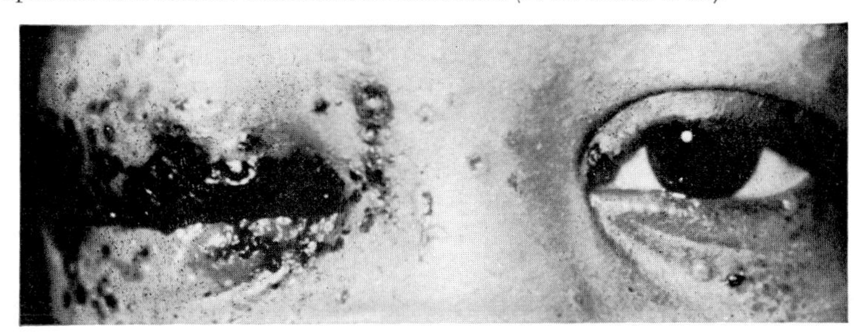

FIG. 45.—The lids show scattered vesicles with a gross hæmorrhagic involvement.

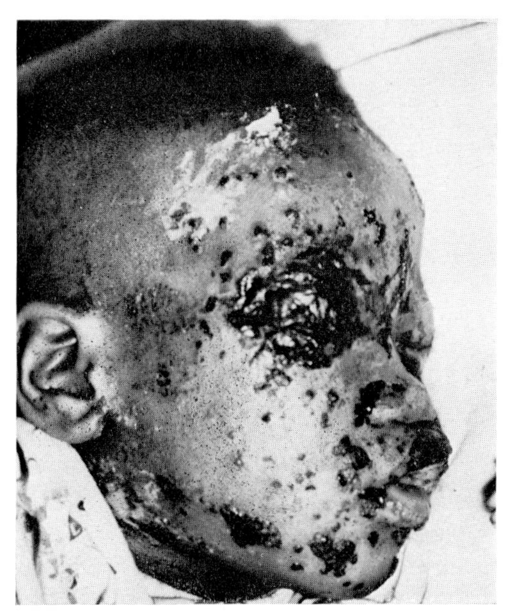

FIG. 46.—At a later stage, the development of Kaposi's varicelliform eruption with secondary hæmorrhages and necrosis of the disseminated lesions in the skin. The patient died. His brother had a similar condition.

doses until the condition is controlled when these should be gradually tapered off. If the lesion has become septic antibiotics may be given systemically if a patch test for sensitivity to them is negative.

The WISKOTT-ALDRICH SYNDROME is interesting in this connection. The condition is characterized by a triad of eczema resembling atopic dermatitis, thrombocytopenia and a liability to recurrent infection (Wiskott, 1937) and is transmitted as a sex-linked recessive character occurring only in males (Aldrich *et al.*, 1954). It is usually apparent within one month of birth and the majority of the children die before the age of two, approximately half the subjects dying from hæmorrhage and half from infections. Podos and his colleagues (1969) reviewed the 80 cases reported up to the time of their paper, and of these 18 had ocular symptoms. Eczema is frequent on the lids, either exudative or dry and exfoliative, which may become purpuric or secondarily infected (Figs. 45–6). Apart from the ecchymoses on the lids, hæmorrhages may occur in the conjunctiva (Dalloz *et al.*, 1965), the periorbita (Huntley and Dees, 1957), the retina (Mills and Winkelmann, 1959) and the vitreous (Root and Speicher, 1963), as well as in practically every organ of the body. The intercurrent infections are due to a remarkable variety of bacteria, viruses, fungi and protozoa, among which the most striking on the lids is the herpes virus, the intensity of the infection tending to result in Kaposi's varicelliform eruption[1]; blepharo-conjunctivitis (Podos *et al.*, 1969) and ulcerative keratitis (Chaptal *et al.*, 1966) have been reported as ocular complications.

The hæmorrhages are due to the lack of platelets resembling thrombocytopenic purpura, but the immunological fault has not been completely elucidated. It seems to result in a defect in the cell-mediated immunity; normal or elevated levels of IgK and IgG have been found, but the isohæmagglutinins are absent from the blood and the IgM is decreased (Stiehm and McIntosh, 1967). The eczema of the lids is susceptible to amelioration but, whatever the treatment, the disease is usually fatal.

Aldrich, Steinberg and Campbell. *Pediatrics*, **13**, 133 (1954).
Chaptal, Royer, Jean *et al.* *Arch. franç. Pédiat.*, **23**, 907 (1966).
Dalloz, Castaing, Nezelof and Seligmann. *Presse méd.*, **73**, 1541 (1965).
Flegel and Plötz. *Derm. Wschr.*, **147**, 215 (1963).
Gelzer and Gasser. *Helv. pædiat. Acta*, **16**, 17 (1961).
Huntley and Dees. *Pediatrics*, **19**, 351 (1957).
Lever. *Histopathology of the Skin*, 4th ed., Phila., 98 (1967).
Mills and Winkelmann. *Arch. Derm.*, **79**, 466 (1959).
Neumann and Winter. *Acta dermato-vener.*, **45**, 272 (1965).
Norman, Lock *et al.* *Brit. med. J.*, **1**, 313 (1962).
Podos, Einaugler, Albert and Blaese. *Arch. Ophthal.*, **82**, 322 (1969).
Root and Speicher. *Pediatrics*, **31**, 444 (1963).
St. Geme, Prince, Burke *et al.* *New Engl. J Med.*, **273**, 229 (1965).
Stiehm and McIntosh. *Clin. exp. Immunol.*, **2**, 179 (1967).
Ten Bensel, Stadlan and Krivit. *J. Pediat.*, **68**, 761 (1966).
Winer and Lipschultz. *Arch. Derm. Syph.* (Chic.), **65**, 270 (1952).
Wiskott. *Mschr. Kinderheilk.*, **68**, 212 (1937).

CONTACT DERMATITIS

CONTACT DERMATITIS (DERMATITIS VENENATA) is a very common condition occurring as *the reaction to some foreign material coming into contact with the skin.* Clinically all stages of eczematous dermatitis are encountered —erythematous, papular, vesicular, pustular; all degrees of severity are seen and, indeed, the reaction occasionally ends in gangrene. The essential symptom is itching, which is usually greater than the objective signs would warrant and may become intolerable. After a single contact the eruption is usually limited and disappears spontaneously, but repeated contacts or constant association with the offending agent may lead to a spread of the reaction far beyond the point of contact and an intensification of the response as the skin becomes progressively more sensitive to the noxa. Some agents are in themselves irritant to all skins, and in these the dermatitis depends on direct chemical trauma (*primary irritant dermatitis*); but more usually the irritant acts with a widely varied intensity on different individuals and not

[1] Gelzer and Gasser (1961), Norman *et al.* (1962), Root and Speicher (1963), St. Geme *et al.* (1965), Ten Bensel *et al.* (1966), Podos *et al.* (1969).

PLATE I

ALLERGIC DERMATITIS (Inst. Ophthal.)

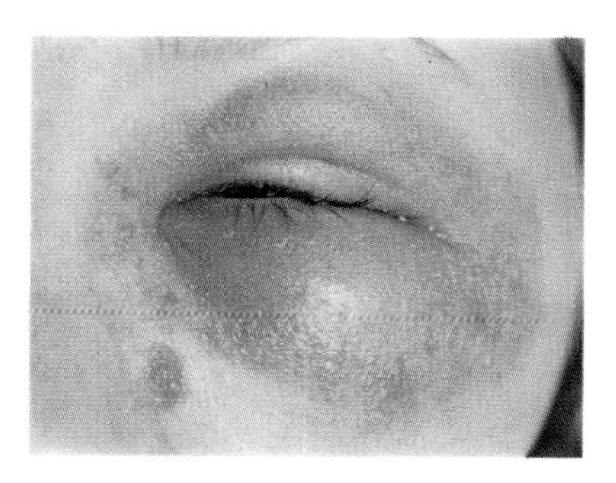

FIG. 1.—A typical case of contact dermatitis.

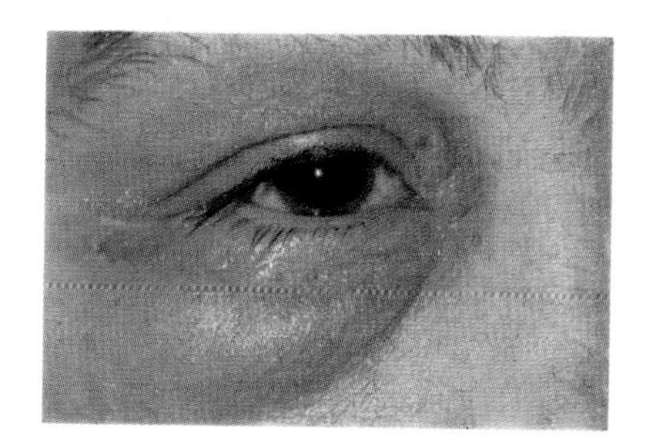

FIG. 2.—Atropine irritation.

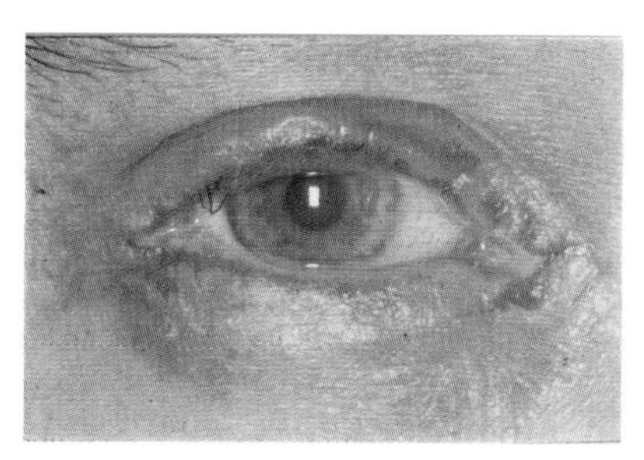

FIG. 3.—Atopic dermatitis with vernal catarrh and corneal ulcers.

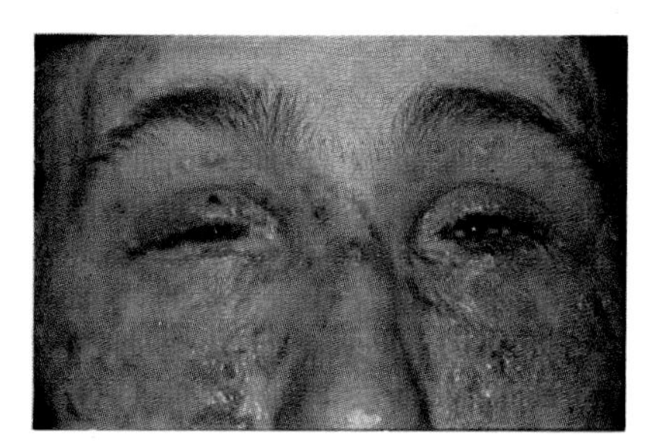

FIG. 4.—Atopic eczema.

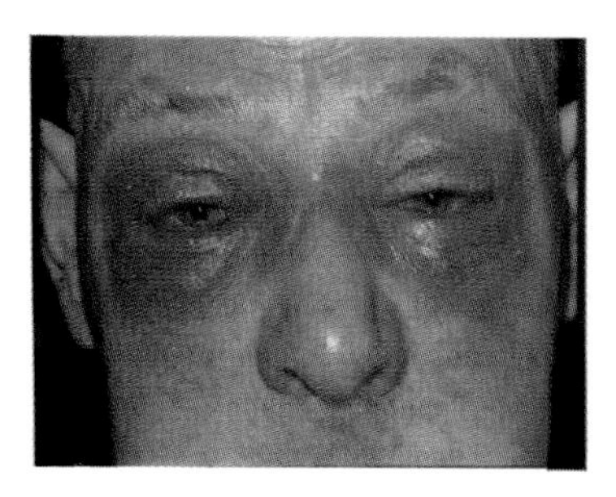

FIG. 5.—Acute allergy to propamidine isethionate (Brolene).

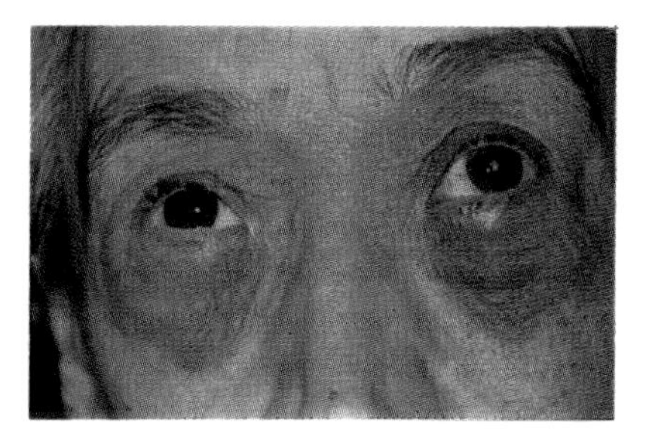

FIG. 6.—Primula dermatitis.

at all on others because of specific hypersensitivity, the reaction being essentially *allergic* in nature. Between the two types of response, however, there is no fundamental difference either clinically or pathologically. The essential difference lies in the condition of hypersensitivity of the tissues, and the clinical picture varies with the reactivity of the individual, and the extent, intensity, duration and frequency of the contact, as well as with the organismal complications superimposed, such as, for example, a seborrhœic dermatitis.

The tissue-response in contact sensitivity belongs to the delayed type of allergy. The agents producing the reaction must be capable of penetrating the unbroken skin, and therefore most are of a relatively small molecular size. These act as haptens, combining with the proteins of the skin to produce conjugates of altered immunological specificity so that they become antigenic to the individual. The antibodies which are formed become fixed in the skin and give rise to the specific reaction on subsequent contact with the hapten-protein conjugate. The nature of the antibodies is not yet known, but although they cannot be passively transferred (Leider and Baer, 1948), they circulate in the blood associated with the lymphocytes (Bergstrand, 1950; Wilhelm *et al.*, 1958; Thies *et al.*, 1965). It is probable that every skin may become sensitized provided a potential allergen is applied sufficiently frequently and in sufficient concentration, but why a peculiar sensitivity develops readily is also unknown: it may appear after a single contact with the antigen, which itself produces no disturbance, but after a latent period varying from 10 to 20 days the entire skin may be sensitized and responds to a second contact, or the reaction may be delayed and suddenly become evident after many years of continual innocuous contact. Once it has developed, however, the allergy may be extreme, so that the most minute traces of the allergen may excite a reaction and rekindle the dormant susceptibility. On the other hand, there is no doubt that if contacts continue, unless the injury is severe, a proportion of patients gradually becomes " hardened " and their sensitivity becomes lessened so that contacts which were initially injurious cease to excite a reaction—an evolution frequently seen in industry (Schwartz, 1940; Peck *et al.*, 1945).

Because they are susceptible tissues, readily capable of becoming swollen and itchy, and any disturbance therein is quickly remarked, the eyelids are frequently the first area to show the signs of an irritative dermatitis to exogenous irritants, exposure to which may be general, and the disturbance may be limited to this area. It is interesting also that the lids are frequently involved although the site of contact has been far afield. A shoe-dye or suspenders may excite a puffiness and itching of the lids as readily as an eyelash dye or a drop of atropine ; and since in the former case the patient seeks medical aid because of his eyes, paying no attention to the irritation of the foot or leg, the recognition of the nature of the condition is frequently difficult. The fact remains, however, that the commonest cause of repeated attacks of puffiness of the lids with itching and sometimes of redness and infiltration is contact dermatitis, and if œdema is marked the condition is allergic.

The ætiology of contact dermatitis is therefore very varied—there is already an almost interminable list of possible excitants, accidentally, medicinally or cosmetically applied, occupationally or seasonally met with, airborne, waterborne or mechanically encountered, a list which is steadily growing from year to year with the advances of synthetic chemistry[1] (Plate I). The most common causes are cosmetics such as face creams, nail polish and dyes for the hair, metals and their salts such as nickel and chromium, topical medications, plastic materials, rubber, dyes applied to furs, leather or fabrics, soaps and detergents and the oleoresins of plants such as poison ivy.

The most common excitants affecting the lids in particular may be noted, and it will be remembered that the eczematous reaction does not differ with different allergens, but varies rather with the intensity of the exposure and the degree of hypersensitivity developed by the patient.

VEGETABLE SUBSTANCES form one of the commonest causes, particular offenders being poison ivy (*Rhus toxicodendron*),[2] primula (*Primula obconica*),[3] poison oak (*Rhus diversilobas*),[4] the ragweeds,[5] and the manzanillo tree (beach apple) (Fig. 47; Plate I, Fig. 6). After contact with the toxic sap of this tree, a member of the Euphorbiaceæ, following a latent period of a few hours the dermatitis around the region of the lids may be severe, the chemotic conjunctiva may protrude between the spastic lids, the cornea usually shows large staining areas which may persist for 5 to 7 days and in neglected cases ulceration may develop; after this period the dermatitis and kerato-conjunctivitis rapidly subside.[6] A generalized rash and a vesicular eruption on the eyelids have been reported in a child who chewed the leaves of poison ivy (Shemeley, 1951). The attacks are usually acute and of short duration, but recurrences are frequent, some individuals being so susceptible that they cannot approach the plant with impunity. At first the attacks are usually seasonal, corresponding to the growing season of the plant, but eventually the affection tends to become perennial, with seasonal exacerbations and remissions in the winter (Shelmire, 1939).

INDUSTRIAL EXCITANTS lead to innumerable occupational dermatoses.[7] The list includes irritation from lacquers, inks (McConnell, 1921), jute (Curjel and Acton, 1924–25), oils such as machine oil or paraffin (Scott, 1922; Peck, 1944), linseed (Vokoun, 1927), turpentine (McCord, 1926), detergents (Morris, 1963), or petroleum (Dutton, 1934). Dyes of all kinds are common offenders, particularly those used in furs (Olson, 1916). Rubber may produce a violent hypersensitivity, usually due to chemicals used in the curing process (Cleveland, 1927; Burrage, 1929; Niles, 1936; Bonnevie and Marcussen, 1944); a primary dermatitis of the lids has been caused by the use of a rubber-covered eyelash curler (E. C. Fox, 1933; Curtis, 1945), by the rubber in an eye-dropper (Bandmann and Hardieck, 1960), or by sponges used for washing or the

[1] Weber (1937), Sulzberger and Finnerud (1938), Hazen (1944), R. L. and R. L. Sutton (1949), Theodore (1958).

[2] McNair (1921–23), Molitch and Poliakoff (1936), Caulfeild (1936–38), and others.

[3] Simpson (1917), Harville (1932–33), Weber (1937), Borrie (1969).

[4] Biederman (1938).

[5] Hannah (1919), Brunsting *et al.* (1934–36), Caulfeild *et al.* (1935–36).

[6] Bodeau (1936), Desoille (1937), Earle (1938), R. Harley (1944), Snow and Harley (1944), Grana (1946), Teulières (1962).

[7] See White (1920), Lane (1922), Cole and Driver (1923), Downing (1935), Sulzberger and Finnerud (1938), Lane *et al.* (1942), Schwartz *et al.* (1947), Borrie (1956).

application of cosmetics (Furman *et al.*, 1950). Similarly a skin allergy may be due to almost every substance used in leather manufacture (Beerman, 1934). "Phosphorus" (P_4S_3) used in the striking surfaces of match boxes sometimes causes sensitization, and since this substance readily volatilizes at a low temperature, the small amount liberated from a match box left in a room may be sufficient to excite a dermatitis particularly affecting the lids in a sensitive person (Martin, 1950). Metals may give rise to similar symptoms—lead, mercury (Ballin, 1933; Kern, 1963), beryllium (van Ordstrand *et al.*, 1945), selenium (Pringle, 1942) or, most dramatically, nickel.

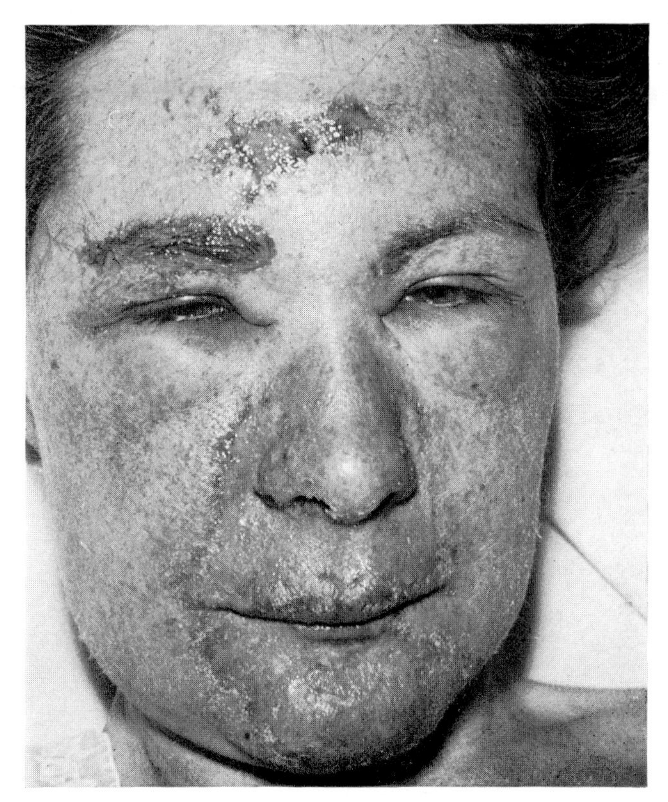

Fig. 47.—Acute Serous Dermatitis.
Due to an allergy to primula (P. Borrie, St. Bartholomew's Hosp.).

NICKEL DERMATITIS. Nickel is one of the few metals to which a definite sensitization can be acquired, and nickel dermatitis, with its picture of an erythemo-vesicular eruption in the acute stage leading to oozing, lichenification and fissuring, is a well-known entity. The possibility of sensitizing guinea-pigs and producing a nickel dermatitis by repeated applications of nickel sulphate was demonstrated by Walthard (1926). Two types are met with clinically. The first reports came from industry and concerned nickel-plating workers whose skin is habitually exposed to strong solutions of the metal[1]; the second type arises as an idiosyncrasy in allergic individuals who

[1] Blaschko (1889), Landsteiner (1924), Schittenhelm and Stockinger (1925), Bulmer and Mackenzie (1926), Jadassohn and Schaaf (1929), Stauffer (1931), Stewart (1933), White (1933), Goldman (1933).

develop sensitivity on occasional contact with nickel-plated articles. All sorts of such articles are on record—rings (Foster and Ball, 1935), suspender clips (Wrong, 1935; Foster and Ball, 1935; Borrie, 1956; Buckley, 1960) (Fig. 48), coins (Rothman, 1931), nickel-plated furniture (Cormia and Stewart, 1935) and, most commonly, spectacles.

The SPECTACLE DERMATITIS caused by nickel spectacle frames makes the subject ophthalmologically interesting. The condition was first noted by Lain

FIGS. 48 and 49.—CONTACT DERMATITIS.

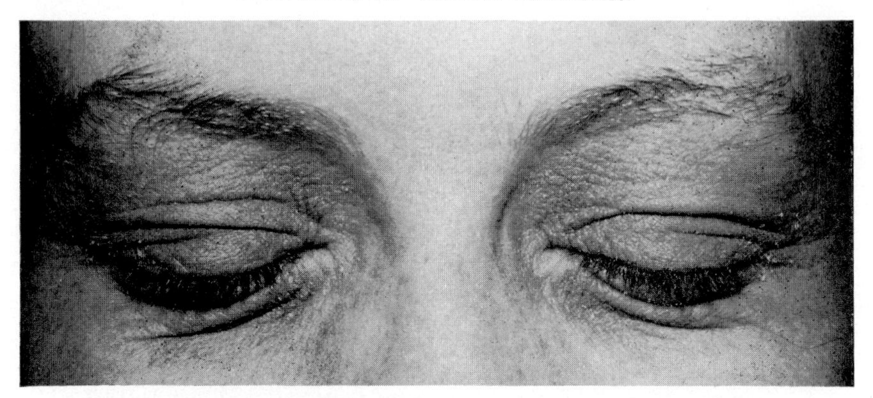

FIG. 48.—Due to sensitivity to the nickel in suspenders (P. Borrie).

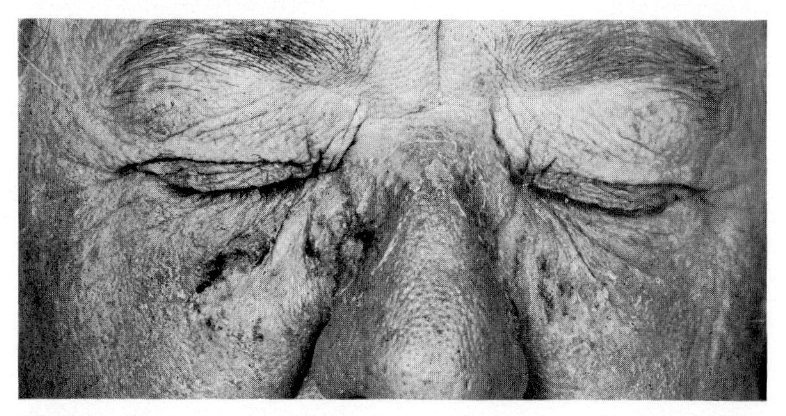

FIG. 49.—Due to contact with plastic spectacles; in the acute stage (D. Calnan).

(1931), and has since been recorded by several authors.[1] The dermatitis affects the areas in contact with the metal—the bridge of the nose, the inner canthus, a horizontal band across the temples, and the post-auricular region—and may be a very acute weeping type of eczema ultimately developing into a chronic infiltrated stage if contacts are maintained. Susceptibility is greater if the nickel content of the frames falls below 15%, in which case a considerable proportion of wearers may be affected. There is no evidence of inherent sensitivity to nickel, for the dermatitis has never been

[1] A. W. and A. W. McAlester (1931), H. Fox (1933), Urbach (1935), Foster and Ball (1935), Cormia and Stewart (1935), Kristjansen (1937), Löwenstein (1938), Taylor *et al.* (1945), and others.

recorded on a first contact; it is presumably brought about by some alteration in the tissue-cells by the nickel ions so that the cells thereafter become hypersensitive and subsequent contacts lead to further changes. Spread of the dermatitis is direct in the skin itself only; nickel dermatitis is therefore a " true " contact dermatitis in contradistinction to others such as ragweed dermatitis which can be induced by pulmonary or alimentary absorption. As a rule the reaction remains local and only in rare instances becomes generalized, a circumstance which is correlated with the fact that the skin hypersensitivity as indicated by the patch test (8% solution of nickel chloride) may be positive locally and negative elsewhere (Cormia and Stewart, 1935).

A dermatitis may also arise from *plastic* (" *horn-rimmed* ") *spectacles* (MacCormac, 1931; Wilde, 1959) (Fig. 49). Sutton (1927), who first noted the condition, attributed the irritation to dyes or faulty curing of the xylonite; Kristjanson (1937) attributed it to artificial resins but it would appear that the fault may lie with the plasticizers or softeners used in manufacture (tricresyl or triphenyl) rather than with the basic material (cellulose acetate) (Berkoff, 1938) or with the dyes used in manufacture (Thistlethwaite, 1943; Gray, 1943). A somewhat similar dermatitis may be traced to the bakelite of a cigar-holder (Sutton, 1927) or of radio headphones (Oelze, 1924). A similar irritation may arise in a socket after wearing a plastic *artificial eye* (MacIvor, 1950), or the use of plastic eye-baths (Quiroga and Guillot, 1955). Smith and Calnan (1966), however, concluded that many cases formerly believed to be an allergic reaction to plastic spectacles occurred in patients already prone to dermatitis so that the condition may depend on a constitutional predisposition.

It is to be noted that chemical solutions used in cleaning spectacle lenses have given rise to dermatitis of the lids (Hollander and Baer, 1935).

CLOTHING DERMATITIS is also of importance for, although the lesion is usually first located at the point of contact, it spreads readily to the hands and then to the eyelids and may become generalized. The most usual excitant is the resinous fabric finish so often found in men's and women's underwear. Apart from synthetic resins, dyes and mordants may give rise to similar irritation (Schwartz *et al.*, 1940; Keil, 1943; Schwartz and Peck, 1945). Soap, starch, bleach and other laundering materials may render previously harmless clothing irritant, while the laundry marking material derived from ral or bella gutti in India is responsible for the virulent " Dhobie mark dermatitis " (Livingood *et al.*, 1943). The dyes, polish, tanning and preservation agents used for shoes may also make these a source of irritation (Shaw, 1944; Peterkin, 1948).

MEDICINAL DERMATITIS forms a further considerable category, of which the examples of most ophthalmological importance are atropine, the sulphonamides, penicillin and other antibiotics, the mercurials and the synthetic local anæsthetics (Figs. 50–1). On this subject a large literatureh as now accumulated—an irritability to mercury (Kesten, 1931; Harper, 1934; Billo, 1941; Samitz, 1944), balsam of Peru (Engelhardt, 1935), picric acid (Pusey and Rattner, 1929; King, 1938), acriflavine (Kiep, 1926), ichthyol (Hugo, 1938), iodoform and iodine (Clifford, 1926; Jacobs and Colmes, 1940), propamidine (Plate I, Fig. 5), adhesive tape (Schwartz and Peck, 1935; Benkwith, 1946), and many others. Of particular ophthalmic interest are the dermatoses set up in association with a conjunctivitis after the local use of physostigmine (Alden and Jones, 1938), pilocarpine (Holmberg, 1955), dionine (Cummer, 1931), scopolamine, homatropine, phenacaine, butacaine (Parkhurst and Lukens, 1939; Lemoine, 1942), procaine (Goodman, 1939; Bandmann and Hardieck, 1960), larocaine (Theodore, 1938), nupercaine (Perera, 1940), paredrine (Laval, 1941), antihistamine drugs (Guerrant and Hollifield, 1951; Swinny, 1951), and particularly the erythemo-vesicular eczema associated with atropine, a reaction probably due to a susceptibility to compounds of the alkaloid and proteins in the tears (Waller, 1934; d'Ermo, 1953) (Plate I, Fig. 2).

Figs. 50 and 51.—Medicinal Dermatitis.

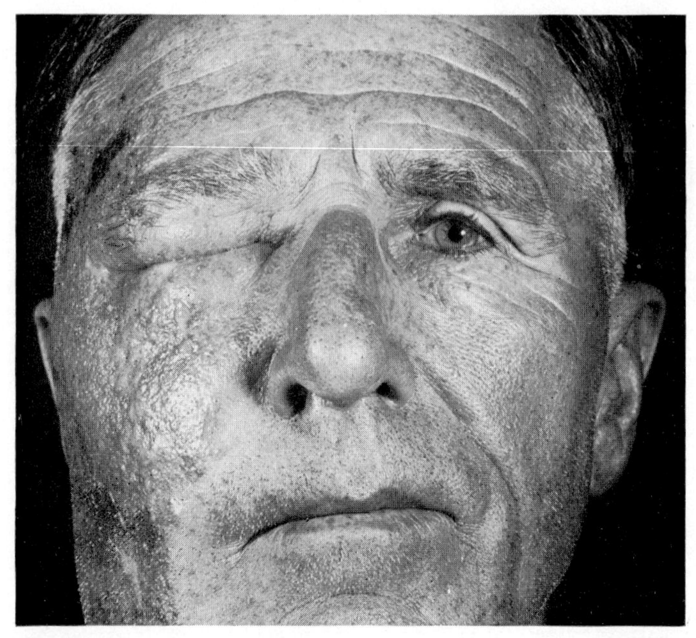

Fig. 50.—Due to the topical application of penicillin ointment after an insect bite (P. Borrie, St. Bartholomew's Hosp.).

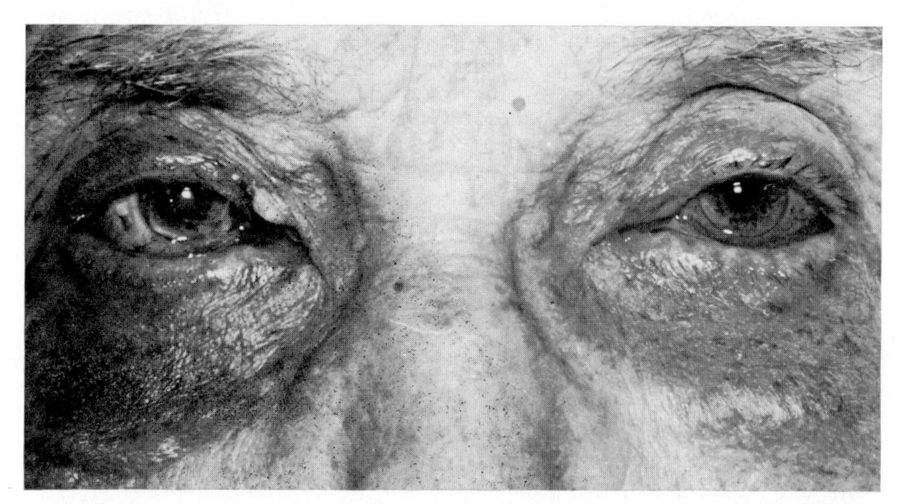

Fig. 51.—Due to the application of an ointment of chloramphenicol (Inst. Ophthal.).

A similar dermatitis, sometimes of considerable severity, may follow the topical use of the *sulphonamides* even in small quantities, either as drops instilled into the conjunctival sac or as ointments around the lids (Alvaro, 1943–45; Sidi and Dobkevitch, 1949; and others). It is noteworthy that the local use of the drugs may give rise to a generalized skin eruption (Shaffer *et al.*, 1943), and that their topical use can induce general sensitivity on a subsequent occasion (Cordes, 1947). Conversely, after previous systemic treatment a severe local reaction resembling the Arthus phenomenon may occur on their local application at a later date (Albright and Seretan, 1948). Of this class of medicament, sodium sulphacetamide is probably the most innocuous, a dermatitis following its use being comparatively rare (Benedict and Henderson, 1947; MacMillan, 1948).

The *antibiotic drugs* give rise to similar reactions. Among these, although penicillin is relatively non-toxic, between 1943 and 1948 more than 350 papers dealt with allergic reactions following its use (E. A. Brown, 1948). Although many of the earlier reactions were due to impurities, some 2·5% of all cases treated with the pure crystalline product

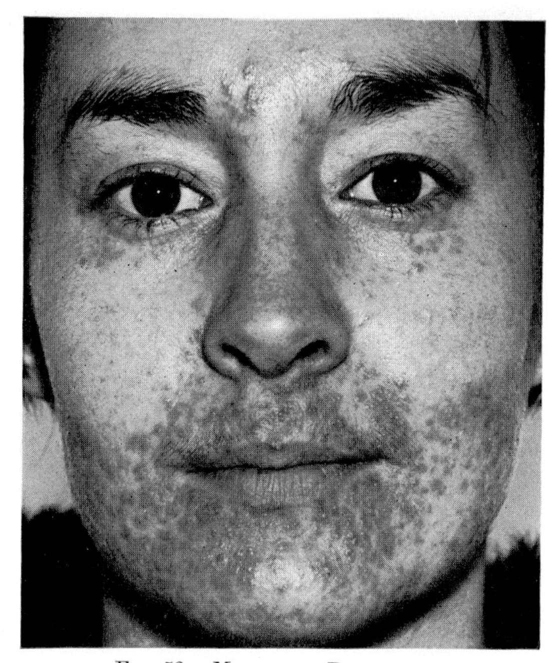

FIG. 52.—MEDICINAL DERMATITIS.

Due to the prolonged application of fluorinated corticosteroids. Note the papules on the lids and the marked perioral dermatitis (A. J. Rook, Addenbrooke's Hosp., Cambridge).

develop symptoms of this type. Mild or severe dermatitis has been recorded by several observers after its topical application to the eye[1] (Fig. 50). As a rule the reaction commences some 5 to 10 days after the initiation of treatment, but appreciable numbers of people are immediately sensitive, even although there is no history of the previous use of the drug; it is possible that these have already been sensitized by an earlier fungus infection (dermatomycosis) (Rostenberg and Welch, 1945; Kolodny and Denhoff, 1946). Nurses and laboratory workers who are in prolonged contact with the drug may also develop a similar dermatitis (Pyle and Rattner, 1944; Gottschalk and Weiss, 1946). Aureomycin (Burstein, 1950), neomycin (Kirton and Munro-Ashman, 1965), chloramphenicol (Fig. 51) and streptomycin may excite a similar

[1] Bellows (1944), Keyes (1944), Barker (1945), Selinger (1945), C. A. Brown (1946), Schultz (1946), Borrie (1956–69), and others.

reaction, and the eczematous dermatitis which may affect the hands and eyelids of nurses and others handling the last drug is so common that they should wear rubber gloves.[1]

The *fluorinated corticosteroids* may give rise to a papular dermatitis usually perioral in distribution but sometimes with papules on the lids (Fig. 52).

Several drugs give rise to a photosensitization of the skin involving an inflammatory reaction: these will be noted subsequently.[2]

COSMETIC DERMATITIS. The use of cosmetics is as old as recorded history—as is seen in the records of Babylon, Egypt, and the biblical story of Jezebel, who painted her face and looked out of her window when Jehu came to Jezreel[3]; and, since their employment has a fundamental biological value, there is no question but that their use will always be widespread so long as man and woman exist. Nevertheless, realization of the dermatological problems raised by their use is of comparatively recent date, and the frequency of cosmetic eruptions is still hardly recognized. Lain (1932), for example, reported that in 1930, 75% of all types of dermatitis of the face and neck in his clinic in the United States of America had this origin (Figs. 53–6).

The most common causes of dermatitis are perfumes and toilet waters, hair tonics, shampoos and setting lotions, mascara, eye-shadow, hair or lash dyes, face creams, bleaches and powders, rouges, lipsticks, deodorants and depilatories (Rattner, 1934; Tulipan, 1938; Garnier and Marshall, 1952) (Figs. 53, 56). In nail lacquer the causal constituents are the solvents and resins and the reaction appears rarely on the fingers, but usually in the places—like the eyelids—where the fingers frequently go[4] (Figs. 54–5). In Hazen's (1944) opinion, it is the most common cause of dermatitis of the lids in women. Hair lotions, tonics, dyes and fixatives, if they do not cause a general dermatitis, are usually effective by contact through pillow-cases; the burning and itching of the lids are thus most evident during the night or in the morning. In this way also it is to be remembered that a woman may affect her sleeping partner. The substances in perfumes and toilet waters to which susceptibility is most usually acquired are the natural essential oils (particularly oil of bergamot), or the alcoholic vehicles. Most face powders and creams are relatively innocuous, but talc, which is itself quite harmless, is seldom free from impurities which may be irritant; moreover, incorporated perfumes (orris, etc.) may produce dermatitis (Roy, 1924). It is sometimes difficult to convince the patient of the cause because frequently the lids alone are affected and not the entire face, but the thinnest and most sensitive part of the skin in the massaged area is the lids and here the trouble tends to start. Moreover, an application which has been used for a long period may suddenly become irritative, sometimes because a sensitivity has developed only after prolonged use, and sometimes because one or more constituents have been changed by the manufacturers.

The most noxious products, however, are the *dyes*, not only hair dyes (Cole, 1927), but also lipstick dye, the reaction of which may travel far afield (Zakon *et al.*, 1947). Those usually employed are aniline, lead and sulphur dyes, silver salts and pyrogallol dyes—all of which may cause irritation, some of them of a dangerous kind. Henna and indigo dyes are harmless in their pure form, but they are rarely sold pure and thus may cause severe reactions (Abramowicz, 1930; Bab, 1936); the most seriously dangerous substance is *paraphenylene-diamine* which may not only bring about a severe dermatitis but also general toxic symptoms. The usual reaction is an accentuation of the wrinkles of the lids with scaling rather than an œdematous swelling (Borrie, 1956) (Fig. 57), but when applied to the brows and lashes it may not only give rise to a

[1] Crofton and Foreman (1948), Dufour (1948), Ricci and Bruna (1948), Charamis (1949), Weekers (1950).

[2] p. 79.

[3] Kings II, ix, 30.

[4] Simon (1943), Dobes and Nippert (1944), Sidi and Leven (1948), Calnan and Sarkany (1958), and others.

FIGS. 53 to 56.—COSMETIC DERMATITIS.

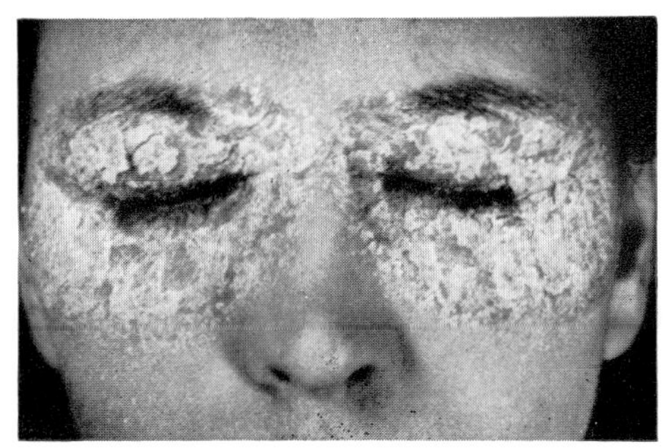

FIG. 53.—Due to an eyelid cream (L. Hollander).

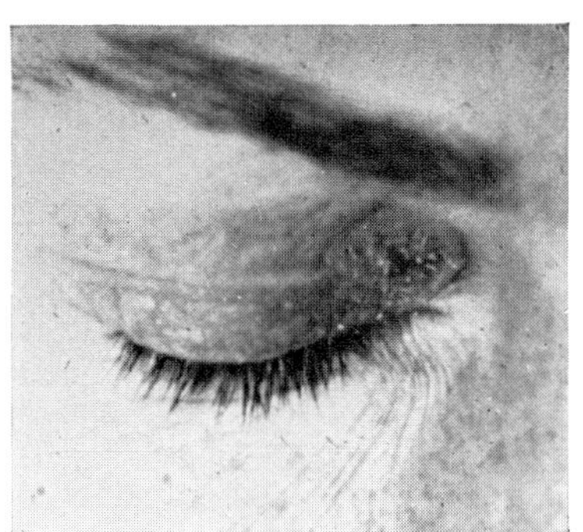

FIG. 54.

FIG. 55.

FIGS. 54 and 55.—Due to nail polish (W. L. Dobes).

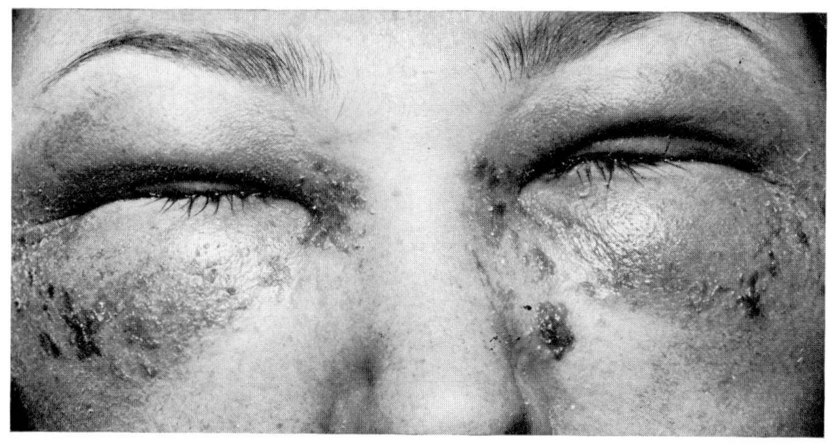

FIG. 56.—Due to eye-shadow (R. J. H. Smith).

violent dermatitis of the lids, but also to a kerato-conjunctivitis of considerable severity involving corneal ulceration with permanent leucomata, cyclitis and secondary glaucoma with disastrous effects on vision[1]; a fatal case has been reported wherein death resulted after gangrene of the lids following the use of an aniline dye containing paraphenylene-diamine on the brows and lashes (Forbes and Blake, 1934). It is quite obvious that if it is decided that the lashes or brows be dyed, a preliminary skin test of sensitization should always be undertaken first in an unimportant part of the body. Once the dermatitis has developed the most effective treatment is to epilate the dyed hair (brows or lashes) immediately.

Diagnosis. The diagnosis of dermatitis venenata is usually easy, but when this has been established, the problem does not end but begins, for unearthing the particular cause is by no means a simple matter. Particularly is this so when involvement of the eyelids occurs as a systemic

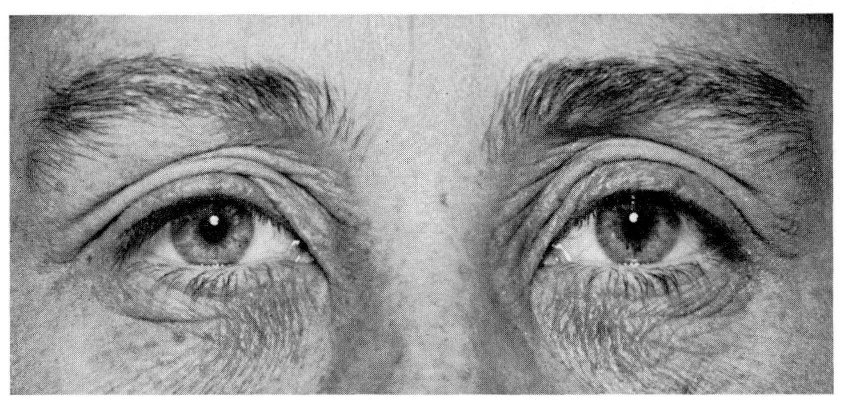

FIG. 57.—COSMETIC DERMATITIS.
Due to the hair dye, paraphenylene-diamine (P. Borrie).

symptom of some distant contact. It is to be remembered that *repeated attacks of swelling and itching of the lids are usually due to dermatitis venenata.* In some cases the trouble is obvious—as in spectacle dermatitis; but more usually the elucidation of a case involves an extensive and tedious investigation into all sorts of unlikely matters concerning the activities and environment of the patient. A rational procedure in the first place, in the event of no obvious environmental cause being found, is to stop all treatment, isolate the patient's skin from all contacts except air, water, cotton or linen, and, if the lesion clears up, to resume the multitudinous contacts of life again one by one until the reaction reappears and the cause can thus be traced (see R. L. and R. L. Sutton, 1949). The diagnosis is, indeed, frequently never established; but if it is suspected, the nature of a particular allergen can be proved by a positive patch test, possibly a flare-up of the

[1] Harner (1933), Bourbon (1933), McCally *et al.* (1933), Jamieson (1933), Greenbaum (1933), Moran (1934), Block (1935), Ramejev (1936), and others.

original sites of dermatitis when the patch reaction occurs and, most important of all, by the cure of the condition when the allergen is completely avoided.

The *patch test*, which is preferably carried out during a period when the dermatitis is relatively quiet, is most effectively and simply done by applying the substance in question to normal skin such as the back, arm or thigh, covering it with linen or cotton which, if the substance is powder or dry particles, should be moistened, and strapping the whole under a thin rubber sheet with adhesive tape. This should be removed in 24 hours when an erythema denotes a positive result; a further examination is, however, advisable in 48 hours (Bloch, 1929; Percival, 1931; Rostenberg and Sulzberger, 1939; Warren, 1943; Henderson and Riley, 1945; Borrie, 1969) (Fig. 58). About 5 hours after the application of the test vasodilatation and œdema appear with migration of lymphocytes into the epidermis carrying with them antibodies; thereafter intercellular vacuoles appear, developing into vesicles in about 8 to 10 hours and within 24 hours into cytoplasmic degeneration of the cells (Fisher and Cooke, 1958; Bandmann, 1960; Miescher, 1961; Flax and Caulfield, 1963).

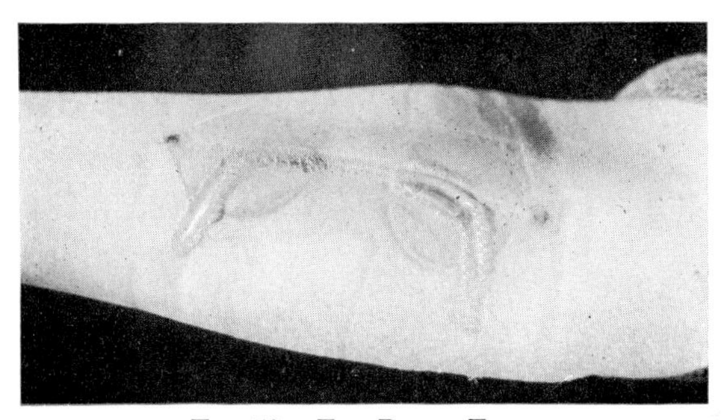

FIG. 58.—THE PATCH TEST.
In a case of spectacle dermatitis (H. R. Vickers).

The essential *treatment* is to remove the cause and give the patient symptomatic relief until the lesion heals. To remove the allergen is sometimes easy but sometimes extremely difficult—either because of its wide dissemination or because the patient's livelihood depends upon his contacts. In this event desensitization may be tried; but it is a long process not by any means universally successful. When an urticarial element is prominent, antihistamine preparations, such as benadryl, may be of value; with its use systemically and locally as an ointment, Fralick and Kiess (1949) found themselves able to control an atropine dermatitis and persist in the use of the drug, but unhappily this does not always occur. As already noted, corticosteroids frequently give immediate and dramatic, although temporary, relief. In all cases, no matter what treatment is adopted, any seborrhœic complication should at the same time receive energetic attention, for this always makes matters worse.

As regards local treatment, none is best, the simplest and mildest applications are the second best; the problem in this disease is of *keeping something away from the skin, not putting it on.* Local applications, if employed at all, should be confined to the blandest available, such as the soothing applications already mentioned[1]; in severe cases the systemic administration of corticosteroids may be useful; while, if chronicity has developed, X-rays may prove helpful in cutting short the irritation. It should be stressed that if the dermatitis clears the allergy remains and the risk of relapses after further contacts with the allergen persists throughout life.

NEURODERMATITIS

Neurodermatitis circumscripta (*chronic lichen simplex*) is a chronic intermittent condition the exacerbations of which are usually associated with stress; it is probable, however, that most cases are due to an unrecognized

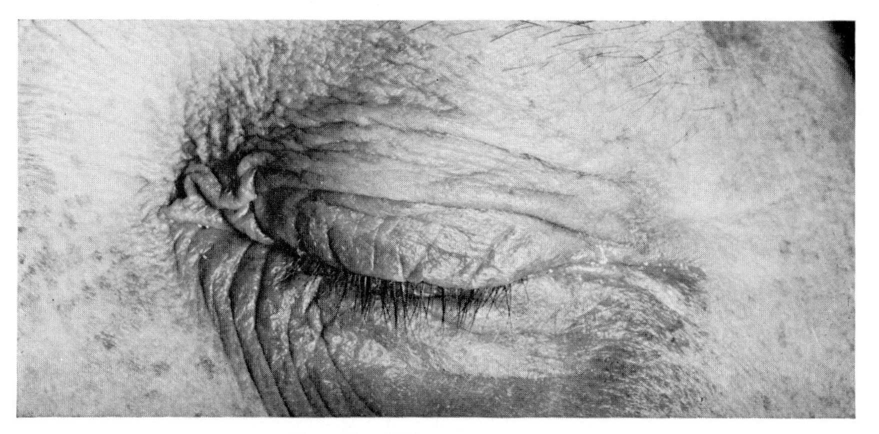

FIG. 59.—NEURODERMATITIS.

There is pronounced irritation accompanied by slight swelling and considerable thickening and pigmentation of the lids (P. Borrie).

contact dermatitis. The eyelids are frequently involved, often with other patches of neurodermatitis elsewhere. There is some swelling and considerable thickening and pigmentation of the skin of the lids but true lichenification does not invariably occur (Fig. 59); irritation is the most prominent symptom. The histopathological picture is that of a chronic dermatitis (Sachs *et al.*, 1946). The condition occurs preferentially in females from the fifth to the seventh decades, all of the patients suffering from anxiety or stress (domestic anxieties, dislike of employment, anxiety about health, and so on). Relief is often obtained by eliminating the cause and by reassurance, with the judicious use of sedatives and amphetamine, while hydrocortisone cream is useful as a topical application to relieve the irritation (Borrie, 1956).

[1] p. 56.

DERMATITIS ARTEFACTA

A dermatitis artefacta of the lids produced by patients who are usually psychopathic is not so common and not nearly so serious as a kerato-conjunctivitis artefacta caused by inserting irritant substances into the conjunctival sac.[1] The diagnosis is frequently difficult since the patient usually strenuously denies the cause of the lesion which varies according to the agent used, whether persistent scratching often to aggravate an existing eruption or the application of irritant fluids. The characteristic features are the absence of the typical appearance of any known disease and its localized configuration depending on the site of the application of the mechanical or chemical irritant used (Fig. 60).

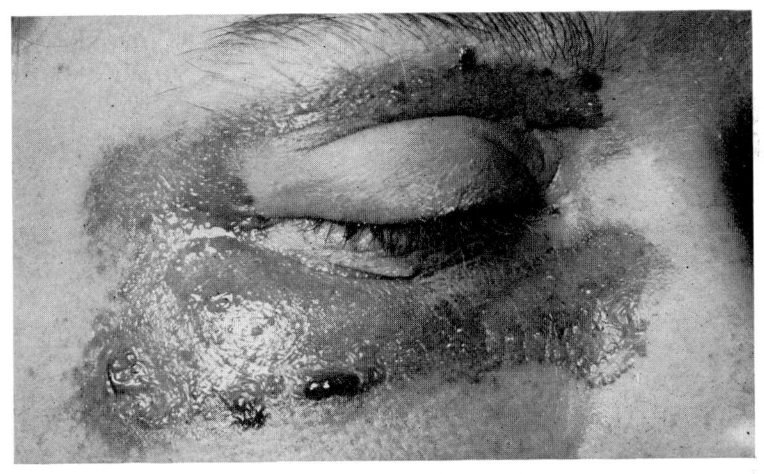

FIG. 60.—DERMATITIS ARTEFACTA.

Note the well-defined margin of the lesion and the sparing of the upper lid itself (P. Borrie, St. Bartholomew's Hospital).

The patients are usually hysterical young women who wish to attract attention to themselves or to escape from an unpleasant situation such as neglect or cruelty at home, in which case the self-infliction sometimes seems to have been done unconsciously, but occasionally it may be produced deliberately for the purpose of malingering; sometimes an attempt is made to prolong a genuine skin eruption with a view to maintaining a sympathy which perhaps has been lacking. The treatment is essentially psychiatric and should be undertaken with sympathetic understanding; to accuse a patient of producing an injury on herself is generally useless since she will tend to react violently against the suggestion and probably continue to mutilate herself elsewhere. Fortunately, in most cases the psychological trouble is of recent origin and can be discovered, and if it is satisfactorily put to rights a cure readily follows.

[1] Vol. XIV, pp. 56, 1016.

Abramowicz. *Klin. oczna*, **8,** 153 (1930).
Albright and Seretan. *Amer. J. Ophthal.*, **31,** 1603 (1948).
Alden and Jones. *Arch. Derm. Syph.* (Chic.), **37,** 82 (1938).
Alvaro. *Arch. Ophthal.*, **29,** 615 (1943).
Amer. J. Ophthal., **28,** 497 (1945).
Bab. *Klin. Mbl. Augenheilk.*, **97,** 391 (1936).
Ballin. *Arch. Derm. Syph.* (Chic.), **27,** 292 (1933).
Bandmann. *Hautarzt*, **11,** 258, 310, 355, 393 (1960).
Bandmann and Hardieck. *Hautarzt*, **11,** 468 (1960).
Barker. *Lancet*, **1,** 177 (1945).
Beerman. *Arch. Derm. Syph.* (Chic.), **29,** 671 (1934).
Bellows. *Amer. J. Ophthal.*, **27,** 1206 (1944).
Benedict and Henderson. *Amer. J. Ophthal.*, **30,** 984 (1947).
Benkwith. *Arch. Ophthal.*, **36,** 620 (1946).
Bergstrand. *Brit. med. J.*, **1,** 89 (1950).
Berkoff. *Arch. Derm. Syph.* (Chic.), **38,** 746 (1938).
Biederman. *New Engl. J. Med.*, **219,** 117 (1938).
Billo. *Amer. J. med. Sci.*, **201,** 756 (1941).
Blaschko. *Dtsch. med. Wschr.*, **15,** 925 (1889).
Bloch. *Arch. Derm. Syph.* (Chic.), **19,** 175 (1929).
Block. *Amer. J. Ophthal.*, **18,** 1052 (1935).
Bodeau. *Arch. Méd. Pharm. nav.*, **126,** 122 (1936).
Bonnevie and Marcussen. *Acta derm.-venereol.* (Stockh.), **25,** 163 (1944).
Borrie. *Brit. J. Ophthal.*, **40,** 742 (1956).
Roxburgh's *Common Skin Diseases*, 13th ed., London, 174 (1969).
Bourbon. *J. Amer. med. Ass.*, **101,** 1559 (1933).
Brown, C. A. *Brit. J. Ophthal.*, **30,** 146 (1946).
Brown, E. A. *Ann. Allergy*, **6,** 723 (1948).
Brunsting and Anderson. *J. Amer. med. Ass.*, **103,** 1285 (1934).
Brunsting and Williams. *J. Amer. med. Ass.*, **106,** 1533 (1936).
Buckley. *J. Irish med. Ass.*, **47,** 98 (1960).
Bulmer and Mackenzie. *J. industr. Hyg.*, **8,** 517 (1926).
Burrage. *J. Amer. med. Ass.*, **92,** 799 (1929).
Burstein. *Amer. J. Ophthal.*, **33,** 973 (1950).
Calnan and Sarkany. *Trans. St. John's Hosp. derm. Soc.*, **40,** 1 (1958).
Caulfeild. *Canad. med. Ass. J.*, **34,** 506 (1936).
J. Allergy, **9,** 535 (1937–8).
Caulfeild, Brown and Waters. *J. Allergy*, **7,** 1 (1935–36).
Charamis. *Brit. J. Ophthal.*, **33,** 714 (1949).
Cleveland. *Canad. med. Ass. J.*, **17,** 695 (1927).
Clifford. *Boston med. Surg. J.*, **195,** 931 (1926).
Cole. *J. Amer. med. Ass.*, **88,** 397 (1927).
Cole and Driver. *J. industr. Hyg.*, **4,** 425 (1923).
Cordes. *Amer. J. Ophthal.*, **30,** 768 (1947).
Cormia and Stewart. *Canad. med. Ass. J.*, **32,** 270 (1935).
Crofton and Foreman. *Brit. med. J.*, **2,** 71 (1948).
Cummer. *Arch. Derm. Syph.* (Chic.), **23,** 68 (1931).
Curjel and Acton. *Indian J. med. Res.*, **12,** 257 (1924–25).
Curtis. *Arch. Derm. Syph.* (Chic.), **52,** 262 (1945).
Desoille. *Encycl. Méd. Chir.*, Paris, 16,037 (1937).
Dobes and Nippert. *Arch. Derm. Syph.* (Chic.), **49,** 183 (1944).
Downing. *J. industr. Hyg.*, **17,** 138 (1935).
Dufour. *Praxis*, **37,** 427 (1948).
Dutton. *Med. Rec.* (N.Y.), **140,** 550 (1934).
Earle. *Trans. roy. Soc. trop. Med. Hyg.*, **32,** 363 (1938).
Engelhardt. *Münch. med. Wschr.*, **82,** 256 (1935).
d'Ermo. *Atti Cong. Soc. oftal. ital.*, **14,** 219 (1953).
Fisher and Cooke. *J. Allergy*, **29,** 411 (1958).
Flax and Caulfield. *Amer. J. Path.*, **43,** 1031 (1963).
Forbes and Blake. *J. Amer. med. Ass.*, **103,** 1441 (1934).
Foster and Ball. *Arch. Derm. Syph.* (Chic.), **31,** 461 (1935).
Fox, E. C. *Arch. Derm. Syph.* (Chic.), **28,** 222 (1933).
Fox, H. *J. Amer. med. Ass.*, **101,** 1066 (1933).
Fralick and Kiess. *Arch. Ophthal.*, **41,** 583 (1949).
Furman, Fisher and Leider. *J. invest. Derm.*, **15,** 223 (1950).
Garnier and Marshall. *S. Afr. med. J.*, **26,** 490 (1952).
Goldman. *Arch. Derm. Syph.* (Chic.), **28,** 688 (1933).
Goodman. *J. invest. Derm.*, **2,** 53 (1939).
Gottschalk and Weiss. *Arch. Derm. Syph.* (Chic.), **53,** 365 (1946).
Grana. *Arch. Ophthal.*, **35,** 421 (1946).
Gray. *Brit. med. J.*, **1,** 648 (1943).
Greenbaum. *J. Amer. med. Ass.*, **101,** 363 (1933).
Guerrant and Hollifield. *Amer. J. Ophthal.*, **34,** 1318 (1951).
Hannah. *J. Amer. med. Ass.*, **72,** 853 (1919).
Harley. *Amer. J. Ophthal.*, **27,** 628 (1944).
Harner. *J. Amer. med. Ass.*, **101,** 1558 (1933).
Amer. J. Ophthal., **17,** 251 (1934).
Harper. *J. Pediat.*, **5,** 794 (1934).
Harville. *J. Allergy*, **4,** 527 (1932–33).
Hazen. *Arch. Derm. Syph.* (Chic.), **49,** 253 (1944).
Henderson and Riley. *J. invest. Derm.*, **6,** 227, 231 (1945).
Hollander and Baer. *Amer. J. Ophthal.*, **18,** 616 (1935).
Holmberg. *Acta ophthal.* (Kbh.), **33,** 371 (1955).
Hugo. *S. Afr. med. J.*, **12,** 763 (1938).

Jacobs and Colmes. *J. lab. clin. Med.*, **26,** 302 (1940).
Jadassohn and Schaaf. *Arch. Derm. Syph.* (Berl.), **157,** 572 (1929).
Jamieson. *J. Amer. med. Ass.*, **101,** 1560 (1933).
Keil. *Arch. Derm. Syph.* (Chic.), **47,** 242 (1943).
Kern. *Ophthalmologica*, **145,** 369 (1963).
Kesten. *Arch. Ophthal.*, **6,** 582 (1931).
Keyes. *J. Amer. med. Ass.*, **126,** 610 (1944).
Kiep. *Trans. ophthal. Soc. U.K.*, **46,** 383 (1926).
King. *W. Virginia med. J.*, **34,** 28 (1938).
Kirton and Munro-Ashman. *Lancet*, **1,** 138 (1965).
Kolodny and Denhoff. *J. Amer. med. Ass.*, **130,** 1058 (1946).
Kristjansen. *Acta derm.-venereol.* (Stockh.), **18,** 519 (1937).
Lain. *J. Amer. med. Ass.*, **96,** 771 (1931).
Sth. med. J., **25,** 718 (1932).
Landsteiner. *J. exp. Med.*, **39,** 631 (1924).
Lane. *Arch. Derm. Syph.* (Chic.), **5,** 589 (1922).
Lane, Dennie, Downing *et al.* *J. Amer. med. Ass.*, **118,** 613 (1942).
Laval. *Arch. Ophthal.*, **26,** 585 (1941).
Leider and Baer. *J. invest. Derm.*, **10,** 425 (1948).
Lemoine. *Arch. Ophthal.*, **28,** 79 (1942).
Livingood, Rogers and Fitz-Hugh. *J. Amer. med. Ass.*, **123,** 23 (1943).
Löwenstein. *Allergische Augenerkrankungen*, Basel (1938).
McAlester, A. W. and A. W. Jr. *Amer. J. Ophthal.*, **14,** 925 (1931).
McCally, Farmer and Loomis. *J. Amer. med. Ass.*, **101,** 1560 (1933).
McConnell. *Publ. Hlth. Rep.* (Wash.), **36,** 979 (1921).
McCord. *J. Amer. med. Ass.*, **86,** 1979 (1926).
MacCormac. *Proc. roy. Soc. Med.*, **24,** 518 (1931).
MacIvor. *Canad. med. Ass. J.*, **62,** 164 (1950).
MacMillan. *Arch. Ophthal.*, **39,** 554 (1948).
McNair. *Arch. Derm. Syph.* (Chic.), **5,** 383, 625 (1921).
Rhus Dermatitis, Chicago (1923).
Martin. *Derm. Wschr.*, **121,** 553 (1950).
Miescher. *Arch. klin. exp. Derm.*, **213,** 297 (1961).
Molitch and Poliakoff. *Arch. Derm. Syph.* (Chic.), **33,** 715 (1936).
Moran. *J. Amer. med. Ass.*, **102,** 286 (1934).
Morris. *Arch. Derm.*, **88,** 220 (1963).
Niles. *N.Y. St. J. Med.*, **36,** 113 (1936).
Oelze. *Derm. Wschr.*, **79,** 997 (1924).
Olson. *J. Amer. med. Ass.*, **66,** 864 (1916).
van Ordstrand, Hughes, de Nardi and Carmody. *J. Amer. med. Ass.*, **129,** 1084 (1945).
Parkhurst and Lukens. *J. Amer. med. Ass.*, **112,** 837 (1939).
Peck. *J. Amer. med. Ass.*, **125,** 190 (1944).
Peck, Gant and Schwartz. *Industr. Med.*, **14,** 214 (1945).
Percival. *Lancet*, **2,** 417 (1931).
Perera. *Arch. Ophthal.*, **24,** 344 (1940).
Peterkin. *Trans. ophthal. Soc. U.K.*, **68,** 353 (1948).
Pringle. *Brit. J. Derm.*, **54,** 54 (1942).
Pusey and Rattner. *Arch. Derm. Syph.* (Chic.), **19,** 917 (1929).
Pyle and Rattner. *J. Amer. med. Ass.*, **125,** 903 (1944).
Quiroga and Guillot. *Rev. argent. Dermatosif.*, **39,** 58 (1955).
Ramejev. *Vestn. Oftal.*, **9,** 509 (1936).
Rattner. *J. Amer. med. Ass.*, **103,** 180 (1934).
Ricci and Bruna. *Boll. Oculist.*, **27,** 330 (1948).
Rostenberg and Sulzberger. *J. invest. Derm.*, **2,** 93 (1939).
Rostenberg and Welch. *Amer. J. med. Sci.*, **210,** 158 (1945).
Rothman. *J. Amer. med. Ass.*, **97,** 336 (1931).
Roy. *J. Amer. med. Ass.*, **82,** 208 (1924).
Sachs, Miller and Gray. *Arch. Derm. Syph.* (Chic.), **54,** 397 (1946).
Samitz. *Arch. Derm. Syph.* (Chic.), **50,** 10 (1944).
Schittenhelm and Stockinger. *Z. ges. exp. Med.*, **45,** 58 (1925).
Schultz. *Arch. Ophthal.*, **35,** 145 (1946).
Schwartz. *Mich. med. Soc. J.*, **39,** 179 (1940).
Schwartz and Peck. *Publ. Hlth. Rep.* (Wash.), **50,** 811 (1935).
J. Amer. med. Ass., **128,** 1209 (1945).
Schwartz, Spolyar, Gastineau *et al.* *J. Amer. med. Ass.*, **115,** 906 (1940).
Schwartz, Tulipan and Peck. *Occupational Diseases of the Skin*, Phila. (1947).
Scott. *Brit. med. J.*, **2,** 381, 1108 (1922).
Selinger. *J. Amer. med. Ass.*, **128,** 437 (1945).
Shaffer, Lentz and McGuire. *J. Amer. med. Ass.*, **123,** 17 (1943).
Shaw. *Arch. Derm. Syph.* (Chic.), **49,** 191 (1944).
Shelmire. *J. Amer. med. Ass.*, **113,** 1085 (1939).
Shemeley. *E.E.N.T. Monthly*, **30,** 543 (1951).
Sidi and Dobkevitch. *Arch. Ophtal.*, **9,** 311 (1949).
Sidi and Leven. *Thérapie*, **3,** 145 (1948).
Simon. *Sth. med. J.*, **36,** 157 (1943).
Simpson. *J. Amer. med. Ass.*, **69,** 95 (1917).
Smith, E. L. and Calnan. *Trans. St. John's Hosp. derm. Soc.*, **52,** 10 (1966).
Snow and Harley. *Arch. Derm. Syph.* (Chic.), **49,** 236 (1944).
Stauffer. *Arch. Derm. Syph.* (Berl.), **162,** 517 (1931).
Stewart. *Arch. intern. Med.*, **51,** 427 (1933).
Sulzberger and Finnerud. *J. Amer. med. Ass.*, **111,** 1528 (1938).

Sutton, R. L. *J. Amer. med. Ass.*, **89**, 1059 (1927).
Sutton, R. L. and R. L. Jr. *Diseases of the Skin*, London (1949).
Swinny. *Ann. Allergy*, **9**, 774 (1951).
Taylor, Fergusson and Atkins. *Brit. med. J.*, **2**, 40 (1945).
Teulières. *Bull. Soc. Ophtal. Fr.*, **62**, 552 (1962).
Theodore. *Arch. Ophthal.*, **20**, 474 (1938). *N.Y. St. J. Med.*, **58**, 2233 (1958).
Thies, Schwarz and Palme. *Arch. klin. exp. Derm.*, **223**, 558 (1965).
Thistlethwaite. *Brit. med. J.*, **1**, 493 (1943).
Tulipan. *Arch. Derm. Syph.* (Chic.), **38**, 906 (1938).
Urbach. *Klinik. u. Therapie d. allergischen Krankheiten*, Wien (1935).
Vokoun. *J. Amer. med. Ass.*, **89**, 20 (1927).
Waller. *Trans. ophthal. Soc. U.K.*, **54**, 96 (1934).
Walthard. *Schweiz. med. Wschr.*, **56**, 603 (1926).
Warren. *Sth. med. J.*, **36**, 435 (1943).
Weber. *Arch. Derm. Syph.* (Berl.), **35**, 129 (1937).
Weekers. *Rev. méd. Liège*, **5**, 320 (1950).
White. *Occupational Affections of the Skin*, London (1920; 1933).
Wilde. *Derm. Wschr.*, **140**, 1089 (1959).
Wilhelm, Sarkany and Calnan. *Trans. St. John's Hosp. derm. Soc.*, **41**, 31 (1958).
Wrong. *Canad. med. Ass. J.*, **32**, 273 (1935).
Zakon, Goldberg and Kahn. *Arch. Derm. Syph.* (Chic.), **56**, 499 (1947).

DERMATITIS MEDICAMENTOSA

DRUG DERMATITIS, *an eruption resulting from the systemic administration of chemicals*—mainly medicinal agents—either by mouth, injection, inhalation or otherwise, in contradistinction to the contact dermatitis which results from their application to the skin, may produce almost any form of dermatitic reaction which, however, is usually of the eczematous type. The urticarial reaction of such drugs has already been noted.[1] A dermatitis is a relatively common phenomenon, one drug producing different reactions in different persons and similar eruptions being caused by dissimilar drugs. They act as allergens (not antigens) with no circulatory antibodies, producing varying degrees of sensitivity, the reaction, of course, being frequently not limited to the skin. Because of its susceptibility, the skin of the eyelids may be the first to show the reaction, and when the dermatitis is general they usually share in it. A patch test is usually positive in susceptible patients and relief follows discontinuance of the drug while the dermatitis is reproduced on re-administration. If no specific remedy exists, antihistaminic drugs (benadryl, antistin, etc.) may bring considerable but only temporary relief, as also may the corticosteroids.

A large number of drugs may be responsible for such reactions. Among the most common and most severe affecting the lids is the reaction excited by *arsenic*, particularly in the form of arsphenamines, characterized typically by an eczematous dermatitis followed by desquamation (exfoliative dermatitis) (Beerman and Stokes, 1941); the associated lesions which may appear in the conjunctiva and the cornea have already been noted.[2] *Gold* (sanocrysin) may give a similar reaction. Both of these are cured by the administration of BAL (dimercaptopropanol), which successfully competes with the vital tissues for the metallic salts to form an inactive compound (Peters *et al.*, 1945; Cohen *et al.*, 1947). An exfoliative dermatitis may also follow the use of *barbiturates* (Winer and Baer, 1941; Moss and Long, 1942), Luminal (Hollander and Baer, 1935), the *sulphonamides* (Livingood and Pillsbury, 1943; Johnson, 1944; and others) and *antibiotics* such as penicillin (Templeton *et al.*, 1947; Farrington *et al.*, 1948; Shaffer, 1948; and others), streptomycin (Steiner and Fishburn, 1947; Lindars

[1] p. 63. [2] Vol. VIII, p. 459.

FIGS. 61 and 62.—BROMODERMA.

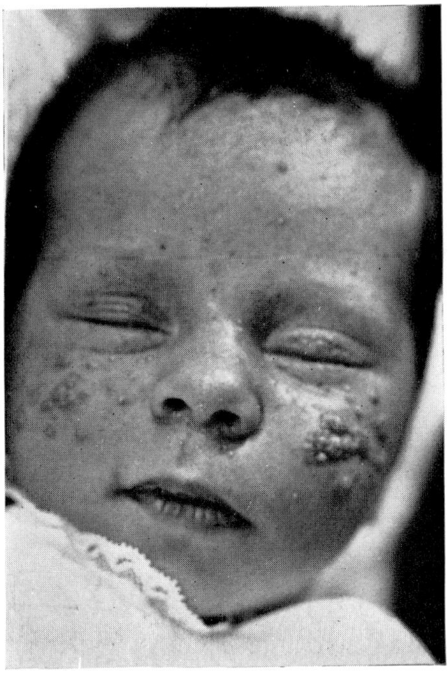

FIG. 61.—Derived from breast-feeding (Dept. of Dermatology, Vanderbilt Clinic, N.Y.).

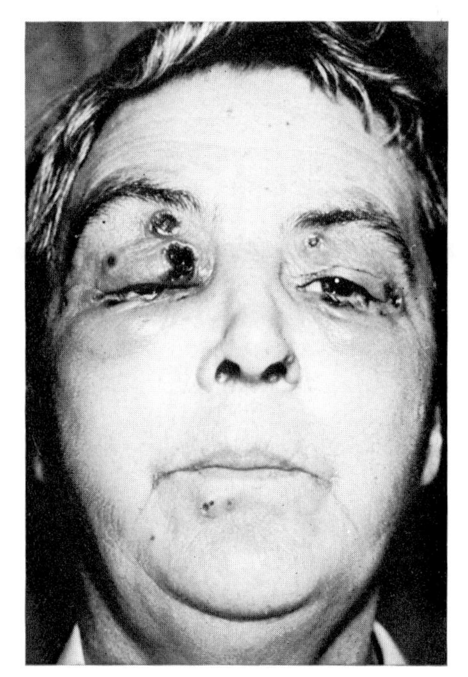

FIG. 62.—A typical case involving the eyelids (N. R. Rowell).

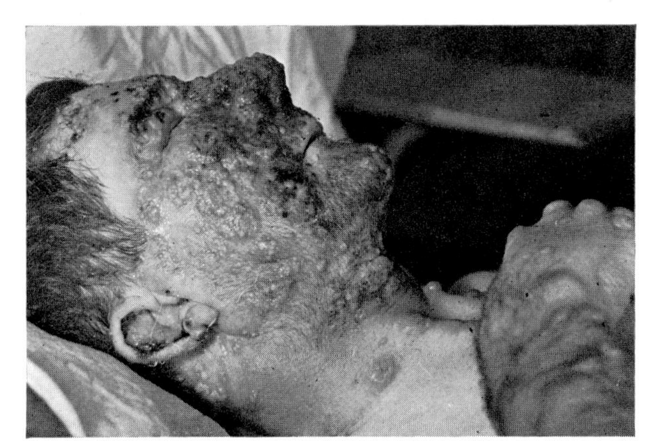

FIG. 63.—DERMATITIS MEDICAMENTOSA: IODODERMA.

A fatal case in a man aged 47 who was given 20 ml. of iodized oil (Lipiodol-Lafay) as an injection into the lung for diagnostic purposes. There was a papulo-pustular eruption over the face involving the eyelids, which spread to the upper parts of the chest and the extensor surfaces of the arms and hands. Death resulted from nephritis and an overwhelming toxæmia (D. W. Goldstein).

1950; Harris and Walley, 1950; and others) and chloramphenicol (Dunphy, 1950); while the furuncular rash or pustules caused by *bromides* (Figs. 61–2) and the generalized *iodine* eruption (urticarial, papular or bullous) may particularly affect the lids (Goldstein, 1936; Cape, 1954) (Fig. 63).

A more severe and extremely toxic reaction results in *toxic epidermal necrolysis* (*Lyell's disease*, 1956) wherein the skin becomes intensely red and exquisitely tender over large areas, this layer becomes necrotic and loosens, large bullæ are formed, and the epidermis peels off in sheets leaving raw weeping surfaces (the *scalded skin syndrome*). The condition is serious with an overall mortality of 25 to 30% (Zak *et al.*, 1964; Bailey *et al.*, 1965) and ocular complications are frequent, sometimes involving symblepharon and leucomata. This syndrome, which is of multiple ætiology, will be described at a later stage.[1]

A large number of drugs has been responsible for this reaction of which Franceschetti and his colleagues (1965) collected 92 cases from the literature. The most common are phenolphthalein (Lang and Walker, 1957; Browne and Ridge, 1961; Rowell and Thompson, 1961; Björnberg *et al.*, 1964), the sulphonamides (Browne and Ridge, 1961), penicillin (Lang and Walker, 1957) and the butazones (Oppel, 1963; Calmettes *et al.*, 1966; Lyell, 1967; Ostler *et al.*, 1970).

Bailey, Rosenbaum and Anderson. *J. Amer. med. Ass.* **191,** 979 (1965).
Beerman and Stokes. *Amer. J. med. Sci.*, **201,** 611; **202,** 606 (1941).
Björnberg, A. and K., and Gisslén. *Acta ophthal.* (Kbh.), **42,** 1084 (1964).
Browne and Ridge. *Brit. med. J.*, **1,** 550 (1961).
Calmettes, Bazex, Salvador and Déodati. *Bull. Soc. Ophtal. Fr.*, **66,** 429 (1966).
Cape. *Brit. med. J.*, **1,** 255 (1954).
Cohen, Goldman and Dubbs. *J. Amer. med. Ass.*, **133,** 749 (1947).
Dunphy. *Arch. Ophthal.*, **44,** 797 (1950).
Farrington, Riley and Olansky. *Sth. med. J.*, **41,** 614 (1948).
Franceschetti, Ricci and Diallinas. *Bull. Soc. franç. Ophtal.*, **78,** 339 (1965).
Goldstein. *J. Amer. med. Ass.*, **106,** 1659 (1936).
Harris and Walley. *Lancet*, **1,** 112 (1950).
Hollander and Baer. *Amer. J. Ophthal.*, **18,** 616 (1935).
Johnson. *J. Amer. med. Ass.*, **124,** 979 (1944).
Lang and Walker. *S. Afr. med. J.*, **31,** 713 (1957).
Lindars. *Lancet*, **1,** 110 (1950).
Livingood and Pillsbury. *J. Amer. med. Ass.*, **121,** 406 (1943).
Lyell. *Brit. J. Derm.*, **68,** 355 (1956); **79,** 662 (1967).
Moss and Long. *Arch. Derm. Syph.* (Chic.), **46,** 386 (1942).
Oppel. *Ber. dtsch. ophthal. Ges.*, **65,** 52 (1963).
Ostler, Conant and Groundwater. *Trans. Amer. Acad. Ophthal.*, **74,** 1254 (1970).
Peters, Stocken and Thompson. *Nature* (Lond.), **156,** 616 (1945).
Rowell and Thompson. *Brit. J. Derm.*, **73,** 278 (1961).
Shaffer. *New Engl. J. Med.*, **238,** 660 (1948).
Steiner and Fishburn. *Arch. Derm. Syph.* (Chic.), **56,** 511 (1947).
Templeton, Lunsford and Allington. *Arch. Derm. Syph.* (Chic.), **56,** 325 (1947).
Winer and Baer. *Arch. Derm. Syph.* (Chic.), **43,** 473 (1941).
Zak, Fellner and Geller. *Amer. J. Med.*, **37,** 140 (1964)

Dermatitis due to Physical Agents

The dermatitis due to *burns* is described at length elsewhere[2] as also is that due to *cold*[3] and *ionizing radiation*.[4] A short note is added here on the reactions due to *light*, particularly short ultra-violet radiation. The pigmentation of sun-tan after repeated exposure to small amounts of light occurs

[1] p. 286.
[2] Vol. XIV, p. 747.
[3] Vol. XIV, p. 776.
[4] Vol. XIV, p. 956.

on the lids as elsewhere, as also its localized deposition as freckles in fair individuals in whom the pigment-forming capacity of the basal layers of the epidermis is confined to small areas.

EPHELIDES

EPHELIDES (FRECKLES), little circumscribed brown spots, seen especially in adolescents and usually following exposure to the sun, are due to deposits of melanin in the deeper epithelial cells (Fig. 64); although not an inflammation they are most conveniently considered here. They may be transient or permanent, and may besprinkle the lids, nose and forehead in profusion during the summer months to disappear gradually as winter develops. There is no elongation of the rete ridges and the number of melanocytes is

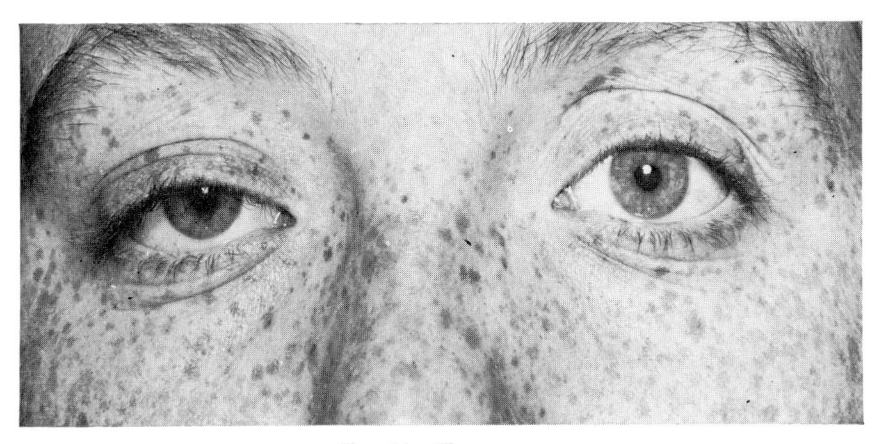

FIG. 64.—FRECKLES.
In a young girl (Inst. Ophthal.).

usually decreased; the melanocytes which are present, however, are larger than usual, are more strongly dopa-positive, with more numerous and longer dendritic processes (Breathnach, 1957). Treatment, if that is necessary, is prophylactic by avoidance of direct sun or the use of a cream which cuts out the ultra-violet rays (disodium naphthal sulphonate). Removal is probably unwise, but can be effected by dabbing on an alcoholic solution[1] which, by setting up an imperceptible exfoliative dermatitis, desquamates the epithelium.

More persistent spots occur in the senile skin (*senile lentigo; benign lentigo*) and malignant changes have been noted in similar lesions (*malignant lentigo* of Hutchinson, 1892–1904; *melanosis circumscripta preblastomatosa*); these are noted elsewhere.[2] In this connection the association of senile keratosis[3] and xeroderma pigmentosum[4] should also be noted.

[1] Mercuric chloride 1, alcohol 25, water 74.
[2] p. 528. [3] p. 404. [4] p. 445.

ACUTE SOLAR DERMATITIS is the result of exposure to light of considerable intensity. It occurs some hours after exposure and is characterized by erythema and œdema with a considerable feeling of burning and tenderness; when it affects the face the œdema of the lids may close the eyes and if the cornea has been exposed a solar keratitis develops.[1] After some days the erythema and swelling subside to be followed by profuse desquamation, and if the affected area is large, headache, malaise and fever may accompany the reaction. Treatment should be prophylactic by protecting the skin with a light-absorbing preparation such as benzophenone cream (Uvistat). When the condition has developed, evaporating lotions such as lead lotion or calamine or hydrocortisone lotion are effective, to be followed by zinc or cold cream. In severe cases systemic corticosteroids reduce the intensity of the skin reactions.

POLYMORPHIC LIGHT ERUPTIONS occur in individuals hypersensitive to light. The cause of such sensitivity is unknown; in some cases there appears to be a genetic factor, in others a metabolic defect such as porphyria, but many cases are unexplained (Storck, 1965). In the milder cases, known as *juvenile spring eruption*, occurring in children in spring and early summer, there is erythema followed by red, œdematous papules and desquamation, while secondary infection may lead to crusting. Sometimes the condition may be more widespread affecting the face, neck, forearms and knees, and persists throughout the summer and even into the winter (*summer eruption*). In adults the polymorphic light reaction is persistent and severe and consists of intensely itchy firm papules on a background of pigmented lichenification.

The more severe cases are known as HYDROA ÆSTIVALE (VACCINIFORME), a condition first described by Bazin (1862), characterized by the formation of conspicuous blisters and crusting followed by scarring. The lesions are confined to the exposed parts—the cheeks, brows, ears and backs of the hands (Fig. 65). As a rule the lids are not markedly affected directly (Crews, 1959), but considerable atrophic stiffening and shrinkage of the tissues may occur so that the palpebral aperture is narrowed and immobile and there is difficulty in opening the eyes (Vollmer, 1903; Linser, 1906; Schmidt-La Baume, 1927); alternatively, the contracture may lead to ectropion (Möller, 1900; Friede, 1921; Wendleberger and Klein, 1938; Slem, 1966). In children the condition tends to resolve at puberty but may persist into adult life. When it does occur in adults it is often more severe and persists indefinitely. The effects of hydroa æstivale on the conjunctiva and cornea have already been noted.[2]

Treatment is as already indicated for solar dermatitis—the avoidance of direct sunlight and the protection of the skin by sun-barrier preparations,

[1] Vol. XIV, p. 922.
[2] Vol. VIII, p. 512.

and the topical use of hydrocortisone lotion supplemented, if necessary, by systemic corticosteroids or antimalarial drugs such as chloroquine (250 mg. daily, slowly decreasing) or hydroxychloroquine and *p*-amino-benzoic acid during the troublesome season (Zarafonetis *et al.*, 1953; Wheeler *et al.*, 1960).

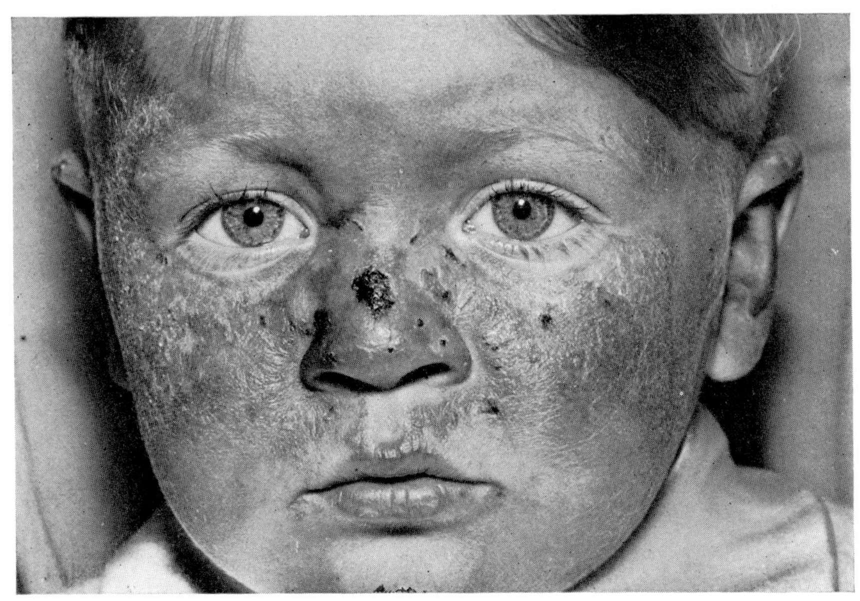

FIG. 65.—POLYMORPHIC LIGHT ERUPTION, JUVENILE TYPE (HYDROA ÆSTIVALE) (P. Borrie, St. Bartholomew's Hosp.).

The conditions of solar elastosis and solar keratosis are discussed at a later stage.

It is to be noted that photosensitivity may be induced by certain drugs, particularly the phenothiazines, some sulphonamides (especially the diuretics and antidiabetics), and certain barbiturates. The rationale of the reaction is obscure but many of these substances are capable of absorbing high-energy photons and are fluorescent, having a molecular structure of three benzene rings arranged linearly (Szent-Gyorgy, 1960).

Bazin. *Leçons théoriques et cliniques sur les affections génériques de la peau*, Paris (1862).

Breathnach. *J. invest. Derm.*, **29,** 253 (1957).

Crews. *Brit. J. Ophthal.*, **43,** 629 (1959).

Friede. *Klin. Mbl. Augenheilk.*, **67,** 26 (1921).

Hutchinson. *Trans. med. Soc. Lond.*, **15,** 472 (1892).

Dtsch. med. Wschr., **30,** 1378 (1904).

Linser. *Arch. Derm. Syph.* (Wien), **79,** 251 (1906).

Möller. *Der Einfluss d. Lichtes auf d. Haut*, Stuttgart, 85 (1900).

Schmidt-La Baume. *Arch. Derm. Syph.* (Berl.), **153,** 368 (1927).

Slem. *Ankara Üniv. Tip. Fak. Göz. Klin. Yill.*, **19,** 129 (1966).

Storck. *Arch. Derm.*, **91,** 469 (1965).

Szent-Gyorgy. *Introduction to Submolecular Biology*, N.Y. (1960).

Vollmer. *Arch. Derm. Syph.* (Wien), **65,** 221 (1903).

Wendleberger and Klein. *Arch. Derm. Syph.* (Berl.), **176,** 522 (1938).

Wheeler, Cawley and Whitmore. *Arch. Derm.*, **82,** 590 (1960).

Zarafonetis, Curtis and Shaw. *J. invest. Derm.*, **21,** 5 (1953).

FIG. 66.—RAYMOND SABOURAUD
[1864–1938].
Seen as a sculptor (from Prof. R. Vanbreuseghem, Antwerp).

DERMATITIS DUE TO INFECTIVE AGENTS

While the pathological school of dermatology may be said to have been founded by Ferdinand von Hebra (Fig. 23) of Vienna, the bacteriological and parasitological period was instituted by RAYMOND JACQUES ADRIEN SABOURAUD [1864–1938] of Paris (Fig. 66). After his graduation he studied histology at the Saint-Louis Hospital and bacteriology at the Pasteur Institute, and so high was his reputation even in his early professional years that in 1900 he was appointed chief of the École Lailler, a combined school and hospital of 300 beds; the revolution he made in the treatment of dermatological diseases allowed him to transfer 150 of these beds back to the Saint-Louis Hospital in four years and today the pursuance of his methods has reduced the number of beds to 20. His whole life was characterized by originality of thought and tenacity of purpose. He established the nature of seborrhœic dermatitis, impetigo contagiosa, ringworm and other mycotic infections, and a host of other conditions about which little or nothing was hitherto known. His application is demonstrated in a work he undertook in five volumes on diseases of the hair: the first three were published between 1902 and 1910 but, writing on a subject until then quite obscure, he spent 18 years of concentrated research before he felt able to publish the remaining two.

A world figure and a prolific writer, Sabouraud was recognized as the Master of dermatology throughout the world and his clinic became the Mecca of that specialty. But his interests were not confined to that discipline. His hobby was gathering rare forms of mushrooms, and in his summer house on the Marne he cultivated unusual species of plants; so effective a sculptor was he (as seen in Fig. 66) that a book of his statues was published; and in his later years he wrote philosophical essays as full of originality and wisdom as were his medical works. He was undoubtedly one of the great French savants of his time.

Dermatitis due to Bacterial Infections

STAPHYLOCOCCAL INFECTIONS

STAPHYLOCOCCAL INFECTIONS of the skin show few topographical characteristics in the lids and can therefore be quickly dismissed. It is to be remembered that this organism is maintained as a saprophyte by most healthy skins. It may be that a decrease in the patient's general resistance may allow such resident organisms to become parasitic, or the local resistance may be lowered by maceration from the secretions of a conjunctivitis or blepharitis, but staphylococci of varying degrees of virulence are so ubiquitous and easily transferred—from one part of a patient's body to another by the hands or clothing (as from the nares, conjunctiva or ears which pyogenic cocci frequently inhabit and from skin lesions elsewhere) or from one person to another by contact, droplet infection or dust—that infection is, in a very definite sense, contagious. Damage to the skin or removal of its superficial layers of epidermis by rubbing or scratching, particularly near a hair follicle, facilitates the invasion of cocci; and because of their exposed situation and the frequency with which they are rubbed by fingers which are never surgically clean, the eyelids are frequent sites of predilection, while it is obvious that the existence of other irritative lesions will predispose to such a sequence.

Staphylococcal lesions may be localized or generalized In the first case they tend to occur in the neighbourhood of hair follicles or sweat glands rather than to give rise to a generalized inflammation, and in type, being usually due to staphylococci which produce necrotizing toxins, they are purulent rather than serous from the start. Their common manifestation is therefore a *pustular folliculitis* (Fig. 67). The resistance of the patient, the virulence of the organism and the site of the infection may be responsible for several types of clinical picture. A superficial intradermal infection leads to an impetigo, and when such an infection of relatively low virulence is preferentially situated at the opening of a pilosebaceous gland in a person showing little protective reaction, an impetiginous folliculitis or a sycosis results. A similar infection by a more virulent organism around the hair follicle in a patient with a better protective reaction is localized and results in the clinical picture of a boil; while a similarly virulent infection when resistance is less may involve the subcutaneous tissues as a carbuncle. It is interesting that from the same patient organisms apparently identical in their reactions can be retrieved from lesions of different types—an impetigo, folliculitis, or a boil.

A FURUNCLE OR BOIL is *an acute, circumscribed infection of the skin by a virulent Staphylococcus aureus, associated with a hair follicle or skin gland resulting in local suppuration and necrosis.* The inflammation commences as a small tender red papule which develops into an indurated nodule; sometimes the lesion absorbs at this stage (" blind boil "), but more usually it progresses to suppuration, and terminates, after a central core of necrosed tissue has been extruded, with the formation of a small depressed scar. The associated œdema may be so intense as to lead to optic atrophy (Wack, 1951). The delicacy of the palpebral skin and the absence of well-developed hair follicles make the occurrence of boils rare in the lids with the exception of the regions of the eyebrows and the lashes where they constitute a hordeolum. In the former site they may cause an intense œdema of the lids; a consideration of their occurrence in the latter will be deferred.[1] Their widespread or frequent recurrence (FURUNCULOSIS) without local cause should call attention to the general health and the lack of immunological resistance to staphylococci.

CARBUNCLES, although rare, may occur in the region of the eyebrows, particularly in aged, debilitated or diabetic patients. The lesion is the same as a furuncle with the exception that the infection is deeper involving the subcutaneous tissues; it is therefore more widespread, develops less rapidly, eventually opens on the surface by several apertures and is followed by more widespread necrosis and sloughing, leaving a ragged ulcer which granulates and heals with considerable scarring. The regional lymph nodes are affected and constitutional symptoms are severe.

IMPETIGO CONTAGIOSA (L. *impetere*, to attack) may be caused by either *Staphylococcus aureus* or the hæmolytic streptococcus; in either case the clinical picture may be much the same, but characteristic differences may exist. There is much confusion in the literature as to the relative prevalence of these two organisms since statistics have usually been based on the findings in particular epidemics which may be caused by one or the other

[1] p. 234.

organism. As a general rule it would seem that in temperate climates staphylococcal or mixed infections are the commonest, while in the tropics streptococci are more frequent (Rook and Roberts, 1972); it has been suggested that the staphylococcus is often a secondary colonizer, but it may well be that the two are synergistic and that either organism can be the primary invader.

The confusion is seen in the following statistics: van Toorn (1961) isolated staphylococci in 89% of his cases and streptococci in 51%; Dillon (1968) found 74% streptococcal infections and 26% staphylococcal; while Swartz and Weinberg (1971) reported 60% staphylococcal, 30% streptococcal and 10% mixed.

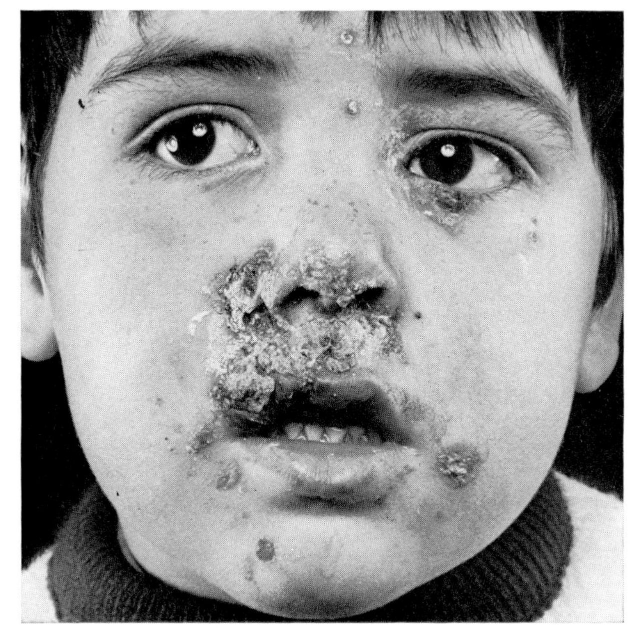

FIG. 67.—IMPETIGO.

The condition affects the lids, the nose and the mouth (St. John's Hospital for Diseases of the Skin).

STAPHYLOCOCCAL IMPETIGO is a superficial, widespread, primary staphylococcal infection of the skin, intra-epidermal in site and characterized by the formation of crops of yellowish blisters which burst at an early stage to leave thin brownish crusts (Fig. 67). Occasionally, spreading centrifugally, they take the form of a ring of vesicles (*impetigo circinata*) (Fig. 68). In other cases the bullæ although localized are larger and more persistent, but when they burst similar crusts are formed (*bullous impetigo*).

In adults a staphylococcal infection may be follicular from the onset, affecting the superficial parts of the hair follicles (*follicular impetigo of Bockhart*, 1887). The lesions are pustular and discrete and may be densely grouped over considerable areas in which crusting becomes extensive. The lids and peri-orbital skin may be involved (Collier, 1970). The scalp is a common site, particularly in children, whence the disorder may spread to the brows and involve the lids in a superficial blepharitis.

Occurring in the first few days of life as a bullous impetigo from external infection (*pemphigus neonatorum*), the blisters in severe cases may cover the entire surface of the body; the face and lids are frequent sites. An exfoliative reaction may be the most prominent feature, due to a group II staphylococcus to which the immunological response is poor (Lorenz, 1968); the condition was produced experimentally in newborn mice by Melish and

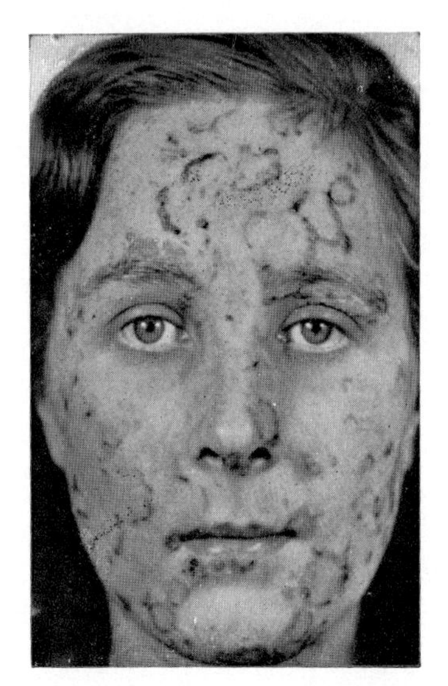

Fig. 68.—Impetigo Circinata.
Of 2 weeks' duration (Sequeira *et al.*, *Diseases of the Skin*, Churchill).

Glasgow (1970). In the GENERALIZED EXFOLIATIVE DERMATITIS of infants (*Ritter's disease*) (Ritter von Rittershain, 1870), the epidermis rapidly and extensively becomes necrotic and exfoliates (Figs. 69–70). In older children, usually under the age of 10 years, the clinical picture is called TOXIC EPIDERMAL NECROLYSIS (*Lyell's disease*, 1956). There are acute symptoms of fever and illness, and the skin over large areas suddenly becomes red and tender, the epidermis loosens, large flaccid bullæ are formed and the epidermis peels off in sheets to leave a raw, weeping and tender surface resembling a second degree burn. Such a condition has aptly been termed the *scalded skin syndrome.*[1] The multiple ætiology and pathology of this condition is described at a later stage as well as the ocular complications.[2]

[1] Lyell (1967), Samuels (1967), Jefferson (1967), Lowney *et al.* (1967), Melish and Glasgow (1970–71), and others.

[2] p. 286.

FIGS. 69 and 70.—GENERALIZED EXFOLIATIVE DERMATITIS (Ritter's disease).

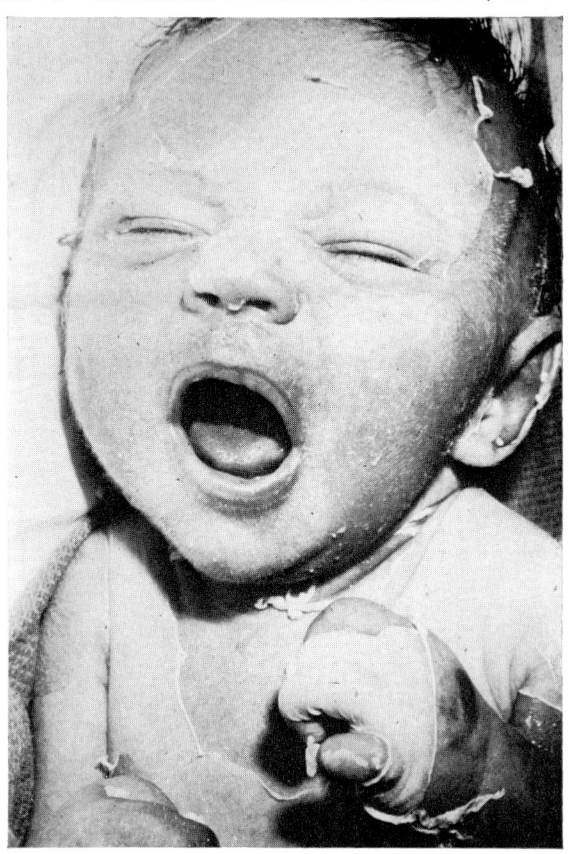

FIG. 69.—The condition in a newborn infant at the height of the exfoliative phase, due to a staphylococcal infection. Recovery followed treatment by methicillin (M. E. Melish and L. A. Glasgow).

FIG. 70.—A similar case in a newborn infant (R. L. and R. L. Sutton).

SYCOSIS (σῦκον, a fig) is a deeper staphylococcal infection introduced into the hair follicles either by scratches from a razor or by inoculation by the finger. The characteristic picture is that of groups of discrete pustules around the hairs surrounded by a swollen œdematous area which eventually becomes heavily crusted by the discharged pus. The disease tends to be progressive and, spreading throughout the beard and moustache area, may attack the eyebrows and lashes, a chronic blepharitis being added to the miseries of a chronically relapsing, oozing, crusted and unshaven face. In old-standing cases the end-result is a smooth, red, atrophied patch in which the hairs are permanently lost (*lupoid sycosis*).

CHANCRIFORM PYODERMA is an uncommon condition, occurring more often in children than adults, which is possibly due to a necrotizing reaction to a strain of *Staph. pyogenes*, usually inoculated by minor trauma (Fig. 71). A solitary sharply defined ulcer occurs near the eyelids with a bright

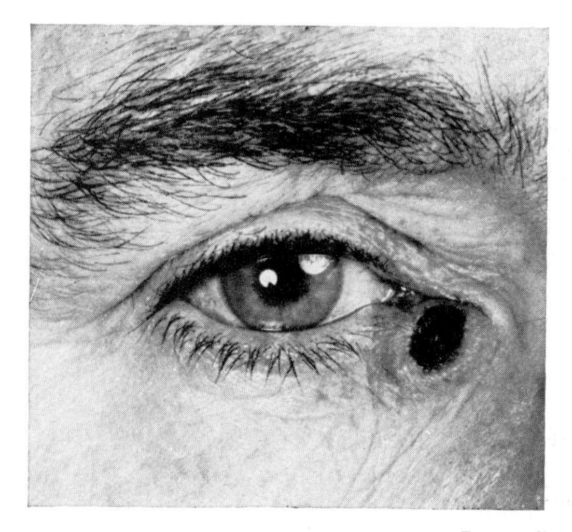

FIG. 71.—CHANCRIFORM PYODERMA AFFECTING THE INNER CANTHUS.

(A. D. Porter, St. John's Hosp. for Diseases of the Skin; from A. Rook's *Textbook of Dermatology*, Blackwell, Oxford).

red areola and an indurated base; the regional lymph nodes are usually enlarged. In the absence of treatment the ulcer tends to persist for several weeks before healing, to leave a superficial scar, but the response to antibiotic treatment is usually not dramatic (Naylor and Rook, 1968).

STAPHYLOCOCCAL SCARLATINA is a further manifestation of infection by this organism in children. The scarlet rash is generalized and is clinically indistinguishable from that of streptococcal scarlet fever. After one or two days cracks appear in the creases of the skin, particularly around the eyes and mouth, followed by the formation of large, thick flakes of skin, a process which becomes generalized leaving healed skin underneath. The infecting agent is *Staph. pyogenes* which produces an erythrogenic toxin, the initial lesion being frequently a conjunctivitis or a similar naso-pharyngeal infection.[1]

INFECTIVE ECZEMATOUS DERMATITIS. As opposed to its usual inflammatory or pyogenic effect, the staphylococcus may evoke an eczematous response, especially in children. It is probably the most common cause of dermatitis of the eyelids (Theodore, 1954–55; 77 out of 238 cases (32%), Borrie, 1956). It is often associated with an infection, either ocular such as blepharitis, conjunctivitis, a stye or a meibomitis, or cutaneous lesions elsewhere such as retro-auricular intertrigo, otitis externa, sycosis barbæ, or

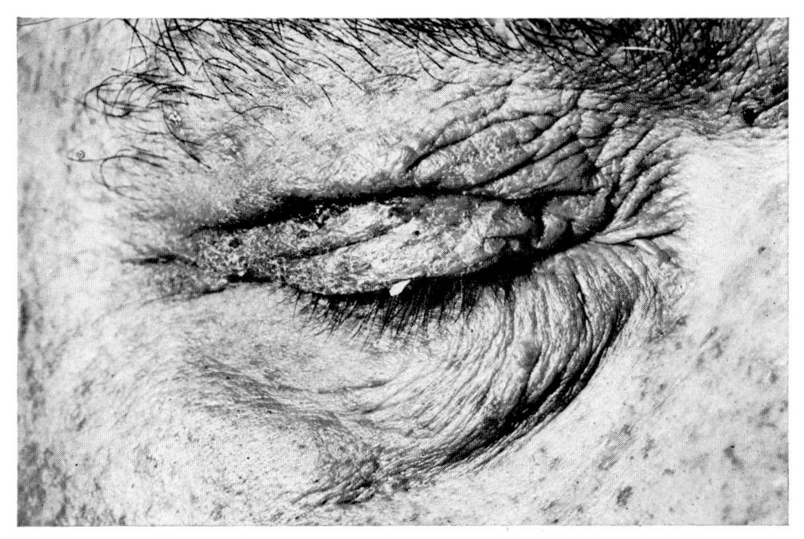

Fig. 72.—Infective Eczematous Dermatitis.
Showing exudation and crusting with swelling of the lids (P. Borrie).

infective conditions of the scalp or anterior nares. The condition is probably an allergic or irritant response to the dermo-necrotizing factor of the staphylococcus combined with an allergy to other products of the bacterium to which the skin of the eyelids is more prone than that of other areas of the body (Hollander and Baer, 1935; Thygeson, 1952). In the lids the reaction frequently starts as a small fissure at one or other of the canthi or in a fold of the upper lid; it is characterized by erythematous, vesicular, scaly or pustular plaques which tend to increase by extension at the periphery or by auto-inoculation (Figs. 72–3). An attack may last for some weeks or months and recurrences are frequent, while rubbing and scratching perpetuate and aggravate the condition. The essential feature in treatment is to

[1] Stevens (1927), Dunnet and Schallibaum (1960), Feldman (1962), Melish and Glasgow (1970–71).

eliminate the infection of the eye or lid-margins[1]; painting the lids, including any fissures in the skin, with 2·0% silver nitrate is often helpful, while a suitable local application is 10% sodium propionate (Theodore, 1954–55). Antibiotics are seldom efficacious but hydrocortisone cream is valuable in the more chronic stages after the bacterial infection has been controlled. Rapid results, however, cannot be expected; it may persist in a single attack up to 6 months, or even 1 to 3 years after treatment has commenced (Borrie, 1956).

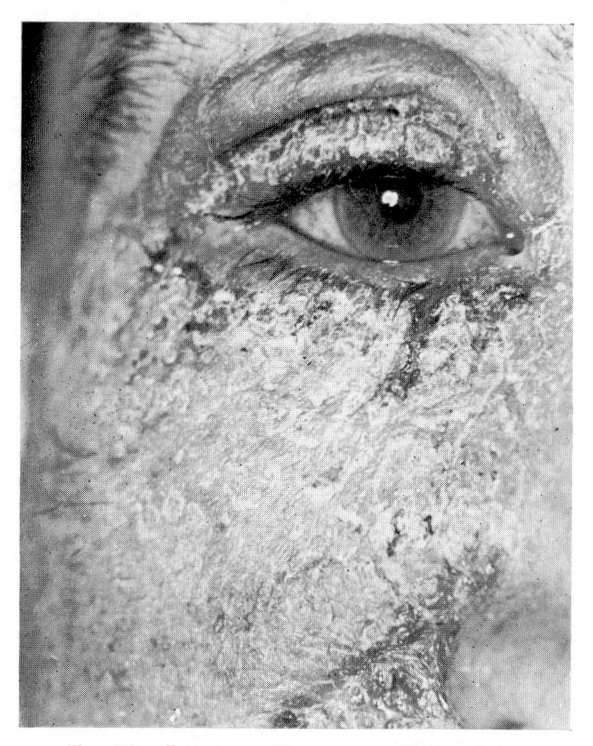

Fig. 73.—Infective Eczematous Dermatitis.

A case in the crusting stage involving large areas of the cheek as well as the lids (Inst. Ophthal.).

It should be noted that a somewhat similar eczematous condition affecting all four lids has been described with an infection elsewhere caused by *Candida albicans* (Ruiz-Moreno, 1947).

The *treatment* of staphylococcal lesions should follow certain general lines. In the first place, if the lesion is at all severe and widespread and particularly if the general condition of the patient is poor, an attack against the organism by systemic antibiotic drugs may be advisable. From the

[1] Vol. VIII, p. 158.

academic point of view, a determination of the sensitivity of the organism involved is a desirable preliminary to treatment, but from the practical point of view when the case is acute and generalized the response to such treatment is frequently so dramatic that diagnosis by therapeutic trial while the bacteriological examination is being carried out may be justifiable. A suitable antibiotic is the potassium salt of phenoxymethyl penicillin (penicillin VK), 250 mg. four times a day. If the organism is resistant to penicillin, novobiocin (0·5 g. 6 hourly) is a useful alternative. In the treatment of acute necrotizing lesions which used frequently to be lethal, Melish and Glasgow (1971) obtained very satisfactory results with methicillin, provided it was given early so that the development of exfoliation was prevented; after the onset of this process, however, antibiotic treatment failed to ameliorate the course of the disease. In these and other acute conditions corticosteroids are contra-indicated since they tend to potentiate the ability of the organism to cause exfoliation.

In the more chronic and localized cases, however, antibiotic treatment may be less satisfactory. These drugs have no effect on the immunological reaction of the patient and when the infection is of some duration and resistance is low, a preliminary response to treatment may be followed by recurrences in which case the organism becomes increasingly resistant. In such cases, antibiotic treatment is frequently inappropriate, or if it is employed initially it should be combined with general steps to increase the health and resistance of the patient, both of which are frequently poor; staphylococcal toxoid, combined perhaps with an autogenous vaccine, used to be advocated, particularly the intracutaneous injection of the toxoid (Hopkins and Burky, 1944), but subsequent experience has led to the abandonment of such measures (Harrison, 1963). Recurrences, particularly after an initial improvement with antibiotic treatment, are generally indicative of a generalized debility and the presence of a persistent local infection such as a staphylococcal infection of the nares, conjunctivitis, blepharitis, pediculosis or scabies, must be cleared up before any staphylococcal lesion will heal.

Local treatment must be varied to meet the type of case. In the more superficial lesions the first essential is the removal of crusts and discharge where these exist (impetigo, sycosis) by oil and saline or hydrogen peroxide, aided if necessary by starch poultices; if vesicles exist their roofs should be excised. Thereafter an antibiotic lotion or cream may be applied, but the effect of this is often disappointing. The choice should be determined by sensitivity tests, but penicillin and streptomycin should generally be avoided owing to the frequency of sensitivity reactions; the tetracyclines and erythromycins should be used sparingly owing to the liability of creating resistance to drugs regularly required for serious systemic infections; neomycin and framycetin are suitable agents and, if these fail, chloramphenicol. Such measures, combined with systemic antibiotic treatment if indicated, are frequently dramatically effective; but when they prove ineffective, the classical applications such as gentian violet (2%), brilliant green (1%), bichloride of mercury (1:10,000) or salicylic acid (1%) in zinc paste may be useful, although perhaps less speedy. The use of corticosteroids is to be avoided.

In the more deeply-seated infections, topical applications of antibiotic drugs or anything else are of little or no value, although systemic treatment with an antibiotic such as penicillin VK may prove effective, particularly in recurring cases. Usually it is best to allow a boil to evacuate itself; where pus is evident, however, and if the boil is deep and painful, it may be cautiously opened—a very localized and superficial incision and swabbing out with pure carbolic, followed by a dry dressing, is the method of choice; hot fomentations are generally to be avoided as they encourage auto-inoculation. Epilation of the affected hair is contra-indicated, since this may give the staphylococcus free access to the subdermal tissues; nor should a boil ever be expressed—" squeeze a pimple and make a boil, squeeze a boil and make a carbuncle ". When a carbuncle fails to abort with antibiotic treatment, conservative treatment usually leads to the best results, but if it becomes fluctuant and is associated with much pain and constitutional disturbance a radical crucial incision may be made, all pockets cleared out and the wound packed.

It is to be remembered that boils on the face are always to be treated seriously, owing to the possibility of the development of an infective venous thrombosis which may become intracranial and fatal. Such a dramatic complication, however, is much more prone to follow a lesion located on the upper lip or nose than on the lids, that is, in the region drained by the angular veins.[1]

Bockhart. *Mh. prakt. Derm.*, **6,** 450 (1887).
Borrie. *Brit. J. Ophthal.*, **40,** 742 (1956).
Collier. *Bull. Soc. Ophtal. Fr.*, **70,** 815 (1970).
Dillon. *Amer. J. Dis. Child.*, **115,** 530 (1968).
Dunnet and Schallibaum. *Lancet*, **2,** 1227 (1960).
Feldman. *New Engl. J. Med.*, **267,** 877 (1962).
Harrison. *Brit. med. J.*, **2,** 149 (1963).
Hinton. *Ann. Surg.*, **85,** 104 (1927).
Hollander and Baer. *Amer. J. Ophthal.*, **18,** 616 (1935).
Hopkins and Burky. *Arch. Derm. Syph.* (Chic.), **49,** 124 (1944).
Jefferson. *Brit. med. J.*, **2,** 802 (1967).
Lorenz. *Wien. klin. Wschr.*, **80,** 240 (1968).
Lowney, Baublis, Kreye *et al.* *Arch. Derm.*, **95,** 359 (1967).
Ludlow. *Med. Times Gaz.*, **5,** 287, 332 (1852).
Lyell. *Brit. J. Derm.*, **68,** 355 (1956); **79,** 662 (1967).
Maes. *Ann. Surg.*, **106,** 1 (1937).
Melish and Glasgow. *New Engl. J. Med.*, **282,** 1114 (1970).
J. Pediat., **78,** 958 (1971).
Naylor and Rook. *Textbook of Dermatology* (ed. Rook *et al.*), Oxford, **1,** 647 (1968).
von Rittershain. *Österr. Jb. Pädiat.*, **1,** 23 (1870).
Roedelius. *Derm. Wschr.*, **78,** 37 (1924).
Rook and Roberts. *Textbook of Dermatology* (ed. Rook *et al.*), 2nd ed., Oxford, **1,** 482 (1972).
Ruiz-Moreno. *Ann. Allergy*, **5,** 132 (1947).
Samuels. *Brit. J. Derm.*, **79,** 672 (1967).
Stevens. *J. Amer. med. Ass.*, **88,** 1956 (1927).
Swartz and Weinberg. *Dermatology in General Medicine* (ed. Fitzpatrick *et al.*), N.Y., 1706 (1971).
Theodore. *Trans. Amer. Acad. Ophthal.*, **58,** 708 (1954).
J. Mt. Sinai Hosp., **21,** 255 (1955).
Thygeson. *Trans. Amer. Acad. Ophthal.*, **56,** 737 (1952).
van Toorn. *Dermatologica*, **123,** 391 (1961).
Wack. *Med. Klin.*, **46,** 865 (1951).

[1] Ludlow (1852), Roedelius (1924), Hinton (1927), Maes (1937), and others.

STREPTOCOCCAL INFECTIONS

In contradistinction to staphylococcal infections, dermatitis due to STREPTOCOCCI is not follicular in type and tends to produce a serous rather than a pustular exudate which, if superficial, appears as blisters and, if introduced into the dermis, gives rise to the serous and cellular exudate of erysipelas. Occasionally a superficial streptococcal dermatitis produces a dry inflammatory patch with fine brawny scales, such as is seen in a scarlatiniform rash, and such patches may be found with the usual vesicular lesion; but these are not typical.

STREPTOCOCCAL IMPETIGO (IMPETIGO CONTAGIOSA of Tilbury Fox, 1864) is an acute dermatitis characterized by the occurrence on localized erythematous areas of discrete, thin-walled superficial vesicles, which become pustular and rupture to form loosely adherent yellow crusts that readily fall off without leaving a scar. A more deeply-seated pustular infection occurs less frequently which invades the cutis and leaves a permanent scar (*ecthyma*). In the early stages the blister may contain pure streptococci, but later usually becomes contaminated with staphylococci, in which case the crusts become thicker and brown. Children are more susceptible than adults and the face, including the brows and lids, is the site of predilection; a streptococcal conjunctivitis or corneal ulcers have been noted as sequelæ (Cunningham, 1907; Krauss, 1908; Hansell, 1912) and a cicatricial ankyloblepharon as an end-result (Bourgeois, 1906). As a primary disease the malady, like streptococcal infections in general, has a peak-incidence in midsummer in Europe (Newman, 1935) and is contagious, contacts being inoculated through any slight abrasion. Occasionally systemic complications such as a nephritis may arise (Kohn, 1921; Sieben, 1922; Sutton, 1934; Callaway and Gilgor, 1967). As a secondary condition it complicates other infections or eczematous states of the skin or such conditions as pediculosis capitis or scabies.

ERYSIPELAS (*St. Anthony's fire*) is *an acute localized inflammation of the skin and subcutaneous tissue, characterized by redness, œdema and induration and associated with constitutional manifestations of fever and toxæmia.* As was originally shown by Fehleisen (1883), it is due to streptococcal infection introduced frequently from an infection of the upper respiratory tract, usually through an abrasion or minute wound, the organism involved being typically the group A hæmolytic streptococcus. The organism is present in the lymphatic channels through which it spreads to neighbouring areas. The lids are rarely primarily affected, although when this does occur the lesion is usually severe (Biermann, 1869); as a rule they are involved secondarily by spread from a neighbouring lesion. The disease is virulently contagious and patients should be isolated.

After an incubation period of 2 to 5 days the illness commences abruptly, usually with a rigor, malaise, headache and sometimes delirium. At the site of infection an area of dermatitis appears; here the skin is hot, red, raised and indurated with a shining glazed surface and a *sharp spreading margin dotted with minute vesicles*—an important diagnostic point. Œdema is marked so that the lids are closed and the whole face so swollen that the features may be scarcely recognizable (Fig. 74). If uninterrupted by treatment, the disease lasts 1 to 3 weeks when involution occurs, the vivid dusky red giving place to a brownish or yellowish patch showing considerable scaly desquamation. As a rule it is self-limited and most healthy adults recover by crisis or lysis, but fatalities may occur in infants or the aged or in those debilitated by other acute or chronic disease or by alcoholism.

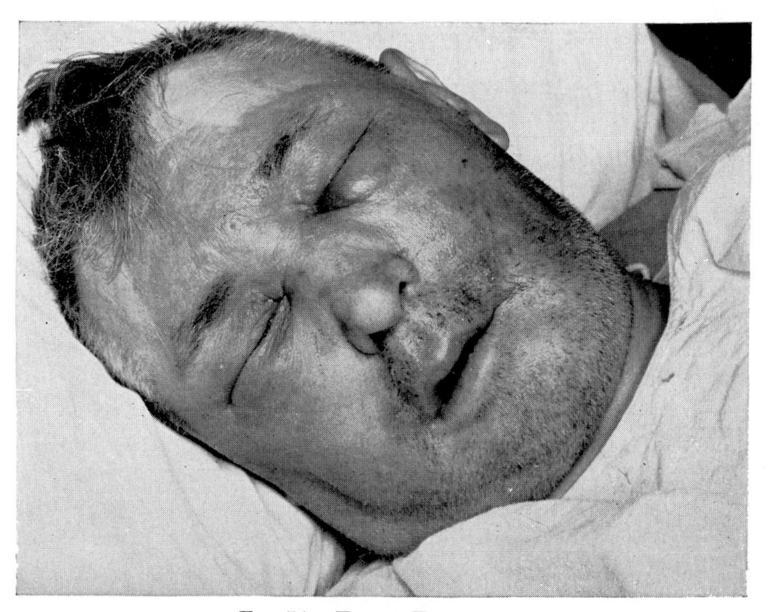

Fig. 74.—Facial Erysipelas.

In the acute stage; note the sharp demarcation of the discoloured, œdematous area (F. H. Top).

Before the use of sulphonamides or penicillin, Spink and Keefer (1936) found a mortality of 16% in 1,400 cases, rising to 39% in the first year of life and to 43% above the age of 70 (Hoyne, 1935, in 1,193 cases). With the newer methods of treatment, however, the lesion rarely spreads after 36 hours and the mortality has been reduced out of all comparison (2·5%, Miller *et al.*, 1945).

Pathologically, throughout the corium there is a fibrinous and leucocytic exudation, with dilated and congested blood and lymph channels and a marked perivascular polymorphonuclear infiltration. Streptococci are distributed throughout the tissues and are found particularly in the lymphatics. Even in severe cases resolution is usually complete without cicatrization.

Local complications, however, may occur: an abscess in the lids themselves or spreading over the surrounding areas, gangrene of the lids (Fig. 75), suppuration in the orbit with extension to the cavernous sinus and meningitis with a fatal termination (Carl, 1884; Strümpell, 1920; and others); or in the eye a suppurative or membranous conjunctivitis, a keratitis sometimes with hypopyon, a uveitis or a panophthalmitis. Superimposed infections, such as a staphylococcal, may produce in the tissues of the eyelids deep necrotizing ulcers which aggravate the prognosis (Stookey *et al.*, 1934). Conversely, the incidence of erysipelas has led to the sudden cure of an already established inflammatory condition, such as lupus, a chronic conjunctivitis or trachoma, or a keratitis, and it has even been claimed to inhibit the growth of a tumour.

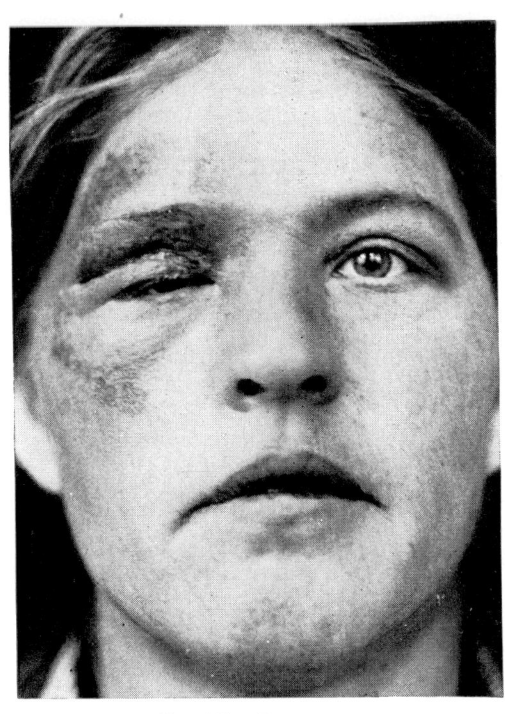

FIG. 75.—ERYSIPELAS.

In the more advanced stage, showing marked œdema and ptosis, with abscess formation and necrosis of the lids (H. Ehlers).

Erysipelas confers no local immunity and recurrences may occur as relatively mild attacks unassociated with acute systemic symptoms (*erysipelas perstans: recurrent erysipelatoid eruption: relapsing streptococcal lymphangitis*). In fact, a previous attack may make the same area liable to a second on very slight provocation (*e.g.*, after the occurrence of atropine irritation, Wolfe and McLeod, 1931). Each attack is characterized by a vivid redness, the area, however, showing no raised advancing border, but only a uniform œdema. It is possible that in many instances the portal for infection of the facial lymphatics is a fissure, for example, in the nostril

(Barber, 1934), and the disease is frequently associated with an impetiginous rhinitis, a nasopharyngeal catarrh or some other such focus of infection, the elimination of which is of essential importance in treatment. In such recurrent cases a permanent indurated thickening of the lid, particularly the upper, may eventually develop, producing a chronic œdema and leading to an elephantiasis due to lymphatic blockage (the *solid œdema* of Hutchinson, 1888) (Critchett, 1899) (Fig. 15). Elsewhere in the body it has been shown by lymphangiography that the recurrences are due to lymphatic deficiency and although this has not been demonstrated on the face it is likely that the cause is similar.

In SCARLET FEVER (SCARLATINA), a generalized streptococcal infection, lid changes are few. The typical rash appears here as elsewhere, and œdema may be considerable in severe cases. As a rarity, however, abscesses (Jackson, 1895) or even partial gangrene used to occur (Saint-Martin, 1884).

The *treatment* of streptococcal infections of the skin should comprise the general maintenance of the patient's resistance by symptomatic stimulatory measures, complete rest and isolation in bed in severe cases with the exhibition of analgesics if necessary, and the control of the infection. The last objective is best obtained by the early use of antibiotics which have made all other methods of treatment for the more serious forms of infection obsolete. With these, systemic administration is indicated, but the keynote of success lies in their early exhibition within the first day or two. The most effective antibiotic is the potassium salt of phenoxymethyl penicillin (penicillin VK) (250 mg. 4 times a day) which should be given immediately and should be continued until some time after the disease has subsided lest relapses occur. While this drug is the most generally effective, other antibiotics such as the tetracyclines may abort an attack. In erysipelas the temperature usually falls within 48 hours of the commencement of antibiotic treatment and the spread of the lesion stops abruptly, but the skin does not return to its normal state until the lapse of some days.

The local treatment of impetiginous lesions should follow much the same lines as those indicated for staphylococcal impetigo—the removal of crusts by bathing with warm water several times daily, the application of warm oil and saline, and the local application of chloramphenicol or bacitracin as an ointment. Penicillin is usually the most effective but should, if possible, be avoided because of the high incidence of sensitivity reactions. The deeper dermal infections (erysipelas), of course, are unaffected by local applications, although hot compresses may be found comforting or, alternatively, cold saline compresses or petrolatum. In recurrent erysipelas penicillin administered systemically frequently aborts a relapse, but is sometimes ineffective. In this event a search should be made for local portals of infection, such as fissures at the canthus, and for streptococcal foci, particularly in the mouth, nose or throat. Disfiguring deformities of the lids may require later treatment by plastic surgery.

Barber. *Trans. ophthal. Soc. U.K.*, **54,** 426, 436 (1934).
Biermann. *Klin. Mbl. Augenheilk.*, **7,** 91 (1869).
Bourgeois. *Un. méd. sci. Nord-Est* (1906). *See Jber. Ophthal.*, **37,** 472 (1906).
Callaway and Gilgor. *Progr. Derm.*, **2,** 11 (1967).
Carl. *Klin. Mbl. Augenheilk.*, **22,** 113 (1884).
Critchett. *Trans. ophthal. Soc. U.K.*, **19,** 7 (1899).
Cunningham. *Brit. med. J.*, **2,** 142 (1907).
Fehleisen. *Die Aetiologie d. Erysipels*, Berlin (1883).
Fox. *Brit. med. J.*, **1,** 467, 495, 553, 607 (1864).
Hansell. *Ophthalmology*, **8,** 180 (1912).
Hoyne. *Med. Rec.* (N.Y.), **141,** 132 (1935).
Hutchinson. *Illust. med. News* (Lond.), **1,** 82 (1888).
Jackson. *Brit. med. J.*, **1,** 641 (1895).
Kohn. *Berl. klin. Wschr.*, **58,** 28, 131 (1921).
Krauss. *Z. Augenheilk.*, **19,** 123 (1908).
Miller, Tucker and Kean. *Sth. med. J.*, **38,** 757 (1945).
Newman. *J. Hyg.*, **35,** 150 (1935).
Saint-Martin. *Bull. Clin. nat. ophtal. Hosp. d. Quinze-Vingts*, **2,** 144 (1884).
Sieben. *Klin. Wschr.*, **1,** 896 (1922).
Spink and Keefer. *J. clin. Invest.*, **15,** 17, 21 (1936).
Stookey, Ferris, Parker *et al.* *J. Amer. med. Ass.*, **103,** 903 (1934).
Strümpell. *Lhb. d. speziellen Pathologie u. Therapie d. inneren Krankheiten*, 22nd ed., **1,** 98 (1920).
Sutton. *Sth. med. J.*, **27,** 798 (1934).
Unna. *Histolog. Atlas z. Pathologie d. Haut*, Leipzig (1898).
Wolfe and McLeod. *J. Amer. med. Ass.*, **97,** 460 (1931).

ANGULAR BLEPHARO-CONJUNCTIVITIS

ANGULAR CONJUNCTIVITIS has already been studied in detail[1] when it was pointed out that the most profound changes occur not in the conjunctiva but in the surrounding skin (Ishihara, 1911). The conjunctival injection is particularly localized near the canthi and is constantly associated with an eczematous condition of the surrounding skin varying in intensity from a slight scurfiness to a well-marked area of redness and maceration associated with considerable itching and smarting.[2] The organism usually inculpated (*Moraxella lacunata* or the diplobacillus of Morax-Axenfeld) is a saprophyte which grows and flourishes more easily in the keratinized epithelial cells of the skin than in the mucous membrane; the former is therefore the principal reservoir of infection, and here it secretes a protein-dissolving enzyme which acts by macerating the epithelium (Lindner, 1921; Pillat, 1922; Howard, 1924; Abe, 1925). Other organisms such as the staphylococcus may well be implicated (Jones *et al.*, 1957).

On the other hand, there is a considerable body of evidence that this disease is associated with a deficiency of vitamin B, particularly pyridoxine (B_6) and that the organism plays a secondary part in the ætiology (Mueller and Vilter, 1950); a similar relationship has been claimed with nicotinic acid (Irinoda, 1953) and riboflavine (Venkataswamy, 1960). Confirmation of this dependence on poor nutrition, however, is lacking (van Bijsterveld, 1973).

Treatment is by zinc as lotions or drops of the sulphate, or ointments of the oxide, or by ionization; zinc is a specific, acting by inhibiting the proteolytic ferment and so rendering the bacillus impotent. Occasionally, however, the disease is very rebellious and may defy treatment for long periods, as in a case reported by Bailliart and Tillé (1935) which, still persisting for more than 10 years, resulted in considerable thickening and ectropion of both upper and lower lids.

[1] Vol. VIII, pp. 79, 184. [2] Vol. VIII, Plate IV, Fig. 3.

Abe. *Nip. Gank. Zas.*, **29,** 339 (1925).
Bailliart and Tillé. *Bull. Soc. Ophtal. Paris,* 157 (1935).
van Bijsterveld. *Amer. J. Ophthal.*, **76,** 545 (1973).
Howard. *Amer. J. Ophthal.*, **7,** 909 (1924).
Irinoda. *Rinsho Ganka*, **7,** 340 (1953).
Ishihara. *Klin. Mbl. Augenheilk.*, **49** (1), 191 (1911).
Jones, Andrews, Henderson and Schofield. *Trans. ophthal. Soc. U.K.*, **77,** 291 (1957).
Lindner. *v. Graefes Arch. Ophthal.*, **105,** 726 (1921).
Mueller and Vilter. *J. clin. Invest.*, **29,** 193 (1950).
Pillat. *Klin. Mbl. Augenheilk.*, **68,** 533 (1922).
Venkataswamy. *J. All-India ophthal. Soc.*, **8,** 33 (1960).

NEISSERIA CATARRHALIS INFECTION

The *Neisseria catarrhalis*[1] may, as a great rarity, excite a dermatitis of the eyelids. Such a case was reported by Nastri (1935) wherein, following vaginal infection transferred at birth through facial excoriations, a widespread dermatitis of the lids was associated with a purulent conjunctivitis.

Nastri. *Boll. Oculist.*, **14,** 674 (1935).

DIPHTHERITIC INFECTION

Skin infection by the DIPHTHERIA BACILLUS (*Corynebacterium diphtheriæ*) is relatively rare, occurring either as a new lesion or contaminating a wound. In the former case the infection commences as a red zone surrounding a clear vesicle which readily ruptures, leaving a central persistent greyish slough which may progress to indolent ulceration and eventually to gangrene of the lids; this is a result of the failure of the blood supply when true membrane formation causes coagulation within the tissues (Thygeson, 1952). Such a primary lesion on the skin of the lids in the absence of conjunctival involvement is very rare (Lichtenstein, 1919; Weinberg, 1921). It has occurred in a patient who had nasal and faucial diphtheria: following a mosquito bite on the upper lid, a deep sore developed (Duke-Elder, 1952). Dermal infections are frequently followed by nerve involvement, and in these cases a severe post-diphtheritic polyneuritis with paresis of the limbs may follow the healing of the ulcer. Diagnosis is made by the identification of the bacillus from the ulcer, and treatment is by the prompt intramuscular or intravenous injection of diphtheria antitoxin (50,000 to 100,000 units) accompanied by the systemic exhibition of penicillin. Antitoxin given early and in adequate amounts is curative, but both it and the antibiotic which, however, has no neutralizing effect on the toxin, are equally advisable (Dodds, 1946).

A more common condition is secondary diphtheritic involvement of the lids from conjunctival infection, a complication which, particularly in infants, may end in *necrosis* or *gangrene* (Steffens, 1900; Schillinger, 1903; Axenfeld, 1907). In a case reported by Naróg (1948), a 3-day-old infant whose mother had a positive Schick reaction, the necrosis was so extensive that the resultant clinical picture resembled a bilateral amputation of the lids at the orbital margins. Coincidental conjunctival and faucial infection is not unusual (Williams, 1943).

[1] Vol. VIII, p. 176.

Axenfeld. *Die Bakteriologie in d. Augenheilkunde*, Jena (1907).

Dodds. *Brit. med. J.*, **2,** 8 (1946).

Duke-Elder. *Text-Book of Ophthalmology*, London, **5,** 4847 (1952).

Lichtenstein. *Klin. Mbl. Augenheilk.*, **63,** 684 (1919).

Naróg. *Klin. oczna*, **18,** 511 (1948).

Schillinger. *Ein weiterer Fall von Lidgangrän mit Diphtheriebazillenbefund* (Diss.), Tübingen (1903).

Steffens. *Klin. Mbl. Augenheilk.*, **38,** 339 (1900).

Thygeson. *Trans. Amer. Acad. Ophthal.*, **56,** 737 (1952).

Weinberg. *v. Graefes Arch. Ophthal.*, **104,** 345 (1921).

Williams. *Brit. med. J.*, **2,** 416 (1943).

ANTHRAX

The *Bacillus anthracis* (ἄνθραξ, coal, carbuncle), a rod-shaped, non-motile organism which forms spores of great vitality capable of survival for several years, was the first micro-organism proved to be the specific cause of an infective disease. It thrives mainly in cattle, sheep, horses and mice, and is found less commonly in carnivora. It is to be noted that the infected animals may be apparently healthy; but infection may be transmitted to man directly, as in farmers or children through close contact with animals (Tahernia, 1967), or more usually in the dust of hides or hair as in shaving brushes. The infection thus frequently occurs in small epidemics (as in the leather industry or in a rug factory, Rundlett, 1948), to prevent which proper preliminary treatment should be given to the hides or other animal products to kill the spores. Transmission from affected patients is rare, but has been reported (Elschnig, 1893). The disease is common in the Middle East, India, Pakistan and South Africa, but is rare in Europe and the U.S.A. The most common site of infection is the skin, and the usual lesion is the *malignant pustule*; more rarely an *anthrax œdema* results, but the two types are often difficult to distinguish. Other avenues of infection are the lungs (*wool-sorters' disease*) and very rarely the intestinal tract. Infection of the skin of the lids is usually conveyed by rubbing with contaminated fingers, the upper and lower lids having approximately an equal incidence (Wilms, 1905; Rosenbach, 1905). All forms of infection may give rise to septicæmia, always of a serious nature and frequently with a fatal termination unless adequately treated at an early stage.

Its occurrence on the lids was noted in the first century of the Christian epoch by Severus, and early cases were recorded by Himly (1843) and Desmarres (1852). Of 352 cases Thielmann (1855) found infection in 2 on the lids, in 3 on the brows; in 1,077 cases Koch (1886) found 282 on the face, of which 10 affected the lids; of 271 cases Iwanowski (1896) found 8 on the lids; of 90 cases Barnshaw and Lovett (1949) found 7 involving the lids, of which 5 were malignant pustule and 2 malignant œdema; and in Iran, H. and M. Farpour (1972) among 300 cases found 47 cases of both types affecting the lids. A pustule on the conjunctiva is excessively rare but has been reported (Knapp, 1876).

ANTHRAX MALIGNANT PUSTULE (*Milzbrand karbunkel*) is the most characteristic form of the disease; it comprises *a papulo-vesicular lesion which progresses to gangrene.* After an incubation period of 1 to 3 days an intensely itchy red papule appears which soon becomes vesicular. Several such lesions may coalesce and spread, and the vesicles, rapidly becoming hæmorrhagic in nature, burst leaving a deep black eschar surrounded by indurated, livid and brawny skin (Figs. 76–9). The surrounding œdema may be intense and may involve the face, neck and trunk—indeed, a lesion in one upper lid has caused swelling of all four lids with bilateral exophthalmos followed by bilateral optic atrophy (Manolescu, 1911). Recovery may take place at this

Figs. 76 to 79.—Anthrax of the Eyelid.

In an agricultural teacher aged 37 who had been vaccinating cattle with a live anthrax vaccine a week previously (N. T. Simmonds).

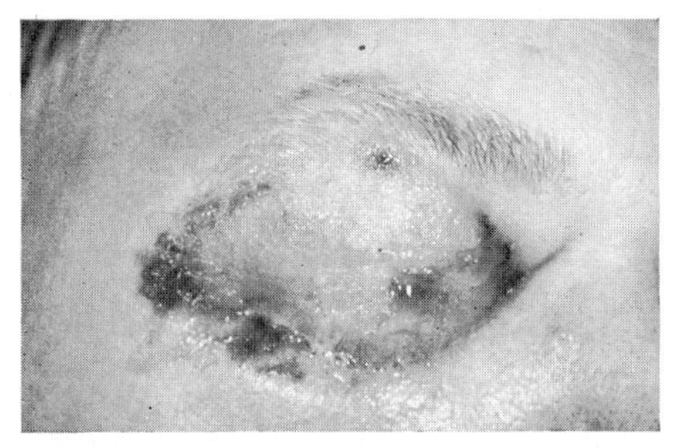

Fig. 76.—The appearance of pustules on the upper lid, five days after hospitalization.

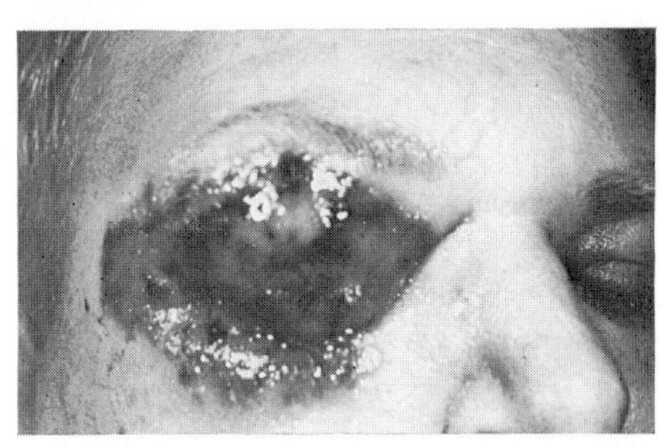

Fig. 77.—Seven days after hospitalization, showing extensive gangrenous ulceration.

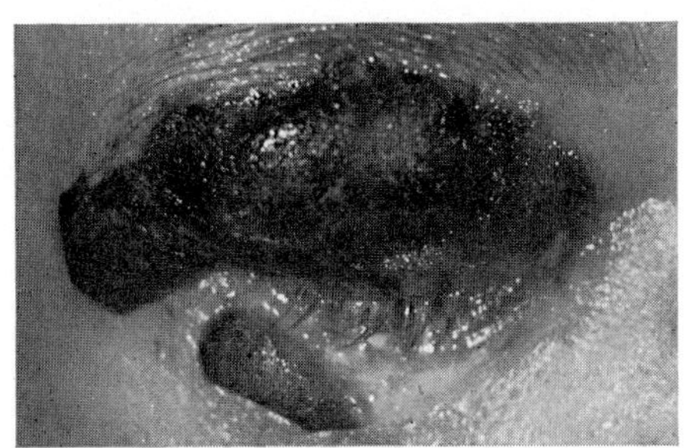

Fig. 78.—Fourteen days after admission to hospital.

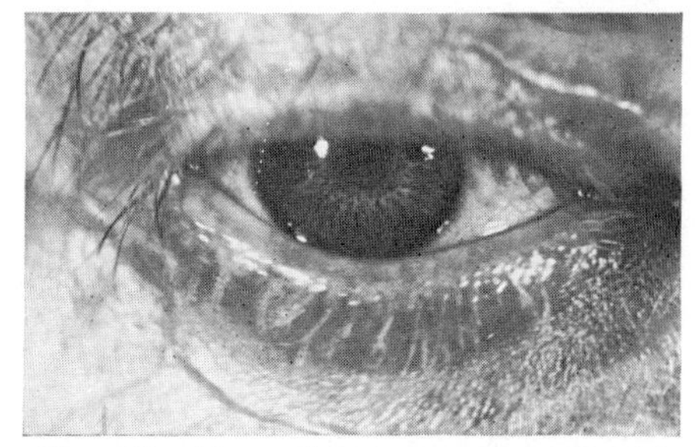

Fig. 79.—The satisfactory end-result with good functioning eyelids and a good cosmetic appearance after treatment with penicillin and sulphadiazine.

stage or, alternatively, satellite lesions may appear usually as the result of scratching so that eventually large areas slough leaving much cicatricial contraction leading to ectropion (Sasso, 1961); sloughing may extend to the periosteum, and panophthalmitis has resulted (Blatt, 1923). Meantime, the regional lymph nodes become swollen and tender, a phenomenon which may become apparent before the pustule itself appears (Praun and Pröscher, 1900). If treatment is not effective the local condition may progress to gangrene while the patient rapidly becomes prostrated, and a general and sometimes fatal septicæmia may ensue.

FIGS. 80 to 82.—MALIGNANT ANTHRAX ŒDEMA.

In a man of 38, an employee in a factory processing goat skins (A. S. Ross and J. S. Shipman).

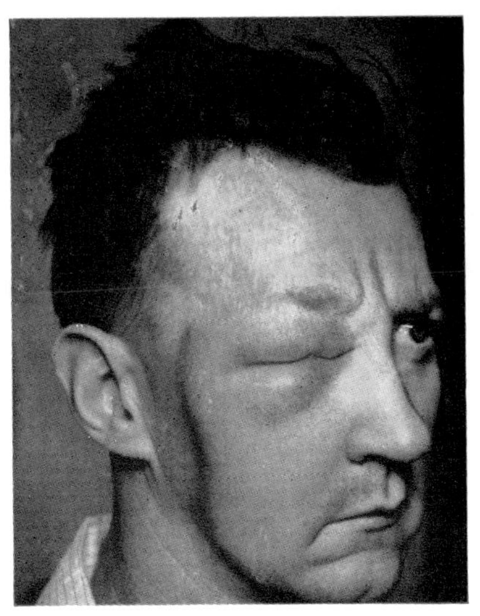

FIG. 80.

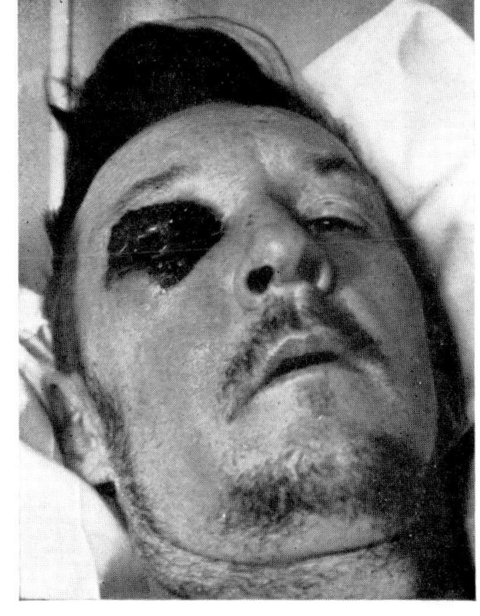

FIG. 81.

FIG. 80.—Early stage of œdema on the second day of the disease.

FIG. 81.—Fully developed disease, showing the formation of an extensive black eschar, 10 days later.

FIG. 82.—The same patient 2 months later after recovery, showing the absence of disfiguring scars after treatment by serum.

FIG. 82.

MALIGNANT ANTHRAX ŒDEMA (*Milzbrandödem; œdème charbonneux ou malin*), a rarer manifestation, appears as an intense widespread œdema.[1]

[1] Després (1886), Moreau (1905), Schmitt (1906), Lundsgaard (1907), Salus (1907), Ross and Shipman (1935), Bernard *et al.* (1956), and others.

The eyelids are the most common site of this clinical type of the infection. Starting in one lid, usually the upper, the doughy reddish swelling rapidly involves the soft tissues of both lids, spreads to the other side and extends to the cheek, the temple, the neck and chest, sometimes involving the buccal mucosa, the tongue and the throat (Debrou, 1865; Ross and Shipman, 1935) (Figs. 80–2). There is no pain, primary pustule or suppuration, but after some days vesicles appear which form extensive black crusts covering a sero-sanguineous fluid. The crusts finally slough off, sometimes leaving scars which may be disfiguring and cause a marked ectropion, sometimes requiring extensive plastic surgery for reconstruction (Bernard *et al.*, 1956). The same acute general symptoms accompany this manifestation of the infection as characterize the malignant pustule.

The *diagnosis* of anthrax may be easily missed largely because the disease is not considered as a possible factor. The absence of pain and of suppuration are important points distinguishing it from such conditions as erysipelas, vaccinia and orbital cellulitis with which it may be confused; a history of working with animals or animal products is significant; while discovery of the characteristic organism in the swab forms conclusive evidence.

The *prognosis* was until comparatively recently always serious. Of 50 cases affecting the lids collected by Morax (1907), 15 died. Among 123 cases in leather workers, Smyth and Bricker (1922) found a death-rate exceeding 21%, but with the efficient use of serum this was considerably reduced, and since the advent of penicillin it became negligible.

The classical methods of *treatment* included early and radical excision of the original lesion with the exhibition of massive doses of iodides. With the introduction of anti-anthrax serum, early and intensive treatment by injections repeated every few hours until the disease was under control, usually led to cure without ill-effects (30 to 100 or even 200 ml. intravenously, intramuscularly or subcutaneously).[1] These methods, however, have now been rendered obsolete by the systemic administration of penicillin, which usually cures the condition rapidly and eliminates the necessity for any other type of treatment (except isolation).[2] Treatment by penicillin G must be intensive (the equivalent of one million units administered immediately followed by a quarter of a million units 6-hourly for 7 days) or chlortetracycline (2,500 mg. daily). As a rule the œdema appears to increase for 2 to 4 days whereafter it begins rapidly to respond; if it does not subside, sulphadiazine (which is also effective) may well be combined with the penicillin (Barnshaw and Lovett, 1949) or the tetracyclines may be used (Gold, 1950–55; Gold and Boger, 1951). With the antibiotic drugs the administration of steroids has been advised (H. and M. Farpour, 1972). In addition to this systemic treatment, it may be well to remove the crusts and provide for

[1] Wilms (1905), J. C. and C. Regan (1919–24), Gold (1935–42), Ross and Shipman (1935), Hodgson (1941), and others.

[2] Murphy *et al.* (1944), Ellingson *et al.* (1946), Sanchez Mosquera (1948), Holgate and Holman (1949), Secret (1952), Huysmans (1959), Simmonds (1960), Tahernia (1967), Genée and Ivandíc (1970).

drainage, in which case bacitracin ointment (500 to 1,000 units per g.) is a suitable application. With such treatment scarring is rarely so severe as to require plastic reconstruction of the lid. A vaccine gives a considerable degree of protection and should be given to those at risk.

Barnshaw and Lovett. *Amer. J. Ophthal.*, **32**, 106 (1949).
Bernard, Maestraggi and Pouillaude. *Pédiat.* (Lyons), **11**, 909 (1956).
Blatt. *v. Graefes Arch. Ophthal.*, **111**, 60 (1923).
Debrou. *Arch. gén. Méd.*, **2**, 403, 422 (1865).
Desmarres. *Hb. d. gesammten Augenheilkunde* (transl. Seitz), Erlangen, 93 (1852).
Després. *France méd.*, **1**, 589 (1886).
Ellingson, Kadull, Bookwalter and Howe. *J. Amer. med. Ass.*, **131**, 1105 (1946).
Elschnig. *Klin. Mbl. Augenheilk.*, **31**, 191 (1893).
Farpour, H. and M. *Bull. Soc. franç. Ophtal.*, **85**, 385 (1972).
Genée and Ivandíc. *Klin. Mbl. Augenheilk.*, **157**, 404 (1970).
Gold. *J. lab. clin. Med.*, **21**, 134 (1935).
Penn. med. J., **40**, 728 (1937).
Arch. intern. Med., **70**, 785 (1942); **96**, 387 (1955).
Amer. J. Med., **8**, 31 (1950).
Gold and Boger. *New Engl. J. Med.*, **244**, 391 (1951).
Himly. *Die Krankheiten u. Missbildungen d. menschlichen Auges*, Berlin, **1**, 204 (1843).
Hodgson. *Lancet*, **1**, 811 (1941).
Holgate and Holman. *Brit. med. J.*, **2**, 575 (1949).
Huysmans. *Ophthalmologica*, **138**, 236 (1959).
Iwanowski. *Vestn. Oftal.*, **13**, 367 (1896).
Knapp. *Arch. Augenheilk. Ohrenheilk.*, **5**, 371 (1876).
Koch. *Milzbrand und Rauschbrand*, Stuttgart; *Dtsch. Chir.*, Lief. ix (1886).
Lundsgaard. *Hospitalstidende*, **50**, 373 (1907).
Manolescu. *Ber. dtsch. ophthal. Ges.*, **37**, 289 (1911).
Morax. *Ann. Oculist.* (Paris), **138**, 338 (1907).
Moreau. *Rev. gén. Ophtal.*, **24**, 193 (1905).
Murphy, La Boccetta and Lockwood. *J. Amer. med. Ass.*, **126**, 948 (1944).
Praun and Pröscher. *Zbl. prakt. Augenheilk.*, **24**, 41 (1900).
Regan, J. C. and C. *Amer. J. med. Sci.*, **157**, 782 (1919); **167**, 255 (1924).
Rosenbach. *Arch. klin. Chir.*, **77**, 715 (1905).
Ross and Shipman. *Amer. J. Ophthal.*, **18**, 641 (1935).
Rundlett. *J. med. Soc. New Jersey*, **45**, 277 (1948).
Salus. *Prag. med. Wschr.*, **32**, 595 (1907).
Sanchez Mosquera. *Arch. Soc. oftal. hisp.-amer.*, **8**, 949 (1948).
Sasso. *Ann. Oculist.* (Paris), **194**, 54 (1961).
Schmitt. *Contribution à l'étude de l'œdème malin charbonneux des paupières* (Thèse), Lyon (1906).
Secret. *Sem. méd. Paris*, **28**, 857 (1952).
Simmonds. *Amer. J. Ophthal.*, **49**, 838 (1960).
Smyth and Bricker. *J. industr. Hyg.*, **4**, 53 (1922).
Tahernia. *Arch. Dis. Childh.*, **42**, 181 (1967).
Thielmann. *Med. Z. Russlands*, **1** (1855).
Wilms. *Münch. med. Wschr.*, **52**, 1100 (1905).

GLANDERS

GLANDERS (*Equinia : Farcy : Rotz : Malleus : Morve*) is *an acute or chronic infection due to the Pfeifferella (Malleomyces) mallei, characterized by vesicular, pustular and carbuncular lesions accompanied by profound systemic involvement and frequently terminating fatally.* The disease is comparatively common in horses, mules and donkeys, rare in cows, and exceptional in man who acquires the infection by association with these animals. Since the coming into force of the Glanders Order in 1908 which enforced the slaughtering of every animal with glanders or giving a positive mallein test, the disease has been eradicated in Britain. Primary infection has been noted in the conjunctiva,[1] and only as great rarities have such primary lesions been described in the human eyelids (Krajewski, 1871; Scheby-Buch, 1878; Neisser, 1892; Filatov, 1907).

ACUTE GLANDERS is characterized by a vesicular or carbuncular lesion,

[1] Vol. VIII, p. 227.

heavily indurated and sloughing. Several such may occur and coalesce to form large gangrenous patches; they may appear on the upper or lower lid or both simultaneously (Krajewski, 1871), and they may eat through the lid substance, resulting in panophthalmitis (Filatov, 1907). The associated lymph nodes are involved and break down ("*farcy buds*"); septicæmic symptoms supervene, a profuse purulent nasal discharge is the rule, and the usual termination is the development of the typhoid state and death.

In the CHRONIC FORM, which may affect the lids or the canthus (Scheby-Buch, 1878; Neisser, 1892), the same local process of ulceration or abscess formation occurs in a milder degree, sometimes associated with metastatic foci. These symptoms may relapse over a period of years and the acute type may supervene at any time, even after apparent recovery. Most of the acute cases and many of the chronic terminate fatally.

The *diagnosis* is made by the superimposition of the nasal involvement and the severe typhoid-like general symptoms on the local lesion, and is confirmed by the microscopic demonstration of the bacillus in the discharge or by the complement-fixation test. *Treatment* is unsatisfactory. The usual surgical measures are to be employed locally, general treatment prescribed as indicated, but specific serum or vaccine therapy has not yielded promising results. Sulphadiazine, however, is very effective (2 to 4 g. immediately, followed by 1 g. every 4 to 6 hours for a minimum of 21 days) (Howe and Miller, 1947; Cravitz and Miller, 1950); of the antibiotics, chloramphenicol, chlortetracycline and streptomycin are the most useful and may be combined with the sulphonamide.

Cravitz and Miller. *J. infect Dis.*, **86**, 52 (1950).
Filatov. *Vestn. Oftal.*, **24**, 195 (1907).
Howe and Miller. *Ann. intern. Med.*, **26**, 93 (1947).
Krajewski. *Klinika* (Poland), **6**, 161 (1871).
Neisser. *Berl. klin. Wschr.*, **29**, 321 (1892).
Scheby-Buch. *Berl. klin. Wschr.*, **15**, 74 (1878).

CHANCROID

CHANCROID (*soft chancre; ulcus molle*) is *a painful ulcer of relatively benign character occurring as a result of infection by Ducrey's streptobacillus* (*Hæmophilus ducreyii*), an organism which probably lives as a saprophyte in the female genital tract and becomes pathogenic when transported into other tissues. The infection is common in areas with low hygienic standards especially in Africa, the Far East and Central and South America; it usually occurs around the genitals owing to venereal inoculation, but extra-genital sores may occur elsewhere as a result of auto-inoculation by the fingers so that multiple lesions may result. We have seen that conjunctival lesions may occur as a rarity[1]; so also may typical lesions appear on the lids (Hirschler, 1866[2]; Bull, 1894; 3 times in 66 cases, Eudlitz, 1897; Bruner, 1912; Appleman and Greenbaum, 1926).

[1] Vol. VIII, p. 210.

[2] In the lid of a surgeon, conveyed whilst syringing a chancre on a patient's penis: it healed under local treatment.

The lesion, which is painful, appears as a reddish macule, which within 24 to 48 hours develops into a pustule. Soon thereafter this breaks down into an ulcer, usually circular in outline with soft edges, a greyish floor bathed in pus, and a surrounding inflammatory areola. The ulcers are auto-inoculable and several may develop which grow by peripheral extension or by coalescence, while the associated lymph nodes become inflamed and may suppurate and break down. Healing usually takes place in a week or more with ordinary hygienic measures, although some cases are more stubborn. Diagnosis is made by smears and culture from the edge of the ulcer.

Treatment in pre-chemotherapeutic days was remarkably effective by moist antiseptic soaks (hydrogen peroxide, pot. permanganate) or gentian violet; vaccine preparations were also advocated (Nicolle, 1923; Nicolle and Durand, 1924; Hunt, 1935; and others); but the sulphonamide drugs usually produce incomparably better results (sulphadiazine or sulphafurazole, 2 g. immediately, followed by 1 g. 6-hourly for 7 days), the condition being usually cured in less than a week (Hutchison, 1938; Batchelor and Lees, 1938; Combes, 1946; and others). Infected lymph nodes when purulent should be opened, but this is required less frequently when chemotherapy is employed; they and the local lesion may be dusted with sulphanilamide powder. It is interesting that, although the organism is penicillin-sensitive *in vitro* (Tung and Frazier, 1945), this antibiotic is without clinical value (Pereyra and Landy, 1944); streptomycin, however, is rapidly curative (Hirsh and Taggart, 1948; Jawetz, 1948; Willcox, 1950).

Appleman and Greenbaum. *Amer. J. Ophthal.*, **9,** 358 (1926).
Batchelor and Lees. *Brit. med. J.*, **1,** 1100 (1938).
Bruner. *Derm. Wschr.*, **54,** 277 (1912).
Bull. *Norsk. Mag. Laeger.*, **9,** 487 (1894).
Combes. *N.Y. St. med. J.*, **46,** 1700 (1946).
Eudlitz. *Arch. gén. Méd.*, **1,** 424 (1897).
Hirschler. *Wien. med. Wschr.*, **16,** 1145, 1161, 1177 (1866).
Hirsh and Taggart. *J. vener. Dis. Inform.*, **29,** 47 (1948).
Hunt. *Proc. Soc. exp. Biol. Med.*, **33,** 293 (1935).
Hutchison. *Lancet*, **1,** 1047 (1938).
Jawetz. *Arch. Derm. Syph.* (Chic.), **57,** 916 (1948).
Nicolle. *Acta derm.-venereol.* (Stockh.), **4,** 353 (1923).
Nicolle and Durand. *Presse méd.*, **32,** 1033 (1924).
Pereyra and Landy. *U.S. Navy med. Bull.*, **43,** 189 (1944).
Tung and Frazier. *Amer. J. Syph.*, **29,** 629 (1945).
Willcox. *Lancet*, **1,** 396 (1950).

Infection of the lids in BRUCELLOSIS occurs as a rarity (Rocco, 1952). Similarly, a primary palpebral lesion appearing in TULARÆMIA giving rise to œdema and swelling of the parotid and submaxillary glands has been reported (Lindsay and Scott, 1951).

Lindsay and Scott. *Canad. J. publ. Hlth.*, **42,** 146 (1951).
Rocco. *Rass. ital. Ottal.*, **21,** 291 (1952).

TUBERCULOSIS

Infection of the lids by the *Mycobacterium tuberculosis* may be either primary or, much more commonly although less frequently than in former years, a reinfection tuberculosis from a focus elsewhere. Since the advent of chemotherapy and antibiotics, these manifestations of the disease are rarely seen.

Primary Tuberculosis of the Lids

PRIMARY TUBERCULOSIS (TUBERCULOUS CHANCRE) is very rare in the skin; it usually occurs in the lung (*Ghon complex*) and appears in previously uninfected individuals who thus do not have an immunity to the organism. It follows that it is typically seen in young children although it may occur in adults. In the skin it appears as an inflammation showing histologically areas of necrosis resulting in ulceration and is associated with involvement of the regional lymph nodes. Tubercle bacilli are present. Some seven cases have been reported in the literature with a chancre of this type on the face; among these, Hallé and Garnier (1930) and Montgomery and Helmholz (1936) recorded cases on the cheek in the region of the lids, while an interesting case was described by Hervouët (1951) on the lid itself, which appeared on the nasal side of the left lower lid of a girl aged one year; two members of the family had died of tuberculosis and the mother had a pulmonary infection so that the lesion was almost certainly caused through kissing. An ulcer formed with a purulent exudate from which the bacilli were recovered and the pre-auricular nodes were swollen. A case was recorded by Scuderi and Cardia (1963) wherein there developed a cutaneous nodular tuberculosis of the lids with regional lymphadenopathy in a woman who had had a plastic operation performed on the lid, with the retention of a catgut suture which the authors concluded was associated with the infection.

Primary tuberculosis may occur in the conjunctiva[1] in which case the lids may become secondarily infected.

Reinfection Tuberculosis

Reinfection tuberculosis of the lids is also rare. The most common lesion is lupus vulgaris; scrofuloderma wherein the lid is infected from diseased tissue underneath is rarer, while tuberculomata may appear on the lids in some cases due to hæmatogenous infection.

LUPUS VULGARIS

LUPUS VULGARIS *is a chronic and slowly progressive tuberculous infection of the skin characterized by soft " apple-jelly like " tubercles*, first adequately described by Robert Willan (1808) (*Willan's disease*). It practically always begins in childhood, sometimes in infancy, the dermal infection arriving in one of four ways: (*a*) direct inoculation into the skin, (*b*) spread from infected mucous membranes, particularly the nasal, (*c*) from some underlying organ, as a lymph node, or (*d*) from the blood-stream in the course of a bacteræmia, a phenomenon particularly evident in the appearance of multiple lesions

[1] Vol. VIII, p. 212.

during a febrile illness when presumably a bacteræmia readily occurs. A primary infection of the eyelids is rare (2 in 374 cases, Bender, 1886; 4 in 137 cases, Sachs, 1886; 2 in 121, Block, 1886; 1 case, Morax and Landrieu, 1913; Hager, 1957); the most usual type is a secondary spread from the nose or cheeks where in many cases the infection is originally derived from bacilli lodging in cracks and fissures in a catarrhal mucosa inside the nostril. A much less common mode of spread is from the conjunctival mucosa.[1] In over 40% of cases the infection is of the bovine type (Lomholt, 1934).

The disease starts with a small nodular tuberculous deposit under the epithelium which pushes its way to the surface to appear as a translucent

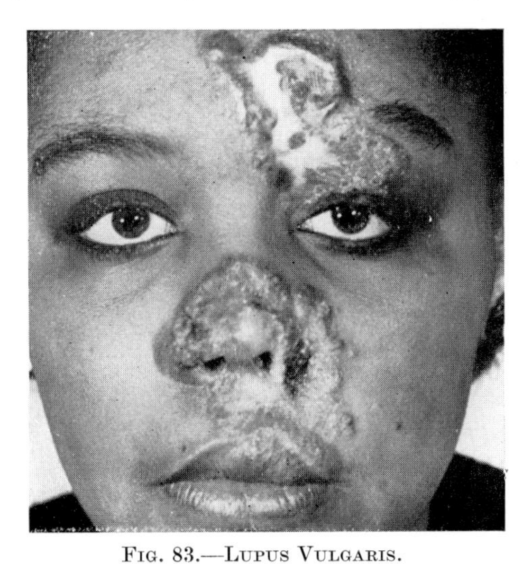

FIG. 83.—LUPUS VULGARIS.
In a relatively early active stage (Dept. of Dermatology, Vanderbilt Clinic, N.Y.).

mass varying in size from a pin-head to a small pea surrounded by an erythematous area. The lesion causes no symptoms. Examined diascopically (*i.e.*, through a thin sheet of glass pressed against the skin to blanch it) it has a characteristic " apple-jelly " appearance. Pathologically it is a typical tubercle with giant cells containing bacilli. As it slowly increases in size and becomes more protuberant, nodules increase in numbers spreading by peripheral deposition. The spread is very slow, but eventually a *lupus-patch* is formed, the irregular periphery of which shows actively spreading changes, while the centre shows retrogressive changes of cicatrization or ulceration which in either case leave a permanent scar (Fig. 83).

This evolution may take place over a period of years or a life-time, so that large areas may become involved, and intermissions and relapses are continually occurring. The ulcers are shallow with sharp margins and a

[1] Vol. VIII, p. 222.

smooth, red, shiny base sometimes studded with readily bleeding granulations. They heal slowly, leading to the development of more scar-tissue, and by interstitial necrosis or secondary infection the scars repeatedly break down. Moreover, in the cicatrized area undestroyed nodules lie buried and constitute foci from which further outbreaks frequently arise. The disease is symptomless, but the destruction of tissue and cicatrization results in the most hideous deformities (Figs. 84–5).

In the usual spread from the nose and face one or all four lids may be involved. The eventual result is almost invariably a high degree of ectropion, and there is on record the complete destruction of all four lids (Laqueur,

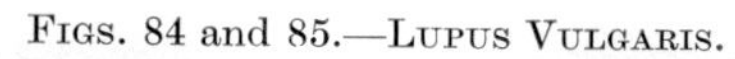

Figs. 84 and 85.—Lupus Vulgaris.

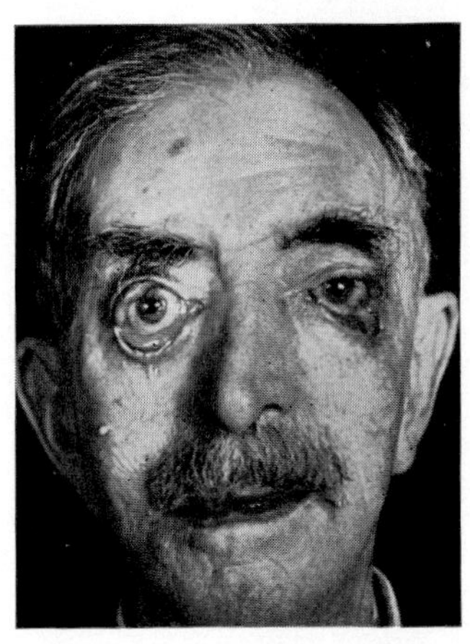

Fig. 84.

Fig. 84.—Lupus of the face which had been active for 65 years, during which time the patient suffered a multitude of treatments. There is bilateral ectropion of a very marked degree (A. MacIndoe).

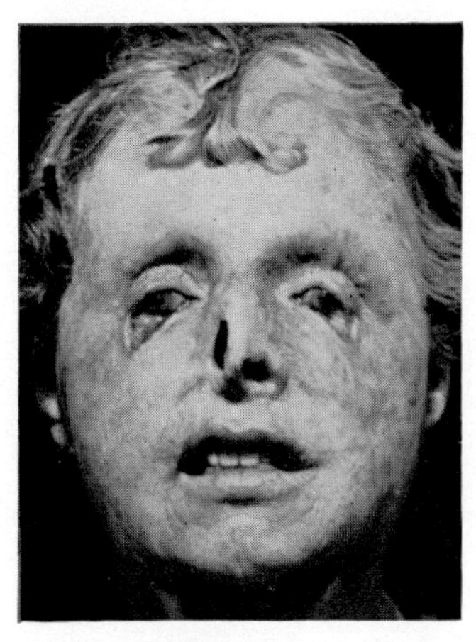

Fig. 85.

Fig. 85.—A still more marked example of the deformities resulting from lupus. The nose has been eaten away to a considerable extent; there is marked bilateral ectropion with exposure keratitis in each eye (T. Pomfret Kilner).

1904). The resultant damage to the eye may be profound: a chronic kerato-conjunctivitis is almost invariable if exposure is marked, the disease may spread to the conjunctiva itself (Arlt, 1864; Alexander, 1875; Pflüger, 1876; and many others), while phthisis bulbi may eventuate (Bentzen, 1904) and enucleation be necessary (Remmlinger, 1898). Finally, carcinomatous changes may develop in long-infected areas of the skin, a complication which may be excited by radiational treatment but is rare in the lids (Capauner, 1901; Lindner, 1914; Hornberger, 1949) (Fig. 86).

The differential diagnosis depends essentially on the long history from childhood, the typical distribution on the face and nose, the presence of apple-jelly nodules as seen by diascopy, and the configuration of the lesion with its active periphery and retrogressive centre in which residual nodules may be noted.

Treatment used to be difficult and tedious, the most eloquent testimony to which is the number of remedies which were advocated. A dramatic therapeutic agent was calciferol (vitamin D_2) given in large doses (100,000–200,000 units daily either as tablets by mouth or by intramuscular injections in an oily solution).[1] A still more effective agent is streptomycin (1 g. daily to twice weekly) combined with isonicotinic acid hydrazide (Isoniazid) (3 to 6 mg./kg. body-weight 2 or 3 times a day) and PAS (4 g. 3 times a day), a

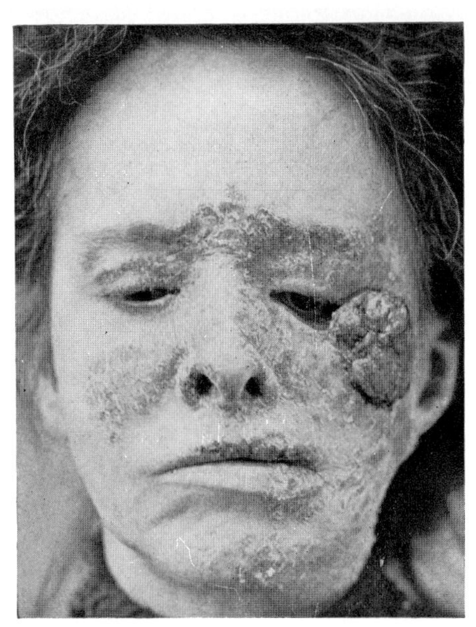

Fig. 86.—Epithelioma on Lupus Vulgaris.
A case of 23 years' duration (Sequeira *et al.*, *Diseases of the Skin*, Churchill).

combination which should be continued for two months after all clinical evidences of the disease have disappeared. Improvement should be evident within a month and small lesions may disappear within 6 to 8 months.

With these drugs local treatment is rarely necessary; a useful adjuvant is ultra-violet light applied through a quartz compressor to render the parts anæmic. This method leaves less obvious scars than do destructive chemical agents pricked into chronic and fibrotic areas which were at one time employed, such as acid nitrate of mercury, silver nitrate or trichloracetic acid. Under an umbrella of Isoniazid, surgical repair can be safely undertaken.

[1] Charpy (1943–46), Dowling (1946), Dowling and Thomas (1946), Gaumond and Grandbois (1947), Hohmann and Beening (1947), Freudenthal (1948), Dawson (1948), Cornbleet (1948), Clarke (1949), Ruiter and Groen (1949), and others.

SCROFULODERMA (TUBERCULOSIS COLLIQUATIVA)

In this variety of tuberculosis the skin becomes secondarily affected from some diseased tissue underneath—a lesion in the orbital bones, the lacrimal sac, the conjunctiva or lymph nodes[1] (Fig. 87). Tuberculous tarsitis which may give rise to a similar lesion is discussed at a later stage.[2] In some of these cases the diagnosis has not been confirmed by the finding of tubercle bacilli but has depended only on the histological appearances. In the typical case a fluctuant purplish swelling appears which, perhaps after a considerable interval of time, breaks down and sloughs to form an ulcer which may be surrounded by granulations and may serve as the mouth of a fistula, discharging more or less continuously. The ulcer has a livid undermined edge and,

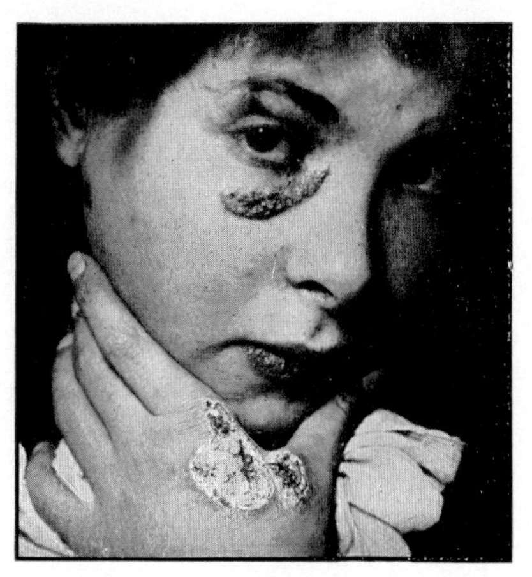

FIG. 87.—SCROFULODERMA.

The lesion on the cheek and lower eyelid resulted from an infection of the skin from a tuberculous abscess of the lacrimal sac; that on the hand was secondary to tuberculosis of the metacarpal bone (D. Paterson and A. Moncrieff, *Diseases of Children*, Arnold).

extending both in depth and area, may cause considerable destruction of the skin of the lids. Tuberculoid structures with caseation are present and the bacilli are usually found. The lesion may persist for years and heal up spontaneously, but usually requires active treatment. The underlying focus should, if possible, be eliminated, the lesion cleaned surgically, while complete healing is effected by streptomycin combined with PAS and Isoniazid (*vide supra*).

A " cold abscess " may develop in the substance of the eyelid without involvement of the skin from a tuberculous lesion in the underlying tissues (as the maxillary antrum, Watrin and Mendelsohn, 1967).

[1] Rollet (1906), von Michel (1908), Wätzold (1912), Salvati (1923), Higab (1939).
[2] p. 237.

TUBERCULOUS GUMMATA are somewhat similar nodular lesions developing in the subcutis, arising from hæmatogenous dissemination and not from an underlying focus. They soften, break down and form indolent ulcers resembling scrofuloderma in appearance (Fig. 88). Indurated subcutaneous nodules which may grow to a considerable size, some of which showed tubercle bacilli histologically and probably come into this category, were described in the lids by Jandot (1906), Rollet (1906), Schöpfer (1928), Gallenga (1930) and Sengupta (1956), and in the region of the brows by Kraus (1905). Treatment is as for lupus vulgaris.

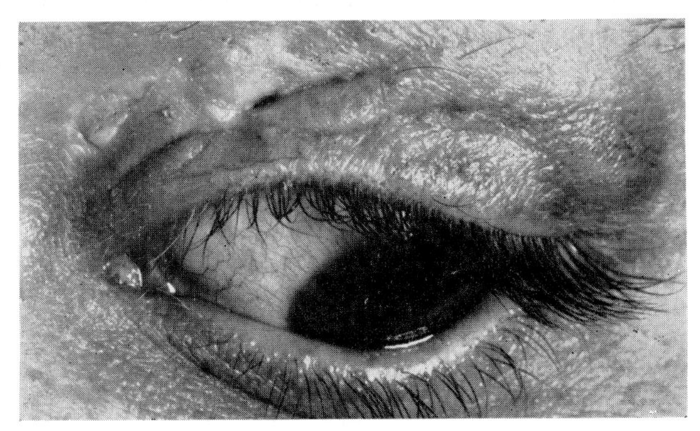

FIG. 88.—TUBERCULOMA.

In a youth of 17. The tuberculoma in the upper fornix had ulcerated through to the skin of the upper lid where a chronic sinus developed (Inst. Ophthal.).

TUBERCULIDES

Several lesions classed as tuberculides by Darier (1896–1903) have given rise to much controversy. They comprise the papulo-necrotic tuberculide, lichen scrofulosorum, erythema induratum (of Bazin), the roseaceous tuberculide (acne rosea lupus), and acne agminata (acnitis, lupus miliaris disseminatus faciei, follicular lupus). In the older literature they were classified as tuberculous, the result of the hæmatogenous dissemination of the bacilli, mainly because of the tuberculoid appearance of the histology of the lesion, sometimes because of the presence of an active focus of tuberculosis elsewhere. In recent years, however, more strict criteria have been applied since it has now become widely accepted that the histological picture does not prove a tuberculous ætiology; this should depend on a hypersensitivity to tuberculin and a favourable response to anti-tuberculous therapy. It is to be noted that in none of them is the tubercle bacillus found. In this respect papulo-necrotic tuberculides are sometimes considered to be tuberculous in nature; the characteristic clinical picture of endarteritis and thrombosis of the dermal vessels has been widely interpreted as a form of vasculitis (Krüger, 1961). Lichen scrofulosorum, a condition of which there are no recent reports in the literature, characterized by indolent papules, may possibly be tuberculous but in many respects resembles sarcoidosis (Thal, 1955; Lever, 1967). Erythema induratum (of Bazin) which usually occurs on the legs of women, and the similar nodular lesion of Wende (1911) which may occur on the face, may sometimes be tuberculous in origin but more frequently present the picture of nodular vasculitis. The rosaceous tuberculide, originally described by Lewandowsky (1917), is now generally accepted as a variant of rosacea (Klingmuller, 1954; Calnan, 1966). Acne agminata (acnitis) is now generally classified as a dermatosis of unknown ætiology and is excluded from the group of tuberculides (Scott and Calnan, 1967; O'Driscoll and Morgan, 1974).[1] This is the only lesion of this class which commonly affects the lids.

Alexander. *Klin. Mbl. Augenheilk.*, **13**, 329 (1875).
Arlt. *Klin. Mbl. Augenheilk.*, **2**, 329 (1864).
Bender. *Dtsch. med. Wschr.*, **12**, 396 (1886).
Bentzen. *Klin. Mbl. Augenheilk.*, **42** (1), 84 (1904).
Block. *Vjschr. Derm.* (Wien), **13**, 201 (1886).
Calnan. *G. ital. Derm.*, **107**, 587 (1966).
Capauner. *Z. Augenheilk.*, **5**, 282 (1901).
Charpy. *Ann. Derm. Syph.* (Paris), **3**, 331, 340 (1943); **4**, 110, 331 (1944).
Lancet, **1**, 400 (1946).
Clarke. *Brit. J. Derm.*, **61**, 409 (1949).
Cornbleet. *J. Amer. med. Ass.*, **138**, 1150 (1948).
Darier. *Ann. Derm. Syph.* (Paris), **7**, 1431 (1896).
Précis de dermatologie, Paris, 547 (1909).
Dawson. *Brit. J. Derm.*, **60**, 164 (1948).
Dowling. *Lancet*, **1**, 590 (1946).
Dowling and Thomas. *Brit. J. Derm.*, **58**, 45 (1946).
Freudenthal. *Brit. J. Derm.*, **60**, 178 (1948).
Gallenga. *Arch. Ottal.*, **37**, 385 (1930).
Gaumond and Grandbois. *Canad. med. Ass. J.*, **56**, 205 (1947).
Hager. *Klin. Mbl. Augenheilk.*, **130**, 704 (1957).
Hallé and Garnier. *Bull. Soc. Pédiat. Paris*, **28**, 191 (1930).
Hervouët. *Bull. Soc. Ophtal. Fr.*, 85 (1951).
Higab. *Bull. ophthal. Soc. Egypt*, **32**, 65 (1939).
Hohmann and Beening. *Ned. T. Geneesk.*, **91**, 78 (1947).
Hornberger. *Strahlentherapie*, **80**, 367 (1949).
Jandot. *La tuberculose nodulaire souscutanée des paupières* (Thèse), Lyons (1906).
Klingmuller. *Derm. Wschr.*, **130**, 1058 (1954).
Kraus. *Arch. Derm. Syph.* (Wien), **74**, 3 (1905).
Krüger. *Arch. klin. exp. Derm.*, **213**, 496 (1961).
Laqueur. *Dtsch. med. Wschr.*, **30**, 190 (1904).
Lever. *Histopathology of the Skin*, 4th ed., Phila., 301 (1967).
Lewandowsky. *Korresp.-Bl. schweiz. Ärz.*, **47**, 1280 (1917).
Lindner. *Klin. Mbl. Augenheilk.*, **53**, 245 (1914).
Lomholt. *Brit. med. J.*, **2**, 291 (1934).
von Michel. *Graefe-Saemisch Hb. d. ges. Augenheilk.*, 2nd ed., Leipzig, **5** (2), 95 (1908).
Montgomery and Helmholz. *Proc. Mayo Clin.*, **11**, 407 (1936).
Morax and Landrieu. *Ann. Oculist.* (Paris), **150**, 266 (1913).
O'Driscoll and Morgan (1974). Personal communication.
Pflüger. *Klin. Mbl. Augenheilk.*, **14**, 162 (1876).
Remmlinger. *Zur Kasuistik d. Tuberkulose d. Bindehaut* (Diss.), Giessen (1898).
Rollet. *Arch. Ophtal.*, **25**, 340 (1905).
Rev. gén. Ophtal., **25**, 385 (1906).
Ruiter and Groen. *Dermatologica*, **99**, 345 (1949).
Sachs. *Vjschr. Derm.* (Wien), **13**, 241 (1886).
Salvati. *Ann. Oculist.* (Paris), **160**, 810 (1923).
Schöpfer. *Klin. Mbl. Augenheilk.*, **81**, 193 (1928).
Scott and Calnan. *Trans. St. John's Hosp. derm. Soc.*, **53**, 60 (1967).
Scuderi and Cardia. *G. ital. Oftal.*, **16**, 372 (1963).
Sengupta. *Indian J. Derm.*, **2**, 38 (1956).
Thal. *Dermatologica*, **111**, 87 (1955).
Wätzold. *Z. Augenheilk.*, **27**, 320 (1912).
Watrin and Mendelsohn. *Bull. Soc. Ophtal. Fr.*, **67**, 1124 (1967).
Wende. *J. cutan. Dis.*, **29**, 1 (1911).
Willan. *On Cutaneous Diseases*, London, **1** (1808).

[1] p. 275.

LEPROSY

The leprosy bacillus of Hansen, the general clinical aspects of leprosy and its effects upon the conjunctiva and cornea have already been discussed,[1] as well as its uveal complications.[2] The clinical aspects of the disease as it affects the lids only will be considered here.

Leprosy (Hansen's disease) attacks the eyebrows and lids with great frequency, and manifestations, usually of a pronounced nature, are evident in this region in more than two-thirds of all cases (Lopez, 1891). There is some geographical variation: thus in Australia leprotic nodules on the lids are comparatively rare (6% of all cases, Gibson, 1950), but in most countries where the disease is endemic lid involvement is so frequent that it is one of the best-known stigmata recognized by physicians and laymen alike.

Loutfy and his colleagues (1937) gave the following incidence of lid affections among patients with leprosy in Egypt:

Loss of cilia or eyebrows	60%
Skin nodules	14%
Paralysis of orbicularis	11%
Paralysis of orbicularis with facial paralysis	5%
Diffuse infiltration of lid	1%
Anæsthetic patches on lid	1%

Emiru (1970) found the following conditions of the lids among 890 patients in Uganda:

	Lepromatous leprosy	Tuberculoid leprosy	Borderline
Madarosis	50	15	8
Lagophthalmos	9	38	3
Entropion	6	3	—
Lacrimal obstruction	—	1	—
Blepharitis	1	1	—

Weerekoon (1972) gave the following statistics from Malaysia in 444 patients: loss of lashes (usually sparse rather than totally denuded), 111 cases; an absence of eyebrows especially over the lateral aspect, 219 cases; involvement of the VIIth nerve with varying degrees of lagophthalmos, 209 cases; nodules on the lids, 7 cases; erythema of the lids, 3 cases; and anæsthetic patches on the lids, 1 case.

Hornblass (1973), among 51 patients in Vietnam, found 27% with paralysis of the VIIth nerve, usually with madarosis; 20% with involvement of the Vth nerve; 16% with madarosis of the eyebrows and 12% with tylosis of the lids.

Three main types of the disease occur: *lepromatous* (*nodular*) *leprosy* in which the typical nodules show a xanthoma-like histological picture, an abundance of bacilli and a negative lepromin (intradermal) test; the *tuberculoid* (*nerve*) *type* attacking particularly the nerves, with a sarcoid-like histological picture, no bacilli and a positive lepromin test; and an *intermediate* (*border-line*) *type* participating in the characteristics of both. The differentiation depends on the immune-status of the patient.

[1] Vol. VIII, p. 844. [2] Vol. IX, p. 285.

The *lepromin test* consists of the injection of 0·1 ml. of an emulsion of killed leprosy bacilli intradermally into the front of the forearm. A positive test consists of the appearance of a small nodule within 48 hours.

In LEPROMATOUS LEPROSY thickening of the supraciliary ridges with loss of the eyebrows is a characteristic early symptom which is followed by the development of the typical nodules, masses of granulation tissue containing the characteristic multinucleated giant " lepra cells " and packed

FIGS. 89 to 91.—NODULAR LEPROSY.

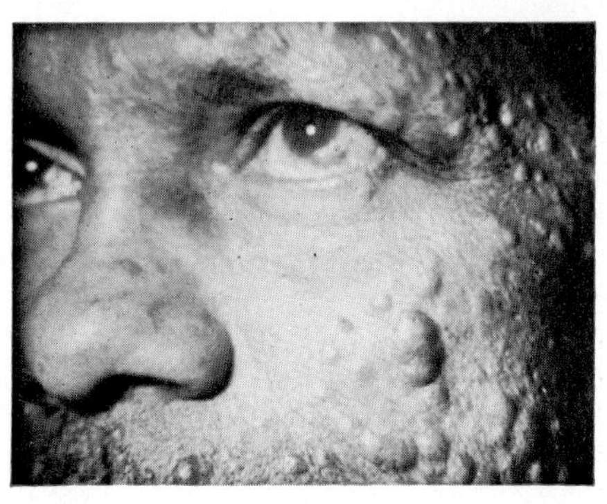

FIG. 89.—Showing multiple nodules on the face and lids and an episcleral nodule (L. Weerekoon).

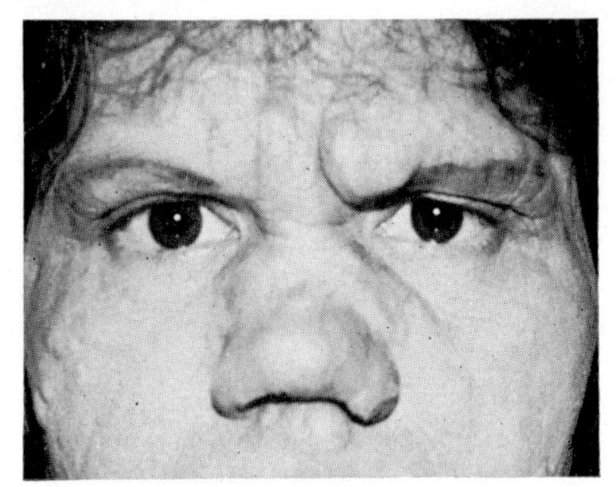

Fig. 90.—The common appearance showing gross nodules particularly around the nose and brows (Inst. Ophthal.).

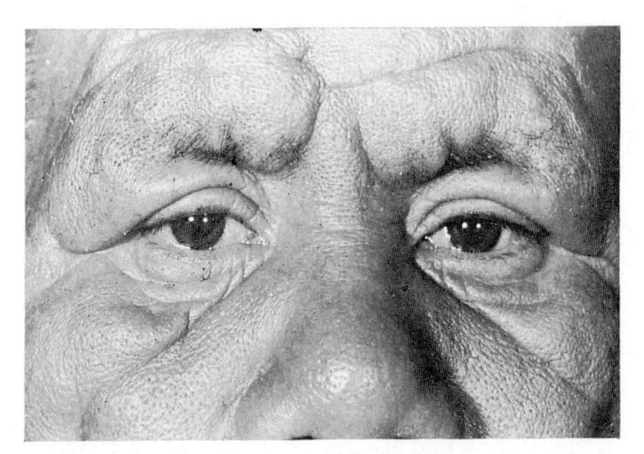

FIG. 91.—Gross nodules around the brows and cheeks with loss of the outer parts of the eyebrows (W. J. Holmes).

with bacilli (Figs. 89–90). Thickening of the lids follows, usually a diffuse infiltration beginning in the intramarginal area, the process frequently extending through the tarsal plate and becoming visible on the conjunctival surface. Starting in the temporal region the eyebrows and lashes tend to turn white and splinter and then to fall out, so that the brows are changed into prominent, irregular, hairless folds and the lid-margins assume the rolled, thickened appearance of tylosis. Eventually, although the nodules on the lids rarely attain a large size or ulcerate but rather resemble chalazia, the forehead, brows and the region of the glabellum become disfigured with irregularly shaped knobs and bosses separated by deep furrows, deformities which, in combination with the thickening and deformation of the lips, nose and ears, produce the hideous " leonine " expression characteristic of these unfortunate people (*elephantiasis græcorum*)[1] (Fig. 91).

It is to be noted that the infection frequently spreads to the lacrimal sac and duct leading to their permanent occlusion.

In the TUBERCULOID TYPE of the disease the ocular adnexa are frequently affected. The characteristic anæsthetic patches, slightly paler in colour than the rest of the skin, are common and appear at an early stage in the lids and brows associated, as elsewhere, with degenerative changes in the brows and lashes leading to alopecia with madarosis (Fig. 92). As a rule no bacilli are found in the lesions. The terminal filaments of the VIIth nerve are commonly involved in a chronic interstitial neuritis, the orbicularis palpebrarum muscle being peculiarly liable to suffer. The consequences are diminished blinking, irregular fibrillation of the muscle and paresis of the lids leading to a paralytic ectropion, epiphora, lagophthalmos and exposure of the globe, the disastrous consequences of which are increased by the frequent loss of corneal sensitivity[2] (Fig. 93). It is interesting that the IIIrd nerve is much less frequently involved so that ptosis is unusual and, since the IVth and VIth nerves are similarly immune, ocular motor disturbances are exceptional, if, indeed, they occur at all. In the later stages of the disease trophic disturbances appear, mutilations of the fingers and face, particularly the nose, being most common, but among which absorption of the tarsal cartilages and even complete disappearance of the lid may be evident (Fig. 94).

Without adequate treatment patients with lepromatous leprosy suffer a gradual progression of the disease, dying of complications within 15 years; in the tuberculoid type the patients tend to recover, sometimes after a protracted course, with a considerable residual nerve damage.

The *treatment* of leprosy was for long a completely baffling problem, and of all the innumerable remedies the somewhat ineffective chaulmoogra oil was the only drug which withstood the test of time. Chemotherapy, however, has transformed a situation hitherto one of the most depressing in medicine, for the derivatives of diamino-diphenyl sulphone[3] have produced therapeutic results far more effective than anything else.[4] High doses used to be admin-

[1] Neve (1900), Pinkerton (1933), Riad (1934), King (1936), Harley (1946), de Carvalho (1948), Aparisi (1950), Weerekoon (1969), Richards and Arrington (1969).

[2] Choyce (1969), Richards and Arrington (1969), Ticho and Ben Sira (1970), Emiru (1970), and others.

[3] Vol. VII, p. 652.

[4] Faget (1947), Sharp and Payne (1948), Lowe and Smith (1949), Molesworth and Naryanaswami (1949), Sloan *et al.* (1950), and others.

FIGS. 92 to 94.—NERVE LEPROSY.

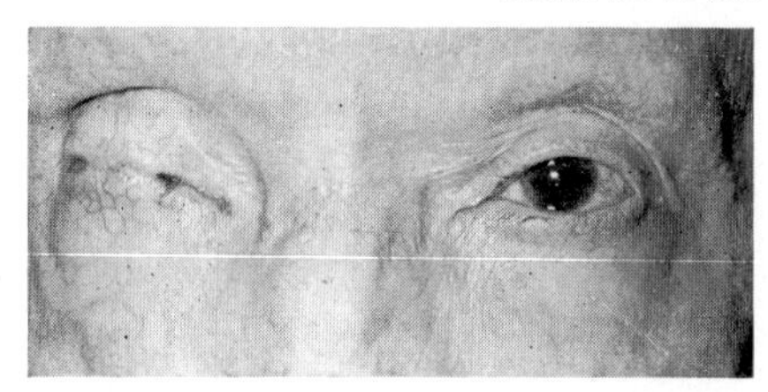

FIG. 92.—Showing absence of the brows and lashes in a native of New Caledonia (J. Beretti).

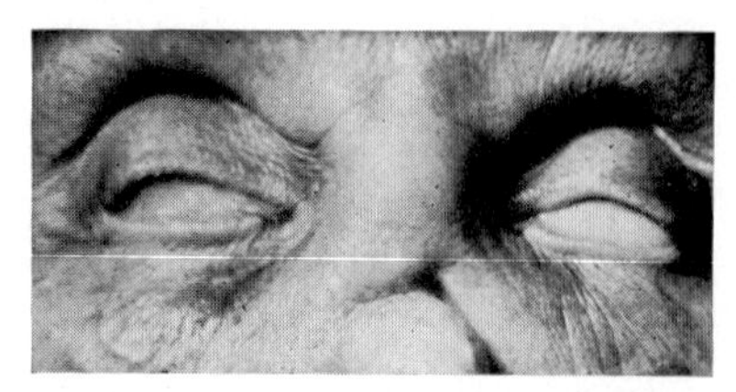

FIG. 93.—Lagophthalmos due to paralysis of the VIIth nerve, in a Malayan (L. Weerekoon).

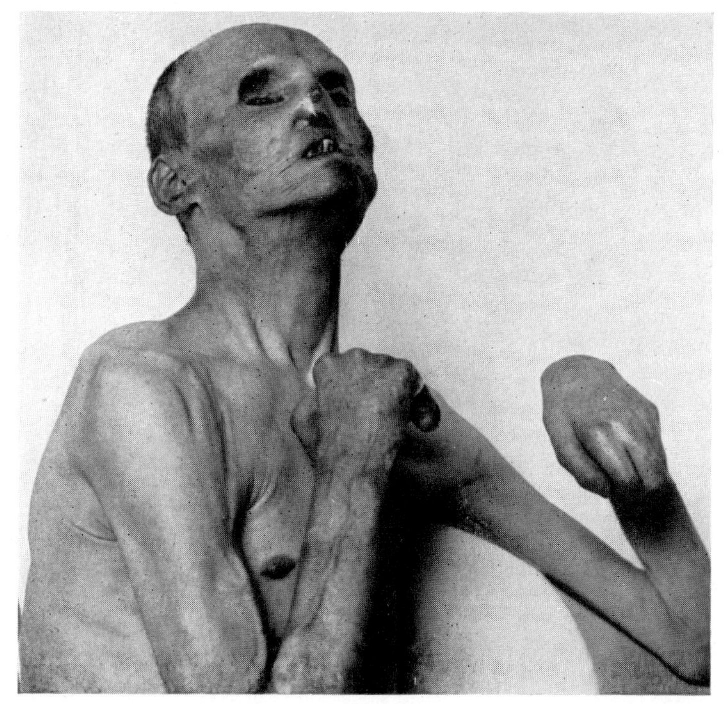

FIG. 94.—Extreme type of nerve leprosy, showing the habitual posture adopted by the patient (Juvenal Esteves).

istered but these have now been diminished; a suitable dose recommended by Pearson and Pettit (1969) is 50 mg. twice weekly. Lowe (1950), after a year's trial using very high doses, found that 70% of patients showed a clinical and 60% a bacteriological improvement, although bacteriologically negative results were rare (3%); for this reason relapses are prone to occur unless the treatment is continued for at least 2 years after all activity has ceased, a process which may take up to 8 years (Erickson, 1950; Doull and Wolcott, 1956). So far as the lids are concerned, surgical procedures are tolerated much better than might be expected and plastic repair for ectropion or exposure keratitis can be successfully undertaken although the results may not be permanent (Pinkerton, 1937; Harley, 1946; Fahim, 1948).

Aparisi. *Arch. Soc. oftal. hisp.-amer.*, **10**, 107 (1950).
de Carvalho. *Rev. bras. Oftal.*, **6**, 31 (1948).
Choyce. *Brit. J. Ophthal.*, **53**, 217 (1969).
Doull and Wolcott. *New Engl. J. Med.*, **254**, 20 (1956).
Emiru. *Brit. J. Ophthal.*, **54**, 740 (1970).
Erickson. *Publ. Hlth. Rep.* (Wash.), **65**, 1147 (1950).
Faget. *Int. J. Leprosy*, **15**, 7 (1947).
Fahim. *Bull. ophthal. Soc. Egypt*, **37**, 116 (1948).
Gibson. *Med. J. Aust.*, **1**, 8 (1950).
Harley. *Amer. J. Ophthal.*, **29**, 295 (1946).
Hornblass. *Amer. J. Ophthal.*, **75**, 478 (1973).
King. *Brit. J. Ophthal.*, **20**, 561 (1936).
Lopez. *Arch. Augenheilk.*, **22**, 318 (1891).
Loutfy, Fahmy and Ismail. *Bull. ophthal. Soc. Egypt*, **30**, 181 (1937).
Lowe. *Lancet*, **1**, 145 (1950).
Lowe and Smith. *Int. J. Leprosy*, **17**, 181 (1949).
Molesworth and Naryanaswami. *Int. J. Leprosy*, **17**, 197 (1949).
Neve. *Brit. med. J.*, **1**, 1153 (1900).
Pearson and Pettit. *Int. J. Leprosy*, **37**, 40 (1969).
Pinkerton. *Laryngoscope*, **43**, 991 (1933). *Amer. J. Ophthal.*, **20**, 715 (1937).
Riad. *Bull. ophthal. Soc. Egypt*, **27**, 79 (1934).
Richards and Arrington. *Amer. J. Ophthal.*, **68**, 492 (1969).
Sharp and Payne. *Int. J. Leprosy*, **16**, 157 (1948).
Sloan, Chung-Hoon, Godfrey-Horan and Hedgcock. *Int. J. Leprosy*, **18**, 1 (1950).
Ticho and Ben Sira. *Brit. J. Ophthal.*, **54**, 107 (1970).
Weerekoon. *Brit. J. Ophthal.*, **53**, 457 (1969); **56**, 106 (1972).

RHINOSCLEROMA

RHINOSCLEROMA, a condition first described by Hebra, is a chronic granulomatous disease affecting essentially the nose and upper respiratory tract causing sclerosis and ultimate deformities; it is due to a short, encapsulated Gram-negative bacillus isolated by von Frisch (1882), the *Klebsiella rhinoscleromatis*. It has a scattered incidence but is found most commonly in south eastern Europe (particularly the Danube basin), North Africa and the West Indies (Pardo-Castello and Martinez Dominguez, 1922). The malady starts with the development of hard, circumscribed plaques, at first subcutaneous but later involving the skin, over which the epidermis becomes a glistening brownish-red with a telangiectatic scaly surface. Over a period of years tumour

FIGS. 95 and 96.—RHINOSCLEROMA.

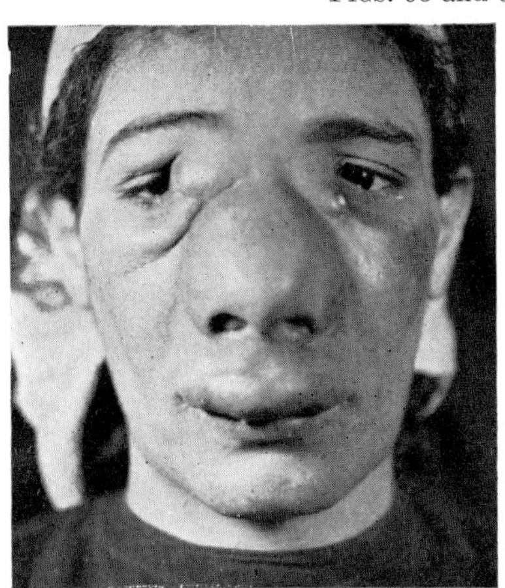

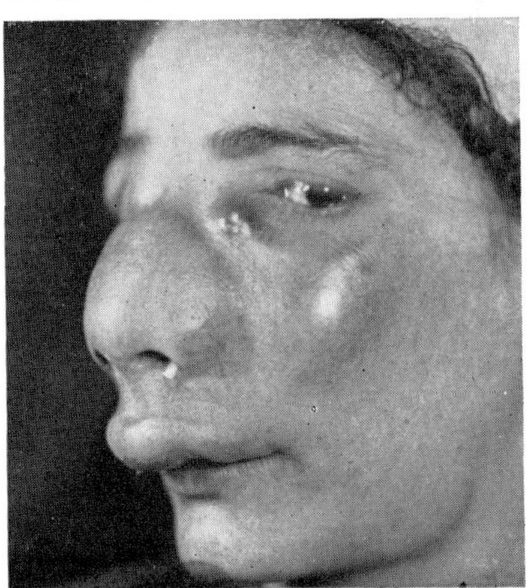

FIG. 95. FIG. 96.

Involving the nose, the upper lip, the lower lids and the region of the internal canthi in a native of Egypt (R. P. Wilson).

masses increase in size irregularly and ultimately shrink producing indurated and cicatricial deformities. Starting usually in the anterior nares and upper lip the disease spreads over the mouth, nose and pharynx, and the infiltration may affect the lower lid and its margin, or spread up the lacrimal passages to the inner canthus, the upper lid and the conjunctiva (Ewetzky, 1898; Gallenga, 1899) (Figs. 95–6).

Histopathologically the tumour is rich in plasma cells together with two characteristic types—the Mikulicz cell, a large rounded cell with an eccentric nucleus containing the organism, and Russell bodies, brilliant bodies without a nucleus formed within a plasma cell as a result of its degeneration. The Mikulicz cells increase until they dominate the histological picture, but they are finally replaced by fibrosis (Kline and Brody, 1949; Convit *et al.*, 1961; Kerdel-Vegas *et al.*, 1963).

The organism is particularly sensitive to streptomycin or chloramphenicol; thereby the proliferative lesions may be arrested (Devine *et al.*, 1947). Surgical excision may be necessary but unless radical is usually followed by recurrences, but although all cases are not responsive, x-rays administered early are palliative and may produce good therapeutic results (Pollitzer, 1906; Alderson, 1932; Chamberlin, 1936; Simpson and Ellis, 1939).

Alderson. *Arch. Derm. Syph.* (Chic.), **26,** 639 (1932).
Chamberlin. *Arch. Otolaryng.*, **23,** 285 (1936).
Convit, Kerdel-Vegas and Gordon. *Arch. Derm.*, **84,** 55 (1961).
Devine, Weed, Nichols and New. *Proc. Mayo Clin.*, **22,** 597 (1947).
Ewetzky. *Beitr. Augenheilk.*, **3** (22), 101 (1898).
von Frisch. *Wien. med. Wschr.*, **32,** 969 (1882).
Gallenga. *Zbl. prakt. Augenheilk.*, **23,** 289 (1899).
Kerdel-Vegas *et al.* *Rhinoscleroma*, Springfield (1963).
Kline and Brody. *Arch. Derm. Syph.* (Chic.), **59,** 606 (1949).
Pardo-Castello and Martinez Dominguez. *Arch. Derm. Syph.* (Chic.), **5,** 478 (1922).
Pollitzer. *Laryngoscope*, **16,** 963 (1906).
Simpson and Ellis. *Arch. Derm. Syph.* (Chic)., **39,** 503 (1939).

GRANULOMA VENEREUM (INGUINALE)

GRANULOMA VENEREUM (GRANULOMA INGUINALE)—an entity established by Conyers and Daniels (1896) and quite separate from lymphogranuloma venereum which is caused by a virus—is an infection common among Negroes in the United States of America and the West Indies but a rare incidental invader of Great Britain and Western Europe. It is a venereal disease affecting both sexes, characterized initially by the appearance of a red papule on the skin which progresses to a chronic granulomatous ulceration without adenopathy and spreads slowly and destructively over the genital and perianal region. There is marked epidermal proliferation and eventually healing is by a thick contracted fibrous scar (Beerman and Sonck, 1952). Extragenital lesions affecting mainly the cheek, lips and mouth constitute about 6% of the case reports in the literature (Greenblatt *et al.*, 1938). Such a case has involved the upper lid of a Negro already venereally infected (Weiner *et al.*, 1943): initially resembling an acute hordeolum (as such it was incised), the lesion progressed as a destructive ulcer covered by a profuse dirty-grey discharge without pain or tenderness until a large portion of the tarsus sloughed away, whereafter it was arrested by treatment (Figs. 97–98).

The lesions are characterized by the presence of intracellular Donovan

FIGS. 97 and 98.—GRANULOMA VENEREUM.

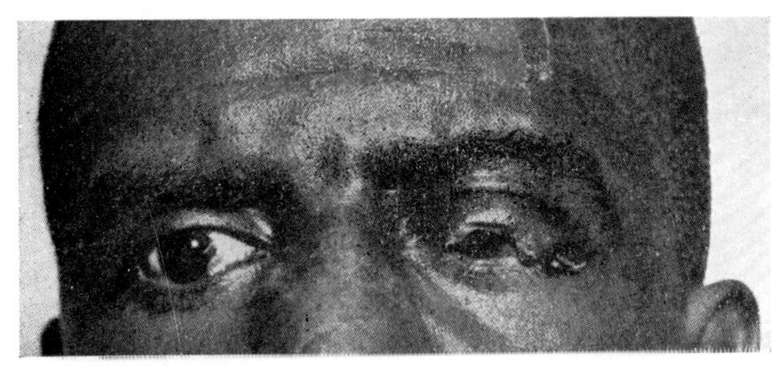

FIG. 97.

In a Negro. In the left upper lid a crescentic area including a large portion of the tarsus has sloughed away (A. L. Weiner, I. E. Gaynon and M. S. Osherwitz).

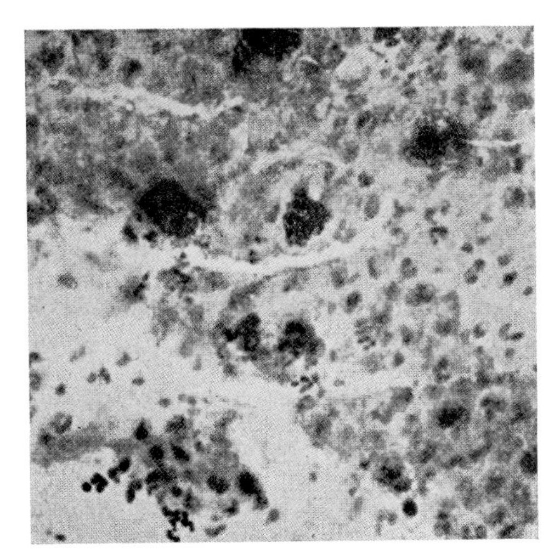

FIG. 98.

Donovan bodies in a histological section from the eyelid of the case shown in Fig. 97.

bodies (1905) which at one time were considered to be protozoa but were cultivated as encapsulated bacteria (*Donovania granulomatis*) by Anderson and her co-workers (1945) and Goldberg and his team (1953); with them successful human inoculations have been reported (Greenblatt *et al.*, 1947). The diagnosis is made by finding these bodies in scrapings of the ulcer. The classical and effective method of treatment has been by antimonial drugs, but these have now been superseded by the much more effectively acting streptomycin[1] or the tetracyclines or chloramphenicol (500 mg. 4 times daily for 14 days),[2] any of which heals the lesions with great rapidity. The surgical correction of cicatricial deformities may be necessary.

[1] Greenblatt *et al.* (1947), Hirsh and Taggart (1948), Kupperman *et al.* (1948), and others.

[2] Hill *et al.* (1949), Robinson *et al.* (1949), Wammock *et al.* (1950).

Anderson, de Monbreun and Goodpasture. *Amer. J. Syph.*, **29,** 165 (1945).
J. exp. Med., **81,** 25 (1945).
Beerman and Sonck. *Amer. J. Syph.*, **36,** 501 (1952).
Conyers and Daniels. *Brit. Guiana med. Ann.*, **8,** 13 (1896).
Donovan. *Indian med. Gaz.*, **40,** 414 (1905).
Goldberg, Weaver and Packer. *Amer. J. Syph.*, **37,** 60 (1953).
Greenblatt, Dienst, Kupperman and Reinstein. *J. ven. Dis. Inform.*, **28,** 183 (1947).
Greenblatt, Dienst, Pund and Torpin. *J. Amer. med. Ass.*, **113,** 1109 (1939).
Greenblatt, Torpin and Pund. *Arch. Derm. Syph.* (Chic.), **38,** 358 (1938).
Hill, Wright, Prigot and Logan. *J. Amer. med. Ass.*, **141,** 1047 (1949).
Hirsh and Taggart. *Amer. J. Syph.*, **32,** 159 (1948).
Kupperman, Greenblatt and Dienst. *J. Amer. med. Ass.*, **136,** 84 (1948).
Robinson, Elmendorf and Zheutlin. *Amer. J. Syph.*, **33,** 389 (1949).
Wammock, Greenblatt and Dienst. *J. invest. Derm.*, **14,** 427 (1950).
Weiner, Gaynon and Osherwitz. *Amer. J. Ophthal.*, **26,** 13 (1943).

BUBONIC PLAGUE. In PLAGUE the skin of the lids, the subcutaneous muscles and the tarsus may show an inflammatory infiltration; in infected animals Mizuo (1910) found capillary lymphatic thromboses rich in *Pasteurella pestis*.

Mizuo. *Arch. Augenheilk.*, **65,** 1 (1910).

ACTINOMYCOSIS

The *Actinomyces israeli* (*bovis*) (ακτῖνος, ray; μύκης, fungus), an organism previously referred to as the *ray fungus* but which should properly be classified with bacteria, has already been described.[1] It is met with naturally on straw, grain or hay, and infects animals in fodder or bedding, and human beings who chew straw or make contact with infected beasts. Infection is conveyed through an abrasion in the skin or mouth, the face and neck being the most common sites, and in the spread of the disease the eyelids may become involved (Partsch, 1893; Janiszewska-Geldner, 1956) (Fig. 99); only rarely are they affected primarily (Gifford, 1921). The lesion is nodular, at first subcutaneous and of wooden hardness, and then as it spreads to the surface it discharges freely through multiple draining sinuses emitting quantities of pus containing green or yellow pinhead granules composed of the organism, the so-called " sulphur granules ". The disease is indolent and intractable, and the lesions multiply, interconnecting with ramifying sinuses, so that the affected area becomes like a rabbit-warren. In addition to the lids the deeper tissues of the orbit may be affected (Wilson, 1937).[2] Castelain (1907) noted an infection of a meibomian gland which appeared as a chalazion. Histopathologically the indurated skin shows many large abscesses containing the yellow sulphur granules and large quantities of the organism (Figs. 100–1); in the later stages fibrosis is widespread. Diagnosis is by microscopic recognition of the organism; and treatment, which used to be very effective by large doses of potassium iodide, is even more so with sulphonamides (Dobson *et al.*, 1941; Dobson and Cutting, 1945), but the massive exhibition of penicillin over a considerable period is the treatment of choice[3]; tetracycline (McVay *et al.*, 1950; Lane *et al.*, 1953) or streptomycin (Torrens and

[1] Vol. VIII, p. 228. [2] p. 910.
[3] McCrea *et al.* (1945), Muskatblit (1947), Lamb *et al.* (1947), Harvey *et al.* (1957), Peabody and Seabury (1960).

FIGS. 99 to 101.—ACTINOMYCOSIS OF THE FACE AND LOWER LID.

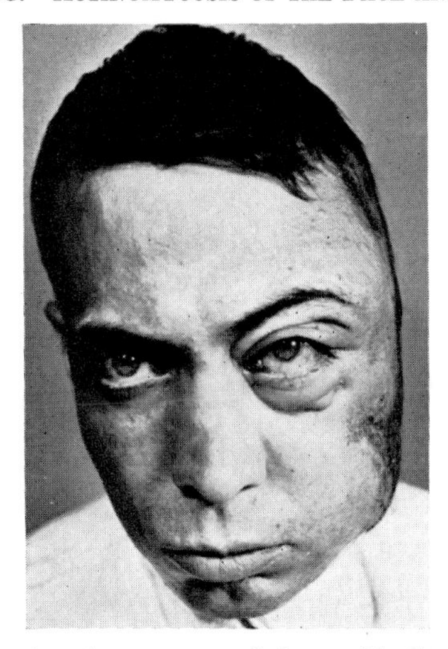

FIG. 99.—Showing the presence of sinuses (A. Gonzales Ochoa).

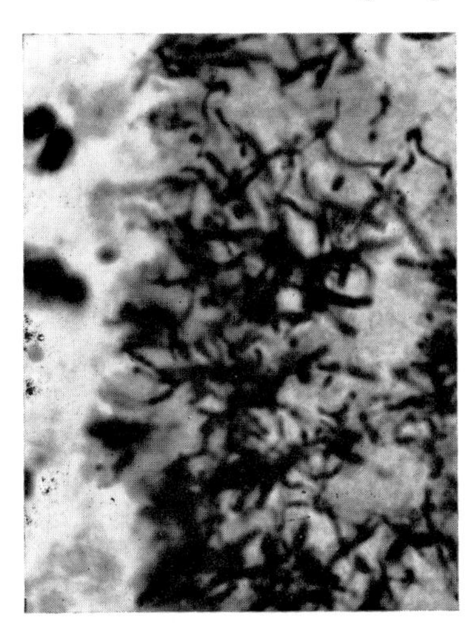

FIG. 100.—A granule of the fungus surrounded by inflammatory exudate.

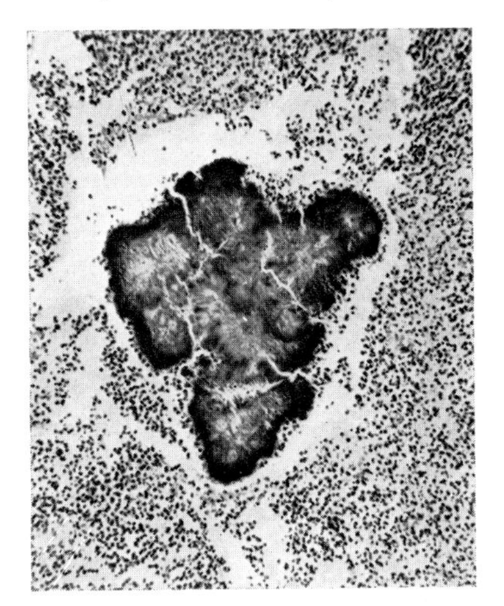

FIG. 101.—A colony stained by Gram's method to show the branching hyphæ (R. D. Baker).

Wood, 1949) is also effective but the administration of these drugs should be accompanied by surgical incision and drainage.

NOCARDIOSIS of the lid is even more rare. A case was reported by Woodruff (1925) which appeared as a nodule in the upper lid.

Castelain. *Ann. Oculist.* (Paris), **138,** 261 (1907).
Dobson and Cutting. *J. Amer. med. Ass.*, **128,** 856 (1945).
Dobson, Holman and Cutting. *J. Amer. med. Ass.*, **116,** 272 (1941).
Gifford. *Amer. J. Ophthal.*, **4,** 1 (1921).

Harvey, Cantrell and Fisher. *Ann. intern. Med.*, **46,** 868 (1957).
Janiszewska-Geldner. *Klin. oczna*, **26,** 221 (1956).
Lamb, Lain and Jones. *J. Amer. med. Ass.*, **134,** 351 (1947).
Lane, Kutscher and Chaves. *J. Amer. med. Ass.*, **151,** 986 (1953).
McCrea, Steven and Williams. *J. lab. clin. Med.*, **30,** 509 (1945).
McVay, Dunavant, Guthrie and Sprunt. *J. Amer. med. Ass.*, **143,** 1067 (1950).
Muskatblit. *Arch. Derm. Syph.* (Chic.), **56,** 706 (1947).
Partsch. *Zbl. prakt. Augenheilk.*, **17,** 161 (1893).
Peabody and Seabury. *Amer. J. Med.*, **28,** 99 (1960).
Torrens and Wood. *Lancet*, **1,** 1091 (1949).
Wilson. *Ann. Rep. ophthal. Lab. Giza*, **12,** 70 (1937).
Woodruff. *Arch. Ophthal.*, **54,** 11 (1925).

ERYSIPELOID

ERYSIPELOID (ERYTHEMA SERPENS) is an acute infection of the skin and deeper tissues caused by the *Erysipelothrix rhusiopathiæ*, an organism belonging to the somewhat loose family of the Actinomycetaceæ. It is a short, non-motile, non-sporing, Gram-negative bacillus widely found in the animal kingdom, living long in putrefying carcases and the cause of swine erysipelas. We have seen[1] that it may cause a severe conjunctivitis among those in close contact with animals, and in people who handle meat, poultry, rabbits or fish it may cause a dermal lesion through the infection of a cut or scratch.

After an incubation period of one to five days after infection the lesion appears as a dull red or purple raised patch, circumscribed but spreading slowly outwards from the scratch to reach a dimension of 2 or 3 inches (Plate II, Fig. 1); this burns or itches without constitutional symptoms. Infection to the face and the region of the eye is usually transferred from a lesion in the hand. The disease is self-limited and usually clears up after 3 or 4 weeks but treatment by penicillin results in its resolution in a few days.

Dermatitis due to Spirochætal Infections

SYPHILIS

The *Treponema pallidum* may affect the lids in any of the three stages of syphilis.

A PRIMARY SYPHILITIC CHANCRE is not a common occurrence on the lids: extragenital chancres are comparatively rare and, among these, lid chancres form a small but definite proportion, taking numerical precedence after the lips, breast, mouth and finger.

Bulkley (1894) analysing 9,058 cases of chancre reported in the literature, Münchheimer (1897) analysing 10,256, and Scheuer (1910) analysing 14,590, all found the lids and conjunctiva to be involved in about 5% of cases. Among these the conjunctiva is more frequently affected than the skin of the lids; thus Maxey (1918), in 89 cases of ocular chancre, found 49 conjunctival, 34 on the lids, usually the lower, 4 in a canthus and 2 of unrecorded location. Again, Zeissl (1877) in 40,000 syphilitics found 8 cases of chancre of the lids, Lesser (1887) out of 201 chancres found 18 extra-

[1] Vol. VIII, p. 235.

genital of which 1 affected an eyelid; Porey-Koschitz (1890), analysing 852 extragenital chancres, found 132 in the ocular region, and Alexander (1895) found 247 cases among 931 extragenital chancres. Pospelow (1889) noted 3 chancres of the lids in a Moscow factory, and Fortuniadès (1890) collected 118 cases. On the other hand, among 16,616 luetics showing 307 extragenital chancres, Wilbrand and Staelin (1897) found none on the lids, and Talbot (1894) out of 434 syphilitic eye affections found only 3 lid chancres. In recent years the lesion has become more rare; thus in Egypt, Sadek (1923) reported 48 cases, but among 7,854,201 new patients attending Egyptian hospitals during the ten years from 1928 to 1937, Ibrahim (1947) could trace records of only 69 chancres of the lid. Sporadic cases, however, are still occasionally reported (Kamel, 1945; Boulach, 1948; Vit, 1952; and others).

Infection may be contracted in a number of ways. Usually it is a contamination from the fingers either obtained venereally or by auto-inoculation from a genital chancre (Holth, 1895). A kiss from a person with an infected mouth, or the curiously unsavoury traditional habit, known even to Christ, of treating ocular disease by spitting on the eyes or removing foreign bodies by licking with the tongue, may infect the lids (Pospelow, 1889; Poljakow, 1893; Shaaban, 1939). Infection has been conveyed by a bite on the lids (Fox and Machlis, 1924), and doctors and nurses dealing with syphilitic patients have not been exempt (Desmarres, 1847; Alexander, 1895; Schreiber, 1924). Finally, an infant may be infected from its mother at or after birth (Little, 1901)—thus a chancre has more than once been recorded on the lid of a 2 months' infant whose mother showed a secondary cutaneous eruption after delivery (Silcock, 1902; Applebaum, 1937), and such cases in older children are on record (14 months, Puig Solanes, 1938; and others).

The infection usually occurs through an abrasion, but its entrance may on rare occasions be facilitated by maceration of the epidermal covering, as in impetigo (von Michel, 1901) or an inflamed meibomian gland (Helbron, 1898). The lesion is usually single and is most frequently situated on the lid-margin. As a rarity a double chancre may occur on the lids of the same eye (Pflüger, 1878; Morel-Lavallée, 1886; Gallemaerts, 1896; Seydel, 1898; Helbron, 1898); one may occur on each side (Igersheimer, 1918); or as a great rarity a triple chancre has been found (Kornacker, 1904). Sometimes the lesion is very extensive, affecting for example, the skin of the lids of both eyes, the bridge of the nose and the bulbar conjunctiva (Oast, 1933). Again, a chancre on the lid may appear simultaneously with one elsewhere (on the chin, Rollet and Genet, 1912).

Clinically the primary chancre may appear as a livid purple patch without ulceration, as an indurated subcutaneous tubercle or, most usually, as an ulcer with a strongly indurated, cartilage-like base (Fig. 102). In its usual marginal situation the commonest picture is that of a V-shaped ulcer extending from the ciliary region down the skin of the lid and encroaching on the tarsal conjunctiva, the whole being covered with a yellow, adherent false membrane. There is always an accompanying conjunctivitis, a considerable œdema of the lids, and the regional lymph nodes—pre-auricular, maxillary and eventually cervical—are invariably affected and considerably swollen. Pain is not a prominent symptom. If the condition is untreated the ulcer resolves in some 5 or 6 weeks, leaving remarkably little scarring or deformity; frequently the only remaining evidence of the chancre is the loss of lashes on the site of ulceration, but even these may grow again.

The diagnosis is suggested by the typical, relatively non-inflammatory, indolent, indurated character of the ulcer, the constant presence of regional adenopathy, the absence of pain, and the lack of response to local medicaments. In suspicious cases it is rendered certain by the demonstration of the treponema by dark-ground microscopical examination, but it must be remembered that positive results are not

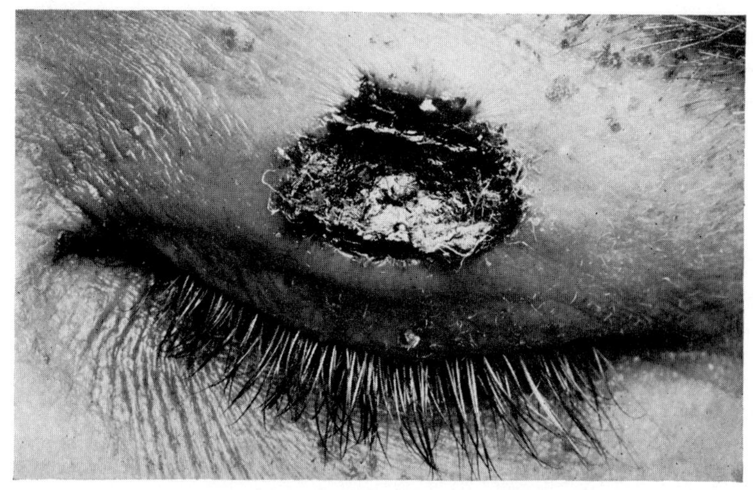

Fig. 102.—Primary Chancre of the Lid.
Occurring on the upper lid of a male aged 20 (Inst. Ophthal.).

always obtained in cases wherein clinical developments or the therapeutic response to anti-syphilitic treatment have established the nature of the case beyond doubt (Del Duca, 1930; Appelbaum, 1937).

Igersheimer (1918) brought forward evidence to suggest that in cases of acquired syphilis when a primary lid chancre has occurred, there is a tendency for the development of unilateral interstitial keratitis at a later date on the same side.

The secondary or early syphilides, appearing as exanthematous syphilitic eruptions, occur on the lids as elsewhere in the body. In general terms, as opposed to the tertiary or late syphilides, they are *superficial in the skin, rapid in development, seldom ulcerate and tend to subside spontaneously*; moreover, they are frequently associated with widely distributed lesions elsewhere and tend to be symmetrical. These manifestations of syphilis are in general either roseolar (erythematous or macular), papular, or mixed in character, but their main characteristics are their pleomorphism and multiformity.

The *macular* or *roseolar syphilide*, the earliest of the exanthematous manifestations, composed of pinkish irregular macules, is a general and symmetrical eruption appearing some 6 weeks after the primary lesion. It shows no peculiarities in the lids, is symptomless and disappears spontaneously in a few weeks with little or no scaling.

The *papular syphilide*, which forms acne-like indurations, is more typical on the lids, the papules varying from pin-head to bean-size or larger. In colour they are usually reddish-brown, with a smooth shiny surface, eventually becoming scaly and occasionally pustulating (Fig. 103). The face is a site of predilection, particularly the forehead, the buccal commissures and margins of the lids. Occasionally the squamous (*papulo-squamous syphilide*) or ulcerative (*papulo-pustular syphilide*) elements are very prominent, the latter occurring particularly at the lid-margin where *condylomata* may appear (Gresser, 1930; Malamud *et al.*, 1952).

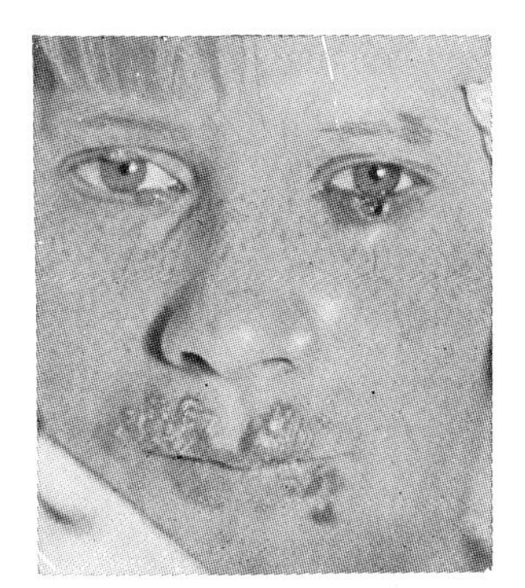

Fig. 103.—Secondary Syphilis.
Showing ulcerating papules of the lids and lips (E. W. Assinder).

Three further features of the exanthematous stage of syphilis should be noted: (1) the occurrence of an *ulcerative blepharitis*, sometimes associated with a cheilitis wherein the lids remain swollen and œdematous, a profuse yellow discharge is prominent and the ulceration if untreated may leave considerable scarring with permanent loss of the lashes; a similar ulceration may occur at a late stage in the disease[1]; (2) a *syphilitic œdema* of the lids; and (3) *syphilitic alopecia* of the brows and lashes which, in contradistinction to the permanent loss following ulcerative blepharitis, is nearly always followed by complete restoration.

None of these lesions, however, is common. Thus Wilbrand and Staelin (1897) found one case showing a squamous syphilide on the upper and lower lid and another with a similar lesion on the brow among 136 secondary syphilitics; while among 196 patients showing generalized specific alopecia they noted loss of the eyebrows in 12 cases, of the lashes in 7, and of both in 4 (*see* Gresser, 1930; Assinder, 1942; and others).

[1] Hirschler(1866), Hutchinson (1888), Fisher (1911), Renard and Halbron (1938) (Fig. 222).

TERTIARY OR LATE SYPHILIDES, appearing 2 to 20 years after the initial infection, are exceedingly rare on the lids (Heckel and Beinhauer, 1925): they occur in only some half per cent of all syphilitics (Renard and Halbron, 1938) or even less than this (0·01%, Narich, 1905). In Austria, Gruder (1898) recorded 2 cases in some 150,000 patients, and in Russia, Lipovskaya (1939) 8 in 219,139. In contradistinction to the secondary exanthematous manifestations of the disease, the late lesions are single or few in number, asymmetrical, attack the deep layers of the skin or the subcutaneous tissues, are sluggish in onset and chronic in course, and they habitually ulcerate. Their onset seems frequently determined by injury which appears to activate latent lues (Gozberk, 1937). In character they are GUMMATOUS[1]: the more superficial, situated in the skin itself, take the form of *nodular* or *tubercular syphilides*, and the deeper ones appear as typical *gummatous syphilides*, both types having a strong tendency to exuberant granulations and deep and destructive ulceration. A further manifestation of late syphilis in the lids is TARSITIS: this will be described subsequently.[2] A gumma on the lid may occur simultaneously with gummata elsewhere; bilateral palpebral gummata are a rarity (Gruder, 1898). Because of their pleomorphism these lesions may present considerable difficulties in clinical diagnosis (*see* Morano, 1947).

The *tubercular* (*nodular*) *syphilide*, which is histologically a gumma of the skin, usually appears in the third or fourth year after infection. The nodule is a smooth, round, reddish-brown elevation with a tendency to form in circinate or serpiginous groups spreading out on the periphery; these develop slowly, may persist for weeks unchanged, and eventually tend to ulcerate leaving atrophic deformed scars. The most frequent site is the lid-margin where the ulcerative destruction may involve both the skin and conjunctival surfaces in this region, eating into the tissues and leaving a serpiginous deformity which may involve the loss of one-half or two-thirds of the lid and large areas of the brow or cheek (Figs. 104–5).[3] At the commencement the lesion may resemble a stye, and the fully developed ulcer is differentiated from the primary lesion by the absence of adenopathy and of the organism on histological examination.

Subcutaneous gummata are also rare. The lesion appears as a rounded circumscribed tumour varying in size from a pea to a walnut, growing rapidly without symptoms. At this stage the gumma may resemble a chalazion, for which it has been mistaken (Hutchinson, 1888; Nazarov, 1939); occasionally two apposing gummata may appear on the upper and lower lids (Postić, 1940; Cassady, 1950) (Plate II, Fig. 2). As the overlying skin is implicated secondarily, it becomes bluish or purple and soon breaks down to form a deep, soft, punched-out ulcer with necrotic edges and a red or greenish granular floor bathed with pus (Ryss-Zalkind, 1933; Lipovskaya, 1939; and others).

[1] Vol. VIII, p. 241.

[2] p. 237.

[3] MacLehose (1902), Aschheim (1903), Rille (1906), McKee (1936), Bushmich (1936), Matteucci (1945), Ellison (1948), and others.

PLATE II

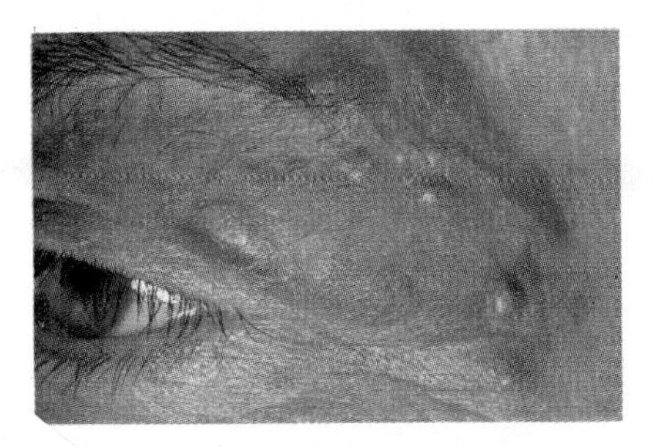

FIG. 1.—Erysipeloid (B. Jay).

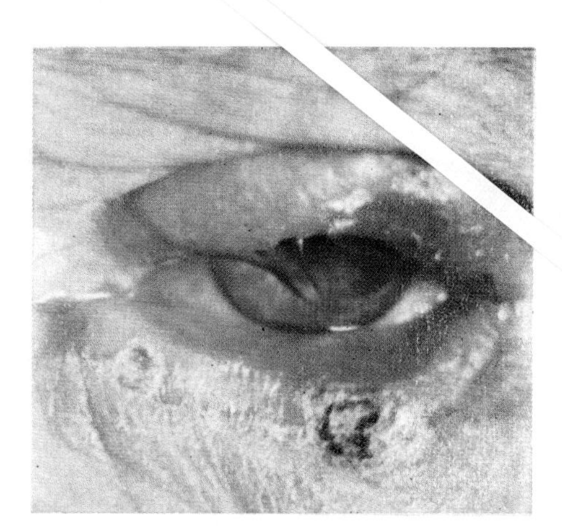

FIG. 2.—Gummatous infiltration of the eyelids with ectropion and ulceration of the lower lid, blepharitis and loss of the lashes (J. V. Cassady).

[*To face p.* 124.

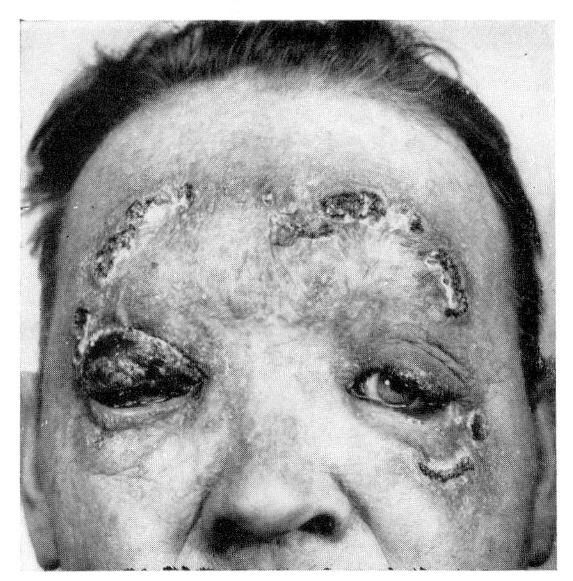

Fig. 104.

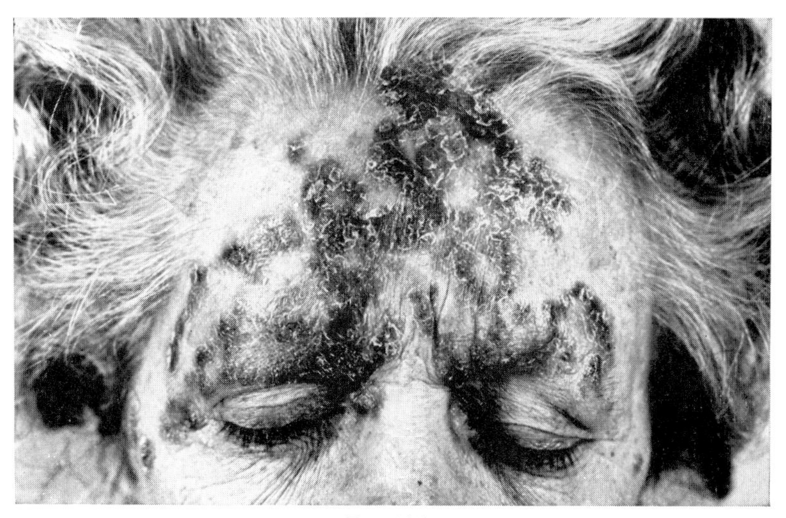

Fig. 105.

Figs. 104 and 105.—Serpiginous Nodulo-squamous Syphilide in Tertiary Syphilis.

(Finsen Inst., Copenhagen, from A. Rook's *Textbook of Dermatology*, Blackwell, Oxford).

As a rule the ulcer is localized but it may extend deeply involving the fibres of the orbicularis palpebrarum muscle (Venco, 1935), or eat through the substance of the lid leaving a gap (Hartridge, 1898), or spread to include the entire lid from the margin to the eyebrow (Wilson, 1930; Girgis, 1939) and

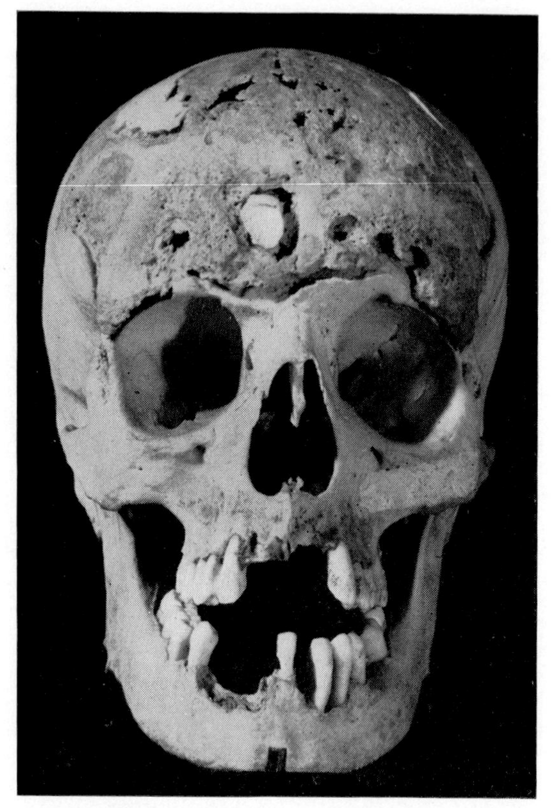

Fig. 106.—Syphilitic Caries.

An advanced case affecting particularly the frontal bone. The sequestra were exposed in ulcers of the scalp and forehead during life and involve the whole thickness of the bone which is perforated in places by caries. There is also extensive necrosis of the bones of the nose and palate (Museum, Royal College of Surgeons).

eventually involve the bone in widespread caries (Fig. 106). Occasionally if untreated, the lesion may become very extensive, resulting in enormous and hideous disfigurement of the entire face.

CONGENITAL SYPHILIS

Congenital syphilis rarely affects the lids, but may do so exceptionally at almost any stage of the disease in infancy, adolescence or even in adult life. As a rule luetic manifestations occur spontaneously, but in a considerable proportion of cases they are preceded by an acute infective disease or traumatism (Gozberk, 1937). Most of the lesions described correspond to those characteristic of the acquired disease: it has, indeed, been well said that pre-natal syphilis can ape any of the characteristics of acquired syphilis and then a great deal more.

Rollin (1935) considered that in congenital syphilitics the palpebral aperture tends to slant downwards and outwards; but this cannot be considered a characteristic. *Rhagades* resembling those found at the angles of the mouth may occur, particularly at the outer commissure. *Papular syphilides* may also appear at the lid-margins and may ulcerate; they may leave thickening, deformity and permanent loss of the cilia (Schreiber, 1924) (Fig. 107). *Gummatous lesions* occur in the later stages, either superficially when large areas at the margin may be destroyed by ulceration, or in the substance of the lids when deforming ulceration may extend over wide areas of the forehead, cheek, nose and temple (Gozberk, 1937; Genet, 1939; and others). In the most severe cases the whole of the lids may be destroyed.

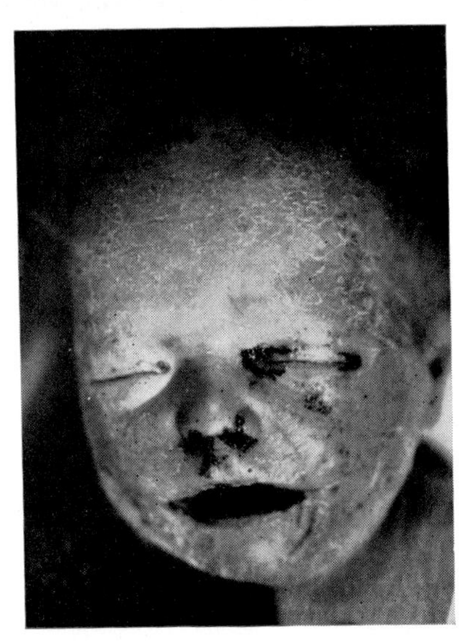

Fig. 107.—Congenital Syphilis.

Showing fissures of the ocular and labial commissures and snuffles (R. L. and R. L. Sutton, *Hb. of Diseases of the Skin*, Mosby).

Colucci (1904) reported an extensive late thickening of the lids due to amyloid degeneration. The occurrence of a gumma on the lid has been noted to initiate an attack of typical syphilitic interstitial keratitis (Lipovskaya, 1939).

The pathology of syphilis has already been discussed.[1] The essential changes are a dense perivascular infiltrate of lymphocytes and plasma cells with marked endarteritis and endophlebitis. In tertiary lesions there is also a tuberculoid infiltrate with epithelioid and giant cells and caseation necrosis (Fig. 108). Treatment conforms with the general systemic treatment of the disease with penicillin as injections of penicillin G or orally

[1] Vol. VIII, p. 241.

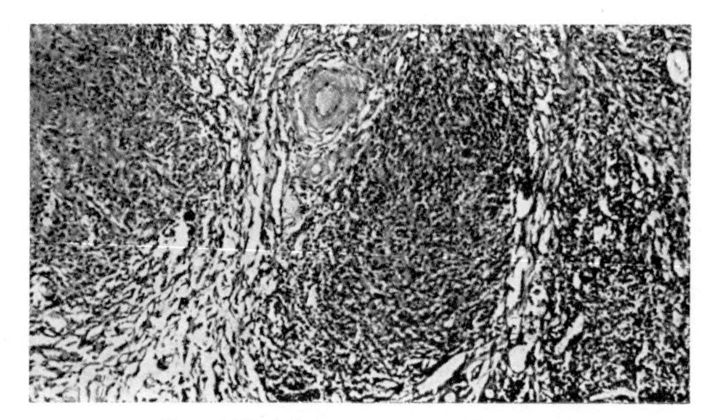

FIG. 108.—GUMMA OF THE EYELID.

In an Egyptian male, aged 30, with an indurated ulcer involving the whole of the central part of the left upper eyelid and eyebrow. The section shows nodules of giant cells and epithelioid cells in the centre and lymphocytes at the periphery, surrounded by fibrous tissue containing vessels with thickened walls and much perivascular infiltration (R. P. Wilson).

as phenoxymethyl penicillin (penicillin V) in large doses. For patients who are allergic to this drug the best alternative is tetracycline or erythromycin. Few diseases, particularly regarding manifestations in the lids, react more readily to treatment.

Alexander. *Neue Erfahrungen ü. luetische Augenerkrankungen*, Wiesbaden (1895).
Appelbaum. *Arch. Ophthal.*, **18**, 920 (1937).
Aschheim. Vossius's *Samml. zwangl. Abhandl. Gebiete Augenheilk.*, **5** (2) (1903).
Assinder. *Brit. J. Ophthal.*, **26**, 1 (1942).
Boulach. *Vestn. Oftal.*, **27** (1), 44 (1948).
Bulkley. *Syphilis in the Innocent*, N.Y. (1894).
Bushmich. *Vestn. Oftal.*, **9**, 65 (1936).
Cassady. *Amer. J. Ophthal.*, **33**, 18 (1950).
Colucci. *Ann. Ottal.*, **33**, 925 (1904).
Del Duca. *Saggi Oftal.*, **5**, 212 (1930).
Desmarres. *Traité théorique et pratique des maladies des yeux*, Paris, 156 (1847).
Ellison. *Trans. ophthal. Soc. U.K.*, **68**, 273 (1948).
Fisher. *Trans. ophthal. Soc. U.K.*, **31**, 268 (1911).
Fortuniadés. *Étude sur le chancre syphilitique des paupières* (Thèse), Paris (1890).
Fox and Machlis. *Amer. J. Ophthal.*, **7**, 701 (1924).
Gallemaerts. *Rev. gén. Ophtal.*, **15**, 518 (1896).
Genet. *Bull. Soc. Ophtal. Paris*, 446 (1939).
Girgis. *Bull. ophthal. Soc. Egypt*, **32**, 50 (1939).
Gozberk. *Ann. Oculist.* (Paris), **174**, 837 (1937).
Gresser. *Amer. J. Ophthal.*, **13**, 886 (1930).
Gruder. *Wien. klin. Wschr.*, **11**, 830 (1898).
Hartridge. *Trans. ophthal. Soc. U.K.*, **18**, 50 (1898).
Heckel and Beinhauer. *Arch. Ophthal.*, **54**, 352 (1925).
Helbron. *Münch. med. Wschr.*, **45**, 663 (1898).
Hirschler. *Wien. med. Wschr.*, **16**, 1145, 1161, 1177 (1866).
Holth. *Arch. Augenheilk.*, **30**, 214 (1895).
Hutchinson. *Roy. Lond. ophthal. Hosp. Rep.*, **12**, 156 (1888).
Ibrahim. *Bull. ophthal. Soc. Egypt* (1947), **40**, 74 (1949).
Igersheimer. *Syphilis u. Auge*, Berlin, 150 (1918).
Kamel. *Bull. ophthal. Soc. Egypt*, **38**, 50 (1945).
Kornacker. *Ueber Initialsklerose der Augenlider* (Diss.), Berlin (1904).
Lesser. *Münch. med. Wschr.*, **34**, 577 (1887).
Lipovskaya. *Vestn. Oftal.*, **15** (2), 88 (1939).
Little. *Trans. ophthal. Soc. U.K.*, **21**, 3 (1901).
MacLehose. *Trans. ophthal. Soc. U.K.*, **22**, 35 (1902).
Malamud, Coll and Foix. *Arch. Oftal. B. Aires*, **27**, 292 (1952).
Matteucci. *Boll. Oculist.*, **24**, 223 (1945).
Maxey. *Amer. J. Ophthal.*, **1**, 13 (1918).
McKee. *Canad. med. Ass. J.*, **35**, 307 (1936).
von Michel. *Arch. Augenheilk.*, **42**, 1, 8 (1901).
Morano. *Boll. Oculist.*, **26**, 449 (1947).
Morel-Lavallée. *Ann. Derm. Syph.* (Paris), **7**, 85 (1886).
Münchheimer. *Arch. Derm. Syph.* (Wien), **40**, 191 (1897).
Narich. *Rev. méd. Suisse rom.*, **25**, 761 (1905).
Nazarov. *Vestn. Oftal.*, **15** (2), 86 (1939).

Oast. *Arch. Ophthal.*, **10,** 704 (1933).
Pflüger. *Ber. Augenklin in Bern für d. Jahr 1877*, 57 (1878).
Poljakow. *Vestn. Oftal.*, **10,** 507 (1893).
Porey-Koschitz. *Die Topographie d. syphilitischen Schankers*, Karkov (1890).
Pospelow. *Arch. Derm. Syph.* (Wien), **21,** 59, 217 (1889).
Postić. *Klin. Mbl. Augenheilk.*, **104,** 319, 433 (1940).
Puig Solanes. *An. Soc. mex. Oftal.*, **12,** 97 (1938).
Renard and Halbron. *Arch. Ophtal.*, **2,** 599 (1938).
Rille. *Münch med. Wschr.*, **53,** 2274 (1906).
Rollot and Genet. *Rev. gén. Ophtal.*, **31,** 145 (1912).
Rollin. *Z. Augenheilk.*, **87,** 104 (1935).
Ryss-Zalkind. *Vestn. Oftal.*, **3,** 55 (1933).
Sadek. *Bull. ophthal. Soc. Egypt*, **16,** 82 (1923).
Scheuer. *Die Syphilis d. Unschuldigen*, Berlin (1910).
Schreiber. *Graefe-Saemisch Hb. d. ges. Augenheilk.*, 3rd ed., *Die Krankheiten d. Augenlider*, Berlin, 123, 143 (1924).
Seydel. *Klin. Mbl. Augenheilk.*, **36,** 117 (1898).
Shaaban. *Bull. ophthal. Soc. Egypt*, **32,** 70 (1939).
Silcock. *Trans. ophthal. Soc. U.K.*, **22,** 35 (1902).
Talbot. *Recherches statistiques sur la syphilis de l'oeil* (Thèse), Paris (1894).
Venco. *Rass. ital. Ottal.*, **4,** 149 (1935).
Vit. *Wien. klin. Wschr.*, **64,** 183 (1952).
Wilbrand and Staelin. *Ueber d Augenerkrankungen in d. Frühperiode d. Syphilis*, Hamburg (1897).
Wilson. *Ann. Rep. ophthal. Lab. Giza*, **5,** 33 (1930).
Zeissl. *Allg. Wien. med. Ztg.*, **22,** 311, 319, 329, 337 (1877).

YAWS (FRAMBŒSIA)

YAWS or FRAMBŒSIA (a raspberry) (known as PARANGI in Sri Lanka, BUBA in the Pacific Islands, COKO in Fiji, PURU in Malaysia) is an endemic disease occurring in North and Central Africa, Madagascar, Asia, the Pacific, Australia and the West Indies, caused by the *Treponema pertenue*—an organism closely related to the *T. pallidum* (Castellani, 1905). The two organisms are morphologically indistinguishable. The *T. pertenue* can be demonstrated on silver impregnation in the first two stages of the disease; it is almost entirely epidermotropic whereas the *T. pallidum* is largely mesodermotropic. The clinical differences between the two diseases, such as the extragenital non-venereal mode of infection in yaws, the absence of hereditary and congenital manifestations, the rarity of mucosal, visceral or nervous lesions, as well as the impossibility of inducing cross immunity in animals between it and syphilis, suggest that they form related but separate entities (Manson-Bahr, 1929; Fox, 1929–43; and others). The Wassermann reaction is positive in both.

The *primary lesion*, communicated by contact, is usually extragenital: it is a moist papule without induration which becomes surrounded by a collection of similar papules which coalesce and ulcerate. The face is a favourite, the lids an occasonal site. In 10 to 12 weeks the *secondary lesions* appear—crops of papules which tend to develop into fungating, granulomatous, raspberry- or cauliflower-like papillomatous masses covered by yellow crusts; again the face is a favourite site and the brows and lids are frequently affected (Fig. 109). Sometimes granulomata appear on the palpebral border of the lids, on occasion covering it with crusts and spreading over to the palpebral conjunctiva (Breda, 1895–1909; Elliot, 1920). As a rule these lesions disappear slowly in about 12 months leaving pigmented areas.

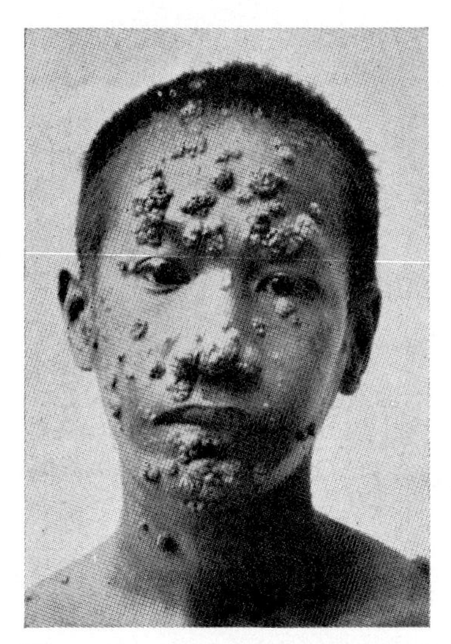

FIG. 109.—YAWS (Armed Forces Institute of Pathology).

FIGS. 110 and 111.—YAWS: GANGOSA.
As seen in natives of the Belgian Congo (J.-M. Habig).

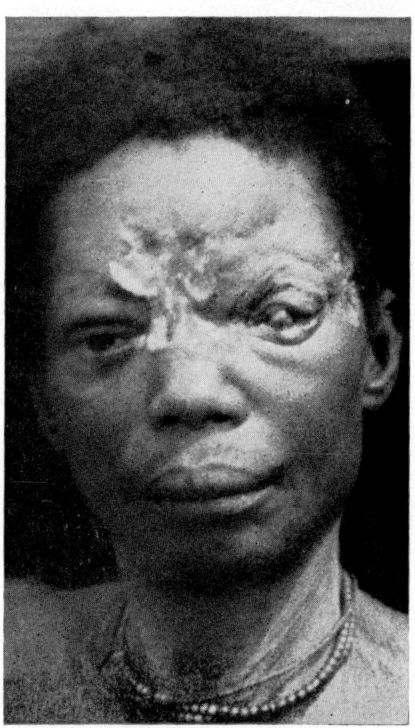

FIG. 110.—A relatively early case showing disappearance of the lids and destruction of the eye.

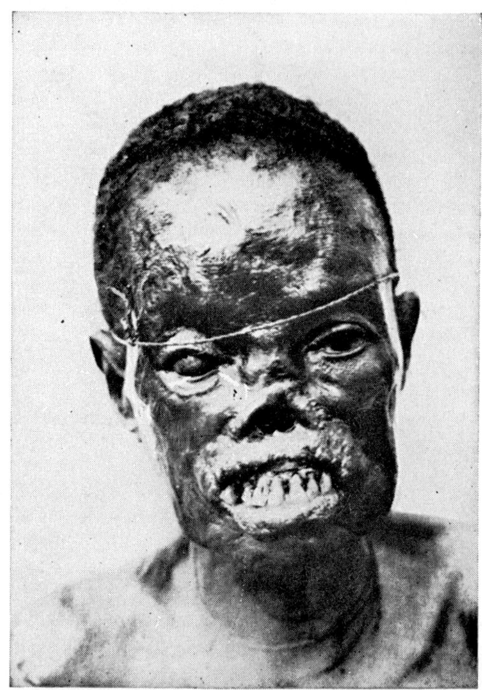

FIG. 111.—An advanced case showing destruction of the lids, particularly of the right eye, the nose and the lips.

It is to be noted that although the papillomatous growths do not overstep the lid-margin, the secondary stage of the disease is frequently associated with a catarrhal conjunctivitis in the secretion from which the treponema has been found (Soeroto, 1927; Hermans, 1928). As a rarity, an interstitial keratitis and a mild iridocyclitis have been noted (Castellani and Chalmers, 1919; Hermans, 1928; Schneider, 1946).

The *tertiary stage*, which is absent in a large percentage of cases, is characterized by the development of gummatous lesions which tend to

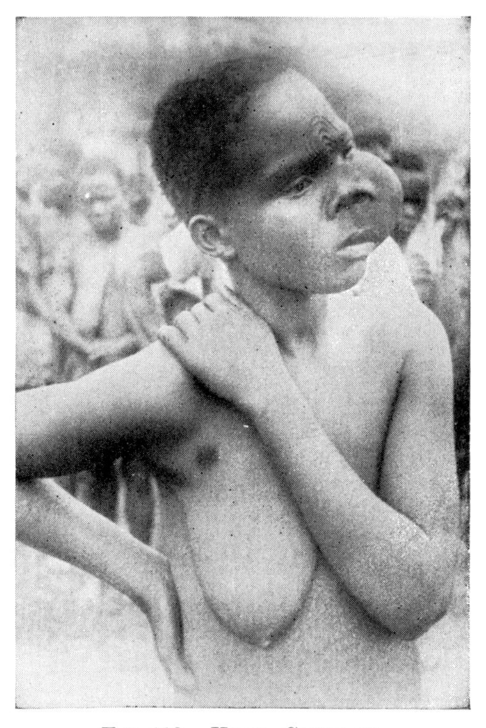

FIG. 112.—YAWS: GOUNDOU.

Showing the large mass growing on the side of the nose and invading the orbit, in a native of the Belgian Congo (J.-M. Habig).

proliferate and ulcerate; they occur in the skin and deeper tissues and bones. Histopathologically in the primary and secondary stages there is acanthosis, papillomatosis and œdema of the epidermis with intra-epidermal micro-abscesses largely composed of neutrophils and a dense infiltrate in the dermis in which plasma cells predominate; unlike syphilis, the vessels show little change (Williams, 1935; Hasselmann, 1952). In the tertiary stage the lesions resemble those of syphilis without the prominent vascular changes (Pardo-Castello, 1939).

Clinically two extreme forms are recognized—gangosa and goundou.

GANGOSA (RHINOPHARYNGITIS MUTILANS) is a virulent form of tertiary yaws which takes the form of an ulcerative rhinopharyngitis. Commencing in the pharynx and soft palate, a progressive ulcerative destruction spreads to the larynx, hard palate, the nose, face and lids, sometimes resulting in very considerable deformity including cicatricial ectropion, lagophthalmos and usually gross corneal damage (Habig, 1949) (Figs. 110–1).

GOUNDOU is a rarer manifestation occurring in Central and South America more often than Africa (Stavaux, 1945; Habig, 1949). Its most characteristic feature is the development of slowly growing bony tumours on the upper portion of the lateral side of the nose, which encroach on the lids and orbit and displace the globe (Fig. 112). This enlargement of the maxillary bone was first described under the title of "the Horned Men of Africa" (MacAlister, 1882), and at one time was considered specific to yaws. This, however, has been questioned and the same clinical picture has been seen in patients in whom there has been no question of this disease being present (Stannus, 1946; Hackett, 1969); the bony growths may well be examples of leontiasis ossea (Byers and Jones, 1969).

The prognosis of yaws is good; and, as in syphilis, treatment is effective by the administration of penicillin (Whitehill and Austrian, 1945; Hill *et al.*, 1946; Dwindelle *et al.*, 1946–47) or the tetracyclines.

Breda. *G. ital. Mal. Vener.*, **30,** 531 (1895).
Mense's *Treatise on Diseases of Tropical Countries*, Turin, **4,** 280 (1909).
Byers and Jones. *Brit. J. Surg.*, **56,** 262 (1969).
Castellani. *Brit. med. J.*, **2,** 1280, 1330, 1438 (1905).
Castellani and Chalmers. *Manual of Tropical Medicine*, 3rd ed., London (1919).
Dwindelle, Rein, Sternberg and Sheldon. *Amer. J. trop. Med.*, **26,** 311 (1946); **27,** 633 (1947).
Elliot. *Tropical Ophthalmology*, London, 497 (1920).
Fox. *Arch. Derm. Syph.* (Chic.), **20,** 820 (1929).
J. Amer. med. Ass., **123,** 459 (1943).
Habig. *Bull. Soc. belge Ophtal.*, No. 92, 284 (1949).
Hackett. *Textbook of Tropical Medicine* (ed. Woodruff), London (1969).
Hasselmann. *Arch. Derm. Syph.* (Chic.), **66,** 107 (1952).
Hermans. *Framboesia tropica* (1928).
Hill, Findlay and MacPherson. *Lancet*, **2,** 522 (1946).
MacAlister. *Trans. roy. Irish Acad.* (1882): *see* Lamprey. *Brit. med. J.*, **2,** 1273 (1887).
Manson-Bahr. *Tropical Medicine*, London (1929).
Pardo-Castello. *Arch. Derm. Syph.* (Chic.), **40,** 762 (1939).
Schneider. *Med. J. Aust.*, **1,** 99 (1946).
Soeroto. *Ned. T. Geneesk.*, **71** (1), 285 (1927).
Stannus. *Trans. roy. Soc. trop. med. Hyg.*, **40,** 219 (1946).
Stavaux. *Rec. Trav. Sci. méd. Congo belge*, No. 3, 177 (1945).
Whitehill and Austrian. *Bull. U.S. Army med. Dept.*, **84** (1945).
Williams. *Arch. Path.*, **20,** 596 (1935).

PINTA

PINTA is a disease common in Central America and Mexico and met with in the tropical areas of South America, the Pacific Islands, the West Indies and tropical Africa, caused by the *Treponema carateum* (*herrejoni*) (Herrejón and Pallares, 1927; Saenz *et al.*, 1940). The primary lesion is a chancre which occurs rarely on the face and is followed in some two months by disseminated plaques (*pintids*) distributed over the skin which eventually result in permanent depigmentation and curiously obvious dyschromic patches in which the lids share (Fig. 113). Active lesions are confined to the skin and lymph nodes. Histopathologically the skin shows a thickening of the epidermis with an intense infiltration of the dermis with lymphocytes and plasma cells; there is usually some perivascular infiltration. The chromatophores initially migrate into the superficial layers of the skin but in the later stages they progressively decrease. The serological tests for

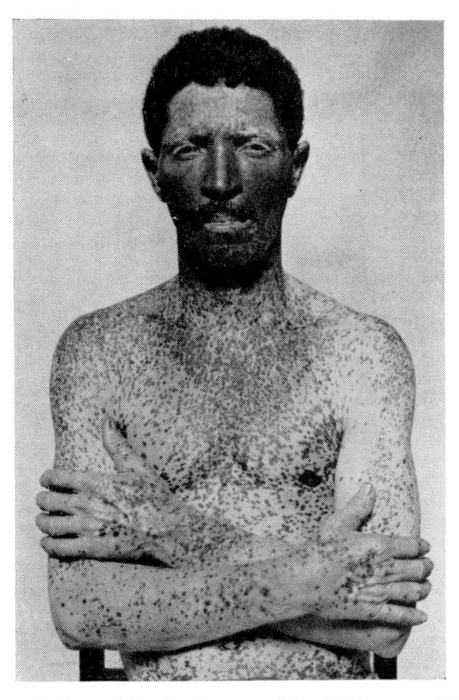

FIG. 113.—PINTA (R. L. and R. L. Sutton, *Hb. of Diseases of the Skin*, Mosby).

syphilis are positive after the primary stage has passed but cardio-vascular and central nervous complications are absent. The disease is not disabling and its worst aspect is the disfigurement.[1]

Treatment, as for syphilis, is by penicillin, which is specific.

Fox. *Arch. Derm. Syph.* (Chic.), **59,** 127 (1949).
Hasselmann. *Arch. klin. exp. Derm.*, **201,** 1 (1955).
Brit. J. vener. Dis., **33,** 5 (1957).
Herrejón and Pallares. *Hosp. Gen.* (Mex.), **2,** 109 (1927).
Latapi and Leon Blanco. *Medicina* (Mex.), **20,** 315 (1940).
Leon Blanco and de Laosa. *Amer. J. Syph.*, **31,** 600 (1947).
Pardo-Castello and Ferrer. *Arch. Derm. Syph.* (Chic.), **45,** 843 (1942).
Saenz, Triana and Armenteros. *Arch. Derm. Syph.* (Chic.), **41,** 463 (1940).

[1] Latapi and Leon Blanco (1940), Pardo-Castello and Ferrer (1942), Leon Blanco and de Laosa (1947), Fox (1949), Hasselmann (1955–57).

RAT-BITE DERMATITIS

The bites of rats have long been known to produce serious systemic effects (Wilcox, 1840), but the establishment of a separate disease-entity is primarily due to the extensive observations of Miyake (1900). RAT-BITE FEVER (SODOKU) is a disease at first thought to be solely due to infection by the *Spirillum minus* (*Spirochæta morsus muris*), a natural parasite in the mouth of the healthy rat (Schottmüller, 1914; Futaki *et al.*, 1916–17; Robertson, 1924); 25% of the rats in London harbour the organism, but the disease is only sporadic in England and is rare except in the Orient and the Pacific coast of America (Mooser, 1924–5; Theiler, 1926). Other animals such as cats or dogs may occasionally serve as vectors (Swyer, 1945). Subsequently many spirillum-free cases, particularly in the United States and Japan, were found to be infected by the *Actinobacillus muris* (*Streptobacillus moniliformis*) (Larson, 1941; Brown and Nunemaker, 1942), an organism extremely common in the nasopharynx of rats (Strangeways, 1933).

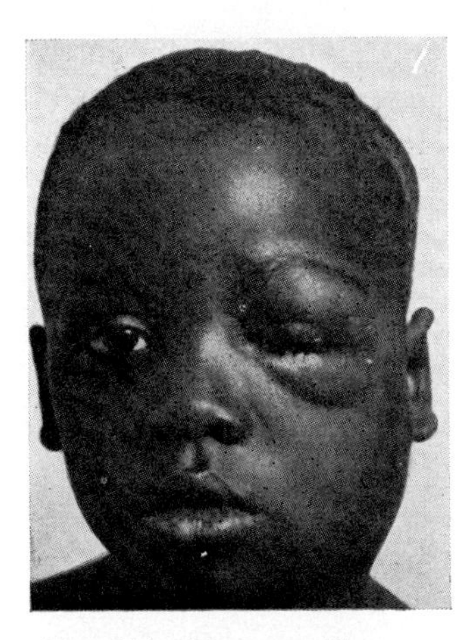

FIG. 114.—RAT-BITE FEVER.

Appearance of the patient 24 hours after admission to hospital. Note the swelling of the anterior cervical lymph nodes (C. Swab).

In both cases the disease is conveyed by a bite, and since this occurs relatively frequently when the victim is asleep and the face is exposed, the region of the lids may be affected (Kusama and others, 1919; Swab, 1930; Monthus *et al.*, 1931; and others). After an incubation period of 10–30 days when the wound of the bite has apparently healed normally, an erythematous, indurated, sclerotic plaque appears without suppuration at the site of inoculation associated with much œdema and a satellite bubo (Fig. 114); this is followed by the appearance of a roseolar eruption accompanied by an intermittent fever of 2- to 7-day periodicity. In the cases due to the actinobacillus the rash is marked and joint involvement is common. Occasionally, prostration, collapse and death may occur. The Wassermann reaction may become positive.

Treatment in spirillar cases may be carried out by the systemic administration of arsenic, the action of which may be dramatic (Hata, 1912;

O'Leary, 1924; Adams, 1925; Leadingham, 1938; and others). The response in actinobacillary cases is less good, but in both types penicillin, streptomycin or the tetracyclines give very effective results (Wheeler, 1945; Lominski *et al.*, 1948); with this treatment the mortality, which used to be considerable (6 to 12%), has become negligible.

Adams. *Arch. Derm. Syph.* (Chic.), **11**, 654 (1925).
Brown and Nunemaker. *Bull. Johns Hopk. Hosp.*, **70**, 201 (1942).
Futaki, Takaki, Taniguchi and Osumi. *J. exp. Med.*, **23**, 249 (1916); **25**, 33 (1917).
Hata. *Münch. med. Wschr.*, **59**, 854 (1912).
Kusama, Kobayashi and Kasai. *J. infect. Dis.*, **24**, 366 (1919).
Larson. *Publ. Hlth. Rep.* (Wash.), **56**, 1961 (1941).
Leadingham. *Amer. J. clin. Path.*, **8**, 333 (1938).
Lominski, Henderson and McNee. *Brit. med. J.*, **2**, 510 (1948).
Miyake. *Mitt. Grenzgeb. Med. Chir.*, **5**, 231 (1900).
Monthus, Favory and Levaditi. *Arch. Ophtal.*, **48**, 493 (1931).
Mooser. *J. exp. Med.*, **39**, 589 (1924); **42**, 539 (1925).
O'Leary. *Arch. Derm. Syph.* (Chic.), **9**, 293 (1924).
Robertson. *Ann. trop. Med. Parasit.*, **18**, 157 (1924).
Schottmüller. *Derm. Wschr.*, **58**, Suppl., 77 (1914).
Strangeways. *J. Path. Bact.*, **37**, 45 (1933).
Swab. *Amer. J. Ophthal.*, **13**, 884 (1930).
Swyer. *Brit. med. J.*, **2**, 386 (1945).
Theiler. *Amer. J. trop. Med.*, **6**, 131 (1926).
Wheeler. *Amer. J. Dis. Child.*, **69**, 215 (1945).
Wilcox. *Amer. J. med. Sci.*, **26**, 245 (1840).

Dermatitis due to Viral Infections

No more suitable photograph could introduce this discussion of viral diseases than that of BARRIE RUSSELL JONES [1921——] (Fig. 115). A New Zealander, he was born at Silverstream and was educated first at Victoria University where he studied chemistry and physics and then at Otago University where he pursued his medical course. After working at the Dunedin Hospital he came to London in 1951 for postgraduate studies, entering Moorfields Eye Hospital in 1952 to become, in 1963, the Professor of Clinical Ophthalmology at this hospital and the Institute of Ophthalmology, the first professorship in this subject to be established in the University of London. An unusually able clinician and a dexterous surgeon, making advances particularly in keratoplasty and the surgery of the lacrimal passages and the orbit, he has rightly earned a world reputation for his sustained and brilliant researches. Among many other subjects these have been concerned mainly with the viruses (considered in the widest sense) and the fungi which affect the eye. His work on the herpes virus, the adenoviruses and vaccinia has considerably advanced our knowledge, but the most fundamental research has been on the *Chlamydia*; he was the first to prove the organismal cause of trachoma and the first to demonstrate its urogenital incidence and the importance of the frequent venereal mode of infection of this disease. This work has led him far afield and his studies, which have taken him and his team on many occasions to countries where trachoma is rife, particularly Iran, have presented much new and valuable information on the clinical aspects, the epidemiology, the natural history and treatment of this most blinding of all diseases.

Certain viral infections which assume an acute exanthematous form involve the eyelids in the same way as the rest of the skin. Although the palpebral lesions usually show no very important features, occasionally events of ophthalmological interest occur.

In MEASLES (MORBILLI), especially when the associated conjunctivitis is acute, a considerable degree of œdema of the lids is not rare, while small tarsal nodules may

FIG. 115.—BARRIE RUSSELL JONES
[1921———].

occur (Morales *et al.*, 1965); exceptionally gangrene of one or more lids has been reported (Saint-Martin, 1884; Fieuzal, 1887; Axenfeld, 1904).

In CHICKENPOX (VARICELLA) the rash may affect the lids severely. This infection will be discussed in association with zoster.[1]

In SMALLPOX (VARIOLA) the rash appears on the lids as elsewhere and runs the usual macular, vesicular and pustular course. Œdema of the lids may be very marked, " blinding " the patient by their swelling. In the hæmorrhagic types the purpuric tendency may show itself first in this region (Gorkom, 1899). Secondary infection has led to abscesses or gangrene (Landesberg, 1874), while pustules on the lid-margins may result in ectropion, trichiasis and madarosis, and even ankyloblepharon (Saxena *et al.*, 1966).

Axenfeld. *Münch. med. Wschr.*, **51,** 779 (1904).
Fieuzal. *Bull. Clin. nat. ophtal. Hosp. Quinze-vingts*, **5,** 198 (1887).
Gorkom. *T. Ned. Indie.*, **39,** 458 (1899). *See Jber. Ophthal.*, **30,** 425 (1899).
Landesberg. *Beitr. z. variolösen Ophthalmie*, Elberfield, 20 (1874).
Morales, Taricco and Phillipi. *Rev. chil. Pediat.*, **36,** 717 (1965).
Saint-Martin. *Bull. Clin. nat. ophtal. Hosp. Quinze-vingts*, **2,** 145 (1884).
Saxena, Garg and Ramchand. *Amer. J. Ophthal.*, **61,** 169 (1966).

Other viral infections may be preferentially localized in the lids: the most common of these are vaccinia, herpes, varicella and zoster, and the viruses causing warts and molluscum contagiosum. Palpebral manifestations of lymphogranuloma venereum, ornithosis and orf may also occur as rarities.

VACCINIA

An accidental infection of the lids with the vaccinia virus, as was first pointed out by Hirschberg (1879), is not very rare: Bedell (1919–20) collected 93 cases from the literature affecting the lids and conjunctiva since Jenner introduced vaccination in 1796, Meder (1919) added 27 additional references, while Atkinson and Scullard (1940) collected a further 91 cases; since then many cases have been reported. Of these, approximately one-third had some ocular involvement. Cases of auto-inoculation have occurred (King and Robie, 1951; Kline, 1951; King and Forrest, 1953; Jones and Al-Hussaini, 1963), but contact contaminations from others is much more common, such as mothers or nurses who are tending a vaccinated child or one child sleeping with another (Brav, 1945; Monod, 1958). Thus of 50 cases of lid vaccinia collected by Pihl (1900) only 8 were examples of self-infection. Occasionally more than one lesion may appear simultaneously (one on the lid, another on the cheek, Frampton and Smith, 1952) (Fig. 118) or the two eyes may be infected together (Eyb, 1949; Taylor, 1957; Ogg, 1963) (Fig. 117).

Two types of lesion occur clinically—a typical vaccination pustule probably determined by the infection of a small abrasion, and vaccinial blepharitis brought about by its simple contact with the lid-margin (Folk and Taube, 1933). The more common is the VACCINIAL PUSTULE marking a site of inoculation on the outer surface of the lid. After an incubation of 3

[1] p. 146.

FIGS. 116 to 120.—VACCINIA.

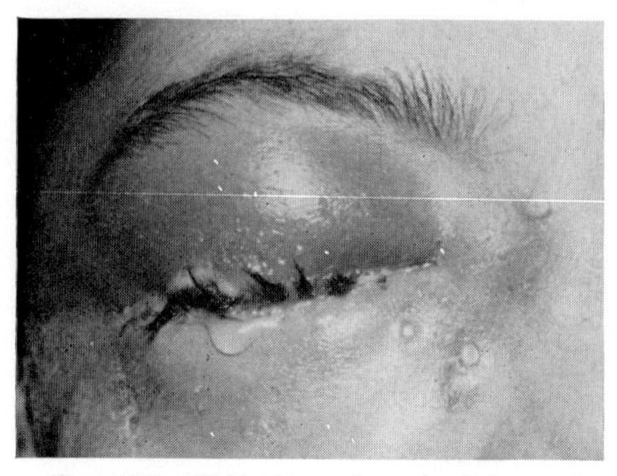

FIG. 116.—Primary ocular vaccinia resulting in a picture resembling orbital cellulitis (B. R. Jones).

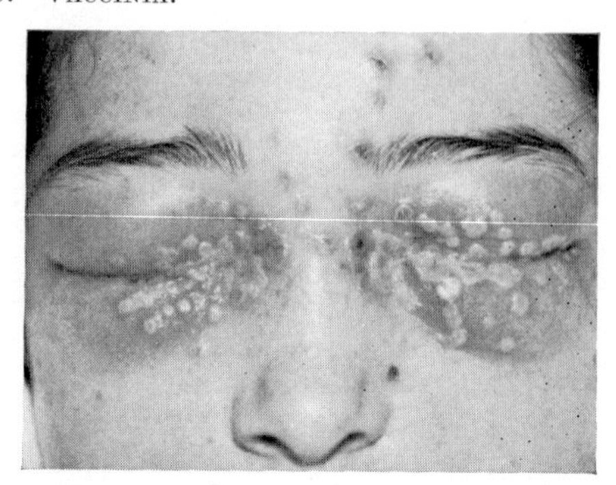

FIG. 117.—Palpebral vaccinia in a girl aged 12 whose father and several girls at school had been vaccinated some days before the condition developed (A. J. Ogg).

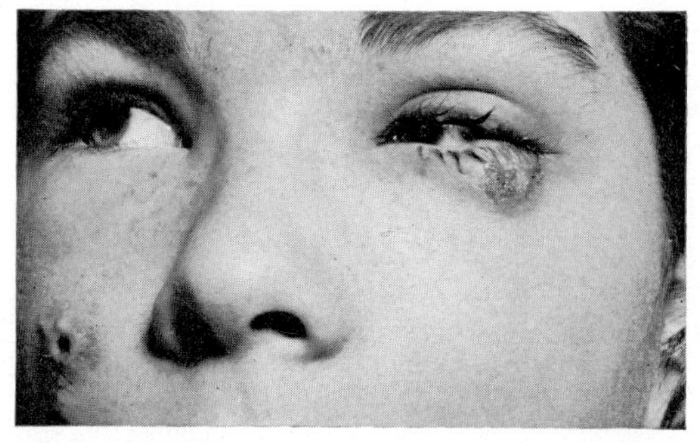

FIG. 118.—Primary vaccinia of the left lower lid and right cheek in an unvaccinated girl aged 14, presumably derived from a friend who had been vaccinated (G. Frampton and C. Smith).

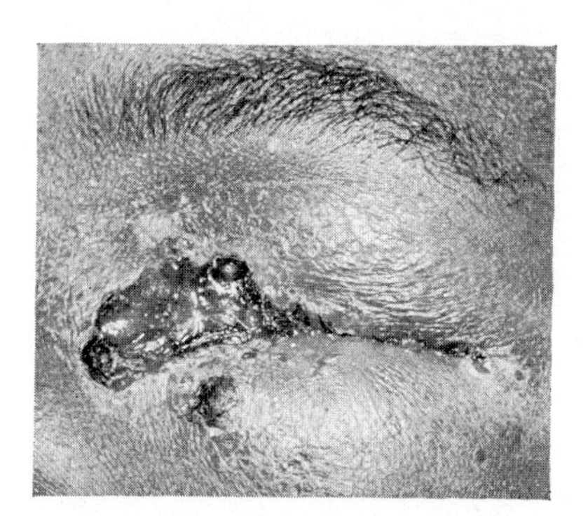

FIG. 119.—The primary lesion of vaccinia in a Negro child (Inst. Ophthal.).

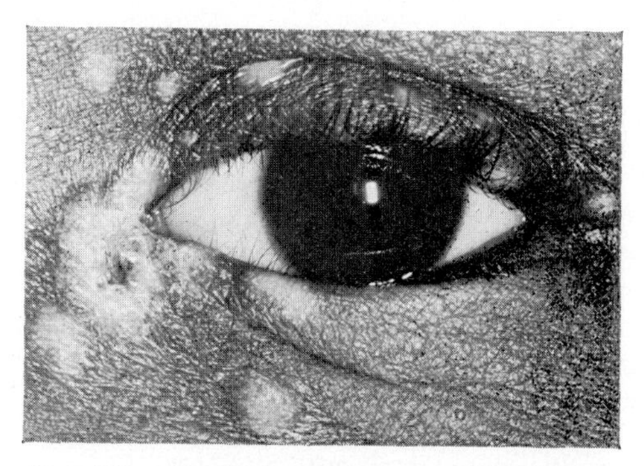

FIG. 120.—Residual scarring of the lids in the case seen in Fig. 119.

days a group of coalescing vesicles appears in a zone of erythema. This is accompanied by a swelling of the lids so intense that the eyes are closed, the pre- and post-auricular lymph nodes become swollen, and by the 8th or 9th day general constitutional symptoms of fever and malaise set in; the condition may be so severe that an orbital cellulitis is simulated (Fig. 116). The vesicular fluid, clear at first, becomes cloudy and purulent, and after the 12th day crusting appears. The ulcer so formed is surrounded by granulations, and the floor covered with a thick, grey, tenacious, necrotic membrane (Figs. 119–20). This eventually falls off to leave a red pitted scar which slowly becomes white and atrophic. As a general rule healing is remarkably good considering the virulence of the initial lesion, especially in those already vaccinated. In these patients, indeed, the lesion may merely assume the form of an itchy papule and develop into a shallow pustule which

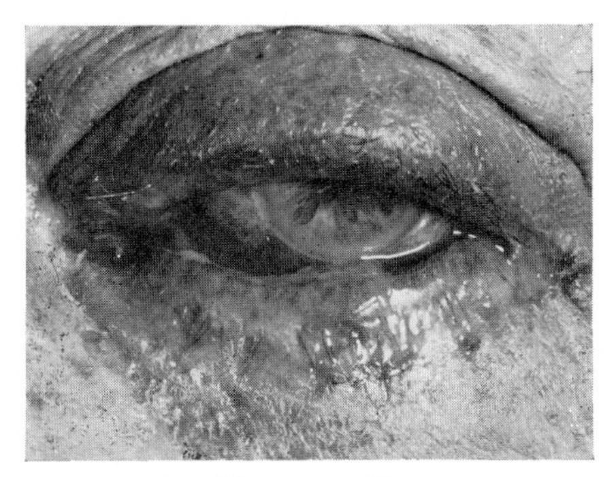

FIG. 121.—VACCINIAL BLEPHARITIS.

In a woman aged 70 showing involvement of the lid-margin and the conjunctiva (Inst. Ophthal.).

dries and crusts with very shallow ulceration. It must be remembered, however, that even although the lid-margin was not involved in the original lesion, serious corneal complications such as vaccinial disciform keratitis may subsequently develop (Perera, 1940).[1]

The second type—VACCINIAL BLEPHARITIS—occurs only on the lid-margins, where ulceration usually of widespread distribution develops and tends to spread over the palpebral conjunctiva (Fig. 121). The ulcers are covered by a thick, tenacious, dirty-grey membrane but, desiccation being prevented by the tears, no scabs are formed; the lashes, however, are frequently lost. Healing usually occurs in about ten days, but a considerable thickening of the lid-margin may persist sometimes for several months and then very gradually disappear (Munns, 1931), or permanent distortion and even symblepharon may result (Bedell, 1920; Gabbay *et al.*, 1970). Frequently, however, especially in non-immune persons, the ulceration extends from the primary scab of inoculation so that large areas may be affected,

[1] Vol. VIII, p. 360.

the opposing surfaces of the upper and lower lids, pressed together by œdematous swelling, become equally involved and spread may occur to the other eye; large areas of the lids may thus appear as one crusted ulcer, and particularly if secondary infection is introduced, necrosis may follow. Moreover, vesiculation may spread to the conjunctiva, the sclera (Bocci, 1947) or, in the form of a marginal infiltrate or a disciform keratitis, to the cornea.[1] Encephalitis is a rare complication.[2]

GENERALIZED VACCINIA may arise in patients with a normal skin in which case crops of lesions appear one or two weeks after vaccination; the spread is hæmatogenous and the lesions are more superficial than in the primary type and heal without scarring. If, however, the patient already

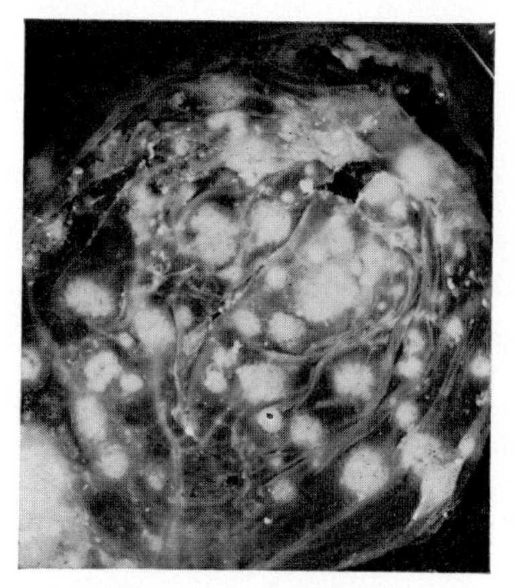

FIG. 122.—VACCINIA.

Semiconfluent vaccinial lesions on the chorio-allantoic membrane of the chick, from the case seen in Fig. 118 (G. Frampton and C. Smith).

suffers from a widespread dermatitis the reaction may be severe (*eczema vaccinatum*) which is a form of Kaposi's varicelliform eruption (Ellis, 1935; Riley and Callaway, 1947).[3] The most effective treatment is systemic cytarabine (2–3 mg./kg./day for 2 or 3 days, Juel-Jensen, 1973). In such persons vaccination should never be performed, particularly in infants in whom a fatal result may ensue (Petersilge and Toomey, 1944).

The vaccinia virus has been cultivated *in vitro* (Rivers, 1931–33; Goodpasture and Buddingh, 1935; Paschen, 1935) (Fig. 122), and the intracellular inclusions—Guarnieri bodies (1893)—containing Paschen bodies which are the ætiological agent, are diagnostic (Paschen, 1932). Electron-microscopic and cultural examination has been used to make a rapid diagnosis (Bybee *et al.*, 1967). Paul's test (1916–17) constitutes a further method of diagnosis: the contents of the vesicles or pustules dried for 3 days on a glass slide, taken up with 50% glycerol in water, are inoculated

[1] Vol. VIII, p. 362. [2] Vol. XII, pp. 125, 750. [3] p. 142.

into the cornea of the rabbit, whereon minute elevations appear in which Guarnieri bodies are found.

Treatment should consist of hyperimmune vaccinial gamma-globulin injected intramuscularly (1,500 mg.) and instilled topically (250 mg. dissolved in distilled water).[1] With this the effect is often dramatic but the virus is not eliminated, so that it should be combined with topical interferon[2] or idoxuridine (IDU)[3] together with vitamin B_7 and oral tetracyclines.[4]

COWPOX

In the literature cowpox and milker's nodules have been described somewhat indiscriminately.[5] It is now generally agreed, however, that the two diseases are distinct and separate from vaccinia, both being caused by poxviruses, the first closely related to the viruses of variola and vaccinia and the second to that of orf.[6] The lesions of cowpox, which may be numerous, consist of papules developing into vesicles which become purulent, usually with central umbilications, followed by crusting affecting the skin of cows particularly the teats and udders; these may spread to milkers' hands which in turn may infect the lids and the eye. Such a case, wherein an ulcerated area developed on the lid-margin in a milker whose herd was suffering from an epidemic of cowpox, was reported by Pfingst (1941).

Atkinson and Scullard. *Arch. Ophthal.*, **23**, 584 (1940).
Bedell. *Trans. Amer. ophthal. Soc.*, **17**, 273 (1919).
Amer. J. Ophthal., **3**, 103 (1920).
Bocci. *Boll. Oculist.*, **26**, 561 (1947).
Brav. *Arch. Ophthal.*, **33**, 67 (1945).
Bybee, Phillips, Ory and Brunschwig. *J. Amer. med. Ass.*, **199**, 126 (1967).
Ellis. *J. Amer. med. Ass.*, **104**, 1891 (1935).
Eyb. *Wien. klin. Wschr.*, **61**, 828 (1949).
Folk and Taube. *Amer. J. Ophthal.*, **16**, 36 (1933).
Frampton and Smith. *Brit. J. Ophthal.*, **36**, 214 (1952).
François, Demolder and Gildemyn. *Bull. Soc. belge Ophtal.*, No. 140, 512 (1965).
Gabbay, Kurz, Henig *et al.* *Ann. Ophthal.* (Chic.), **2**, 686 (1970).
Goodpasture and Buddingh. *Amer. J. Hyg.*, **21**, 319 (1935).
Gregoratos. *Ann. Ophthal.* (Athens), **9**, 93 (1972).
Guarnieri. *Arch. ital. Biol.*, **19**, 195 (1893).
Hirschberg. *Arch. Augenheilk.*, **8**, 187 (1879).
Jack and Sorenson. *Arch. Ophthal.*, **69**, 730 (1963).
Jones and Al-Hussaini. *Trans. ophthal. Soc. U.K.*, **83**, 613 (1963).
Jones, Galbraith and Al-Hussaini. *Lancet*, **1**, 875 (1962).
Juel-Jensen. *Brit. J. Hosp. Med.*, **10**, 402 (1973).
Kaposi. *Pathologie u. Therapie d. Hautkrankheiten*, Wien, 483 (1887).
Kaufman, Nesburn and Maloney. *Virology*, **18**, 567 (1962).
King and Forrest. *J. Amer. med. Ass.*, **153**, 31 (1953).
King and Robie. *Amer. J. Ophthal.*, **34**, 339 (1951).
Kline. *Amer. J. Ophthal.*, **34**, 342 (1951).
Madden. *Trans. ophthal. Soc. U.K.*, **89**, 955 (1969).
Meder. *Veröffentl. a. d. Geb. d. Med.*, **9**, 405 (1919).
Moffatt. *Brit. J. Ophthal.*, **36**, 211 (1952).
Monod. *Bull. Soc. Ophtal. Fr.*, 66 (1958).
Munns. *Amer. J. Ophthal.*, **14**, 1037 (1931).
Ogg. *Brit. J. Ophthal.*, **47**, 123 (1963).
Paschen. *Zbl. Bakt., Abt.* 1, **124**, 89 (1932).
IX Cong. int. Derm., Budapest, **2**, 225 (1935).
Paul. *Wien. med. Wschr.*, **66**, 861 (1916).
Dtsch. med. Wschr., **43**, 900, 1415 (1917).
Perera. *Arch. Ophthal.*, **24**, 352 (1940).
Petersilge and Toomey. *Arch. Pediat.*, **61**, 455 (1944).
Pfingst. *Amer. J. Ophthal.*, **24**, 257 (1941).
Pihl. *Klin. Mbl. Augenheilk.*, **38**, 454 (1900).
Riley and Callaway. *J. invest. Derm.*, **9**, 321 (1947).
Rivers. *J. exp. Med.*, **54**, 453 (1931); **58**, 635 (1933).
Taylor. *Brit. J. Ophthal.*, **41**, 243 (1957).

[1] Moffatt (1952), Taylor (1957), Jones *et al.* (1962), Ogg (1963), Madden (1969).
[2] Vol. VII, p. 655. [3] Vol. VIII, p. 251.
[4] Kaufman *et al.* (1962), Jones and Al-Hussaini (1963), Jack and Sorenson (1963), François *et al.* (1965), Gabbay *et al.* (1970), Gregoratos (1972).
[5] As in Vol. VIII, p. 366. [6] p. 158.

HERPES (SIMPLEX : FEBRILIS)

The virus of herpes has been described in an earlier Volume.[1] In all Western countries it is a common cause of infection; thus it has been found that in the U.S.A. 60% of individuals before the age of 5 and 90% by the age of 15 have been infected as indicated by the presence of neutralizing antibodies in the serum (Buddingh *et al.*, 1953). The mode of infection is obvious for the vesicular fluid of a herpetic lesion is heavily packed with organisms as also are the nasal secretion and stools of a large proportion of the population (Rake, 1957). Related viruses are found in many animals; the organism causing herpes in man is *Herpesvirus hominis*.

The life-cycle of the virus is important. A primary infection is frequently acquired in youth, sometimes at birth from the vulva of the mother, in which case the disease is systemic and severe and sometimes fatal. More usually, however, the child is protected for the first few months of life by circulating maternal antibodies, but after the age of 6 months the baby is liable to infection, usually by being kissed by a relative, frequently in the region of the eyelids. If he escapes this peril the adolescent runs a similar risk by being kissed by a lover. There are thus two peaks of herpetic infection, the first few years of life (after the first 6 months) and between the ages of 16 and 25 years. Once the infection is acquired the virus remains harmless and latent at the site of the original infection, but the symbiosis has a persistent habit of breaking down with recurrences of the lesion by some extraneous factor, frequently of an insignificant nature—a coryza or a febrile attack of any kind especially involving the upper respiratory tract, systemic fevers, antityphoid inoculation (Moro, 1955), or even psychological disturbances.[2]

The *primary infection* which may affect the lids or lid-margins may be subclinical but may be associated with an acute illness (Gundersen, 1936; Maumenee *et al.*, 1945). The usual lesion is a crop of vesicles of pinhead size on the lids, particularly the lower, or the lips, on a swollen, œdematous and slightly erythematous base, typically unilateral in distribution. At first filled with clear yellow fluid they dry to form yellowish-brown crusts which drop off leaving no scars in about 7 days (Figs. 123–4). In the primary infection the attack may be of some severity and associated with fever. The eye may also be affected with an acute follicular conjunctivitis and a keratitis with regional lymphadenopathy. As a rule the condition is benign, but occasionally a generalized severe infection develops—Kaposi's varicelliform eruption.

KAPOSI'S VARICELLIFORM ERUPTION (ECZEMA HERPETICUM OR VACCINATUM) is a generalized vesico-pustular lesion occurring with a primary infection by the virus of herpes or vaccinia in patients with a pre-existing dermatosis, usually atopic in nature. It may take the form of a severe

[1] Vol. VIII, p. 337. [2] Vol. VIII, p. 313.

and sometimes fatal illness, particularly in young children but also occurs in young adults (Braley, 1957). An eruption closely resembling smallpox suddenly appears, affecting particularly the face and head and frequently involving a heavy infection of the lids, associated with fever and constitutional symptoms; initially erythematous and vesicular, the lesions rapidly

FIGS. 123 and 124.—PRIMARY HERPES.

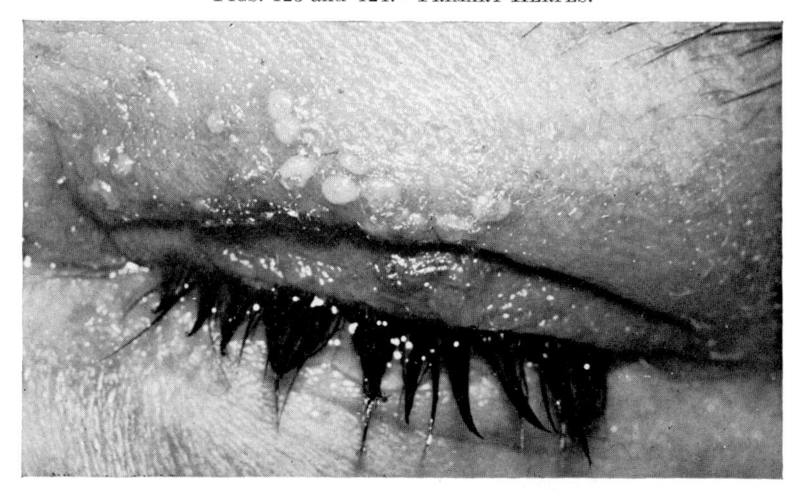

FIG. 123.—Affecting the upper lid, in the vesicular stage (B. Jay).

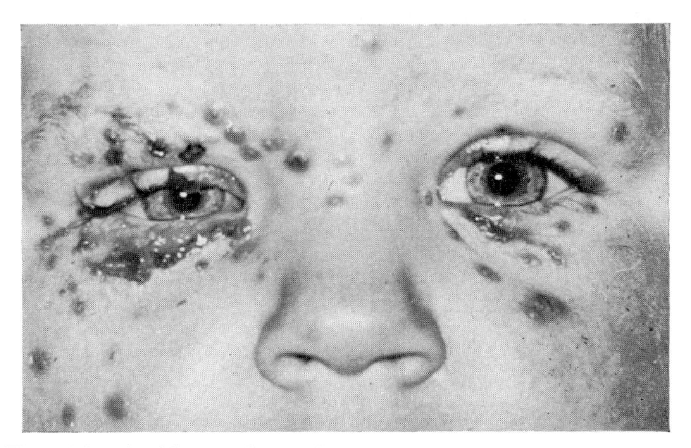

FIG. 124.—A widespread eruption in a young child (Inst. Ophthal.).

become pustular and desiccate leaving considerable scarring[1] (Figs. 125–6). In very young infants complications such as hepatitis and encephalitis may develop, sometimes with a fatal result (Zuelzer and Stulberg, 1952; France and Wilmers, 1953), and in adults, although rarely, a meningo-encephalitis

[1] Juliusberg (1898), Brown (1934), McLachlan and Gillespie (1936), Barton and Brunsting (1944), and others.

occasionally involving neurological and mental sequelæ which again may be fatal (Whitman *et al.*, 1946; Ross and Stevenson, 1961). The malady was originally described by Kaposi (1887) as a varicelliform eruption, being a complication of infantile eczema, and for long it was considered to be identical with generalized vaccinia, the widespread eruption occurring after the vaccination of persons already suffering from some widespread form of dermatitis. The virus of herpes, however, has been demonstrated in the vesicles in many cases.[1] A complicating bacterial infection frequently by

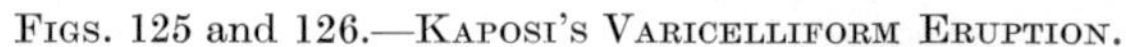
FIGS. 125 and 126.—KAPOSI'S VARICELLIFORM ERUPTION.

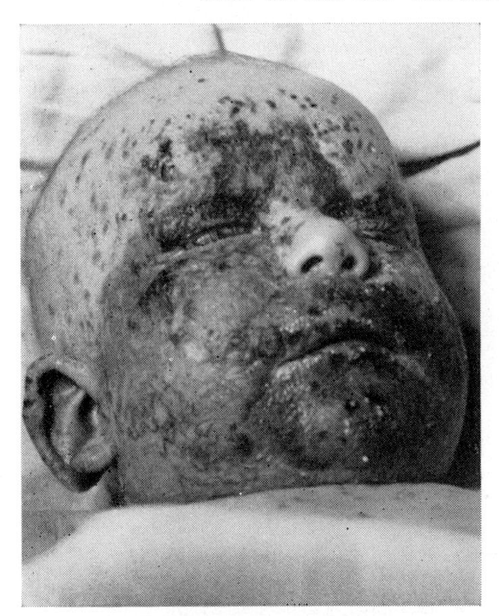

FIG. 125.—In the acute stage (T. L. Chester-Williams).

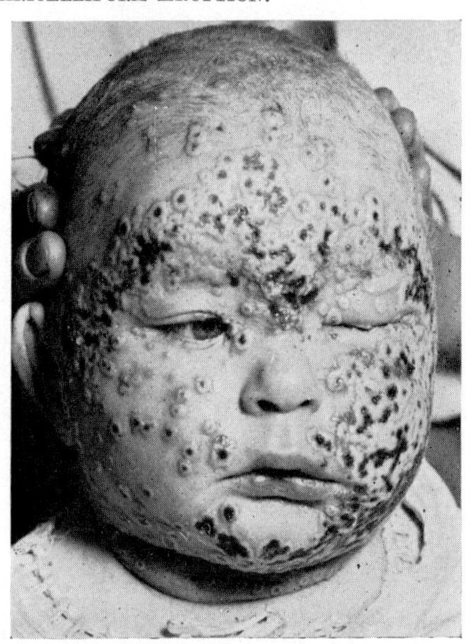

FIG. 126.—In the subacute stage (W. H. Brown).

Pseudomonas pyocyanea may lead to a fatal termination from septicæmia. The condition may also develop in the Wiskott-Aldrich syndrome[2] in which case it is also fatal.

Recurrent herpes is a more common affection, the most serious aspect of which is the frequent occurrence of ocular lesions.[3] On the lids recurrent outcrops of vesicles appear which behave as those of the primary affection and progress to the formation of eschars (Figs. 127–8). These recurrences are most serious when they involve the lid-margins when pustulation and recurrent ulceration may develop, in which case the most feared complication is a keratitis (Fig. 129). A staphylococcal infection may lead to the formation of an ulcerated eschar (Nauheim and Sussman, 1971) (Fig. 130).

[1] Barton and Brunsting (1943), Wenner (1944), Lynch *et al.* (1945), Lynch and Steves (1947), Ruchman *et al.* (1947), and others.
[2] p. 57. [3] Vol. VIII, p. 313.

FIGS. 127 to 130.—RECURRENT HERPES.

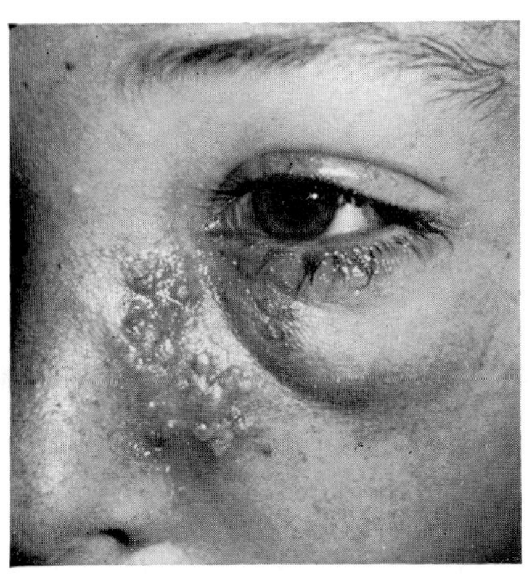

FIG. 127.—Vesicular eruption in the lower lid and nose (J. S. Nauheim and W. Sussman).

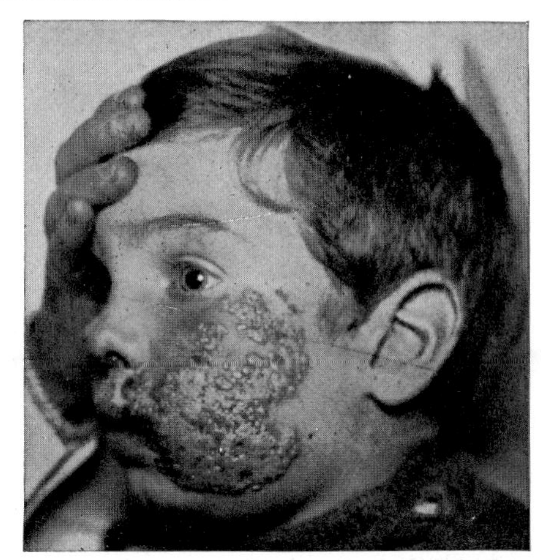

FIG. 128.—Vesicular eruption affecting the lower lid and cheek (D. Paterson and A. Moncrieff).

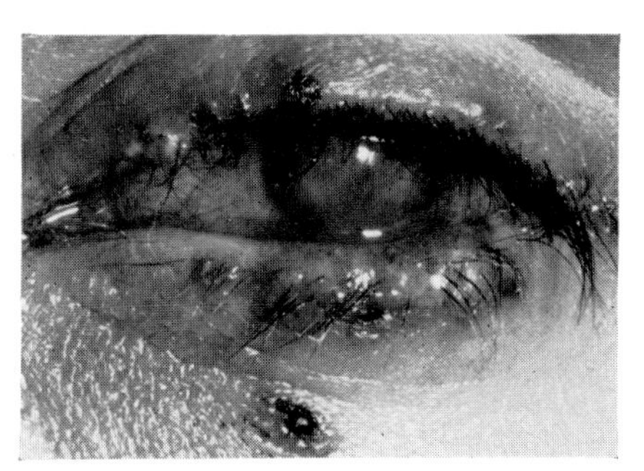

FIG. 129.—The eruption affects the lower lid-margin; there is also a severe keratitis (Inst. Ophthal.).

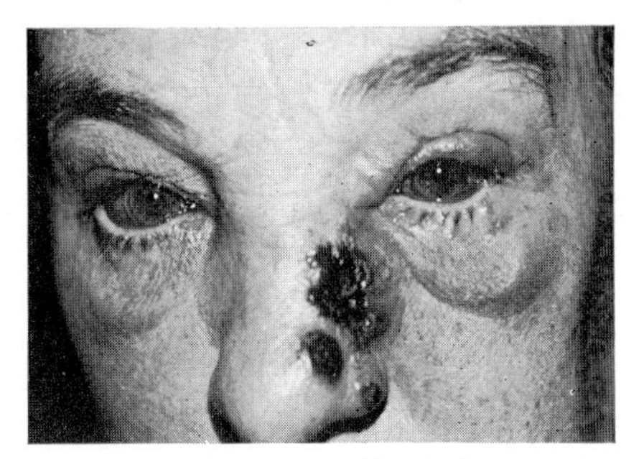

FIG. 130.—Herpes complicated by a staphylococcal infection. There is an ulcerated eschar on the side of the nose from which the herpes virus and staphylococci were cultured (J. S. Nauheim and W. Sussman).

Palpebral herpes is benign and despite recurrences the prognosis is good if the eye itself is not affected. The best symptomatic *treatment* in the early stages is the application of alcoholic solutions and in the later stages soothing lotions (1 in 500 aluminium acetate). In a generalized infection these applications should be given and stimulatory and sedative measures taken; an antibiotic such as tetracycline is indicated if secondary infections develop (Baer and Miller, 1949; Bereston and Carliner, 1949; Bookman, 1950); these, however, have no effect on the virus.

Specific treatment is by idoxuridine (IDU) which may be applied with a spray-gun or injected percutaneously into the lesion (which is painful) or applied with a paint-brush as a 5% suspension in dimethyl sulphoxide

(DMSO) (Juel-Jensen and MacCallum, 1965; MacCallum and Juel-Jensen, 1966; Juel-Jensen, 1973). An alternative drug is cytarabine (cytosine arabinoside) (aqueous solution of 3 mg./kg./day by intravenous injection) which arrests the replication of the virus and has been found to have beneficial effects in severe attacks of both primary and recurrent herpes (Juel-Jensen, 1973). It is to be hoped that other specific drugs, such as vidarabine, may become available (Juel-Jensen, 1973). Corticosteroids are contra-indicated for herpes simplex.

Baer and Miller. *J. invest. Derm.*, **13**, 5 (1949).
Barton and Brunsting. *Proc. Mayo Clin.*, **18**, 199 (1943).
Arch. Derm. Syph. (Chic.), **50**, 99 (1944).
Bereston and Carliner. *J. invest. Derm.*, **13**, 13 (1949).
Bookman. *J. Allergy*, **21**, 68 (1950).
Braley. *Amer. J. Ophthal.*, **43** (2), 105 (1957).
Brown. *Brit. J. Derm.*, **46**, 1 (1934).
Buddingh, Schrum, Lanier and Guidry. *Pediatrics*, **11**, 595 (1953).
France and Wilmers. *Lancet*, **1**, 1181 (1953).
Gundersen. *Arch. Ophthal.*, **15**, 225 (1936).
Juel-Jensen. *Brit. med. J.*, **1**, 406 (1973).
Juel-Jensen and MacCallum. *Brit. med. J.*, **1**, 901 (1965).
Juliusberg. *Arch. Derm. Syph.* (Wien), **45**, 21 (1898).
Kaposi. *Pathologie u. Therapie d. Hautkrankheiten*, Wien, 483 (1887).
Lynch, Evans, Bolin and Steves. *Arch. Derm. Syph.* (Chic.), **51**, 129 (1945).
Lynch and Steves. *Arch. Derm. Syph.* (Chic.), **55**, 327 (1947).
MacCallum and Juel-Jensen. *Brit. med. J.*, **2**, 805 (1966).
McLachlan and Gillespie. *Brit. J. Derm.*, **48**, 337 (1936).
Maumenee, Hayes and Hartman. *Amer. J. Ophthal.*, **28**, 823 (1945).
Moro. *Ann. Ottal.*, **81**, 169 (1955).
Nauheim and Sussman. *Trans. Amer. Acad. Ophthal.*, **75**, 1236 (1971).
Rake. *Amer. J. Ophthal.*, **43** (2), 113 (1957).
Ross and Stevenson. *Lancet*, **2**, 682 (1961).
Ruchman, Welsh and Dodd. *Arch. Derm. Syph.* (Chic.), **56**, 846 (1947).
Wenner. *Amer. J. Dis. Child.*, **67**, 247 (1944).
Whitman, Wall and Warren. *J. Amer. med. Ass.*, **131**, 1408 (1946).
Zuelzer and Stulberg. *Amer. J. Dis. Child.*, **83**, 421 (1952).

VARICELLA—ZOSTER

The two diseases, varicella and zoster, are caused by the same virus, the *V-Z virus*, the characteristics of which have been described in a previous Volume.[1] Varicella is a mild but highly infectious febrile disease occurring particularly in children, transmitted by droplet infection, the port of entry being the upper respiratory tract. Zoster, on the other hand, tends to attack older people in whom the dorsal root ganglia of the cord or the extra-medullary ganglia of the cranial nerves, both sensory and motor, become infected and inflamed and the disease is manifested by the occurrence of lesions, at first vesicular and eventually necrotic, in the area of distribution of the nerve. How the virus reaches these ganglia is unknown, but zoster may develop by contact with patients with varicella in epidemics. The late and localized characteristics of zoster are difficult to explain; it may be that this manifestation of the infection is due to a partial immunity caused by a previous attack of varicella or that it is due to the activation of the virus previously existing in a latent state. It used to be suggested that zoster existed in two forms, an idiopathic (epidemic) form eventually shown to be of viral origin, and a symptomatic form due to the involvement of the

[1] *Herpesvirus varicellæ* or the *varicella-zoster virus*, Vol. VIII, p. 336.

ganglion in some infective, neoplasic or traumatic disturbance; it is likely, however, that the second type results from the activation of a latent virus in the ganglion and that the two are identical.

VARICELLA (*chickenpox*) is characterized by a cutaneous rash which may be papular or vesicular and occasionally pustular. The rash is usually widespread, but it may attack the lids, particularly their margins as well as the conjunctiva (Griffin and Searle, 1953); it has been reported as appearing first at this site (Comby, 1884). The lesions may become pustular and develop into excavated ulcers with swollen, dark red margins, associated with a considerable reaction and swelling of the lids (Hilbert, 1902; Pearson, 1903; Raybaud *et al.*, 1947; Aubineau, 1947). The lesions usually regress and the only available treatment is by bland lotions, but cases of gangrene of one or both lids have been reported (Horner, 1889; Römer, 1900; Rolleston, 1909; Wintersteiner, 1909).

ZOSTER (ζωστήρ, a girdle) (*shingles, zona*) is of ophthalmological importance when the gasserian ganglion is involved; the first division of the trigeminal is usually affected, the second more rarely, and both on occasion (Figs. 131–3), while the simultaneous involvement of all three divisions has been recorded (O'Neill, 1945; Garrett, 1958). Bilateral cases are extremely rare (Veasey, 1919). The disease may occur sporadically, but sometimes appears in a mild epidemic form, often associated with chicken-pox. It typically attacks adults and the aged, but occasionally occurs in the young: in an infant at birth whose mother was affected in pregnancy (Feldman, 1952), an infant of 20 hours (Counter and Korn, 1950), a child of 1 year (Poulsen, 1955), 2 years (Garrett, 1958), 4 years (Aubertin *et al.*, 1950; Schwartz, 1972), 5 years (Safar, 1949; Malik *et al.*, 1964), 7 years (Kielar *et al.*, 1971) and 8 years (Birks, 1963). Some of the cases in very young patients may well have been due to the occurrence of varicella in the mother during her pregnancy (at the age of 18 months, McKendrick and Raychoudhury, 1972; Lewkonia and Jackson, 1973).

In common with other viral infections, zoster may assume very grave and occasionally fatal aspects in patients with a deficiency or absence of gamma globulin in the serum (Keidan and Mainwaring, 1965; Kielar *et al.*, 1971) (Fig. 134); this occurs particularly in those suffering from malignant reticuloses. In cases with this immunological defect the zoster may occur in childhood and is frequently associated with gangrene and intra-ocular complications involving the loss of vision and even of the eye (Doden, 1960; Blodi, 1968).

The lesion presents a characteristic picture, first described and accurately detailed by Jonathan Hutchinson (1864–69) and later reviewed with all its complications by Scheie (1970). The onset is sudden, introduced by fever, prostration and neuralgic pain over the distribution of the nerve. Sometimes simultaneously with the pain but usually 3 or 4 days later a blushing of the skin is followed by a marked œdema over the affected area, swelling of the

FIGS. 131 to 134.—OPHTHALMIC ZOSTER.

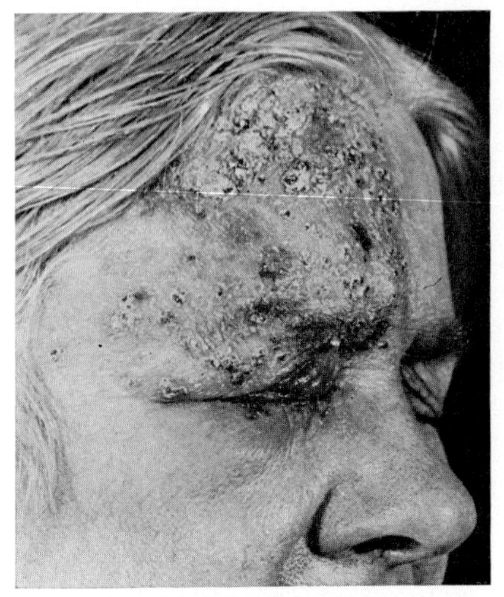

FIG. 131.—Affecting the ophthalmic division of the trigeminal (Inst. Ophthal.).

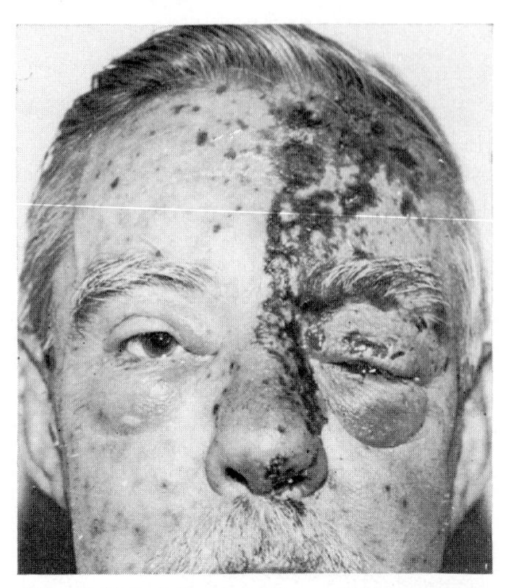

FIG. 132.—Affecting the ophthalmic division of the trigeminal with involvement of the naso-ciliary branches (Inst. Ophthal.).

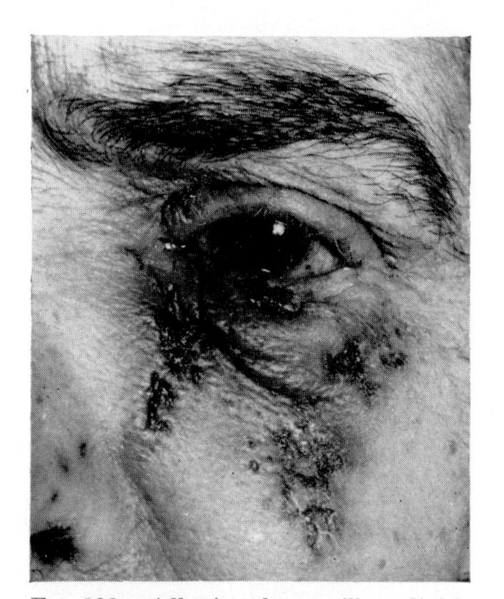

FIG. 133.—Affecting the maxillary division of the trigeminal (Inst. Ophthal.).

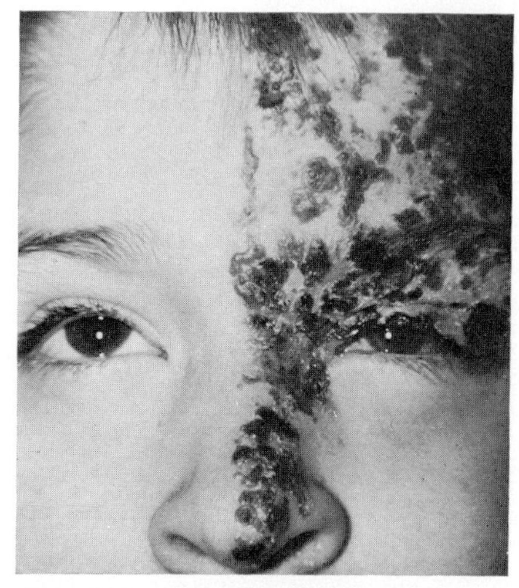

FIG. 134.—Severe lesions occurring in a child aged 7 with IgA deficiency (R. A. Kielar *et al.*).

lids, and the appearance of vesicles filled with clear fluid in which the virus is abundant in the first 12 hours to become scanty within 24 hours; these rapidly become turbid and yellow and within a short time burst, forming scabs which on separating leave deeply pitted scars; a permanent epicanthoid lesion on the lids has been observed (Goldsmith, 1964). Only occasionally are the typical lesions absent or replaced by a seborrhœic eruption (Ross, 1949; Michel and Chauvire, 1952; Lewis, 1958). A widespread concurrent varicelliform eruption may appear as a rarity (Vernier and Fouche, 1951; Glotzer, 1952; Garrett, 1958).

The most clamant symptom is pain which, although sometimes mild and transient, is usually severe, burning and of a neuralgic nature associated with an exquisite tenderness of the skin and in the scalp in lesions of the first division of the trigeminal nerve, going on continuously during the day and night for weeks or months, aggravated by every contact or draught of air. With the subsidence of the eruption it usually subsides, but it may persist for several years and so severely as to lead to suicide. The affected area of the skin is frequently left with diminished sensitivity and numbness which, when the neuralgia persists, produces the distressing condition of anæsthesia dolorosa.

When the naso-ciliary branch of the trigeminal nerve is involved ocular complications are more common than otherwise. These are dealt with elsewhere—conjunctivitis and keratitis,[1] uveitis,[2] retinitis,[3] optic neuritis and atrophy,[4] pupillary disturbances and ocular palsies[5]; Bell's palsy may also occur.[6] The unusual coexistence of zoster and herpes febrilis affecting the lip and the cornea has been reported (Acers and Vaile, 1967), but a dendritic-like ulcer may be due to the virus of zoster alone (Pavan-Langston and McCulley, 1973).

Treatment. Once the virus has ceased to replicate and done permanent damage, no specific chemotherapy is of value. If, however, a lesion on the skin is treated at a sufficiently early stage, good results have been obtained by a strong solution (35 to 40%) of the antiviral agent *idoxuridine* in dimethyl sulphoxide applied continuously for 3 or 4 days on a piece of lint re-wetted daily and kept in place by a bandage such as netelast. If this treatment were instituted sufficiently early Juel-Jensen and his colleagues (1970) found that in the majority of cases the pain had gone within a week and late sequelæ were rare. Similar effective results have been claimed to follow the daily intravenous injection of another antiviral agent, *cytarabine* (3 mg./kg. on the first day followed by 2 mg./kg. for another 3 days) (Juel-Jensen, 1973), or by continuous intravenous infusion for 5 days (Pierce and Jenkins, 1973). During this phase injections of vitamin B_{12} have been advocated by some writers.[7]

If the early stages are past, rest in bed and attempts to relieve the pain are essential; salicylates or codeine may be effective, the injection of posterior pituitary extract has been recommended, and high doses of corticosteroids have given favourable results (Scheie and Alper, 1955; Scheie and McLellan, 1959; Elliott, 1964; Freiwald, 1968; Scheie, 1970), but these have not been invariable (Nover, 1970). In the later stages drugs such as carbamazepine

[1] Vol. VIII, p. 340. [2] Vol. IX, p. 345. [3] Vol. X, p. 269.
[4] Vol. XII, p. 126. [5] Vol. XII, pp. 620 and 750. [6] Vol. XII, p. 923.
[7] Vol. VII, p. 709.

or even morphia preparations may be necessary for the relief of the pain. Many types of dressing have been suggested for the dermal lesions, such as a collodion paint in the early stages or dry dusting powders (zinc oxide, bismuth, sulphanilamide); antibiotics are of little or no value. The pain is equally recalcitrant and measures such as deep x-ray therapy over the gasserian region (McCombs *et al.*, 1940), ganglionectomy or section of the root of the nerve, or even medullary tractotomy whereby the spinal trigeminal tract is sectioned (White and Sweet, 1955) are by no means invariably effective. It is important to persuade the patient to resume his normal life as soon as he is able.

Acers and Vaile. *Amer. J. Ophthal.*, **63**, 992 (1967).
Aubertin, Pesme and Rivière. *Arch. franç. Pédiat.*, **7**, 547 (1950).
Aubineau. *Bull. Soc. Ophtal. Fr.*, 91 (1947).
Birks. *Brit. J. Ophthal.*, **47**, 60 (1963).
Blodi. *Amer. J. Ophthal.*, **65**, 686 (1968).
Comby. *Progr. Méd.*, **12**, 39 (1884).
Counter and Korn. *Arch. Pediat.*, **67**, 397 (1950).
Doden. *Klin. Mbl. Augenheilk.*, **136**, 230 (1960).
Elliott. *Lancet*, **2**, 610 (1964).
Feldman. *Arch. Dis. Childh.*, **22**, 126 (1952).
Freiwald. *Med. Times* (N.Y.), **96**, 43 (1968).
Garrett. *Amer. J. Ophthal.*, **46**, 741 (1958).
Glotzer. *N.Y. St. J. Med.*, **52**, 3033 (1952).
Goldsmith. *Amer. J. Ophthal.*, **57**, 843 (1964).
Griffin and Searle. *Lancet*, **2**, 169 (1953).
Hilbert. *Zbl. prakt. Augenheilk.*, **26**, 39 (1902).
Horner. Gerhardt's *Hb. d. Kinderkrankheiten*, Tübingen, **5** (2), 220 (1889).
Hutchinson. *Roy. Lond. ophthal. Hosp. Rep.*, **4**, 189 (1864); **5**, 191, 331 (1866); **6**, 46, 181, 263 (1867–69).
Juel-Jensen. *Brit. med. J.*, **1**, 406 (1973).
Juel-Jensen, MacCallum, Mackenzie and Pike. *Brit. med. J.*, **4**, 776 (1970).
Keidan and Mainwaring. *Clin. Pediat.*, **4**, 13 (1965).
Kielar, Cunningham and Gerson. *Amer. J. Ophthal.*, **72**, 555 (1971).
Lewis. *Brit. med. J.*, **2**, 418 (1958).
Lewkonia and Jackson. *Brit. med. J.*, **3**, 149 (1973).
McCombs, Tuggle and Guion. *Amer. J. med. Sci.*, **200**, 803 (1940).
McKendrick and Raychoudhury. *Scand. J. infect. Dis.*, **4**, 23 (1972).
Malik, Sood, Gupta, S. B. and D. K. *Orient. Arch. Ophthal.*, **2**, 92 (1964).
Michel and Chauvire. *Bull. Soc. franç. Derm. Syph.*, **59**, 303 (1952).
Nover. *Klin. Mbl. Augenheilk.*, **156**, 305 (1970).
O'Neill. *Arch. Ophthal.*, **33**, 237 (1945).
Pavan-Langston and McCulley. *Arch. Ophthal.*, **89**, 25 (1973).
Pearson. *Brit. med. J.*, **1**, 1492 (1903).
Pierce and Jenkins. *Arch. Ophthal.*, **89**, 21 (1973).
Poulsen. *Acta med. scand.*, **151**, 131 (1955).
Raybaud, Farnarier, Garrouste and Payan. *Rev. Oto-neuro-ophtal.*, **19**, 366 (1947).
Römer. *Samml. zwangl. Abhandl. Geb. Augenheilk.*, **3** (4), 12 (1900).
Rolleston. *Med. Chron.* (Manchester), **49**, 215 (1909).
Ross. *Arch. Ophthal.*, **42**, 808 (1949).
Safar. *Wien. klin. Wschr.*, **61**, 191, 413 (1949).
Scheie. *Trans. ophthal. Soc. U.K.*, **90**, 899 (1970).
Scheie and Alper. *Arch. Ophthal.*, **53**, 38 (1955).
Scheie and McLellan. *Arch. Ophthal.*, **62**, 579 (1959).
Schwartz. *Amer. J. Ophthal.*, **74**, 142 (1972).
Veasey. *Trans. Amer. ophthal. Soc.*, **17**, 297 (1919).
Vernier and Fouche. *Bull. Soc. franç. Derm. Syph.*, **58**, 62 (1951).
White and Sweet. *Pain. Its Mechanisms and Neurosurgical Control*, Springfield (1955).
Wintersteiner. *Z. Augenheilk.*, **21**, 268 (1909).

VERRUCA

VERRUCÆ (*warts*) are *small, auto-inoculable epidermal and papillary growths caused by a filtrable virus.* The condition has long been considered contagious, its infectivity being first proved by inoculation by Variot (1894) and Jadassohn (1896); the filtrability of the infective agent was demonstrated by Ciuffo (1907) and the virus fully studied by Wile and Kingery (1919) and Goodman and Greenwood (1934).

After an incubation period of from one to five and sometimes twenty months (Templeton, 1935), a wart begins as a tiny grey epithelial thickening which grows into an irregular, greyish-yellow, horny excrescence. These may be single or multiple, usually discrete but sometimes coalescing; their appearance on the skin of the lids is due to infection carried by the fingers from elsewhere and in this locality they give rise to no symptoms.

On the margins of the lids, however, where warts are found between the lashes, particularly on the upper lids, the conjunctival epithelium may be affected and ocular symptoms arise somewhat akin to those seen in molluscum contagiosum (de Rötth, 1933–39; Vito, 1936; Turtz, 1942) (Figs. 135–6). The reaction here, however, differs from that in the skin. It would appear that without the protection of a horny epidermal layer and exposed to the stronger immunological attack of the conjunctival tissues, the virus cannot

Figs. 135 and 136.—Warts at the Lid Margin.

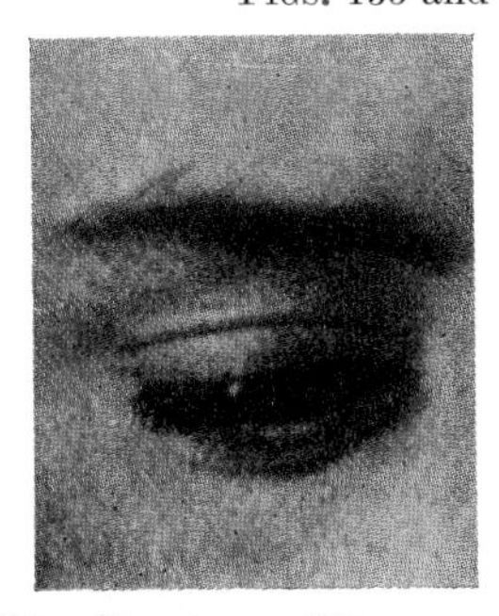

Fig. 135.—Showing a filiform wart in the middle of the upper lid margin in a boy aged 13 with a 6 months' history.

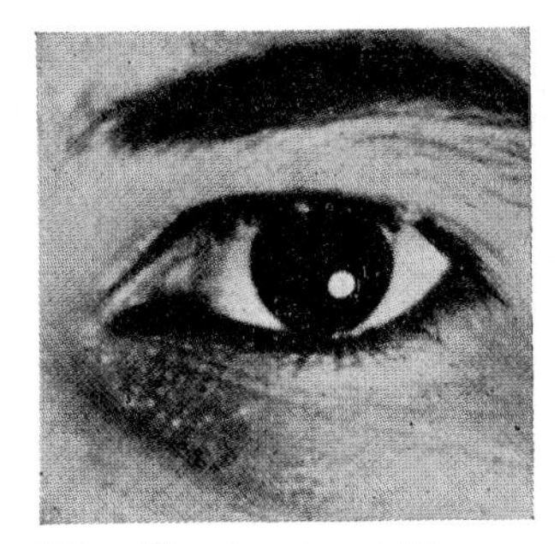

Fig. 136.—Showing two filiform warts at the margin of the upper lid in a woman aged 21 with a 3 months' history (A. de Rötth).

survive the very prolonged incubation period; moreover, infected mucosal cells are probably cast off so that proliferating excrescences are not produced. At the same time, after constant infection by a marginal wart an irritative *verrucose conjunctivitis*[1] develops of a subacute nature, with little discharge and usually characterized by a smooth conjunctival surface, although follicles may occasionally be present. Similarly a *verrucose keratitis* has been reported: it involves initially the epithelial layer only, producing a superficial punctate keratitis, repeated transient attacks recurring with considerable pain and photophobia. In more severe cases the deeper tissues are involved and an infiltration with new vessels occurs, the condition resembling a rosacea keratitis. Finally, actual ulceration may appear. No inclusions have been found in the cells (de Rötth, 1939) and the clinical picture suggests that the reaction is allergic in nature since both the conjunctivitis and the keratitis disappear on removal of the warts.

[1] Vol. VIII, p. 375.

In form, warts on the skin of the lids may be round excrescences (VERRUCA VULGARIS), flat and sessile (VERRUCA PLANA), digitate, composed of several projections with horny caps grouped on a narrow base (VERRUCA DIGITATA) or filiform, slender and threadlike, covered with apparently normal skin (VERRUCA FILIFORMIS): the last type is found very commonly on the lids. Multiple common warts are seen in Fig. 137.

VERRUCA ACUMINATA (CONDYLOMATA ACUMINATA, SECONDARY PAPILLOMATA) are prolifically growing warts composed of closely aggregated, tufted projections which are found near muco-cutaneous junctions in moist localities. The most usual site is round the vulva or underneath the foreskin where they may be very exuberant (venereal wart); but as a rarity they may occur at the lid-margin where they project as pink, fleshy masses among the lashes (Hirschler, 1866).

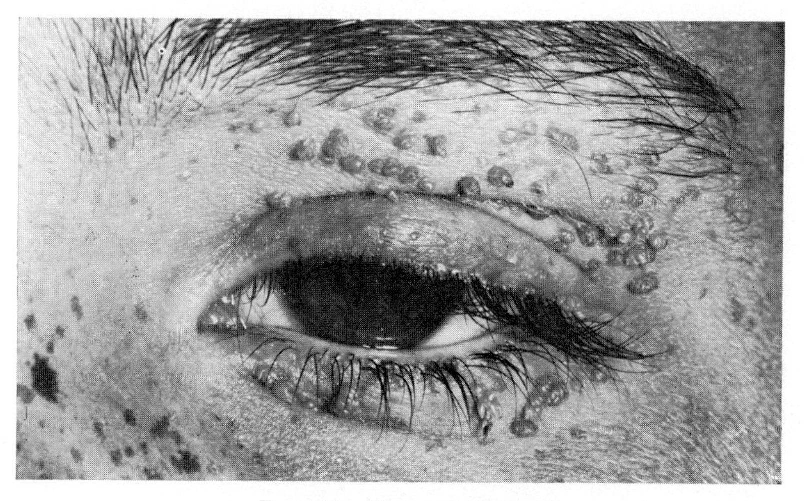

FIG. 137.—MULTIPLE WARTS.
In a boy aged 10 (Inst. Ophthal.).

It is probable that the different clinical types of human wart are due to the same infective agent, its results being modified by its surroundings. Morphologically dissimilar lesions often coexist in the same patient, and a filtered extract of a wart of one type may give rise to an apparently different type if inoculated into a different epithelial surface (von Rooyen and Rhodes, 1940). Brain (1937) was unable to demonstrate any antibody formation, although the possibility that immunity reactions may follow inoculations was suggested by the experiments of Findlay (1930).

Histopathologically, in verruca vulgaris the papillomatous growths show acanthosis, hyperkeratosis and some parakeratosis, the cells being often arranged in tiers overlying the crests of the papillomatous elevations in the stratum malpighii (Fig. 138). In the upper layers of the stratum malpighii and the granular layer there are large vacuolated cells in the nuclei of which the virus is found in early lesions, occasionally replacing the nuclear material (Blank *et al.*, 1951; Almeida *et al.*, 1962). Rounded eosinophil bodies may also be found in the nuclei and cytoplasm of the epidermal cells; these were at one time considered to be viral inclusion bodies (Lipschütz, 1931; Strauss *et al.*, 1950), but electron microscopy has demonstrated that they are degeneration products (Almeida *et al.*, 1962). In verucca plana there is no papillomatosis, but the cells of the stratum malpighii and the granular layer are enlarged and vacuolated. In the condylomatous type there may be a considerable downgrowth of the epidermis into the underlying tissue, producing a picture which may mimic a squamous-celled carcinoma (Machacek and Weakley, 1960).

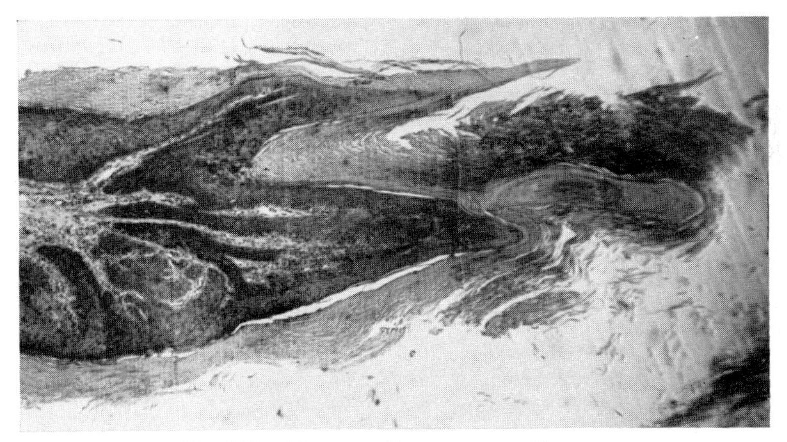

Fig. 138.—Verruca Filiformis of Eyelid.

There is acanthosis and hyperkeratosis resulting in the formation of an elongated finger-like process which projects from the surface (H. and E. × 30) (N. Ashton).

The *treatment* of warts in the region of the eye is best by surgical excision or scraping with a sharp spoon and touching the base with iodine or silver nitrate. Other serviceable methods which may be used in localities away from the lid-margin are cauterization by the actual cautery, fulguration, or destruction by such caustics as trichloracetic or salicylic acid, zinc chloride or the silver nitrate pencil in repeated applications; the pencil of carbon dioxide snow gives excellent therapeutic results. Ionization with an anodal zinc needle thrust through the base of the wart is also effective. The application of Cantharon (0·7% cantharidin in equal parts of acetone and collodion) which can be repeated at 8-day intervals if necessary was recommended by Bock (1965) if the warts are outside the line of the lashes. When the causative warts have been removed any ocular symptoms immediately clear up in the absence of permanent corneal damage.

It is important to remember that podophyllin, a very efficient therapeutic agent for acuminate warts, is dangerous to use near the eye as it excites a serious conjunctivitis.[1]

Almeida, Howatson and Williams. *J. invest. Derm.*, **38,** 337 (1962).
Blank, Buerk and Weidman. *J. invest. Derm.*, **16,** 19 (1951).
Bock. *Amer. J. Ophthal.*, **60,** 529 (1965).
Brain. *Brit. med. J.*, **2,** 1064 (1937).
Ciuffo. *G. ital. Mal. vener.*, **42,** 12 (1907).
Findlay. *A System of Bacteriology* (*Med. Res. Cncl.*), London, **7,** 248 (1930).
Goodman and Greenwood. *Arch. Derm. Syph.* (Chic.), **30,** 659 (1934).
Hirschler. *Wien. med. Wschr.*, **16,** 1161 (1866).
Jadassohn. *Vhdl. dtsch. derm. Ges.*, **5,** 497 (1896).
Lipschütz. Jadassohn's *Hb. d. Haut- u. Geschlechtskrankheiten*, Berlin, **11** (1) (1931).
Machacek and Weakley. *Arch. Derm.*, **82,** 41 (1960).
van Rooyen and Rhodes. *Virus Diseases of Man*, Oxon (1940).
de Rötth. *Klin. Mbl. Augenheilk.*, **91,** 196 (1933).
Arch. Ophthal., **21,** 409 (1939).
Strauss, Bunting and Melnick. *J. invest. Derm.*, **15,** 433 (1950).
Templeton. *Arch. Derm. Syph.* (Chic.), **32,** 102 (1935).
Turtz. *Amer. J. Ophthal.*, **25,** 452 (1942).
Variot. *J. Clin. Thér. inf.*, **2,** 529 (1894).
Vito. *Boll. Oculist.*, **15,** 627 (1936).
Wile and Kingery. *J. Amer. med. Ass.*, **73,** 970 (1919).

MOLLUSCUM CONTAGIOSUM

MOLLUSCUM CONTAGIOSUM is *a mildly contagious disease of the skin characterized by the appearance of small globular, umbilicated epithelial tumours* (*water warts*) (Figs. 139–40). That the malady is infective, caused by a filtrable virus, was shown by Marx and Sticker (1902) and Juliusberg (1905), and confirmed by Wile and Kingery (1919); it has been classified with the poxviruses.

The disease usually occurs in the young, but has been seen in those over middle age (63 years, Lauber, 1913; 64, Fleischer, 1932). Infection is by direct spread or by fomites and the incubation period has been estimated as 35 days (Findlay, 1930). The lesions first appear as small pimples and remain discrete, but as a rule they become multiple, spreading widely over

FIGS. 139 and 140.—MOLLUSCUM CONTAGIOSUM (Inst. Ophthal.).

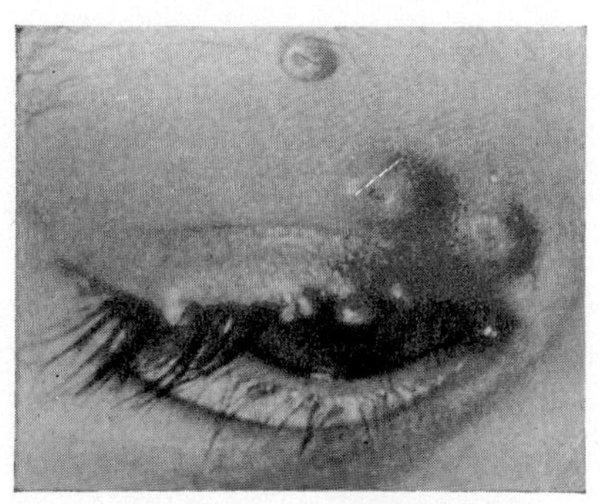

FIG. 139.—In a boy aged 2.

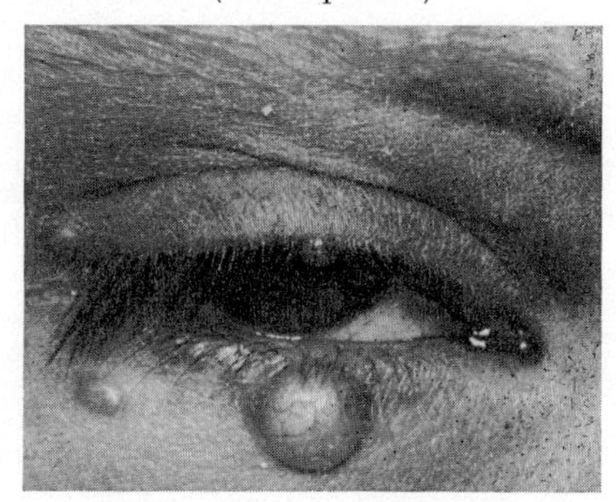

FIG. 140.—In a girl aged 8.

[1] Vol. XIV, p. 1192.

affected areas. Their numbers may be very large, particularly on the face, breasts, genitalia and the inner surfaces of the thighs: thus, in addition to multitudes scattered over the rest of the body, Ebert (1871) noted 15 on the left and 16 on the right lids in a 4-year-old girl, and de Vincentiis (1877) 19 on the upper, 11 on the lower right lid, and 5 on the upper left in a 15-year-old girl; the caruncle has also been affected (Vannas and Lapinleimu, 1967). Auto-inoculation undoubtedly occurs, and contact inoculation is seen in the infection of corresponding sites on each lid of the same eye. The lesions develop slowly from the size of a pin-point and may remain small, hidden by the lashes or under masses of desquamating epithelium, but usually enlarge to become globular in shape with a broad base and a flattened umbilicated top in the centre of which a tiny dark spot becomes apparent. This is said to resemble the hole in a pearl shirt-button and is important diagnostically. Occasionally they reach a giant size on the lid (0·5 cm. in diameter, Lugossy, 1948; occupying the entire lower lid, Maastricht and Gomperts, 1950; on the upper lid, Svestuchina, 1951; 16 mm., Ricci, 1958; 12 × 10 mm., Einaugler and Henkind, 1968). At first firm and solid, they slowly soften, and on being squeezed firmly they exude a waxy-looking material resembling sebum. Sometimes they burst and disappear spontaneously; alternatively they may persist for many months giving rise to no symptoms if they affect the outer skin of the lids only; but sometimes they break down and suppurate, finally healing without scarring (Julianelle and James, 1943; Saubermann, 1948; Mathieu and Henry, 1948; and others).

Histologically the mollusc is formed by hyperplasia of the cells of the rete malpighii whereby the epidermis grows down into the dermis as multiple bean-shaped bodies[1] (Figs. 141–2). The tumour, which is enveloped in a fibrous tissue capsule, is divided somewhat after the manner of an orange by septa into lobules, each of which is formed by a closely packed mass of rete cells. Among these epithelial cells a number suffers a peculiar degeneration, becoming greatly enlarged and assuming a round or oval shape, their nuclei being pushed to the side and destroyed and their cytoplasm being replaced by the molluscum body.

It is interesting that these *molluscum bodies* were at one time regarded as the parasites which cause the malady. This body, an eosinophilic structure, was studied by Goodpasture and King (1927), and by van Rooyen (1938–39); it grows from an elementary body and may eventually form a large oval structure as much as 37μ in length. van Rooyen has shown that the molluscum body is surrounded by a membrane and that it can, by micro-dissection, be removed complete from the cell; it consists of innumerable elementary bodies lying in a protein matrix. The elementary bodies measure about 0·35μ in diameter when calculated by micro-metric extinction methods after staining by a mordant (van Rooyen, 1938; Rake and Blank, 1950; Banfield *et al.*, 1951; Nasemann and Stanka, 1959).

Animal inoculation experiments have failed (Findlay, 1930; Goodpasture and Woodruffe, 1931; van Rooyen and Rhodes, 1940), and specific agglutinins were not

[1] Neisser (1888), Macallum (1892), Kingery (1920), Mescon *et al.* (1954), Lutzner (1963), and others.

FIGS. 141 and 142.—MOLLUSCUM CONTAGIOSUM.

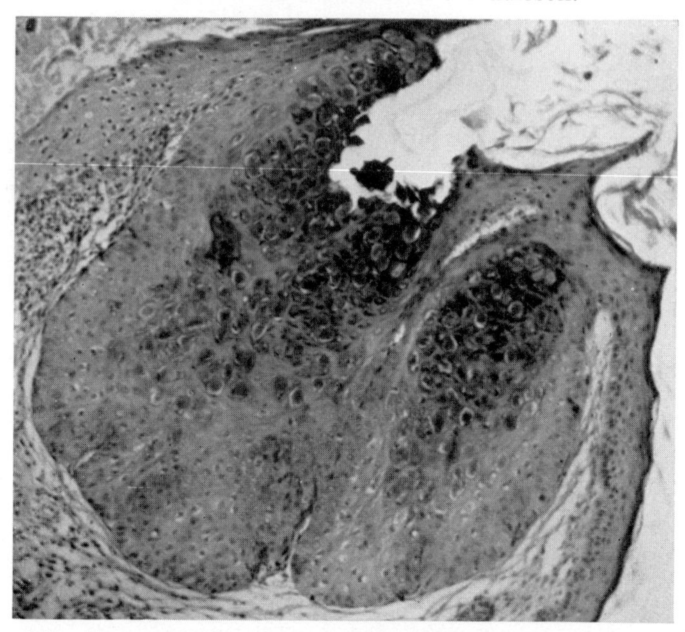

FIG. 141.—Low-power view of part of the tumour showing a lobulated mass of hyperplasic epidermis extending into the dermis. Many inclusion bodies may be seen and they become more numerous in the centre of the lesion. The central indentation on the surface is filled with keratinized epithelial cells containing inclusion bodies (H. and E. × 95) (Norman Ashton).

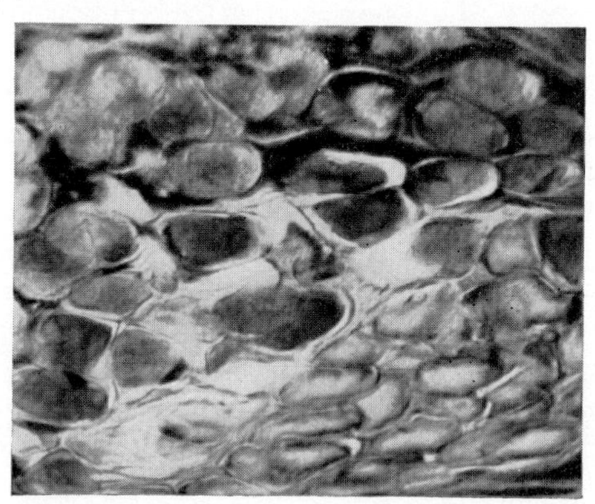

FIG. 142.—The eosinophilic inclusion bodies are highly refractile and oval in shape. The nuclei of the epithelial cells are pushed to one side (H. and E. × 185) (Norman Ashton).

demonstrated in human sera by van Rooyen and Rhodes (1940). The rarity of antibodies may be due to the fact that the virus is confined to the superficial epithelium and does not gain access to the general circulation and the reticulo-endothelial system; Mitchell (1953), however, demonstrated complement-fixing antibodies in the serum of some patients.

The significance of molluscum at the lid-margins and sometimes some distance away (Curtin and Theodore, 1955) is greatly increased by the incidence of ocular complications, due probably to the toxin of the virus, perhaps acting allergically (Bardelli, 1929); the virus itself does not generally invade the conjunctival or corneal epithelium (Thygeson, 1943), and nodules on the conjunctiva are exceptional (Redslob, 1927). A *molluscum conjunctivitis*,[1] differing from the simple conjunctivitis produced by warts, is typically a follicular disease with considerable papillary hypertrophy; it is relatively common and has been mistaken for trachoma, a mistake frequently facilitated by the presence of pannus.[2] A *molluscum keratitis*, on the other hand, is more rare, assuming the form of a superficial punctate keratitis (Nover, 1950; Jones, 1962), phlyctenular keratitis (Elschnig, 1897), a peripheral infiltration (Cavara, 1924) or a corneal ulcer (H. and S. R. Gifford, 1921; Offret and Duperrat, 1938; Mathieu and Henry, 1948). Apart from the clinical appearance of these lesions, their toxic or allergic nature is suggested by the negative response produced by direct inoculation and by their rapid subsidence if the causative lesions on the lid-margins are removed (Thygeson, 1943; Postić, 1959).

Treatment is usually by incising the tumours with a sharp-pointed knife, squeezing out the contents, and swabbing the cavity with carbolic acid, alcohol or tincture of iodine (see Appelmans *et al.*, 1960). Alternatively, simple incision, making sure that the cavity fills with blood-clot, is usually sufficient; fulguration is also effective (Goodman, 1935). There is no tendency to relapse if all the lesions are adequately dealt with, and Singh and Grover (1958) reported the disappearance of all the nodules in a multiple infection after the removal of two.

Appelmans, Michiels and Jamotton. *Bull. Soc. belge Ophtal.*, No. 125, 1018 (1960).
Banfield, Bunting, Strauss and Melnich. *Proc. Soc. exp. Biol. Med.*, **77**, 843 (1951).
Bardelli. *Boll. Oculist.*, **8**, 817 (1929).
Cavara. *Boll. Oculist.*, **3**, 1 (1924).
Atti Accad. Fisiocritici Siena, **3**, 673 (1929).
Curtin and Theodore. *Amer. J. Ophthal.*, **39**, 302 (1955).
Ebert. *Jb. Kinderheilk.*, **3**, 152 (1871).
Einaugler and Henkind. *J. pediat. Ophthal.*, **5**, 201 (1968).
Elschnig. *Wien. klin. Wschr.*, **10**, 943 (1897).
Dtsch. med. Wschr., **34**, 1531 (1908).
Arch. Ophthal., **51**, 237 (1922).
Findlay. *A System of Bacteriology* (*Med. Res. Cncl.*), London, **7**, 248 (1930).
Fleischer. *Klin. Mbl. Augenheilk.*, **88**, 13 (1932).
Gifford, H. and S. R. *Arch. Ophthal.*, **50**, 227 (1921).
Goodman. *Brit. J. Derm.*, **47**, 413 (1935).
Goodpasture and King. *Amer. J. Path.*, **3**, 385 (1927).
Goodpasture and Woodruffe. *Amer. J. Path.*, **7**, 1 (1931).
Jones. *Int. Ophthal. Clin.*, **2**, 591 (1962).
Julianelle and James. *Amer. J. Ophthal.*, **26**, 565 (1943).
Juliusberg. *Dtsch. med. Wschr.*, **31**, 1598 (1905).
Kingery. *Arch. Derm. Syph.* (Chic.), **2**, 144 (1920).
Lauber. *Z. Augenheilk.*, **30**, 246 (1913).
Klin. Mbl. Augenheilk., **52**, 284 (1914).
Lee. *Arch. Ophthal.*, **31**, 64 (1944).
Lugossy. *Ophthalmologica*, **115**, 187 (1948).
Lutzner. *Arch. Derm.*, **87**, 436 (1963).
Maastricht and Gomperts. *Amer. J. Ophthal.*, **33**, 965 (1950).
Macallum. *J. cutan. Dis.*, **10**, 93 (1892).
McCulloch. *Arch. Ophthal.*, **28**, 362 (1942).
Marx and Sticker. *Dtsch. med. Wschr.*, **28**, 893 (1902).
Mathieu and Henry. *Bull. Soc. Ophtal. Paris*, 449 (1948).
Mescon, Gray and Moretti. *J. invest. Derm.*, **23**, 293 (1954).
Mitchell. *Brit. J. exp. Path.*, **34**, 44 (1953).
Mütze. *Arch. Augenheilk.*, **33**, 302 (1896).
Arch. Ophthal., **26**, 15 (1897).
Nasemann and Stanka. *Derm. Wschr.*, **140**, 747 (1959).
Neisser. *Vjschr. Derm.* (Wien), **15**, 553 (1888).

[1] Vol. VIII, p. 376.
[2] Mütze (1896–97), Elschnig (1897–1922), H. and S. R. Gifford (1921), Cavara (1924–29), Nichelatti (1928), Bardelli (1929), Tzanck *et al.* (1938), McCulloch (1942), Lee (1944), Curtin and Theodore (1955), and others.

Nichelatti. *Lettura Oftal.*, **5,** 222 (1928).
Nover. *Klin. Mbl. Augenheilk.*, **117,** 302 (1950).
Offret and Duperrat. *Arch. Ophtal.*, **2,** 993 (1938).
Postić. *Bull. Soc. franç. Ophtal.*, **72,** 539 (1959).
Rake and Blank. *J. invest. Derm.*, **15,** 81 (1950).
Redslob. *Bull. Soc. Ophtal. Paris*, 315 (1927).
Ricci. *Ann. Ottal.*, **84,** 522 (1958).
van Rooyen. *J. Path. Bact.*, **46,** 425 (1938); **49,** 345 (1939).
van Rooyen and Rhodes. *Virus Diseases of Man*, Oxon (1940).
Saubermann. *Ophthalmologica*, **116,** 121 (1948).
Singh and Grover. *J. All-India ophthal. Soc.*, **6,** 37 (1958).
Svestuchina. *Vestn. Oftal.*, **30** (5), 41 (1951).
Thygeson. *Arch. Ophthal.*, **29,** 488 (1943).
Tzanck, Offret and Duperrat. *Bull. Soc. franç. Derm. Syph.*, **45,** 261 (1938).
Vannas and Lapinleimu. *Acta ophthal.* (Kbh.), **45,** 314 (1967).
de Vincentiis. *Movim. med.-chir.* (1877). *See Jber. Ophthal.*, **8,** 346 (1877).
Wile and Kingery. *J. cutan. Dis.*, **37,** 431 (1919).

ORF (ECTHYMA CONTAGIOSUM)

ORF is a contagious, pustular dermatitis occurring in sheep caused by a poxvirus which may be transmitted to man[1] (Fig. 143); the virus concerned has been isolated from man by tissue-culture (Nagington and Whittle, 1961). In man the initial lesion, which may affect the lids, is a large, dark red papule with pronounced central umbilication, firm in consistency and practically painless, which may grow to about 3 cm. in size. The lesion is single and usually heals without treatment in a few weeks and seems to confer an immunity since the infection does not recur. If secondary infection develops a purulent exudate appears and granulations may complicate the picture; in this event after excision or cauterization the lesion shrivels up in a few weeks.

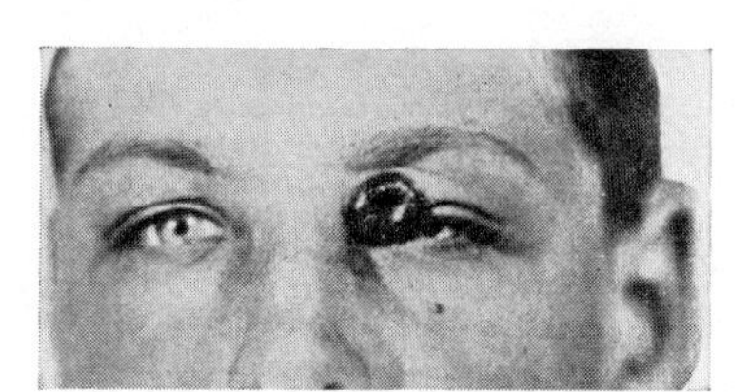

FIG. 143.—ORF (G. A. G. Peterkin).

MILKER'S NODULES

This disease, sometimes known as *pseudo-cowpox* or *paravaccinia*, is frequently confused with cowpox although it is caused by a different poxvirus related to that causing orf. The lesion occurs on the udders of cows as a hemispherical cherry red papule without umbilications, essentially comprised of endothelial proliferation, the cells containing inclusion bodies; thence they may spread to milkers and affect their hands. A case causing a chalazion-like nodule on the margin of the lid with lymphadenopathy was reported by Ludwig (1936) in a milkmaid; such cases must be very rare indeed.

Aynaud. *Ann. Inst. Pasteur*, **37,** 408 (1923).
Blakemore, Abdussalam and Goldsmith. *Brit. J. Derm.*, **60,** 404 (1948).
Kingery and Dahl. *Arch. Derm. Syph.* (Chic.), **51,** 359 (1945).
Lloyd, Macdonald and Glover. *Lancet*, **1,** 720 (1951).
Ludwig. *Arch. Augenheilk.*, **109,** 346 (1936).
Lyell and Miles. *Brit. med. J.*, **2,** 1119 (1950).
Nagington and Whittle. *Brit. med. J.*, **2,** 1324 (1961).
Peterkin. *Brit. J. Derm.*, **49,** 492 (1937).
Wallace. *Brit. J. Derm.*, **59,** 379 (1947).

[1] Aynaud (1923), Peterkin (1937), Kingery and Dahl (1945), Wallace (1947), Blakemore *et al.* (1948), Lyell and Miles (1950), Lloyd *et al.* (1951).

LYMPHOGRANULOMA VENEREUM

LYMPHOGRANULOMA VENEREUM (INGUINALE), an infectious disease transmitted by sexual contact caused by a virus of the PLT group, occasionally affects the region of the eye, usually by venereal contact through the fingers. Conjunctival lesions are the most common, corneal involvement is more rare,[1] but more widespread lesions may appear consisting of proliferations and granulomata involving stenosis of the lacrimal puncta and dacryo-adenopathy associated with a massive and characteristic palpebral œdema which may become consolidated owing to lymphatic blockage (Curth *et al.*, 1940–42) (Fig. 144). In the early stages treatment is effective by the sulphonamides or the broad-spectrum antibiotics, but in the later stages when widespread sclerosis

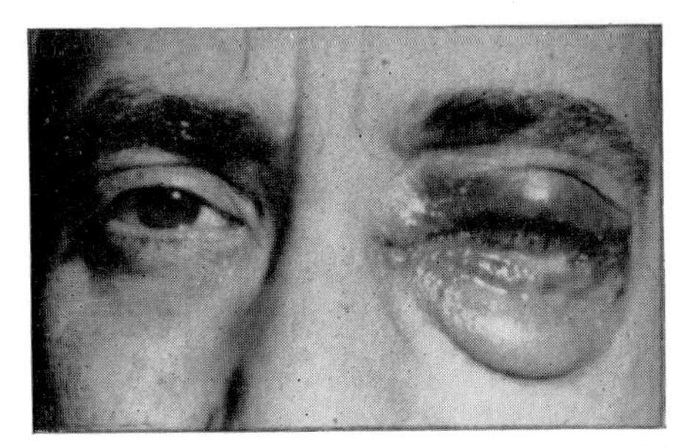

FIG. 144.—LYMPHOGRANULOMA VENEREUM (P. Thygeson).

has occurred, the tissue-changes do not resolve. In such a case wherein the conjunctiva, the lids and the lacrimal gland had been involved for 10 years, Ruíz Barranco and Díaz Rodríguez (1961) found that non-specific measures such as phenylbutazone or protein shock were the most helpful.

Curth, W. and H., and Sanders. *J. Amer. med. Ass.*, **115**, 445 (1940); **118**, 973 (1942).
Ruíz Barranco and Díaz Rodríguez. *Arch. Soc. oftal. hisp.-amer.*, **21**, 501 (1961).

ORNITHOSIS

This viral disease, derived mainly from poultry and pigeons, called psittacosis when occurring in parrots, has been reported as an influenzal type of infection causing a dacryo-adenitis followed by a painful hard œdema of the upper lid with marginal corneal ulcers and scleritis (Kloucek, 1969). The lesion occurred in a woman who worked on a poultry farm. Histological examination of the tissues of the lid showed a chronic perivascular round-cell infiltration mainly in the subcutis.

Kloucek. *Cs. Oftal.*, **25**, 226 (1969).

Dermatitis due to Rickettsiæ

A few diseases due to Rickettsiæ,[2] a group of organisms lying mid-way between bacteria and viruses and conveyed by the bites of ticks or mites, can affect the lids or the neighbouring region of the face where the primary lesion may be incidentally localized. Most of the ticks and mites which act as vectors have a wide geographical distribution and they produce a number of somewhat similar clinical syndromes of the tick-typhus type.

[1] Vol. VIII, p. 305.
[2] Vol. VIII, p. 244; Vol. IX, p. 326.

Some of the ticks are derived from animals while others are found in grass and other vegetation. In all cases the Rickettsiæ are parasitic in the vascular endothelium, causing widespread lesions characterized by a proliferating angiitis throughout most organs of the body, leading to proliferating perivascular nodes, necroses, thromboses and hæmorrhages.

Four groups of Rickettsiæ are usually recognized: the typhus group, causing epidemic typhus, murine typhus and the Brill-Zinsser disease; the spotted fever group, causing Rocky Mountain spotted fever, fièvre boutonneuse (Marseilles fever), South African tick-bite fever and rickettsialpox; the tsutsugamushi group, causing scrub typhus in Malaysia, tsutsugamushi fever in Japan and mite typhus in Indonesia, mainly Sumatra; and Q fever. Among these the following affections of the lids have been recorded.

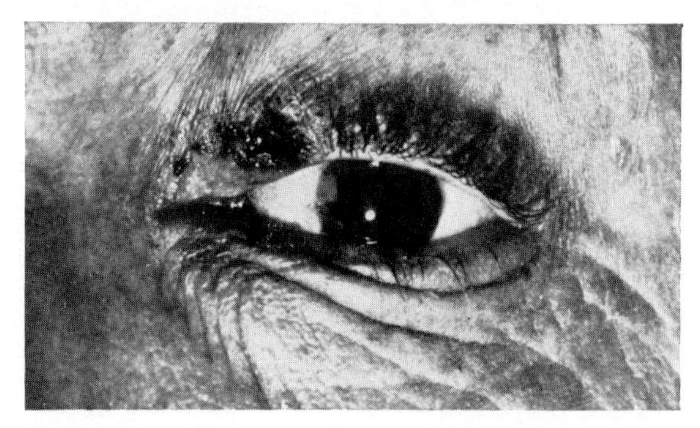

Fig. 145.—Scrub Typhus.

The primary lesion appearing as an eschar resulting from a mite bite involving the upper lid (Harold Scheie).

BOUTONNEUSE FEVER (ESCARRO NODULAIRE: MARSEILLES FEVER), encountered on the Mediterranean littoral, transmitted by *R. conori* conveyed by a tick from a dog, has a primary lesion consisting of small vesicles or weals with a central hæmorrhagic punctum, followed by a fever and a skin eruption; the palpebro-conjunctival reaction may be severe (Conor and Bruch, 1910; Moutinho, 1947; Blatt, 1961).

SOUTH AFRICAN TICK-BITE FEVER, caused by the bites of larval ticks from tall grass, has a similar initial sore with a black necrotic centre and an associated adenitis; it may be followed by a generalized eruption and considerable fever (Pijper and Crocker, 1938).

TSUTSUGAMUSHI FEVER, found in Japan, is caused by *R. nipponica*, and SCRUB TYPHUS, found in Malaysia and the East Indies, by the probably identical organism, *R. orientalis*. The primary lesion, an ulcer with a black eschar and adenopathy, may occur on the lids, and in the febrile stage conjunctivitis and a puffy œdema of the lids are common (Ahlm and Lipshutz, 1944; Scheie, 1947) (Fig. 145).

Q FEVER (" Query fever "),[1] due to *Coxiella burneti*, a rickettsia-like organism, when the infection is in the region of the eye may be associated with gangrene of the lids (Fig. 146) (Calmettes *et al.*, 1960; Perdriel *et al.*, 1961). Cattle and sheep act as carriers of the disease which is of widespread distribution particularly in rural communities.

[1] Vol. VIII, p. 247.

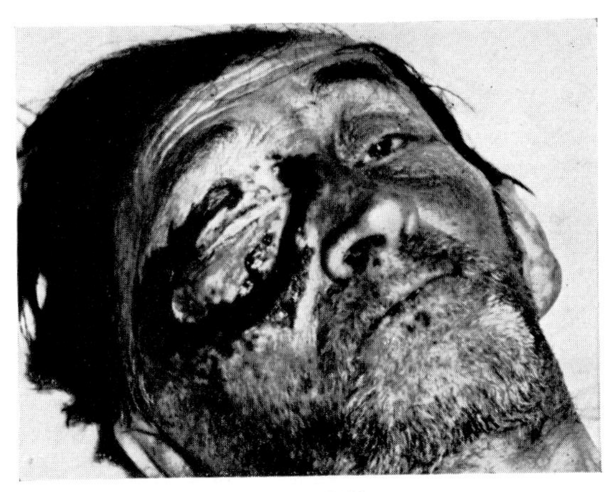

Fig. 146.—Q Fever.

In a man aged 64. To show the widespread ulceration in the region of the lids with pre-auricular adenopathy (L. Calmettes, F. Déodati and P. Bec).

The *treatment* of the rickettsial diseases until recently was unsatisfactory and largely symptomatic, but antibiotics such as chloramphenicol (Smadel and Jackson, 1947) and the tetracyclines (Gear and Harington, 1949; Perdriel *et al.*, 1961) are rapidly curative; this is not so in the case of Q fever.

Ahlm and Lipshutz. *J. Amer. med. Ass.*, **124**, 1095 (1944).

Blatt. *Riv. ital. Tracoma*, **13**, 217 (1961).

Calmettes, Déodati and Bec. *Bull. Soc. franç. Ophtal.*, **73**, 509 (1960).

Conor and Bruch. *Bull. Soc. Path. exot.*, **3**, 492 (1910).

Gear and Harington. *S. Afr. med. J.*, **23**, 507 (1949).

Moutinho. *Bol. Soc. port. Oftal.*, **5**, 163 (1947).

Perdriel, Michel, Guyard and Hoel. *Ann. Oculist.* (Paris), **194**, 957 (1961).

Pijper and Crocker. *S. Afr. med. J.*, **12**, 613 (1938).

Scheie. *Trans. Amer. ophthal. Soc.*, **45**, 637 (1947).

Smadel and Jackson. *Science*, **106**, 418 (1947).

Dermatitis due to Fungal Infections

As would be appropriate in many parts of this *System* we are introducing this subject with the photograph of JULES FRANÇOIS [1907——] (Fig. 147). A native of Gingelom in Belgium, he studied medicine at the University of Louvain and in Paris, and in 1942 became associate professor and since 1948 has been professor of ophthalmology at the University of Ghent. His clinical abilities are exceptional; he has described a large number of syndromes and diseases including the dyscephalic syndrome, the dermo-chondro-corneal syndrome, the malformative syndrome with cryptophthalmos, otomandibular dysostosis, sex-linked myopic chorioretinal heredo-degeneration, fundus flavimaculatus (with Franceschetti), pseudo-papillitis vascularis and several forms of corneal degeneration, as well as furthering the elaboration of new techniques such as electro-oculography, the determination of dark adaptation and the culture of hyalocytes for the replacement of the vitreous. All this is contained in more than 1,200 papers and the astonishing number of 25 books written by himself or with others, among them on gonioscopy (1948), heredity in ophthalmology (1958), congenital cataract (1959), toxoplasmosis and its ocular manifestations (1963), the chorioretinal heredo-degenerations (1963), chromosomal aberrations in ophthalmology (1972), in addition to a review of ocular mycoses (1968), the subject of our present

FIG. 147.—JULES FRANÇOIS
[1907——].

study. These remarkable contributions to our specialty have been amply rewarded: he has been presented with the Favoloro, Donders, von Helmholtz and Gonin Medals, as well as the Medal of the American Society of Contemporary Ophthalmology, honorary degrees from nine universities, the honorary membership of more than 40 ophthalmological or medical societies and of several national academies; he is a past-president of the Royal Academy of Medicine of Belgium. His international activities are exemplified in his holding the Secretaryship of the European Ophthalmological Society and the Presidency of the International Council of Ophthalmology.

The general nature of fungi has already been discussed in another Volume[1]; it will be remembered that they form a sprawling, inchoate group probably comprising more species than are found among green plants. Of these only a few are met with in the eyelids.

It is interesting, however, that fungi are frequently present on the lid-margins (and in the conjunctivæ[2]), the incidence depending largely on environmental factors. Many varieties of saprophytic organisms are found both in the healthy and diseased lid-margins; they are particularly prevalent among agricultural workers and, judging by the demonstration of different species on periodic examinations of the same patients, the contamination would appear to be transient and by chance, derived frequently from fungi present on plants or in the soil, the organisms being air-borne or carried on the hands. Thus in the Indian school-children in California and Arizona, Olson (1969) found an incidence of 41% and 86%.

SPOROTRICHOSIS

SPOROTRICHOSIS (σπόρος, seed; θρίξ, τριχός, a hair) is a disease characterized by granulomatous formations and multiple abscesses in the skin. It is due to a fungus, the *Sporothrix*[3] found on plants, green vegetables (cabbage, lettuce) and grass, whence it is implanted into the human skin through an abrasion or injury. Described by Schenck (1898) in America and de Beurmann (1909) in Europe, it is frequently known as *S. schenkii-beurmanni*.

Infection of the lids is rare; the earliest reports were by Danlos and Blanc (1907), de Beurmann and Gougerot (1907) and Morax and Carlotti (1908), while summaries of the reported cases have been made by Toulant (1913), Wilder and McCullough (1914), Bedell (1914), Gifford (1922), Hill (1930), Gordon (1947), de Rezende (1962) and François and Elewaut-Rysselaere (1968).

The lids may be primarily affected—they have been successfully inoculated by subcutaneous injection of the fungus into a dog (Fischer-Galati, 1914): they may be involved with the lacrimal sac or the orbit (Fage, 1908; Bonnet, 1909; and others) or participate in a facial (Bolanos and Trejos, 1943) or systemic sporotrichosis (King, 1927). When the eyelid is infected one or more indolent nodules usually appear subcutaneously which are hard,

[1] Vol. VIII, p. 385. [2] Vol. VIII, p. 146. [3] Vol. VIII, p. 386.

movable and painless; occasionally this lesion remains indefinitely but usually the skin becomes involved and, turning purple, eventually breaks down to form an indolent ulcer or an abscess discharging yellow pus (Figs. 148–9; Plate III). A sinus formed in this way may simulate a lacrimal fistula (Morax, 1913): several such abscesses may coalesce to form large broken-down ulcers, or adjacent abscesses may communicate with each other through tortuous sinuses. Secondary nodules may appear along the efferent lymphatics and the regional lymph nodes may become swollen and tender; but although there is some fever, pain and systemic disturbance are slight. Early lesions show the histological picture of a non-specific infiltrate composed of neutrophils, lymphocytes, plasma cells and histiocytes (Fetter and Tindall, 1964); in the later stages multiple abscesses and tuberculoid formations occur in the dermis (Carr *et al.*, 1964; Maberry *et al.*,

FIGS. 148 and 149.—OCULAR SPOROTRICHOSIS.
In a 48-year-old woman (D. M. Gordon).

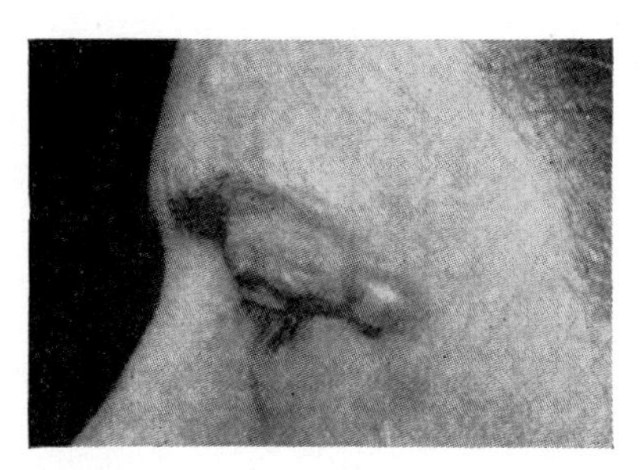

FIG. 148.—The upper lid on the 39th day of the disease.

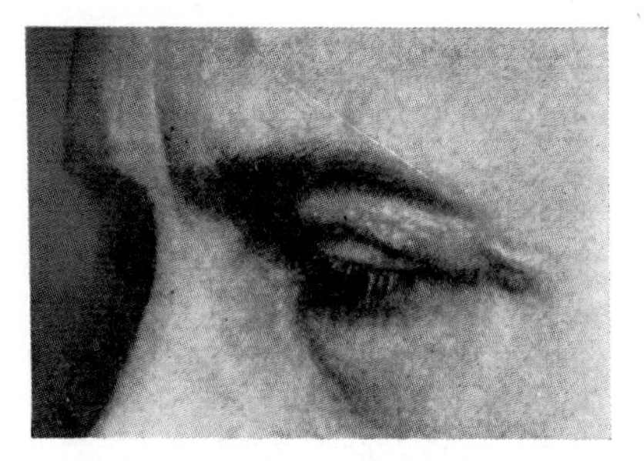

FIG. 149.—The involution of the lesions after 5 weeks' iodide therapy.

1966). The disease shows little tendency to spontaneous resolution and its spread may involve the orbit and its bony walls as well as the globe (Aloin and Vallin, 1920; Hill, 1930); only occasionally does generalized hæmatogenous dissemination end fatally.

In Hill's (1930) case the lesion, starting as a small nodule on the upper lid which was opened with the diagnosis of chalazion, progressed over a period of 5 years to involve the frontal and malar bones as well as the globe with the formation of a hypopyon ulcer. The orbit was exenterated and the antral wall removed, but after 3 months a nodule had to be removed from the canine fossa. Five months later the patient was cured. In this case the response to iodides was poor.

The diagnosis is by recognition of the fungus, and treatment is generally effective with iodine locally and systemically as potassium iodide. Although sulphonamides may be effective against the fungus *in vitro* (Noojin and Callaway, 1944), they are not usually so clinically; better results have been obtained with hydroxystilbamidine or amphotericin B.

PLATE III

Fungal Infections

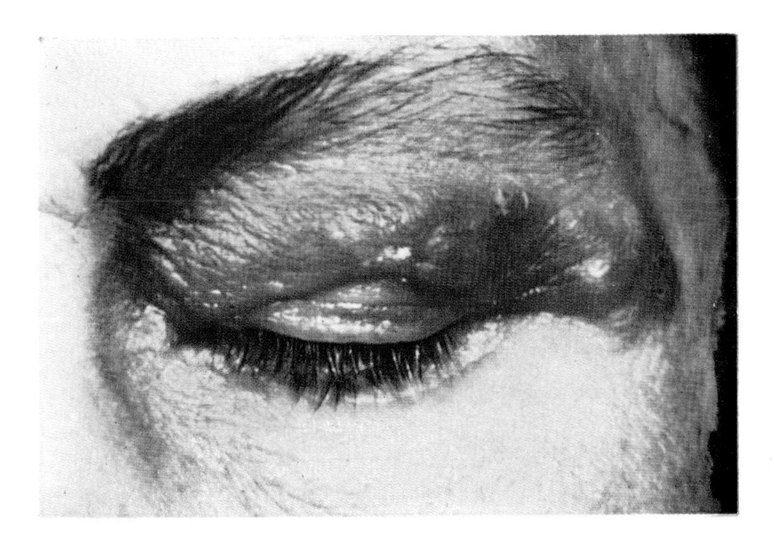

Fig. 1.—Ocular sporotrichosis.
The same case as in Figs. 148–9 (D. M. Gordon).

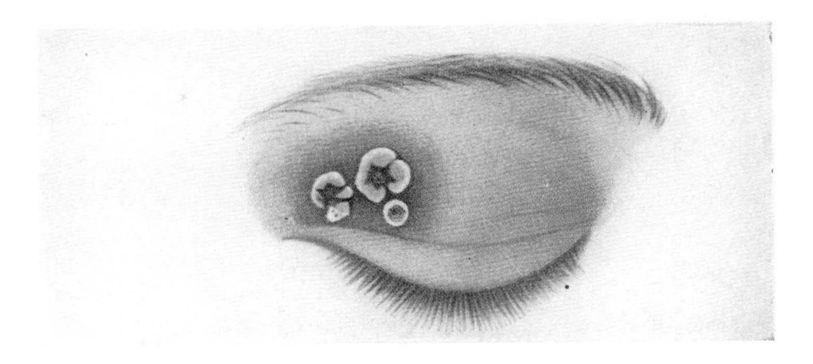

Fig. 2.—Favus.
Treacher Collins's classical illustration of favus of the upper lid. The yellow scutulæ, each with a central depression, are well shown on an erythematous patch.

Aloin and Vallin. *Lyon méd.*, **129,** 859 (1920).
Bedell. *Trans. Amer. ophthal. Soc.*, **13,** 720 (1914).
de Beurmann and Gougerot. *Bull. Soc. méd. Hôp. Paris*, **24,** 302, 309 (1907); **26,** 1046 (1909).
Bolanos and Trejos. *Rev. méd. Costa Rica*, **5,** 369 (1943).
Bonnet. *Lyon Chir.*, **2,** 515 (1909).
Carr, Storkan, Wilson and Swatek. *Arch. Derm.*, **89,** 124 (1964).
Danlos and Blanc. *Bull. Soc. méd. Hôp. Paris*, **24,** 1450 (1907).
Fage. *Progr. méd.*, **24,** 248 (1908).
Fetter and Tindall. *Arch. Path.*, **78,** 613 (1964).
Fischer-Galati. *v. Graefes Arch. Ophthal.*, **87,** 122 (1914).
François and Elewaut-Rysselaere. *Bull. Soc. belge Ophtal.*, No. 148, 1 (1968).
Gifford. *Arch. Ophthal.*, **51,** 540 (1922).
Gordon. *Arch. Ophthal.*, **37,** 56 (1947).
Hill. *Trans. Amer. Acad. Ophthal.*, **35,** 128 (1930).
King. *Sth. med. J.*, **20,** 541 (1927).
Maberry, Mullins and Stone. *Arch. Derm.*, **93,** 65 (1966).
Morax. *Ann. Oculist.* (Paris), **141,** 321 (1909); **149,** 183 (1913).
Morax and Carlotti. *Ann. Oculist.* (Paris), **139,** 418 (1908).
Noojin and Callaway. *Arch. Derm. Syph.* (Chic.), **49,** 305 (1944).
Olson. *Arch. Ophthal.*, **81,** 351 (1969).
de Rezende. *Acta XIX int. Cong. Ophthal.* (New Delhi), **1,** 347 (1962).
Schenck. *Bull. Johns Hopk. Hosp.*, **9,** 286 (1898).
Toulant. *Essai sur la sporotrichose oculaire* (Thèse), Paris (1913).
Wilder and McCullough. *J. Amer. med. Ass.*, **62,** 1156 (1914).

NORTH AMERICAN BLASTOMYCOSIS

In NORTH AMERICAN BLASTOMYCOSIS (BLASTOMYCETIC DERMATITIS, Gilchrist, 1896) (βλαστός, a bud; μύκης, a fungus), caused by a yeast-like fungus, *Blastomyces dermatitidis* (Gilchrist and Stokes, 1898), the infection may be cutaneous, pulmonary or disseminated. It occurs on the North American continent and has also been culturally identified in Africa (Emmons *et al.*, 1964). The cutaneous form of the disease is found chiefly among labourers on the land, producing an indolent, warty, papillomatous lesion (Moore, 1945); as a rule the general health of the patient is not disturbed unless pyæmia develops. The lids are frequently affected, sometimes primarily but more often in the spread of an infection from the cheek,[1] or they may also be involved in systemic blastomycosis (Noojin and Praytor, 1951) (Figs. 150–3). The lesion begins as a small red papule or nodule which slowly becomes pustular and crusts. On the crusts being removed, an elevated irregular base stands out surrounded by a red areola, and this gradually spreads and becomes warty and papillomatous. Sometimes these growths become large and cauliflower-like with irregular clefts bathed in pus, while in the surrounding areola are numerous subepithelial yellow points which on pricking prove to be minute abscesses containing masses of the fungus. Sometimes these lesions break down to form crater-like ulcers with hard raised edges.

Histopathologically there is a non-specific inflammatory infiltrate and in it are numerous organisms, mainly in a budding state (Fig. 154). Meantime, with periods of comparative quiet the disease progresses peripherally while the central area tends to show healing by destructive scarring, leading at times to great deformity and ectropion of the lids (Figs. 152–3). So long as the condition is confined to the skin the symptoms are mild; the patches are not

[1] Ricketts (1901), Gilchrist (1902), Pusey (1903), Wood (1904), Wilder (1904), Jackson (1915), Forsberg and Stern (1929), McKee (1930), Pichette (1949), Blodi and Huffman (1958).

FIGS. 150 to 153.—BLASTOMYCOSIS.

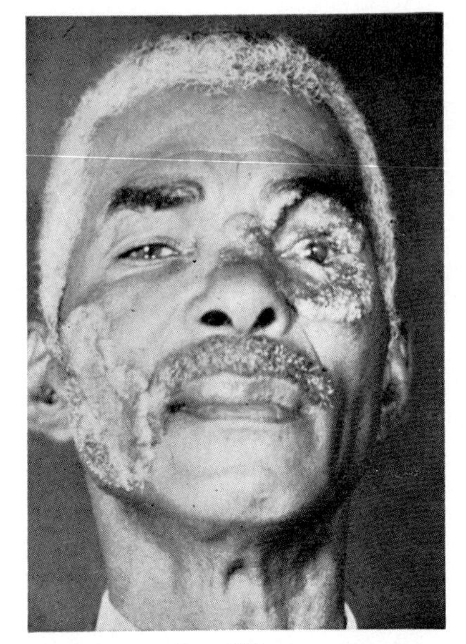

FIG. 150.

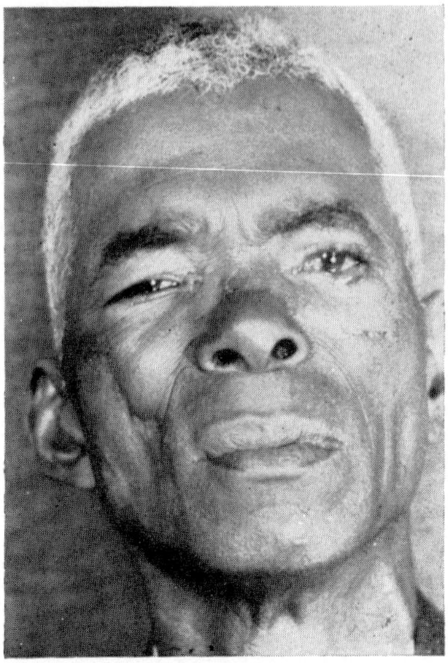

FIG. 151.

FIGS. 150 and 151.—A typical case affecting the face and eyelids (Fig. 150). The appearance of the healed cicatrized lesion after the exhibition of potassium iodide by mouth and local x-ray treatment is seen in Fig. 151 (W. L. Dobes).

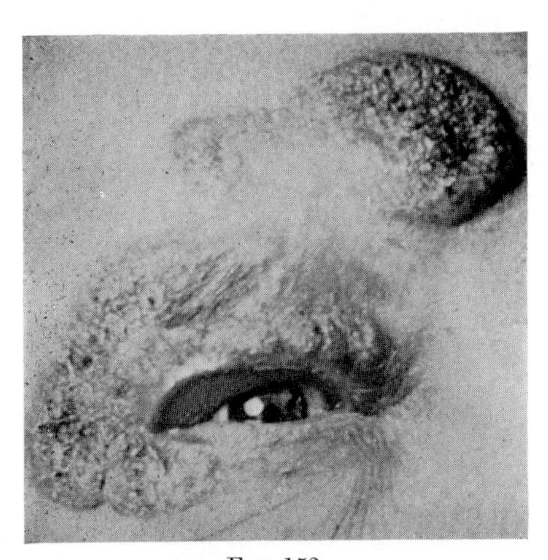

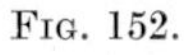

FIG. 152.

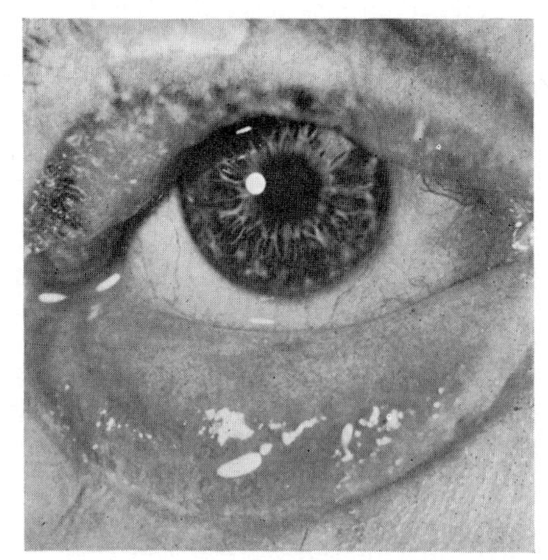

FIG. 153.

Affecting all four lids of a man aged 57 (Fig. 152); the lesions had been present for 20 years. The condition was treated with stilbamidine isethionate; one year later there was marked cicatricial ectropion which was surgically repaired (Fig. 153) (F. C. Blodi and W. C. Huffman).

sensitive to pressure, pain is minimal and the general health is maintained. If, however, the disease becomes disseminated, generalized symptoms become acute and the prognosis without effective treatment is poor.

Diagnosis is effected by recognition of the fungus. Treatment used to be by heroic doses of iodide (up to 400 g. per day of potassium iodide, or in small doses by mouth combined with sodium iodide intravenously). The introduction of stilbamidine and the less toxic 2-hydroxystilbamidine (225 mg. in saline infused intravenously over a 3-hour period each day for 30 days until a total dose of 8 to 12 g. has been administered, Schoenbach *et al.*, 1952) has revolutionized the prognosis, as also has the fungicide, amphotericin B. The reconstruction of the ectropionized lids may require extensive skin-grafting.

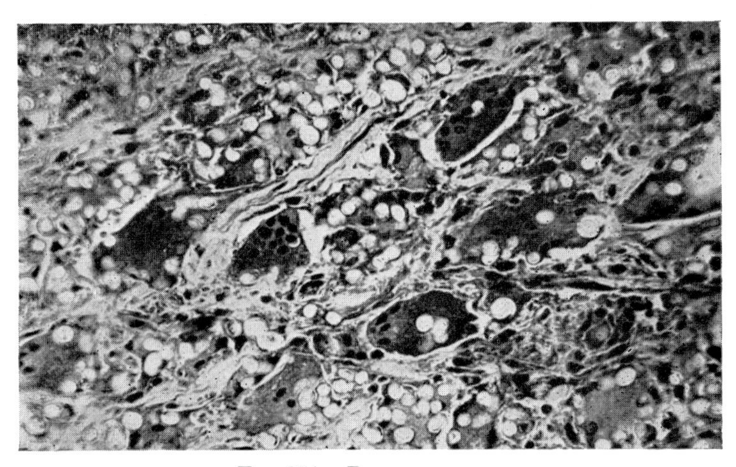

FIG. 154.—BLASTOMYCOSIS.

Chronic granulomatous inflammatory tissue due to blastomycosis. Photomicrograph shows many spherical, refractile blastomycetes, some of which have been engulfed by giant cells (N. Ashton).

Blodi and Huffman. *Arch. Ophthal.*, **59**, 459 (1958).
Emmons, Murray, Lurie *et al.* *Sabouraudia*, **3**, 306 (1964).
Forsberg and Stern. *Journal-Lancet*, **49**, 498 (1929).
Gilchrist. *Johns Hopk. Hosp. Bull.*, **1**, 269 (1896).
Brit. med. J., **2**, 1321 (1902).
Gilchrist and Stokes. *J. exp. Med.*, **3**, 53 (1898).
Jackson. *J. Amer. med. Ass.*, **65**, 23 (1915).
McKee. *Arch. Ophthal.*, **3**, 301 (1930).
Moore. *J. invest. Derm.*, **6**, 149 (1945).
Noojin and Praytor. *J. Amer. med. Ass.*, **147**, 749 (1951).
Pichette. *Trans. Canad. ophthal. Soc.*, **12**, 111 (1949).
Pusey. *J. cutan. Dis.*, **21**, 223 (1903).
Ricketts. *J. med. Res.*, **6**, 373 (1901).
Schoenbach, Miller and Long. *Ann. intern. Med.*, **37**, 31 (1952).
Wilder. *J. Amer. med. Ass.*, **43**, 2026 (1904).
Wood. *Ann. Ophthal.*, **13**, 92 (1904).

CRYPTOCOCCOSIS (TORULOSIS: EUROPEAN BLASTOMYCOSIS)

The *Cryptococcus neoformans* may cause a disseminated infection, in which the uvea may share[1] as well as the meninges and lungs, in which case the disease is frequently fatal, or it may be purely cutaneous. The lids may be involved in either

[1] Vol. IX, p. 400.

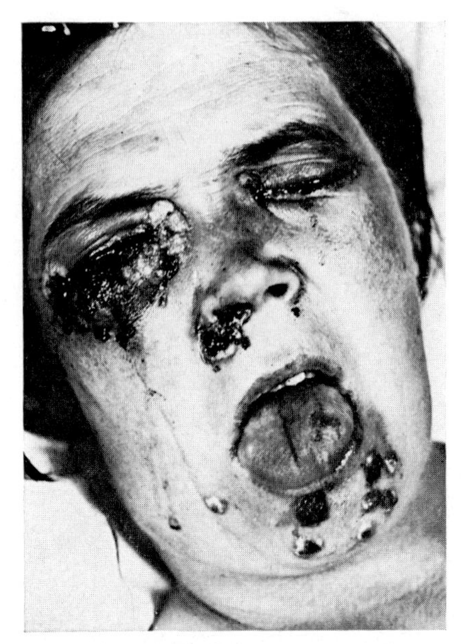

FIG. 155.—CRYPTOCOCCOSIS.

In a woman aged 37 who had nasal catarrh for a month and had developed symptoms of cryptococcal dissemination for a week. In both eyes but more markedly in the right she developed a metastatic ophthalmitis and episcleritis (W. Kreibig).

form as multiple ulcers in the generalized type (Wilson and Plunkett, 1965) or localized nodular or ulcerating lesions on the lids whence the infection may spread to the face (Dósa, 1938; Kreibig, 1940; Fedukowicz, 1963) (Fig. 155). Treatment is by amphotericin B which is successful in some 80% of cases.

Dósa. *Orv. Hetil.*, **82,** 815 (1938).
Fedukowicz. *External Infections of the Eye*, N.Y. (1963).
Kreibig. *Klin. Mbl. Angenheilk.*, **104,** 64 (1940).
Wilson and Plunkett. *The Fungus Diseases of Man*, Calif. (1965).

PARACOCCIDIOIDOMYCOSIS

PARACOCCIDIOIDOMYCOSIS (PARACOCCIDIOIDAL GRANULOMA; SOUTH AMERICAN BLASTOMYCOSIS) was long considered identical with coccidioidal granuloma but is now known to be due to a different but related organism, *Paracoccidioides brasiliensis* (Jordan and Weidman, 1936). The disease occurs particularly in Brazil, and being of vegetable origin the fungus affects preferentially the mouth and exposed parts of the face from which it spreads by continuity to the lids or by dissemination and, unless adequately treated, is frequently fatal (Harrell and Curtis, 1959; Wilson, 1961; de Rezende, 1962; Furtado, 1963). The lesion starts as a furuncle which ulcerates, becomes verrucose and pustulates. Treatment is by amphotericin B.

Furtado. *Derm. tropica*, **2,** 27 (1963).
Harrell and Curtis. *Amer. J. Med.*, **27,** 750 (1959).
Jordan and Weidman. *Arch. Derm. Syph.* (Chic.), **33,** 31 (1936).
de Rezende. *Acta XIX int. Cong. Ophthal.* (New Delhi), **1,** 342 (1962).
Wilson. *Arch. intern. Med.*, **108,** 292 (1961).

COCCIDIOIDOMYCOSIS

COCCIDIOIDAL GRANULOMA, due to infection by the *Coccidioides immitis*[1] is somewhat rare in the lids (Fig. 156). The disease occurs typically in California, Texas and Mexico (*valley* or *desert fever*, Dickson, 1937). The cutaneous lesions may be primary, but more frequently they are secondary to a generalized infection; in this case the disease may remain virtually asymptomatic or attack the lungs, the internal organs and the central nervous system when it may be rapidly fatal.[2]

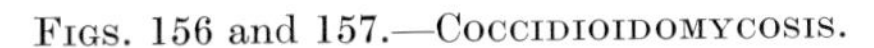

FIGS. 156 and 157.—COCCIDIOIDOMYCOSIS.

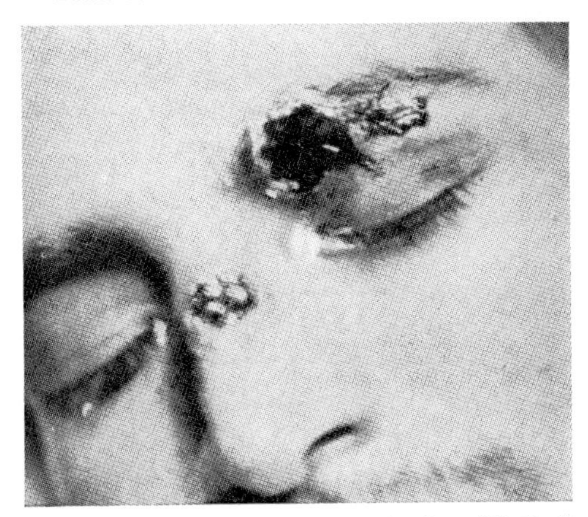

FIG. 156.—Coccidioidal granulomata of the face (H. P. Jacobson).

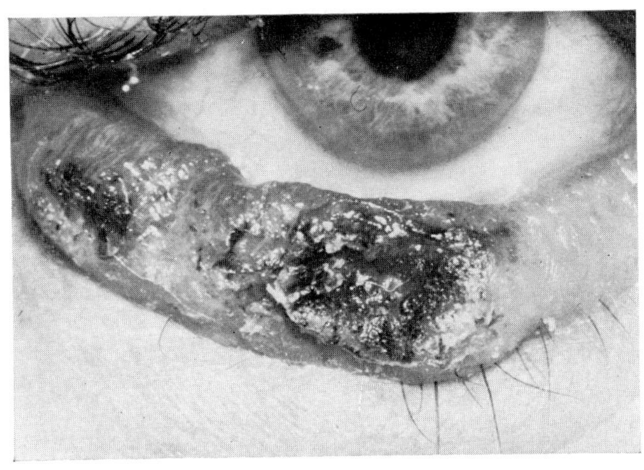

FIG. 157.—In a Californian man aged 73 who developed generalized coccidioidomycosis. The granuloma is seen in the lower lid (A. Ray Irvine).

[1] Vol. VIII, p. 390.
[2] Farnell and Starr (1917), Zeisler (1932), Smith *et al.* (1946), Trowbridge (1952), Irvine (1968).

The initial cutaneous lesion is a dusky red nodule which ulcerates and exudes pus rich in the fungus, and eventually develops into an ulcerating and purulent granulomatous mass somewhat resembling blastomycosis, and frequently involving considerable destruction of tissue (Figs. 157–8). In other cases, particularly when of metastatic origin, subcutaneous lesions occur, appearing as flaccid tumours, commencing as cellular infiltrations and ending as abscesses with sinus-formation or breaking down in deep ulceration like a gumma. Finally, involvement of the lymph nodes may initiate lesions resembling scrofuloderma[1] (Jacobson, 1928–30).

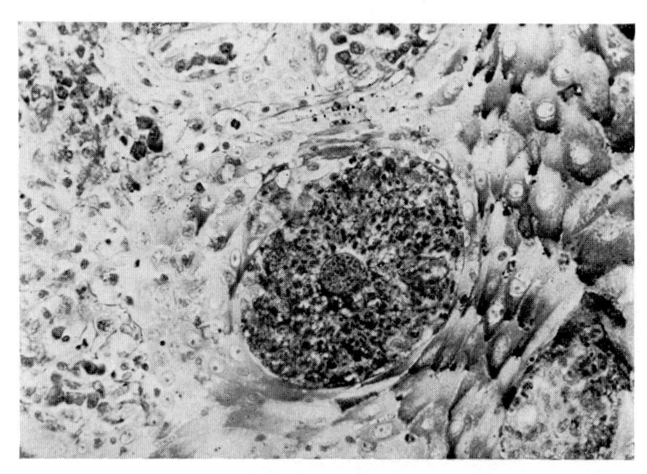

FIG. 158.—COCCIDIOIDAL GRANULOMA.

From the case seen in Fig. 157. Section from the lid showing a typical spherule in the centre of a micro-abscess (× 100) (A. Ray Irvine; photographed by his "Man Friday", Zolton Yuhasz).

Treatment has been unsatisfactory and many remedies were proposed until the advent of chloramphenicol or, more effective, amphotericin B which has received favourable reports (Cohen and Miller, 1950; Skipworth *et al.*, 1960; Faulkner, 1962; Irvine, 1968).

Cohen and Miller. *Ann. West. Med. Surg.*, **4,** 342 (1950).

Dickson. *Arch. intern. Med.*, **59,** 1029 (1937). *Calif. West. Med.*, **47,** 151 (1937).

Farnell and Starr. *Bost. med. surg. J.*, **176,** 771 (1917).

Faulkner. *Amer. J. Ophthal.*, **53,** 822 (1962).

Irvine. *Trans. Amer. Acad. Ophthal.*, **72,** 751 (1968).

Jacobson. *Arch. Derm. Syph.* (Chic.), **18,** 562 (1928); **21,** 790 (1930).

Smith, Beard, Rosenberger and Whiting. *J. Amer. med. Ass.*, **132,** 833 (1946).

Skipworth, Bergin and Williams. *Arch. Derm.*, **82,** 605 (1960).

Trowbridge. *Trans. Pac. Cst. oto-ophthal. Soc.*, **32,** 229 (1952).

Zeisler. *Arch. Derm. Syph.* (Chic.), **25,** 52 (1932).

MYCETOMA

MYCETOMA (MADUROMYCOSIS), caused by various species of fungi or actinomycetes, has only rarely been found on the lids, but was reported from Egypt (Aldridge and

[1] p. 108.

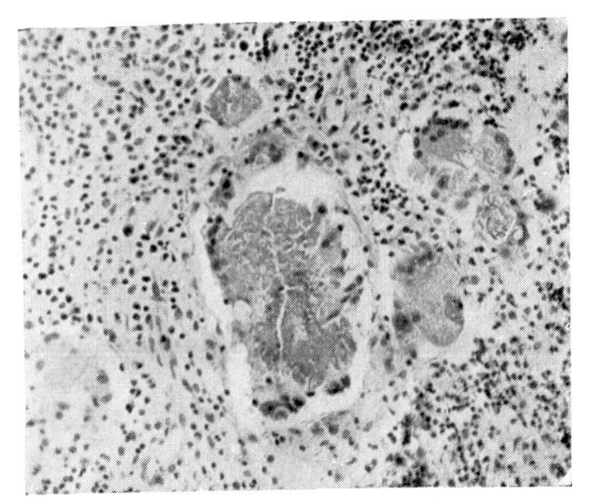

FIG. 159.—MYCETOMA OF THE EYELID.

In a Nubian girl aged 4 with a tumour of the left upper eyelid of 2 months' duration. The section of the granuloma shows the grains of the fungus (× 200) (J. S. Aldridge and R. Kirk).

Kirk, 1940). Occurring as a lump on the upper lid, the tissue showed the usual histological changes of chronic inflammation associated with fungal infection—plasma cell and leucocytic infiltration, giant-cell formations and fibrosis with the presence of irregularly distributed grains of the fungus (Fig. 159). Sulphonamides (Dixon, 1941; Calero, 1947), penicillin (Twining *et al.*, 1946) and the tetracyclines have been claimed to exercise some beneficial effect although the therapeutic action is apparently not dramatic. The treatment is thus unsatisfactory, but some forms respond to amphotericin B.

Aldridge and Kirk. *Brit. J. Ophthal.*, **24,** 211 (1940).
Calero. *Arch. Derm. Syph.* (Chic.), **55,** 761 (1947).
Dixon. *Virg. med. Mthly.*, **68,** 281 (1941).
Twining, Dixon and Weidman. *Bull. U.S. Army Med. Rep.*, **46,** 417 (1946).

RHINOSPORIDIOSIS

The *Rhinosporidium seeberi* (Seeber, 1900) or *kinealii* (O'Kinealy, 1903) (ῥίς, ῥινος nose; σπόρος, seed), a member of the Phycomycetes which has not yet been isolated in culture, is not commonly found in man—cases have been reported mainly in India, South Africa and sporadically in the United States (Ashworth, 1923; Karunaratne, 1936; Kaye, 1938; Sharma *et al.*, 1958). The great majority of lesions occurs in the anterior nares, but sporadic cases have been reported occurring in the pharynx, larynx, ear, lacrimal sac, conjunctiva[1] and the skin; some 4% of the lesions occur on the skin of the lids (Kuriakose, 1963). It is probable that the cutaneous lesions are inoculations from foci located elsewhere (Allen and Dave, 1936).

The lesions, which are painless, appear as masses of exuberant villous polypi, pedunculated or sessile, sometimes of enormous dimensions but showing no infiltration at the base (Figs. 160–3). They are composed of granulation tissue with plasma cells and giant cells in which are cysts of mucoid matter (sporangia) containing the fungus spherules; these morulæ can sometimes be seen discharging in enormous masses into the surrounding granulation tissue (Fig. 164). Treatment is by excision at the base, after ligation, followed by cauterization of the site by 2·0% silver nitrate;

[1] Vol. VIII, p. 391.

FIGS. 160 to 163.—RHINOSPORIDIOSIS.

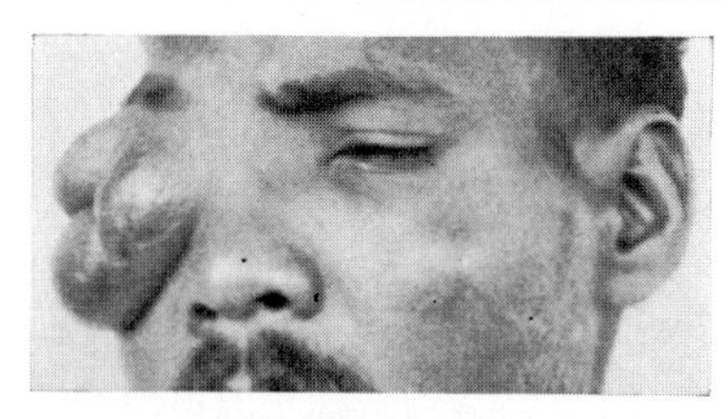

FIG. 160.—In an Indian with a large mass involving the whole of the right lower lid, measuring 8 cm. at its largest diameter and entirely covering the globe (K. D. Sharma *et al.*).

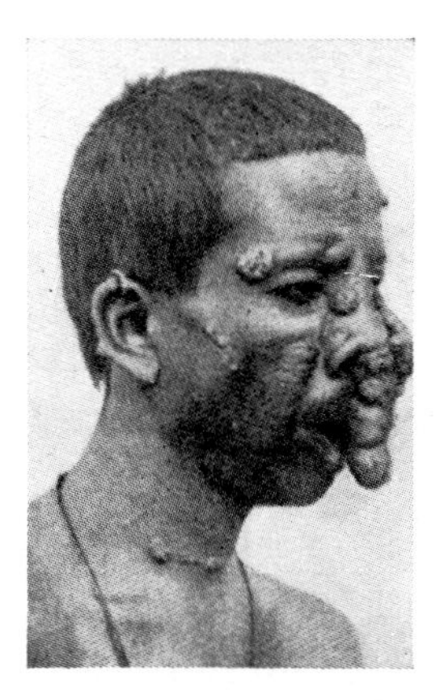

FIG. 161.

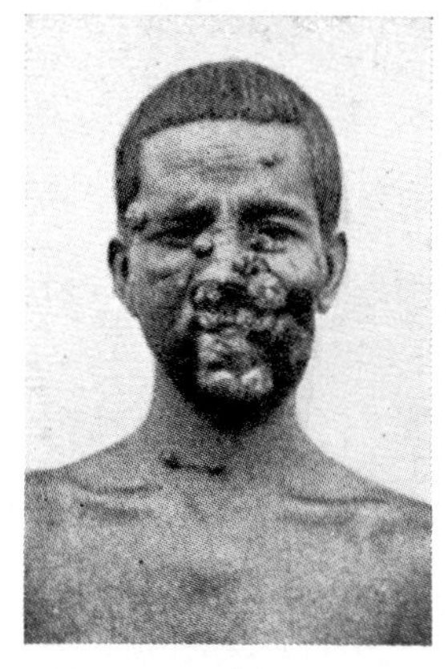

FIG. 162.

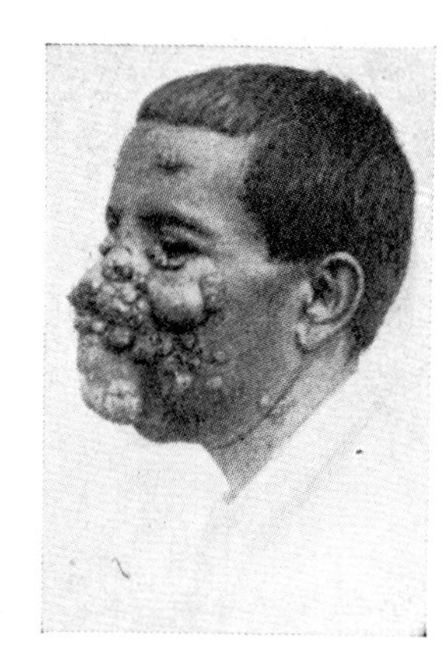

FIG. 163.

FIGS. 161 to 163.—In an Indian man showing exuberant lesions on the lids, nose and face (F. R. Allen and M. L. Dave).

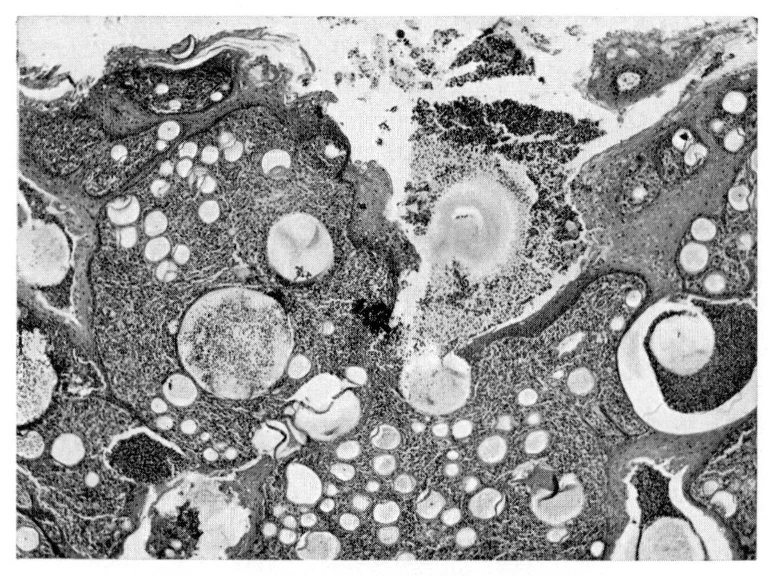

FIG. 164.—RHINOSPORIDIOSIS OF THE LID.

From a polyp on the inner surface of the lid. The section shows a mature sporangium which has burst and is discharging its spores (H. Kaye).

if this is not practicable, painting the lesions with 2·0% tartrated antimony solution is effective when continued for some weeks.

Allen and Dave. *Ind. med. Gaz.*, **71**, 376 (1936).
Ashworth. *Proc. roy. Soc. Edinb.*, **53**, 301 (1923).
Karunaratne. *J. Path. Bact.*, **42**, 192 (1936).
Kaye. *Brit. J. Ophthal.*, **22**, 449 (1938).
Kuriakose. *Brit. J. Ophthal.*, **47**, 346 (1963).
O'Kinealy. *Proc. laryng. Soc. Lond.*, **10**, 109 (1903).
Seeber. *Tesis Univ. Nac. de B. Aires* (1900).
Sharma, Shrivastav and Agarwal. *Brit. J. Ophthal.*, **42**, 572 (1958).

ASPERGILLOSIS

Although corneal lesions are common[1], infection of the lids by this genus of organism, usually by *Aspergillus fumigatus*, is rare, but granulomatous lesions of a chronic type have been reported wherein the fungus has been isolated (Bennett *et al.*, 1962; Timm, 1963; Harrell *et al.*, 1966) (Figs. 165–7). Treatment is difficult and resort has been made to excision followed by skin-grafting; Harrell and his co-workers, however, reported the disappearance of an isolated lesion on the lid after the local injection of amphotericin B.

FIGS. 165 to 167.—ASPERGILLOSIS OF THE EYELID (E. R. Harrell, J. R. Wolter and R. F. Gutow).

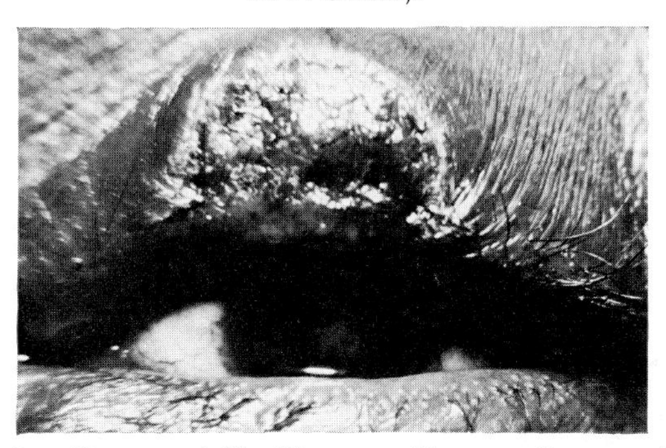

FIG. 165.—In a Negro aged 27. The aspergilloma on the upper lid before treatment.

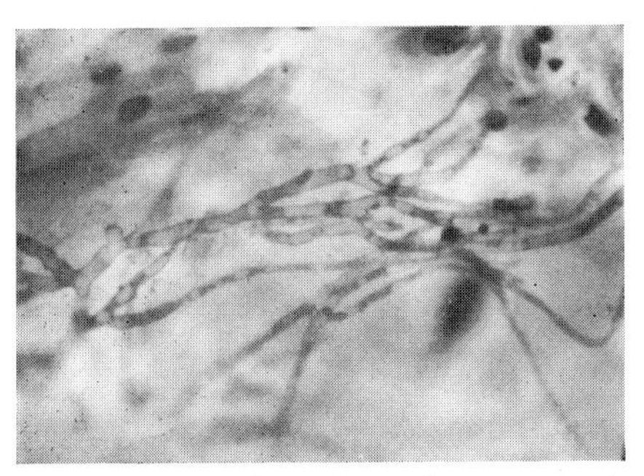

FIG. 166.—A biopsy from the case seen in Fig. 165 showing the hyphal growth of the fungi (PAS stain).

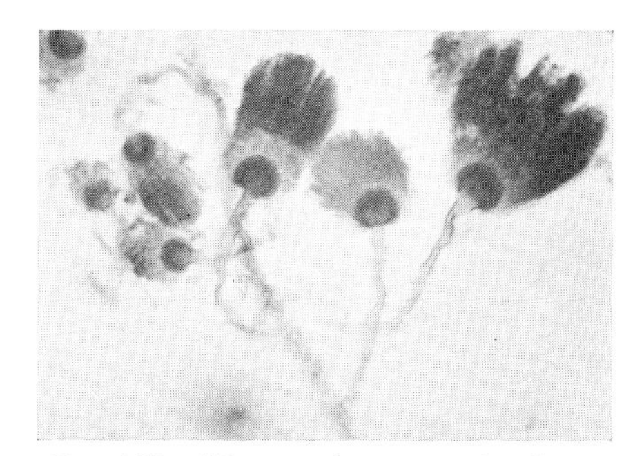

FIG. 167.—Microscopic preparation from a culture showing groups of spores arising from the tips of sterigmata typical of *Aspergillus fumigatus* (× 800).

[1] Vol. VIII, p. 795.

Cases of blepharitis have been reported with infection by *A. niger* by Rosenvold (1942), Ibrahim *et al.* (1964) and Mostafa (1966). Nystatin given systemically has cleared up the condition.

Bennett, Kirby and Blocker. *Plast. reconstr. Surg.*, **29**, 684 (1962).
Harrell, Wolter and Gutow. *Arch. Ophthal.*, **76**, 322 (1966).
Ibrahim, El-Gammal, Mostafa and Zaki. *Bull. ophthal. Soc. Egypt*, **57**, 93 (1964).
Mostafa. *Amer. J. Ophthal.*, **62**, 1204 (1966).
Rosenvold. *Amer. J. Ophthal.*, **25**, 588 (1942).
Timm. *Ophthalmologica*, **146**, 250 (1963).

CANDIDIASIS (MONILIASIS; THRUSH)

The Candida form a heterogeneous group of fungi of which the commonest member is *Candida* (*Monilia* or *Oidium*) *albicans* which causes THRUSH.[1] The disease occurs essentially in children and debilitated adults in whom the mouth is the common site of infection, whence it may be carried to the face, including the lids and conjunctiva,[2] the perineum, vulva, thighs, interdigital

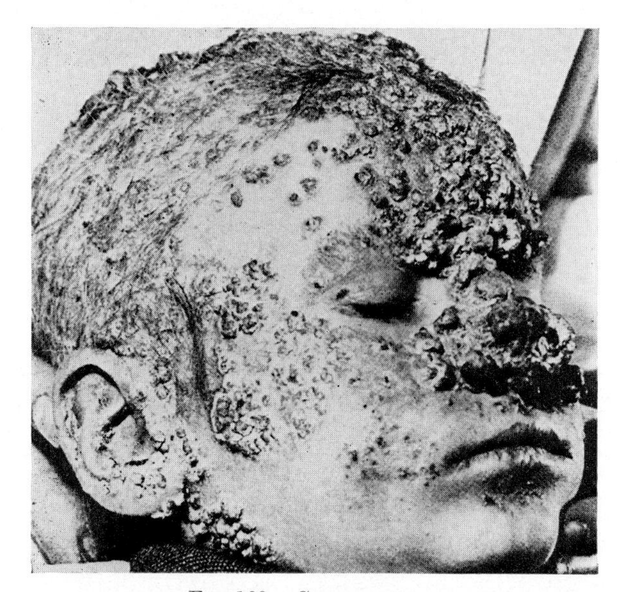

FIG. 168.—CANDIDIASIS.

A generalized case affecting the face and cranium (after J. Alkiewicz; J. François and M. Elewaut-Rysselaere).

spaces, nails and breasts. Occasionally it becomes generalized and even fatal (Fig. 168). A pseudo-tumour of the lids has been reported in a case of generalized candidiasis (Black and Eddy, 1937). Primary infection of the lids is rare (Tanaka, 1952; François and Elewaut-Rysselaere, 1968); secondary infection from the skin and conjunctiva is more common (Erdös,

[1] It is interesting that the fungus causing thrush was the first important human pathogen to be seen; it was observed by Langenbeck in 1839 and named *Oidium albicans* by Charles Robin in 1843. Schönlein discovered the organism causing favus in the same year as Langenbeck's observation.

[2] Vol VIII, p. 394.

1935; Privat and Benne, 1962). A case has been reported in a boy aged 13 after prolonged systemic treatment by antibiotics (Vozza and Bagolini, 1964). In the mouth there is a white, slightly adherent, membranous deposit on a reddened base; on the lids the milder cases show scaly, grey or reddish, definitely marginated lesions, while in the more severe cases vesicles develop which rupture leaving oozing areas and small pustules enclosed by undermined epidermis (Greenbaum and Klauder, 1922; Shelmire, 1925).

The diagnosis is confirmed by the culture of the fungus from the scales. Topical treatment to a lesion of the lids is by nystatin, while a generalized infection is best treated by this drug or amphotericin B. Coincident treatment of the stomatitis and the general condition is, of course, of the utmost importance and cessation of any antibiotic treatment if this is practicable.

Black and Eddy. *J. lab. clin. Med.*, **22**, 584 (1937).
Erdös. *Szemészet*, **70**, 82 (1935).
François and Elewaut-Rysselaere. *Bull. Soc. belge Ophtal.*, No. 148, 1 (1968).
Greenbaum and Klauder. *Arch. Derm. Syph.* (Chic.), **5**, 332 (1922).
Privat and Benne. *Bull. Soc. franç. Derm. Syph.*, **69**, 142 (1962).
Shelmire. *Arch. Derm. Syph.* (Chic.), **12**, 789 (1925).
Tanaka. *Acta Soc. ophthal. jap.*, **56**, 653 (1952).
Vozza and Bagolini. *Boll. Oculist.*, **43**, 433 (1964).

TINEA: RINGWORM

RINGWORM affecting the skin of the lids is caused by a considerable variety of fungi (Dermatophytes) which can be conveniently divided into two types—small- and large-spored. Each of these has many sub-varieties.

In the *Microspora* the mycelium is irregularly jointed and branching with fusiform swellings, and the " spores " are round or oval and do not hang together in

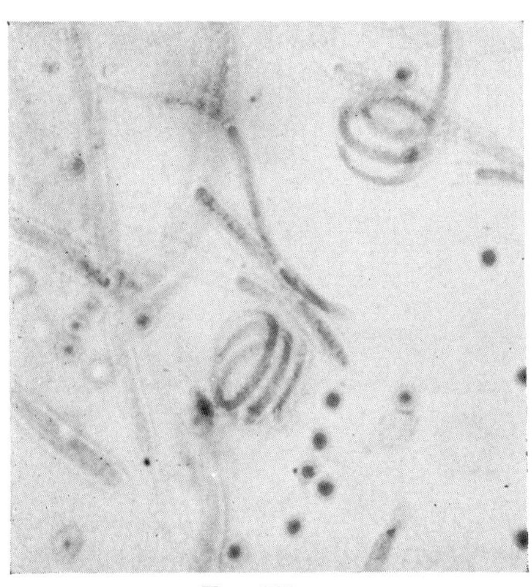

FIG. 169.

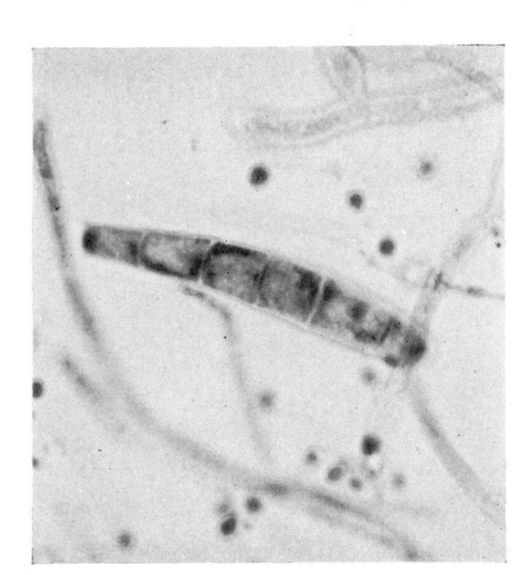

FIG. 170.

FIGS. 169 and 170.—*TRICHOPHYTON MENTAGROPHYTES.*

Showing coiled hyphæ (Fig. 169) and a macroconidium and microconidia (Fig. 170) (H. B. Ostler, M. Okumoto and C. Halde).

chains. The most common variety met with in children is *Microsporum audouini* which only affects human beings or *M. canis* derived from dogs or cats. These organisms are responsible for most cases of ringworm of the scalp.

The *Trichophyta* or *Macrospora*, on the other hand, have hyphæ divided into short segments of equal length and the spores are large, squarish and hang together in chains; these are responsible for much of the ringworm of the glabrous skin in persons of all ages and for most cases of adult ringworm in all sites; certain types (*T. mentagrophytes* and *T. verrucosum*) are derived from horses and cattle (see Weidman, 1927; White, 1927; Blaisdell, 1930; Shaw, 1934; and others) (Figs. 169–70).

A third variety, *Epidermophyton*, which infests humans, is found on the skin of the groin and feet.

A related organism is *Trichophyton* (*Achorion*) *schoenleini* causing TINEA FAVOSA (FAVUS); it is a mycelium of rod-like segments some of which bear spores. PITYRIASIS (TINEA) VERSICOLOR, characterized by brawny scales and affecting mainly the upper trunk, is caused by *Malassezia furfur* and has been associated with the *Pityrosporum orbiculare*.[1]

The most common involvement of the lids in ringworm is TINEA CORPORIS (CIRCINATA) (L. *tinea*, a worm) or TRICHOPHYTOSIS (*θρίξ*, a hair; *φυτόν*, a plant); this usually affects the lids secondarily by spread from the face or occurs in association with patches elsewhere on the body as on the feet. The infection starts as a red spot from which centrifugal extension spreads peripherally as a squamous, fawn-coloured plaque, while the central area heals (Fig. 171). This mode of spread leads to a ringed shape and, as the lesions continually multiply, coalescence of neighbouring lesions results in a polycyclic formation. The infection is superficial and the inflammatory reaction is slight although pustular points may occur. Alternatively, a widespread lesion resembling a pyoderma may involve the brow and both eyelids although the lashes, the hair of the eyebrows and the conjunctiva may escape (Ostler *et al.*, 1971) (Fig. 173). The formation of actual ulcers on the lids is rare but may occur (Wirtz, 1922). The hairs of the eyebrows and cilia, however, may be lost, the lids are usually œdematous and red, particularly at the ciliary margin and sometimes round the orbital rim, and the proximal lymph nodes may be enlarged.[2]

Ringworm of the hairy regions attacks the lids more rarely but the hairs of the brows and lashes may be involved in ringworm of the scalp (TINEA CAPITIS) or of the beard (TINEA BARBÆ).[3] The infection, however, is relatively rare (1 case of infection of the lashes in 560 cases of infection of the scalp; Montgomery and Walzer, 1942). There is a folliculitis of the hair of the brows and lashes, leading in the latter case to a purulent blepharitis, involving in the early stages some œdema of the lids and in the later some

[1] p. 267.

[2] Snell (1902), Lesczynski (1923), Louste and Baranger (1924), Pais (1924), Ballarini (1925), Behdjet (1927), Pignot (1930), Ruete (1930), and others.

[3] Mibelli (1894), Voerner (1901), Lourier and Zwitkis (1928), Elschnig (1928), Arievic (1930), Davidson and Gregory (1932), Weidman (1937), Costa (1943), Silvers (1944), Muskatblit and Targan (1945), Franks *et al.* (1950), Franks and Mandel (1950), Hopkins and Krause (1950), Mariotti (1951), Latte and Piredda (1954), Pimentel and Machado (1956), Bonanni (1957), Lisch (1958), D. H. Hoffmann (1965).

FIGS. 171 to 173.—RINGWORM INFECTIONS OF THE LIDS.

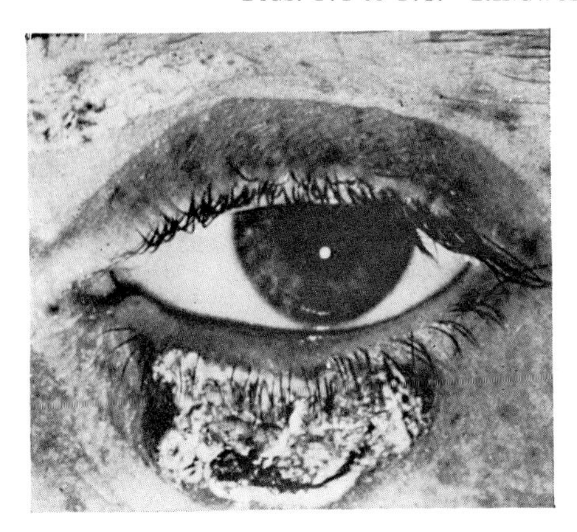

FIG. 171.—Infection of the lower lid and under part of the eyebrow by *Trichophyton quinckeanum* (D. H. Hoffmann).

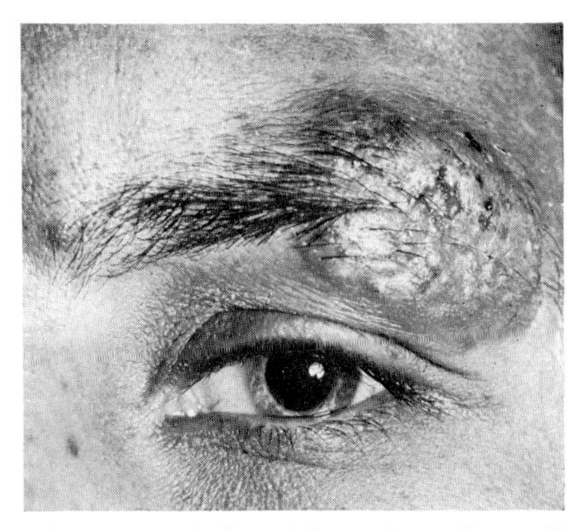

FIG. 172.—Kerion of the eyebrow due to *T. verrucosum*. Note the temporal loss of the eyebrow (P. Borrie, St. Bartholomew's Hosp.).

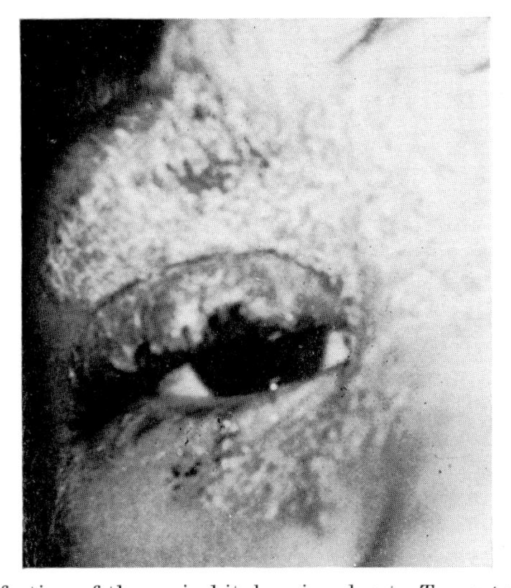

FIG. 173.—Infection of the periorbital region due to *T. mentagrophytes* (see Figs. 169–170) (H. Bruce Ostler *et al.*).

scarring and loss of the lashes, a condition which tends to progress indefinitely unless the somewhat rare diagnosis is remembered. Children are more susceptible than adults probably because of the higher concentration of fungistatic free fatty acids in the hair of the latter (Rothman *et al.*, 1946–47). In the type of trichophytous infection caught from horses and cattle in adults a *kerion* (κηρίον, honeycomb) is commonly seen (Fig. 172). This is a localized, painful, elevated, boggy, erythematous tumefaction with deep-seated pustules on its surface.

The fungi are present in the horny layer of the skin and in the follicles in and around the hairs (Graham *et al.*, 1964); the dermis contains no fungi in either tinea capitis or barbæ, but shows a perifollicular infiltrate with giant cells.

Diagnosis depends on the demonstration of the fungi, which are readily found in the enlarged, whitish bulbs of the broken hairs which are easily extracted or in the scales at the periphery of the lesions; in preparing material for examination a 20% solution of potassium hydroxide is employed which clears but does not disintegrate the hair or the cells when the fungus is seen as a network of lines cutting across the outlines of the cells. It is interesting that hairs infected with microspora fluoresce with ultra-violet light filtered through a glass containing nickel oxide (Wood's light) (Margarot and Deveze, 1925; Kinnear, 1931; Muskatblit and Targan, 1945).

The *treatment* of ringworm is usually easy. The classical method of painting with tincture of iodine night and morning for 3 days is efficacious but in the region of the lids a salicylic ointment is safer: a suitable preparation is Whitfield's ointment (benzoic 6% and salicylic acid 3%). This may well be combined with griseofulvin taken systemically which has given excellent results (François, 1968; Lejman and Bogdaszewska-Czabanowska, 1968). When the brows and lashes are involved a preliminary epilation may be advisable.

It is noteworthy that treatment with corticosteroid ointments may not only increase the severity but change the character of the disease (Ives and Marks, 1968); as in other mycotic infections, these drugs are to be avoided despite their undoubted ability to give prompt temporary relief of symptoms by suppression of the inflammation.

TINEA FAVOSA or FAVUS (L. *favus*, a honeycomb) is not common on the lids.[1] It may affect the glabrous skin and be generalized; most usually it affects the scalp, particularly in children; it has been recorded as affecting one eyelid alone (Greenbaum, 1924). The disease is contagious and is derived from animals, particularly mice, dogs, cats and kids.

The lesion starts usually in the upper lid as a small red pimple with a scaly surface situated at a hair follicle and pierced centrally by a hair; the hyphæ and a few spores are present in the epidermis within and around the hairs. A yellow point appears which assumes the typical crust or *scutula*, a sulphur-yellow saucer-shaped mass, depressed in the centre where the hair, if not already destroyed, projects. The scutula consists of keratinized and parakeratotic cells, plasma and inflammatory cells heavily intermingled with hyphæ and spores which become degenerated in the centre of the lesion, while the dermis underneath shows an inflammatory infiltrate with plasma cells and giant cells of the foreign-body type. The mass is fixed in the skin by a thin cuticle, and usually becomes surrounded by satellite formations

[1] Narkiewicz-Jodko (1870), Schiess-Gemuseus (1873), MacHardy (1885), Pergens (1897), Gloor (1898), Libman (1898), Schmidt-Rimpler (1901), Collins (1903), Pegoraro (1904), van der Wijk (1904), Brault (1912), E. Hoffmann (1913), Truffi (1914), Lévy-Franckel and Offret (1921), Greenbaum (1924), Stargardt (1925), Ballarini (1925), Pignot (1930), Kerl (1930), Fazakas (1937), Chan (1951), and others.

(Plate III). It is to be noted that in favus of the scalp the lashes may become infected with the fungus (Arcoleo, 1871). In a case reported by Chan (1951) wherein the lids and face were affected, the conjunctiva and the cornea as a superficial punctate keratitis were also involved.

Treatment of favus on the lid by the classical method was thorough washing and the application of an ointment of copper sulphate or mercury oleate. In the usual case when the scalp is infected, this area, of course, must receive primary attention. More recently, however, griseofulvin has given excellent results after the epilation of infected hairs.

Arcoleo. *Res. Clin. Ottal. Palermo*, 269 (1871). *See Jber. Ophthal.*, **2,** 372 (1871).
Arievic. *Derm. Wschr.*, **90,** 683 (1930).
Ballarini. *Arch. ital. Derm.*, **1,** 76 (1925).
Behdjet. *Derm. Wschr.*, **85,** 1028 (1927).
Blaisdell. *New Engl. J. Med.*, **202,** 1059 (1930).
Bonanni. *G. ital. Derm.*, **98,** 667 (1957).
Brault. *Bull. Soc. franç. Derm. Syph.*, **19,** 215 (1912).
Chan. *Chin. med. J.*, **69,** 262 (1951).
Collins. *Trans. ophthal. Soc. U.K.*, **23,** 1 (1903).
Costa. *Arch. Derm. Syph.* (Chic.), **48,** 65 (1943).
Davidson and Gregory. *Canad. med. Ass. J.*, **27,** 485 (1932).
Elschnig. *Klin. Mbl. Augenheilk.*, **80,** 246 (1928).
Fazakas. *Klin. Mbl. Augenheilk.*, **99,** 106 (1937).
François. *Bull. Soc. belge Ophtal.*, No. 148, 208 (1968).
Franks and Mandel. *Arch. Derm. Syph.* (Chic.), **62,** 708 (1950).
Franks, Mandel and Sternberg. *Arch. Derm. Syph.* (Chic.), **62,** 54 (1950).
Gloor. *Arch. Augenheilk.*, **37,** 358 (1898).
Graham, Johnson, Burgoon and Helwig. *Arch. Derm.*, **89,** 528 (1964).
Greenbaum. *Amer. J. Ophthal.*, **7,** 6 (1924).
Hoffmann, D. H. *Adv. in Ophthal.*, **16,** 63 (1965).
Hoffmann, E. *Dtsch. med. Wschr.*, **39,** 93 (1913).
Hopkins and Krause. *Amer. J. Ophthal.*, **33,** 1793 (1950).
Ives and Marks. *Brit. med. J.*, **3,** 149 (1968).
Kerl. *Zbl. Hautkrankh.*, **32,** 180 (1930).
Kinnear. *Brit. med. J.*, **1,** 791 (1931).
Latte and Piredda. *Boll. Oculist.*, **33,** 761 (1954).
Lejman and Bogdaszewska-Czabanowska. *Hautarzt*, **19,** 264 (1968).
Lesczynski. *Ann. Derm. Syph.* (Paris), **4,** 536 (1923).
Lévy-Franckel and Offret. *Bull. Soc. franç. Derm. Syph.*, **28,** 437 (1921).
Libman. *Arch. Ophthal.*, **27,** 173 (1898).
Lisch. *Klin. Mbl. Augenheilk.*, **133,** 729 (1958).
Lourier and Zwitkis. *Derm. Wschr.*, **87,** 1019 (1928).
Louste and Baranger. *Bull. Soc. franç. Derm. Syph.*, **31,** 143 (1924).
MacHardy. *Trans. ophthal. Soc. U.K.*, **5,** 42 (1885).
Margarot and Deveze. *Bull. Soc. Sci. méd. biol. Montpellier*, **6,** 375 (1925).
Mariotti. *Ann. Ottal.*, **77,** 38 (1951).
Mibelli. *Ann. Ottal.*, **23,** 368 (1894).
Montgomery and Walzer. *Arch. Derm. Syph.* (Chic.), **46,** 40 (1942).
Muskatblit and Targan. *Arch. Derm. Syph.* (Chic.), **52,** 116 (1945).
Narkiewicz-Jodko. *Klin. Mbl. Augenheilk.*, **8,** 78 (1870).
Ostler, Okumoto and Halde. *Amer. J. Ophthal.*, **72,** 934 (1971).
Pais. *Rinasc. med.*, **1,** 425 (1924).
Pegoraro. *Arch. Ottal.*, **11,** 311 (1904).
Pergens. *Klin. Mbl. Augenheilk.*, **35,** 241 (1897).
Pignot. *Bull. Soc. franç. Derm. Syph.*, **37,** 1109 (1930).
Pimentel and Machado. *Arch. bras. Oftal.*, **19,** 295 (1956).
Rothman, Smiljanic, Shapiro and Weitkamp. *J. invest. Derm.*, **8,** 81 (1947).
Rothman, Smiljanic and Weitkamp. *Science*, **104,** 201 (1946).
Ruete. *Z. ärztl. Fortbldg.*, **27,** 649 (1930).
Schiess-Gemuseus. *Klin. Mbl. Augenheilk.*, **11,** 211 (1873).
Schmidt-Rimpler. *Dtsch. med. Wschr.*, **27,** 16 (1901).
Shaw. *J. lab. clin. Med.*, **20,** 113 (1934).
Silvers. *Arch. Derm. Syph.* (Chic.), **49,** 436 (1944).
Snell. *Trans. ophthal. Soc. U.K.*, **22,** 36 (1902).
Stargardt. *Klin. Mbl. Augenheilk.*, **75,** 244 (1925).
Truffi. *G. ital. Mal. vener.*, **55,** 330 (1914).
Voerner. *Klin. Mbl. Augenheilk.*, **39** (2), 871 (1901).
Weidman. *Arch. Derm. Syph.* (Chic.), **15,** 415 (1927).
Appleton's *System of Medicine*, London, **10,** 133 (1937).
White. *Arch. Derm. Syph.* (Chic.), **15,** 387 (1927).
van der Wijk. *Ned. T. Geneesk.*, **48,** 906 (1904).
Wirtz. *Klin. Mbl. Augenheilk.*, **68,** 384 (1922).

Fig. 174.—Sir William Boog Leishman
[1865–1926].
(The Wellcome Institute of the History of Medicine.)

Dermatitis due to Protozoa

A suitable introduction to this section is the photograph of SIR WILLIAM BOOG LEISHMAN [1865–1926] (Fig. 174), one of the several officers in the medical service of the British Army to make outstanding contributions to tropical medicine; he eventually became Director General of the Royal Army Medical Corps. In addition to his work on leishmaniasis, he demonstrated the effect of serum on stimulating the phagocytic activity of leucocytes, evolved the preparation of the anti-typhoid vaccine and introduced the stain for blood and organisms which bears his name. He identified the protozoon parasite which caused kala-azar in a post-mortem smear at Dum Dum near Calcutta in 1900, and published his work in 1903 describing the organism as possibly a trypanosome. Later in the same year Major Charles Donovan [1883–1951] found the same bodies taken in life from splenic punctures in Madras and they were named *Leishmania donovani*. In the same year the parasite was found in a Delhi boil by James Homer Wright [1870–1928] and was named *Leishmania tropica*, an organism which had in fact been described five years earlier by a Russian, Petr Fokich Borovsky [1863–1932], but his writing in Russian was not generally appreciated. The organism was cultivated in 1904 by Sir Leonard Rogers [1868–1962], an Anglo-Indian doctor who made many contributions to tropical medicine and showed the development of the rounded organisms into flagellates, while in 1911 Charles Morley Wenyon [1878–1948] demonstrated that the sandfly, *Phlebotomus*, was the transmitter of cutaneous leishmaniasis.

LEISHMANIASIS

LEISHMANIASIS is a granulomatous type of infection conveyed by parasitic flagellates, essentially an infection of lower animals conveyed to man. Although recognized towards the end of last century by several workers, our real knowledge of these parasites arises from the work of Leishman and Donovan in 1903. They are nucleated ovoid bodies with an active flagellum which generally live in endothelial cells or macrophages (Fig. 175). They may affect the skin (oriental sore in the Orient; American leishmaniasis in South America) or the viscera (kala-azar); but in each case the infection is conveyed by the bites of flies (*Phlebotomi*). It is probable that the original source is usually rodents and that the sand-fly is the common vector transmitting the infection to man (Kojeonikov, 1941; Hoare, 1944).

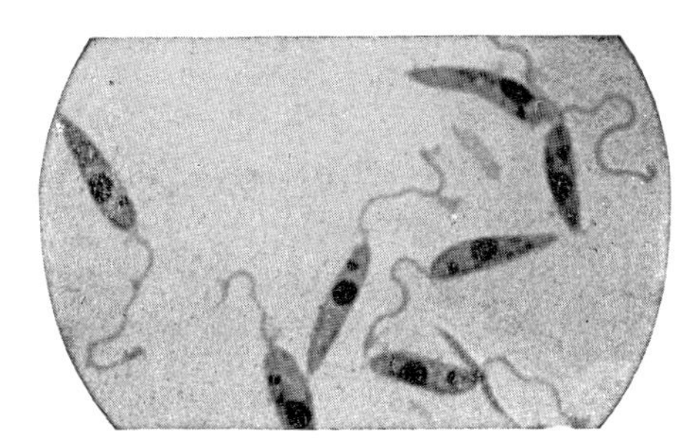

FIG. 175.—*LEISHMANIA DONOVANI.*
The flagellated form seen in culture (U.S. Army Medical Museum).

This was proved by the production of an oriental sore in a human volunteer after an incubation period of 84 days by inoculating cutaneous scarifications with the macerated pulp of a *Phlebotomus* collected at Biskra (*see* Adler and Theodor, 1925–26).

CUTANEOUS LEISHMANIASIS

(*a*) ORIENTAL SORE (ALEPPO, DELHI, BAGHDAD or BISKRA BOIL; BISKRA BUTTON; DERMATITIS ULCEROSA CIRCUMSCRIPTA; LEISHMANIASIS CUTANEUM EXTERNUM) is a disease caused by the *Leishmania tropica* (Wright, 1903–4), which takes the form of a granuloma on the face or other exposed parts of the body. It is met with particularly in central and north Africa, western and southern Asia and Russia where it is endemic, and to a less extent in Sicily and southern Italy (d'Amico, 1931), and occurs in dogs and other lower animals as well as in man. It is to be noted that it is seen in European emigrants to America (Cordero, 1949). It is a mild disease, confined to the skin, responding readily to treatment, never fatal and although multiple lesions are the rule (usually 3 to 8), one attack usually confers immunity (Fox, 1931; Goodall, 1937; and others). For this reason it is the custom amongst some Arabs to infect their children in some convenient spot to forestall more disfiguring lesions in later life. The immunity, however, is not invariably absolute: El Kattan (1936), for example, found subsequent recurrences in some 6% of cases in Egypt.

Two clinical types are recognized—a non-ulcerating, dry or papular type wherein cutaneous nodules appear which run a chronic course,

FIGS. 176 to 178.—LEISHMANIASIS (G. Morgan).

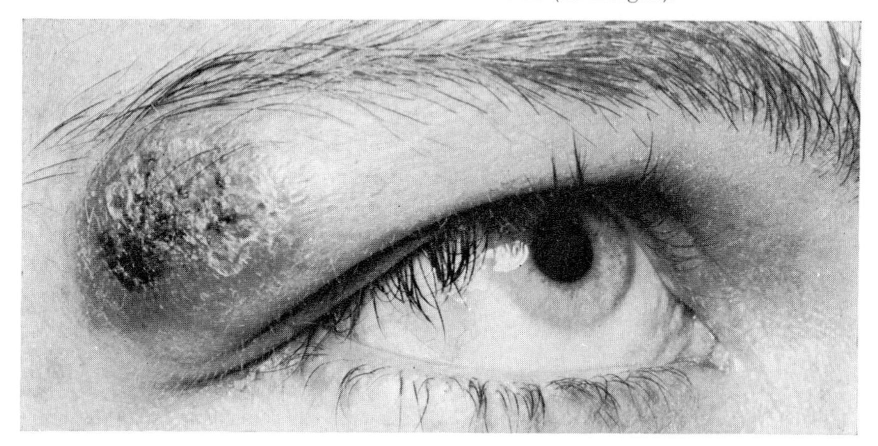

FIG. 176.—Oriental sore in an 18-year-old English girl after a holiday in Majorca.

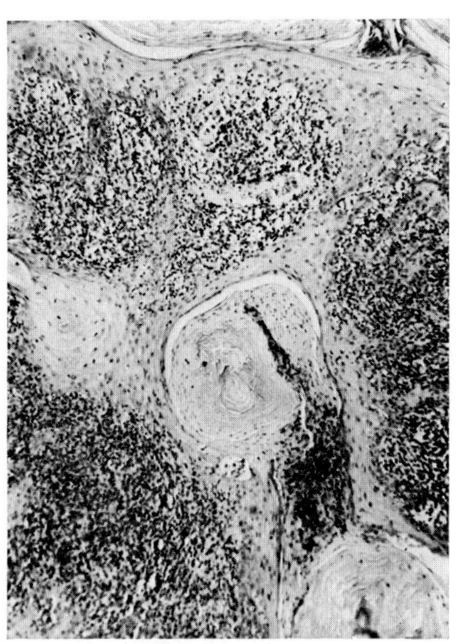

FIG. 177.—The lesion in the case illustrated in Fig. 176 showing irregular acanthosis and a marked inflammatory infiltrate in the epidermis and dermis (H. & E.; × 55).

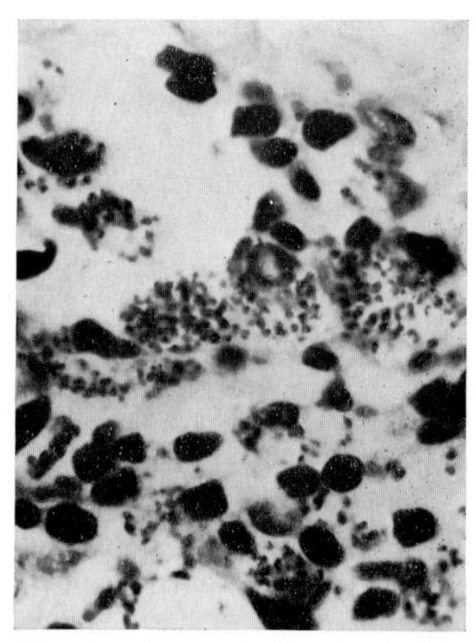

FIG. 178.—Intra- and extra-cellular Leishman-Donovan bodies in the dermis (Giemsa; × 910).

and the more common ulcerating or moist type which runs a more acute course and breaks down relatively early to form an open sore (Thomson and Balfour, 1910). Such a sore may occur on the lids, although these are not one of the most common sites, occurring in about 2 to 5% of cases[1] (Fig. 176). In this connection the lesion may simulate a dacryocystitis (Motolese, 1962). The latent period is long—some months to some years—so that it may appear some time after the patient has left the endemic area—frequently a cause of diagnostic confusion. The individual lesion is a red or purple papule exuding a serous fluid, which crusts while an ulcer develops underneath. Pain and adenopathy are absent unless secondary infection occurs. The course is slow and indolent and, if untreated, spontaneous resolution occurs in about a year's time with the formation of a characteristic indelible scar which may lead to ectropion. It is interesting that the non-ulcerative, dry type is common in towns, while the classical ulcerative type, corresponding to the American *espundia*, is usually found in Russia, Asia and Africa in country or desert dwellers.

Pathologically the disease affects the corium primarily and the epidermis breaks down secondarily. In the former tissue a non-distinctive collection of plasma and round cells with epithelioid cells showing a perivascular grouping is distributed diffusely and focally, among which the demonstration of the specific parasites with Giemsa's staining is diagnostic (*Leishman-Donovan bodies*) (Riehl, 1886; Kuhn, 1897; Weidman, 1937; Morgan, 1965; and others) (Figs. 177–8).

Various specific *diagnostic* tests have been suggested—an intradermal vaccine test (Dostrovsky, 1935), a formol-gel serological reaction (Napier, 1922), the Brahmachari reaction wherein the patient's serum precipitates when mixed with two parts of distilled water (Brahmachari, 1922; Gradwohl, 1938) and the Chopra antimony reaction, wherein the serum diluted 10 times with distilled water becomes cloudy on the addition of an antimonial preparation (4% urea stibamine) (Napier, 1928; Gradwohl, 1938). The most efficient test is the intradermal leishmanin (Montenegro) test wherein 0·1 to 0·2 ml. of a flagellate suspension of antigen is injected intradermally; a positive response is the appearance of a papule or nodule surrounded by an erythematous halo which is obtained in some 92% of cases of oriental sore and 98% of cases of American leishmaniasis (Battistini, 1945; Fasel and Gradow, 1951).

(*b*) AMERICAN LEISHMANIASIS (ESPUNDIA; UTA; MUCO-CUTANEOUS LEISHMANIASIS) is a more severe, resistant malady which spreads more extensively than the corresponding disease found in the Old World. It is a fly-borne cutaneous infection by the *Leishmania brasiliensis*, an organism morphologically identical with that of oriental sore (Strong *et al.*, 1913), the infected material reaching the lid from naso-mucosal lesions by way of the naso-lacrimal duct (Machado *et al.*, 1958). It is endemic in the great rubber district of South America comprising large areas of Brazil, Bolivia, Peru, Ecuador and Colombia, and occurs sporadically from Mexico to the

[1] Willemin (1854), Marzinowsky and Bogrow (1904), Marzinowsky (1907), Bettmann (1907), Dostrovsky (1925), Feigenbaum (1927), Miterstein (1935), Kamel (1945), Ricca (1947), Scuderi (1947), di Ferdinando (1948), Agnello (1948–9), Shusha (1952), Theodorides and Christides (1952), Pestre (1955), Rende (1956), Perelli and Cesa (1960), Morgan (1965), and others.

northern Argentine particularly in richly forested regions (Marback, 1953). After an incubation period of from 1 to 3 months, one or more erythematous patches appear on an exposed part such as the face (Fig. 179); in a few days a pustule forms which breaks down into an ulcer—the *espundic chancre*—covered with a thick crust of proliferating granulations, surrounded by œdema and associated with lymphangitis. There is usually a small number of ulcers (one to three) but sometimes 200 or more disseminated sores occur and along the course of the inflamed lymphatics ulcerating tumours may form. The ulcers are slow and indolent, lasting 6 to 8 months or 20 years, but eventually heal leaving a characteristic, permanent, star-shaped scar. At a varying period after the appearance of the primary ulcers, lesions develop

FIGS. 179 and 180.—AMERICAN LEISHMANIASIS.

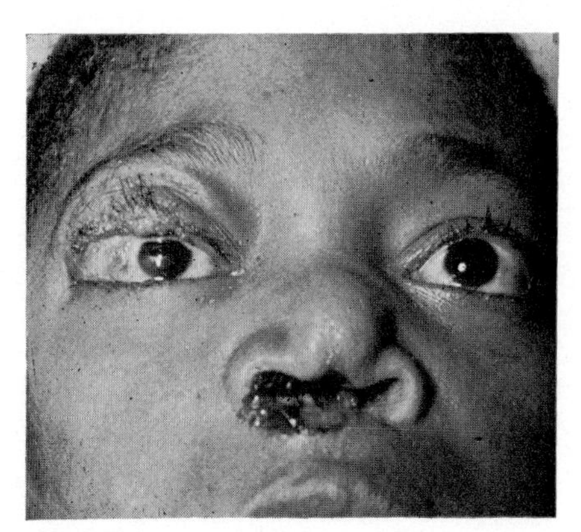

FIG. 179.—In a Brazilian, affecting the lids and nostrils (H. Marback).

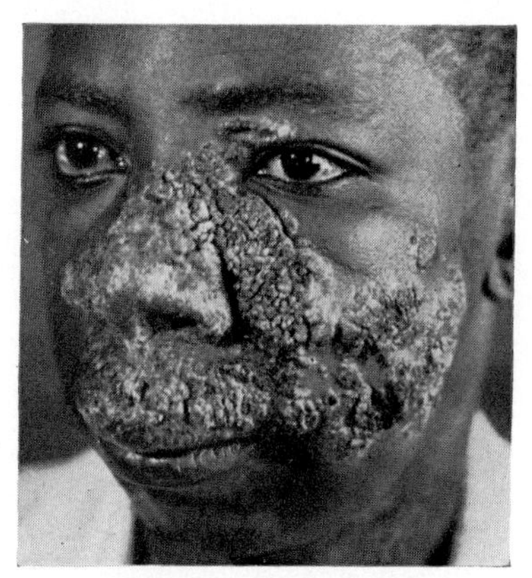

FIG. 180.—A rare type of the verrucose variety in a Brazilian (R. L. and R. L. Sutton).

in the mucous membrane in some 10 to 20% of cases. If they appear immediately after the primary stage they are called *secondary manifestations* (UTA); if they are delayed for some years they are designated *tertiary* (ESPUNDIA) (Weidman, 1937). In either case they are most common in the nose; here and in the mouth ulcers develop producing large vegetating sanguineous granulations, destroying the cartilage and bone in the region of the face. At this stage the lids may again be heavily involved in the widespread destruction (Fig. 180).

A solid œdema of the lids with the formation of new connective tissue in their substance may be caused mechanically by leishmaniasis of the nose (Busacca and Maia, 1935).

Pathologically the lesion resembles oriental sore with the addition of lymphatic involvement (Evans, 1938). The Montenegro test has been devised in diagnosis; and the parasite is seen histologically in the lesions.

Prophylaxis consists in avoiding forests as much as possible, for since the disease is transmitted by forest flies, infection is avoided by camping in clearings far from trees.

VISCERAL LEISHMANIASIS : KALA-AZAR

KALA-AZAR (DUM-DUM FEVER) is essentially a visceral leishmaniasis caused by *L. donovani*; like all leishmanial infections it is transmitted by flies, and is indigenous in India, China and the Sudan. The irregular fever and leucopenia are accompanied by lesions in the spleen, liver and large intestine. In the course of the disease, which is long and insidious, cutaneous lesions containing parasites may develop; among the sites attacked are the face and eyelids. Here depigmented zones appear, nodules, papillomatous lesions, or areas resembling xanthomata associated with considerable thickening of the tissues.

The *treatment* of all forms of leishmaniasis is slow and difficult. Suggested measures for local application have included destruction by CO_2 snow, cleansing with alcohol after curettage (Agnello, 1937) or diathermic coagulation (Kamel, 1943; Agnello, 1948–49). The most effective treatment, however, is by the quinquavalent or pentavalent derivatives of antimony which may be given systemically by intramuscular or intravenous injection and, if necessary, injected in quantities of about 2 ml. at points around the periphery into the base of the sore at weekly intervals. The more useful preparations are ethylstibamine and sodium stibogluconate. In extensive cases X-radiation has given good results. When the condition is generalized and in kala-azar the pentavalent antimony compounds such as ethylstibamine injected intravenously or the less toxic sodium antimony gluconate or stibamine glucoside are the most useful medicaments but the disease is frequently refractory.[1] If antimonial drugs are not effective a useful alternative (or addition) is one other of the diamidines such as stilbamidine. The most satisfactory method of handling the disease, however, is prophylactic—the wholesale systematic poisoning of the burrows of the rodents in endemic areas, a policy which has reputedly been pursued with success in Russia (Latyshev and Kruikova, 1941; Kojeonikov, 1941), or the elimination of infected dogs and rodents and measures against the sandfly vector by filling up cracks in walls, removing rubbish and vegetation near living quarters and spraying with insecticides such as DDT.

Adler and Theodor. *Ann. trop. Med. Parasit.*, **19**, 365 (1925); **20**, 175, 355 (1926).
Agnello. *Rass. ital. Ottal.*, **6**, 303 (1937).
Ann. Ottal., **74**, 129 (1948).
G. ital. Oftal., **2**, 231 (1949).
d'Amico. *Ann. Ottal.*, **59**, 347 (1931).
Battistini. *Rev. Med. exp.*, **4**, 116 (1945).
Bettmann. *Munch. med. Wschr.*, **54**, 289 (1907).
Brahmachari. *Ind. J. med. Res.*, **10**, 492 (1922).

[1] Castellani (1916–23), Dostrovsky (1925), Goodall (1937), Holmes (1937), Kamel (1945), Most and Lavietes (1947), Snow *et al.* (1948), Cordero (1949), Ceccarini (1949), Donatelli (1949), Morgan (1965), and others.

Busacca and Maia. *Folia ophthal. orient.*, **1**, 372 (1935).
Castellani. *Brit. med. J.*, **2**, 552 (1916); **1**, 283 (1923).
Ceccarini. *Ann. ital. Derm.*, **4**, 490 (1949).
Cordero. *Prensa méd. argent.*, **36**, 1775 (1949).
Donatelli. *Ann. ital. Derm.*, **4**, 502 (1949).
Donovan. *Brit. med. J.*, **2**, 79, 1401 (1903).
Dostrovsky. *Arch. Schiffs- u. Tropenhyg.*, **29**, 101 (1925).
Ann. trop. Med., **29**, 123 (1935).
El Kattan. *Bull. ophthal. Soc. Egypt*, **28**, 12 (1936).
Evans. *Brit. J. Derm.*, **50**, 17 (1938).
Fasel and Gradow. *Arch. Derm. Syph.* (Chic.), **64**, 487 (1951).
Feigenbaum. *Ber. dtsch. ophthal. Ges.*, **46**, 395 (1927).
di Ferdinando. *G. ital. Oftal.*, **1**, 359 (1948).
Fox. *Arch. Derm. Syph.* (Chic.), **23**, 480 (1931).
Goodall. *Ind. med. Gaz.*, **72**, 3 (1937).
Gradwohl. *Clinical Laboratory Methods and Diagnosis*, 2nd ed., St. Louis, 1485 (1938).
Hoare. *Trop. Dis. Bull.*, **41**, 331 (1944).
Holmes. *J. roy. Army med. Corps*, **69**, 258 (1937).
Kamel. *Bull. ophthal. Soc. Egpyt*, **36**, 75 (1943).
Rep. Mem. ophthal. Lab. Giza, **14**, 124 (1945).
Kojeonikov. *Problems of Cutaneous Leishmaniasis*, Ashkhabad, 75, 93, 127, 169 (1941).
Kuhn. *Virchows Arch. path. Anat.*, **150**, 372 (1897).
Latyshev and Kruikova. *Problems of Cutaneous Leishmaniasis*, Ashkhabad, 55 (1941).
C.R. Acad. Sci. Moscow, **30**, 90 (1941).
Leishman. *Brit. med. J.*, **1**, 1252; **2**, 1326 (1903).
Machado, N. and J., and Moura. *Rev. bras. Oftal.*, **17**, 279 (1958).
Marback. *Lesões oculares da leishmaniose tegumentar americana*, Bahia (1953).
Marzinowsky. *Z. Hyg. Infekt.-Kr.*, **58**, 327 (1907).
Marzinowsky and Bogrow. *Virchows Arch. path. Anat.*, **178**, 112 (1904).
Miterstein. *Folia ophthal. orient.*, **1**, 383 (1935).
Morgan. *Brit. J. Ophthal.*, **49**, 542 (1965).
Most and Lavietes. *Medicine* (Balt.), **26**, 221 (1947).
Motolese. *Acta XIX int. Cong. Ophthal.*, New Delhi, **2**, 1084 (1962).
Napier. *Ind. J. med. Res.*. **9**, 830 (1922).
Ind. med. Gaz., **63**, 687 (1928).
Perelli and Cesa. *Riv. ital. Tracoma*, **12**, 22 (1960).
Pestre. *Algérie méd.*, **59**, 589 (1955).
Rende. *Riv. ital. Tracoma*, **8**, 148 (1956).
Ricca. *Ann. Ottal.*, **73**, 115 (1947).
Riehl. *Arch. Derm. Syph.* (Wien), **13**, 805 (1886).
Scuderi. *Rass. ital. Ottal.*, **16**, 335 (1947).
Shusha. *Bull. ophthal. Soc. Egypt*, **42**, 239 (1952).
Snow, Satulsky and Kean. *Arch. Derm. Syph.* (Chic.), **57**, 90 (1948).
Strong, Tyzzer, Brues *et al.* *J. Amer. med. Ass.*, **61**, 1713 (1913).
Theodorides and Christides. *Bull. Soc. héllen. Ophtal.*, **20**, 10 (1952).
Thomson and Balfour. *J. roy. Army med. Corps*, **14**, 1 (1910).
Weidman. Appleton's *System of Medicine*, London, **5**, 289 (1937).
Willemin. *Gaz. méd. Paris*, **9**, 200, 228, 252 (1854).
Wright. *J. med. Res.*, **10**, 472 (1903).
J. cutan. Dis., **22**, 1 (1904).

TRYPANOSOMIASIS

The trypanosomes, flagellated protozoa, are inoculated into man through the bites of flies, causing SLEEPING SICKNESS in Africa (*T. gambiense* and *T. rhodesiense* carried by tsetse flies of the genus *Glossina palpalis*) and CHAGAS'S DISEASE in South America (*T. cruzi* transmitted by blood-sucking bugs of the family Triatomidæ). The initial lesion is cutaneous and in Chagas's disease frequently involves the eyelids. It is raised, reddened with a dark centre and is associated with regional adenopathy, regressing slowly after one or two weeks. In the early stages of the disease a common feature is the development of tense urticarial swellings of the lids with a tendency to recur intermittently; they are frequently unilateral, often localized to the outer part of the lower lid in Africa (Fig. 181), affecting both lids in America (Fig. 182); the combination of unilateral œdema of the lids and inflammation of the lacrimal gland is known as *Romaña's sign*.[1] The swellings may also

[1] Daniels (1911), Urrets Zavalia and de Anquin (1939), Toulant (1950), Borzone (1951), Trindade *et al.* (1955), Paez Allende (1964), Cordero Moreno and Echerlo Salcedo (1971).

occur in the terminal stages of the disease, 3 or 4 days before death (Broden, 1904). The œdema has been said to be toxic in nature caused by the presence of the parasite in the blood, but is more probably an anaphylactic phenomenon due to the acquirement of sensitivity to the proteins of dead organisms (Habig, 1949); it gives a characteristic appearance to the face which in endemic areas is diagnostic.

The general symptoms are frequently profound with fever and prostration, complicated by meningo-encephalitic and cardiac manifestations. The *treatment* of the African type of the disease is by melarsoprol (a pentavalent arsenical compound) or suramin sodium followed by tryparsamide;

Figs. 181 and 182.—Trypanosomiasis.

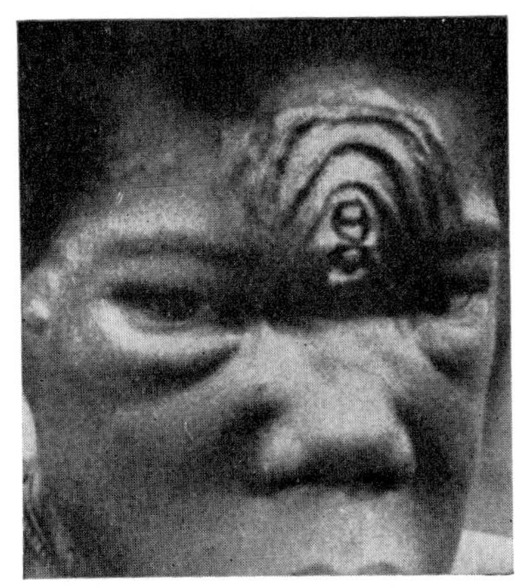

Fig. 181.—African trypanosomiasis showing the typical anaphylactic type of palpebral œdema, in this case affecting both lower lids. The markings on the forehead are artificial (J.-M. Habig).

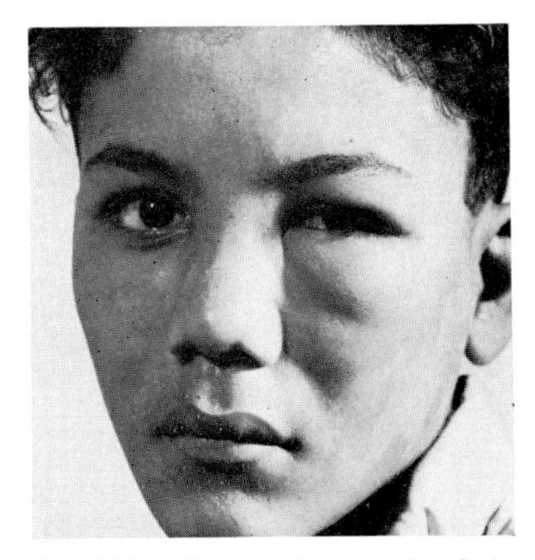

Fig. 182.—Chagas's disease. South American trypanosomiasis, showing unilateral œdema of the lids with congestion of the conjunctiva and swelling of the pre-auricular nodes (R. Cordero Moreno and E. Echerle Salcedo).

Chagas's disease is less amenable to treatment, but drugs such as primaquine, pentaquine and Spirotrypan have been recommended. For particulars of the use of these drugs text-books of tropical medicine should be consulted.

Borzone. *Sem. méd. B. Aires*, **51,** 812 (1951).

Broden. *Z. ang. Mikr.*, Leipzig, **10,** 35 (1904).

Cordero Moreno and Echerle Salcedo. *Arch. Soc. españ. Oftal.*, **31,** 289 (1971).

Daniels. *J. trop. Med.*, **14,** 161 (1911).

Habig. *Bull. Soc. franç. Ophtal.*, **62,** 383 (1949).

Paez Allende. *Arch. Oftal. B.Aires*, **39,** 191 (1964).

Toulant. *Manifestations oculaires d. trypanosomiases humaines*, Paris (1950).

Trindade, Brant and Lopez. *Rev. bras. Malar.*, **7,** 371 (1955).

Urrets Zavalia and de Anquin. *Arch. Oftal. B.Aires*, **14,** 836 (1939).

Dermatitis due to Metazoa

Infestations of the lids by animal parasites are not very common, nor do they show many peculiarities there; they may therefore be dismissed rapidly.

NEMATODES

Most of the round and thread worms affect the lids only secondarily by toxic or allergic effects; the exception to this general rule is the Filaria which may invade the lids and cause considerable local disturbances therein.

ASCARIASIS. The *Ascaris lumbricoides*, the white round worm which may inhabit the gut in enormous numbers, particularly of children, may give rise to a toxic swelling and urticaria of the lids in infested persons; eight cases of unilateral œdema of the lids were reported by Gogina (1964) which he attributed to a migration of the larvæ into the tissues; a toxic-allergic sero-fibrinous choroiditis may also occur. Such attacks of œdema may be recurrent, being related to the life-cycle of the parasite (Sakić, 1964; Castellazzo, 1967). The body-fluid of the worm is extremely poisonous (Weinberg and Julien, 1911), and if directly rubbed on the lids it produces a violent conjunctivitis and urticaria, an accident which may occur to butchers or pathologists (Snell, 1906; Bäumler, 1907; Dorff, 1912; Craig and Faust, 1951; Kozakiewicz and Tarzynska, 1962). In treatment, piperazine and its numerous derivatives have been found to be both effective and safe. A recent promising alternative is mebendazole (Chaia and Da Cunha, 1971).

ENTEROBIASIS. The *Enterobius (Oxyuris) vermicularis*, another intestinal white worm, produces similar toxic œdemas and urticarias of the lids in infested persons; it also occasionally figures in the ætiology of blepharitis (Nègre, 1929), conjunctivitis and keratitis.[1]

ANKYLOSTOMIASIS. Both the *Ankylostoma duodenale* and the *Necator americanus* also cause toxic dermatoses with swelling of the lids (Laignier-Terrasse, 1932). The former variety is largely subtropical in distribution; the latter is prevalent in the equatorial parts of Africa and America.

TRICHINIASIS (TRICHINOSIS). *Trichinella spiralis* is a common intestinal parasite usually acquired by eating inadequately cooked pork in the United States, Europe and China. The larvæ migrate into the muscles and become encysted, giving rise to the same toxic palpebral symptoms as those parasites already mentioned although to a greater degree—œdema of the lids sometimes with an erythematous rash, accompanied on occasion by chemosis, subconjunctival petechiæ and photophobia (Drake *et al.*, 1935; Pierose and Butt, 1945; M. and L. Croll, 1952; Edwards, 1954; Schoop *et al.*, 1961).[2]

GNATHOSTOMIASIS. Infection of the ocular region by the *Gnathostoma* is rare. If an infection near the lids occurs the palpebral œdema is intense (Sen and Ghose, 1945). The lids may be primarily invaded (Okabe and Kuwahara, 1955).

FILARIASIS

The thread worms and their life-histories have already been discussed in detail[3] where it was shown that filariasis is endemic in almost all tropical countries (Fig. 183).

Loa (Filaria) loa, which is found in tropical West Africa[4] and gives rise to *Loaiasis*, pursues an erratic course in subcutaneous tissues, its peregrinations being marked by the occasional appearance of transient CALABAR SWELLINGS resembling angio-œdema; these probably represent an ana-

[1] Vol. VIII, pp. 401, 417.
[2] Vol. IX, p. 460.
[3] Vol. VIII, p. 401.
[4] Vol. VIII, p. 402.

phylactic reaction of the host to the loa antigen (Lloyd and Chandra, 1933) (Fig. 184). When they appear in the lids they tend to give rise to intense irritation and much œdema (*ambulant œdema*). These swellings may be of considerable size reaching, for example, from the lid-margin to the brows, and disappear as rapidly as they appear. Permanent hypertrophy of the tissues owing to prolonged irritation and lymphatic blockage is rare in the lids.

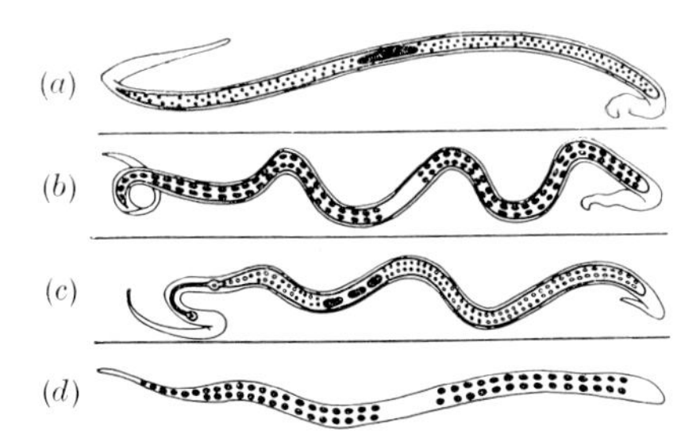

FIG. 183.—DIAGRAMMATIC DRAWINGS OF HUMAN PATHOLOGICAL FILARIÆ.

The gross anatomical structure of the main pathogenic filariæ affecting man. (a) *Wuchereria bancrofti;* (b) *Loa loa;* (c) *Wuchereria malayi;* (d) *Onchocerca volvulus* (after M. Gentilini).

The worm, which on occasion may grow to a great length (6·5 cm., Pacalin, 1930; 6 cm., Nolasco, 1947), may travel very actively, leaving the subcutaneous tissue of the lids for the conjunctiva, disappearing into the orbit, flitting over the bridge of the nose to the lids of the other eye or running away across the cheek. This may cause intense irritation and considerable mental distress, symptoms dramatically described in an autobiographical case by Johnstone (1947). As a general rule, heat, as on sitting in front of a fire, entices the parasite to the surface, while cold drives

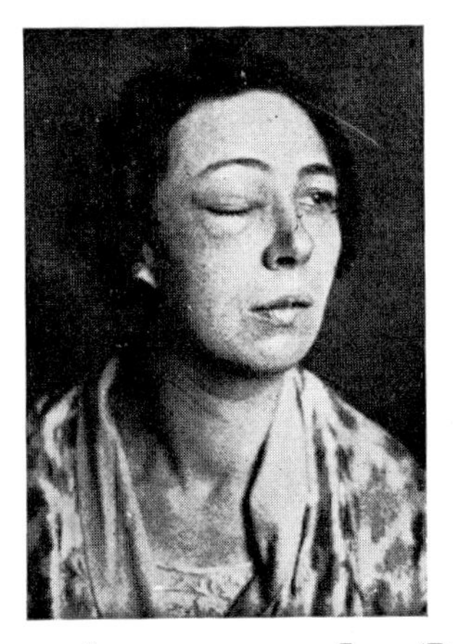

FIG. 184.—CALABAR SWELLING OF THE LIDS (P. Manson-Bahr).

it into the deeper tissues.[1] One fact which must be remembered is that the filariæ may remain in the body for many years so that the patient may have travelled far and long from the infested region of Africa before the symptoms appear (17 years, Manson-Bahr, 1945; 14 years, Laveran, 1916; 11 years, Pacalin, 1930), a circumstance which often leads to difficulties in diagnosis. The only effective treatment is to wait until the worm can be felt in the subcutaneous tissue, whereupon it is seized by a subcutaneous stitch, immobilized by subcutaneous topical anæsthesia and extracted surgically. The site of election for excision is from the conjunctiva,[2] but excision from the lids when the worm is seen under the thin skin is possible although difficult (Gifford and Konné, 1943; Johnstone, 1947; Castresana y Guinea, 1949). It is to be remembered, however, that even if one worm is removed in an infected person from the region of the lids others may appear either here (4, Johnstone, 1947) or elsewhere in the body. Treatment by diethylcarbamazine (Hetrazan) (2 mg. per kg. body-weight 3 times a day for 7 to 14 days) kills both the microfilariæ and the adult female worms.[3]

Wuchereria bancrofti, which is found particularly in sub-tropical south-eastern Asia, produces enormous œdematous swellings of the lids (Leber, 1914) or indurated masses extending down the cheek (Badir, 1956; Mohamed, 1959); in such a case Badir (1956) found a chronic inflammatory granuloma with lymphocytes, eosinophils and foreign-body giant cells, swarming with microfilariæ. It is probable that the suffusions are due to the toxic action of the secretions of the microfilariæ. It is this parasite which produces *filarial elephantiasis* by blocking the lymphatics.[4] Hetrazan is rapidly lethal to the microfilariæ.

DRACONTIASIS. The *Dracunculus* (*Filaria*) *medinensis* or *Guinea worm*[5] found in the north and west of Africa and southern Asia may produce an intense œdema of the lids associated with general urticaria in infested subjects (Bartet, 1908). It may, however, invade the lids themselves, where it excites an intense irritation and causes a nodular swelling or an abscess (Laignier-Terrasse, 1932; Reddy, 1966). Treatment again is by surgical excision of the worm. The natives remove it by winding it round a stick, a few centimetres each day, the operation being completed within 10 to 14 days; niridazole is the effective drug (Rousset, 1952).

ONCHOCERCIASIS. The life-history of the *Onchocerca* (*Filaria*) *volvulus*, found in central and western Africa and central America, as well as the enormous ocular damage it causes in these regions known as "river blindness" have already been described.[6] Nodules (CERCOMATA) containing the adult worms appear in the subcutaneous tissues (on the head, including the eyelids,

[1] Morton (1877), Argyll Robertson (1895–97), Sorel (1911), Laveran (1916), Rousseau (1919), Elliot (1920), Villard (1923), Volmer (1926), Pacalin (1930), Laignier-Terrasse (1932), Clothier (1943), Harley (1958), and others.

[2] Vol. VIII, p. 405.

[3] Shookhoff and Dwork (1949), Esteban (1950), Ridley and Anderson (1950), Schneider (1951), Brumpt (1952), Hawking (1955).

[4] Elephantiasis lymphangioides or arabicum (p. 356).

[5] Vol. VIII, p. 414.

[6] Vol. VIII, p. 406; Vol. IX, p. 444.

in Central America (Fig. 185); below the shoulders in Africa), and from these, microfilariæ pervade the skin in enormous numbers (Ridley, 1945; Rodger, 1957) (Fig. 188). The skin infested in this way suffers a severe and intensely itchy dermatitis resembling erysipelas in the acute phase with intense photophobia and blepharospasm, and resulting in œdema and hyperpigmentation in the chronic stage (Figs. 186–7). Pathologically there is considerable

FIGS. 185 to 187.—ONCHOCERCIASIS.

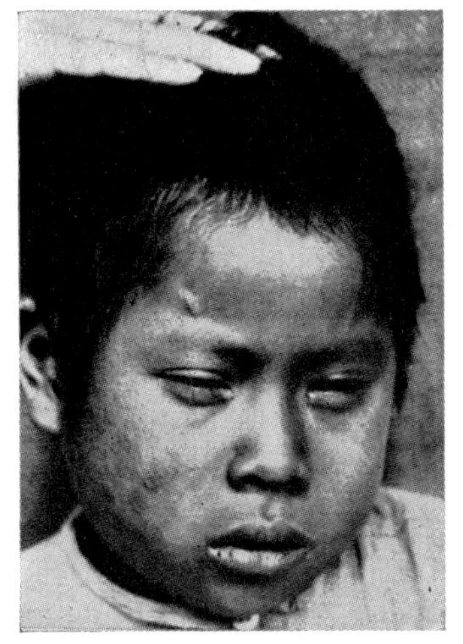

FIG. 185.—The cercomata on the lids and forehead (R. L. and R. L. Sutton).

FIG. 186.—Photophobia and blepharospasm in an acute case (M. Puig Solanes).

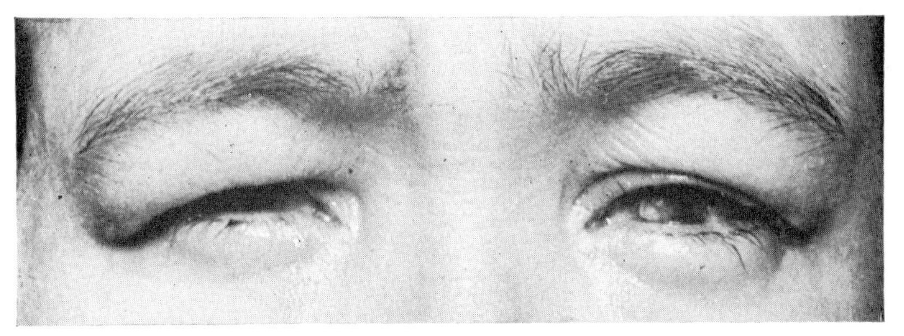

FIG. 187.—Palpebral œdema in a chronic case (M. Puig Solanes).

perivascular proliferation and cellular infiltration in the corium into which the microfilariæ burrow (Figs. 188–9). In other cases an œdema of the lids and orbit involving proptosis and occasionally papillœdema may represent, as do calabar swellings, an anaphylactic reaction to the toxins of the dead parasites (Scott, 1944; Ridley, 1945).

In association with ocular onchocerciasis a permanent swelling of the upper lid may be prominent, colourfully described by Owen and Henessey

(1932) as "bung eye". This has usually been described as œdema, but Rodger (1960) showed that in some cases at any rate the cause was bilateral lobulated lipomata lying deep in the dermis, a lipomatous change frequently found in onchocercomata elsewhere in the body. In the dermis which is œdematous microfilariæ are found in considerable numbers in the lipomata. It would seem that the toxins of the parasite act as a stimulant to the fibroblasts and the adipose tissue with the subsequent development of a lipoma.

Treatment is difficult. Surgical excision of the cercomata wherever they appear should be practised (Strong *et al.*, 1934; Puig Solanes *et al.*, 1949).

FIGS. 188 and 189.—SKIN NODULES IN ONCHOCERCIASIS DUE TO *O. VOLVULUS* (H. Ridley).

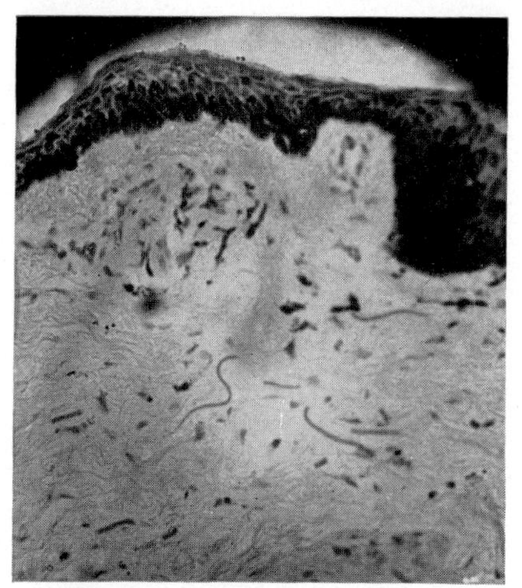

FIG. 188.—A section of the skin through a nodule showing numerous microfilariæ.

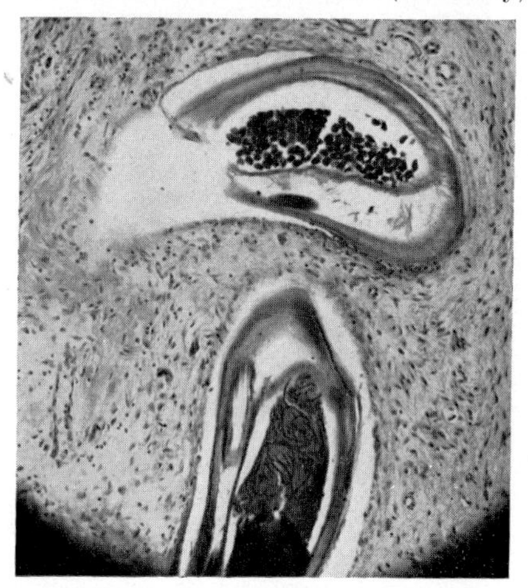

FIG. 189.—Section of a cercoma showing eggs and microfilariæ in an adult worm.

Diethylcarbamazine (Hetrazan) (1 to 2 mg. per kg. body-weight 3 times a day for 2 to 3 weeks) is effective against the microfilariæ but not against the adult worm, while suramin (5 g. over 3 weeks) given intravenously kills the adults and the microfilariæ die in several months. In both cases, however, marked allergic reactions including ocular and orbital manifestations occur from the rapid destruction of the parasites; these may be lessened by systemic corticosteroids given the day before treatment is commenced and during the first four days of treatment, accompanied by an antihistaminic drug.[1]

DIROFILARIASIS. The Dirofilariæ play a small part in ocular disease. *Dirofilaria conjunctivæ* most commonly affects the conjunctiva where small subconjunctival cysts are formed,[2] but subcutaneous cysts also appear (de Meillon and Gillespie, 1943; Lopez-Neyra and Rubio, 1951); the occurrence of subcutaneous nodules in the lids

[1] Ashburn *et al.* (1949), Mazzotti (1951), Hawking (1952–56), Nelson (1955), Conran and Waddy (1956).

[2] Vol. VIII, p. 416.

containing *D. immitis* was noted by Carter (1958). The *D. repens* has been reported to form a subcutaneous nodule in the lid in Russian patients (Skriabin *et al.*, 1930; Malayan, 1968; Gogina, 1970).

THELAZIASIS. The *Thelazia callipœda* (*Filaria circumocularis*) is a nematode encountered usually in animals which may invade the conjunctival sac of man.[1] A case was recorded by Howard (1927) from China wherein a wart-like tumour of some years' duration on the lower lid was shown after excision to be infested with Thelazia larvæ, and similar cases have been seen in Japan (Mori *et al.*, 1969).

Ashburn, Burch and Brady. *Bull. Ofic. san. panam.*, **28,** 1107 (1949).
Badir. *Öst. ophthal. Ges.*, **2,** 95 (1956).
Bäumler. *Arch. Augenheilk.*, **57,** 69 (1907).
Bartet. *Bull. Soc. Path. exot.*, **1,** 330 (1908).
Brumpt. *C.R. Soc. Biol.* (Paris), **146,** 209 (1952).
Carter. *Amer. J. Ophthal.*, **45,** 853 (1958).
Castellazzo. *Minerva oftal.*, **9,** 146 (1967).
Castresana y Guinea. *Arch. Soc. oftal. hisp.-amer.*, **9,** 293 (1949).
Chaia and Da Cunha. *Fôlha méd.*, **63,** 67 (1971).
Clothier. *Clinics* (Phila.), **2,** 875 (1943).
Conran and Waddy. *J. trop. Med.*, **59,** 52 (1956).
Craig and Faust. *Clinical Parasitology*, 5th ed., London, 380 (1951).
Croll, M. and L. *Amer. J. Ophthal.*, **35,** 985 (1952).
Dorff. *Klin. Mbl. Augenheilk.*, **50** (2), 670 (1912).
Drake, Hawkes and Warren. *J. Amer. med. Ass.*, **105,** 1340 (1935).
Edwards. *Trans. ophthal. Soc. U.K.*, **74,** 495 (1954).
Elliot. *Tropical Ophthalmology*, London, 161 (1920).
Esteban. *Arch. Soc. oftal. hisp.-amer.*, **10,** 616 (1950).
Gifford and Konné. *Arch. Ophthal.*, **29,** 578 (1943).
Gogina. *Oftal. Zh.*, **19,** 341 (1964); **25,** 465 (1970).
Harley. *Amer. J. Ophthal.*, **45,** 901 (1958).
Hawking. *Brit. med. J.*, **1.** 992 (1952).
Pharmacol. Rev., **7,** 279 (1955).
Trop. Dis. Bull., **53,** 829 (1956).
Howard. *Amer. J. Ophthal.*, **10,** 807 (1927).
Johnstone. *Lancet*, **1,** 250 (1947).
Kozakiewicz and Tarzynska. *Klin. oczna*, **32,** 55 (1962).
Laignier-Terrasse. *Nemathelminthes et platyhelminthes de l'appareil oculaire humain*, Paris (1932).
Laveran. *Bull. Soc. Path. exot.*, **9,** 436 (1916).
Leber. *v. Graefes Arch. Ophthal.*, **87,** 541 (1914).
Lloyd and Chandra. *Ind. J. med. Res.*, **20,** 1197 (1933).
Lopez-Neyra and Rubio. *Medicamenta*, **9,** 291 (1951).
Malayan. *Vestn. Oftal.*, **81,** 88 (1968).
Manson-Bahr. *Tropical Diseases*, London (1945).
Mazzotti. *Amer. J. trop. Med.*, **31,** 628 (1951).
de Meillon and Gillespie. *S. Afr. med. J.*, **17,** 5 (1943).
Mohamed. *Bull. ophthal. Soc. Egypt*, **52,** 425 (1959).
Mori, Okuda and Shiragami. *Folia ophthal. jap.*, **20,** 639 (1969).
Morton. *Amer. J. med. Sci.*, **74,** 113 (1877).
Nègre. *Bull. Soc. Ophtal. Paris*, 726 (1929).
Nelson. *E. Afr. med. J.*, **32,** 413 (1955).
Nolasco. *Bol. Soc. port. Oftal.*, **5,** 216 (1947).
Okabe and Kuwahara. *J. clin. Ophthal.*, **9,** 432 (1955).
Owen and Henessey. *Trans. roy. Soc. trop. Med. Hyg.*, **25,** 267 (1932).
Pacalin. *Arch. Ophtal.*, **47,** 108 (1930).
Pierose and Butt. *Calif. West. Med.*, **62,** 174 (1945).
Puig Solanes, Noble and Fonte. *Amer. J. Ophthal.*, **32,** 1207 (1949).
Reddy. *Proc. All-India ophthal. Soc.*, **23,** 325 (1966).
Ridley. *Brit. J. Ophthal.*, Monog. Suppl. 10 (1945).
Ridley and Anderson. *Brit. J. Ophthal.*, **34,** 688 (1950).
Robertson, Argyll. *Trans. ophthal. Soc. U.K.*, **15,** 137 (1895); **17,** 227 (1897).
Rodger. *Trans. ophthal. Soc. U.K.*, **77,** 267 (1957).
Amer. J. Ophthal., **49,** 560 (1960).
Rousseau. *Bull. Soc. Path. exot.*, **12,** 35 (1919).
Rousset. *Bull. méd. Afr. occid. franç.*, **9,** 351 (1952).
Sakić. *Lijecn. Vjesn.*, **86,** 173 (1964).
Schneider. *Acta tropica*, **8,** 345 (1951).
Schoop, Lieb, Lamina and Hiemisch. *Klin. Mbl. Augenheilk.*, **139,** 433 (1961).
Scott. *Brit. med. J.*, **1,** 553 (1944).
Sen and Ghose. *Brit. J. Ophthal.*, **29,** 618 (1945).
Shookhoff and Dwork. *Amer. J. trop. Med.*, **29,** 589 (1949).
Skriabin, Althausen and Shulman. *Trop. med. vet.* (Moskva), **8,** 9 (1930).

[1] Vol. VIII, p. 416.

Snell. *Trans. ophthal. Soc. U.K.*, **26,** 8 (1906).
Sorel. *Bull. Soc. Path. exot.*, **4,** 205 (1911).
Strong, Sandground, Bequaert and Ochoa. *Harv. Inst. trop. Biol. Med.*, No. 6 (1934).
Villard. *Bull. Soc. Ophtal. Paris*, 94 (1923).
Volmer. *Klin. Mbl. Augenheilk.*, **76,** 807 (1926).
Weinberg and Julien. *C.R. Soc. Biol.* (Paris), **70,** 337 (1911).

PLATYHELMINTHES

The flat worms play little part in the pathology of the lids, being found there merely by accident. Two Trematodes (flukes) and three Cestodes (tape-worms) are to be noted.

SCHISTOSOMIASIS (BILHARZIASIS). During the period of systemic invasion by the *Schistosoma* (*hæmatobium, mansoni* or *japonicum*)[1] urticaria and œdema of the lids may occur (Meira *et al.*, 1951; Somerset, 1962). One instance has been recorded in Egypt of a lump in the lower lid of a child proving on excision to be a mass of bilharzial ova embedded in granulation tissue (Sobhy Bey, 1928). Treatment is by trivalent antimony compounds (potassium antimony tartrate, sodium antimonyl gluconate, or stibophen).

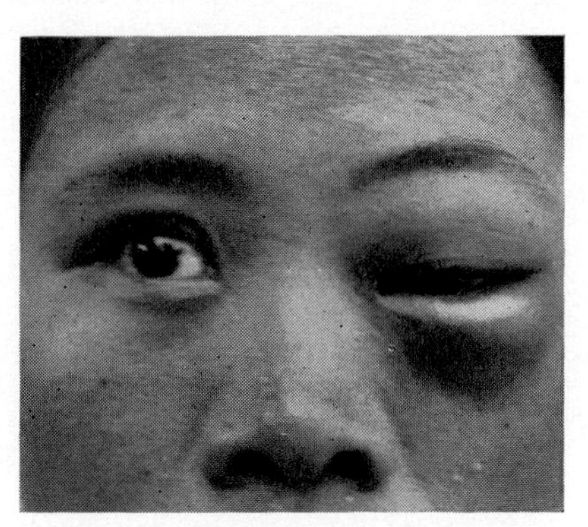

FIG. 190.—SPARGANOSIS.

In an Annamite girl from Hanoi (R. L. and R. L. Sutton, *Hb. of Diseases of the Skin*, Mosby).

DISTOMIASIS. Cases have been recorded wherein the *Distoma ringeri* (*Paragonimus westermani*) (the lung-fluke of Japan and the Far East), usually a parasite in the lungs, has been found encysted in the lids (Elliot, 1920).

SPARGANOSIS. The *Sparganum mansoni* (*Diphyllobothrium mansoni*), a Cestode which has already been described,[2] may be found in the lids, particularly in South East Asia; more frequently it is found in the orbit into which the parasite penetrates. Its frequent occurrence in the ocular region in this part of the world is said to be a direct infection from the application of newly killed frogs as a therapeutic measure in ocular disease. In the lid, usually the outer part of the upper, a tumour appears surrounded by much œdema (Fig. 190); it is diffuse, non-cystic and not adherent to the skin, and on excision is found to contain one or more parasites.[3] Treatment is by excision of the nodule after killing the parasite by an injection of ethyl alcohol and procaine (Houdemer *et al.*, 1934) or by intravenous arsenic injections (Keller, 1937).

[1] Vol. VIII, p. 418. [2] Vol. VIII, p. 420.
[3] Motais (1920–29), Motais (fils) (1921), Collin (1930), Shen (1950), Craig and Faust (1951), Chen and Chou (1971).

ECHINOCOCCOSIS. The localization of a hydatid cyst[1] in the lids is exceptionally rare (upper lid, Orloff, 1908).

CYSTICERCOSIS. The *Cysticercus cellulosæ*, while being relatively frequently localized under the conjunctiva and in the orbit, is rarely found in the subcutaneous tissues of the lids (Sichel, 1847; Hirschberg, 1870–92; Rosenzweig, 1938; Liu, 1957; Reddy and Satyendran, 1964; Wolpiuk, 1967; Jampol *et al.*, 1973). Gros (1871) noted one under the eyebrow. The characteristics are those of similar cysts elsewhere[2]; and treatment is by excision.

Chen and Chou. *Trans. ophthal. Soc. Sineca*, **10,** 123 (1971).
Collin. *Bull. Soc. franç. Ophtal.*, **43,** 395 (1930).
Craig and Faust. *Clinical Parasitology*, 5th ed., London, 561 (1951).
Elliot. *Tropical Ophthalmology*, London, 152 (1920).
Gros. *Gaz. Hôp. Paris*, **23,** 469 (1871).
Hirschberg. *Arch. Augenheilk.*, **1** (2), 138 (1870).
Zbl. prakt. Augenheilk., **3,** 172 (1879).
Berl. klin. Wschr., **29,** 359 (1892).
Houdemer, Dodero and Cornet. *Ann. Oculist.* (Paris), **171,** 311 (1934).
Jampol, Caldwell and Albert. *Arch. Ophthal.*, **89,** 319 (1973).
Keller. *Arch. Ophtal.*, **1,** 779 (1937).
Liu. *Chin. J. Ophthal.*, **7,** 262 (1957).
Meira, Behmer and Bloise. *Rev. Med. Cirug.* (S. Paolo), **11,** 169 (1951).
Motais, F. *Bull. Soc. Path. exot.*, **13,** 215 (1920).
Bull. Soc. med.-chir. Indochine, **7,** 363 (1929).
Motais (fils). *Ann. Oculist.* (Paris), **158,** 329 (1921).
Orloff. *Ann. Oculist.* (Paris), **139,** 117 (1908).
Reddy and Satyendran. *Amer. J. Ophthal.*, **57,** 664 (1964).
Rosenzweig. *Vestn. Oftal.*, **13,** 270 (1938).
Shen. *Nat. med. J. China*, **36,** 351 (1950).
Sichel. *Ann. Oculist.* (Paris), **18,** 223 (1847).
Sobhy Bey. *Ann. Oculist.* (Paris), **165,** 675 (1928).
Somerset. *Ophthalmology in the Tropics*, London, 129 (1962).
Wolpiuk. *Klin. oczna*, **37,** 577 (1967).

ARTHROPODA

In this connection three parasites are of interest—two varieties of lice which cause pathological changes and the *Demodex folliculorum*, a saprophytic mite which may sometimes intensify inflammatory changes.

PEDICULOSIS

LICE (PEDICULI) may attack the eyebrows or lashes. Of the three varieties of blood-sucking lice parasitic on man, the head-louse, *Pediculus humanus capitis*, is found in this situation rarely, the body-louse, *Pediculus humanus corporis* (*vestimentorum*), never, and the crab-louse, *Phthirus pubis*, most commonly. They are found only in persons of unclean habits or those who have been in intimate contact with them.

The *Pediculus humanus capitis* or head-louse is found on the lids only on rare occasions in infestation of the scalp, in which case the brows and lashes may harbour lice and their ova—PEDICULOSIS PALPEBRARUM (Bock, 1892; Ammann, 1897; Hill, 1919; Kumer, 1922). The louse is a small ashen-grey creature (the female 2·5 × 1 mm. and the male somewhat smaller) and the minute ova are attached to the hairs about $\frac{1}{8}$ in. from the skin (Fig. 191). The lice bite to suck blood and this leads to irritation, scratching and secondary infection; in the scalp in milder cases this results in the development of tell-tale excoriations, and in advanced conditions the

[1] Vol. VIII, p. 421. [2] Vol. VIII, p. 423.

hair becomes matted into a fœtid thicket with blood-stained semi-dried pus in which the parasites crawl, while the lymph nodes are enlarged and impetiginous sores break out on the face. Similarly in the lids a persistent blepharitis is the rule which may be associated with marked irritation, follicular conjunctivitis (Goldenberg, 1887) and phlyctenular keratitis (Herz, 1886; and others). Excoriations of the skin of the lids may lead to the formation of abscesses (Rodger and Farooqi, 1959).

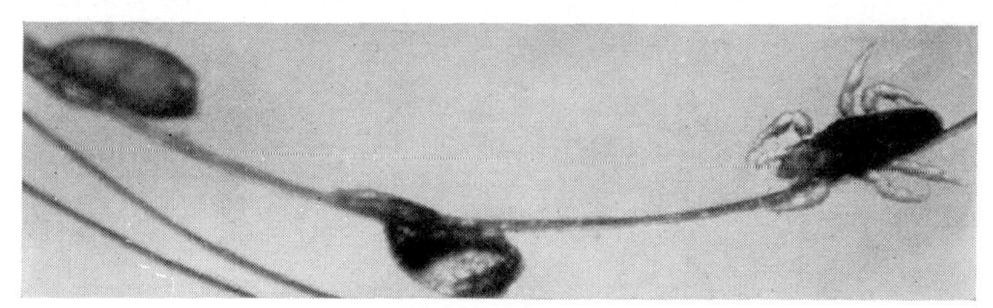

FIG. 191.—The head-louse with nits attached to the hair (R. L. and R. L. Sutton, *Hb. of Diseases of the Skin*, Mosby).

PHTHIRIASIS PALPEBRARUM

Infection of the lashes by the crab-louse (Figs. 192–5), a creature usually more at home around the genital organs, is commoner than by the head-louse but is nevertheless rare: its occurrence on the lashes and brows has in fact been recorded from very early times.

It was noted by Celsus,[1] Scarpa (1801), Chelius (1839), Harkness (1859), Steffen (1866) and others; the work of Jullien (1891–92) fully established the nature of the condition. Several subsequent reports of interest have appeared.[2]

The parasite probably reaches the region of the eyes by transmission by the hand from the pubic hair, and in this way one eye only may be affected (Friedman and Wright, 1934): only occasionally are the lashes infested without pubic involvement (Bab, 1937). The louse, or several of them (17, Boase, 1945), may be seen gripping on to the roots of the lashes or brows with its claws (Fig. 194) while the ova adhering to the hairs may be deposited in great numbers like a heavy, dark powdering (Figs. 195–6). The lice themselves may be difficult to see, but the tell-tale nits on the lashes are readily visible. A blepharitis is not by any means invariable for lice are often found on normal palpebral margins, but frequently itching, scratching and rubbing lead to a true inflammation which may be very persistent and intense: a severe conjunctivitis simulating trachoma has been reported

[1] c. 25 B.C. to A.D. 50. See Hirschberg, *Graefe-Saemisch Hb. d. ges. Augenheilk.*, 2nd ed., Leipzig, **12,** Kap. 23, 258 (1899).

[2] Villard (1908), Passera (1910), Gabriélidès (1911), Brault and Montpellier (1914), Freischmidt (1919), Oppenheim (1922), Waddy (1922), Brusselmans (1923), Friedman and Wright (1934), Goldman (1948), Boase (1945), Chandler (1955), Bose (1955), Beasley (1964), Moses (1965), Schulze (1968), Velasquez (1968), Wolter and Burnes (1970), and Verdy (1971).

FIGS. 192 and 193.—PHTHIRIASIS PALPEBRARUM.

Classical drawings by the ophthalmologist, J. F. Streatfeild, at Moorfields Eye Hospital in 1859.

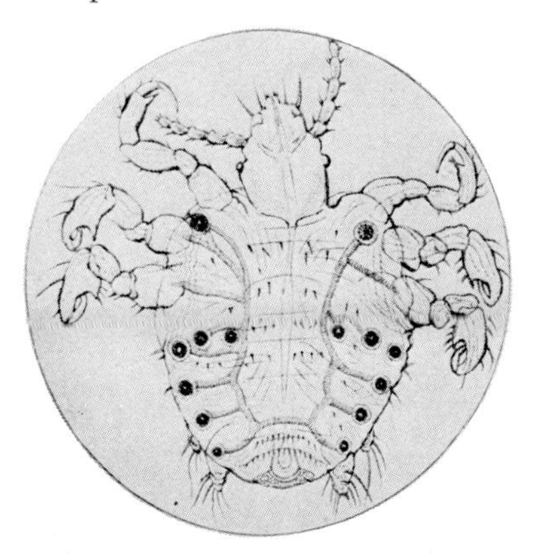

FIG. 192.—The parasite mounted in glycerin (× 80).

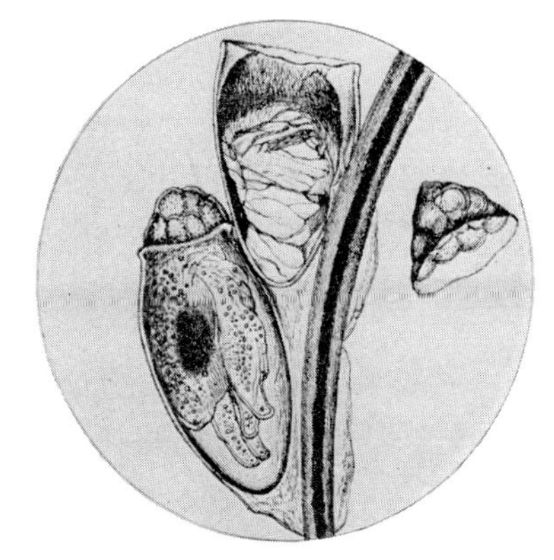

FIG. 193.—An egg containing the fœtal louse and an egg-case with its lid detached showing the puckered internal membrane (× 130).

FIGS. 194 and 195.—PHTHIRIASIS PALPEBRARUM.

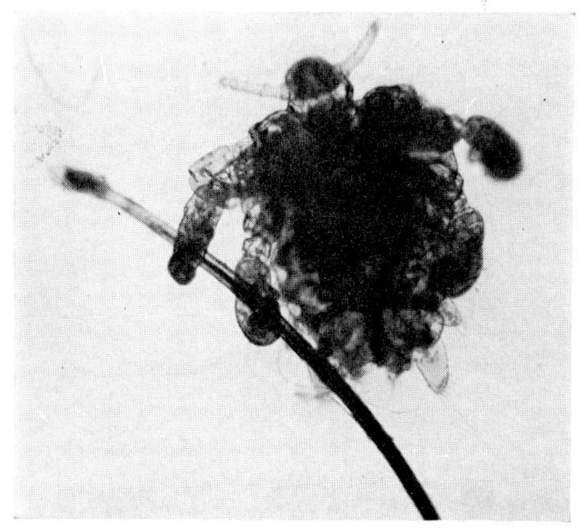

FIG. 194.—A specimen of the crab-louse, *Phthirus pubis*, on the eyelash from a 21-year-old female hippie student (J. R. Wolter *et al.*).

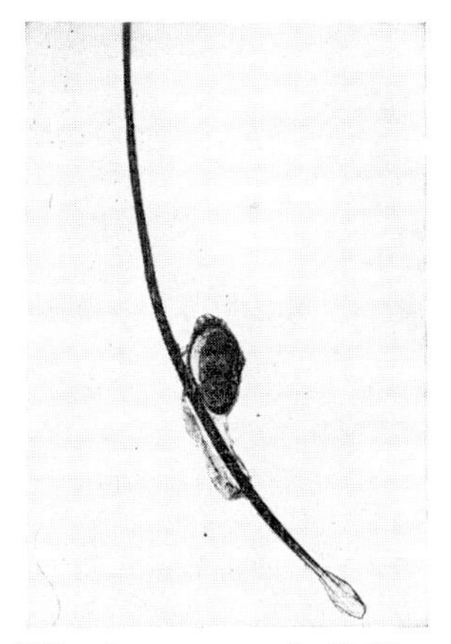

FIG. 195.—An ovum of *Phthirus pubis* attached to the lash from the lower lid. The ovum contains an embryo; the operculum is intact (R. Friedman and C. S. Wright).

(Sabata, 1932). The infestation is most commonly met with in children, particularly the inhabitants of crowded slums or in neglected country areas: but even among these not many cases are encountered in England or the United States. Ruata (1904) found the infestation common in West Africa, Boase (1945) in East Africa, and Rodger and Farooqi (1959) in India.

The *treatment* of pediculosis in this region is the removal of the parasites by forceps (a procedure which may necessitate an anæsthetic in young children, Goldman, 1948) and the application of yellow oxide of mercury ointment (Perlman *et al.*, 1956); equally good results may be obtained with drops or ointment of physostigmine (1%) or other anticholinesterase drugs (tetra-ethyl-pyro-phosphate) (Cogan and Grant, 1949), pyrethrum ointment (Roy and Ghosh, 1944) or gamma benzene hexachloride (1% cream); epilation or shaving is rarely necessary; but the efficient delousing of the initial nidus in the head or genitalia by such preparations as DDT (dichloro-diphenyl-trichloroethane), benzyl benzoate or one of its many preparations, must be thorough (Frazer, 1946; Eddy, 1946; and others).

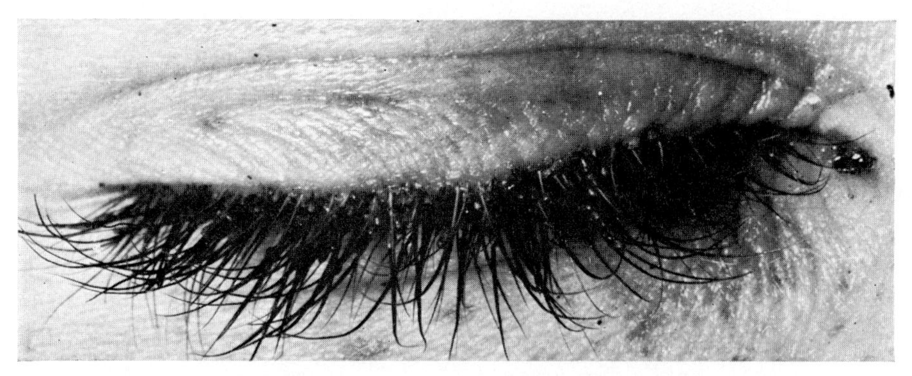

FIG. 196.—LICE ON THE LASHES.
The eggs of the parasite (nits) are seen as minute white dots (Inst. Ophthal.).

DEMODICIDOSIS, the infection of the follicles of the cilia with *Demodex folliculorum*, will be discussed in the section dealing with blepharitis.[1]

Ammann. *Klin. Mbl. Augenheilk.*, **35,** 308 (1897).
Bab. *Klin. Mbl. Augenheilk.*, **98,** 81 (1937).
Beasley. *J. Fla. med. Ass.*, **51,** 533 (1964).
Boase. *E. Afr. med. J.*, **22,** 271 (1945).
Bock. *Zbl. prakt. Augenheilk.*, **16,** 260 (1892).
Bose. *Amer. J. Ophthal.*, **39,** 211 (1955).
Brault and Montpellier. *Bull. Soc. franç. Derm. Syph.*, **25,** 78, 206 (1914).
Brusselmans. *Vlaam. geneesk. T.*, **4,** 553 (1923).
Chandler. *Introduction to Parasitology*, 9th ed., N.Y., 605 (1955).
Chelius. *Hb. d. Augenheilkunde*, Stuttgart (1839).
Cogan and Grant. *Arch. Ophthal.*, **41,** 627 (1949).
Eddy. *J. invest. Derm.*, **7,** 85 (1946).
Frazer. *Brit. med. J.*, **2,** 263 (1946).
Freischmidt. *Derm. Wschr.*, **69,** 517 (1919).
Friedman and Wright. *Arch. Ophthal.*, **11,** 995 (1934).
Gabriélidès. *Ann. Oculist.* (Paris), **146,** 34 (1911).
Goldenberg. *Berl. klin. Wschr.*, **24,** 866 (1887).
Goldman. *Arch. Derm. Syph.* (Chic.), **57,** 274 (1948).
Harkness. *Roy. Lond. ophthal. Hosp. Rep.*, **2,** 125 (1859).
Herz. *Klin. Mbl. Augenheilk.*, **24,** 418 (1886).
Hill. *Brit. med. J.*, **2,** 136 (1919).
Jullien. *Ann. Oculist.* (Paris), **106,** 450 (1891).
Bull. Soc. franç. Derm. Syph., **2,** 457 (1891).
Ann. Derm. Syph. (Paris), **3,** 158 (1892).
Kumer. *Zbl. Haut- u. Geschl.-Kr.*, **6,** 330 (1922).
Moses. *Amer. J. Ophthal.*, **60,** 530 (1965).
Oppenheim. *Zbl. Haut- u. Geschl.-Kr.*, **6,** 332 (1922).
Passera. *G. Med. milit.* (Rome), **58,** 803 (1910).
Perlman, Fraga and Medina. *J. Pediat.*, **49,** 88 (1956).
Rodger and Farooqi. *Brit. J. Ophthal.*, **43,** 676 (1959).

[1] p. 226.

Roy and Ghosh. *Bull. entomol. Res.*, **35,** 231 (1944).
Ruata. *Ann. Ottal.*, **33,** 445 (1904).
Sabata. *Oftal. Sborn.*, **7,** 282 (1932).
Scarpa. *Saggi di osservazioni e d'esperienze sulle principali malattie degli occhi*, Pavia (1801).
Schulze. *Klin. Mbl. Augenheilk.*, **152,** 823 (1968).
Steffen. *Klin. Mbl. Augenheilk.*, **4,** 43 (1866).
Velasquez. *J. Parasit.*, **54,** 1140 (1968).
Verdy. *Méd. trop.*, **31,** 463 (1971).
Villard. *Arch. Ophtal.*, **28,** 496 (1908).
Waddy. *Med. J. Aust.*, **1,** 127 (1922).
Wolter and Burnes. *J. pediat. Ophthal.*, **7,** 5 (1970).

PALPEBRAL MYIASIS

We have already discussed at length the question of ocular myiasis wherein the tissues are invaded and destroyed by the larvæ of flies[1]; it will be remembered that in larval conjunctivitis an attack may be made upon the lids and in destructive myiasis the entire orbit may be invaded, the globe and the lids being destroyed.[2] A primary palpebral myiasis is much more rare. This may assume two forms. Sometimes, particularly in the case of *Dermatobia*, the larvæ are deposited in the skin and remain

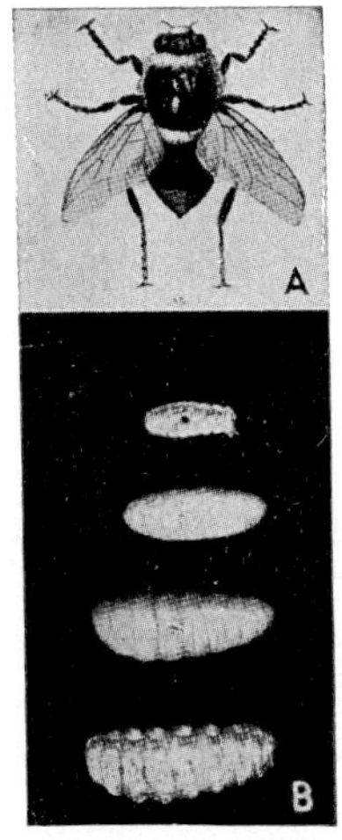

FIG. 197.—*HYPODERMA BOVIS*.

A, the fly; B, the larvæ (M. C. Ennema).

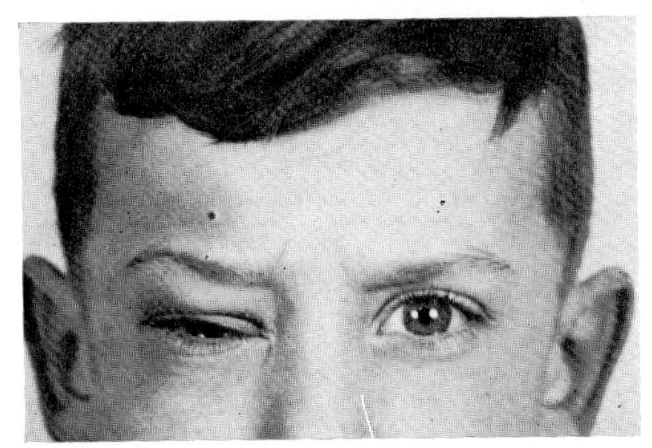

FIG. 198.—PALPEBRAL MYIASIS.

Infestation of the brow by *Hypoderma lineata* in a boy aged 7. Pus was expressible from a small hole above the centre of the right eyebrow which was swollen and indurated; pressure resulted in the extrusion of a larva. Subsequent surgical exploration revealed nothing abnormal (A. Smith and D. P. Greaves).

to give rise to a furuncular lesion (FURUNCULAR MYIASIS) or may even cause destruction of the lids (Cordero-Moreno, 1973). In this type of infection there may be much pain and itching with considerable surrounding œdema, and on opening the swelling the larvæ are seen (Gradenigo, 1894; Lagleyze, 1914; Vianna, 1951; Pacurariu and Budai, 1957; Padrón, 1957). Such a localized lesion may on rare occasions be due to the larvæ of *Hypoderma lineata* (the warble or gadfly which commonly attacks cattle) (Fig. 197), in which case there is much œdema and swelling; local and general symptoms may be absent but are occasionally severe. On the summit of the swelling a hole representing the point of exit is evident from which serum or pus may exude, but on expression of the larva resolution is rapid (Style, 1924; Smith and Greaves, 1946) (Fig. 198). *Hypoderma bovis*, which invades the eyeball itself, has given rise to a migratory

[1] Vol. VIII, p. 426; Vol. IX, p. 490. [2] p. 931.

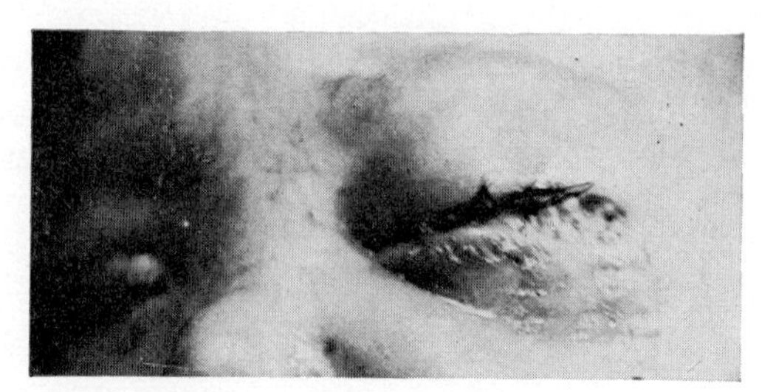

FIG. 199.

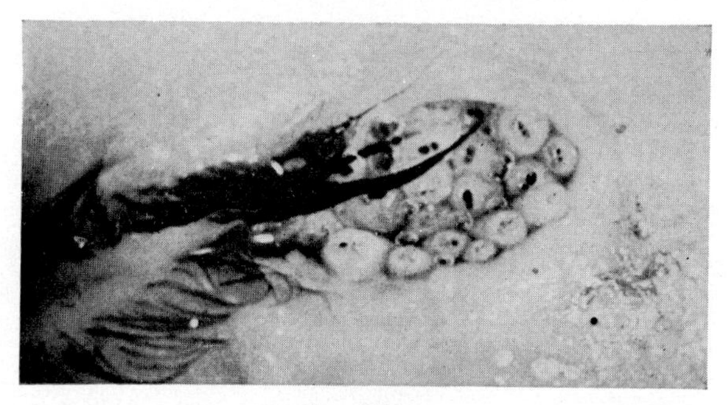

FIG. 200.

FIGS. 199 and 200.—OCULAR MYIASIS.

In a 2-year-old Egyptian boy. Both lids were swollen with an extensive ulcer at the outer canthus. At the base of the ulcer the extremities of the larvæ of *Wohlfahrtia magnifica* are seen as rounded bodies with pits bearing short black lines, their movements being associated with the bubbling of air. Twenty-two such larvæ were removed (A. Kamel).

swelling in the lids and conjunctiva from which the larva has been extracted (Bronner and Callot, 1955; Suárez-López, 1965). The *Wohlfahrtia magnifica* (flesh-fly) which habitually infests barns and cow-sheds, may similarly attack the lids (Figs. 199–200). Infestation by the house-fly (*Musca domestica*) is very rare (Teodorescu *et al.*, 1956).

Alternatively, the larva migrates, causing a linear lesion either tortuous or straight, bright red at the advancing edge and pink behind (CREEPING MYIASIS: DERMATOMYIASIS LINEARIS) (Somerset, 1962). This form of myiasis is caused by members of the Œstridæ group, usually the *Hypoderma* in Europe and North America and *Dermatobia* in South America. The treatment is incision and cleaning of the wound with chloroform water (Stewart and Boyd, 1934), and antibiotics to control secondary infection.

Bronner and Callot. *Bull. Soc. Ophtal. Fr.*, 688 (1955).

Cordero-Moreno. *Amer. J. Ophthal.*, **75**, 349 (1973).

Gradenigo. *XI int. Cong. Med.*, Rome, **6**, 40 (1894).

Lagleyze. *Arch. Oftal. hisp.-amer.*, **14**, 556 (1914).

Pacurariu and Budai. *Oftalmologia* (Buc.), **1**, 69 (1957).

Padrón. *Rev. Oftal. Venezol.*, **2**, 123 (1957).

Smith and Greaves. *Brit. med. J.*, **2**, 120 (1946).

Somerset. *Ophthalmology in the Tropics*, London, 149 (1962).

Stewart and Boyd. *J. Amer. med. Ass.*, **103**, 402 (1934).

Style. *Brit. med. J.*, **1**, 1086 (1924).

Suárez-López. *Arch. Soc. oftal. hisp.-amer.*, **25**, 100 (1965).

Teodorescu, Bobulescu and Furtunescu. *Oftalmologia* (Buc.), **2**, 60 (1956).

Vianna. *Arch. bras. Oftal.*, **14**, 47 (1951).

POROCEPHALIASIS

The larva of the arachnid, *Porocephalus* (*Armillifer*) *armillatus*, may be found in the conjunctiva,[1] in the anterior chamber and even in the posterior segment[2]; it has also been reported in the skin of the eyelid of a Liberian boy (Neumann and Gratz, 1962). There was a painless swelling near the inner canthus on the lower lid which was found to contain pus in which lay the larva.

Neumann and Gratz. *Amer. J. Ophthal.*, **54,** 305 (1962).

BITES AND STINGS: THE SPIT OF SNAKES

The stinging arthropods—bees, wasps, hornets, scorpions and some ants—show no predilection for the eyelids but, being an exposed part and being clothed by thin skin, they receive their full share of stings and the effects are more obvious and dramatic than elsewhere. The sting—a modified ovipositor in the female—usually lies in the abdomen and is typically provided with toothed lancets as well as poison glands. When a honey-bee stings, the stinging apparatus with the tip of the abdomen is left in the wound; bumble-bees do not lose their sting and can thus attack repeatedly. Wasps and hornets are more formidable than bees; the sting of the wasp contains 2% histamine and large quantities of hyaluronidase. Stinging ants, too, are not to be despised and they occur in the tropics in hosts which have to be seen to be believed: one type of South American ant can cause a swelling 15 cm. across (Weber, 1939).

On the lids a sting may give a violent reaction associated, if the stings are multiple, with much pain, œdema and sometimes erythema and, as a rarity, gangrene (Pes, 1904); an indurated lump is usually evident at the site of the wound. If the sting is apparent it should be lifted or scraped out rather than pinched, a procedure rarely successful because of the barbs and inadvisable since it is apt to spread the poison further afield in the tissues. Occasionally a sting has travelled through the lid and injured the conjunctiva or eyeball (Hilbert, 1904; Kraupa, 1911; Orendorff, 1911; Norris, 1944): a case has been recorded wherein it remained in the lid 7 years and then injured the eye (Dorff, 1921). The sting of a bee has even caused an osteoperiostitis (Kalt and Lemaitre, 1923).

Bites by insects may be equally painful. Many *ants* bite rather than sting, particularly the North American varieties; they make a wound with their strong mandibles and pass venom into it from poison glands. The little red ants of India, which in some regions seem to be everywhere in untold numbers, are similarly vicious and, running over the face during sleep, may inflict very painful bites on the lids involving enormous swellings (Elliot, 1920). They are hard to find for they lie hidden in the folds of the swollen lid; and harder to detach, for so tight is their grip that the body is often pulled away leaving the head biting into the tissue from which it

[1] Vol. VIII, p. 431. [2] Vol. IX, p. 494.

eventually sloughs out. In Ruata's (1904) experience, bites of this type are a commonplace on the lids of sleeping infants in the West Indies.

Mosquito bites are also very evident on the lids. Individuals differ in their susceptibility and some seem to develop immunity. A poison is injected from the salivary glands which causes red, itchy weals (Hecht, 1929–33). The swelling they may cause on the lids may lead to complete closure of the eye (*bung eye*) with marked chemosis of the conjunctiva, a reaction which gives the patient much anxiety.

Among Arachnids, *ticks* (*Ixodidæ*) are also poisonous owing to their salivary secretion; man is not their natural prey but they may leave their usual host (cattle or other wild or domestic animals) and attack him for blood-sucking purposes. If they choose the lids, with some species little

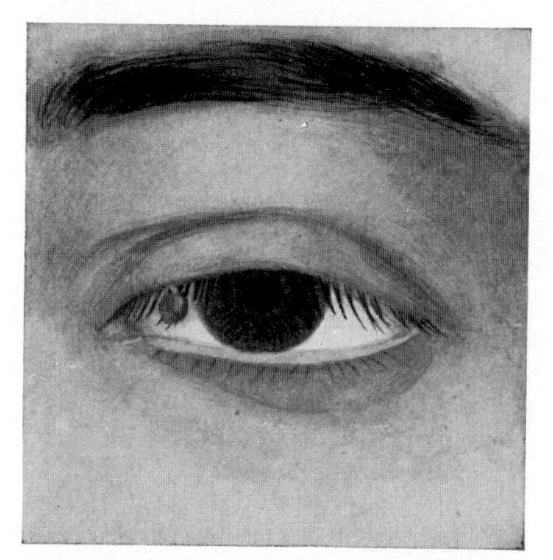

FIG. 201.—A TICK ON THE UPPER EYELID.
(*Dermacentor auratus* nymph) (E. Kirwan).

reaction may occur (Kirwan, 1935) (Fig. 201); with others great swelling and œdema may result, so great that the tick is frequently hidden in the folds (Misar, 1959). Once found it may be hard to destroy even although it is bloated with blood, and attempts to remove it may result in the head being left behind so that a suppurating sore develops. Such an accident has led to the development of a conjunctival cyst (Cosmettatos, 1912). The larva of a tick has caused a small bluish tumour on the lower lid (O'Rourke, 1963). A drop of oil or cocaine, however, makes the tick loosen its hold in which case an urticarial itchy weal is left. Of greater importance than the actual bite, of course, is the danger of many species transmitting other diseases (*e.g.*, the *Treponema duttoni* causing African relapsing fever, the pasteurella of tularæmia or the rickettsia of Rocky Mountain spotted fever, etc.).

Spider bites, because of the venom injected, may give rise to toxic symptoms of considerable severity (*araneidism*), the site becoming swollen and sometimes purpuric: occasionally general symptoms of poisoning and collapse ensue. Immediate incision and suction may be tried, while an anti-venom, if given in time, may be useful (Kirby-Smith, 1945).

Chigoe bites occur on occasion on the lids. They are due to the sand-flea (*Tunga penetrans*) found in some parts of the tropics particularly of America and Africa. The pregnant female burrows under the skin sucking blood for several days and reaching the size of a mistletoe berry, in the meantime causing ulceration and abscesses. She should be removed by a blunt needle before she lays her multitudes of eggs; in case these may be deposited, great care should be taken not to burst her abdomen in the process. If removal is difficult the animal should be touched with phenol—for if she dies *in situ* she will slough out in the ensuing abscess.

The *harvest* (*chigger*) *mite* (*Trombicula autumnalis* in Europe; *T. irritans* in America), a minute spider-like creature which abounds in the autumn fields and may serve as a vector of scrub typhus in endemic areas, produces an ulcer which becomes black and necrotic associated with an indurated erythematous eruption. In Africa, Boase (1947) found the lashes powdered with the minute red larvæ of the mite.

Centipedes and *caterpillars*, if they are disturbed crossing the face, may cause an urticarial eruption or a weal on the lids sometimes of considerable severity—by digging in their claws in the case of centipedes, by leaving spines or hairs behind in the case of caterpillars.

Bed-bug bites frequently occur on the lids, the *Cimex lectularius* (or *C. rotundalius* in the tropics) preferring to attack exposed parts of the body. The lesion produced is a large, intensely itchy urticarial weal, probably due to irritating fluid injected by the bug as it bites to increase the flow of blood. Frequently the lesions are multiple and linear.

Leech bites are common in tropical countries where warmth and excessive moisture produce a dank and luxuriant vegetation. In such swamps leeches may abound in clouds, and they cling to man with a leech-minded tenacity and ferocious determination. The face and eyelids may be attacked, resulting in œdema and swelling. Fortunately they drop off as soon as they are engorged with blood; or they can readily be removed by a strong saline solution.

The *treatment* of all these bites varies with the customs of different lands. The first essential, as has already been noted, is to get rid of the attacker or its sting. To relieve the subsequent irritation of stings weak ammonia or iodine is as good an application as any; for bites a remedy which gives rapid relief from the intense itching in most cases is a paint of ichthyol and ether.[1] In them all, of course, the best treatment—if it can be done—is prevention.

BEETLES of the genus *Pæderus* (Fig. 202) found in South America, Africa and South-East Asia, give rise to a vesicular *ophthalmo-dermatozoosis* by the action of a noxious substance secreted by the insect as a defence-mechanism. The lesion may

[1] Ichthyol 4 gm., methylated ether 2 ml.

appear on any part of the exposed skin but the region of the eyelids and the contiguous parts of the face are commonly affected. Initially there is œdema and vesicles frequently develop with a narrow yellowish line running along the centre which appears as if the surface cells had been coagulated (Fig. 203) (Somerset, 1961; Mukherjee and Ahmed, 1966; Ahmed and Roy, 1969). The conjunctiva may also be affected.[1] In India the lesion is called " spider-lick " although the irritant substance is excreted by a beetle mainly by the hinder end of the abdomen; its chemical nature is unknown. The usual history is that the insect is brushed away and no disturbance results for a period of 48 hours; recovery is uninterrupted unless secondary infection develops. The lesion should not be confused with the round, circumscribed bullæ caused by the secretions of vesicant beetles of the family Meloidæ (Swarts and Wanamaker, 1946).

FIGS. 202 and 203.—" SPIDER-LICK " (E. J. Somerset).

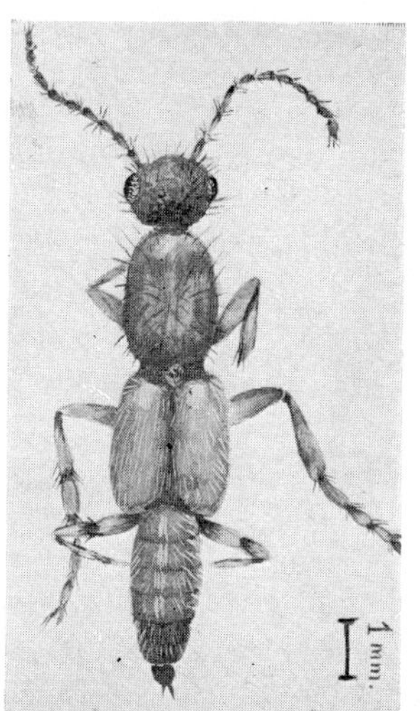

FIG. 202.—*Pœderus fuscipes* (7 mm. long).

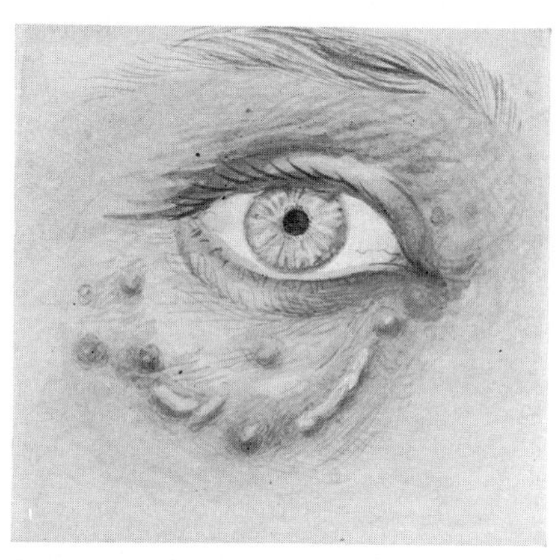

FIG. 203.—The typical lesion on the eyelids.

SPITTING SNAKES, such as the black-necked cobra (*Naja nigricollis*) met with in Africa, are apt to eject their venom for some 8 to 12 feet with great force and accuracy, and usually select the face of an intruder as a target. If the eyes are involved an intense burning pain immediately blinds the patient and blepharospasm and œdema of the lids rapidly appear, associated with an acute conjunctivitis. Unless actual damage is done to the cornea, in which tissue a bullous, exfoliative keratitis may be caused, the condition subsides within a week with atropine and bathing with saline (Zanettin, 1935; Pergola, 1942; Ridley. 1944; Somerset, 1962). In severe cases anti-venine, produced from the serum of horses which have received graduated doses of venom, may be helpful.

Ahmed and Roy. *J. All-India ophthal. Soc.*, **17**, 145 (1969).
Boase. *E. Afr. med. J.*, **24**, 363 (1947).
Cosmettatos. *Clin. Ophtal.*, **4**, 572 (1912).
Dorff. *Klin. Mbl. Augenheilk.*, **67**, 256 (1921).
Elliot. *Tropical Ophthalmology*, London (1920).
Hecht. *Derm. Wschr.*, **88**, 793, 839 (1929); **96**, 588 (1933).
Hilbert. *Wschr. Therap. Hyg. Auges*, **7**, 201 (1904).
Kalt and Lemaitre. *Bull. Soc. Ophtal. Paris*, 359 (1923).
Kirby-Smith. *Sth. med. J.*, **38**, 696 (1945).

[1] Vol. VIII, p. 430.

Kirwan. *Brit. J. Ophthal.*, **19,** 659 (1935).
Kraupa. *Zbl. prakt. Augenheilk.*, **35,** 321 (1911).
Misar. *Wien. klin. Wschr.*, **71,** 712 (1959).
Mukherjee and Ahmed. *J. All-India ophthal. Soc.*, **14,** 178 (1966).
Norris. *Brit. J. Ophthal.*, **28,** 139 (1944).
Orendorff. *Ophthal. Rec.*, **20,** 242 (1911).
O'Rourke. *Brit. med. J.*, **2.** 544 (1963).
Pergola. *Boll. Soc. ital. Med. Igiene trop.*, **1,** 80 (1942).
Pes. *Z. Augenheilk.*, **12,** 438 (1904).
Ridley. *Brit. J. Ophthal.*, **28,** 568 (1944).
Ruata. *Ann. Ottal.*, **33,** 445 (1904).
Somerset. *Brit. J. Ophthal.*, **45,** 395 (1961). *Ophthalmology in the Tropics*, London, 20 (1962).
Swarts and Wanamaker. *J. Amer. med. Ass.*, **131,** 594 (1946).
Weber. *Science*, **89,** 127 (1939).
Zanettin. *Arch. ital. Sci. med. colon.*, **16,** 856 (1935).

INFLAMMATIONS OF THE LID-MARGINS

BLEPHARITIS

JOHN DALRYMPLE [1803–1852] (Fig. 204) is a suitable introduction to the section of this volume dealing with blepharitis. In his classical work, *The Pathology of the Eye* (1852), he was one of the early writers not only to describe diseases of the eye with great accuracy but to illustrate them lavishly with painted plates which up to his time had never been equalled in their artistic merit and the faithfulness with which the conditions were depicted. He gave minute descriptions of blepharitis in its various forms and discussed its sequelæ (" lippitudo ") and also dealt with warts, horns, syphilitic blepharitis, lupus (which was said not to affect the eyelids), molluscum contagiosum (known as " tumor glandiformis "), infestation with lice, and many other conditions affecting the lid-margins and the eyes. His father was a general surgeon who devoted time to ocular pathology, and with him John Dalrymple served his apprenticeship in ophthalmology and pathology. He commenced work at Moorfields Eye Hospital in 1832, and also joined the Saunderian Institution where he studied anatomy and pathology; as a result he wrote his book *The Anatomy of the Human Eye* (1834). His writings were by no means limited to ophthalmology for he published several papers on general pathology and natural history. In his later years his health broke down and he retired from active work at Moorfields in 1849, and died just after his classical work on pathology was published in 1852.

BLEPHARITIS (MARGINAL or CILIARY BLEPHARITIS) (βλέφαρον, an eyelid), a subacute or chronic inflammation of the margins of the lids, is an extremely common disease. As the meeting place of the skin and conjunctiva this region tends to share in the affections of either, while the complexity of its follicular and glandular structures provides ready access and prolonged hospitality to intruding organisms. The ætiology is unusually complex and treatment may be exceptionally difficult and protracted.

General Clinical Picture

Apart from specific infections which will be discussed separately, the pathological changes in inflammations of the lid-margin as a rule conform to one of two general types, the first of which may sometimes merge into the second: *simple squamous blepharitis*, a superficial, non-destructive dermatitis, and *follicular* (*purulent*) *blepharitis*, wherein the follicles and glandular structures are affected producing either a pustular or ulcerative condition. This differentiation, however, although clinically useful, does not rigidly indicate the organism involved (Galin, 1962).

FIG. 204.—JOHN DALRYMPLE
[1803–1852].

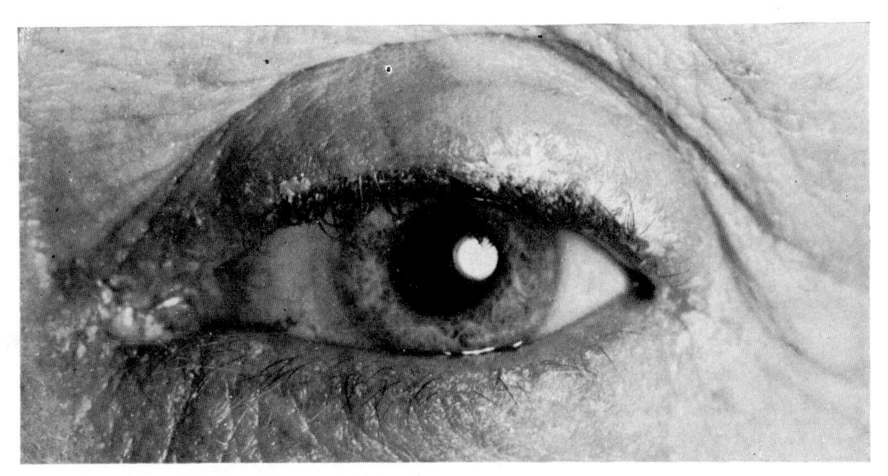

FIG. 205.—SQUAMOUS BLEPHARITIS (Inst. Ophthal.).

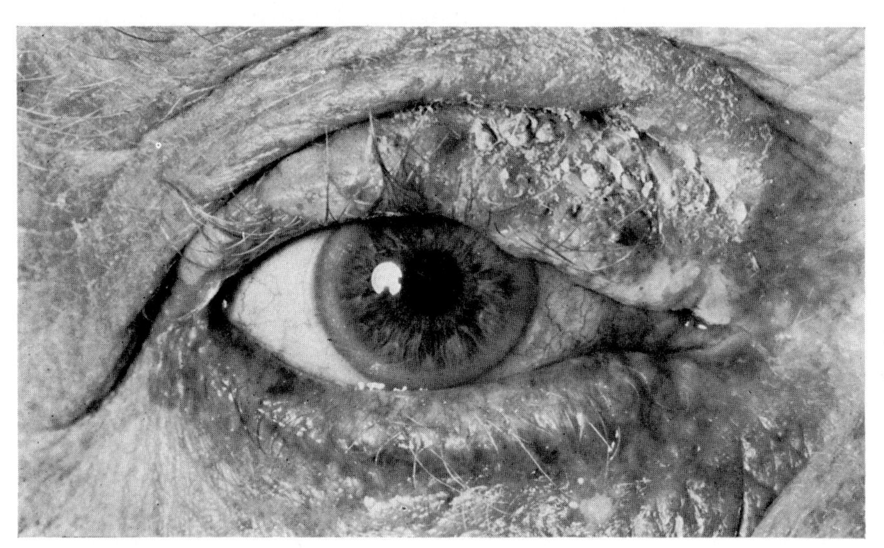

FIG. 206.—BLEPHARITIS WITH HYPERKERATOTIC CHANGES (Inst. Ophthal.).

SQUAMOUS BLEPHARITIS is essentially an eczema-like cutaneous inflammation, characterized pathologically by *parakeratosis* (*desquamation*) *and acanthosis* (*hypertrophy*) *of the epidermis, and by congestion, œdema and infiltration of the dermal tissues* (Plate IV, Fig. 1; Figs. 205–6). Clinically the initial symptom is hyperæmia of the lid-margin. In the normal lid-margin the anterior (cutaneous) zone, which is a continuation of the skin, is separated by a fine linear trace of a greyish tone (the intermarginal linear zone) from the posterior (tarsal) zone. In the early stages of inflammation the anterior zone under high magnification shows fine vascularization, the linear zone disappears, and the tarsal zone becomes red, vascular and irregularly granular. Associated with these changes a bran-like scaling of the epidermis (*pityriasis*) occurs. The hyperæmia gives an unpleasant, bleary and swollen appearance to the lids, and the scales, at first loose and fine like dandruff, may become coarser and adherent with secretion to form yellow crusts. The presence of a smooth hyperæmic bed without ulceration when these scales are removed differentiates this type of blepharitis from the follicular type. As time goes on, however, the margins become permanently swollen and thickened, and the posterior edge of the palpebral margin, losing its sharp angulation and therefore its perfect coaptation with the globe, allows the tears to accumulate so that epiphora results. The continual wetting of the skin of the lids associated with constant rubbing of the eyes and the spreading of secretion from the inflamed margins frequently give rise to eczema followed by contraction so that an eversion of the lid-margin ensues; this in the lower lids gradually develops into an ectropion. The eversion of the lacrimal punctum thus induced adds to the epiphora and the increased rubbing of the lids, usually with a soiled handkerchief or dirty fingers, perpetuates and adds to the infection, so that a vicious circle tends to be set up. Meantime the lashes are damaged and frequently lost owing to hyperkeratosis at the infundibulum and neck of the follicles, but the alopecia is usually partial and may be temporary unless the condition is of very long standing; moreover, in the absence of deeper inflammation, distortion of the lashes does not result (Fig. 207).

FOLLICULAR BLEPHARITIS (PUSTULAR OR ULCERATIVE BLEPHARITIS) (Plate IV, Figs. 2 and 3), a deeply seated and purulent infection, is a more serious condition wherein the inflammatory symptoms are more pronounced and permanent sequelæ result; it is less common than the squamous variety and is becoming more rare. Pathologically it is characterized by *a suppurative inflammation of the ciliary follicles and the associated glands of Zeis and Moll.* In the initial stages the suppuration is confined to the hair follicles; intra-follicular abscesses are formed which appear as tiny pustules developing on the margin of the lids (*pustular blepharitis; folliculitis interna* of Herzog, 1904). A further stage is an extension of the inflammation outside the hair-follicle into the surrounding connective tissue where a *peri-folliculitis* appears (*folliculitis externa* of Herzog, 1904). In the third stage a peri-follicular abscess is formed resulting in considerable destruction and ulceration (*ulcerative blepharitis*); the final stage is that of healing by scarring involving deformity of the lid-margin and destruction of the lashes. The first stage is typified in impetiginous infections, the remaining two are seen in sycotic infections and, when extending to the sebaceous glands of Zeis, constitute a hordeolum.[1]

Clinically the lid-borders are red and inflamed and the lashes matted with yellow crusts which, when they are removed, reveal not merely the

[1] p. 234.

PLATE IV

BLEPHARITIS (Inst. Ophthal.)

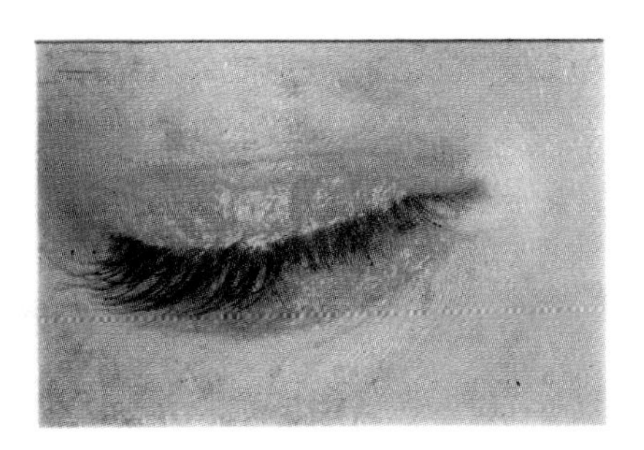

FIG. 1.—Squamous blepharitis.

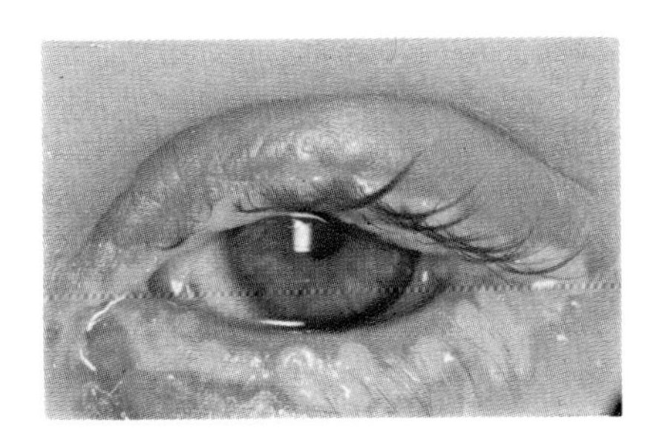

FIG. 2.—Purulent blepharitis.

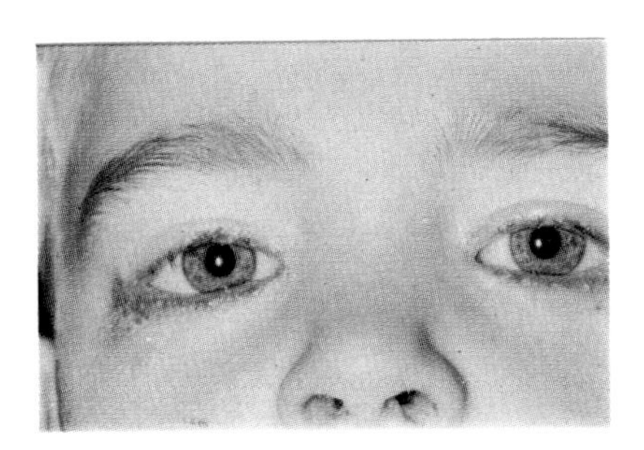

FIG. 3.—Ulcerative blepharitis.

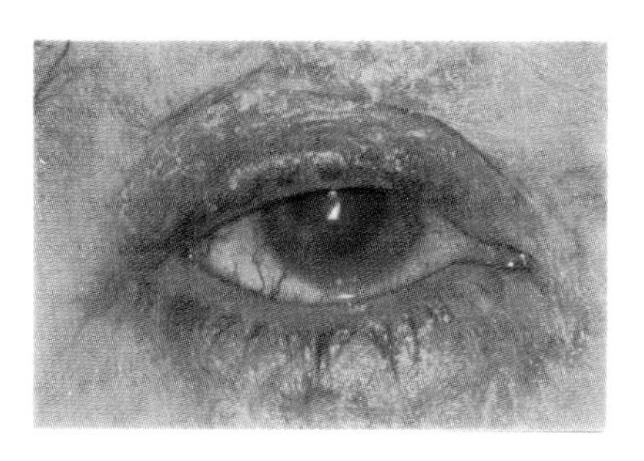

FIG. 4.—Rosacea blepharitis.

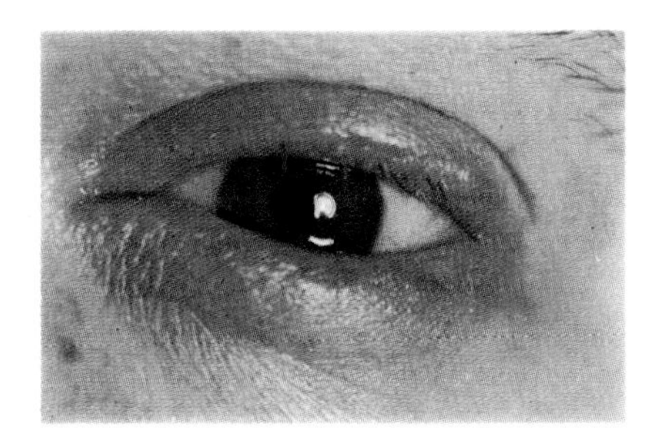

FIG. 5.—Chronic staphylococcal blepharitis.

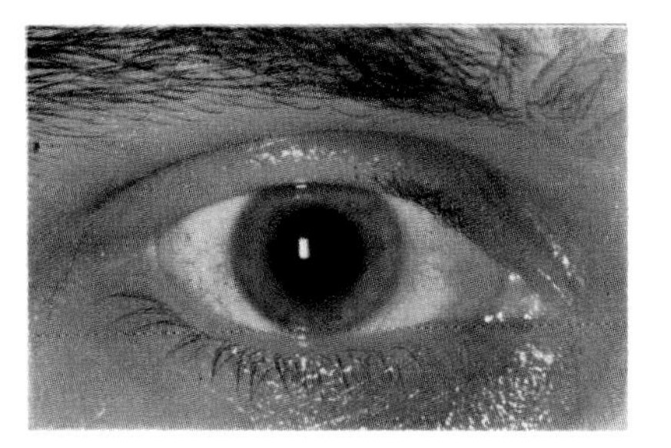

FIG. 6.—Madarosis after blepharitis.

hyperæmia of the squamous type, but suppurative processes in the tissues. Scattered over the lid-margin small yellow elevations are seen each pierced by a cilium, each a tiny abscess of the underlying hair-follicle and the gland associated with it. As the perifollicular tissues become involved an ulcer is formed which heals with the formation of a scar injuring or destroying the cilium. As the process spreads, abscesses, ulcers and scars multiply, and more and more lashes are damaged by suppuration and scarring, until

FIGS. 207 and 208.—ALOPECIA FOLLOWING BLEPHARITIS.

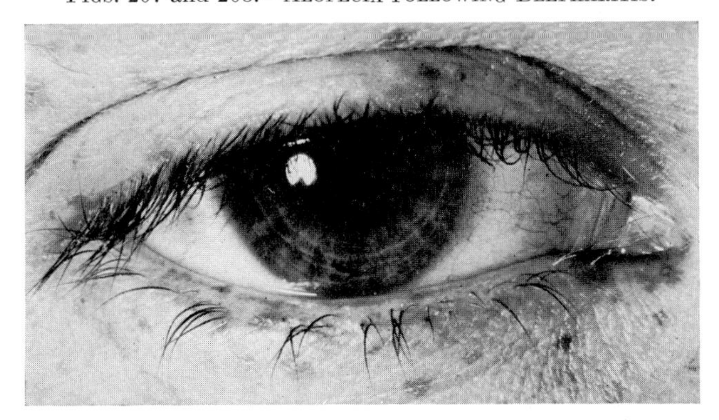

FIG. 207.—With partial loss of the lashes (Inst. Ophthal.).

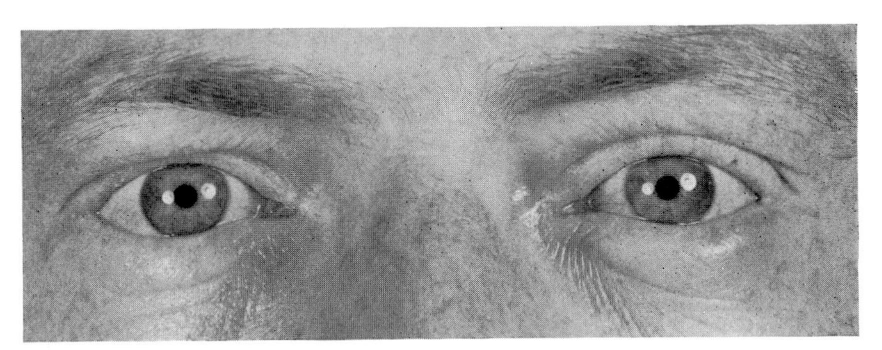

FIG. 208.—With complete loss of the lashes (B. Jay).

only isolated groups remain glued together into tufts by secretion; frequently these are stunted, deformed and pulled out of alignment by cicatricial contracture so that they project onto the cornea (trichiasis).[1] In the ulcerated region the papillæ become hypertrophied and may even form painful warty excrescences which readily bleed. At the same time the continued irritation of the surrounding skin by discharge, epiphora and mechanical rubbing may set up an eczematous dermatitis with subsequent contraction leading to ectropion and eversion of the border of the lower lid,

[1] p. 383.

while a chronic conjunctival catarrh is invariable. As these changes progress the marginal region becomes hypertrophied, rounded and swollen, bordered by a raised rim of fleshy-looking conjunctiva, so that the thickened mis-shapen borders droop in consequence of their own weight (*hypertrophic blepharitis; tylosis: τύλος*, a lump). Thus the process goes on, sometimes for many years, sometimes all through life, and eventually on the thickened and red lid-margin only a few scattered, abortive hairs are left (*madarosis*, *μαδαρός*, bald)—a most unsightly condition—until finally, when all have been destroyed and no more follicles remain to harbour suppuration, the disease tends automatically to die out leaving a permanently deformed lid-border (Fig. 208; Plate IV, Fig. 6).

The *complications* of blepharitis may therefore be summarized thus:

(1) Chronic conjunctivitis—a constant concomitant.

(2) Destruction of the lashes. In the squamous type *alopecia* of the lashes results if of long duration; in the follicular type distortion of the follicles leading to *trichiasis* or destruction of the lashes (*madarosis*).

(3) Permanent reddening of the lids.

(4) Hypertrophy, thickening and deformity of the lid-border (*tylosis*).

(5) Eversion of the border of the lid developing into a post-inflammatory *ectropion*.

(6) The trichiasis may produce corneal complications, ulcers and nebulæ with permanent visual impairment.

Symptoms are usually not prominent, the patient frequently coming for treatment for cosmetic reasons long after the disease has taken hold, the complaint being of the appearance of the red-rimmed eyes and the scantiness of lashes. On the other hand, with relatively mild objective signs, symptoms of heat, itching and grittiness may be prominent and sometimes actual pain. Frequently there develops an increased sensitivity of the eyes, which tire readily, are intolerant of light, dust or heat, and persistently weep. A relatively constant feature is difficulty in opening them in the morning when the lids are glued together. In the more severe cases, of course, symptoms of pain, blepharospasm and photophobia are severe, particularly when the disease of the ciliary margin is complicated by an eczema of the skin and conjunctival and corneal involvement.

The *diagnosis* is usually simple. A matting of the lashes by conjunctival discharge is the only point to be remembered. In all cases the scales and discharge must be carefully wiped away and the lid-margins examined with a loupe or a slit-lamp—if the underlying skin is normal, conjunctivitis only is present; if it is red, squamous blepharitis; and if pustules or ulcers are seen, follicular blepharitis is diagnosed.

General Ætiology

As is the case in most diseases, the ætiology of blepharitis is rarely simple. Considered in their widest sense, the factors determining the occurrence of marginal inflammation can be divided into two main categories—those responsible for creating a suitable soil for its development, and those which act as excitants.

Predisposing Factors. To a large extent the same factors which lead to marginal hyperæmia predispose to the establishment of inflammation, since prolonged or intermittent congestion prepares the way for further pathological developments. These have already been discussed,[1] but the more important must be summarized here.

(*a*) *External irritants*, such as dust, wind, cold, heat and other radiation not only in warm climates but associated with industrial processes, smoke, chemicals, and polluted or hygienically bad atmospheres, all play a role of considerable importance. To these must be added the common habit of rubbing the eyes with dirty fingers or soiled handkerchiefs which not only irritates the lids mechanically but also frequently infects them directly by conveying infection from extraneous sources or from other parts of the body as the nose, the scalp or cutaneous lesions elsewhere. This is a habit common to all, but particularly amongst children especially at the time of the first dentition when salivation is profuse and the fingers rarely leave the mouth.

(*b*) *Eyestrain* in its widest sense—refractive errors, accommodative strain, muscular imbalance, over-work of a fine nature, particularly in conditions of poor or ill-placed lighting, and loss of sleep—has long been given an important place in the ætiology of blepharitis; indeed, it used to be considered by some clinicians as the most common effective cause not only of ciliary hyperæmia but even of inflammation.

This view was first strongly advocated by St. John Roosa (1876) and Keyser (1877) in America who contended that even ulcerative blepharitis could be caused in this way, particularly by hypermetropia (Hall, 1882; Richet, 1885; Clarke, 1894; Hotz, 1899). Winselmann (1899) and others published statistics supporting the same contention by the claim that most sufferers had considerable refractive errors. It is to be remembered, however, that among children, the most common sufferers from blepharitis, hypermetropia is the rule, and that others have found no significant difference in refraction between the blepharitic and the normal nor any tendency for uniocular blepharitis in anisometropia (Schreiber, 1924; Somerset, 1939). It is obvious that the first view is exaggerated, but on the other hand it is possible that, while eyestrain cannot reasonably be arraigned as a primary cause of blepharitis, its presence may, by inducing continuous reflex hyperæmia, aid and abet other causes which, acting themselves, might have been relatively ineffective, and that, for the same reason, the relief of the strain by spectacles and wise ocular hygiene may be followed by improvement in a certain number of cases.

(*c*) *Constitutional factors* form the third category of important predisposing causes which in many instances are hereditary. These are many and various. In the first place the *nature of the skin* has an influence, particularly in squamous blepharitis of seborrhœic origin; blonds with fair skins are more susceptible than brunettes, and Bassin and Skerlj (1937) found its incidence twice as great with blue or grey eyes than with brown. Sex has no significant influence. *Metabolic disturbances* are frequently in

[1] p. 6.

evidence—over-indulgence in food, particularly of a starchy nature, or stimulants, dyspepsia, faulty elimination, renal or hepatic insufficiency, gout or diabetes are all of importance. The influence of the *endocrine system*, particularly the thyroid and the gonads, has been cited, although with little evidence, in view of the prevalence of blepharitis at the critical sexual epochs. *Toxic factors* have been considered as of ætiological importance, particularly foci of infection in the teeth, tonsils, nose and sinuses (Serov and Ovchinnikova, 1971). This is seen particularly in children, among whom blepharitis commonly accompanies chronic catarrh of the nasal mucosa, nasal obstruction and enlarged adenoids, and it is significant that the same organism, frequently a hæmolytic staphylococcus, may be found both in the nares or upper respiratory tract and on the margins of the lids (Prigal, 1953). Finally, *allergic reactivity* is frequently of importance, the lid-margin sharing sometimes even more markedly than the skin in the tendency, sometimes hereditary, to eczema due to a host of causes, some of which predispose to the persistence of a chronic or recurrent inflammation while others, whether irritant factors in the environment, constituents of the diet, or contact materials, either industrial, accidental or cosmetic, may provoke an acute crisis. Allergy to micro-organisms should not be forgotten, particularly hypersensitivity to the exotoxin of a staphylococcus which need not be present in the conjunctiva but frequently inhabits the nasal and nasopharyngeal mucosa.

On such a background is blepharitis prone to flourish, given a specific cause. The main factors, which may operate sometimes singly and sometimes in combination, may be classified thus:

I. *Associated inflammations* which affect the lid-margins by direct spread, either cutaneous, conjunctival or lacrimal.

II. *Localized Marginal Inflammations.*

(*a*) *Simple blepharitis*, the basis of which in the great majority of cases is either seborrhœa or rosacea. Here the infective element is secondary and saprophytic while the terrain is important, a fact first clearly pointed out by Cuénod (1894). In this connection an excessive population of *Demodex folliculorum* should be remembered.[1] Unless infective complications occur the clinical picture is of the squamous type (*blepharitis sicca* or *oleosa*).

(*b*) *Infective Blepharitis.* Here the infective element is the more important, being usually staphylococcal (Jones *et al.*, 1957), most often situated in or around the roots of the lashes (Norn, 1970), and the clinical picture is of the follicular type varying from a simple superficial folliculitis to an ulcerative sycosis or acute furunculosis. In Japan an organism, the *Micrococcus conjunctivæ* has been isolated in many cases (Mitsui *et al.*, 1951; Hasegawa, 1952). The flora obtained by Galin (1962) from cultures of lid-scrapings in the USA were as follows:

[1] p. 226.

Staph. aureus (coag. pos.)	10	Budding yeast	10
α streptococcus	4	Mixed Staph. aureus (coag. pos.) and budding yeast	14
β streptococcus	1	Pseudomonas	1
Pneumococcus	1	Normal flora	8
Proteus	1		

(c) *Irritative blepharitis*, due to physical or chemical irritants or allergy thereto. The clinical picture may be either of the squamous or follicular type.

III. *Inflammations of a specific infective nature*, frequently associated with systemic infections—the eruptive fevers, vaccinia, tuberculosis, syphilis, leprosy, viral infections, fungal infections, parasites, and so on. The clinical picture in this type varies considerably with the exciting agent.

Of all these causes *the commonest are dirt and the staphylococcus in children, seborrhœa in adolescents, with rosacea and allergic sensitization added in adult life.*

General Treatment

The treatment of blepharitis is frequently a difficult business, for with its complicated ætiology many factors have to be taken into consideration and controlled. Moreover, the constitutional basis which is frequently present tends to chronicity and recurrences unless the therapeutic measures are as fundamental and far-reaching as the origins of the disease, while the anatomical complexity of the ciliary region harbours infection indefinitely unless the local condition is tackled with persistent and thorough care. In most cases treatment must embrace general measures of health and hygiene and local therapeutic procedures, as well as a specific attack against particular organisms.

General constitutional measures must be determined by an assessment of the patient as a whole, his hereditary and previous personal history. In many cases, particularly in children, the general health requires attention, frequently by a complete change from surroundings which may be unhygienic and a life which may be physically trying, to a holiday environment in the country, the mountains or by the sea, where cleanliness, open air exercise and rest are easy to find. Of great importance is diet; irregular feeding, particularly with a carbohydrate diet, or over-feeding should be replaced by a sensibly balanced intake together with regulated intestinal hygiene. To this may be added abundance of fruit juices and vitamins, particularly A, B and D as in cod-liver oil or brewers' yeast. Underlying sources of focal infection especially in the mouth or nasopharynx, and particular diseases such as anæmia, gout or diabetes may also require attention. Sources of allergy should also be remembered, particularly in recurrent cases; the environment at work or at home should be examined from this point of view and appropriate measures taken either by eliminating the offending agent, changing the employment or inducing desensitization.

Sensitivity to articles in the diet should not be forgotten in this connection; care in this respect regarding particular items (for example, shell-fish, fruit) may change the course of the disease, and weaning or a change of milk may in itself suffice to clear up an obstinate blepharitis in a suckling.

Local sources of irritation come next for review. If a marked refractive error or heterophoria complicates the situation, particularly if eyestrain seems an obvious factor as at school or in occupations involving prolonged close or detailed work, spectacles should be prescribed and more than usual care taken with the efficiency of the illumination. In predisposed persons extremes of heat or cold, the irritation of wind or dust, vitiated smoky atmospheres or heated air either over-dry as with excessive central heating or in forges and foundries, or over-damp as in certain tropical climates or industrial processes, should be avoided; when that ideal course is impossible, such protection as may be obtained by dark glasses or goggles should be considered. Contamination of the lids themselves by rubbing should be guarded against, particularly in children and in trades where oily or gritty dirt abounds. Frequently of equal importance is contamination of the lids in susceptible persons by cosmetic creams, powders, lash-blacks and dyes. These should be discontinued or examined for sensitivity and if necessary changed. If cosmetic colouring of the brows and lashes is a necessity the safest material is non-irritating lamp-black; even such an apparently innocuous powder as mascara contains grains and crystals which are mechanically and chemically irritating and may perpetuate a blepharitis which in itself would be of minor importance; and the favourite face-powder which has succeeded innocuously and happily in its function for years may eventually and surprisingly have suddenly determined a state of sensitivity.

Associated lesions should also receive attention; it is useless to treat a blepharitis as an isolated entity in the presence of a seborrhœic or lousy scalp, a rosacea of the face or widespread dermatitis, a conjunctivitis or persistent nasal infection. This last local source should be investigated with particular care.

Local treatment consists first and foremost in the maintenance of cleanliness with constant removal of scales, exudates and crusts. Without this being done scrupulously, determinedly and persistently, every case of blepharitis tends to become chronic long after its original cause has disappeared. In blepharitis sicca this is relatively easy, but it is to be remembered that the fine bran-like scales return rapidly after they have been removed. They are best removed by a pledget of cotton-wool soaked in oil (olive oil, cod-liver oil, paraffin); a very fine comb was suggested by J. W. Smith (1938). In the greasy type of squamous blepharitis a non-greasy fat-dissolving substance is best; a tooth-pick swab soaked in benzene, tincture of iodine or weak silver nitrate is useful. In pustular and ulcerative conditions cleansing of the lid-margins is more difficult and must be preceded by a thorough removal of the discharge and softening of the crusts by hot fomentations, washing with warm water and soap, lathered if necessary into the lashes with the eyes closed, or by the application of olive oil or an ointment; the crusts should then be removed by a pledget of cotton soaked in oil or hydrogen peroxide, stroked in the direction of the lashes until they and the ciliary border are completely exposed. This process may be difficult to do thoroughly and at the same time gently, particularly in children since it may be painful, but it should be persisted in night and morning, for only by the complete removal of the crusts can medicaments reach the diseased areas, and only by great care is hæmorrhage avoided which itself not only prevents access to the follicles but also forms an ideal nidus for further bacterial growth.

Once the ciliary region has been cleaned the next step is the removal of dammed-up secretion or pus. In the seborrhœic types of the disease the removal of excessive secretion in the glands by tarsal massage frequently produces astonishing results; the process should be repeated as often as possible until the condition has resolved. In the ulcerative types follicular abscesses are opened by the epilation of infected cilia, using a cilium forceps with broad rounded ends. This should be done thoroughly and drastically every day. Under the magnification of the loupe an infected cilium looks dark and thick near its root, and when epilated the root comes away encased in a cylindrical mass of necrotic material; if it is not removed this acts as a source of infection, a pustule forms, the lash is permanently lost, and the resulting perifollicular ulcer causes scarring and mutilation. It is wiser, therefore, to be prodigal and remove too many lashes than too few, and in severe and spreading cases the safest plan is to epilate the lashes completely, a process which should occupy several sittings and which may have to be repeated several times. Indeed, in these cases, it is well to keep repeating total epilation time after time until the lids appear normal; if the follicles are prevented in this way from being destroyed the lashes will always grow again. It is important to remember that the most effective single measure in the treatment of blepharitis is good lid-hygiene.

The ciliary margins having been cleaned and buried infection drained, there remains the application of healing medicaments or antibiotic drugs. It is to be remembered that all applications if used habitually must be non-irritative; moreover their reaction varies considerably with each patient and with the same patient in an inconstant way so that changes are frequently desirable.

Lotions have been employed from time immemorial; and while they are gratefully accepted, their use should not be overdone lest the skin is macerated. They should always be bland, for unless definite infection is present, even mild drugs such as mercury oxycyanide are not well borne. Sodium borate or bicarbonate lotions[1] are ideal, or even saline.

Ointments are frequently combined with lotions, being applied parti-

[1] A bland isotonic lotion—borax, 0·75; sodium bicarbonate, 0·75; aq. ad 100.

cularly at night to avoid the lids sticking together. Again they should be non-irritant, and the common but irritating yellow oxide of mercury frequently does more harm than good and sometimes reacts virulently (Kesten, 1931). Mercury, indeed, is rarely indicated; zinc and ichthyol will usually be found more satisfactory. In many cases, however, particularly of the seborrhœic type when the skin is already greasy, the lids are intolerant of ointments, becoming red and macerated under them. A keratolytic agent such as salicylic acid (2%) or ammoniated mercury (5%) has been advised to increase the penetrance of antibacterial drugs (Galin, 1962).

The best way to apply an ointment is for the patient to rub it into the closed palpebral fissure with his (washed) finger after the scales and crusts have been freshly removed. In severe cases, however, a more prolonged and penetrating action is obtained by spreading it thickly on a swab which is bandaged over the eye and left in place through the night.

Maceration and fissures of the skin resulting from epiphora and discharge, and sometimes from wrong treatment, are most easily cured under an eschar precipitated by the application of 10% silver nitrate; this is not removed for some days and may be followed by the application of a refrigerating paste.[1]

In pustular or ulcerative conditions antiseptics or antibiotics should be applied to the diseased areas as soon as the margins are cleaned. Many drugs have been employed, the oldest being tincture of iodine in 5% solution, silver nitrate as a paint or ointment (Bargy, 1949) or the silver nitrate stick delicately applied; swabbing with protargol or other silver preparations used to be widely practised. Thereafter the margins were covered by a salve of zinc, boracic, lanolin and paraffin or by olive or cod-liver oil. An alternative type of application, more easily applied and usually more effective, is one of the antiseptic aniline dyes. Flavine was one of the earliest used (Lawson, 1919); brilliant green (Medvedeff and Euffa, 1929; Tulipan, 1936; Franks, 1941) and malachite green (Aleeva, 1931) have been widely employed as also has gentian violet in alcoholic solution (Doherty, 1940) and mixtures of these (liquor tinctorium). These aniline preparations are sometimes exceedingly effective in both the squamous and ulcerative types of the disease.

In infective conditions, however, most of these applications have been replaced by one or other of the bacteriostatic drugs incorporated in an ointment. The first of these to be widely used were the sulphonamides, sodium sulphacetamide being particularly applicable because of its relative non-irritancy. To a large extent these in their turn have been replaced by antibiotics, the particular drug being ideally chosen by the results of sensitivity tests. These drugs have undoubtedly revolutionized the treatment of acute infective conditions of the lids and rendered antiseptic applications obsolete. It is to be remembered, however, that in chronic infections their effect is less dramatic and if such chronic infections are recurrent, the organism may develop resistance. Moreover, in seborrhœic

[1] Ichthyol, 0·1, zinc oxide, 1·0, bismuth subnitrate, 1·0, ung. aq. rosæ, 10·0.

states and conditions wherein infection is by a non-susceptible organism or the inflammation is due to the exotoxin of an organism elsewhere (as a nasopharyngeal staphylococcus), they are useless. It is to be remembered also that not infrequently they excite sensitivity reactions. Finally, even in those cases wherein they are most obviously indicated, their use must be combined with adequate ætiological measures together with the maintenance of local cleanliness or their effect is almost certain to be merely temporary. There is a definite place for such drugs in the treatment of blepharitis, but their indiscriminate use will result in as many failures as cures. The most useful will be indicated when we are discussing the various infections. In chronic conditions Thygeson (1969) advised a combined corticosteroid-antibiotic preparation; more rapid results have been claimed with such combined treatment (Aragones, 1973). After such treatment for a considerable time an unhappy complication is a super-infection with *Candida*; such an event should be monitored by scrapings and cultures and treated with ointments of nystatin or amphotericin B.

Immunization in infective conditions also has a place in therapy, particularly against the staphylococcus and especially in those cases—which are not rare—wherein a hypersensitivity to this organism exists. The use of an autogenous vaccine has had its advocates; immunization with staphylococcal toxoid sometimes combined with an autogenous vaccine has been said to be more effective and has produced good results in chronic cases when all local methods have failed (Thygeson, 1938–69), a view, however, not universally maintained. Such treatment, of course, must be supplemented by efficient local treatment and be persisted in, and its results assessed bacteriologically: if after the course of injections a culture of the lid-margins shows no infection with the *Staphylococcus aureus*, the probability is that further relapses will not occur: if, however, there is symptomatic and not bacteriological cure, a relapse can be expected within six months.

Finally, *plastic surgery* may be required to alleviate a deformity of the lids. Thickening and tylosis may be improved by repeated massage of the lid with yellow oxide of mercury ointment. Trichiasis should be treated by electrolysis if few lashes are involved. An eversion of the lid may be required in entropion and an inversion in ectropion and to relieve epiphora. To mitigate the cosmetic defect of the loss of the lashes many expedients have been tried, from applying cosmetics or tattooing a line of indian ink to simulate the cilia, to the grafting of extraneous hairs into the lid-margin.[1]

Clinical Types of Blepharitis

BLEPHARITIS OF CUTANEOUS ORIGIN

As would be expected most of the types of dermatitis affecting the lids are liable to spread to their margins, whether they be of irritative, allergic or

[1] p. 381.

infective origin. These have already been discussed both as to their ætiology, clinical course and treatment and need detain us little here. It is to be remembered, however, that owing to their vascularity and complex folliculo-glandular structure, the margins of the lids tend to become even more heavily affected than the palpebral skin and inflammatory processes tend to linger here longer—too often indefinitely.

Any dermatitis may act in this way producing sometimes a squamous and sometimes a follicular inflammation.

The many forms of *eczema* frequently attack the lid-margins even although the adjacent skin of the lids is unaffected.

Sycosis is usually spread from a corresponding staphylococcal infection of the chin and produces a typical and recalcitrant suppurative ciliary folliculitis.

Impetigo is usually carried by the fingers from the face, lips, nose or ears, the coccal infection producing a rapidly developing superficial folliculitis.

Erysipelas rarely leaves the lid-margins intact. A chronic squamous blepharitis complicated by loss of the lashes, frequent relapses and styes is the most usual manifestation; but if the lid-margin is directly affected, necrosis may result with subsequent deformity, ectropion and trichiasis.

The more chronic dermatoses may also be particularly evident in the ciliary margin, both the proliferating exfoliative type of lesion as *psoriasis*, *pityriasis rubra pilaris*, *dyskeratosis follicularis*, *lupus erythematosus*, or *dermatitis exfoliativa*, and the rare hereditary *keratosis follicularis spinulosa decalvans*, and the vesicular type of lesion as *pemphigus* or *epidermolysis bullosa*.

BLEPHARITIS OF CONJUNCTIVAL ORIGIN

Inflammatory affections of the conjunctiva rarely leave the lid-margins unscathed. The involvement may take the form of direct spread of a conjunctival inflammation to the ciliary margin, irritation from infective conjunctival discharges, or irritation from lacrimal disease.

The direct spread of conjunctival infection occurs in some degree in every acute and chronic conjunctivitis, in which case the posterior lip of the palpebral margin first becomes red and swollen (*blepharitis marginalis posterior* of Aubaret, 1930): eventually the glueing together of the lids and lashes particularly in the morning is due to blepharitis as well as conjunctival secretion. Here again the infection may so entrench itself that the ciliary inflammation may linger after the conjunctival involvement has cleared and may eventually dominate the clinical picture. On the other hand, it is to be remembered that in many cases the inflammatory reaction here is less virulent than in the conjunctiva, a fact perhaps due to inhibition of organisms which thrived in the alkaline conjunctival secretion when they reach the acid reaction of the skin (Gowen, 1937).

Three conditions are of particular importance:

In *angular conjunctivitis* due to infection by the diplobacillus of Morax-Axenfeld the associated *angular blepharitis* dominates the clinical picture, and the ciliary region as well as the skin of the lids forms the main reservoir of re-infection. It will be remembered that this organism thrives more easily in keratinized squamous cells than in those of the conjunctival mucosa and here produces its peculiar proteolytic enzymes in greatest abundance. The resulting clinical picture is that of vivid œdematous eczema with parakeratosis of the epidermal cells involving their intumescence and maceration, and leading to the formation of fissures and rhagades. The treatment of this condition has already been described in a previous Volume.[1]

Trachoma also affects the ciliary margin with great constancy, leading to chronic inflammation of the ciliary follicles and thickening, scarring and distortion of the palpebral margins with subsequent entropion and intractable trichiasis from such cilia as survive.

Phlyctenular kerato-conjunctivitis is also constantly associated with a blepharitis, probably not so much by direct spread, but owing to the fact that the palpebral margins are simultaneously affected by the allergic factors determining the disease.

Infection from a conjunctival discharge assists the direct spread of conjunctival inflammation to set up a blepharitis, a factor frequently accentuated by spastic closure of the lids and the irritation caused by rubbing them convulsively with the fingers or fists or burrowing the head into pillows or clothes which are soiled with discharge. Naso-lacrimal infections whether catarrhal or purulent also lead to an obstinate *lacrimal blepharitis*, the sequence being aided by the accompanying epiphora with its macerating effect on the epidermis.

SEBORRHŒIC BLEPHARITIS

The seborrhœic state[2] is the commonest basis of all types of blepharitis. Seborrhœic blepharitis starts typically in adolescence and persists into late adult life in a patient with a scurfy head, acne, seborrhœic dermatitis of the back and chest and, particularly in the male, with a tendency to superficial staphylococcal infections on the face which become manifest on the lids as styes and chalazia. The tendency is undoubtedly hereditary.

The inflammation superimposed upon the seborrhœic state has been associated with the presence of two yeast-like fungi, *Pityrosporum ovale*, isolated by Moore and his colleagues (1935–36) and the *P. orbiculare* (Fig. 209), isolated by Gordon (1951), both of which have been cited as the cause of seborrhœic blepharitis as well as seborrhœic dermatitis (Gots *et al.*, 1947; Keddie and Shadomy, 1963).[3] It has been demonstrated that both organisms are found in the skin of the lids and lid-margins, but *P. orbiculare* is more common on the margins than on the skin of the face or scalp. At the same time, the frequency of these organisms in the normal skin is very high (Martin-Scott, 1952; Spoor *et al.*, 1954): 100% in seborrhœic and 85% of non-seborrhœic scalps (Portnoy, 1950), 98% in seborrhœic and 58% in normal subjects on the lid-margins (Thygeson and Vaughan, 1954), 100% in seborrhœic blepharitis and 93% on normal lid-margins (Parunovic and Halde, 1967), 100% in seborrhœic blepharitis and 50% in subjects with normal lid-margins (Uchida *et al.*, 1969). It is generally accepted that the organism is saprophytic and not parasitic, but it would seem possible that the seborrhœic skin with its excess of sebum allows it to flourish so that it may assume more than its normal saprophytic role while, owing to the damage it inflicts upon the horny layer, the way is paved for the entrance of staphylococci and other organisms. At the present time its role as a causal factor is by no means clear.

[1] Vol. VIII, p. 79. [2] p. 29. [3] p. 267.

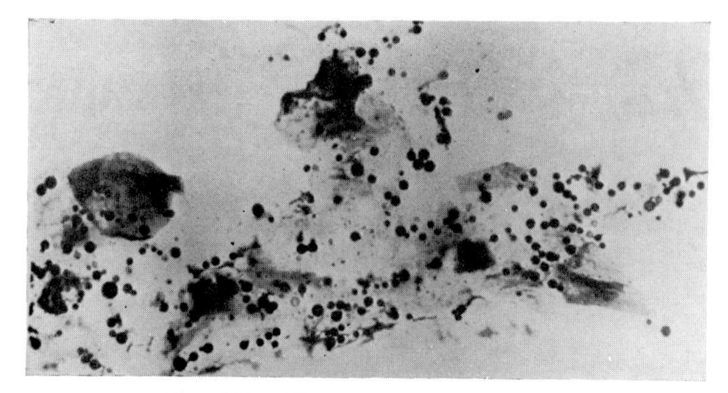

FIG. 209.—*PITYROSPORUM ORBICULARE.*

The scale from a seborrhœic lid-margin showing abundant organisms (A. Parunovic and C. Halde).

The blepharitis may assume one of two forms, the dry form (*blepharitis sicca*) or the oily (*blepharitis oleosa*); the first is the more frequent, the more persistent and resistant to treatment.

In BLEPHARITIS SICCA the disease is at first manifested only by the appearance of fine, dry, grey scales clinging loosely to the lashes associated with symptoms of itching and burning (*ciliary pityriasis*) (Fig. 210). Sometimes the scales are scanty and at other times all the lashes are powdered heavily with them, but in either case they are renewed as soon as they are removed. The skin itself may appear normal merely showing a " dandruff of the lids " corresponding to the accompanying dandruff of the scalp and brows, but evidences of inflammation tend to flare up on the slightest irritation by wind, dust or smoke. This condition may persist unchanged or with intermissions until late adult life when it tends to disappear, leaving as evidence of its former presence a partial alopecia of the lashes corresponding to the baldness of the scalp; alternatively the character of the disease may change and the lids become greasy and permanently red.

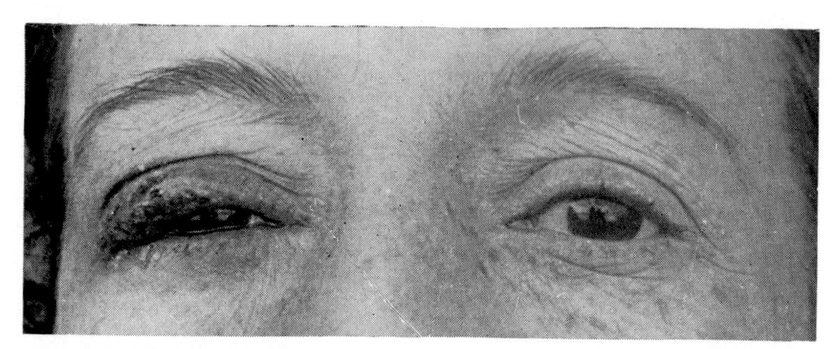

FIG. 210.—SEBORRHŒIC BLEPHARITIS (W. J. O'Donovan and I. C. Michaelson).

In BLEPHARITIS OLEOSA the margins of the lids assume a persistent redness and the scales thicken and become greasy, while an associated dermatitis[1] is usually evident elsewhere. The exfoliating epithelium combining with dried secretion forms crusts which are glued like yellow wax to the margins of the lids and lashes. Itching and irritation become prominent and rubbing tends to aggravate the inflammation and add infection to it so that an intense dermatitis may develop associated with a chronic conjunctivitis. The infection is usually staphylococcal, setting up a generalized follicular eruption (*acne blepharitis*); occasionally the glands associated with the follicles are involved and styes and infected chalazia develop in crops. Finally the lid-margins become ulcerated, encrusted and thickened, and there is a progressive loss of the lashes resulting eventually in alopecia.

Pathologically the condition is associated typically with over-secretion of the glands of Zeis, and occasionally the accumulation of epithelial debris and sebum may produce plugs of greasy material in the follicular pore of the lash; this may progress to the formation of an actual *ciliary comedo*. At other times the meibomian glands seem especially implicated, when a *meibomian blepharitis* develops,[2] masses of sebaceous material being expressed therefrom on tarsal massage.

The treatment of seborrhœic blepharitis should be undertaken on the lines indicated,[3] first tackling the general condition as thoroughly as possible, then other affected areas such as the scalp and brows, and then the lids: the latter are rarely cured alone. Moreover, as we have already pointed out, treatment of the general condition must last for years or throughout life. For the local condition the maintenance of cleanliness and the application of selenium sulphide as an ointment are frequently effective. In high concentrations this substance causes corneal ulcers in rabbits and a keratitis in man (Wong *et al.*, 1956); Cohen (1954) found concentrations up to 25% safe, but most clinicians prefer a strength of 0·5% which rarely gives rise to a transient keratitis that clears on cessation of the drug (Bahn, 1954; Wong *et al.*, 1956). The ointment should be rubbed into the lid-margins after removal of the scales and not inserted into the lower conjunctival cul-de-sac. Arruga (1959) combined it with hydrocortisone and chloramphenicol.[4] On the other hand, it has been claimed that an ointment of ammoniated mercury (5%) is equally effective (Wong *et al.*, 1956).

ROSACEA BLEPHARITIS

ROSACEA is a common disease the ætiology and general features of which have been described in a previous Volume.[5] The essential feature is a paralytic dilatation of the superficial blood vessels leading to erythema and telangiectasia; this may develop into the formation of papules and pustular acneiform lesions which heal without scarring; a third form is associated

[1] p. 267. [2] p. 241. [3] p. 29.
[4] Selenium disulphide 1 g., hydrocortisone 0·25 g., chloramphenicol 0·25 g., non-greasy excipient *ad* 100 g.
[5] Vol. VIII, p. 534.

with gross glandular hyperplasia as is seen in rhinophyma.[1] It affects the exposed parts of the face but not the eyelids themselves apart from their margins (Fig. 211; Plate IV, Fig. 4). There is nothing specific in the histopathology, and the ætiology is unknown. A host of theories has been suggested from time to time which have been listed in connection with rosaceal conjunctivitis, keratitis and episcleritis,[2] but for any of them there

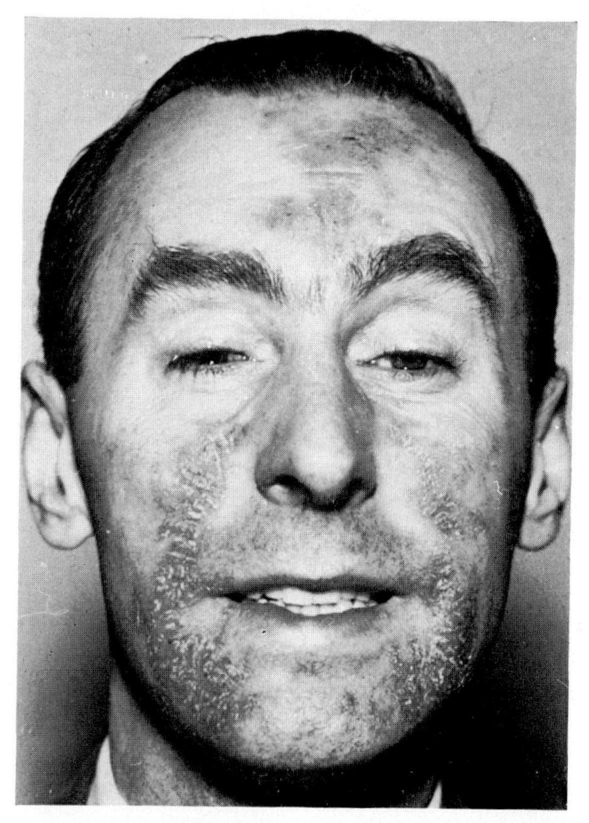

Fig. 211.—Rosacea.
A typical facial case with blepharo-conjunctivitis (Inst. Ophthal.).

is no conclusive evidence; it is interesting that, despite the erythema, the vascular reactions are essentially normal (Borrie, 1955) and the presence of *Demodex folliculorum* (S. and S. Ayres, 1961; Ayres and Mihan, 1967) or of *Corynebacterium acnes* in the majority of the pustules (Marples and Izumi, 1970) is probably fortuitous.

A mild blepharitis or frequently a blepharo-conjunctivitis appears in the course of most cases of rosacea; it is commoner but much less important

[1] p. 32; Fig. 21. [2] Vol. VIII, p. 537.

than the corneal complications which may arise. Its significance lies in the fact that, in addition to the occurrence of chalazia with which it is frequently associated, it is in a large number of cases the forerunner of keratitis. The blepharitic tendency occurs essentially in seborrhœic people; indeed, it may be said to take the place of acne vulgaris in such persons. It starts with hyperæmia of the lid-margins which persists with intermissions until a state of chronic stasis and congestion is reached. Such a state leads eventually to disturbances of secretion and congestion of the sebaceous glands of the lid-border, and at this stage frank symptoms of blepharitis appear. Its most characteristic symptom is its obstinacy; the persistent erythema, scale-formation and the development of pustules and furuncles lead eventually to permanent thickening, hypertrophy and loss of the lashes (Fig. 212).

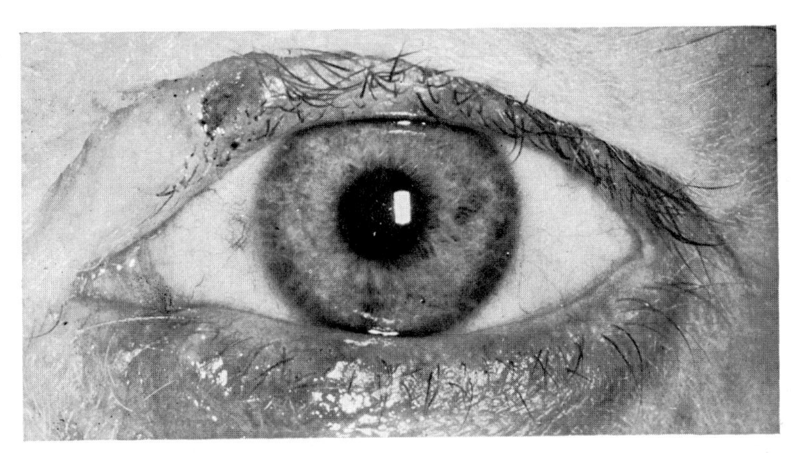

Fig. 212.—Rosacea Blepharitis (Inst. Ophthal.).

No specific *treatment* exists for this type of blepharitis, but measures should be taken to control the disease on the face. For this reason cold and exposure to the sun should be avoided and the diet regulated, avoiding highly seasoned foods, alcohol and other stimuli exciting vasodilatation. The element of acne is best treated with sulphur ointment, and systemic tetracyclines which also appear to diminish the erythematous element in rosacea are of value (Sneddon, 1966). For the lids themselves the establishment of cleanliness is essential, any pustules should be eliminated and dark glasses are usually welcomed. If an infective element is absent a stimulatory ichthyol and zinc ointment is often effective, if seborrhœa is evident selenium disulphide, and if cultures show an infection an appropriate antibiotic. The condition, however, is very resistant to treatment and, although it can be ameliorated, the patient must frequently resign himself to live with red-rimmed eyes for many years.

FOLLICULAR IMPETIGINOUS BLEPHARITIS is rare; it is coccal in nature and usually accompanies impetigo elsewhere on the face. The lesion is a rapidly developing superficial folliculitis. At the roots of the lashes purulent bullæ or pustules are seen which dry into soft yellow crusts under which the epidermis is merely eroded. With appropriate treatment—careful removal of the crusts and the application of a cream of penicillin or other antibiotics—the condition heals up leaving no trace.

FOLLICULAR SYCOTIC (STAPHYLOCOCCAL) BLEPHARITIS

This is the usual type of infective blepharitis, occurring commonly in children, frequently persisting with innumerable relapses all through life, and habitually complicating the more simple forms of marginal inflammation, such as seborrhœic blepharitis; it is typically due to the mannitol-positive coagulase-positive *Staphylococcus aureus* although a rare agent is a hæmolytic *Staphylococcus albus* (Plate IV, Fig. 5). The great majority of cases of chronic sycosis is complicated by blepharitis; and conversely it would appear that the chronic staphylococcal blepharitis of childhood may give rise to a sycosis commencing in the central part of the upper lip and spreading therefrom (Sabouraud, 1922). In these cases the infection from the lids may be carried by the tears down the naso-lacrimal duct to set up a chronic rhinitis and a pustular infection of the follicles of the nasal vestibule; the final stage begins with the growth of the moustache, the rhinorrhœa giving rise to a pustular folliculitis of the moustache hairs (*sycosis sous-narinaire*). In the reverse sense, a staphylococcal infection in the nose may easily be transferred to the lids by the fingers or a handkerchief.

Sycotic blepharitis is a serious and exceedingly intractable disease characterized over periods of many years by continual relapses, each of which shows the four stages of follicular inflammation—follicular abscess, peri-follicular infiltration and abscess formation, ulceration, and healing by scarring of the lids with deformity or destruction of the lashes. Unless effective treatment is undertaken it tends to persist and spread until no lash follicles remain, by which time the lid-margin is persistently red, hypertrophied and deformed (*tylosis*), trichiasis has injured the cornea or madarosis has supervened. Quite a number of these persistent and recurrent cases shows an allergy or hypersensitivity to the organisms.

Occasionally a staphylococcal infection may assume a more acute and severe form and a cellulitis followed by a purulent inflammation involves the glands associated with the follicles. When such a condition is generalized the picture of *furunculous blepharitis* is produced when the entire lid becomes swollen, soggy and œdematous with pus-points showing at several places along the margins while pain and general malaise are prominent; the local manifestation of the same lesion is the external hordeolum or stye.

Systemic spread of the infection is a great rarity; thus B. Berger (1962) reported a case of a boy aged 7 who for 2 months suffered from a purulent blepharitis which gave rise to intense swelling of the lids, an orbital abscess, cavernous sinus thrombosis, a meningo-encephalitis and a staphylococcal pyæmia which proved fatal.

Several complications occur, the most common of which is a secondary chronic conjunctivitis; the inflammation is of the papillary type and is probably not due to invasion by the organism but to the liberation of its toxins (Thygeson, 1937); an epithelial keratitis affecting the lower part of the cornea is also a frequent accompaniment[1] and a phlyctenular keratitis with a pannus, again most marked inferiorly (Thygeson, 1969). In addition, recurrent hordeola, meibomitis and infectious eczematoid dermatitis may complicate the clinical picture.

The *treatment* of follicular blepharitis must be carried out with painstaking thoroughness or the condition may never fully resolve. The general measures already outlined[2] must be followed and adverse constitutional and hygienic influences counteracted. An accompanying staphylococcal infection of the nasopharynx should receive particular attention. Local treatment should maintain the most careful cleanliness and meticulous and gentle removal of crusts and scales, the radical epilation of diseased lashes, and the application of one or other of the anti-staphylococcal antibiotic drugs. Unfortunately, even if the drug is shown to be active against the organism, a dramatic cure is rare, perhaps because of deep-seated infection in the glands of the lids, sometimes because the treatment is complicated by irritation or contact sensitization to the drug. *Bacterial and not symptomatic cure is the ideal:* disappearance of the inflammatory symptoms should be followed after an interval by a bacteriological examination not only of the lid-margin but of the roots of an epilated lash and, if the organism is still present, immunization with toxoid and autogenous vaccine should be considered (Thygeson, 1938–69). In cases of hypersensitivity to the staphylococcus, however, it may be a very difficult and lengthy matter to establish a satisfactory and lasting immunity. Because of the complications arising when the infection has become established, the importance of early treatment cannot be over-estimated, particularly in children.

IRRITATIVE AND ALLERGIC BLEPHARITIS

As we have seen, physical, chemical and allergic irritants are frequent predisposing factors in the ætiology of blepharitis; if, however, they are sufficiently potent, they may set up a primary marginal inflammation. Usually this is of the squamous type, sometimes it is ulcerative. Thus radiation such as heat (in furnace workers, glass-blowers, etc.), or ultra-violet rays (in welders or in snow-blindness) or dust (in furriers) or chemical fumes may produce a chronic hyperæmia and squamous blepharitis without the aid of any adjuvant factors. Allergic sensitivity accounts for a larger group, and here again the squamous type of inflammation preponderates (Fig. 213). In this type a whole galaxy of substances is involved—vegetable or animal allergens, medicaments or cosmetics—which have already been

[1] Vol. VIII, p. 157. [2] p. 213.

discussed in dealing with contact dermatitis.[1] Special mention, however, should be made of cosmetics applied particularly to the lid-margins (mascara, lash dyes), the virulent effects of which, as we have already noted, may include an ulcerative or necrotic or even gangrenous inflammation involving permanent damage not only to the lids but also to the eye.

The *treatment* of all these conditions is, of course, avoidance of the irritant, either by protective measures as by goggles for radiation, or by its elimination from the environment; when this is not possible desensitization may be tried. Locally the blandest sedative applications should be used or—frequently better—none at all, and measures adopted to avoid complicating infection.

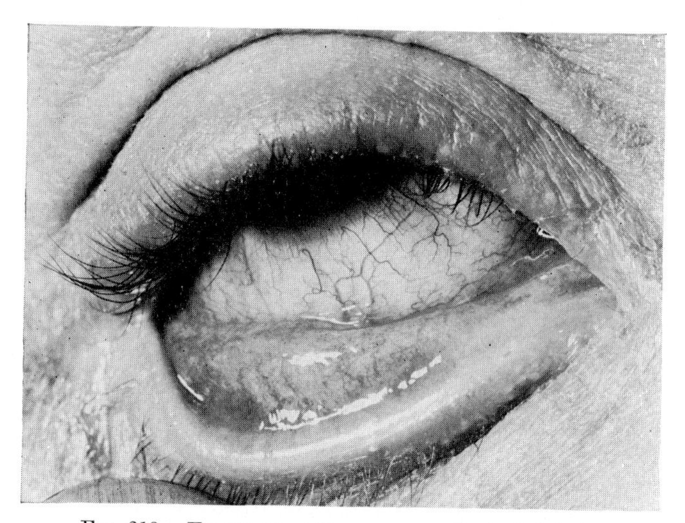

FIG. 213.—ECZEMATOUS BLEPHARITIS (Inst. Ophthal.).

DEMODECTIC BLEPHARITIS

A blepharitis associated with the presence of the *Demodex folliculorum* (*blepharitis acarica*, Raehlmann, 1899) is probably more common than is generally supposed.

Demodex, a mite discovered by E. Berger (1845), can infest many mammals and the species found in the human skin, the *Demodex folliculorum*, is found in sebaceous follicles anywhere in the body in adults all over the world with a predilection for the nose, cheek, the external auditory meatus, the forehead and the margins of the lids. On squeezing the follicles on the nose, Du Bois (1910) found no mites in children under the age of 5, but they were present in 50% of those between 5 and 10 years of age and universally in those over 25. On epilated lashes Coston (1967) found none below the age of 16 but they were present in 25% of adults between the ages of 20 and 83, while Norn (1971) found them in practically all middle-aged and old subjects. In heavily infected people, 3 to 5 mites are frequently found clinging to every epilated lash;

[1] p. 58.

up to 25 have been counted on one lash, and others are presumably left behind in the follicle.

The mite itself is a small, transparent, vermiform creature, the male being 28μ in length and the female somewhat shorter. It has a well-differentiated head and thorax bearing four paired appendages which permit active movement and the abdomen is distinguished by annular marks (Fig. 214). A more stumpy variant, the caudate or short type (*D. folliculorum brevis*), was originally described by Erasmus Wilson (1844) (Fig. 215); it resides in the sebaceous gland while the more common type inhabits the follicle itself (English, 1971; Desch and Nutting, 1972).

FIGS. 214 to 216.—*DEMODEX FOLLICULORUM* (F. P. English).

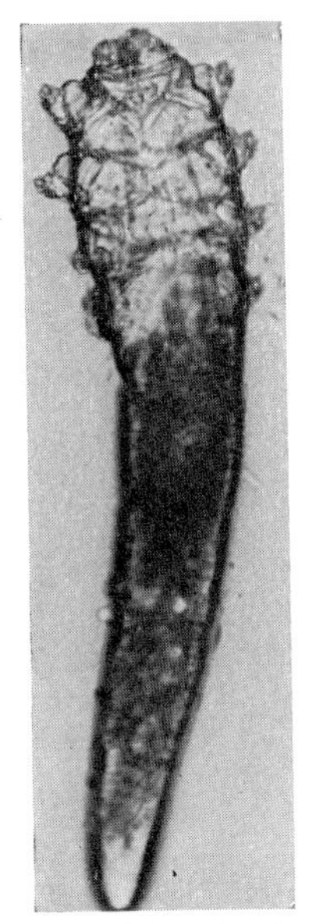

FIG. 214.—Ventral view of the adult (× 400).

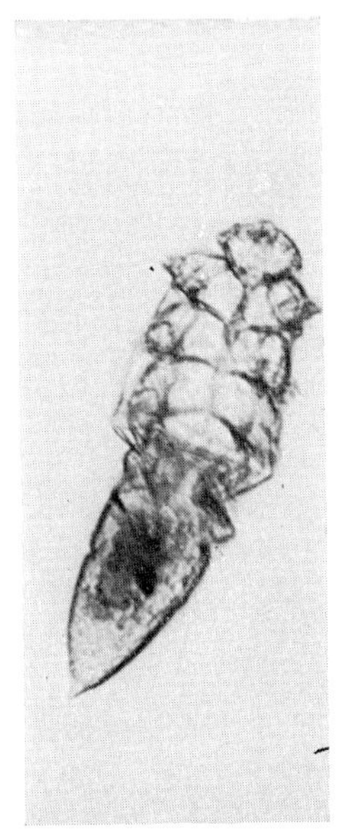

FIG. 215.—Ventral view of the caudate (short) type (× 250).

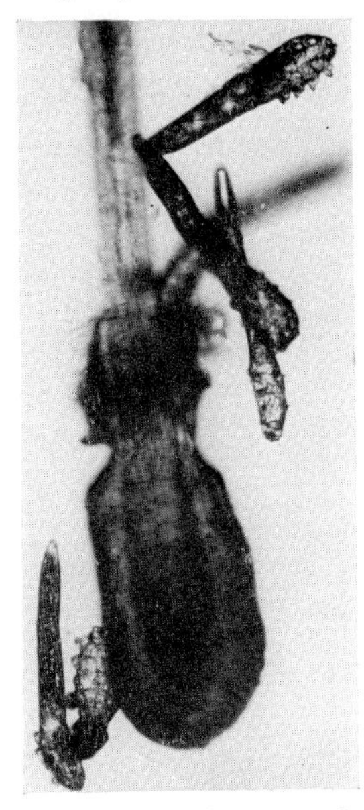

FIG. 216.—The egg, the immature and the adult stages. The short variety is located near the bulb of the eyelash (× 60).

The life-history of the mites is interesting, particularly their voracious appetite and urgent sex-life. They live in the follicle with their heads towards its base, feeding on sebum (Fuss, 1933–35) (Fig. 216), work their way to the opening of the follicle where copulation takes place; the gravid female returns to the follicle where the eggs are deposited and the larvæ are hatched, pass through two metamorphoses to form a deutonymph which crawls about over the surface of the skin in the hours of darkness to enter a follicle where it transforms into an adult. After a total life of about a fortnight the mite dies, its body lying at and blocking the opening of the follicle (Spickett, 1961).

The part these parasites play in disease of the lids, particularly blepharitis, has been questioned. The fact that they are found so frequently in the follicles of the cilia and in the meibomian glands of normal individuals (64%, Joers, 1899; 50%

Gmeiner, 1908), becoming more numerous as age advances (Oyenard, 1908), led most early investigators to conclude that they had little or no pathological effect.[1] It had, however, been contended by Whitfield (1920), Ayres (1930–63) and Lawrence (1935) and later by Morgan and Coston (1964) that the parasite was the causative agent of pityriasis folliculorum and the papulo-pustular type of rosacea, and in recent years considerable agreement has been reached that it may participate in the ætiology of blepharitis.[2]

Figs. 217 to 219.—Demodectic Blepharitis.

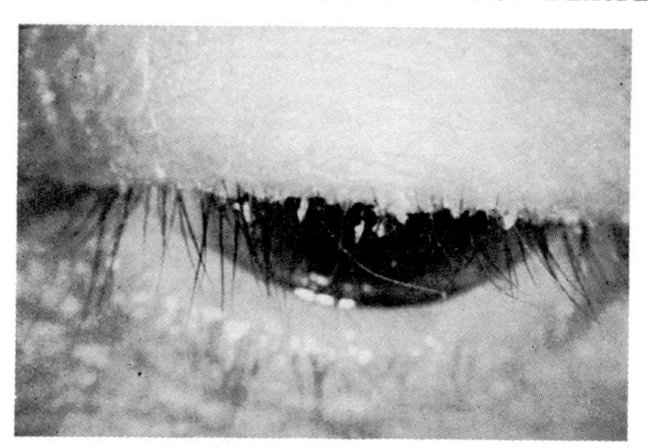

Fig. 217.—Showing crusting and cuffing of the lashes (C. McCulloch).

Fig. 218.—Slit-lamp view showing collaring of the lashes (T. O. Coston).

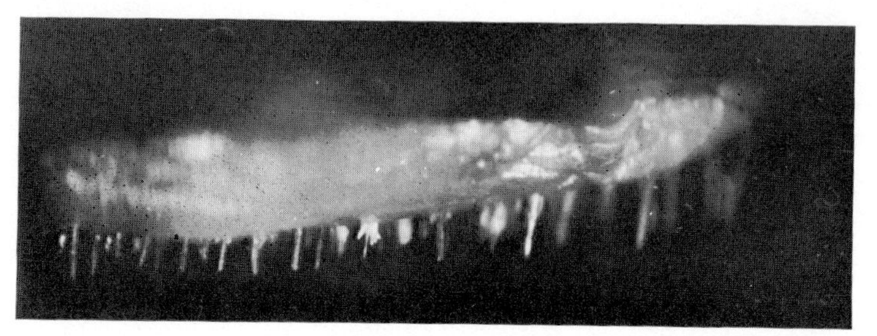

Fig. 219.—The mites are seen with the slit-lamp partly emerging in the central area around a lash after a cotton applicator saturated with ether is massaged along the lid-margin to clean away the waxy fæces (T. O. Coston).

A *blepharitis acarica* on observation with the unaided eye shows few characteristics except perhaps the presence of dry bran-like scales on the lid-margins. With the slit-lamp, however, waxy debris is seen at the base of the lashes and around them, forming a cuff (Figs. 217–8); this, which represents accumulations of the excreta of the parasite, is diagnostic. The lower lid is usually the most heavily involved. Erythema may develop, and a mucoid discharge and eventually the lashes may be distorted and lost.

[1] Carron du Villards (1855), Majocchi (1879–1900), Burchardt (1884), Stieda (1890), Raehlmann (1899), Joers (1899), Hunsche (1900), Herzog (1904–9), Gmeiner (1908), and others.

[2] Post and Juhlin (1963), Ayres and Mihan (1967), Coston (1967), S. Smith and McCulloch (1969), English (1969), English *et al.* (1970), Jacobson (1971).

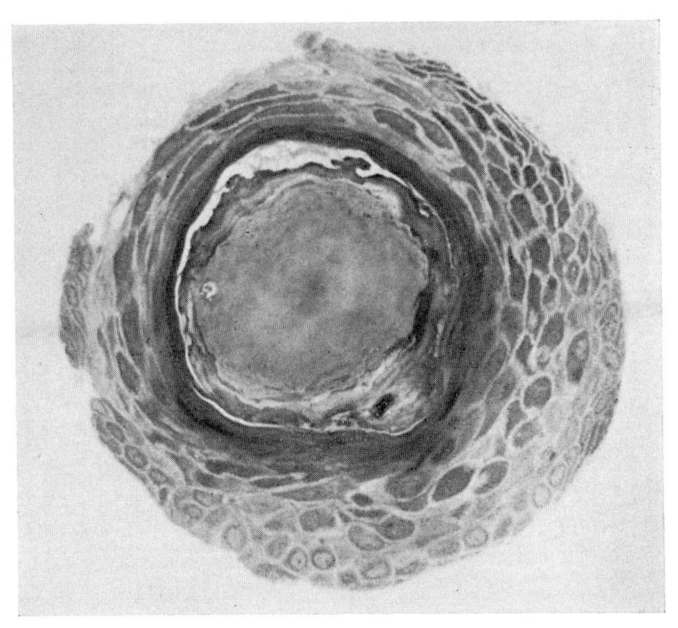

Fig. 220.—Demodectic Blepharitis.

Cross-section of an eyelash; a parasite is adherent and strictly localized to one side of the bulb. There is marked intercellular œdema with teasing out of the cellular components associated with a localized area of hydrops and hyperproduction of keratin (F. P. English).

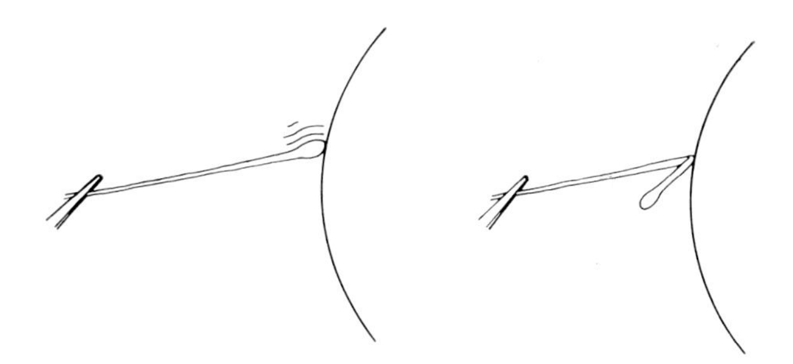

Fig. 221.—A Healthy and a Diseased Cilium.

The cilium is held by a pair of forceps and briskly tapped end-on against the anæsthetized cornea. A healthy eyelash (on the left) displays a vibrant character and rapidly assumes its normal configuration. In demodectic infestation the eyelash (seen on the right) shows no resilience but crumples like a piece of moistened blotting paper (F. P. English).

The appearance of pustules and a lesion resembling rosacea is unusual (Ayres and Mihan, 1967). When the lid-margin is stimulated with an irritating agent such as ether, there is an incomplete evacuation of the parasites, the heads remaining in the ducts of the follicles and the tails, as seen with the slit-lamp, protruding from them like fine bristles (Fig. 219). The diagnosis is verified by epilating some lashes and demonstrating the mites microscopically on a slide in a drop of peanut oil under a cover-slip. If several are found clinging to one lash, the follicles of the cilia are over-populated (Coston, 1967).

The lashes are easily pulled out and are swollen and soggy in consistency; there is marked intercellular œdema in the portion just distal to the bulb and an increase in the production of keratin (Fig. 220). When a normal lash is tapped end-on against a solid surface it displays a vibrant quality and resumes its normal configuration; an infested lash, on the other hand, crumples as if it were a piece of blotting paper (Fig. 221). Around the follicle there may be a mild hyperkeratosis and an infiltration of lymphocytes.

The pathological role played by the parasite itself has not been proved. Electron-microscopic observation, however, has shown that bacteria lie on its surface (English *et al.*, 1970). Apart from the irritation it causes, the acarid may act as a vector of organisms, and in its migratory movements from the depth of the follicle to the surface of the skin it may serve as a reservoir of infection leading to recurrent attacks of inflammation which topical therapy can only temporarily abate. If it is a pathogen, it is extremely benign, effective more by virtue of its numbers than by its presence.

Treatment should therefore be directed to the elimination of the parasite. The method suggested by Coston (1967) is as follows. A drop of anæsthetic is placed in the eye and a cotton applicator soaked in ether is massaged briskly along the dermal side of the lid-margins. The ether cleans away the excreta and as the tails of the mites begin to emerge from the follicles as demonstrated with the slit-lamp (Fig. 219), the lid-margin is massaged with an ointment such as ammoniated mercury (1 to 3%), yellow oxide of mercury (1%), selenium sulphide or sodium sulphacetamide (10%); this application should be repeated twice daily for several weeks. This usually decreases the irritation by lowering the population of mites, but if treatment is discontinued for a long period recurrences are common. This is only to be expected in the case of a parasite omnipresent on man, but when once the population has been reduced to a non-irritative level, it is good practice to wash the face with soap and water rather than to rely on cleansing creams which serve as an excellent culture-medium for the organism (Fakes, 1965).

BLEPHARITIS DUE TO SPECIFIC INFECTIONS is common but since most of them have already been discussed at length, a short recapitulation only is necessary here.

EXANTHEMATOUS BLEPHARITIS. The *eruptive fevers* are almost invariably accompanied by inflammation of the lid-margins, a generalization peculiarly applicable to *measles*. In this disease, associated with the conjunctivitis which characterizes the first stage and may precede symptoms, a blepharitis appears, and when the eruption occurs the lid-margins are red, swollen and inflamed. The affection is of the squamous type without specific characteristics unless staphylococcal infection supervenes, but it may become persistent and obstinate, outlasting the measles, and may precede other lesions such as phlyctenular kerato-conjunctivitis, eczema or impetigo of the lids. All the other exanthemata may produce a similar mild blepharitis—influenza, chickenpox, typhoid, typhus and so on. More destructive, however, are the lesions of smallpox (*variolar blepharitis*). In the pre-vaccination era this relatively common condition was pustular and ulcerative, the lids being markedly swollen with their margins caked with pus and crusts in the acute stage and thickened and covered with granulations in the chronic. Permanent scarring, trichiasis and madarosis were frequent sequels leaving considerable deformity of the ciliary margin, blepharophimosis and symblepharon.

A DIPHTHERITIC BLEPHARITIS is a rarity but has been reported in the absence of other manifestations in the conjunctiva, nose and throat as a moist, eczematous condition interrupted by inflamed hypertrophic areas showing the specific organism on culture (Friedmann, 1931). A similar infection may be transmitted to a contact; treatment is by antitoxin and penicillin.

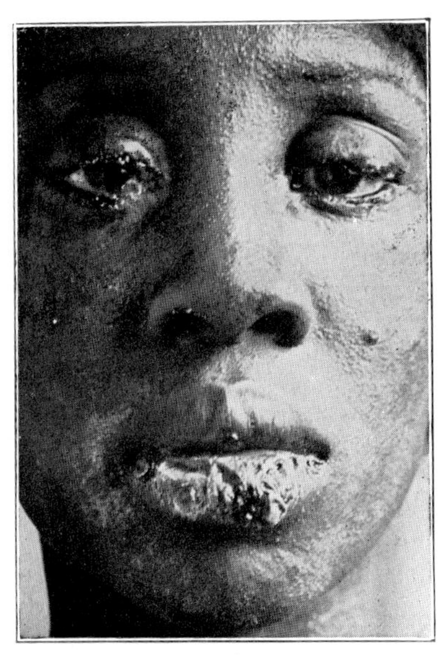

FIG. 222.—SECONDARY SYPHILIS.
With blepharitis, cheilitis and iritis (J. A. Fordyce and G. M. Mackee).

SYPHILITIC BLEPHARITIS. In syphilis a primary lesion may appear on the lid-margin, mucous patches or condylomata may occur, and gummata may be located here either in tertiary or congenital syphilis. In this connection, however, the most apposite manifestation is the syphilitic ulcerative blepharitis of Hutchinson (1888) with its associated scarring and madarosis (Fig. 222).

TUBERCULOUS BLEPHARITIS. Tuberculosis, apart from its common influence as a predisposing factor in the ætiology of blepharitis, is occasionally complicated by lesions on the ciliary border, usually as a direct inoculation from neighbouring disease such as lupus vulgaris of the skin, scrofuloderma or tuberculous rhinitis.

LEPROTIC BLEPHARITIS. Leprosy almost invariably affects the lid-margins. Alopecia of the lashes is an early sign, and the margins themselves, usually swollen, reddened and thickened, frequently become deformed with infiltrated nodules which may break down into abscess-formation and ulceration resulting sometimes in complete disorganization of the lid-border and even of large areas of the substance of the lid itself.

VIRAL BLEPHARITIS. Viral infections of the lid-margins have already been noted—*herpes*, *vaccinia*, *verruca* and *molluscum contagiosum*.

MYCOTIC BLEPHARITIS. Fungal infections as a rule affect the lid-margins secondary to involvement of the palpebral skin, causing blepharitis only incidentally; their incidence depends largely on the individual's environment, such as agricultural workers without good hygienic conditions, and their contamination is usually transitory and incurred by chance (Olson, 1969). Of specific infections, the destructive lesion of *sporotrichosis* may spread to the border of the lid where small yellow abscesses and ulceration may occur (Morax and Carlotti, 1908). *Blastomycosis* may produce small abscesses round the lashes (Gilchrist, 1902; Wood, 1904). The most common infection, however, is *tinea* (*trichophytosis*), which causes a purulent and ulcerative blepharitis and *favus* which may involve the lid-margin incidentally; both of these have already been described. An infection by *Candida* may develop especially after prolonged treatment of an infective blepharitis by antibiotics and steroids.

PARASITIC BLEPHARITIS. Animal parasites rarely cause marginal inflammation with one exception—the *Phthirus pubis* or crab-louse, the presence of which is frequently associated with an irritative blepharitis. The *Pediculus capitis* is seldom seen on the lashes, but a persistent blepharitis is constantly associated with its infestation of the head.

Aleeva. *Russ. oftal. J.*, **13,** 420 (1931).
Aragones. *Ann. Ophthal.* (Chic.), **5,** 49 (1973).
Arruga. *Ann. Oculist.* (Paris), **192,** 528 (1959).
Aubaret. *Arch. Ophthal.*, **3,** 762 (1930).
Ayres, S. *Arch. Derm. Syph.* (Chic.), **21,** 19 (1930).
Calif. Med., **98,** 328 (1963).
Ayres, S., and S. *Arch. Derm.*, **83,** 154 (1961).
Ayres, S., and Mihan. *Arch. Derm.*, **95,** 63 (1967).
Bahn. *Sth. med. J.*, **47,** 749 (1954).
Bargy. *Concours méd.*, **71,** 1745 (1949).
Bassin and Skerlj. *Klin. Mbl. Augenheilk.*, **98,** 314 (1937).
Berger, B. *Nord. Med.*, **68,** 1376 (1962).
Berger, E. *C.R. Acad. Sci.* (Paris), **20,** 1506 (1845).
Borrie. *Brit. J. Derm.*, **67,** 73 (1955).
Burchardt. *Zbl. prakt. Augenheilk.*, **8,** 229 (1884).
Carron du Villards. *Ann. Oculist.* (Paris), **33,** 241; **34,** 65 (1855).
Clarke. *Ophthal. Rev.*, **13,** 345 (1894).
Cohen. *Amer. J. Ophthal.*, **58,** 560 (1954).
Coston. *Trans. Amer. ophthal. Soc.*, **65,** 361 (1967).
Cuénod. *Bactériologie et parasitologie clinique des paupières* (Thèse), Paris (1894).
Desch and Nutting. *J. Parasit.*, **58,** 169 (1972).
Doherty. *Amer. J. Ophthal.*, **23,** 567 (1940).
Du Bois. *Ann. Derm. Syph.* (Paris), **1,** 188 (1910).
English. *Med. J. Aust.*, **1,** 1359 (1969).
Trans. Aust. Coll. Ophthal., **2,** 89 (1970).
Brit. J. Ophthal., **55,** 742, 747 (1971).
English, Iwamoto, Darrell and DeVoe. *Arch. Ophthal.*, **84,** 83 (1970).
Fakes. *St. George's Hosp. Gaz.*, **50,** 8 (1965).
Franks. *Arch. Ophthal.*, **25,** 334 (1941).
Friedmann. *Dtsch. med. Wschr.*, **57,** 1542 (1931).
Fuss. *Ann. Derm. Syph.* (Paris), **4,** 1053 (1933); **6,** 326 (1935).
Galin. *Arch. Ophthal.*, **67,** 746 (1962).
Gilchrist. *Brit. med. J.*, **2,** 1321 (1902).
Gmeiner. *Arch. Derm. Syph.* (Berl.), **92,** 25 (1908).
Gordon. *J. invest. Derm.*, **17,** 267 (1951).
Gots, Thygeson and Waisman. *Amer. J. Ophthal.*, **30,** 1485 (1947).
Gowen. *Ill. med. J.*, **71,** 216 (1937).
Hall. *Med. Rec.* (N.Y.), **21,** 399 (1882).
Hasegawa. *Acta Soc. ophthal. jap.*, **56,** 9 (1952).
Herzog. *Z. Augenheilk.*, **11,** 151, 245, 342; **12,** 180 (1904).
v. Graefes Arch. Ophthal., **69,** 492 (1909).
Hotz. *Klin. Mbl. Augenheilk.*, **37,** 485 (1899).
Hunsche. *Münch. med. Wschr.*, **47,** 1563 (1900).
Hutchinson. *Roy. Lond. ophthal. Hosp. Rep.*, **12,** 156 (1888).
Jacobson. *Trans. Amer. Acad. Ophthal.*, **75,** 1242 (1971).
Joers. *Dtsch. med. Wschr.*, **25,** 220 (1899).

Jones, Andrews, Henderson and Schofield. *Trans. ophthal. Soc. U.K.*, **77,** 291 (1957).
Keddie and Shadomy. *Sabouraudia*, **3,** 21 (1963).
Kesten. *Arch. Ophthal.*, **6,** 582 (1931).
Keyser. *Phila. med. Times*, **7,** 266 (1877).
Lawrence. *Arch. Derm. Syph.* (Chic.), **32,** 633 (1935).
Lawson. *Trans. ophthal. Soc. U.K.*, **39,** 109 (1919).
Majocchi. *Atti Accad. med. Roma*, **5,** 43 (1879).
Ann. Ottal., **29,** 493 (1900).
Marples and Izumi. *J. invest. Derm.*, **54,** 252 (1970).
Martin-Scott. *Brit. J. Derm.*, **64,** 257 (1952).
Medvedeff and Euffa. *Russ. oftal. J.*, **10,** 723 (1929).
Mitsui, Hinokuma and Tanaka. *Amer. J. Ophthal.*, **34,** 1579 (1951).
Moore. *Arch. Derm. Syph.* (Chic.), **31,** 661 (1935).
Moore, Kile, Engman, M. F. and M. F., *Arch. Derm. Syph.* (Chic.), **33,** 457 (1936).
Morax and Carlotti. *Ann. Oculist.* (Paris), **139,** 418 (1908).
Morgan and Coston. *Sth. med J.*, **57,** 694 (1964).
Norn. *Acta ophthal.* (Kbh.), **48,** 237 (1970); Suppl. 108 (1970).
Dan. med. Bull., **18,** 14 (1971).
Olson. *Arch. Ophthal.*, **81,** 351 (1969).
Oyenard. *Arch. Oftal. hisp.-amer.*, **8,** 97 (1908).
Parunovic and Halde. *Amer. J. Ophthal.*, **63,** 815 (1967).
Portnoy. *Med. Ill.* (Lond.), **4,** 1 (1950).
Post and Juhlin. *Arch. Derm.*, **88,** 298 (1963).
Prigal. *N.Y. St. J. Med.*, **53.** 1327 (1953).
Raehlmann. *Klin. Mbl. Augenheilk.*, **37,** 33 (1899).
Richet. *Recueil Ophtal.*, **7,** 321 (1885).
Roosa. *V int. Cong. Ophthal.*, N.Y., 137 (1876).
Sabouraud. *Entrétiens dermatologiques*, Paris (1922).
Schreiber. *Graefe-Saemisch Hb. d. ges. Augenheilk.*, 3rd ed., *Die Krankheiten d. Augenlider*, Berlin, 308 (1924).
Serov and Ovchinnikova. *Oftal. Zh.*, **26,** 297 (1971).
Smith, J. W. *Arch. Ophthal.*, **20,** 658 (1938).
Smith, S., and McCulloch. *Canad. J. Ophthal.*, **4,** 3 (1969).
Sneddon. *Brit. J. Derm.*, **78,** 649 (1966).
Somerset. *Brit. J. Ophthal.*, **23,** 205 (1939).
Spickett. *Parasitology*, **51,** 181 (1961).
Spoor, Traub and Bell. *Arch. Derm. Syph.* (Chic.), **69,** 323 (1954).
Stieda. *Zbl. prakt. Augenheilk.*, **14,** 193 (1890).
Thygeson. *Arch. Ophthal.*, **18,** 373 (1937); **20,** 271 (1938).
Amer. J. Ophthal., **68,** 446 (1969).
Thygeson and Vaughan. *Trans. Amer. ophthal. Soc.*, **52,** 173 (1954).
Tulipan. *Arch. Derm. Syph.* (Chic.), **33,** 349 (1936).
Uchida, Hirai, Ohshima and Kameyama. *Rinsho Ganka*, **23,** 221 (1969).
Whitfield. *Proc. roy. Soc. Med.*, **13,** *Sect. Derm.*, 102 (1920).
Wilson. *Phil. Trans.*, **134,** 305 (1844).
Winselmann. *Klin. Mbl. Augenheilk.*, **37,** 240 (1899).
Wong, Fasanella, Haley *et al. Arch. Ophthal.*, **55,** 246 (1956).
Wood. *Ann. Ophthal.*, **13,** 92 (1904).

Inflammations of the Glands of the Lid-margin

The glands of Moll—modified apocrine glands—are rarely infected alone, probably because of the length and narrowness of their ducts, but when those opening separately on the ciliary margin are involved the clinical picture of hidradenitis arises. The glands of Zeis—large acinous glands of the usual sebaceous type—on the other hand, are frequently inflamed, producing as a rule a more violent inflammation than the analogous lesion in the skin.

HIDRADENITIS (SUDORIPAROUS BLEPHARITIS, Aubaret, 1929–30) is an acute or subacute infection of the glands of Moll of little consequence, associated with a generalized infection of the ciliary border. The infection of the orifices of the glands is seen clinically by the appearance of minute, red inflammatory papules along the ciliary margin which develop rapidly and disappear usually without suppuration or ulceration. Treatment is on the general lines for blepharitis.[1]

[1] p. 213.

An interesting case was reported by Sachs and Gordon (1967) wherein a woman aged 40 had a generalized suppurative inflammation affecting the apocrine glands in the axillæ, perineum, the retro-auricular skin, the mammary glands and the glands of Moll. For 15 years abscesses occurred in one or other of these sites, most cultures revealing *Proteus*. On the lid-margins a tenacious mucopurulent discharge was present, crusting the lashes, which evidently came from the openings of the glands of Moll along the lash-lines and the roots of the cilia. Treatment proved very difficult but transient relief was obtained by the topical application of polymyxin, neomycin and gramicidin hourly for 5 days.

In inflammations of the perifollicular tissues of the lid-margin, the glands of Moll are, of course, frequently involved. Thus in acute conditions such as the cellulitis associated with a hordeolum, these sweat glands are implicated; and in chronic infiltrations they may be seriously involved, as in trachoma when hyperplasia and cystic formation result[1] or in tuberculosis of the lids wherein the glands may be invaded by typical tuberculous granulomatous tissue (Moauro, 1891).

HORDEOLUM (EXTERNUM)

A HORDEOLUM (L. *hordeolum*, barley) or STYE (A.S., *steigan*, to rise), is *a localized suppurative inflammation at the lid-margin commencing in connection with a ciliary follicle and involving particularly the associated gland of Zeis* (ZEISIANUM). It corresponds to a furuncle of the skin.

Ætiologically a stye is an infection by the staphylococcus, usually the *Staphylococcus aureus*, which may attain a considerable degree of virulence. This may arise as a pure infection, in association, for example, with a follicular blepharitis,[2] but very commonly it complicates acne vulgaris, of which it forms a typical pustule. In this event, particularly in young adults, styes are frequently recurrent, appearing in crops with tantalizing persistence (*acne ciliaris*). Apart from this, their recurrence is usually associated with some constitutional cause, such as debility, focal infections, a badly controlled carbohydrate diet or diabetes (Hiwatari and Kito, 1955). In these cases, indeed, the actual staphylococcal infection is secondary in importance to the general condition (Herxheimer, 1907; and others).

Clinically a stye usually starts with an œdema of the lids which in some cases may be circumscribed and in others, particularly if near the outer canthus where the inflammatory reaction blocks the venous return, may be intense and widespread involving both lids to such an extent as to suggest an orbital cellulitis, accompanied, perhaps, by conjunctival chemosis. It may happen that the swelling is so severe as to mask the actual point of inflammation. Here a swollen, red indurated area appears, sometimes of considerable size, just on the lid-margin over which the skin is tense and exquisitely tender (Fig. 223). Pathologically it represents a purulent infection of the follicle and its associated gland with a cellulitis of the surrounding dense connective tissue. In the matter of a few days a yellow point of pus appears on the lid-margin around the roots of a lash at the most prominent part of the swelling; this accumulates until the skin gives way and

[1] Vol. VIII, p. 278. [2] p. 208.

pus escapes, sometimes with a central slough, whereupon the swelling subsides and after the lapse of a further few days the spot, partially denuded of lashes, is marked only by a cicatrix. Spread of the infection, however, is common, either by direct inoculation of the opposite lid-margin or of other follicles some distance away, and it is not unusual for several styes to be present simultaneously, all in different stages of evolution. Occasionally the inflammation aborts and an indurated area is left.

The symptoms are frequently out of all proportion to a lesion which is usually regarded as a very trifling ailment. At first there is a feeling of fullness or heaviness and heat but soon the swelling and induration in the dense, richly innervated tissues of the ciliary margin give rise to acute pain, while the inflamed spot itself is markedly tender to the touch. Fortunately it is

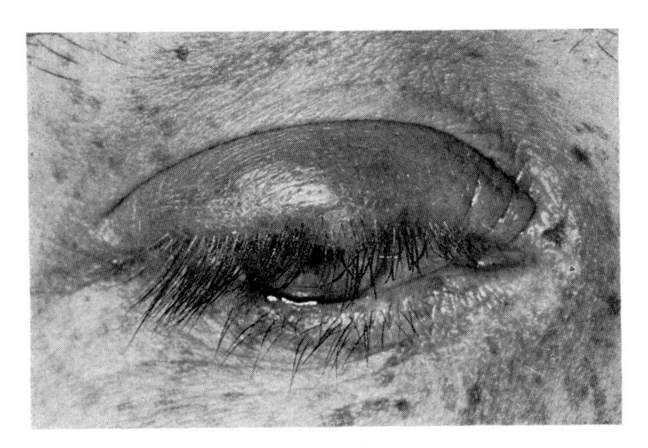

Fig. 223.—Acute Hordeolum.

At a relatively early stage showing diffuse œdema of the upper lid (Inst. Ophthal.).

usually not of long duration; as pus forms, the pain, at first sharp and cutting, becomes dull and throbbing, and the moment pus escapes it immediately subsides.

Complications are few. In the early stages the severity of the pain may disturb the sleep and feverishness may occur, the general symptoms of absorption being probably due to the great vascularity of the lid and its constant mobility. In debilitated and cachectic individuals a spreading cellulitis may occur particularly if the stye is near the inner canthus; an orbital thrombophlebitis (Mylius, 1925) or cellulitis (Green, 1926; Caramazza, 1938) has followed which has caused death from meningitis or cavernous sinus thrombosis (Lesniowski, 1895; Guth, 1898; Eagleton, 1935; Vail, 1949) and an acute and fatal staphylococcal septicæmia has been reported (Vlacek, 1935).

The *diagnosis* is usually easy and apparent, the only difficulty arising when the extent of the œdema obscures the causative lesion and simulates an orbital cellulitis or a dacryocystitis. In all such cases, however, careful palpation of the lid-margin reveals the point of tenderness and indicates the cause. A ciliary chalazion, a tuberculoma, a syphilide or a gumma, an early carcinoma or a mycotic infection may on rare occasions lead to confusion.

The *treatment* of an acute stye[1] is on general surgical principles: locally, hot fomentations or compresses until the abscess points, followed by an evacuation of the pus and subsequent control of any blepharitis which may be present. A stye should never be opened until pus points; sometimes epilation of the affected lash suffices to evacuate the pus—there need be no hesitation in doing this as the lash is to be lost in any event—but usually opening with a sharp, narrow knife is more satisfactory, a procedure, however, which should be delayed until pus is near the surface when the layer of dead epidermis should be lifted up rather than incised. This should be a painless operation. After opening, the lid-margin should never be squeezed. Subsequent treatment designed to prevent recurrences depends on the local and general condition. An antibiotic such as neomycin and gramicidin applied to both the nose and the eyes is frequently effective (Copeman, 1958), otherwise the treatment should be as for staphylococcal blepharitis.[2]

Aubaret. *Bull. Soc. franç. Ophtal.*, **42**, 1 (1929).
Arch. Ophthal., **3**, 762 (1930).
Caramazza. *Riv. Oto-neuro-oftal.*, **15**, 1 (1938).
Copeman. *Lancet*, **2**, 728 (1958).
Eagleton. *Arch. Ophthal.*, **14**, 1 (1935).
Green. *Amer. J. Ophthal.*, **9**, 34 (1926).
Guth. *Prag. med. Wschr.*, **23**, 25 (1898).
Herxheimer. *Dtsch. med. Wschr.*, **33**, 1481 (1907).
Hiwatari and Kito. *Acta Soc. ophthal. jap.*, **59**, 587 (1955).
Lesniowski. *Ann. Oculist.* (Paris), **113**, 71 (1895).
Moauro. *Ann. Ottal.*, **20**, 324 (1891).
Mylius. *Z. Augenheilk.*, **56**, 302 (1925).
Klin. Mbl. Augenheilk., **74**, 781 (1925).
Sachs and Gordon. *Arch. Ophthal.*, **77**, 635 (1967).
Vail. *Postgrad. Med.*, **5**, 439 (1949).
Vlacek. *Cs. Oftal.*, **2**, 203 (1935).

DEEP INFLAMMATIONS OF THE LIDS

TARSITIS

Owing to its dense, fibrous, avascular structure, primary inflammations of the tarsal connective tissue are rare: indeed, with the exception of the chronic granulomata, all inflammations therein are secondary to inflammatory processes originating in the skin and subcutaneous tissues or the conjunctiva.

Among the *acute inflammations*, erysipelas, furuncle, lid-abscess or gangrene or a very severe conjunctivitis such as that due to the gonococcus may spread to the tarsus, infiltrating it with inflammatory cells, softening it and sometimes destroying it. Among the *chronic inflammations* the most important is TRACHOMA[3] which if of any severity infiltrates the tarsus, at

[1] " It is commonly said that the eye-lid being rubbed by the tail of a black cat would do it much good if not entirely cure it, and having a black cat, a little before dinner I made a trial of it, and soon after dinner I found my eye-lid much abated of the swelling and almost free of pain. I cannot therefore but conclude it to be of the greatest service to a Stiony on the eye-lid. Any other cat's tail may have the above effect in all probability—but I did my eye-lid with my own black Tom-cat's tail." (Parson Woodforde, 1791.)

[2] p. 225. [3] Vol. VIII, p. 278.

first along the superior and inferior vascular arcades (Wolfring, 1868; Greeff, 1902; W. and M. Goldzieher, 1906) and then diffusely through the connective tissue of the tarsus (Junius, 1902), thickening, softening and deforming it. It is the weight and thickening which contribute to the ptosis characteristic of this disease and the softening and deformation which are potent factors in inducing entropion and trichiasis. The tarsal glands are also affected, becoming infiltrated, strangulated, atrophied and eventually cystic (Fig. 407) (Zykulenko, 1935; Pulvertaft, 1935; MacCallan, 1936). In this way, it is to be noted, the tarsus frequently serves as a reservoir for re-infection in cases of persistent disease. In VERNAL CATARRH,[1] also, the infiltration may invade the deeper layers of the tarsus leading to thickening and hyaline infiltration.[2] Other deeply penetrating conjunctival inflammations, such as PEMPHIGUS, may also involve the tarsus.

Of the infective granulomata, tuberculosis, leprosy and syphilis have a place in the pathology of the tarsus.

TUBERCULOSIS has not been proved to originate primarily in the tarsus although the connective tissue and glands may be heavily infiltrated from neighbouring disease (Rollet, 1905; Aurand, 1909). A tuberculous lesion may occur as a rarity by hæmatogenous spread. Thus in his experiments observing the effect of the injection of tubercle bacilli into the blood-stream, Stock (1907) produced chalazion-like tuberculomata in the upper lid, and Palermo (1896) induced tuberculosis of the tarsus resulting in ulceration and death by direct inoculation in rabbits. In a case recorded by Belz (1948) all four tarsal plates were swollen, in the lower lids over their entire area, in the upper with discrete, chalazion-like masses: the skin and the conjunctiva were normal and adenopathy was marked; tracheo-bronchial nodes were seen radiologically, skin tests for tuberculosis were positive, and although the case was not proven bacteriologically, an excised tarsus on histological examination showed a typical tuberculous picture without caseation. Several other cases have been reported, the literature concerning which was summarized by Gát and his colleagues (1962) and Serifoglu (1966).

TULARÆMIA, affecting the tarsal conjunctiva,[3] may invade the tarsal plate giving rise to chalazion-like swellings (Vail, 1926; Francis, 1942).

SYPHILITIC TARSITIS

Tarsitis is one of the rarest of syphilitic infections: Vancea (1927) was able to find only 25 cases in the literature. Three main types occur—(1) a diffuse gummatous infiltrate which may affect one or all four lids (Fig. 224), (2) a marginal type characterized by ulceration leading to deformity of the lid-margin with loss of the lashes, and (3) a circumscribed nodular lesion

[1] Vol. VIII, p. 475.

[2] Rschanitzin (1905), Goldzieher (1906), Thaler (1906), Axenfeld (1907), Schieck (1907), and others.

[3] Vol. VIII, p. 204.

resembling a chalazion, one or more of which may occur on the same lid or on more than one (Fig. 225). The upper lids are the more commonly affected.[1]

Clinically the condition is a tertiary infection, although it may occur precociously in the secondary stage; Igersheimer (1918), indeed, considered the infiltration to be of the papular type. It affects both children as a hereditary manifestation (*e.g.*, in a 9-months-old child, Stern, 1892) and adults with acquired syphilis, and may occur at an earlier stage of the disease (2 to 12

FIGS. 224 and 225.—SYPHILITIC TARSITIS.

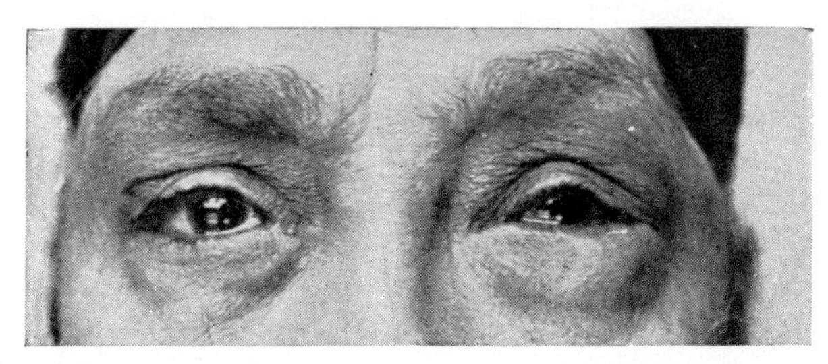

FIG. 224.—In a 65-year-old Egyptian woman. There is thickening of all four lids with bilateral ptosis (M. Khalil).

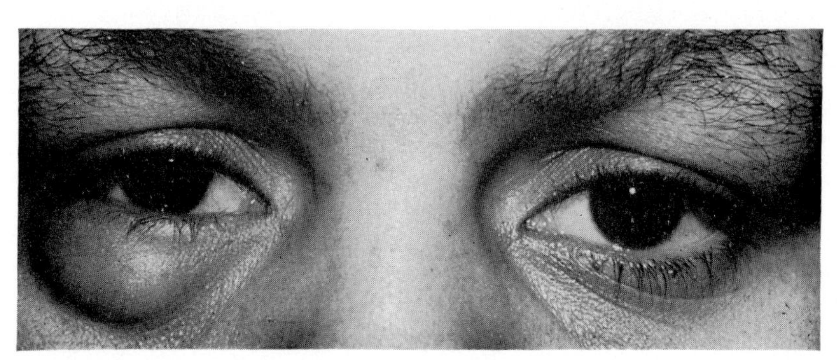

FIG. 225.—In a 33-year-old Negress affecting the right lower lid. The swelling was of one month's duration (L. Turner).

months) than gummata elsewhere (Heckel and Beinhauer, 1925; Turner, 1959). It is of gradual development, without pain, appearing as an indolent swelling over which the skin is at first freely movable and eventually becomes tense, reddened and violaceous. The pre-auricular node is swollen. Meantime the tarsus, or its swollen part, becomes stony-hard and ungainly in form, the entire lid becomes enlarged and, in the upper lid, ptosis is marked and eversion of the lid impossible. Indeed, even when the lower lid alone is involved the swelling may be so great that the eye may be opened only with

[1] Fuchs (1878), Rogman (1898), Reiner (1898), Bull (1899), Basso (1900), Le Roux (1926), Vancea (1927), Whiting (1932), Khalil (1937), Cornet (1937), Renard and Halbron (1938), Turner (1959), and others.

FIGS. 226 and 227.—SYPHILITIC TARSITIS (L. Turner).

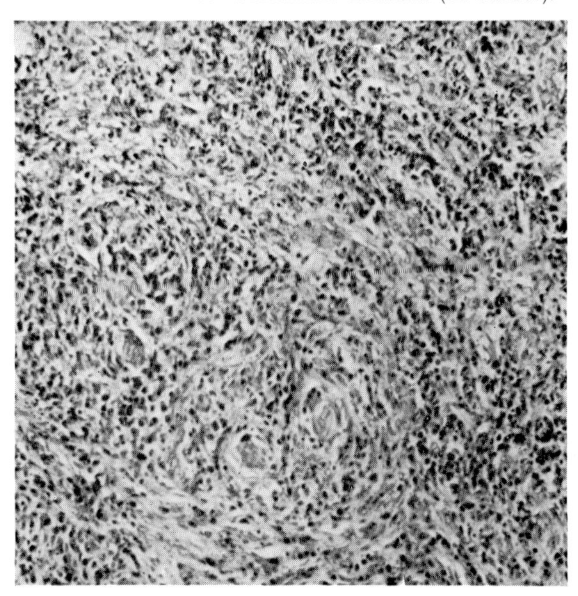

FIG. 226.—The granulomatous lesion with proliferation of fibrous tissue, endothelial swelling and a diffuse infiltration with inflammatory including multinucleated cells (× 162).

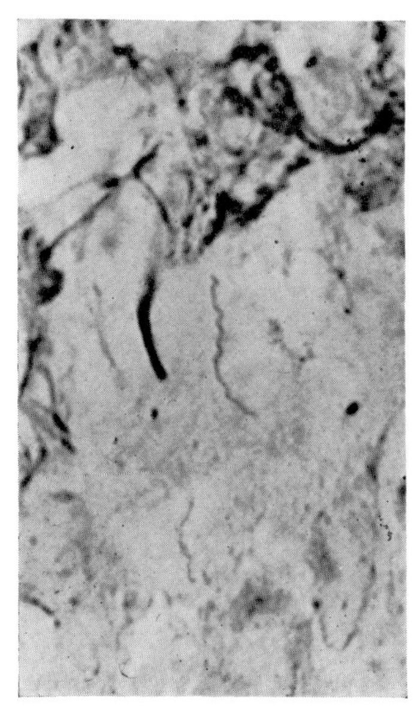

FIG. 227.—A spirochæte is in focus in the centre of the picture while others are slightly out of focus (Dieterle stain; × 1,788).

difficulty and the patient can scarcely get about. The conjunctiva is usually normal, but in cases of long duration may show changes suggestive of trachoma while in those of 7 or 8 years' duration, epidermoid or xerotic changes may appear. Occasionally indolent and painless ulcers develop, either on the skin or on the conjunctival surface (Le Roux, 1926). As a rule, however, when untreated the swelling subsides very slowly until the tarsus resumes its former size or becomes smaller through atrophy.

As a rarity, anomalous forms have been met with. Vancea (1927) described a *polypoid form* in which a pedunculated tumour rose from the conjunctival surface. A more common form attacks the ciliary margin which involves loss of the lashes and dense scarring (Yamaguchi, 1905) or a gross deformity with the appearance of a colobomatous defect (Druais, 1905).

When the tissue is incised it is found to be hard, lardaceous and avascular and does not bleed. Pathological studies, which have been conducted by several authors,[1] show the ordinary syphilitic infiltration of round and plasma cells, extreme endarteritis, and hyaline and sometimes even calcareous degeneration (Figs. 226–7).

The diagnosis is made by serological tests but is sometimes difficult clinically, particularly in the nodular form which may simulate the common chalazion. The general appearance may make the diagnosis clear, but in the event of an incision being made under this impression, the hard lardaceous consistency without central softening will arouse suspicion, and if histological examination is made, perhaps to eliminate an adenoma, the diagnosis will be apparent. In the ulcerative stage, tuberculosis and epithelioma may be simulated; while in the diffuse type a diagnosis of amyloid tumour has suggested itself (Yamaguchi, 1905).

Treatment is by standard anti-syphilitic measures.

Aurand. *Clin. Ophtal.*, **1,** 587 (1909).
Axenfeld. *Zbl. allg. Path. path. Anat.*, **18,** 813 (1907).
Bull. Soc. franç. Ophtal., **24,** 1 (1907).
Basso. *Ann. Ottal.*, **29,** 645 (1900).
Belz. *Bull. Soc. Ophtal. Paris*, 580 (1948).
Bull. *Med. News* (N.Y.), **74,** 609 (1899).
Cornet. *Rev. int. Trachome*, **14,** 264 (1937).
Druais. *Trib. méd.* (Paris), **37,** 229 (1905).
Francis (quoting Blum). *Arch. Ophthal.*, **28,** 711 (1942).
Fuchs. *Klin. Mbl. Augenheilk.*, **16,** 21 (1878).
Gát, L., Pintér and Gát, G. *Ophthalmologica*, **144,** 175 (1962).
Goldzieher, W. *Klin. Mbl. Augenheilk.*, **44** (2), 521 (1906).
Goldzieher, W. and M. *v. Graefes Arch. Ophthal.*, **63,** 287 (1906).
Greeff. Orth's *Lhb. d. spec. path. Anat.*, Berlin, Erg. **1** (2), 32 (1902).
Heckel and Beinhauer. *Arch. Ophthal.*, **44,** 644 (1925).
Igersheimer. *Syphilis u. Auge*, Berlin, 165 (1918).
Ischreyt. *Tarsitis luetica*, Berlin (1906).
Junius. *Z. Augenheilk.*, **8,** 77 (1902).
Khalil. *Brit. J. Ophthal.*, **21,** 648 (1937).
Le Roux. *Arch. Ophtal.*, **43,** 530 (1926).
MacCallan. *Brit. med. J.*, **1,** 635 (1936).
Palermo. *Ann. Ottal.*, **25,** 481, 559 (1896).
Pulvertaft. *Rev. int. Trachome*, **12,** 19 (1935).
Reiner. *Beitr. Augenheilk.*, **3** (23), 225 (1898).
Renard and Halbron. *Arch. Ophtal.*, **2,** 599 (1938).
Rogman. *Ann. Oculist.* (Paris), **120,** 89 (1898).
Rollet. *Arch. Ophtal.*, **25,** 340 (1905).
Rschanitzin. *Vestn. Oftal.*, **22,** 746 (1905).
Sabbadini. *Saggi Oftal.*, **4,** 561 (1929).
Schieck. *Klin. Mbl. Augenheilk.*, **45** (1), 449 (1907).
Serifoglu. *Ankara Numune Hast. Bült.*, **6,** 298 (1966).
Stern. *Casuistik d. syphilitischen Erkrankung d. Tarsus nebst einem Beitrag* (Diss.), Würzburg (1892).
Stock. *v. Graefes Arch. Ophthal.*, **66,** 1 (1907).
Thaler. *Z. Augenheilk.*, **16,** 16 (1906).
Turner. *Amer. J. Ophthal.*, **47,** 389 (1959).
Vail. *Arch. Ophthal.*, **55,** 236 (1926).
Vancea. *Arch. Ophtal.*, **44,** 644 (1927).
Whiting. *Trans. ophthal. Soc. U.K.*, **52,** 140 (1932).
Wolfring. *v. Graefes Arch. Ophthal.*, **14** (3), 159 (1868).
Yamaguchi. *Arch. Augenheilk.*, **51,** 8 (1905).
Zykulenko. *Folia ophthal. orient.*, **1,** 366 (1935).

[1] Rogman (1898), Reiner (1898), Basso (1900), Yamaguchi (1905), Ischreyt (1906), Igershiemer (1918), Sabbadini (1929), Whiting (1932), Turner (1959) and others.

INFLAMMATIONS OF THE TARSAL GLANDS

Inflammations of the tarsal glands may be divided into two types—those depending primarily on secretory anomalies wherein infection is an incidental factor, and those depending essentially on infection. The first class can be divided into two groups depending on whether the inflammation is generalized throughout the tarsus (*seborrhœic meibomitis*) or mainly localized in one or more glands when the most important feature is retention (*chalazion*); the second is most easily divided clinically into acute and chronic inflammations (*hordeolum internum* and *chronic infective meibomitis*).

SEBORRHŒIC MEIBOMITIS

SEBORRHŒIC MEIBOMITIS (CHRONIC MEIBOMITIS, Gifford, 1921) is a chronic affection, particularly common in persons over middle age, wherein a true inflammatory process is superimposed upon a seborrhœa of the meibomian glands.[1] The clinical picture was recognized by the early clinicians and corresponds to the *ophthalmia tarsi* of Mackenzie (1835) or the *puriform palpebral flux* of Scarpa (1801). We have already seen that in simple meibomian seborrhœa the only symptom is the accumulation of a white frothy secretion along the lid-margins collecting particularly at the canthi. This hypersecretion of the meibomian glands which occurs with blepharitis is not a cause but a symptom (Indeikin, 1961); the foam-like secretion found at the outer palpebral canthus usually consists of fat and fatty epithelial scales originating from the lid-margin (Norn, 1963). On the development of inflammation more obvious symptoms of irritation appear. The lids become thickened and the glands may be seen through the conjunctiva clothing the tarsus, while a yellowish oily, serous or purulent fluid can be expressed from the ducts in considerable quantity (Cowper, 1922). In the secretion no significant organismal infection has been found (staphylococcus, Doyne, 1910; *Coryne. xerosis*, Gifford, 1921; Mohammed, 1937). Complications, however, are the rule, due to some extent to irritation caused by the lipids in the secretion—almost invariably a seborrhœic squamous blepharitis,[2] a chronic conjunctivitis meibomiana[3] (Elschnig, 1908; Roesen, 1928) and sometimes a keratitis the most significant feature of which is recurrent marginal ulcers (keratitis meibomiana, Filatov, 1930). In the *triad comprised of chronic conjunctivitis, chronic blepharitis and meibomitis*, the meibomitis is too frequently forgotten in the clinical picture and in treatment, with unfortunate results for it is the cause of the other two and is responsible for maintaining the vicious circle. Eventually considerable hypertrophy may occur in the tarsus with permanent thickening of the lids, and occlusion of the ducts may lead to the development of chalazia which may appear in crops over periods of years unless efficient treatment is carried out.

[1] p. 33. [2] p. 219. [3] Vol. VIII, p. 80

Treatment should consist essentially of tarsal massage[1] which should be continued until the condition clears up; a useful adjunct is suction by means of a medicine dropper along the lid-margin (Lemere, 1944). This should be combined with the use of a bicarbonate lotion as well as general measures to combat the seborrhœic tendency.[2] In the worst cases it may be necessary to split the tarsus into two leaves and scrape the glands with a spoon (Filatov, 1922). Couzi (1949) recommended sealing the orifices of the tarsal glands by electro-coagulation.

Couzi. *Ann. Oculist.* (Paris), **182,** 761 (1949).
Cowper. *Amer. J. Ophthal.*, **5,** 25 (1922).
Doyne. *Trans. ophthal. Soc. U.K.*, **30,** 85 (1910).
Elschnig. *Dtsch. med. Wschr.*, **34,** 1133 (1908).
Filatov. *Klin. Mbl. Augenheilk.*, **69,** 657 (1922); **84,** 380 (1930).
Gifford. *Amer. J. Ophthal.*, **4,** 489, 566 (1921).
Indeikin. *Oftal. Zh.*, No. 1, 32 (1961).
Lemere. *Arch. Ophthal.*, **31,** 95 (1944).
Mackenzie. *Practical Treatise on the Diseases of the Eye*, 2nd ed., London, 149 (1835).
Mohammed. *Bull. ophthal. Soc. Egypt*, **30,** 127 (1937).
Norn. *Acta ophthal.* (Kbh.), **41,** 531 (1963).
Roesen. *Klin. Mbl. Augenheilk.*, **81,** 855 (1928).
Scarpa. *Saggi di osservazioni e d'esperienze sulle principali malattie degli occhi*, Pavia (1801).

CHALAZION

A CHALAZION (χαλάζιον, a small hail-stone) is a *chronic inflammatory granuloma caused primarily by the retention of the secretion of a tarsal gland.* It was formerly regarded as a meibomian cyst analogous to the atheromata of the sebaceous glands of the skin,[3] but while retention cysts of the meibomian glands do occur,[4] they are very rare; a truer analogy is with acne rosacea of the skin wherein the sebaceous glands play a part comparable to that of the meibomian glands in the development of a chalazion.

The clinical picture is very characteristic (Figs. 228–9). Typically, a chalazion starts as a hard, circumscribed, painless tumour which grows slowly and indolently without marked symptoms, and is felt under the normal skin of the lid like a pellet. It occurs more frequently in the upper lid than the lower, and more commonly in adults than in the young. It may be single, more than one lid may be affected, while in the same lid more than one may coincide; frequently chalazia occur in crops particularly in young people, especially the seborrhœic who for some time have had a chronic meibomitis with blepharitic and conjunctival inflammation (the *conjunctivite à chalazion* of Dianoux, 1891) or in old people, particularly sufferers from acne rosacea of long standing.

A chalazion may appear after a blepharitic or a local tarsal infection, but more usually its onset is slow and gradual without apparent cause. Over the little hard tumour the skin is at first freely movable and normal, but as growth continues it may become stretched and vivid. Underneath, the conjunctiva is velvet-like, red and swollen, but as the tarsus becomes thinned by pressure, a bluish-grey spot is seen on the conjunctiva when the lid is everted. The tumour may remain indefinitely without growing; in some

[1] p. 34. [2] p. 29.
[3] p. 395. [4] p. 398.

FIGS. 228 to 230.—CHALAZIA.

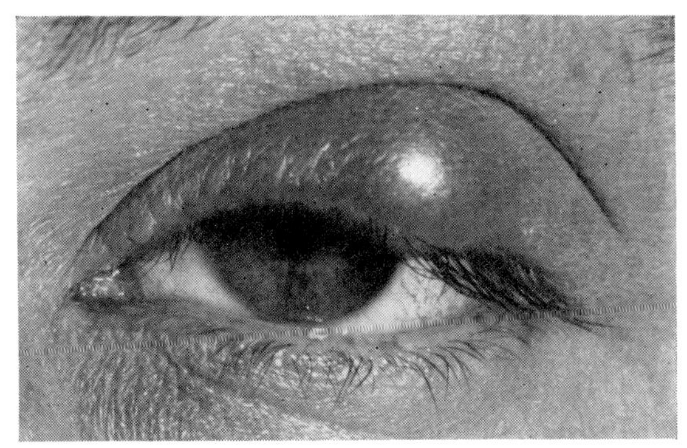

FIG. 228.—Chalazion in the upper lid (Inst. Ophthal.).

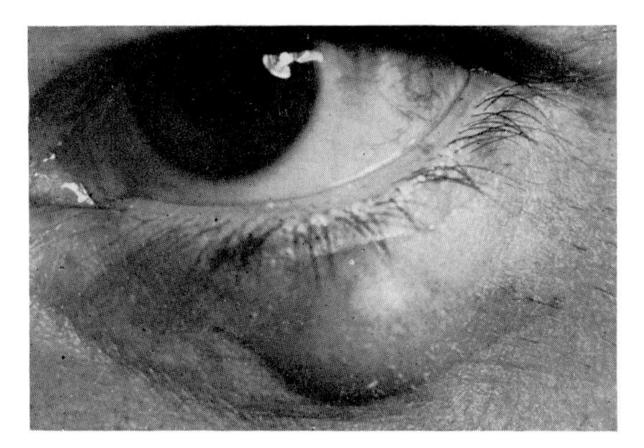

FIG. 229.—Chalazion in the lower lid in a cystic state (P. MacFaul).

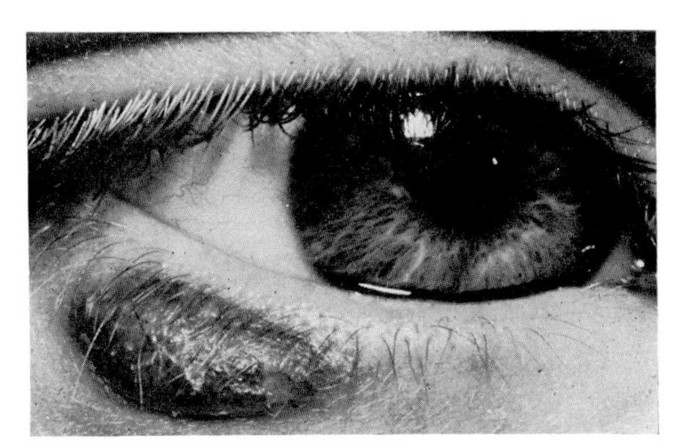

FIG. 230.—Recurrent chalazion in the lower lid (P. Watson, Addenbrooke's Hosp., Cambridge).

early cases it may be absorbed and disappear; occasionally secondary infection and suppuration occur which may even lead to a purulent tenonitis (Chadwick, 1963) or an orbital abscess (Myska and Uher, 1961), exceptionally to calcification (Searle, 1952; Llopis Rey, 1957) or ossification (Franklin and Cordes, 1924; Rizzini, 1953; Gemolotto, 1960; Nath *et al.*, 1966); but more usually it increases slowly in size and becomes softer in consistency until it eventually bursts on the conjunctival surface. Here, after the central softened portion has escaped as a viscid, gelatinous mass, fleshy granulations form excrescences which may remain protruding from the conjunctiva for some considerable time; it usually requires many months for the tumour to disappear completely. Alternatively and more rarely it may perforate the tarsus in front in which case the granulation tissue grows into the subcutaneous layer beyond the lid-border forming a soft, flat tumour which may remain unchanged for months (Fig. 230) (CHALAZION EXTERNUM of Lagrange, 1889, as opposed to the more usual CHALAZION INTERNUM); to it the skin, discoloured a reddish-blue, becomes adherent and may even break down so that a sinus is formed. Again, a chalazion may develop on the duct of a tarsal gland in which case it projects like a nipple from the ciliary border, being flattened out on its posterior side by pressure on the globe (MARGINAL CHALAZION, Erdmann, 1905). In cases wherein multiple chalazia appear the entire tarsus may be affected and two or more tumours may become confluent so that the whole lid forms nodular projections and is thickened to such an extent that eversion is difficult; the skin remains free although it may be discoloured, but the tarsal conjunctiva is coarse and velvety, uneven and nodular, and the clinical picture may suggest a tarsitis or a neoplasm.

Symptoms are few; indeed, the patient may be ignorant of its presence until it has reached a considerable size. The usual complaint is cosmetic, but some heaviness and discomfort may be present; occasionally there may be some inflammatory concomitants but these are usually insignificant; if, however excrescences develop on the conjunctival surface, the discomfort may be accentuated. It is interesting that the pressure of the tumour may alter the refraction and cause some astigmatism which occasionally may be considerable but disappears on removal of the granuloma (Ormond, 1921; Safar, 1947; Casanovas, 1949; Asseman *et al.*, 1965; and others).

The *pathology* of chalazion shows it to be a peculiar inflammation of a meibomian gland producing granulation tissue rich in giant cells (Fig. 231): for this reason de Vincentiis (1875) called it a GRANULOMA GIGANTO-CELLULARE CAPSULATUM. Fuchs (1878) considered that the process started as a desquamative catarrh with proliferation of the epithelium of the acinus of the gland so that the central cells degenerate and break down into granular debris among which are found fatty droplets. The surrounding tissue of the tarsus becomes densely infiltrated with leucocytes and the fixed cells proliferate, a process which becomes predominant so that both the acini and the peri-acinous tissue are ultimately lost in an undifferentiated mass of granulations surrounded by a dense ring of fibrous tissue. The granulation tissue exhibits the usual cytological characteristics, containing plasma, epithelioid and lymphoid cells with the addition of giant cells and a considerable amount of fibrosis. In the middle of the infiltrated parts spherules of cells may be isolated by connective-tissue fibres among which drops of fat are found extracellularly as well as within the giant cells so that the picture of a lipogranuloma is produced. Bakker (1947) found cholesterol esters present. As the process evolves the peripheral fibrous tissue becomes compressed to form a dense capsule, and retrogressive changes occur centrally. Here the

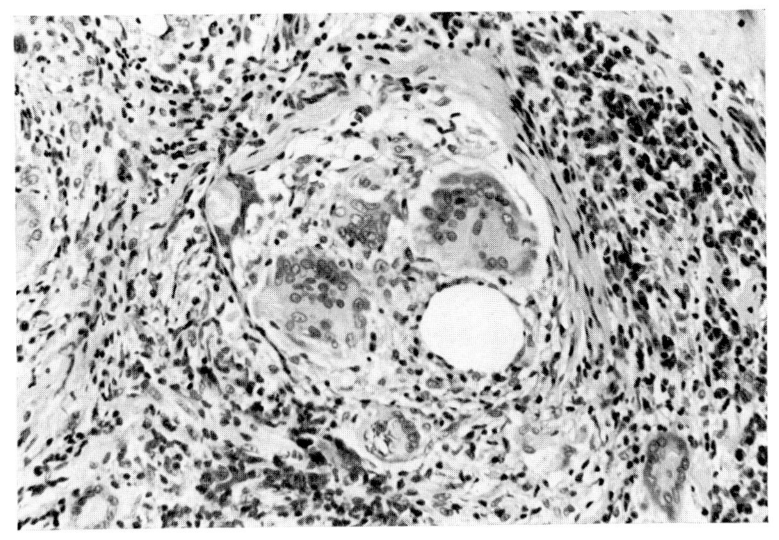

Fig. 231.—Chalazion.

The chronic inflammatory tissue contains a cluster of foreign-body giant cells and a lipid globule (H. & E.; × 186) (N. Ashton).

fibres become hyaline and fuse into a fluid or gelatinous homogeneous mass, the cells become vacuolated and disappear and the entire contents may eventually become liquefied. Sometimes fibrosis is very prominent when the nodule is almost of cartilaginous consistency with a dense fibrous capsule and an interior made up of interlacing fibrous septa in which globules of glairy fluid are held. Sometimes the necrotic process is predominant in which case a thin fibrous sac containing a glairy fluid forms the traditional meibomiam "cyst" comparable to a cold abscess. Depending on which of the two processes is more evident any intermediate stage between these two extremes may result (Parsons, 1904; Lowenstein, 1913–14; del Monte, 1916–17; Schall, 1926; Levaditi, 1934; and others); in recurrent cases heterotopic ossification may result (Nath *et al.*, 1966) (Figs. 366–7).

In a marginal chalazion Erdmann (1905) found the same cellular elements heaped up free from the acini as a granulomatous mass surrounded by fibrous tissue.

The *pathogenesis* of chalazion has long been a matter of dispute—a curious fact in a lesion so common and accessible. The early writers considered it to be an aborted internal hordeolum (*hordeolum induratum*) (Arlt, 1858) or a *retention cyst* of a meibomian gland (de Wecker, 1867). With the demonstration of the granulomatous nature of the tumour by de Vincentiis (1875), however, ideas changed, but there remained some controversy whether glandular changes were primary (Fuchs, 1878) or secondary to changes in the surrounding tissues. Deyl (1893) suggested the theory that the mass was due to the obstruction of the secretion of the gland; Deutschmann (1891) considered it a chronic hyperplasic folliculitis and perifolliculitis, while Sabrazès and Lafon (1908) compared it with acne, a view strenuously opposed by Buri (1912). On the other hand, Palermo (1896) considered it a *foreign-body granuloma due to chemical* (or physical) *irritation* and the more recent views follow from this, based on the work of Schall (1926), Levaditi (1934) and Hagedoorn (1935). The retained meibomian secretion, particularly its lipid constituents, has an irritative effect which excites a foreign-body reaction producing granulomatous tissue rich in giant cells. A similar reaction from chemical irritation can be produced by the injection of vernix caseosa or sterile mutton fat (Schall, 1926) or yellow wax (Levaditi, 1934) into the tissues of rabbits and guinea-pigs, a process comparable to the fat-necrosis or lymphogranulomatosis which accompanies diseases of the pancreas (Hagedoorn, 1935) or to the foreign-body reaction sometimes seen around sebaceous cysts in the skin. In all these cases the histological picture is the same and the surrounding tissue tends to become progressively affected by the diffusion of lipolytic ferments, a process which accounts for the subcutaneous spread of a chalazion externum.

That a chalazion is a chronic *low-grade infection* is a view which has had many advocates and many types of organism have been implicated—the tubercle bacillus, cocci, the *Demodex folliculorum*, and even protozoa. The large literature can be found in the monograph by Schreiber (1924). It is true that a chalazion may contain several types of organism, but it is equally true that these are incidental rather than causative. In 80 cases Abboud and his colleagues (1968) found that it was associated with a deficiency in the serum of vitamin A.

The *diagnosis* of a chalazion is usually easy. The more widespread types may simulate a tarsitis, and occasionally the question of a neoplasm (adenoma, carcinoma and, very occasionally, sarcoma) may arise. In the latter case suspicions are usually aroused when the tumour is incised or by its *recurrence thereafter*, in which case the diagnosis is made by histological examination (Hughes, 1932; Jaensch, 1933; Lotin, 1937; Engelhart, 1962; and others). *A chalazion which persistently recurs after adequate surgical treatment should always be subjected to this examination with a view to eliminating malignancy.* Other less frequent confusing conditions include tuberculomata (Zubczewska, 1962), tularæmia, gummata, chronic suppurative meibomitis or mycotic infections.

Treatment. When small and causing no symptoms, a chalazion may be left alone, the precaution being taken of massaging the lids to empty other glands as a prophylactic measure against their involvement. If the tumour is of any size, however, an operation is required. If the granuloma is cystic, incision and vigorous curettage under infiltration anæsthesia is usually sufficient; if secondary infection has occurred and pus is evacuated, little or no curettage should be done; if incision shows the growth to be largely fibrous it should be bisected and each half dissected out. A marginal chalazion should be snipped off and the duct (and gland if necessary) slit up and curetted; while granulations growing on the conjunctival surface should be cut away and scraped. With multiple chalazia no hesitation should be felt in a widespread careful dissection; fibrous tissue heals gaps in the tarsal plate with no deformity. In every case the tarsal glands should be thoroughly massaged to prevent recurrences in other glands, and in all cases an adequate portion of the diseased tarsus should be removed lest recurrences necessitate further operative interference.

Abboud, Osman and Massoud. *Exp. Eye Res.*, **7**, 388 (1968).
Arlt. *Die Krankheiten d. Auges*, Prague (1858).
Asseman, Corbel and Taine. *Bull. Soc. Ophtal. Fr.*, **65**, 148 (1965).
Bakker. *Ned. T. Geneesk.*, **91**, 461 (1947).
Buri. *Beitr. Augenheilk.*, **8**, 207 (1912).
Casanovas. *Arch. Soc. oftal. hisp.-amer.*, **9**, 23 (1949).
Chadwick. *Brit. J. Ophthal.*, **47**, 364 (1963).
Deutschmann. *Beitr. Augenheilk.*, **2**, 109 (1891).
Deyl. *Ueber d. Ätiologie d. Chalazion*, Prague (1893).
Dianoux. *Arch. Ophtal.*, **11**, 302 (1891).
Engelhart. *Praxis*, **51**, 170 (1962).
Erdmann. *Arch. Augenheilk.*, **51**, 171 (1905).
Franklin and Cordes. *J. Amer. med. Ass.*, **82**, 519 (1924).
Fuchs. *v. Graefes Arch. Ophthal.*, **24** (2), 121 (1878).
Gemolotto. *Boll. Oculist.*, **39**, 309 (1960).
Hagedoorn. *Amer. J. Ophthal.*, **18**, 424 (1935).
Hughes. *Trans. ophthal. Soc. U.K.*, **52**, 557 (1932).
Jaensch. *Klin. Mbl. Augenheilk.*, **90**, 598 (1933).
Lagrange. *Arch. Ophtal.*, **9**, 226 (1889).
Levaditi. *C.R. Soc. Biol.* (Paris), **115**, 1592 (1934).
Llopis Rey. *Arch. Soc. oftal. hisp.-amer.*, **17**, 449 (1957).
Lotin. *Vestn. Oftal.*, **10**, 891 (1937).
Löwenstein. *Klin. Mbl. Augenheilk.*, **51** (2), 597 (1913).
Z. Augenheilk., **30**, 450 (1913).
v. Graefes Arch. Ophthal., **87**, 391 (1914).
del Monte. *Arch. Ottal.*, **21**, 83 (1914); **23**, 84 (1916).
Ann. Oculist. (Paris), **154**, 607 (1917).
Myska and Uher. *Cs. Oftal.*, **17**, 40 (1961).
Nath, Rahi, A. and S. *Arch. Ophthal.*, **75**, 642 (1966).
Ormond. *Brit. J. Ophthal.*, **5**, 117 (1921).
Palermo. *Ann. Ottal.*, **25**, 481, 559 (1896).
Parsons. *Pathology of the Eye*, London, **1**, 10 (1904).
Rizzini. *G. ital. Oftal.*, **6**, 607 (1953).
Sabrazès and Lafon. *Sem. méd.* (Paris), **28**, 541 (1908).
Safar. *Wien. klin. Wschr.*, **59**, 484 (1947).
Schall. *v. Graefes Arch. Ophthal.*, **117**, 662 (1926).
Schreiber. *Graefe-Saemisch Hb. d. ges. Augenheilk.*, 3rd ed., *Die Krankheiten d. Augenlider*, Berlin, 395 (1924).
Searle. *J. roy. Army med. Corps*, **98**, 66 (1952).
de Vincentiis. *Della struttura e genesi del chalazion*, Naples (1875).
de Wecker. *Traité théorique et pratique des maladies des yeux*, Paris (1867).
Zubczewska. *Klin. oczna*, **32**, 127 (1962).

HORDEOLUM INTERNUM (ACUTE MEIBOMITIS)

A HORDEOLUM INTERNUM (MEIBOMIANUM) is *an acute infection of a meibomian gland,* usually by a staphylococcus, which progresses to suppuration. Like a hordeolum externum it is a suppurative inflammation of a sebaceous gland, analogous to acne of the skin; but in so far as the tarsal glands are larger and more securely encased in dense fibrous tissue than the glands of Zeis, the internal hordeolum is usually a more violent and prolonged inflammation than the more common stye. A suppurating chalazion presents much the same clinical picture, but its symptoms are usually less acute.

Clinically the disease starts with pain and inflammatory œdema which at first may be widespread over the whole lid; usually appearing in the upper lid, the œdema may spread to the lower and even to the bulbar conjunctiva, and may be so severe as to make it impossible to open the eye (Darabos, 1966). A localized spot of tenderness, however, indicates the site

Figs. 232 and 233.—Acute Internal Meibomitis (Inst. Ophthal.).

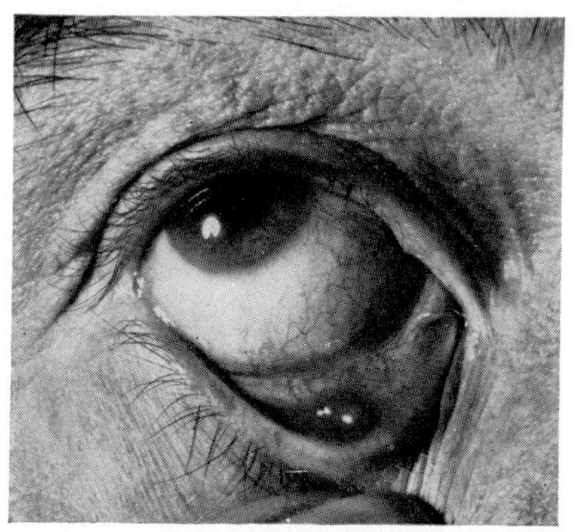

Fig. 232.—The small white point (barely seen) just visible at the lid-margin to the right of the swelling in the lower lid indicates the mouth of the gland exuding pus.

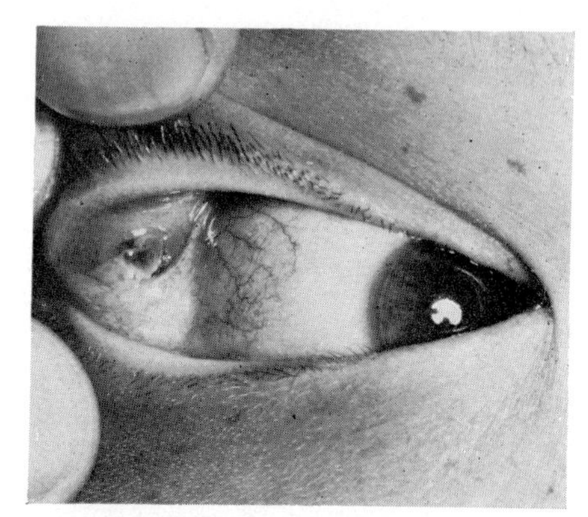

Fig. 233.—An internal hordeolum affecting the upper lid.

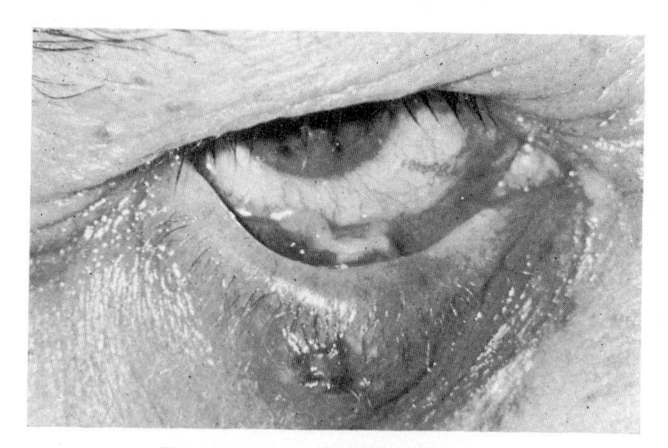

Fig. 234.—Meibomian Abscess.

The lesion has broken out on the surface of the skin; note the pus in the conjunctival sac (P. MacFaul).

of the inflammation; on the ciliary margin the orifice of the affected gland may be seen swollen and enlarged, and if the lid is everted a yellow area shining through the coarsened and inflamed conjunctiva sooner or later indicates the site of pus (Figs. 232–3). As suppuration progresses the symptoms of pain, swelling and tenderness tend to increase, the pre-auricular node is usually involved, and considerable general malaise may develop. Eventually the abscess perforates, in most cases through the conjunctiva; occasionally it discharges through the orifice of the gland; and very rarely perforation occurs through the skin of the lid (Fig. 234).

A hordeolum internum usually occurs singly; but in debilitated people or in the presence of persistent blepharitic or conjunctival infection, several may appear in succession persistently over long periods. Occasionally they appear in crops simultaneously when the condition may be very distressing: in one case Schreiber (1924) noted 24 such abscesses situated in the four lids. In such cases a considerable necrosis of the tarsal tissue may result with multiple perforations through the conjunctiva (TARSITIS NECROTICANS of Mitvalský, 1897), and even when the process does not become so extensive subsequent scarring may cause considerable embarrassment. Very rarely the disease has been fatal owing to the spread of infection into the orbit and the development of cavernous sinus thrombosis (Vittadini, 1933; Gözberk, 1938).

Pathologically the condition is a suppurative adenitis and peri-adenitis. The organism involved is almost invariably the staphylococcus, but others such as pneumobacilli (Priouzeau, 1898) have been described.

Treatment should consist of persistent hot fomentations until pus can be localized on the conjunctival surface, whereupon an incision is made and hot bathings continued until the inflammation has subsided. Any associated infection should, of course, be vigorously attacked, and if recurrences persist or multiple abscesses occur, the general health should receive attention as well as the immunological response to the staphylococcus. Antibiotic drugs may abort a lesion or prevent recurrences.

Darabos. *Ophthalmologica*, **151,** 477 (1966).
Gözberk. *Ann. Oculist.* (Paris), **175,** 159 (1938).
Mitvalský. *Zbl. prakt. Augenheilk.*, **21,** 47, 73 (1897).
Priouzeau. *Ann. Oculist.* (Paris), **119,** 126 (1898).
Schreiber. *Graefe-Saemisch Hb. d. ges. Augenheilk.*, 3rd ed., *Die Krankheiten d. Augenlider*, Berlin, 392 (1924).
Vittadini. *Boll. Oculist.*, **12,** 683 (1933).

CHRONIC SUPPURATIVE MEIBOMITIS

A *chronic suppurative inflammation of the tarsal glands* is not common. A chronic meibomitis, wherein inflammatory symptoms supervene on a seborrhœic condition, may become purulent in which case pus can be massaged from the ducts; here the infection is also usually staphylococcal (Doyne, 1910). Occasionally a suppurating meibomian gland, instead of running an acute course, quietens down and may remain for some months, usually causing a velvety papillary hypertrophy in the adjacent tarsal conjunctiva (Natanson, 1907; Schreiber, 1924). More rarely still, a chronic suppurative condition may be associated with tuberculosis (Rollet, 1905; Aurand, 1909; Zubczewska, 1962), an actinomyces (Castelain, 1907), or a fungal infection such as by *Acrostalagmus cinnabarinus* (Fazakas, 1934) or *Acrothecium hominis* (Fazakas, 1936–38). In these cases the tarsus becomes thickened, the ducts of the tarsal glands dilate and from their openings a sero-purulent fluid containing granules can be expressed, or a chalazion-like tumour may develop discharging pus through fistulæ into the conjunctival sac.

A chronic infective meibomitis may result secondarily from adjacent inflammation, such as a conjunctivitis, particularly due to diphtheria (Igersheimer, 1907) or trachoma (Ginsberg, 1903; Zykulenko, 1935; Pulvertaft, 1935).

Aurand. *Clin. Ophtal.*, **1,** 587 (1909).
Castelain. *Ann. Oculist.* (Paris), **138,** 261 (1907).
Doyne. *Trans. ophthal. Soc. U.K.*, **30,** 85 (1910).
Fazakas. *Magy. orv. Arch.*, **35,** 337 (1934). *Klin. Mbl. Augenheilk.*, **96,** 401 (1936); **101,** 387 (1938).
Ginsberg. *Grundriss d. path. Histologie d. Auges*, Berlin (1903).
Igersheimer. *v. Graefes Arch. Ophthal.*, **67,** 162 (1907).
Natanson. *Klin. Mbl. Augenheilk.*, **45** (1), 529 (1907).
Pulvertaft. *Rev. int. Trachome*, **12,** 19 (1935).
Rollet. *Arch. Ophtal.*, **25,** 340 (1905).
Schreiber. *Graefe-Saemisch Hb. d. ges. Augenheilk.*, 3rd ed., *Die Krankheiten d. Augenlider*, Berlin, 393 (1924).
Zubczewska. *Klin. oczna*, **32,** 127 (1962).
Zykulenko. *Folia ophthal. orient.*, **1,** 366 (1935).

CHAPTER III

DERMATOSES OF UNKNOWN OR VARIED ÆTIOLOGY

THIS Chapter contains a large number of diseases affecting the skin, some of them of multiple but most of them of unknown ætiology. In many of them the implication of the lids is incidental and relatively rare, but nevertheless important and interesting. They form a somewhat untidy group, often described with a confusing and duplicated terminology, difficult to arrange in a logical sequence because of our ignorance of their ætiology; we have therefore followed the example of Robert Willan in classifying them according to their objective appearances.

ROBERT WILLAN [1757–1812] (Fig. 235) may well be considered as the Father of Modern Dermatology. He was a Yorkshire Quaker who first divided cutaneous diseases into eight classes for which he suggested a nomenclature based on clinical appearances which is largely retained today: papular, squamous, exanthematous, bullous, vesicular, pustular, tubercular and macular. For his work in systematizing dermatology he was awarded the Fothergillian Gold Medal in 1790. His classical work, *On Cutaneous Diseases*, was published in four parts between 1796 and 1808 and after his death was completed by his pupil, Thomas Bateman, in the *Delineations of Cutaneous Diseases*. Willan's book contains original descriptions with illustrations of many diseases—ichthyosis, pityriasis, prurigo—while other conditions were clearly defined and differentiated—the various types of eczema, psoriasis, sycosis, impetigo, urticaria, tinea versicolor, lupus vulgaris, and others.

Dermatoses characterized by Anomalous Keratinization

A heterogeneous group of diseases which may affect the skin of the eyelids has as its main feature a disturbance of the process of keratinization associated with the production of defective keratin (parakeratosis) when the horny cells retain their nuclei, with a premature keratinization of the cells (dyskeratosis), or with an excessive thickening of the horny layer (hyperkeratosis). Our knowledge of the biochemistry of keratinization or of the dynamics of the metabolism of the skin is not sufficiently advanced to allow the presentation of a logical classification of these dermatoses characterized by abnormal or excessive keratinization; some of them appear to be primary defects, others secondary to other pathological processes; many of them are probably manifestations of a widespread disturbance of cellular metabolism and few of them are necessarily related. The most common affecting the lids are the ichthyosiform dermatoses wherein the scaling is generalized, the follicular keratoses wherein the abnormality of keratinization is focal, and acanthosis nigricans wherein the hyperkeratosis takes on a proliferative warty configuration.

FIG. 235.—ROBERT WILLAN
[1757–1812].
(From the Wellcome Institute of the History of Medicine.)

THE ICHTHYOSIFORM DERMATOSES

The ichthyosiform dermatoses (ἰχθύς, a fish) are a group of hereditary disorders characterized by dryness and roughness of the skin with an excessive accumulation of epidermal scales varying in size from the fine bran-like scaling seen in xeroderma to immense flakes two or three millimetres in thickness so that the " harlequin fœtus " appears to be encased in armour. These conditions were first described by Willan (1808), while Brocq (1902) described several forms, but the most useful contribution to our knowledge was derived from the work of Wells and Kerr (1965–66) who discussed the clinical and pathological aspects of the various types on a basis of their genetic transmission (Table I). The most common types are ichthyosis vulgaris which may be either dominantly transmitted or sex-linked, non-bullous ichthyosiform erythroderma (lamellar ichthyosis) and bullous ichthyosiform erythroderma (epidermolytic hyperkeratosis).

TABLE I

GENETIC CLASSIFICATION OF ICHTHYOSIS
(after Wells and Kerr, 1965–66, and Jay *et al.*, 1968)

MODE OF INHERITANCE	PHENOTYPE	SYNONYMS
AUTOSOMAL DOMINANT	(1) Ichthyosis vulgaris (2) Bullous ichthyosiform erythroderma (3) Ichthyosis hystrix gravior	Ichthyosis nitida or simplex; xeroderma Bullous ichthyosiform or epidermolytic hyperkeratosis Porcupine man
AUTOSOMAL RECESSIVE	(1) Non-bullous ichthyosiform erythroderma (2) Lamellar ichthyosis (3) Sjögren-Larsson syndrome (4) Refsum's syndrome	Ichthyosis congenita, larvata or inversa; harlequin fœtus Collodion baby
X-LINKED RECESSIVE	Sex-linked ichthyosis	Ichthyosis serpentina or vulgaris

The ætiology of these genetically determined diseases is quite unknown. The principal determining factor is a cellular hyperplasia leading to an increased formation of cells in the corneal layer and an increased rate of their transit through the epidermis giving rise to profuse scaling. In all forms the trans-epidermal loss of water is increased (Frost *et al.*, 1968) so that the stratum corneum becomes dehydrated, a tendency which is lessened by humidity or immersing the skin in water when it becomes fairly smooth until dehydration reappears. In some forms an abnormality in the lipid metabolism has been found (Finkelstein and Cass, 1967).

DOMINANT ICHTHYOSIS VULGARIS is a relatively common condition in both Caucasians and Negroes with an incidence in Berkshire, England, of about 1/5,300 (Wells and Kerr, 1966) (Fig. 236). It usually appears between

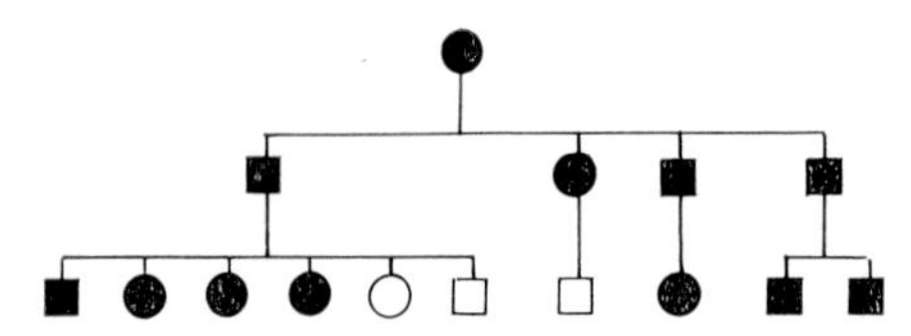

FIG. 236.—DOMINANT ICHTHYOSIS VULGARIS.
Autosomal dominant inheritance (after A. Touraine).

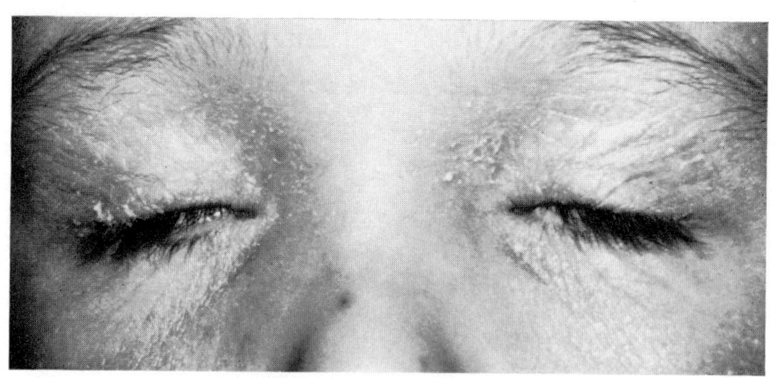

FIG. 237.—DOMINANT ICHTHYOSIS VULGARIS.
Showing involvement of the eyelids and scales on the lashes (B. Jay *et al.*).

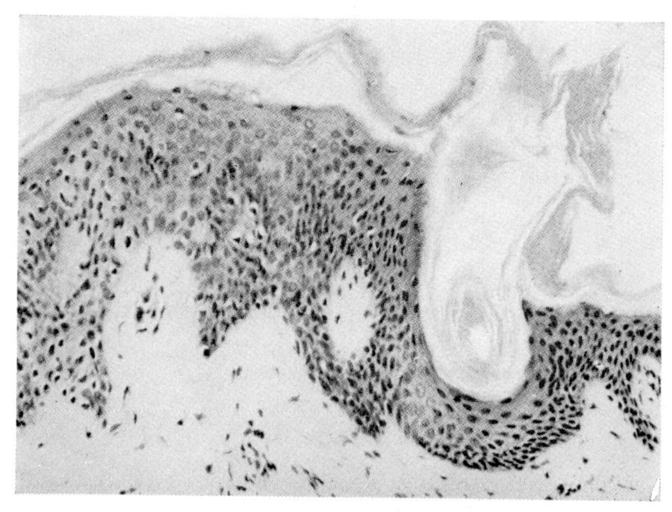

FIG. 238.—DOMINANT ICHTHYOSIS.
Showing hyperkeratosis and an absence or thinning of the granular layer (R. S. Wells and C. B. Kerr).

the ages of 1 and 4 years whereafter it tends to become progressively more severe but often shows spontaneous improvement in later life. Histologically there is a mild hyperkeratosis and the granular layer is reduced or absent (Fig. 238). In the milder cases (*xeroderma*) the symptoms are few and may be limited to a dryness of the skin and a fine scaling which in the more severe cases may be relatively profuse. The face and lids may be affected and the scales may accumulate on the lashes (Fig. 237).

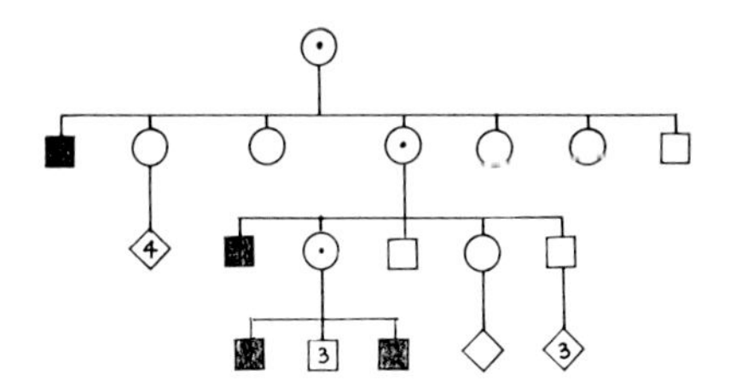

FIG. 239.—SEX-LINKED ICHTHYOSIS.
Sex-linked recessive inheritance (after P. J. Waardenburg, 1946).

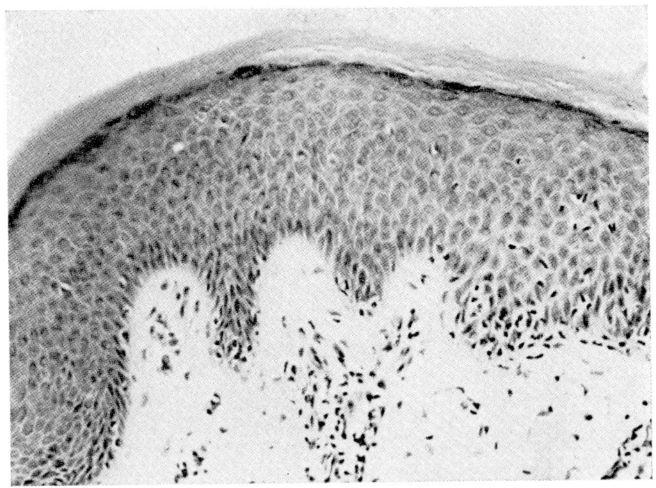

FIG. 240.—SEX-LINKED ICHTHYOSIS.
Showing hyperkeratosis and increased thickness of the granular layer; acanthosis and perivascular infiltrate (R. S. Wells and C. B. Kerr).

SEX-LINKED ICHTHYOSIS VULGARIS is a more rare and more severe disease affecting males and transmitted by normal females, with an incidence in Berkshire, England, of about 1/6,000 males (Fig. 239.) It starts very early in infancy with the formation of large dark scales nearly universal in distribution and persisting throughout life. Histologically there is hyperkeratosis with an increase in the horny and granular layers associated with a vascular infiltrate mainly of lymphocytes (Fig. 240). The lids are often heavily affected and the lashes are loaded with scales.

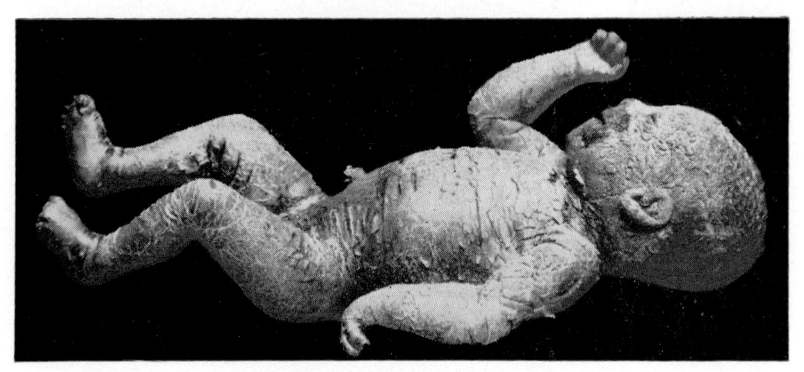

FIG. 241.—HARLEQUIN FŒTUS.

Congenital non-bullous ichthyosiform erythroderma (Edith Potter, *Pathology of the Fetus and the Infant*, Year Book Med. Publishers).

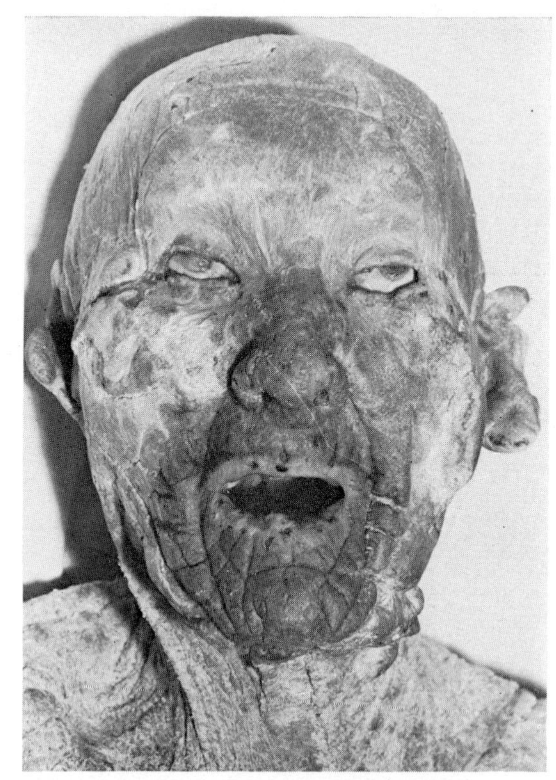

FIG. 242.—NON-BULLOUS ICHTHYOSIFORM ERYTHRODERMA.

In a man aged 69 (A. J. Rook, *Textbook of Dermatology*, Blackwell, Oxford).

BULLOUS ICHTHYOSIFORM ERYTHRODERMA (EPIDERMOLYTIC HYPERKERATOSIS) is a severe condition transmitted by an autosomal dominant gene. It is present at birth or in the first days of life, when crops of bullæ appear associated with erythema and desquamation; at a later stage hyperkeratotic changes appear and persist throughout life which in the less severe cases may be of normal expectancy but in severe cases death may occur in childhood. The hyperkeratosis is marked and in the granular layer there are keratohyalin-like granules. The lids may be affected with profuse scaling on the lashes.

NON-BULLOUS ICHTHYOSIFORM ERYTHRODERMA, a severe disease transmitted as an autosomal recessive character, affects the fœtus (*ichthyosis fœtalis*); the infant is born, usually prematurely, encased in an armour of

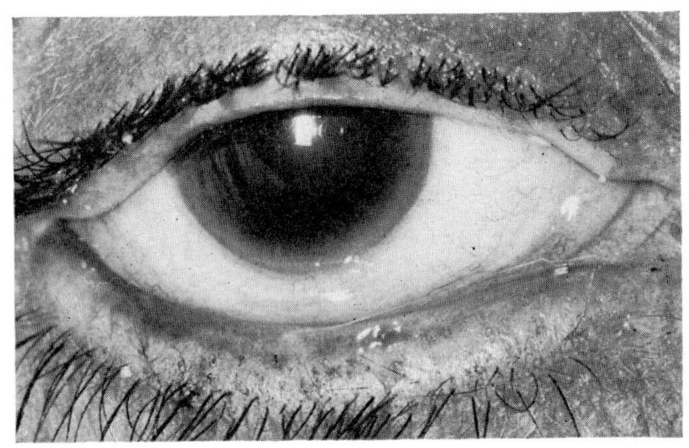

FIG. 243.—ECTROPION IN NON-BULLOUS ICHTHYOSIFORM ERYTHRODERMA (B. Jay *et al.*).

polygonal plates which are desquamated in sheets (*harlequin fœtus*) (Fig. 241). The histopathology resembles that of the sex-linked type and the sweat glands are hypoplasic or absent (Butterworth and Strean, 1962). The hands and feet resemble claws, the face is grotesquely deformed and, in addition to the scaling, ectropion which may be marked is a frequent complication. Few children survive more than a few days, but in more moderate cases erythema and scaling may persist until old age (Fig. 242).

LAMELLAR ICHTHYOSIS is a less severe form of the autosomal recessive type; since the report of Ballantyne (1895) who described 33 cases, Lentz and Altman (1968) collected 103 cases from the literature, 33% of which had bilateral ectropion. The child is born with an appearance suggesting that he is encased in a tight red collodion membrane (*collodion baby*) comprised of the thickened horny layers of the epidermis. After a few days fissuring of the membrane occurs and it gradually peels off leaving the face and the lids free, although for a time their motion is stiff; eventually they move normally and their eversion disappears (Figs. 244–5) (Mortada, 1966; Shapiro and Soentgen, 1969; Rose, 1971).

The most common gross defect in the lids in non-bullous ichthyosiform erythroderma is the occurrence of ectropion due to dryness and contraction (Fig. 243). This usually affects the lower lids but may lead to contracture of the upper, resulting in damage to the cornea through lagophthalmos.[1] For this a skin-graft is the most suitable treatment (Sondermann, 1923;

Figs. 244 and 245.—Lamellar Ichthyosis (A. Mortada).

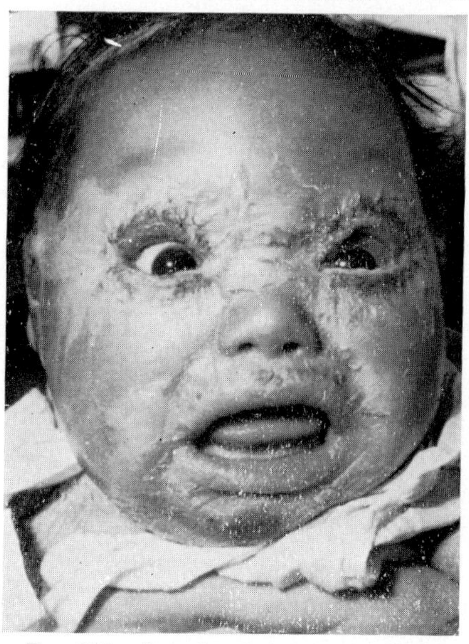

Fig. 244.—In a female infant aged 4 days. There is a contracting membrane covering the face (and body) causing bilateral ectropion of the upper and lower lids.

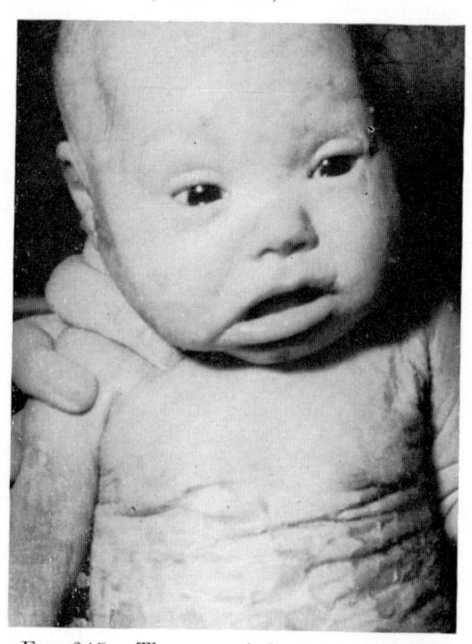

Fig. 245.—The same infant 4 weeks later showing absence of the membrane on the face and self-cure of the ectropion of the lids.

Elschnig, 1923; Hudson, 1926; Hill and Rodrigue, 1971; Shindle and Leone, 1973). The corneal complications—multiple erosions,[2] cornification of the epithelium and stromal opacities[3]—have been described elsewhere, as well as the granular pigmentation in the periphery of the retina.[4]

The *treatment* of these conditions is purely symptomatic and frequently unrewarding; the ideal is to live in a warm moist climate, and if that is impossible an atmosphere of high humidity and periodic exposures to mild

[1] Arnold (1834), Caspary (1886), Riecke (1923), Kaulich (1906), Ischreyt (1908), Sondermann (1923), Contino (1924), Hudson (1926), von Jarmersted (1928), Siemens (1928–29), Simizu (1934), Sannicandro (1936), Shimkin (1945), Bloom and Goodfried (1962), Mortada (1966), Jay *et al.* (1968), Lentz and Altman (1968), Sever *et al.* (1968), Rose (1971), and others.

[2] Vol. VIII, p. 694. [3] Vol. VIII, p. 548. [4] Vol. III, p. 883.

degrees of ultra-violet light may be helpful. Many local applications have been tried but the most effective is simple petrolatum which retards the rate of the loss of water and is rendered more effective by the addition of a keratolytic agent such as salicylic acid which helps to remove the scales[1]; vitamin A acid cream (0·1%) is frequently effective (Frost and Weinstein, 1969). In the more severe types of ichthyosiform erythroderma corticosteroids may be of value (Hanssler, 1957), and in the bullous forms antibiotics may be useful to prevent infection (Siemens and van der Neut, 1961); in the former methotrexate has given satisfactory results (Esterly and Maxwell, 1968).

An ichthyosiform eruption may occur in certain hereditary syndromes.

Refsum's syndrome (1945–46) has already been described[2]; it is essentially due to an anomaly in lipid metabolism transmitted by an autosomal recessive gene and is characterized by a chronic polyneuritis with progressive pareses, nerve-deafness, cataract and an atypical pigmentary retinal degeneration. The ichthyosis which shows no specific features usually develops in childhood.

The *Sjögren-Larsson syndrome* (1957),[3] also transmitted by an autosomal recessive gene, is a combination of oligophrenia and spastic paraplegia, frequently complicated by cataract and macular degeneration in the retina. Ichthyosiform changes are common and may occasionally take the form of ichthyosiform erythroderma from birth (*xerodermal idiocy*).

Rud's syndrome (1927),[4] probably transmitted by an autosomal recessive gene and characterized by hypogonadism, oligophrenia, infantilism, epilepsy and pigmentary retinopathy, is accompanied by ichthyosiform erythroderma.

Conradi's disease (1914) (chondrodystrophia congenita punctata),[5] a condition showing widespread mesodermal and ectodermal anomalies including cataract, involves a dryness of the skin and a universal or patchy erythema and scaling from birth.

An *acquired ichthyosis*, taking the form of ichthyosis vulgaris, may occur in certain malignant diseases, particularly Hodgkin's disease (Bureau *et al.*, 1958; Stephens and Rhodes, 1964; Balabanov and Andreev, 1966), lymphosarcoma and mycosis fungoides (Borda *et al.*, 1956). Similar changes have been noted in some 10% of patients with leprosy in South Africa (Schulz, 1965).

Arnold. *Med. Korresp.-Bl. d. Württemb. ärztl. Vereins*, No. 21 (1834).
Balabanov and Andreev. *Hautarzt*, **17**, 252 (1966).
Ballantyne. *Diseases and Deformities of the Foetus*, Edinb. (1895).
Bloom and Goodfried. *Arch. Derm.*, **86**, 336 (1962).
Borda, Stringa, Abulafia and Villa. *Arch. argent. Derm.*, **6**, 47 (1956).
Brocq. *Ann. Derm. Syph.* (Paris), **3**, 1 (1902).
Bureau, Picard and Barrière. *Ann. Derm. Syph.* (Paris), **85**, 30 (1958).
Butterworth and Strean. *Clinical Genodermatology*, Baltimore (1962).
Caspary. *Vjschr. Derm. Syph.*, **13**, 3 (1886).
Conradi. *Jb. Kinderheilk.*, **80**, 86 (1914).
Contino. *Ann. Ottal.*, **52**, 153 (1924).
Elschnig. *Klin. Mbl. Augenheilk.*, **71**, 155 (1923).
Esterly and Maxwell. *Pediatrics*, **41**, 120 (1968).
Finkelstein and Cass. *Nature* (Lond.), **216**, 717 (1967).
Frost and Weinstein. *J. Amer. med. Ass.*, **207**, 1863 (1969).
Frost, Weinstein, Bothwell and Wildnauer. *Arch. Derm.*, **98**, 230 (1968).
Hanssler. *Dtsch. med. Wschr.*, **82**, 1733 (1957).
Hill and Rodrigue. *Canad. J. Ophthal.*, **6**, 89 (1971).
Hudson. *Proc. roy. Soc. Med.*, **19**, 11 (1926).
Ischreyt. *Petersburg. med. Wschr.*, **33**, 687 (1908).
von Jarmersted. *Z. Augenheilk.*, **66**, 408 (1928).

[1] Glycerini amyli, adip. lanæ hydros., ptes aeq. with 2% acid salicyl.
[2] Vol. X, p. 488; Vol. XI, p. 194.
[3] Vol. III, p. 1131; Vol. XI, p. 207.
[4] Vol. III, p. 1130.
[5] Vol. III, p. 1118; Vol. XI, p. 206.

Jay, Blach and Wells. *Brit. J. Ophthal.*, **52,** 217 (1968).
Kaulich. *Z. Augenheilk.*, **15,** 375 (1906).
Lentz and Altman. *Arch. Derm.*, **97,** 3 (1968).
Mortada. *Ophthalmologica*, **152,** 68 (1966).
Refsum. *Nord. Med.*, **28,** 2682 (1945).
Acta physiol. scand., Suppl. 38, 1 (1946).
Riecke. *Münch. med. Wschr.*, **70,** 379 (1923).
Rose. *Brit. J. Ophthal.*, **55,** 750 (1971).
Rud. *Hospitalstidende*, **70,** 525 (1927).
Sannicandro. *Arch. ital. Derm.*, **12,** 84 (1936).
Schulz. *Brit. J. Derm.*, **77,** 151 (1965).
Sever, Frost and Weinstein. *J. Amer. med. Ass.*, **206,** 2283 (1968).
Shapiro and Soentgen. *Postgrad. Med.*, **45,** 216 (1969).
Shimkin. *Brit. J. Ophthal.*, **29,** 363 (1945).
Shindle and Leone. *Arch. Ophthal.*, **89,** 62 (1973).
Siemens. *Arch. Derm. Syph.* (Berl.), **156,** 624 (1928); **158,** 111 (1929).
Siemens and van der Neut. *Hautarzt*, **12,** 186 (1961).
Simizu. *Acta Soc. ophthal. jap.*, **38,** 451 (1934).
Sjögren and Larsson. *Acta psychiat. neurol. scand.*, **32,** Suppl. 113 (1957).
Sondermann. *Klin. Mbl. Augenheilk.*, **70,** 180 (1923).
Stephens and Rhodes. *Trans. St. John's Hosp. derm. Soc.*, **50,** 58 (1964).
Touraine. *L'hérédité en médecine*, Paris (1955).
Waardenburg, Franceschetti and Klein. *Genetics and Ophthalmology*, Oxford, **1,** 220 (1961).
Wells and Kerr. *Arch. Derm.*, **92,** 1 (1965).
Brit. med. J., **1,** 947 (1966).
J. invest. Derm., **46,** 530 (1966).
Willan. *On Cutaneous Diseases*, London (1808).

THE FOLLICULAR KERATOSES

This group of dermatoses is characterized by focal disturbances of keratosis which involve particularly but not exclusively the pilosebaceous follicles.

DYSKERATOSIS FOLLICULARIS (the *Darier-White disease*), simultaneously described by Darier (1889) and White (1889–91), is a rare condition of widespread and symmetrical distribution which may occur on the lids, either on the skin of the upper lid (Rusch, 1921) or on the lid-margins (Broich, 1908; Löhe, 1921; Wright, 1963). It generally begins in early life, particularly affecting males, and is transmitted as an autosomal dominant (Svendsen and Albrectsen, 1959) (Fig. 246). The eruption consists of small reddish papules at the pilosebaceous orifices; these may enlarge to form papillomatous growths and may ulcerate or suppurate and occasionally may develop into pseudo-carcinomatous hyperplasia suggesting a squamous-celled carcinoma (Charache, 1937; Beerman, 1949), but in no case has malignancy been reported. Histologically hyperkeratosis and parakeratosis are prominent; in the stratum corneum small dark cells with grain-shaped pyknotic nuclei (" grains ") are found, and in the granular layer large cells with a homogeneous basophilic dyskeratotic cytoplasmic mass surrounded by a clear halo (" corps ronds ") (Figs. 247 and 248). These were mistaken by Darier for psorosperms, whence the old misnomer sometimes applied to the condition, *pseudospermosis follicularis* (*vegetans*). The disease is chronic and progressive and treatment is unsatisfactory. It has been associated with disturbed vitamin A metabolism (Peck *et al.*, 1941–43) and good results have occasionally (but not invariably) been reported with the prolonged administration of this vitamin in large doses.

Conjunctival and corneal changes may also occur (Jaensch, 1927; Keim, 1930; Brünauer, 1931).[1]

[1] Vol. VIII, p. 560.

FIGS. 246 to 248.—DARIER'S DISEASE (J. C. Wright).

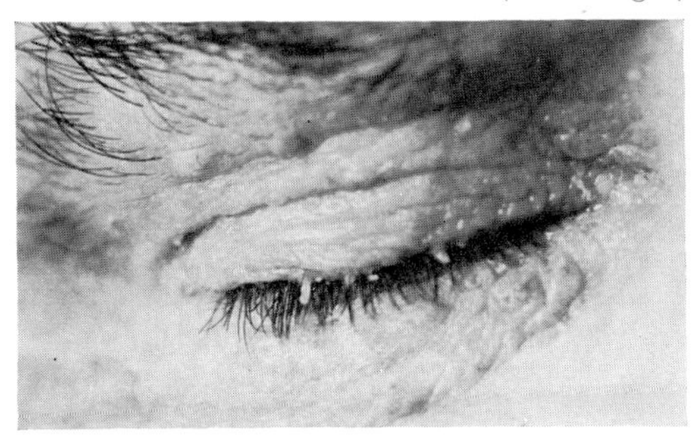

FIG. 246.—Multiple keratotic horn-like growths on the lid-margins.

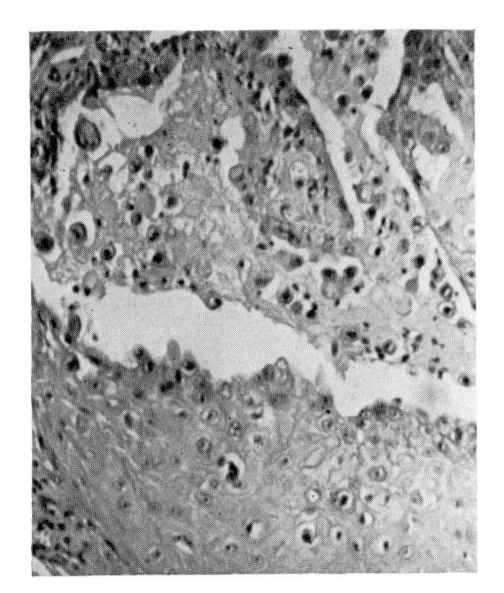

FIG. 247.—Section of biopsy showing hyper- and para-keratosis.

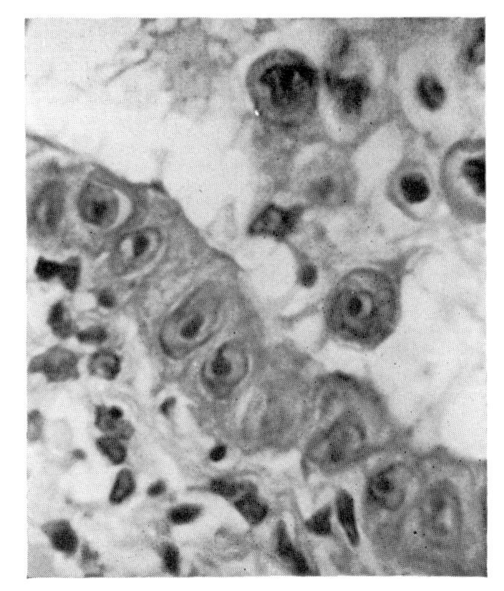

FIG. 248.—" Corps ronds " in the granular layer and grains in the stratum corneum.

PITYRIASIS RUBRA PILARIS. This rare disease is characterized by the widespread appearance of follicular papules crowned by horny spines which coalesce into red scaly plaques showing a clinical resemblance to psoriasis although the histopathology is different, consisting of follicular and diffuse hyperkeratosis with focal parakeratosis. The papules are found mainly on the extensor surfaces of the limbs, especially on the dorsum of the hands, but most of the surface of the body may be affected. The eruption is vividly red and the desquamation may be profuse. On the lid-margin papules may appear (Török, 1897; Etzine and Ross, 1958), and the formation of plaques on the lids may lead to ectropion and conjunctivitis with loss of the lashes and brows (du Castel and Kalt, 1900; Yano, 1921). We have already seen that typical mother-of-pearl patches may appear on the conjunctiva, and that corneal and intra-ocular complications may eventuate.[1]

The disease has been associated with a deficiency of vitamin A and this has been advocated for treatment (Weiner and Levin, 1943; Waldorf and Hambrick, 1965), but on this there is no universal agreement. Methotrexate has also given good results (Brown and Perry, 1966; Weinstein, 1971).

KERATOSIS PILARIS ATROPHICANS describes a group of diseases of varied ætiology characterized by a disorder of keratinization affecting the hair follicles succeeded by atrophy. Two types affect the lids, both of which have a hereditary tendency.

KERATOSIS PILARIS DECALVANS (KERATOSIS FOLLICULARIS SPINULOSA DECALVANS), a very rare hereditary disease described by Siemens (1925–26) and termed ICHTHYOSIS FOLLICULARIS by Laméris (1905), begins on the face in infancy; it is characterized by keratosis of the follicles in the hairy regions, resulting in sparseness of the lashes and disappearance of the outer part of the eyebrows, ectropion and corneal opacities with

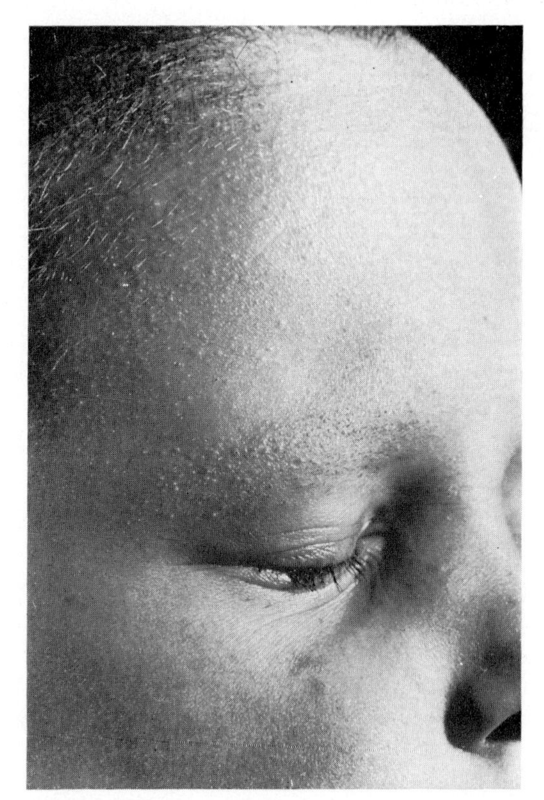

FIG. 249.—KERATOSIS PILARIS ATROPHICANS FACIEI.

Horny plugs have destroyed the eyebrows (A. J. Rook, *Textbook of Dermatology*, Blackwell, Oxford).

[1] Vol. VIII, p. 560.

the development of pannus (Fig. 385) (Franceschetti *et al.*, 1956–57; Zeligman and Fleisher, 1959; Adler and Nyhan, 1969). In the cornea the epithelium and superficial layers of the stroma may be involved.[1]

KERATOSIS PILARIS RUBRA ATROPHICANS FACIEI (*Ulerythema ophryogenes*) also commences at birth or in early infancy. The erythema and the appearance of small horny plugs are restricted to the outer halves of the eyebrows extending medially and destroying the hair-follicles (Davenport, 1964) (Fig. 249).

ACANTHOSIS NIGRICANS, a rare condition already noted in association with its conjunctival and corneal manifestations,[2] also occurs on the lids. It is characterized in the skin by a reddish brown pigmentation and hyperkeratosis with symmetrically distributed papillary growths ranging in size from small elevations giving the skin a velvety texture to large warty growths. It occurs in three forms. The *benign* form, often showing an autosomal

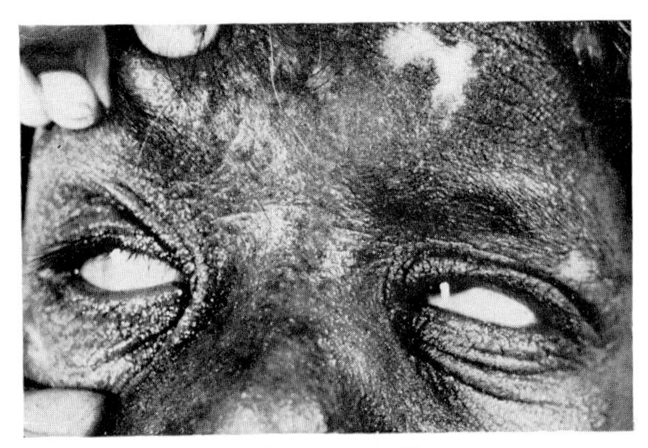

FIG. 250.—ACANTHOSIS NIGRICANS.

Benign form in a boy aged 6. There is thickening and pigmentation of the skin of the lids, but no pigment in the conjunctiva (P. A. Lamba and S. Lal).

irregular dominant heredity, usually develops in early infancy; it progresses slowly, tends to get worse at puberty and thereafter remains stationary or regresses (Lamba and Lal, 1970) (Fig. 250). Treatment is ineffective, but surgical excision may solve the problem of disfigurement. The second form, known as *pseudo-acanthosis nigricans*, occurs in states of obesity or in association with endocrine disorders such as various pituitary syndromes; the condition tends to resolve if the nutritional obesity is made to disappear or is reversible when the endocrine disturbance is controlled.

The *malignant* form which occurs typically in adults is invariably associated with advancing malignant disease in some internal organ. The lesions in the skin are more severe and extensive and are steadily progressive. No treatment is effective; the removal of the causal tumour may be followed by partial regression which, however, is not always maintained (Thorn, 1944; Kierland, 1947; El Nasr and El Zawahry, 1950; and others).

[1] Vol. VIII, p. 557. [2] Vol. VIII, p. 550.

Particularly marked localization around the lid-margins of the multiple little tumours, usually associated with loss of the brows and cilia, has been noted by several observers.[1]

Adler and Nyhan. *J. Pediat.*, **75,** 436 (1969).
Beerman. *Arch. Derm. Syph.* (Chic.), **60,** 500 (1949).
Birch-Hirschfeld and Kraft. *Klin. Mbl. Augenheilk.*, **42** (1), 232 (1904).
Bogrow. *Arch. Derm. Syph.* (Wien), **94,** 271 (1909).
Broich. *Vhdl. dtsch. derm. Ges.*, Frankfurt, **10,** 110 (1908).
Brown and Perry. *Arch. Derm.*, **94,** 636 (1966).
Brünauer. *Hb. d. Haut- u. Geschlectskr.*, (ed. Jadassohn), Berlin, **8** (2), 210 (1931).
du Castel and Kalt. *Ann. Derm. Syph.* (Paris), **1,** 1228 (1900).
Charache. *Arch. Derm. Syph.* (Chic.), **35,** 480 (1937).
Darier. *Ann. Derm. Syph.* (Paris), **10,** 597 (1889); **4,** 865 (1893).
Davenport. *Arch. Derm.*, **89,** 74 (1964).
Ebling and Rook. *Textbook of Dermatology* (ed. Rook *et al.*), Oxford, **2,** 1059 (1968).
El Nasr and El Zawahry. *J. roy. Egypt. med. Ass.*, **33,** 81 (1950).
Etzine and Ross. *Med. Proc.*, **4,** 208 (1958).
Franceschetti, Jaccottet and Jadassohn. *Ophthalmologica*, **133,** 259 (1957).
Franceschetti, Rossano, Jadassohn and Paillard. *Dermatologica*, **112,** 512 (1956).
Hallopeau, Jeanselme and Meslay. *Ann. Derm. Syph.* (Paris), **4,** 876 (1893).
Jaensch. *Klin. Mbl. Augenheilk.*, **78,** 96 (1927).
Keim. *Arch. Derm. Syph.* (Chic.), **22,** 573 (1930).
Kierland. *J. invest. Derm.*, **9,** 299 (1947).
Küttner. *Mitt. Grenzgeb. Med. Chir.*, **39,** 276 (1926).
Kuznitzky. *Arch. Derm. Syph.* (Wien), **35,** 3 (1896).
Lamba and Lal. *Dermatologica*, **140,** 356 (1970).
Laméris. *Ned. T. Geneesk.*, **49** (2), 1524 (1905).
Löhe. *Derm. Z.*, **34,** 72 (1921).
Mateos. *Arch. Soc. oftal. hisp.-amer.*, **18,** 1043 (1958).
Newman and Carsten. *Arch. Ophthal.*, **90,** 259 (1973).
Peck, Chargin and Sobotka. *Arch. Derm. Syph.* (Chic.), **43,** 223 (1941).
Peck, Glick, Sobotka and Chargin. *Arch. Derm. Syph.* (Chic.), **48,** 17 (1943).
Rusch. *Arch. Derm. Syph.* (Berl.), **137,** 58 (1921).
Siemens. *Arch. Rassen- u. Ges.-biol.*, **17,** 47 (1925).
Arch. Derm. Syph. (Berl.), **151,** 384 (1926).
Svendsen and Albrectsen. *Acta derm. venereol.* (Stockh.), **39,** 256 (1959).
Thorn. *J. Amer. med. Ass.*, **125,** 10 (1944).
Török. *Arch. Derm. Syph.* (Wien), **38,** 464 (1897).
Mh. prakt. Derm., **25,** 31 (1897).
Waldorf and Hambrick. *Arch. Derm.*, **92,** 424 (1965).
Weiner and Levin. *Arch. Derm. Syph.* (Chic.), **48,** 288 (1943).
Weinstein. *Dermatology in General Medicine* (ed. Fitzpatrick *et al.*), N.Y., 277 (1971).
White. *J. cutan Dis.*, **7,** 201 (1889); **9,** 13 (1891).
Wright. *Amer. J. Ophthal.*, **55,** 134 (1963).
Yano. *Dtsch. med. Wschr.*, **47,** 652 (1921).
Zeligman and Fleisher. *Arch. Derm.*, **80,** 413 (1959).

The Erythemato-squamous Dermatoses

Several dermatoses which may affect the eyelids are characterized by erythema and scaling; all of them are of unknown ætiology. The most common which interest us are psoriasis, parapsoriasis, seborrhœic dermatitis and erythroderma (exfoliative dermatitis).

PSORIASIS

PSORIASIS is a common skin disease of unknown origin characterized by sharply defined flat patches of red colour covered, sometimes thickly and sometimes sparsely, with diagnostic silvery scales; it is of an obstinate chronic nature and indefinitely recurrent habit, with a hereditary tendency

[1] Darier (1893), Hallopeau *et al.* (1893), Kuznitzky (1896), Birch-Hirschfeld and Kraft (1904), Bogrow (1909), Küttner (1926), Mateos (1958), Ebling and Rook (1968), Newman and Carsten (1973), and others.

and a predilection to appear first towards the end of the second decade. The lesions begin as minute papules, each capped by a scale, which eventually spread in various forms—guttate, nummular, annular or gyrate—until large plaques may be formed which may amalgamate into immense continuous sheets. The sites of election are the extensor surfaces of the limbs, particularly at the elbows and knees, but the disease may become generalized over practically the entire body and present the clinical picture of general exfoliative psoriasis. An intractable itchiness is a distressing symptom.

Histopathologically the main features are a regular elongation of the rete ridges, an elongation and capillary dilatation of the papillæ with œdema and an inflammatory infiltrate, and a parakeratosis of the cells of the horny layer. In the exfoliative type there is an increase in mitotic activity in the lower layers of cells of the rete ridges and a great increase in the production of the epidermal cells which cannot mature because of their rapid production and are cast off in quantity (Burks and Montgomery, 1943; Van Scott and Ekel, 1963; Weinstein and Van Scott, 1965; Hashimoto and Lever, 1966).

The face and lids are rarely affected, but lesions occasionally occur in this site, particularly when the scalp is involved; these atypical cases may be familial[1] (Figs. 251–2). As a rule the site of election is the lid-margin; the eyebrows may also be affected and in severe and neglected cases the typical picture of PSORIASIS VEGETANS RUPIOIDES, wherein the crusts attain an excessive thickness, may appear on the lids (Schmidt-Labaume, 1927). There may be considerable destruction of tissue even when the disease is relatively localized; thus in one of a family of four cases reported by Kogan (1922) wherein the elbows, knees and lids only were affected, the lid-margins were destroyed with widespread loss of the cilia to such a degree that the eyes could not be closed. It is also to be noted that conjunctival and corneal manifestations may ultimately lead to phthisis bulbi, the ocular involvement sometimes spreading from plaques on the skin of the lids and sometimes appearing primarily.[2]

The ætiology of the disease is unknown and, unfortunately, no specific *treatment* is available. The standard topical application, chrysarobin, is now replaced by its less toxic derivative, anthralin (Ingram, 1953; Weigand and Everett, 1967). Alternatively, a corticosteroid ointment may be applied to the affected skin; a suitable preparation is Betnovate (bethamethasone) cream with 2% salicylic acid which enhances penetration and facilitates the dissolution of scales. *N*-crotonoyl-*N*-ethyl-*o*-toluidine (Eurax) is useful to relieve the itching. A cytostatic drug such as azathioprine (Imuran) given systemically may be of value, particularly in exfoliative cases, but during its administration frequent blood-counts are advisable to ward against any depression of the bone-marrow. In generalized cases a systemic corticosteroid

[1] Hutchinson (1890), Pergens (1901), Brünauer (1921), Weil (1927), Dreyer (1927), Tóth (1949), Kaldeck (1953), Manna and Jankowski (1966).

[2] Vol. VIII, p. 556.

is useful in suppressing the disease but, not being curative, side-effects should be guarded against in the treatment of a disease of this chronic nature. In serious generalized cases a carcino-chemotherapeutic agent such as methotrexate may be taken—but again in small doses and under supervision (Rees *et al.*, 1967; Coe and Bull, 1968).

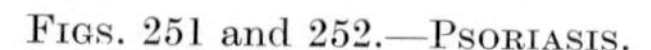

FIGS. 251 and 252.—PSORIASIS.

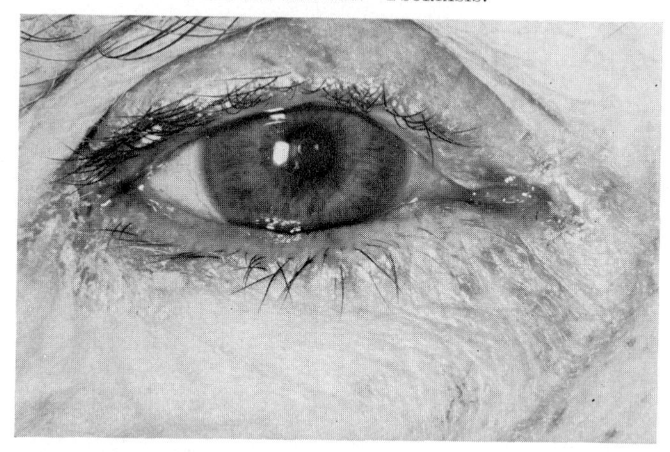

FIG. 251.

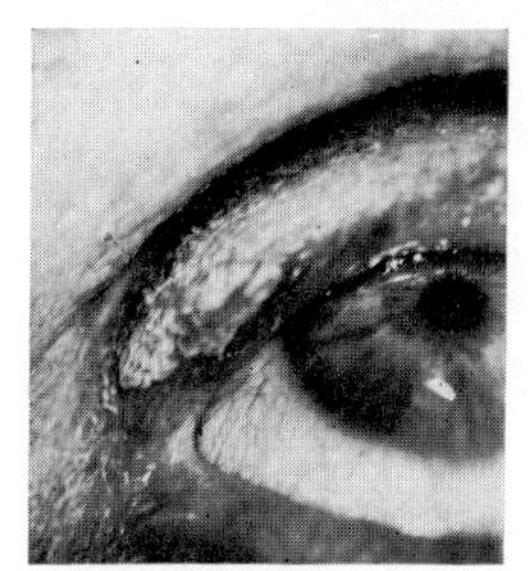

FIG. 252.

FIG. 251.—In an adult, affecting both lids and the outer canthus (Inst. Ophthal.).

FIG. 252.—In a patient aged 56 (Z. Tóth).

PARAPSORIASIS

This is a term introduced by Brocq (1902) to define a group of dermatoses of unknown ætiology characterized by an eruption of brownish-red plaques, each covered by a scale, occurring mainly on the trunk; they evolve slowly and may remain unchanged for decades. In the (possibly infective) condition, *guttate parapsoriasis* (or *pityriasis lichenoides*) the numerous lesions appear in crops, beginning as erythematous patches, slowly become scaly and persist for some time. The disease occurs in both adolescents and adults and the eruption is of wide distribution; it has only rarely been reported as affecting the lids (Friede, 1920). In this case, in the skin of the lids pin-sized reddish lesions were numerous, each covered with an adherent scale, which when scraped off left a red purpuric spot (Fig. 253). In Bantu patients the face has been

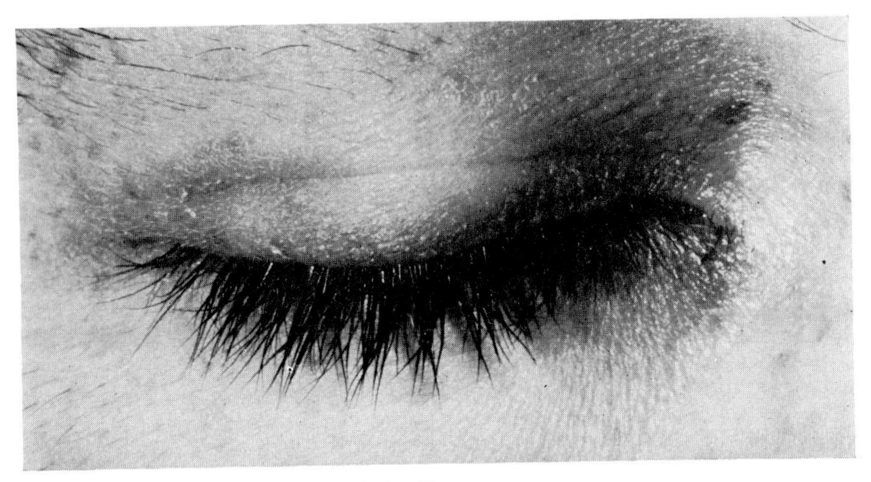

FIG. 253.—PARAPSORIASIS.
Pityriasis lichenoides chronica (P. Watson, Addenbrooke's Hospital, Cambridge).

reported as being particularly affected (Marshall, 1956). No specific treatment is available although ultra-violet light, natural or artificial, may be of value; in acute cases a short course of corticosteroids may be justifiable.

SEBORRHŒIC DERMATITIS

SEBORRHŒIC DERMATITIS, a somewhat loose and unsatisfactory term, is a common acute or subacute inflammation of the outer layers of the skin, characterized essentially by rounded or circinate lesions covered with yellow, greasy scales. The most common form affects the scalp (PITYRIASIS CAPITIS (πίτυρον, bran); DANDRUFF), whence it may spread to the forehead, brows, upper lids and lashes; sometimes, however, it affects the region of the lids without involvement of the scalp, in which case the diagnosis is less obvious. The naso-labial folds are also frequently involved, as are the sides of the neck. Other parts of the body may be attacked—the interscapular and sternal areas and the axillary and genito-crural regions. It is a frequent complication of acne vulgaris and contact eczematous dermatitis.

Ætiologically the disease is due to an eczematous condition superimposed upon a seborrhœic state of the skin[1]; whether a specific infection is present is not yet clear. Unna (1887), who first defined the condition as a clinical entity, concluded it was due to a " morococcus ", which later was identified with various types of staphylococci. A yeast-fungus originally described as a microsporon by Malassez in 1874, interpreted as the *bottle bacillus* by Sabouraud (1902), was cultivated as the *Pityrosporum ovale* by Moore and his colleagues (1935–36). These workers concluded that the pityrosporum was the cause of seborrhœic dermatitis, while Gots and his co-workers (1947) found the organism in abundance. At a later date a somewhat similar organism, the *P. orbiculare*, was isolated by Gordon (1951). We have seen, however, in discussing seborrhœic blepharitis[2] that these organisms while being almost universally

[1] p. 29. [2] p. 219.

found in seborrhœa are present almost as frequently on normal skin and normal lid-margins. Their pathogenicity is not yet proved and it should be remembered that a dermatitis of the sicca type may occur in the absence of sebaceous overactivity.

The onset of the disease is insidious and at first it may appear as a little unobtrusive scaling; but as time progresses this becomes more pronounced and a sharply marginated patch is formed, yellowish at the centre and reddish at the periphery, the whole being covered by scales. Sometimes the area is dry and the scales brawny and abundant (PITYRIASIS SICCA), a picture which, when of marked degree, may resemble psoriasis; at other times it may be moist and oily with a raised papular rash (PITYRIASIS STEATOIDES); occasionally, when superimposed infection is greater, oozing and crusting may develop with swelling of the pre-auricular nodes. The course of the malady is extremely chronic and patches may remain for months without much spread and showing no signs of clearing, while the lids become swollen.

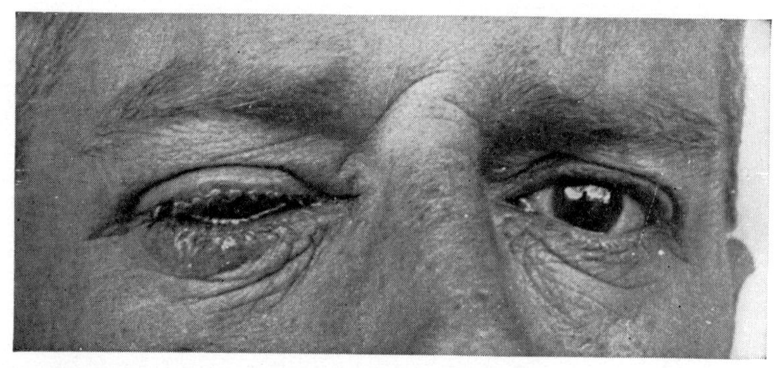

FIG. 254.—SEBORRHŒIC DERMATITIS (W. J. O'Donovan and I. C. Michaelson).

Itching may be considerable, especially when the eyebrows and lashes are involved. Sometimes it is associated with a punctate kerato-conjunctivitis (O'Donovan and Michaelson, 1946; Wright and Meger, 1962) (Fig. 254).

As a rule the involvement of the ocular area is secondary to disease of the scalp and forehead, when the brows and lashes are almost constantly crowded with dry, dirty white scales (Fig. 255). Patches may appear on the skin of the lids by actual spread or they may appear there independently of disease of the scalp. It is important to remember, particularly in adolescents, that a marked and recalcitrant blepharitis[1] may persist while any trouble in the scalp is quite inconspicuous—a circumstance which gives rise to diagnostic difficulties.

LEINER'S DISEASE (1908) is usually regarded as a generalized form of seborrhœic dermatitis occurring in infants; it develops suddenly, generally between the second and fourth months but sometimes in the second week of life. Scaling on the face is fine and bran-like, the scalp and eyebrows are heavily crusted, and over the body the

[1] p. 219.

scales are large, grey and opaque. Secondary infection is common and keratomalacia may result (Leiner, 1908; Krämer, 1921; Salvioli, 1924; Katayama, 1931; Crotty, 1955); without adequate treatment the disease is potentially fatal (10% of cases). Treatment should concentrate on any accompanying metabolic disturbance with regulation of the fluid intake and of the temperature in an incubator. Secondary infection should be controlled by antibiotics and large doses of vitamin B, including biotin, should be prescribed. Steroids should not be administered as a routine but only if the condition is not controlled by these measures.

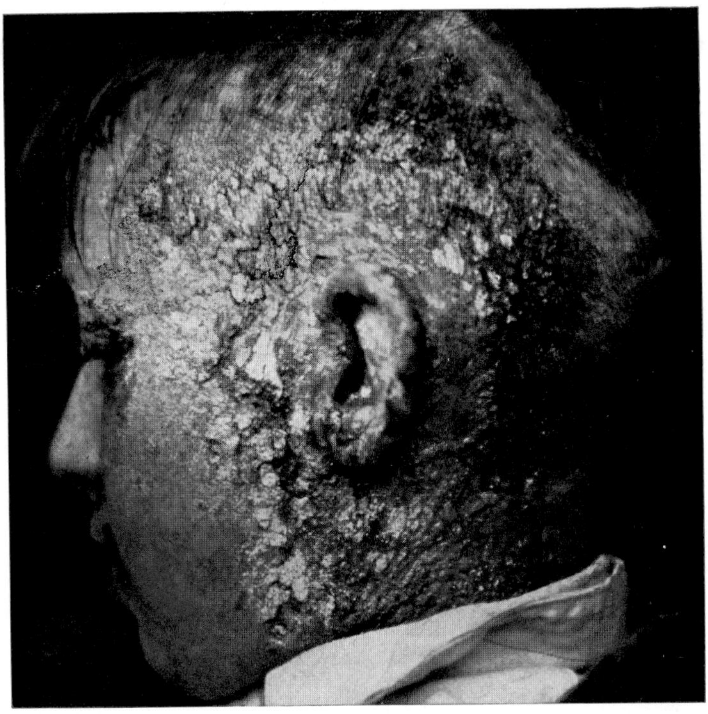

Fig. 255.—Seborrhœic Dermatitis.
Severe eczematous type (A. C. Roxburgh, *Common Skin Diseases*, Lewis, London).

It is obvious that since the organisms associated with seborrhœic dermatitis are almost universal and possibly not causative, *treatment* should be directed not to them but particularly to the morbid state of the host which encourages any change from saprophytism to parasitism. The general treatment, metabolic and dietetic, of the seborrhœic state has already been described; next in importance should come an attack on the dandruff of the scalp and the prevention of secondary infection. For local treatment to the lids sulphur is specific, either as a powder, a salve or a lotion, while topical corticosteroids are probably the most effective preparation. To the scalp, sulphur and salicyclic acid are effective, but selenium sulphide is particularly useful in the form of a shampoo in seborrhœa sicca but less so in the oleosa type (Slinger and Hubbard, 1951; Bahn, 1954; Slepyan, 1952; S. and S. Ayres, 1954; and others). Topical steroid therapy has been claimed to be effective. When inflammatory complications are marked and secondary bacterial infections are present it is advisable to add antibiotics to the topical preparations (neomycin, polymyxin, etc.) and for secondary fungal infections, nystatin. Preliminary cleansing treatment with soothing applications may be necessary (carbolized calamine lotion; salicylic acid and mercuric chloride lotion), and in very recalcitrant cases X-ray therapy may be useful (Hickey, 1950; and others).

ERYTHRODERMA (EXFOLIATIVE DERMATITIS)

ERYTHRODERMA is an inflammatory and persistent process of generalized distribution; the dermis is œdematous with an inflammatory exudate which penetrates the epidermis; this layer is spongiotic and the stratum corneum thickened and parakeratotic. When there is much desquamation the term *exfoliative dermatitis* is applied.

As a rule the condition is *secondary* to other dermal or systemic diseases: eczema of various types such as contact dermatitis, psoriasis, ichthyosiform erythroderma, pemphigus foliaceus and lichen planus are the most common. It also occurs following the administration of drugs, particularly organic arsenic, gold and mercury and sometimes penicillin and the barbiturates. It may appear in the course of leucæmia (1 to 2% of all cases) and other lymphomatous diseases (Abrahams *et al.*, 1963). Adam (1968) in 176 patients found 36% with a previous history of a dermatosis, 10% due to malignancy, 11% to drugs, while the remainder had no significant history. Occasionally the condition is secondary to metabolic disturbances such as steatorrhœa (Shuster, 1968).

A *primary* type or, at any rate, a type in which no ætiology can be found, also occurs usually in middle life or old age. Clinically the disease starts in a patchy fashion and the erythema extends to cover the whole body-surface in a few days or weeks. This is followed by copious scaling, sometimes fine and bran-like but often large, sometimes reaching a length of several inches, and the skin underneath is palpably thickened, red, dry and shiny, differing from the moist surface in eczema. At a later stage the erythema may give place to hyperpigmentation. The nails and hair may be shed and occasionally the eyelids are particularly affected, becoming reddened and thickened and finally shrunken and atrophic, resulting in ectropion, conjunctivitis, keratitis and symblepharon (Schreiber, 1924). Irritation is sometimes severe but a generalized feeling of tightness is more usual.

Systemic disturbances are sometimes profound (Zoon and Mali, 1957; Fox *et al.*, 1965). There is an intermittent fever reaching 102 ° or 103 °F. The immense flow of blood through the skin which may be increased by as much as 66% at rest leads to a disturbance of the body-temperature, hypothermia with a compensatory hypermetabolism, and cardiac failure (Mali, 1952; Krook, 1960–61). The loss of protein in the profuse scaling may reach 9 g/m^2 of skin each day (Freedberg and Baden, 1962) which partly accounts for the loss of plasma albumin, and the plasma γ-globulins are increased, possibly due to abnormal immunological responses. Secondary infections in the skin and respiratory complications provide unfortunate hazards.

The disease may last for months or years and may have a fatal outcome, depending on the patient's age and general resistance and on the cause of the condition. Thus the erythroderma due to drugs has a relatively good prognosis in a healthy patient if the drug is immediately withdrawn. Most cases, however, should receive the most intensive treatment in hospital with careful surveillance over the protein and electrolyte balance, the temperature and the circulatory condition. Topical treatment is of little value although preparations such as zinc cream, calamine liniment or an oily lotion such as equal parts of arachis oil and lime water may be soothing. The only effective agents to suppress the erythema and avoid its serious metabolic consequences are systemic steroids given early and in large doses; a dose equivalent to 20 mg. prednisone, which may be doubled after 3 days if necessary, is usually suitable and if effective it can be slowly reduced after 10 days although a continued maintenance dosage may be required for several months or even for some years. Apart from the usual metabolic complications which may result, this treatment may entail disadvantages in the encouragement of infection. If this develops antibiotics will be required. For the metallic drugs BAL is appropriate treatment.

Abrahams, McCarthy and Sanders. *Arch. Derm.*, **87,** 96 (1963).

Adam. *Canad. med. Ass. J.*, **99,** 661 (1968).

Ayres, S. and S. *Arch. Derm. Syph.* (Chic.), **69,** 615 (1954).

Bahn. *Sth. med. J.*, **47,** 749 (1954).

Brocq. *Ann. Derm. Syph.* (Paris), **3,** 433 (1902).

Brünauer. *Zbl. Haut- u. Geschl.-Kr.*, **3,** 427 (1921).

Burks and Montgomery. *Arch. Derm. Syph.* (Chic.), **48,** 479 (1943).

Coe and Bull. *J. Amer. med. Ass.*, **206,** 15 (1968).

Crotty. *Arch. Derm.*, **71,** 587 (1955).

Dreyer. *Zbl. Haut- u. Geschl.-Kr.*, **21,** 135 (1927).

Fox, Shuster, Williams *et al.* *Brit. med. J.*, **1,** 619 (1965).

Freedberg and Baden. *J. invest. Derm.*, **38,** 277 (1962).

Friede. *Z. Augenheilk.*, **44,** 253 (1920).

Gordon. *J. invest. Derm.*, **17,** 267 (1951).

Gots, Thygeson and Waisman. *Amer. J. Ophthal.*, **30,** 1485 (1947).

Hashimoto and Lever. *Derm. Wschr.*, **152,** 713 (1966).

Hickey. *J. Maine med. Ass.*, **41,** 135 (1950).

Hutchinson. *Arch. Surg.*, **2,** 160 (1890).

Ingram. *Brit. med. J.*, **2,** 591 (1953).

Kaldeck. *Arch. Derm. Syph.* (Chic.), **68,** 44 (1953).

Katayama. *Acta derm.* (Kyoto), **17,** 235, 394 (1931).

Kogan. *Klin. Verein. Ophthal. Kiev* (1922). See *Zbl. ges. Ophthal.*, **9,** 166 (1923).

Krämer. *Wien. med. Wschr.*, **71,** 1063 (1921).

Krook. *Acta derm. venereol.* (Stockh.), **40,** 142 (1960); **41,** 443 (1961).

Leiner. *Arch. Derm. Syph.* (Wien), **89,** 163 (1908).

Malassez. *Arch. Physiol. norm. path.* (Paris), **1,** 451 (1874).

Mali. *Dermatologica*, **104,** 19 (1952).

Manna and Jankowski. *Klin. oczna*, **36,** 371 (1966).

Marshall. *S. Afr. med. J.*, **30,** 210 (1956).

Moore. *Arch. Derm. Syph.* (Chic.), **31,** 661 (1935).

Moore, Kile and Engman. *Arch. Derm. Syph.* (Chic.), **33,** 457 (1936).

O'Donovan and Michaelson. *Brit. J. Ophthal.*, **30,** 193 (1946).

Pergens. *Z. Augenheilk.*, **5,** 14 (1901).

Rees, Bennett, Maibach and Arnold. *Arch. Derm.*, **95,** 2 (1967).

Sabouraud. *Maladies du cuir chevelu*, Paris (1902).

Salvioli. *Sperimentale*, **78,** 5 (1924).

Schmidt-Labaume. *Zbl. Haut- u. Geschl.-Kr.*, **22,** 29 (1927).

Schreiber. *Graefe-Saemisch Hb. d. ges. Augenheilk.*, 3rd ed., *Die Krankheiten d. Augenlider*, Berlin, 102 (1924).

Shuster. *N. Y. J. Med.*, **68,** 3160 (1968).
Slepyan. *Arch. Derm. Syph.* (Chic.), **65,** 228 (1952).
Slinger and Hubbard. *Arch. Derm. Syph.* (Chic.), **64,** 41 (1951).
Tóth. *Klin. Mbl. Augenheilk.*, **114,** 562 (1949).
Unna. *J. cutan. Dis.*, **5,** 449 (1887).
Van Scott and Ekel. *Arch. Derm.*, **88,** 373 (1963).
Weigand and Everett. *Arch. Derm.*, **96,** 554 (1967).
Weil. *Zbl. Haut- u. Geschl.-Kr.*, **22,** 31 (1927).
Weinstein and Van Scott. *J. invest. Derm.*, **45,** 257 (1965).
Wright and Meger. *Amer. J. Ophthal.*, **53,** 686 (1962).
Zoon and Mali. *Arch. Derm.*, **75,** 573 (1957).

Papular Dermatoses

Dermatoses wherein papular lesions predominate are typified in lichen and the lichenoid eruptions; into this category also comes acne agminata.

LICHEN PLANUS

LICHEN PLANUS (LICHEN RUBER PLANUS) was first described by Erasmus Wilson (1869) and since his paper appeared little of significance has been said or done. It may be a chronic localized disease usually occurring in adults, but occasionally takes on acute generalized characteristics. Numerous small flat-topped polygonal papules appear, a millimetre or two in diameter and usually of a bluish-pink colour. Sometimes the papules are discrete but they may coalesce to form scaly plaques which may have a mottled appearance, some areas being white and others violet (the striæ of Wickham); annular lesions also occur and, more rarely, bullæ. Histopathologically there is an increase in the horny layer of the epidermis and of the granular layer which shows an irregular acanthosis, a necrosis of the basal cell layer and an intense infiltration of the dermis with lymphocytes and histiocytes (Summerly and Wilson-Jones, 1964; Ryan, 1966). The acute form of the disease often runs a relatively short course but the chronic form progresses slowly for months or years. The itching may be intense or very slight.

Lichen planus may be found on any site of the skin and on mucous membranes such as in the mouth, affecting the tongue and cheeks, and the penis or vulva; the involvement of the conjunctiva is noted elsewhere.[1] It rarely attacks the lids, so rarely, indeed, that they have often been considered immune. The skin of the lids, however, has on occasion been affected (Jarisch, 1900) (Fig. 256). Michelson and Laymon (1938) saw five cases with lid involvement and described three types of lesion: the typical glistening, lilac-coloured papules with filigree scaling coalescing to form rough scaly patches such as are found elsewhere on the body, annular whitish plaques such as are frequently seen on the glans penis, and a type which would seem to be peculiar to the eyelids. The first two types were associated with lesions elsewhere; the third type (seen in two patients) affected the eyelids only and therefore might easily be misdiagnosed. The lesion resembled erythema ab igne, the four lids being covered with a dark sepia retiform eruption without elevated papules, identical with the melanotic

[1] Vol. VIII, p. 559.

areas seen in the later stages of regression of lichen. Histological investigation showed the typical band-like infiltration of lichen planus; but the lesions were without symptoms, constituting a cosmetic deformity only.

The ætiology is unknown, but it has been associated with emotional stimuli or shock; it became frequent during the Second World War. In treatment an immense number of preparations has been tried including arsenic, mercury, bismuth, anti-malarials, antibiotics and others, but none has been found to alter the course of the disease (Samman, 1961). Systemic corticosteroids relieve the symptoms in many cases, but cure seems to depend mainly on natural evolution.

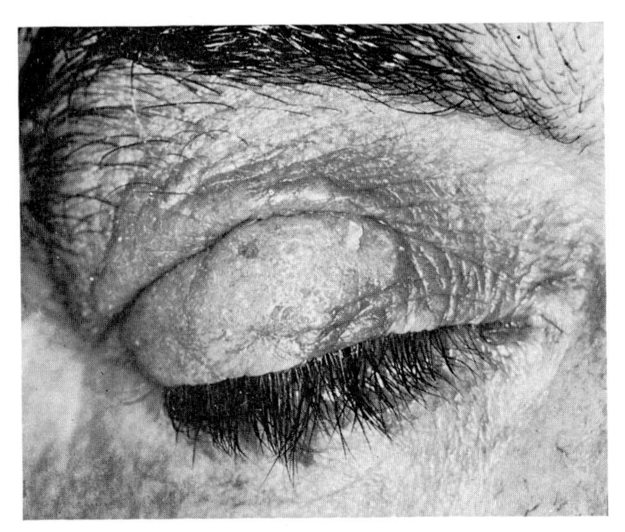

Fig. 256.—Lichen Planus (Inst. Ophthal.).

LICHENOID ERUPTIONS consisting of aggregations of flat-topped, violaceous, shiny papules may be caused by some of the irritant substances employed in colour photography or by certain drugs such as quinine, mepacrine, chloroquine, arsphenamine, gold, certain phenothiazines and the thiazide diuretics.

LICHEN SCLEROSUS ET ATROPHICUS is an uncommon disease of unknown ætiology first described by Hallopeau (1887–98); he considered it a rare sclerosing form of lichen planus and it has been compared with scleroderma, but it is now generally recognized as a separate entity (Kyrle, 1925). It commences in the skin as a crop of small shiny papules showing prominent orifices of the ducts of the pilosebaceous and sweat glands, usually arranged in plaques, followed in the later stages by atrophy and shrinkage. It usually occurs in adult women anywhere on the body, but the most common site is the vulva. Laws and Katz (1968) reported a case wherein the lids were affected without other involvement elsewhere: it concerned a woman aged 31 who presented with a white plaque-like thickening on the medial third of one upper lid; apart from an occasional slight itchiness the lesion was symptomless except that about once each month it became swollen and puffy and exuded fluid as if a blister had burst.

FIGS. 257 to 259.—LICHEN SCLEROSUS ET ATROPHICUS (H. W. Laws and F. Kalz).

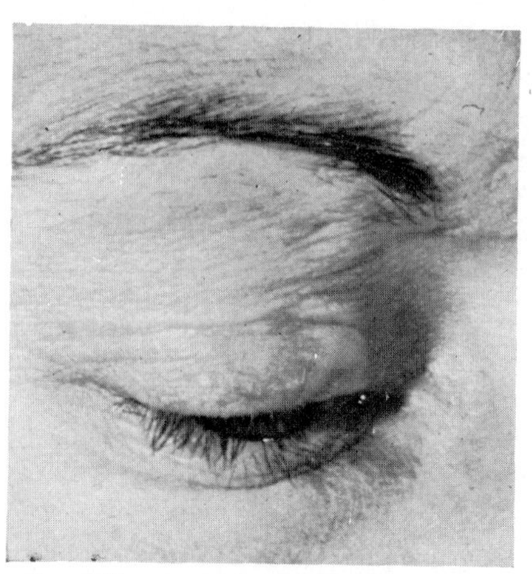

FIG. 257.—An early case in a woman aged 31 with a lesion on the medial third of the upper lid.

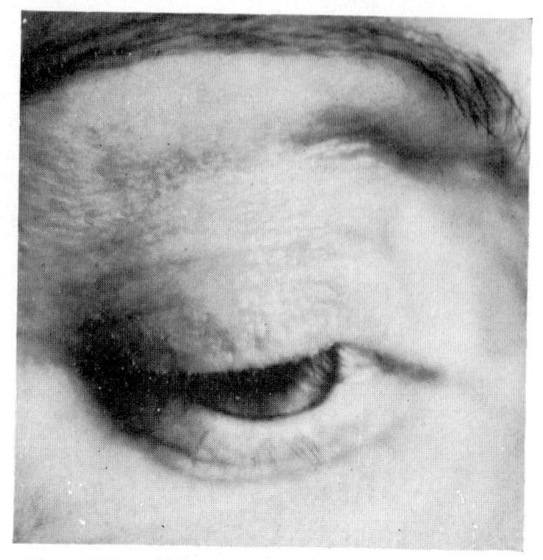

FIG. 258.—The same patient 3 years later, showing scarring and notching of the lid-margin with loss of the lashes.

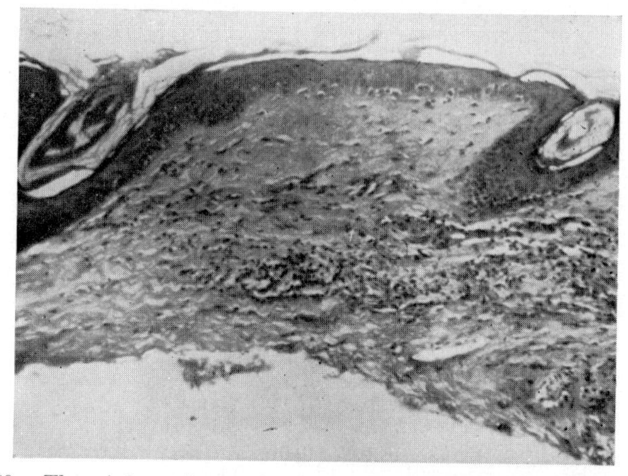

FIG. 259.—There is hyperkeratosis with keratotic plugging. A thin rete malpighii showing focal areas of degeneration. The subepithelial zone shows œdema, homogenization of collagen fibres with few cellular elements; below this is a band of cellular infiltrate consisting of lymphocytes intermingled with histiocytes.

Two years later a second lesion appeared above the first, extending to the eyebrow. Thereafter the condition took on the form of a symptomless, quiet retracted scar with generalized contraction towards its centre and notching of the lid-margin with loss of the lashes (Figs. 257–8). A biopsy showed the typical appearance of lichen sclerosus et atrophicus: an atrophic epidermis with hyperkeratosis and follicular plugging, underneath which was a thick zone of hyalinized collagen with a band-shaped lymphocytic infiltrate in the mid-cutis (Fig. 259). There is no effective treatment.

ACNE AGMINATA

ACNE AGMINATA (L. *agmen, agminis,* army, crowd) was the term first used by Radcliffe-Crocker (1903) to describe a papular eruption of the face which he considered to be the same clinical entity as *disseminated miliary lupus of the face, disseminated follicular lupus* (Fox, 1878) and *acnitis* (Barthélemy, 1891). For a considerable time these conditions were considered to be tuberculides[1] which were presumed to be due to the hæmato-

FIGS. 260 and 261.—ACNE AGMINATA (T. O'Driscoll and G. Morgan).

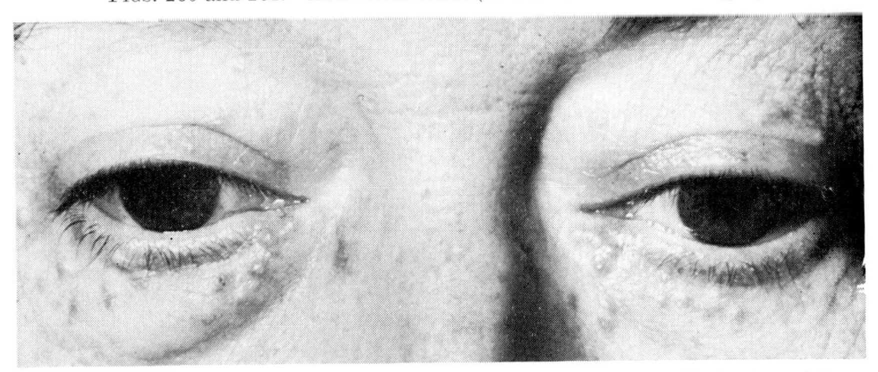

FIG. 260.—In a woman aged 60 years who gave a history of irritation of the eyes of six months' duration; the papular nodules are obvious mainly on the lower lids towards the medial side.

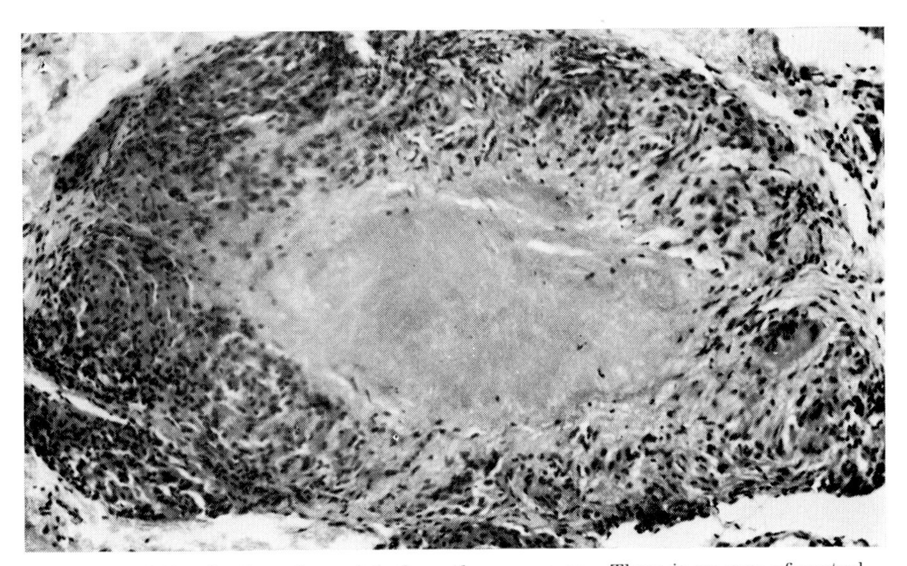

FIG. 261.—Section of a nodule from the same case. There is an area of central necrosis surrounded by a mass of epithelioid cells with giant cells (H. & E.; ×135).

[1] p. 109.

genous dissemination of bacilli from a tuberculous focus elsewhere in the body analogous in a sense to miliary tuberculosis. In recent years, however, although the histological findings are typical of a tuberculoid type of granuloma with caseation, it has become clear that the presence of an active focus of tuberculosis, a hypersensitivity to tuberculin and the response to anti-tuberculous therapy are exceptional rather than the rule; moreover, tubercle bacilli are not found in the lesions. It is now generally accepted both by histologists (Lever, 1967) and dermatologists (Wilkinson, 1972) that the condition is a clinical entity of unknown ætiology; moreover, it is now recognized that the typical histological picture of a tuberculoid granuloma with caseation does not imply a tuberculous ætiology.

The clinical picture has been well documented[1] (Fig. 260). The disease generally attacks patients in mid-adult life, more frequently males than females. It is characterized by the appearance of indolent papules or nodules, a few millimetres in diameter, giving the impression of being deeply situated in the skin. The papules frequently have a reddish-brown tinge and there is a tendency for their apices to crust. Their distribution is mainly confined to the face. Scott and Calnan (1967) found them in the following sites in 17 patients: eyelids 12, cheeks 12, upper lip 11, nose 8, chin 8, forehead 6, with one case showing lesions on the neck and another on the arms. This distribution corresponds to the findings of most dermatologists. The papules are indolent and chronic and may well last several years leaving small irregular pigmented scars. Histologically the lesion consists of a granuloma in the dermis composed of epithelioid cells and multinucleated giant cells similar to those seen in tuberculous lesions with an area of central necrosis (Fig. 261), the granulomata being frequently surrounded by a mantle of lymphocytes (Scott and Calnan, 1967; Lever, 1967; O'Driscoll and Morgan, 1974; and many others). There is no specific treatment; anti-tuberculous therapy is without effect even when carried out for long periods. The tendency is towards eventual spontaneous cure which may be hastened by corticosteroid ointments or other non-specific applications.

Barber. *Trans. ophthal. Soc. U.K.*, **54**, 446 (1934).
Barthélemy. *Arch. Derm. Syph.* (Wien), **2**, 1 (1891).
Fox. *Lancet*, **2**, 35, 75 (1878).
Hallopeau. *Un. méd. Paris*, **43**, 729 (1887).
Ann. Derm. Syph. (Paris), **20**, 447 (1889); **27**, 57 (1896); **29**, 358 (1898).
Jarisch. *Die Hautkrankheiten*, Wien (1900).
Kyrle. *Vorlesunger ü. Histobiologie d. menschlichen Haut u. ihre Erkrankungen*, Berlin (1925).
Laws and Katz. *Canad. J. Ophthal.*, **3**, 39 (1968).
Lever. *Histopathology of the Skin*, 4th ed., Phila., 301 (1967).
Michelson and Laymon. *Arch. Derm. Syph.* (Chic.), **37**, 27 (1938).
Montgomery. *Arch. Derm. Syph.* (Chic.), **42**, 639 (1940).
O'Driscoll and Morgan (1974). Personal communication.
Ormsby and Montgomery. *Diseases of the Skin*, 8th ed., London (1954).
Peck. *Arch. Derm. Syph.* (Berl.), **158**, 545 (1929).
Radcliffe-Crocker. *Diseases of the Skin*, 3rd ed., London (1903).
Ryan. *Brit. J. Derm.*, **78**, 403 (1966).
Samman. *Trans. St. John's Hosp. derm. Soc.*, **46**, 36 (1961).

[1] Radcliffe-Crocker (1903), Peck (1929), Barber (1934), Montgomery (1940), Ormsby and Montgomery (1954), Scott and Calnan (1967).

Scott and Calnan. *Trans. St. John's Hosp. derm. Soc.*, **53,** 60 (1967).
Summerly and Wilson-Jones. *Trans. St. John's Hosp. derm. Soc.*, **50,** 157 (1964).
Wilkinson. *Textbook of Dermatology* (ed. Rook *et al.*), 2nd ed., Oxford, 629 (1972).
Wilson. *J. cutan Med.* (Lond.), **3,** 117 (1869).

The Bullous Dermatoses

Several conditions in which a prominent feature is the formation of vesicles or bullæ are met with in the lids. The bullæ may be of two types

Fig. 262.—Cutaneous Pemphigus.

Note that the pemphigoid bulla is located intra-epidermally and is formed by separation of the epithelial cells and rupture of the intercellular bridges in a process of acantholysis (W. F. Lever).

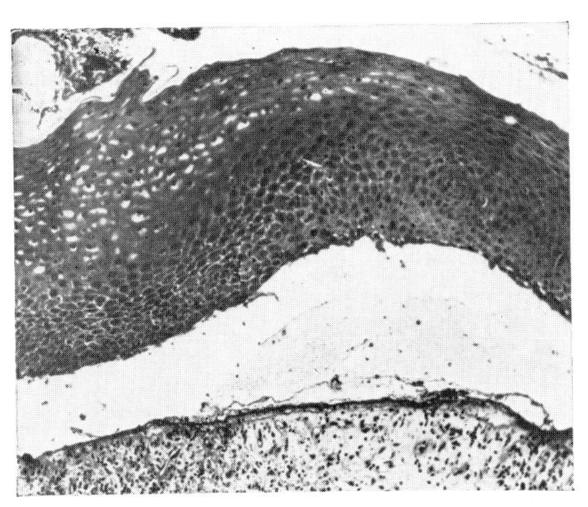

Fig. 263.—Mucosal Pemphigoid.

A bulla from the mucous membrane of the mouth; it is located in the submucosal layer leaving the mucosa intact but detached from the submucosa. There is no acantholysis (W. F. Lever).

(Lever and Talbott, 1942; Klauder and Cowan, 1942; Lever, 1953–65). In pemphigus they are formed within the epidermis by a rupture of the intercellular bridges and a separation of the epithelial cells, a process known as *acantholysis*. In a second type the bullæ are subepidermal leaving the epidermis intact but detaching it from the dermis (Figs. 262–3); this group includes pemphigoid, dermatitis herpetiformis, epidermolysis bullosa, erythema multiforme and toxic epidermal necrolysis. These diseases are generally said to be of " toxic " origin, a viral cause has sometimes been suggested, a relationship with the connective-tissue diseases or auto-immunity has been cited; but the ætiology of them all is obscure. Similarly, no specific treatment is available for any of them and resort is usually made to blanket their effects by corticosteroids.

PEMPHIGUS

Pemphigus is a rare disease of unknown ætiology specifically characterized by the formation of bullæ within the epidermal layer of the skin

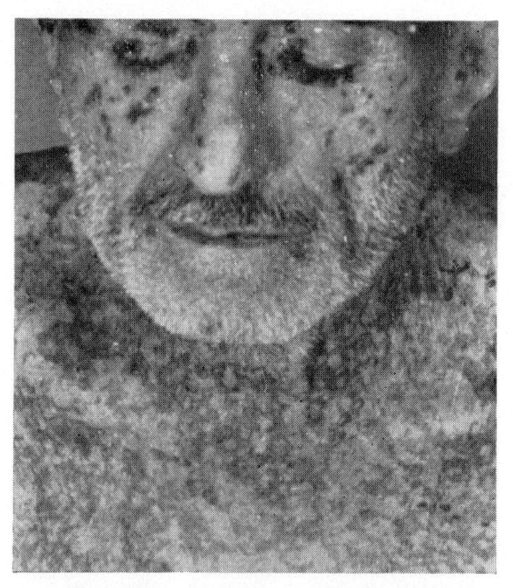

FIG. 264.—PEMPHIGUS VULGARIS.

A generalized eruption in the subacute stage of pemphigus involving the lids (Dept. of Dermatology, Vanderbilt Clin., N.Y.).

(acantholysis) (Lever and Talbott, 1942; Lever, 1953). As it affects the lids it occurs in several forms, all of which are discussed with their corneo-conjunctival complications in a previous Volume.[1]

PEMPHIGUS VULGARIS usually affects older people between 50 and 70 years of age. Crops of bullæ appear on the apparently normal skin and mucous membranes particularly in the mouth, without preceding erythema,

[1] Vol. VIII, p. 498.

to dry up leaving an erythematous scaly patch without scarring; these are succeeded by fresh crops of bullæ, a sequence which may persist for years with a consequent deterioration of the general health until, in the absence of treatment, death supervenes, frequently from septic infection. In the general eruption the lids may be affected (Fig. 264).

Figs. 265 and 266.—Foliaceous Pemphigus.
(The illustrations are from Brazil.)

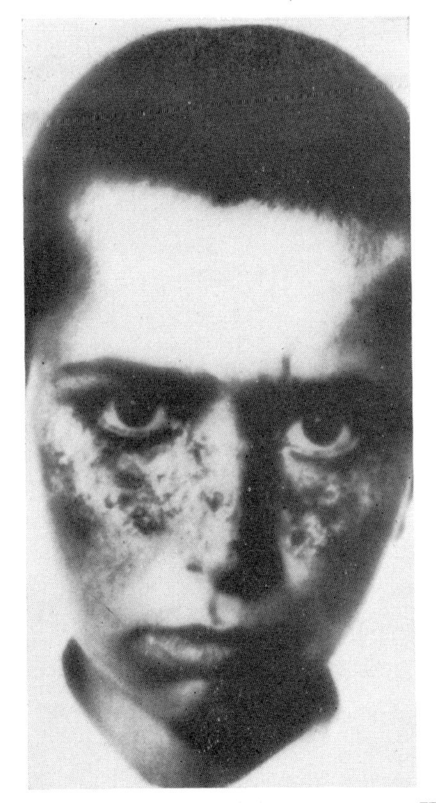

Fig. 265.—The pre-foliaceous stage (H. Marback).

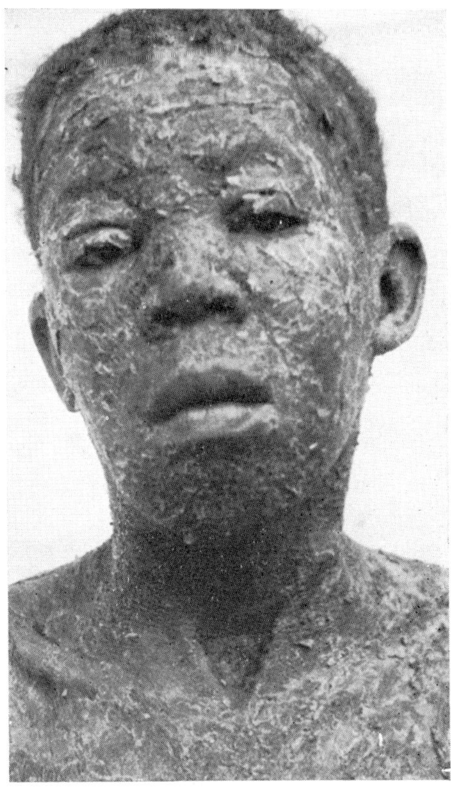

Fig. 266.—A fully developed case (Hilton Rocha).

PEMPHIGUS FOLIACEUS (*Cazenave's disease*, 1844) is rarely seen in Europe but has a high incidence in Brazil. It begins with a generalized erythema followed by the formation of bullæ after which there is a generalized exfoliation of the skin in which the entire surface of the body may eventually be involved; the face gives the appearance of a grotesque mask as if covered with flour and the lids are always implicated (Amêndola, 1949–63) (Figs. 265–6). Some patients recover but others sink into a state of cachexia and apathy and die within a few years.

In PEMPHIGUS VEGETANS the denuded erosions left by the bursting of the bullæ develop florid granulations at their edges often exuding serum or pus, which eventually become hyperkeratotic and fissured.

PEMPHIGUS ERYTHEMATOSUS (*Senear-Usher syndrome*, 1926) is a variant of the foliaceous type wherein the erythematous phase is prolonged, the bullæ are small and the exfoliative condition resembles seborrhœic dermatitis. It is a chronic and prolonged disease, death being usually delayed for many years.

ACUTE PEMPHIGUS[1], described by Pernet and Bulloch (1896) as a rapidly fatal bullous disease seen in a butcher, is now considered as a form of severe erythema multiforme (Lever, 1965) owing to the fact that the bullæ are subepidermal (Vilanova and Piñol Aguade, 1953; Geerts and Dupont, 1955).

The most effective *treatment* for pemphigus is corticosteroids which, when given in large doses, have reduced the mortality rate from almost 100% within 14 months (Lever, 1965) to about 40%.

PEMPHIGOID

BULLOUS PEMPHIGOID may occur in children but is usually a disease of the elderly from 60 to 80 years of age wherein subepidermal bullæ develop (Lever, 1953–67). In the epidermis which forms the roof of the vesicles

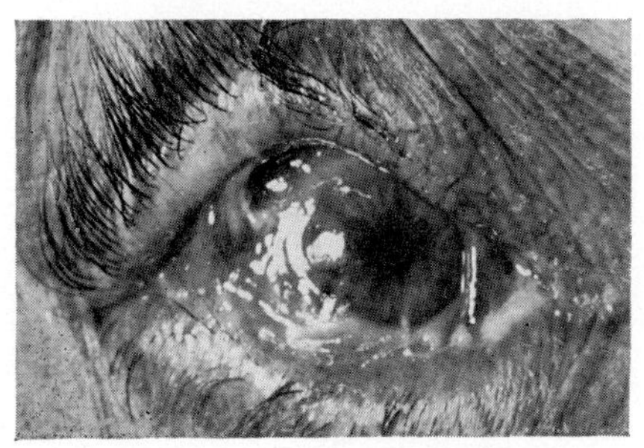

FIG. 267.

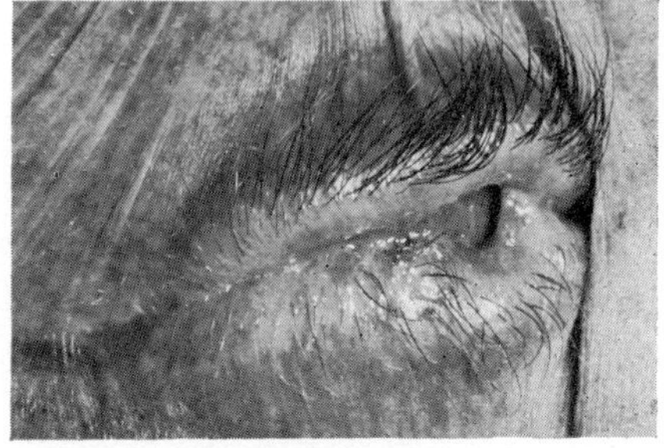

FIG. 268.

FIGS. 267 and 268.—BENIGN MUCOSAL PEMPHIGOID.
The two eyes of the same patient (Inst. Ophthal.).

acantholysis is absent and the primary pathological changes take place in the dermis. The condition usually starts as a localized crop of bullæ, often on an erythematous area and frequently on the legs, whereafter a generalized eruption occurs extending over most of the body, the blisters being tense and sometimes hæmorrhagic. The mucous membranes are usually spared except in the mouth, and the lids are only incidentally affected. The condition usually runs a chronic course extending over years with spontaneous remissions and relapses and the prognosis in elderly debilitated

[1] Vol. VIII, p. 500.

people is often poor. Treatment by steroids, however, has considerably lowered the mortality.

BENIGN MUCOSAL PEMPHIGOID (CICATRICIAL PEMPHIGOID) has been fully discussed in a previous Volume.[1] In this condition the bullæ are subepidermal but the destructive process is the invasion of the submucous tissue by newly formed connective tissue leading, in the eye, to essential shrinkage of the conjunctiva. Cutaneous lesions are usually sparse and localized but may be extensive and the lids may be involved (Figs. 267–8); lesions of this type occurring in two young siblings were reported by François and his colleagues (1972). A case described by Tripodi (1971) which involved symblepharon showed poliosis of the lashes and areas of vitiligo on the lids and elsewhere.

DERMATITIS HERPETIFORMIS

DERMATITIS HERPETIFORMIS (*Duhring-Brocq*) is a somewhat uncommon benign but chronic recurrent disease of unknown ætiology occurring usually between the ages of 20 and 50, characterized by the appearance of successive crops of vesicles on an erythematous base distributed widely and irregularly over the surface of the body. The eruption is pleomorphic and symmetrical, consisting of macules, papules, or bullæ, but is usually papulo-vesicular and gives rise to intense itching; scratching frequently leads to the development of secondary infection. The vesicles are at first multilocular but later become unilocular and are associated with an infiltration of neutrophils and eosinophils, sometimes with micro-abscesses (Piérard, 1963; MacVicar *et al.*, 1963). They are subepithelial and may remain for a considerable time and tend to recur at long intervals so that the disease may last some 10 or 15 years with remissions.[2] Meantime, the patient's health may remain good apart from loss of sleep and secondary infection as the result of scratching. The lids may be incidentally involved and the conjunctiva rarely.[3] It differs from pemphigus in the polymorphism of the eruption arising from erythematous (not normal) skin, the intense itching and the maintenance of good general health.

Treatment is by dapsone (diamino-diphenyl sulphone, usually 100 mg. daily) and since it is suppressive and not curative it must be continued indefinitely in maintenance doses but in large quantities the drug may induce a hæmolytic anæmia in patients with a deficiency of glucose-6-dehydrogenase (Smith and Alexander, 1959; Evans and Fraser, 1963). An alternative effective drug is sulphapyridine (1 to 2 g. daily in divided doses); systemic corticosteroids are of little value (Evans and Fraser, 1963) but topical steroid cream may allay the irritation. Antibiotics may be required to combat secondary infection. The most important factor, however, is a gluten-free diet.

ERYTHEMA MULTIFORME

ERYTHEMA MULTIFORME (EXUDATIVUM) (Hebra, 1866) has its main ophthalmological interest in the lesions it causes in the conjunctiva, in which connection it has been fully described in another Volume.[4] In the skin it is characterized by an eruption which may take many forms usually consisting

[1] Vol. VIII, p. 502.

[2] Degos and Civatte (1961), Alexander (1961–63), Björnberg and Hellgren (1962), Bolgert and Chastanet (1963), Pegum and Mares (1963), and others.

[3] Vol. VIII, p. 515.

[4] Vol. VIII, p. 517.

of sharply defined erythematous patches with an œdematous exudate, distributed symmetrically, occurring preferentially on the hands, forearms and the back of the neck but often more extensively and frequently involving the mucous membranes. In the more simple type red or violaceous macules and papules appear, frequently taking the form of circular lesions resembling targets with a central purpuric spot surrounded by erythematous rings (*erythema* or *herpes iris*). The typical patches may appear on the lids, but here, as elsewhere on the skin, they are evanescent and without complications (Koke, 1941). When the exudation is profuse,

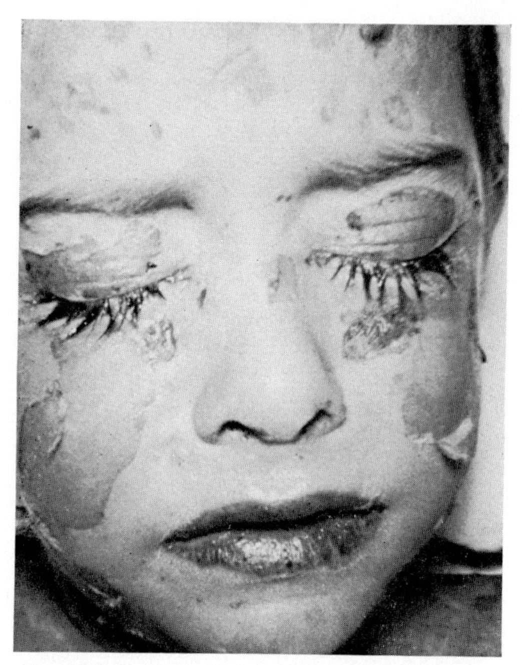

FIG. 269.—ERYTHEMA MULTIFORME EXUDATIVUM.
In a boy aged 6 during the acute phase of the disease (H. L. Gould).

vesicular or bullous lesions are formed in which case the eruption on the skin is extensive and the mucosæ are heavily involved (*erythema vesiculorum* or *bullosum*); when the exudation is intense it may become blood-stained (*erythema purpuricum*). This severe type of the disease accompanied by marked systemic symptoms is frequently referred to as the *Stevens-Johnson syndrome* (1922) which is characterized by lesions on the skin, the conjunctiva, the mouth, the genitalia and elsewhere. The lids may be involved in the general eruption and when the conjunctiva is implicated the lid-margins or their entire structures may participate and share in the typical hæmorrhagic crusting (Fig. 269). The involvement of one eye only is exceptional (Edmund, 1935; Shapiro, 1969).

The disease is usually seen in children and young men and is generally ushered in by malaise and fever; it becomes fully established within 24 hours and reaches its maximum within three weeks. The maculo-papular type is relatively benign and usually rapidly disappears, but in the vesiculo-bullous type the illness is severe, and while individual lesions may resolve in some weeks, fresh ones tend to appear in the following months or even years, while pulmonary and renal complications may supervene.

The pathology differs from other types of erythematous eruption and consists of a serous infiltration of the subepithelial layers with leucocytes followed by lymphocytes and histiocytes, accompanied by vascular dilatation leading to diapedesis and the extravasation of blood (Ackerman *et al.*, 1971). The roofs of the bullæ are formed by an intact epithelium, which frequently suffers complete necrosis. Similar changes occur in the conjunctiva and have been found on autopsy in the mucous membranes of the gut and the respiratory tract.

In a series of 81 cases reported by Ashby and Lazar (1951) the following involvements occurred: mouth 100%, eyes 91%, skin 83%, male genitalia 57%, anal mucosa 5%, bronchitis 6%, pneumonia 23%.

The ætiology is obscure. In approximately half the cases the attacks appear to be precipitated by some toxic factor but in the remainder no cause can be traced, and in both types the identity of the lesions suggests a similar ætiology, perhaps a reaction of hypersensitivity to many agents. The most common precipitating factors are viral infections, particularly herpes, mumps, variola, vaccinia and poliomyelitis; indeed, the virus of herpes has been isolated from the bullæ (Foerster and Scott, 1958; Schmidt, 1961; Pandi, 1964; Shelley, 1967), as also has the *Mycoplasma pneumoniæ* (Sieber *et al.*, 1967), but whether these organisms are causative or precipitating factors is unknown. A very similar lesion is produced by a number of drugs: sulphonamides, sulphones, antibiotics such as penicillin, barbiturates, phenylbutazone, bromides, iodides, trimethadione, coal-tar derivatives and others.[1] The list of reported causes, however, which includes radiotherapy, is large (Löffler *et al.*, 1969).

The *treatment* should first include a detailed search for a precipitating cause and if such can be found it should be dealt with; otherwise only symptomatic measures can be applied. Systemic steroids sometimes have a repressive effect in the more severe cases but this is by no means invariable, and antibiotics should be given in cases wherein secondary infection develops. Topical steroid lotion, with or without an antibiotic, is all that can be used to relieve the bullous condition on the skin.

EPIDERMOLYSIS BULLOSA

EPIDERMOLYSIS BULLOSA is a relatively rare hereditary condition wherein subepithelial bullæ develop, occasionally spontaneously but usually after minor trauma, sometimes in great numbers. Three (and sometimes four) types are recognized (Touraine, 1942).

EPIDERMOLYSIS BULLOSA SIMPLEX, with an autosomal dominant inheritance, usually appears within the first year of life. The bullæ are determined by trauma, usually affecting the hands and feet owing to the

[1] See Vol. VIII, p. 518.

child crawling, and heal without scarring, leaving no permanent changes; the mucous membranes are affected rarely and mildly. The condition usually persists throughout life but may improve at puberty.

DYSTROPHIC (HYPERPLASIC) EPIDERMOLYSIS BULLOSA, also determined by a regular dominant gene, may appear in infancy but its onset may be delayed until puberty. The bullæ, which are usually traumatic but may be spontaneous in origin, heal leaving scars which may be keloidal and mutilating (Fig. 270). The mucosæ are affected in 20% of cases and other anomalies may be present such as ichthyosis, keratosis pilaris and hypertrichosis. The vesicular eruption on the conjunctiva and cornea has been described elsewhere.[1]

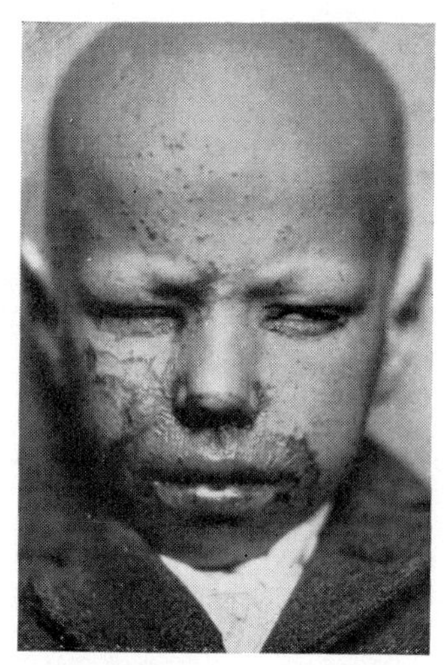

FIG. 270.—EPIDERMOLYSIS BULLOSA.
A hereditary case (R. L. and R. L. Sutton).

POLYDYSPLASIC EPIDERMOLYSIS BULLOSA, inherited as a recessive trait, appears at birth or early infancy, wherein the bullæ, which may be large and hæmorrhagic, heal to leave atrophic scars sometimes with cicatrization.[2] The mucosæ are usually affected and stricture of the œsophagus is a common complication; survival beyond the second decade is unusual. A rapidly fatal type (LETHAL EPIDERMOLYSIS BULLOSA) may be a severe variant of this form; at birth or shortly thereafter the epidermis is shed in large sheets, the mucosæ are extensively involved including the upper respiratory tract and survival is impossible.

When the conjunctiva is affected the lids may be involved in a recalcitrant blepharitis followed by scarring which may lead to ankyloblepharon,[3]

[1] Vol. VIII, p. 524. [2] Vol. III, p. 1133.
[3] Linser (1907), Sakaguchi (1915), Tobias (1928), Paufique *et al.* (1964), Cordella and Peralta (1966).

and in the severe forms of the disease bullæ on the skin of the lids may lead to extensive scarring resulting in ectropion and lagophthalmos (Figs. 271–2). If this occurs the only effective treatment is early grafting and, if the skin of the other lid is not available, split-thickness skin gives the best results; to prevent subsequent cicatrization and ectropion owing to the widespread

Figs. 271 and 272.—Epidermolysis Bullosa (M. Cordella).

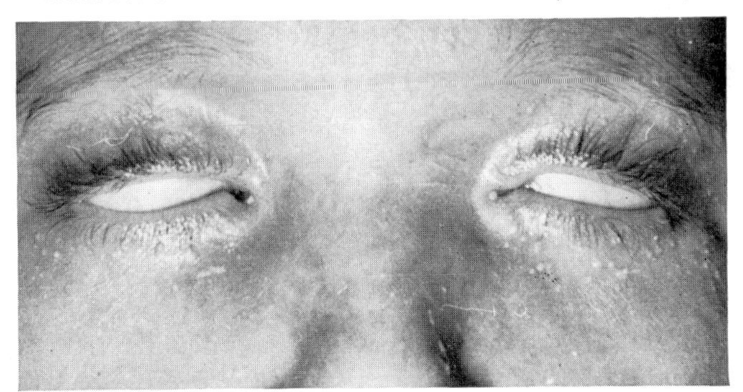

Fig. 271.—The polydysplasic form, showing the lesions on the eyelids and lagophthalmos with inability to close the eyes.

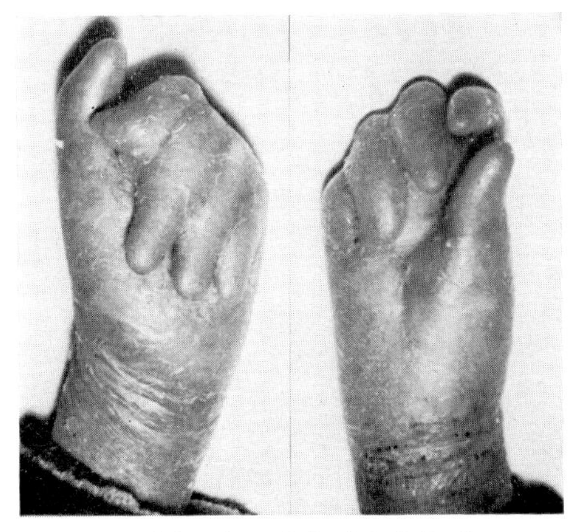

Fig. 272.—The deformity of the hands in the same case.

involvement of the skin, partial tarsorrhaphies at the medial and lateral thirds of the lids are advisable (Hill and Rodrigue, 1971). General treatment is unsatisfactory. Corticosteroids may be life-saving combined with antibiotics to control secondary infection, while topical steroids may be of some value (Severin and Farber, 1967), but these are merely temporary expedients. It is essentially a disease which should call for the avoidance of having other children in the family.

TOXIC EPIDERMAL NECROLYSIS (*Lyell's Disease*)

This is an acute condition first adequately described by Alan Lyell (1956) of Aberdeen, and Lang and Jean Walker (1956) of South Africa, characterized by widespread necrosis of the epidermis as the result of a severe toxæmia. The disease starts with acute symptoms of fever and malaise; the skin over large areas suddenly becomes erythematous and the epidermis separates easily on pressure by the finger (Nikolsky's sign, 1896); thereafter the epidermis loosens with the formation of large flaccid bullæ which rapidly disintegrate so that the sodden necrotic epidermis peels off in large sheets leaving a raw, oozing and exquisitely tender surface resembling a superficial scald (Fig. 273). For this reason the condition has been called the *scalded skin syndrome*; how appropriate is the name is shown in one of the early cases occurring in the infant of a Zulu mother who was accused of boiling her infant preparatory to a cannibalistic orgy and was only saved from the appropriate punishment by Jean Walker's (1959) diagnosis. The toxæmia is usually short-lived, all the damage being done in the first few days, and if and when recovery starts there may be no scarring on the skin unless the condition has been unusually severe (Beare, 1962). Histologically there is complete necrosis of the epidermis with few changes in the dermis (Ostler *et al.*, 1970) (Fig. 274). When the disease is generalized and involves more than one half of the body-surface, the prognosis is grave, especially in infants; localized forms are more benign and recurrences, which are common (29 cases in 121, Franceschetti *et al.*, 1965), are usually less severe.

Implication of the lids, conjunctiva and cornea may occur producing a picture somewhat resembling Stevens-Johnson's syndrome, which may result in symblepharon (Saraux *et al.*, 1965; Pesch and Wessing, 1967; Lorenz, 1968; Reich and Vogel, 1969); the abolition of lacrimal secretion (Oppel, 1963), loss of the corneal epithelium (Ostler *et al.*, 1970), ulceration of both corneæ, complete leucomata (Franceschetti *et al.*, 1965; Kaluzny and Stankiewicz-Strózewska, 1969; Schum, 1971) or perforation of the globe may occur (Björnberg *et al.*, 1964).

The differential diagnosis of a lesion presenting patches of erythema with bullæ and necrosis should include exfoliative dermatitis, erythema multiforme which particularly attacks the mucosæ before the rather scanty eruptions appear on the skin, and pemphigus which is gradual in onset.

The ætiology is varied, but the condition probably represents a hypersensitive state with numerous precipitating factors aided in some cases by a defective immunological response (a low level of IgG, Lorenz, 1968). Most cases in adults have followed the administration of drugs (92 out of 121 cases, Franceschetti *et al.*, 1965); these have already been discussed,[1] and among them phenolphthalein is prominent (Figs. 275–6). Other cases, especially those seen in children under 10 years of age, are due to

[1] p. 76.

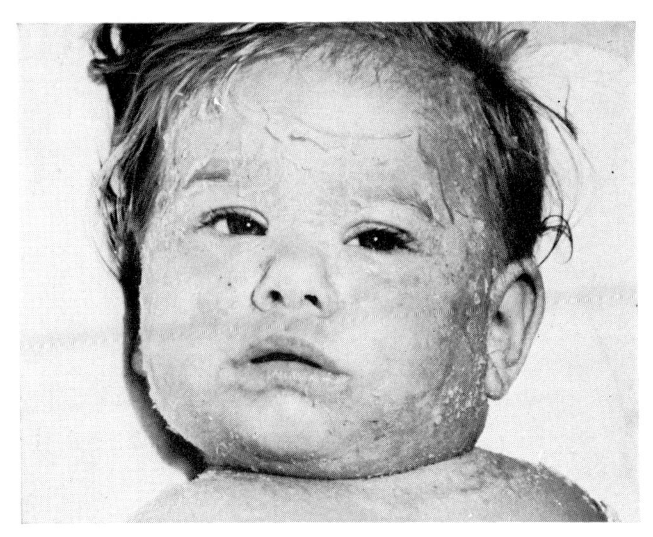

Fig. 273.—Toxic Epidermal Necrolysis.

In a baby girl aged 10 months with the exfoliated form of the scalded skin syndrome, at the stage of secondary desquamation due to a staphylococcal infection. The exfoliated areas have dried and large thick flakes have appeared particularly around the eyes and mouth (M. E. Melish and L. A. Glasgow).

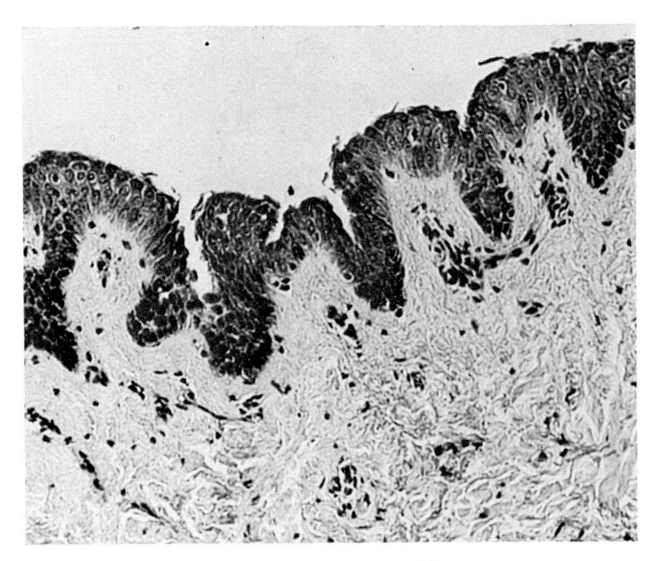

Fig. 274.—Toxic Epidermal Necrolysis.

A fatal case in a woman aged 26. Section of the skin showing the loss of the epidermis and the characteristic absence of inflammatory cells from the dermis (H. B. Ostler *et al.*).

FIG. 275.—TOXIC EPIDERMAL NECROLYSIS.

In a woman aged 42 who had been taking phenolphthalein in small doses intermittently for a year but immediately before her illness she had ingested 1·3 g on a single occasion. The skin lesions healed after intensive treatment with corticosteroids but the patient died after 9 months from septicæmia and toxæmia (A. and K. Björnberg and H. Gisslén).

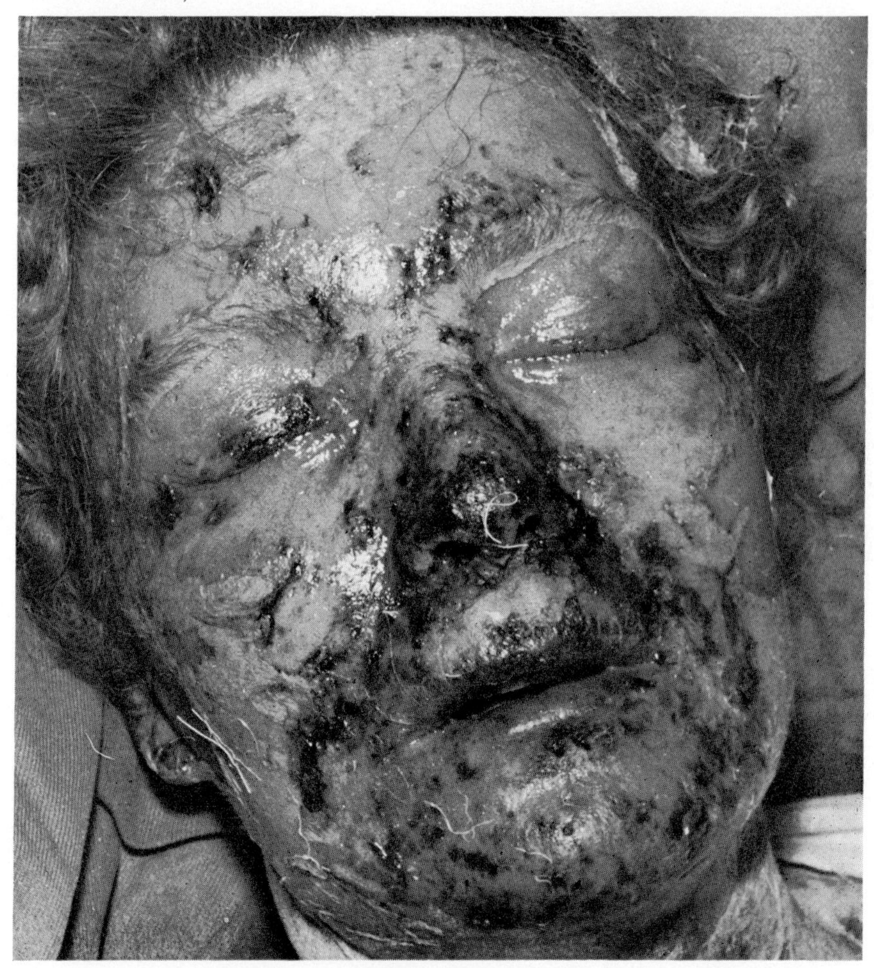

FIG. 275.—Seven days after the beginning of the disease (see also Fig. 276).

infection with Group II staphylococci; these also have been noted.[1] It is interesting that the reaction has been produced in newly born mice by injecting this organism (Melish and Glasgow, 1970). Measles (Lorenz and Lazarini, 1960) and radiotherapy (Garnier, 1962) have been cited as antecedents, while in others no precipitating cause has been found (Rook, 1957; Walker, 1962; Bonnet and de Marigny, 1970).

[1] p. 84.

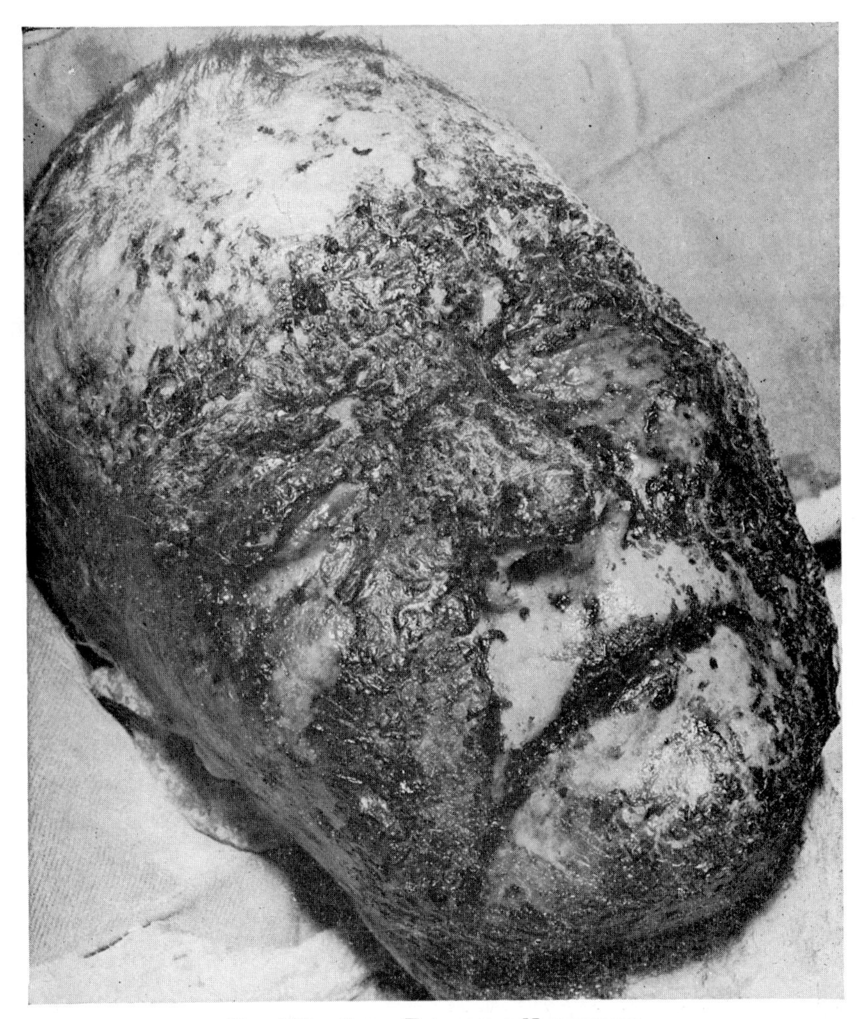

FIG. 276.—TOXIC EPIDERMAL NECROLYSIS.
The same case as Fig. 275; 14 days after the beginning of the disease.

The *treatment* and prognosis depend on the ætiology. In children with infection by staphylococci, most of which are resistant to penicillin, a penicillinase-resistant anti-staphylococcal agent such as methicillin is the drug of choice (Melish and Glasgow, 1971); corticosteroids are contra-indicated. If the antibiotic is given early with adequate attention to the stability of temperature and fluid balance, the prognosis is relatively good with a mortality rate varying between 1 and 7% (Lyell, 1967; Lowney *et al.*, 1967; Samuels, 1967). In adults in whom the ætiology is multiple, including drugs, the treatment should be similar to that for severe burns with particular attention to the fluid balance. If corticosteroids are given systemically before epidermal necrosis occurs, the course of the disease may be curtailed and recurrent attacks modified. In these cases, however, the prognosis is relatively grave, a mortality-rate of 25 to 30% being commonly reported (Zak *et al.*, 1964; Bailey *et al.*, 1965).

Ackerman, Penneys and Clark. *Brit. J. Derm.*, **84,** 554 (1971).
Alexander. *Brit. J. Derm.*, **73,** 267 (1961); **75,** 289 (1963).
Amêndola. *Amer. J. Ophthal.*, **32,** 35 (1949).
Arch. bras. Oftal., **26,** 29 (1963).
Ashby and Lazar. *Lancet*, **1,** 1091 (1951).
Bailey, Rosenbaum and Anderson. *J. Amer. med. Ass.*, **191,** 979 (1965).
Beare. *Arch. Derm.*, **86,** 638 (1962).
Björnberg, A. and K., and Gisslén. *Acta ophthal.* (Kbh.), **42,** 1084 (1964).
Björnberg, A., and Hellgren. *Dermatologica*, **125,** 205 (1962).
Bolgert and Chastanet. *Ann. Derm. Syph.* (Paris), **90,** 259 (1963).
Bonnet and de Marigny. *Bull. Soc. Ophtal. Fr.*, **70,** 1235 (1970).
Cazenave. *Ann. Mal. de Peau*, **1,** 208 (1844).
Cordella and Peralta. *Minerva oftal.*, **8,** 155 (1966).
Degos and Civatte. *Brit. J. Derm.*, **73,** 295 (1961).
Edmund. *Acta ophthal.* (Kbh.), Suppl. 7 (1935).
Evans and Fraser. *Trans. St. John's Hosp. derm. Soc.*, **49,** 108 (1963).
Foerster and Scott. *New Engl. J. Med.*, **259,** 473 (1958).
Franceschetti, Ricci and Diallinas. *Bull. Soc. franç. Ophtal.*, **78,** 339 (1965).
François, Pierard, Coppieters and van Daele. *Bull. Soc. belge Ophtal.*, No. 162, 783 (1972).
Garnier. *Presse méd.*, **70,** 1437 (1962).
Geerts and Dupont. *Arch. belges Derm.*, **11,** 181 (1955).
Gould. *Amer. J. Ophthal.*, **70,** 37 (1970).
Hebra. *Diseases of the Skin*, London (1866).
Hill and Rodrigue. *Canad. J. Ophthal.*, **6,** 89 (1971).
Kaluzny and Stankiewicz-Strózewska. *Klin. oczna*, **39,** 337 (1969).
Klauder and Cowan. *Amer. J. Ophthal.*, **25,** 643 (1942).
Koke. *Arch. Ophthal.*, **25,** 78 (1941).
Lang and Walker. *S. Afr. med. J.*, **30,** 97 (1956).
Lever. *Medicine* (Balt.), **32,** 1 (1953).
Pemphigus and Pemphigoid, Springfield (1965).
Histopathalogy of the Skin, 4th ed., Phila., 122 (1967).
Lever and Talbott. *Arch. Derm. Syph.* (Chic.), **46,** 348, 800 (1942).
Linser. *Arch. Derm. Syph.* (Berl.), **84,** 369 (1907).
Löffler, Jenny and Zimmermann. *Dermatologica*, **138,** Suppl., 5 (1969).
Lorenz. *Wien. klin. Wschr.*, **80,** 240 (1968).
Lorenz and Lazarini. *Arch. Kinderheilk.*, **163,** 48 (1960).
Lowney, Baublis, Kreye *et al. Arch. Derm.*, **95,** 359 (1967).
Lyell. *Brit. J. Derm.*, **68,** 355 (1956); **79,** 662 (1967).
MacVicar, Graham and Burgoon. *J. invest. Derm.*, **41,** 289 (1963).
Melish and Glasgow. *New Engl. J. Med.*, **282,** 1114 (1970).
J. Pediat., **78,** 958 (1971).
Nikolsky. *Univ. Isviestiya*, Kiev (1896).
Oppel. *Ber. dtsch. ophthal. Ges.*, **65,** 52 (1963).
Ostler, Conant and Groundwater. *Trans. Amer. Acad. Ophthal.*, **74,** 1254 (1970).
Pandi. *Brit. med. J.*, **1,** 746 (1964).
Paufique, Ravault, Moulin and Montibert. *Bull. Soc. Ophtal. Fr.*, **64,** 215 (1964).
Pegum and Mares. *Brit. J. Derm.*, **75,** 123 (1963).
Pernet and Bulloch. *Brit. J. Derm.*, **8,** 157 (1896).
Pesch and Wessing. *Ophthalmologica*, **153,** 81 (1967).
Piérard. *Ann. Derm. Syph.* (Paris), **90,** 121 (1963).
Reich and Vogel. *Münch. med. Wschr.*, **111,** 2258 (1969).
Rook. *Arch. belges Derm.*, **13,** 391 (1957).
Sakaguchi. *Arch. Derm. Syph.* (Berl.), **121,** 379 (1915).
Samuels. *Brit. J. Derm.*, **79,** 672 (1967).
Saraux, Chigot, Chassagne and Devaux. *Bull. Soc. Ophtal. Fr.*, **65,** 915 (1965).
Schmidt. *Acta derm.-venereol.* (Stockh.), **41,** 53 (1961).
Schum. *Fortschr. Med.*, **89,** 277 (1971).
Senear and Usher. *Arch. Derm. Syph.* (Chic.), **13,** 761 (1926).
Severin and Farber. *Arch. Derm.*, **95,** 302 (1967).
Shapiro. *Amer. J. Ophthal.*, **67,** 369 (1969).
Shelley. *J. Amer. med. Ass.*, **201,** 153 (1967).
Sieber, John, Fulginiti and Overholt. *J. Amer. med. Ass.*, **200,** 79 (1967).
Smith, R. and Alexander. *Brit. med. J.*, **1,** 625 (1959).
Stevens and Johnson. *Arch. Derm. Syph.* (Chic.), **24,** 526 (1922).
Tobias. *Arch. Derm. Syph.* (Chic.), **18,** 224 (1928).
Touraine. *Ann. Derm. Syph.* (Paris), **2,** 309 (1942).
L'hérédité en médecine, Paris (1955).
Tripodi. *Boll. Oculist.*, **50,** 26 (1971).
Vilanova and Piñol Aguade. *Ann. Derm. Syph.* (Paris), **80,** 574 (1953).

Walker. *Brit. med. J.*, **2,** 1094 (1959).
Med. Proc. (South Africa), **8,** 208 (1962).

Zak, Fellner and Geller. *Amer. J. Med.*, **37,** 140 (1964).

Granulomatous Lesions

PYODERMA VEGETANS

PYODERMA VEGETANS is a non-specific reaction of the skin characterized by a chronic granulomatous growth with epithelial hyperplasia and often the formation of pustules, ulcers, multiple abscesses and fistulæ. Such a condition was first described by Hallopeau (1898) as *pyodermite végétante*, classified by some authorities as a type of pemphigus vegetans, and was fully annotated by F. P. McCarthy (1949). A similar lesion may occur

FIGS. 277 and 278.—PYODERMA VEGETANS (W. Leydhecker and O. E. Lund).

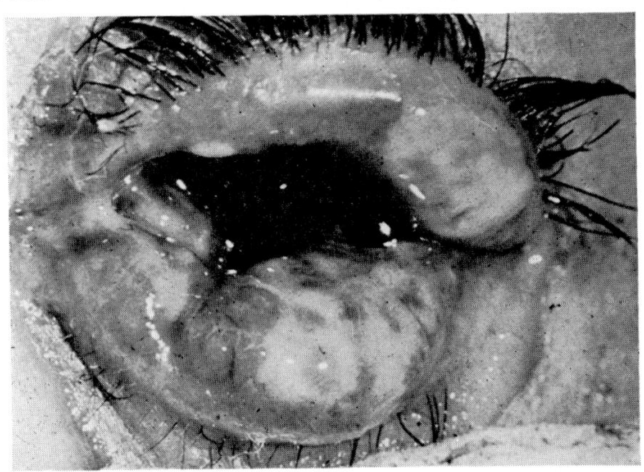

FIG. 277.—In a patient aged 11, showing marked cobblestone swelling of the conjunctiva and much secretion.

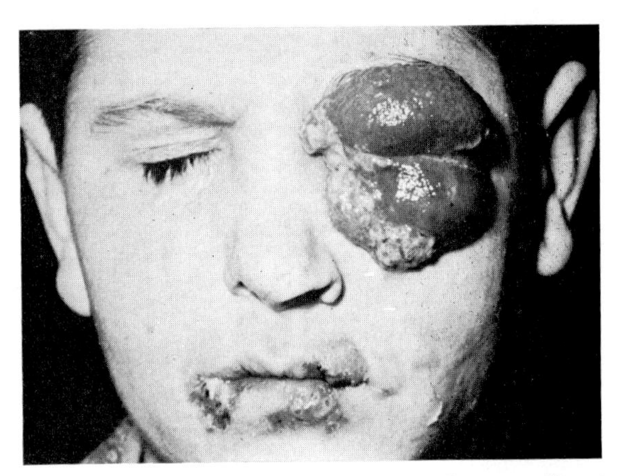

FIG. 278.—Exacerbation of the lesion after radiotherapy. Exenteration of the orbit was performed.

in the mouth (*pyostomatitis vegetans*) wherein multiple miliary abscesses develop in the mucosæ including the tongue and palate, followed by proliferative changes producing a verrucose appearance.

There are no specific histological or bacterial findings and there is no specific therapy. Several organisms have been isolated but the peculiar reaction cannot be attributed to any of them; possibly it may be due to some abnormality in the host. It may occur in apparently healthy individuals but has on occasion been associated with such conditions as ulcerative colitis (Brunsting and Underwood, 1949; Forman, 1966), or malnutrition and alcoholism (Melczer, 1959). Most cases, however, are inexplicable.

Leydhecker and Lund (1962) described a case in a boy: at 5 years of age his mouth and nose were affected and although this resolved at the age of 12, gross follicular changes appeared on the conjunctiva associated with keratitis (Fig. 277). After irradiation the inflammatory changes increased markedly so that the lids appeared as a tumour the size of a mandarin protruding from the orbit necessitating exenteration (Fig. 278); histological examination revealed an uncharacteristic proliferating inflammatory mass. Cases involving the lids in older patients were reported by Azzolini (1951) affecting the lateral canthus, and Woillez and his colleagues (1964) affecting the upper lid. No specific treatment is available, but the administration of broad-spectrum antibiotics should be tried in full doses. A careful examination should also be made to exclude any underlying disease.

GRANULOMA ANNULARE

GRANULOMA ANNULARE is a chronic but benign granulomatous disorder of the dermis or subcutaneous tissues of unknown ætiology; it may perhaps be significant that a certain proportion of patients has obvious or latent diabetes (Rhodes *et al.*, 1966). Histopathologically it is characterized by partial or complete necrosis of the cells and connective tissue typically in the mid-cutis surrounded by a heaped-up palisade of histiocytes sometimes with islands of epithelioid cells and occasional giant cells (Prunty and Montgomery, 1942; Gray *et al.*, 1965) (Fig. 280).

Clinically the lesion consists of a firm nodule which slowly undergoes central involution and extends peripherally to form a ring which may be oval, circular or irregular of infiltrated papules varying in diameter from 1 to 3 cm. The colour varies from ivory to violet and symptoms are absent except for slight itching. Such a lesion may slowly progress for several months or years, extending at the periphery and healing in the centre; it never ulcerates but tends to disappear spontaneously in time without leaving a scar. These occur most frequently on the backs of the hands and on the extensor surfaces, but may occur anywhere; an affection of the lids is very rare but has been reported in a boy aged 7 years who had a lesion in the upper lid with the typical histology, as well as granulomata on his left hand and foot (Ziemssen, 1958) (Figs. 279–80).

Treatment may be unnecessary in view of the tendency to spontaneous recovery, but a mildly irritant ointment (such as salicylic acid), the application of carbon dioxide snow or the intralesional injection of dilute solutions of hydrocortisone acetate may be of value; it is interesting that the removal of a small fragment, which may be undertaken for diagnosis, sometimes makes the entire lesion resolve. The existence of any tendency for diabetes should be explored and treated.

FIGS. 279 and 280.—GRANULOMA ANNULARE (C. H. Ziemssen).

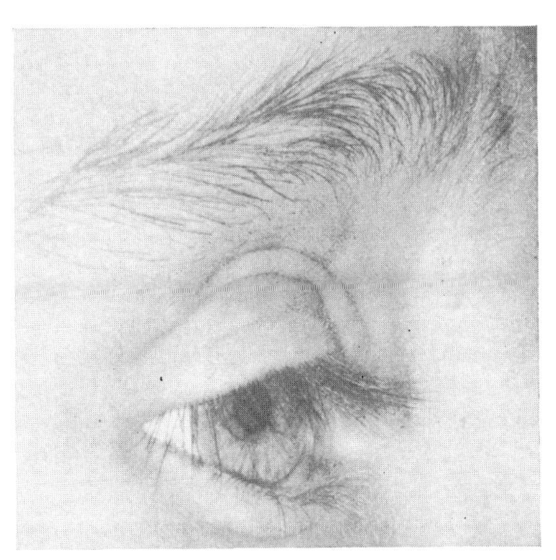

FIG. 279.—In a boy aged 7 years with a lesion on the right upper lid.

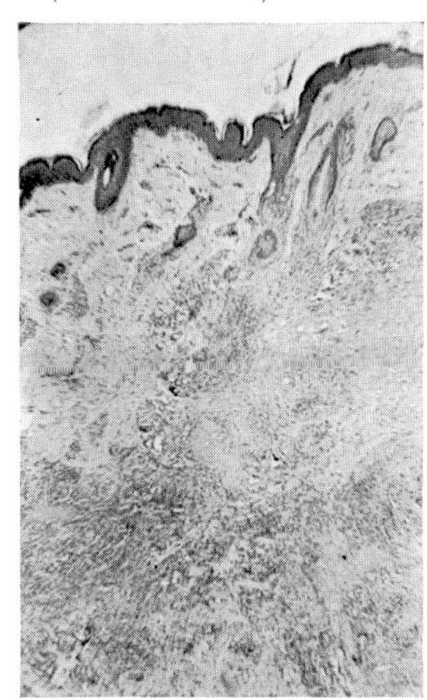

FIG. 280.—The granulomatous tissue in the skin with central necrosis, epithelioid cells and lymphocytes, and perivascular infiltrate.

GRANULOMA FACIALE

GRANULOMA FACIALE WITH EOSINOPHILIA is a rare benign condition of unknown ætiology unassociated with systemic disorders occurring preferentially in the middle aged and elderly. Histopathologically it consists of a dense and circumscribed granulomatous infiltrate in the upper half of the dermis in which eosinophils and neutrophils predominate, associated with vasculitis; in the later fibrotic stages of the disease lymphocytes, plasma cells and fibroblasts become numerous (Lever *et al.*, 1948; P. L. McCarthy, 1958; Johnson *et al.*, 1959).

Clinically the lesions, which may be single or multiple, are circumscribed and nodular, soft and elastic in consistency and often brownish or purplish in colour with telangiectasias. They are asymptomatic although some itching may be present. Ulceration does not occur but spontaneous resolution is rare. The face is the site of predilection and a typical lesion has been described in the upper eyelid which underwent a fluctuating course and caused sufficient swelling to obstruct the vision (Marshall and Pepler, 1953).

The diagnosis can be made only by histological examination. It must be distinguished from eosinophilic granuloma of the bone, classified as a histiocytosis X of which the characteristic feature is large monocytoid histiocytes,[1] and differentiated from erythema elevatum diutinum in which there are many fibrinoid changes around the small blood vessels and numerous polymorphs suffering fragmentation.

For *treatment* surgical excision may be advisable, but a good result has been reported following the intralesional injection of triamcinolone (Wilkinson, 1968).

[1] p. 304.

Azzolini. *G. ital. Oftal.*, **4,** 81 (1951).
Brunsting and Underwood. *Arch. Derm. Syph.* (Chic.), **60,** 161 (1949).
Forman. *Trans. St. John's Hosp. derm. Soc.*, **52,** 139 (1966).
Gray, Graham and Johnson. *J. invest. Derm.*, **44,** 369 (1965).
Hallopeau. *Arch. Derm. Syph.* (Wien), **43,** 289; **45,** 323 (1898).
Johnson, Higdon and Helwig. *Arch. Derm.*, **79,** 42 (1959).
Lever, Lane, Downing and Spangler. *Arch. Derm. Syph.* (Chic.), **58,** 430 (1948).
Leydhecker and Lund. *Klin. Mbl. Augenheilk.*, **141,** 595 (1962).
McCarthy, F. P. *Arch. Derm. Syph.* (Chic.), **60,** 750 (1949).
McCarthy, P. L. *Arch. Derm.*, **77,** 458 (1958).
Marshall and Pepler. *S. Afr. med. J.*, **27,** 448 (1953).
Melczer. *Minerva derm.*, **34,** 308 (1959).
Prunty and Montgomery. *Arch. Derm. Syph.* (Chic.), **46,** 394 (1942).
Rhodes, Hill, Ames *et al.* *Brit. J. Derm.*, **78,** 532 (1966).
Wilkinson. *Textbook of Dermatology* (ed. Rook *et al.*), Oxford, **1,** 444 (1968).
Woillez, Béal and Asseman. *Bull. Soc. Ophtal. Fr.*, **64,** 676 (1964).
Ziemssen. *Klin. Mbl. Augenheilk.*, **133,** 562 (1958).

CHAPTER IV

THE LIDS IN SYSTEMIC DISEASE

WE ARE introducing this Chapter with the photograph of one of the greatest English physicians and pathologists—SIR WILLIAM WITHEY GULL [1816–1890] (Fig. 281). In the context of this Chapter, along with Thomas Addison he was the first to describe xanthomata, under the name "vitiligoidea",[1] and to define myxœdema.[2] These were by no means his only contributions to medicine; he was one of the first to note the implication of the posterior spinal roots in locomotor ataxia, described intermittent hæmoglobinuria and demonstrated that disease of the blood vessels in chronic nephritis was not limited to the kidneys. These and many more arose from his remarkable insight into medicine and the assiduous work he carried out at Guy's Hospital in London where he spent his professional life. He affected a therapeutic nihilism: "the road to medical education is through the pathological Hunterian museums and not through an apothecary's shop". In appearance he was Napoleonic, in manner somewhat magisterial and imperious, in disposition witty, genial and attractive, and he was adored equally by his patients and his pupils.

Lipid-storage Diseases (Lipoidoses)

The LIPID-STORAGE DISEASES constitute an ill-defined group of conditions characterized by an abnormal accumulation of lipids in the tissues. Their general pathology has already been described in this *System*,[3] as well as their manifestations on the outer eye,[4] the uvea[5] and the retina.[6] Here we shall concern ourselves only with their effects on the lids.

It will be remembered that these conditions can be conveniently classified into four groups:

(*a*) ESSENTIAL HYPERLIPÆMIC XANTHOMATOSIS (the *Bürger-Grütz disease*), in which there is an increased neutral fatty content of the blood, and eruptive xanthomata may appear on the lids.

(*b*) ESSENTIAL HYPERCHOLESTEROLÆMIC XANTHOMATOSIS, characterized by an increase in the total cholesterol in the serum, wherein xanthomata occur in various tissues including the lids as xanthelasma palpebrarum.

(*c*) SECONDARY XANTHOMATOSIS may be associated with a raised content of lipids (cholesterol and/or triglycerides) in the serum due to hepatic, renal, pancreatic or metabolic disease (diabetes mellitus, myxœdema or von Gierke's disease); the sudden development of eruptive xanthomata is common.

(*d*) The HISTIOCYTOSES comprising a number of disorders showing no specific alteration in the level or pattern of the circulating lipids, but characterized by the accumulation of lipid material of various types

[1] *Guy's Hosp. Rep.*, London, **7**, 265 (1851); **8**, 149 (1852–53).
[2] *Trans. clin. Soc. Lond.*, **12**, 180 (1873–74).
[3] Vol. VII, p. 145.
[4] Vol. VIII, p. 1079.
[5] Vol. IX, p. 656.
[6] Vol. X, p. 458.

Fig. 281.—Sir William Withey Gull
[1816–1890].
(Courtesy of the Wellcome Institute of the History of Medicine.)

in histiocytic cells in many tissues. Into this category come juvenile xanthogranuloma and those diseases designated by Lichtenstein (1953) as *histiocytosis X*—the Hand-Schüller-Christian disease, the Letterer-Siwe disease and eosinophilic granuloma.

The pathology of these lesions is very similar. Owing to deficient oxidation there is an accumulation of lipids in the tissues, an unmasking of the invisibly fine fatty emulsion normally present in the cells and its deposition in visible form, a process which excites a foreign-body reaction involving the mobilization of histiocytic cells, the proliferation of fibrous tissue and the appearance of giant cells. Initially this reaction is defensive and protective, for the histiocytes ingest the cholesterol ester and become distended as *foam cells* (*xanthoma cells*) in which the nuclei lie mainly in the periphery, but eventually the reparative processes may go awry and end by becoming proliferative (Figs. 282–3). To some extent the pathology has affinities with an atheromatous plaque. The foam cells are essentially perivascular in their arrangement, particularly in a xanthelasma. Occasionally multi-nucleated giant cells with a lipid cytoplasm and a ring of nuclei (*Touton cells*) may be found, but in xanthelasmata they are relatively rare (Fig. 284). These aggregations of cells are found in the dermis; the epidermis is usually thinned through pressure, but in its basal layer intra- and extra-cellular lipids may be found.[1] In xanthelasmata Kahán and his colleagues (1967) found that the lipid preferentially deposited was cholesterol esterified with the more saturated fatty acids.

XANTHOMATOSIS

Xanthomatosis may occur in the lids in two forms, both having the same pathology—xanthoma planum (xanthelasma) and xanthoma tuberosum.

XANTHELASMA PALPEBRARUM

The common lesion on the lids is XANTHELASMA PALPEBRARUM (XANTHOMA PLANUM). Several large series of cases have been reported (369, Grönvall, 1953; 896, Pedace and Winkelmann, 1965; 141 cases, Chhetri *et al.*, 1967); it usually affects females, occurring principally in persons in the sixth decade (Grönvall, 1947) although it may be found at any age (congenital, Barlow, 1884; in a child, Poensgen, 1885; from 20 to 84, Hutchinson, 1871; and many others). It may occur as an isolated phenomenon in the skin in the absence of any systemic abnormality (the normo-cholesterolæmic type), or as a dermal expression of hypercholesterolæmia. It is therefore important to determine whether such a lesion is localized and unimportant or associated with a disturbance of the lipid metabolism, and at the same time

[1] Geber and Simon (1872), Villard (1903), Birch-Hirschfeld (1904), Mawas (1914), Weber (1924), Wile *et al.* (1929), Schaaf and Werner (1930), Heath (1933), Sená and Cerboni (1948), Lussier and Grenier (1967), Anderson (1969), Pelfini *et al.* (1970), Moro (1972).

FIGS. 282 and 283.—XANTHOMA.

Section of a nodule from the lids from the case seen in Fig. 288.

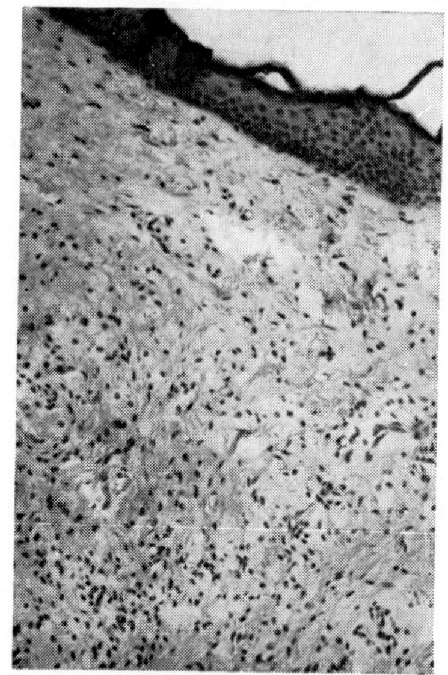

FIG. 282.

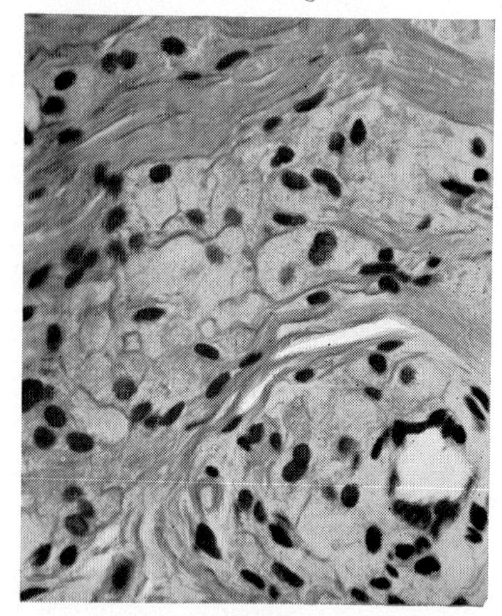

FIG. 283.

FIG. 282.—Shows the normal epidermis and the immense accumulation of xanthoma cells (×150). Fig. 283 shows these cells with clear cytoplasm in which a fine honeycomb striation may be seen; the nuclei stain darkly and lie at the periphery (H. & E.; ×570) (J. A. Sená and F. C. Cerboni).

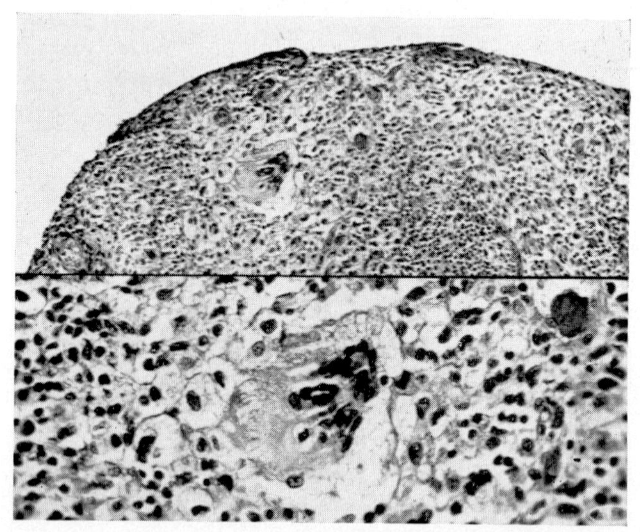

FIG. 284.—JUVENILE XANTHOGRANULOMA.

Histology of the lesion on the skin, showing the histiocytic infiltration and a prominent Touton giant cell (T. E. Sanders).

arteriosclerosis and diabetes should be excluded since xanthelasmata may be associated with occlusive vascular disease of the coronary vessels or the arteries of the extremities.[1]

The lesion itself is very characteristic (Figs. 285–6). It is a rounded or oval chamois-coloured flat plaque, soft and velvety in texture and with a smooth surface which can be resolved on magnification into minute yellow granules. The plaque lies embedded in the corium and may be slightly

FIGS. 285 and 286.—XANTHELASMA PALPEBRARUM.

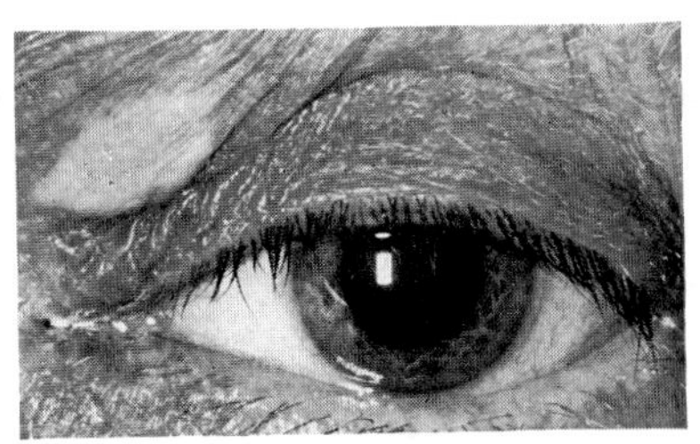

FIG. 285.—In the medial aspect of the upper lid (Inst. Ophthal.).

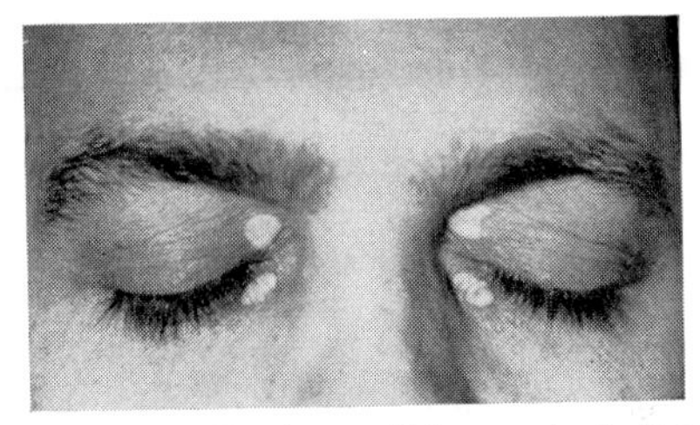

FIG. 286.—In the medial aspect of all four lids (R. Friedman and C. S. Wright).

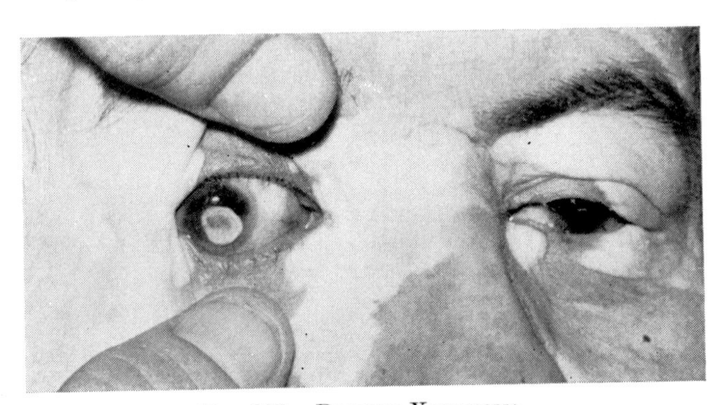

FIG. 287.—DIFFUSE XANTHOMA.

Showing widespread involvement of the lids and surrounding skin and a xanthomatous lesion of the right cornea (H. Rattner).

raised, while over it the epidermis is thin and perhaps a little wrinkled. It usually starts at the inner canthus of the upper lid and sooner or later becomes symmetrical. The lower lid may then become involved and numerous plaques may form laterally in slow progression which may coalesce to form large areas surrounding the palpebral aperture. Symptoms are absent and the only disability is cosmetic.

The histopathology of the lesion shows the typical accumulation of lipid-laden histiocytic cells converted into xanthoma cells; these are

[1] Fagge (1873), Montgomery (1937–40), Stecher and Hersh (1944), Jones and Robertson (1948), Kornerup (1948), Mishpel and Freeman (1966), and others.

strikingly perivascular in their arrangement and Touton giant cells and fibrosis are unusual. Why the site of predilection is the inner half of the lids is not clear. Wolff (1951), however, pointed out that the skin of the inner portion of the lids differs from that of the outer in that it lacks fully developed sebaceous glands; in their place are numerous unicellular sebaceous glands lying in the deep layer of the epidermis. He suggested that xanthelasma palpebrarum is in reality a neoplasia of these unicellular structures, a suggestion which may explain the common occurrence of the lesion at this site.

The *treatment* of xanthelasma consists of its removal when cosmetic considerations dictate. Simple excision may well be successful provided removal is complete, but the deposition has a tendency to recur after some years and if the lesion is extensive and excision is to be repeated, subsequent contracture may have unfortunate results. Alternative methods which are often more satisfactory are destruction of the xanthomatous masses by the application of 50% to 75% trichloracetic acid (Lussier and Grenier, 1967), the galvanic needle, carbon dioxide snow or cryosurgery (Custovic and Streiff, 1971). Other therapeutic expedients which have been recommended include x-rays (Schindler, 1910), the monopolar high-frequency current (Williams, 1929; Silvers, 1935), ironing the lesions out with the actual cautery so that they slough leaving a smooth, flat and pliable scar (R. L. and R. L. Sutton, 1949), or light-coagulation (Rydberg, 1964).

An ERUPTIVE XANTHOMA is a somewhat similar yellow plaque with an identical histological appearance arising from an inflammatory base; it develops suddenly and occurs usually in association with similar plaques elsewhere in patients with essential hyperlipæmia or diabetes when the level of serum lipids is rising and is frequently accompanied by similar dispositions in the cardiovascular system, the lymph nodes and internal organs (Dunphy, 1950). They may also be a feature of the secondary xanthomatoses as in hepatic or renal disease.

DIFFUSE XANTHOMA PLANUM is a rare condition occurring in the middle-aged or elderly wherein the xanthelasma becomes widespread, involving not only all four lids but large areas of the face and forehead, the scalp, neck and limb-folds (Fig. 287). The primary change appears to be a proliferation of histiocytes with a secondary accumulation of lipids and it is associated with normal serum lipids (Altman and Winkelmann, 1962; Cairns, 1963). Many cases have been associated with multiple myeloma or leucæmia, while others have occurred in essential hypercholesterolæmia and biliary cirrhosis.

XANTHOMA TUBEROSUM

The nodular type of xanthoma is comparatively acute in its onset and may be widespread and generalized, affecting not only the lids but also particularly the joints of the extremities. The most common site is on the inner half of the upper lid, but occasionally a single nodule may appear as a

FIGS. 288 to 291.—XANTHOMA TUBEROSUM.

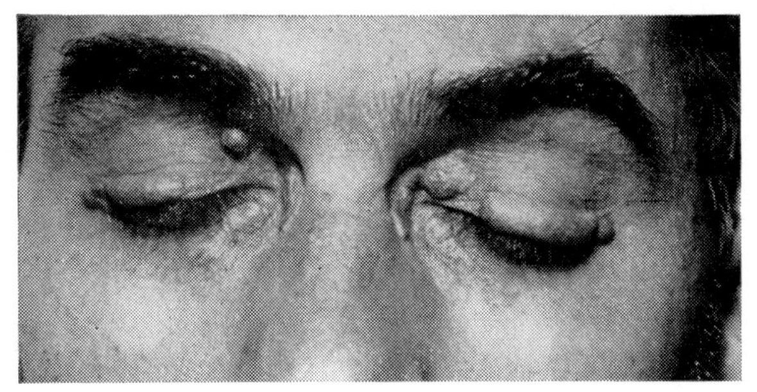

FIG. 288.

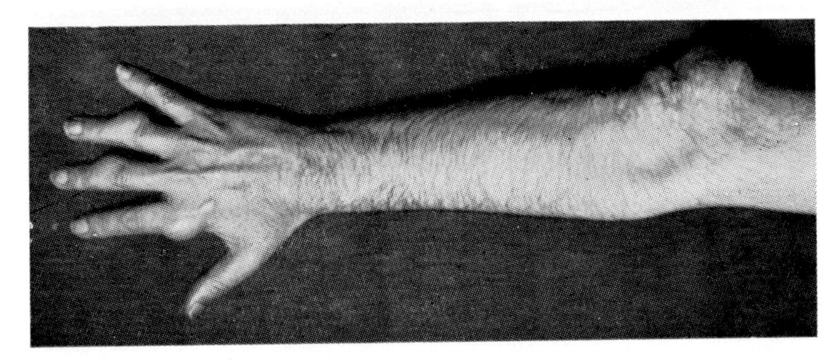

FIG. 289.

FIGS. 288 and 289.—A hereditary case in a man aged 45 showing hard yellow masses around the lids (Fig. 288) and nodules on the elbow and at the metacarpal and phalangeal joints (Fig. 289) (J. A. Sená and F. C. Cerboni).

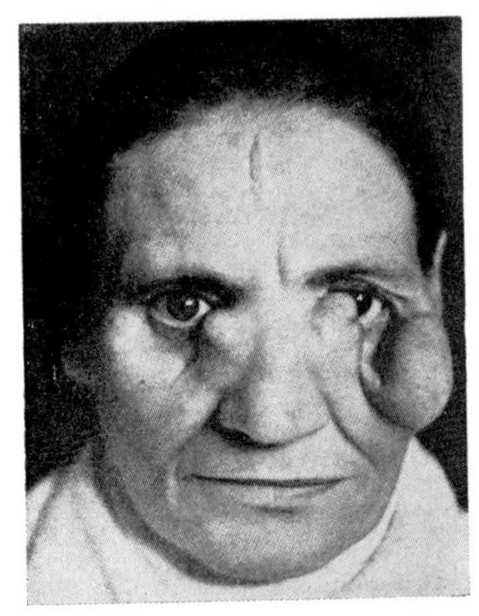

FIG. 290.—An extreme case of xanthoma tuberosum in a woman aged 59. The masses at the corners of the eyes appeared 10 years previously and the patient suffered only cosmetic inconvenience. Operative removal yielded a good cosmetic result (J. Strandberg).

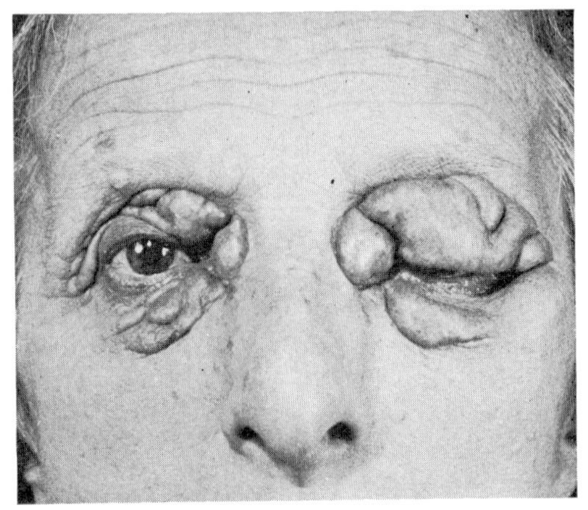

FIG. 291.—In a woman aged 60, showing marked yellowish growths on all four lids of 10 years' duration. The patient had hypercholesterolæmia and cerebello-pyramidal symptoms. The lesions were excised with a good result (F. Moro).

small firm tumour at the ciliary margin (Reese, 1936) (Figs. 288–9). The lesions may be numerous forming solid, reddish-yellow papules, 0·5 to 1·0 cm. in diameter, with a yellow apex somewhat resembling a pustule. On the other hand, they may reach a considerable size (Pajtás, 1957; Roe, 1968; Moro, 1972) (Figs 290–1); thus in a case associated with hepatic disease the xanthoma prevented easy opening of the lids (Bayer, 1969). They develop slowly, may be itchy or tender and occasionally involute and disappear, particularly if the underlying metabolic fault is alleviated. In default of this the only effective treatment is excision.

These lesions occur in hypercholesterolæmic or hyperlipæmic patients or as a secondary xanthomatosis. Out of 26 cases Serpell (1957) found that 18 were secondary to systemic diseases.

XANTHOMA DISSEMINATUM is a rare histiocytic proliferative disorder characterized by widespread cutaneous xanthomata, often involving the

FIGS. 292 and 293.—XANTHOMA DISSEMINATUM (S. D. Liebman *et al.*).

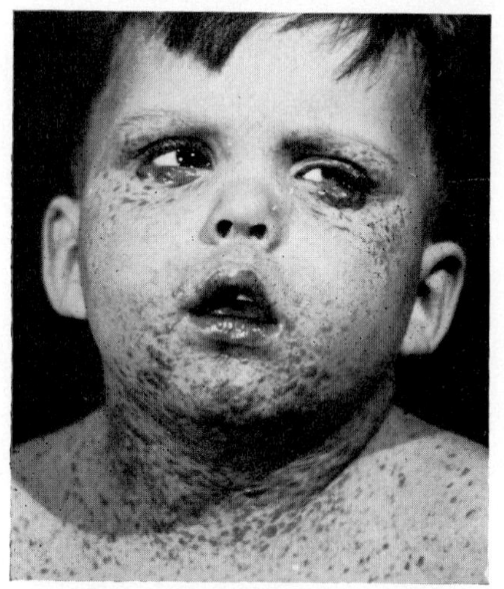

FIG. 292.—A boy aged $3\frac{10}{12}$ years; the xanthomata on the skin and eyelids showing maximal activity.

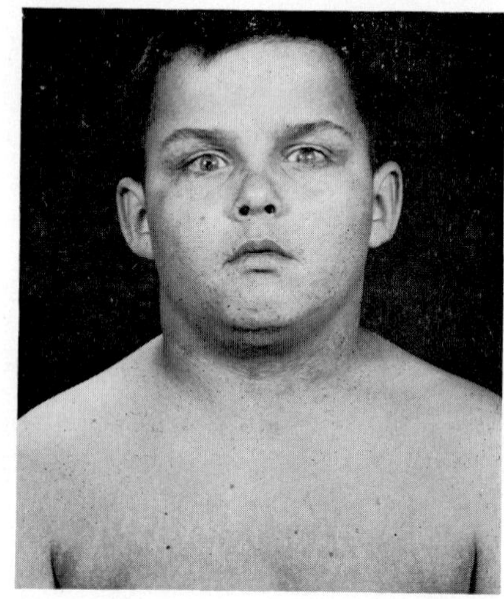

FIG. 293.—At the age of $11\frac{0}{12}$ years; the cutaneous lesions are now completely involuted and have remained so up to the age of 15.

lids, frequently associated with diabetes insipidus but without a disturbance of the lipid metabolism. It occurs usually between 5 and 20 years of age, and is most frequent in males. The prognosis is good for the condition usually resolves spontaneously with a disappearance of the xanthomata and the diabetes. The cutaneous lesions are reddish-yellow papules or nodules, appearing in great numbers usually arranged symmetrically over the face and eyelids, the trunk and the limbs. The pathological picture is identical with that of a tuberous xanthoma with histiocytic proliferation undergoing lipid impregnation to form xanthoma and Touton cells. A typical case was reported by Liebman and his colleagues (1966) wherein the disease appeared at the age of 2 years with involvement of the lids and xanthomatous invasion of the cornea; by the age of $4\frac{1}{2}$ years the dermal lesions were resolving (Figs. 292–3).

XANTHOMA DIABETICORUM is a more eruptive deposition of the tuberous type associated with diabetes mellitus or biliary cirrhosis in the presence of hyperlipæmia (Thannhauser, 1940; Hartmann, 1943–48; and others). The deposition appears suddenly in crops in the lids and elsewhere and may disappear spontaneously, sometimes to appear again. Lewis (1950–52)

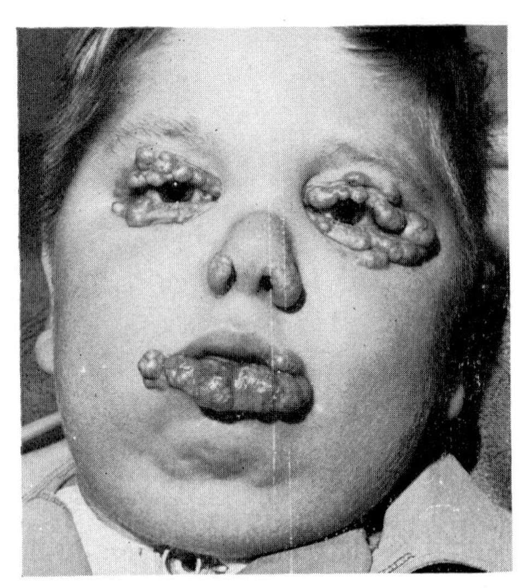

FIG. 294.—DISSEMINATED LIPOGRANULOMATOSIS.

A case of Farber's disease with generalized lesions showing gross involvement of the eyelids, conjunctiva and lips. The disease presented at the age of 7 days and the boy died at the age of 16 years (R. Zetterström).

described an interesting case associated with biliary cirrhosis wherein xanthomatous masses in the fundus caused a retinal detachment, and Davidson and his co-workers (1951) a case complicated by a lipid keratopathy. The condition is rare, generally affecting males; its local pathology resembles that of the familial type; and it can to some extent be controlled by dietetic or insulin treatment to check the hyperlipæmia (Garb, 1943).

DISSEMINATED LIPOGRANULOMATOSIS OF FARBER (1952) is a fatal generalized disturbance of the lipid metabolism with a slowly progressive course. Granulomatous infiltrations of histiocytic cells containing lipids (ceramides) and carbohydrates appear in the periarticular tissues, the throat, lungs and various viscera (Farber *et al.*, 1957). Multiple lesions appear subcutaneously sometimes on the lids and sometimes subconjunctivally (Fig. 294) (Zetterström, 1958; Samuelsson and Zetterström, 1971). The retina may also be affected.[1] Treatment is ineffective and the prognosis is poor.

THE HISTIOCYTOSES

In several histiocytic proliferative disorders, all of unknown ætiology, the deposition of lipids may occur in cutaneous or systemic tissues. Pathologically these lesions may show an evolution in three stages—the proliferation of histiocytes, granulomatous formation and xanthomatous infiltration with lipids. Three of them were classified by Lichtenstein (1953) together as *histiocytosis X:* the Hand-Schüller-Christian disease, the Letterer-Siwe disease, and eosinophilic granuloma of bone. It was pointed out in a previous Volume[2] that pathologically these were all histiocytic proliferations with or

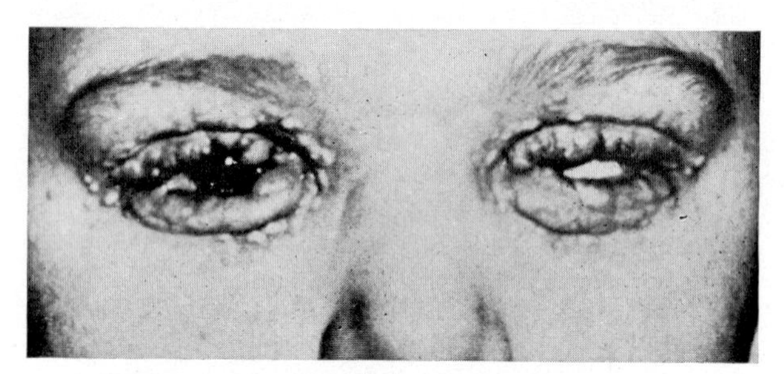

Fig. 295.—The Hand-Schüller-Christian Disease.
In a boy aged 7 years, showing gross involvement of the lids (J. and E. Bacskulin).

without eosinophils which, according to circumstances, may appear in acute, chronic or abortive forms and may even evolve from one type to the other. The manifestations of these conditions in the lids are relatively rare; we shall deal with their characteristics more fully among diseases of the orbit[3]. It may be noted, however, that the Letterer-Siwe disease is an acute condition occurring in infants typical of the histiocytic stage, eosinophilic granuloma of bone of the granulomatous aspect and the Hand-Schüller-Christian disease of the xanthomatous phase.

In the HAND-SCHÜLLER-CHRISTIAN DISEASE (1893–1921/1915/1919–20), a chronic disseminated disease, lipid infiltration is a constant finding and from the ophthalmological point of view affects the orbit.[4] Cutaneous manifestations are in the form of xanthomatous lesions which are present in only about one-third of the cases; the eyelids may occasionally be involved, exceptionally to a gross extent (Fig. 295).[5]

[1] Vol. X, p. 488. [2] Vol. VII, p. 151.
[3] p. 968. [4] p. 969.
[5] Wheeler (1931), Sosman (1932), Rogers (1934), Walsh (1947), Bauer (1958), Altman and Winkelmann (1963), Blodi (1964), J. and E. Bacskulin (1965), François and Bacskulin (1967).

In the LETTERER-SIWE DISEASE (1924/1933–49), a much more serious acute disseminated condition, lethal in 90% of cases, the same type of lesion occurs but the lipidosis is secondary. Cutaneous manifestations take the form of a papular eruption which may occur on the lids (François and Bacskulin, 1967).

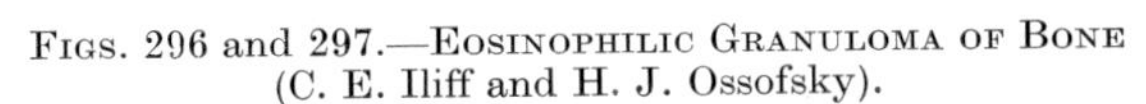

FIGS. 296 and 297.—EOSINOPHILIC GRANULOMA OF BONE (C. E. Iliff and H. J. Ossofsky).

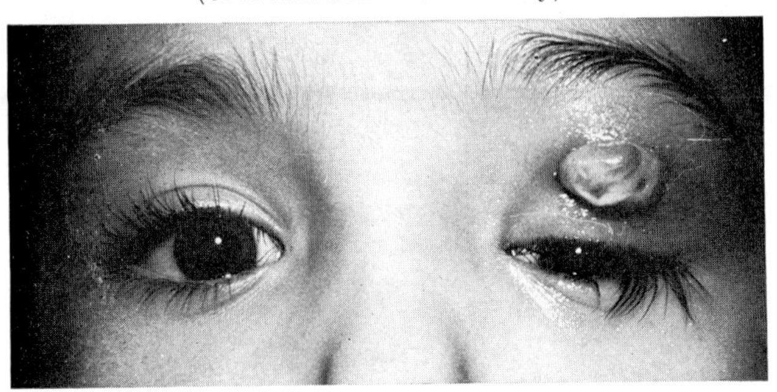

FIG. 296.—In a boy aged 7 years. A lesion on the lid after an attempted incision and drainage with negative results. The mass was removed and the orbit irradiated and the patient remained well during a follow-up period of 15 years. Radiography revealed a large defect in the frontal bone; there were no other skeletal lesions.

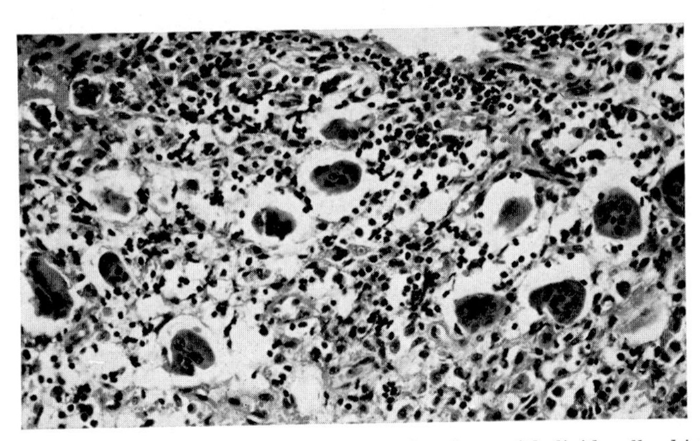

FIG. 297.—Section from the same case showing epithelioid cells, histiocytes, giant cells and clumps of eosinophils (×220).

In EOSINOPHILIC GRANULOMA OF BONE, a more chronic and benign disease of relatively good prognosis, which involves mainly the skeleton and more rarely the skin and the orbit, small yellow papules may appear on the lids (McCraney and Falk, 1958; Mumford, 1958; Iliff and Ossofsky, 1962) (Figs. 296–7).

In the more atypical forms of histiocytosis X with wide systemic evidences which cannot be classified under any particular one of these three defined syndromes,

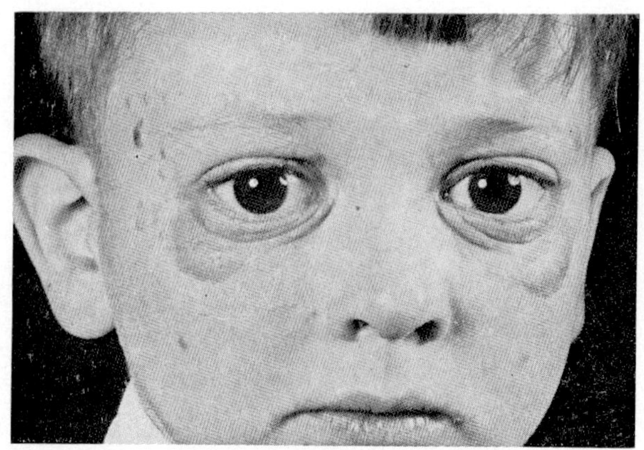

Fig. 298.—Histiocytosis X.

A boy aged 3 with atypical histiocytosis X, showing xanthomata in the lids and face; he also had papulæ on the neck, arms and buttocks, an enlarged spleen and liver and pancytopenia (J. S. Pegum and P. Wallis).

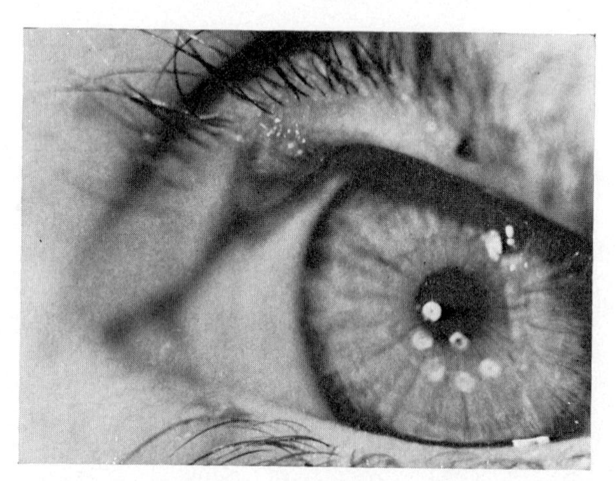

Fig. 299.—Chronic Disseminated Histiocytosis X.

In a boy aged 14 showing a tumour on the upper lid (W. C. Edwards and R. E. Reed).

papular and plaque-like xanthomata have been observed in the eyelids, the forehead, the cheek and elsewhere (Pegum and Wallis, 1962) (Fig. 298), while a histiocytic tumour on the lid-margin was described in a somewhat similar case by Edwards and Reed (1969) (Fig. 299).

An interesting condition was described as SINUS HISTIOCYTOSIS WITH MASSIVE LYMPHADENOPATHY by Rosai and Dorfman (1969) occurring in young children and characterized histologically by a proliferation of histiocytes at first within the sinusoids of lymphoid tissue but eventually leading to destruction of the normal architecture of the follicles associated with a similar invasion of the lymph nodes. Codling and his collaborators (1972) described a familial case of this type which is described with diseases of the orbit.[1]

JUVENILE XANTHOGRANULOMA

This rare disease occurring in infants and young children has long been known to dermatologists by whom it was described as *xanthoma* (*multiplex*) (Jackson, 1890; Adamson, 1905; MacLeod, 1907); at a later date it was called *nævo-xantho-endothelioma* by McDonaugh (1909–12) and eventually, in an intensive study of 53 cases, JUVENILE XANTHOGRANULOMA by Helwig and Hackney (1954). The ætiology is unknown, and the histopathological features are those of a typical lipoidal histiocytosis with a close resemblance to the members of the histiocytosis X group of diseases. The benign nature of the condition, however, and the lack of any tendency to appear in inter-

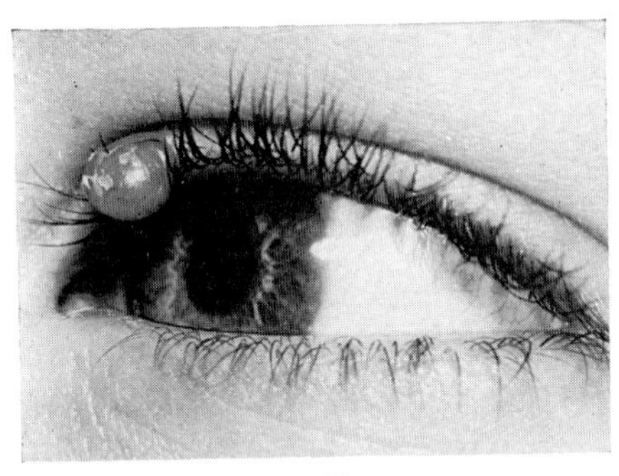

FIG. 300.—JUVENILE XANTHOGRANULOMA.
A solitary lesion on the upper lid (B. Daicker).

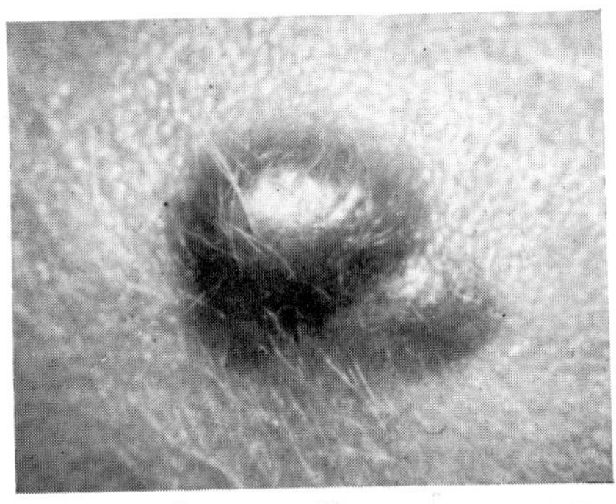

FIG. 301.—JUVENILE XANTHOGRANULOMA.
A skin lesion on the forehead (T. E. Sanders).

mediate forms characteristic of this group justify the general agreement to regard it as a nosological entity (Sanders, 1962).

The disease is essentially dermatological in its incidence, but in a previous Volume[2] we have seen that tumours may occur in the iris and ciliary body which grow rapidly with a hæmorrhagic tendency so that the eye may be enucleated owing to the development of secondary glaucoma; lesions may also occur in the epibulbar tissues and the orbit (Zimmerman, 1965).[3] As a general rule the disease appears in infancy, usually before the age of 6 months, but it has occurred in adults. The age incidence of lesions of the lids is slightly higher: in 12 cases reported by Zimmerman (1965) the age distribution was 2 in the first year of life, 7 between 17 months and 5 years of age, the remaining patients being 11, 19 and 30 years old.

The most common involvement of the lids is a single isolated lesion unassociated with other dermatological implications which, indeed, is often

[1] p. 974. [2] Vol. IX, p. 656. [3] p. 974.

FIGS. 302 and 303.—JUVENILE XANTHOGRANULOMA (L. E. Zimmerman).

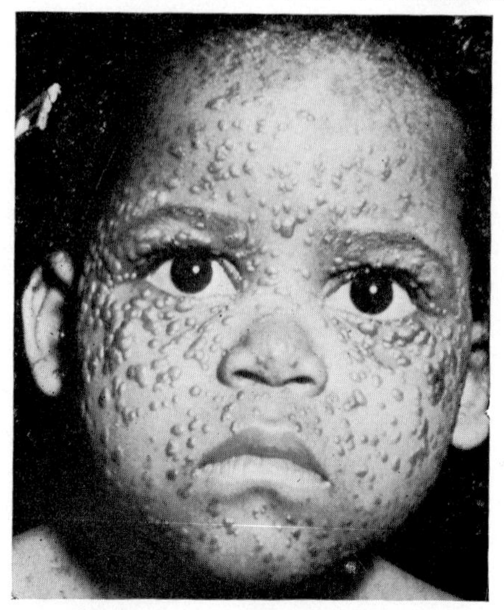

FIG. 302.—In a child aged 2½ years, with extensive involvement of the face and thorax.

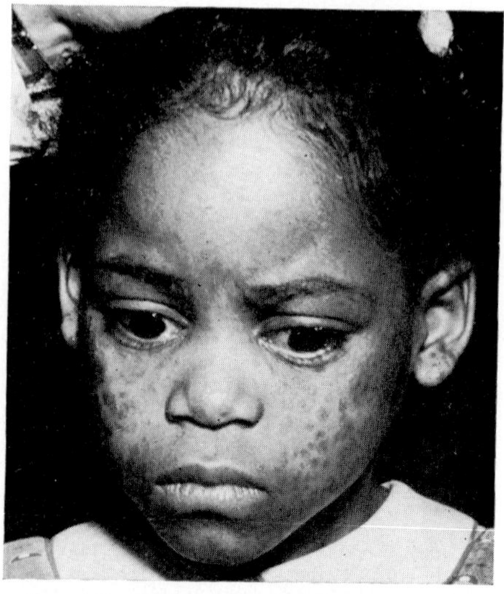

FIG. 303.—The same patient aged 6 years; spontaneous regression of the lesions began at the age of 4.

FIGS. 304 and 305.—JUVENILE XANTHOGRANULOMA OF THE EYELID (B. Daicker).

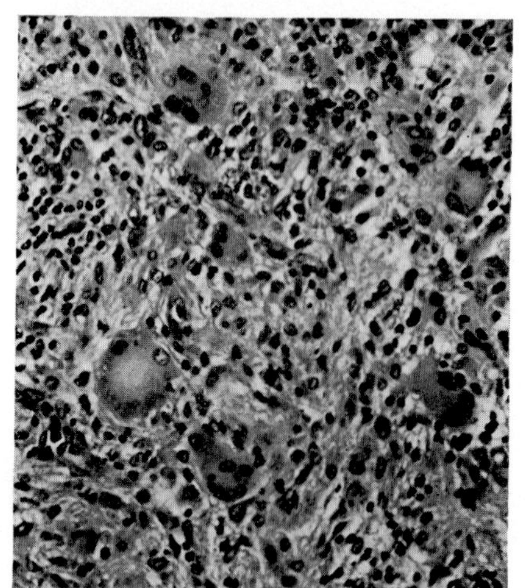

FIG. 304.—General histological view of the lesion seen in Fig. 300.

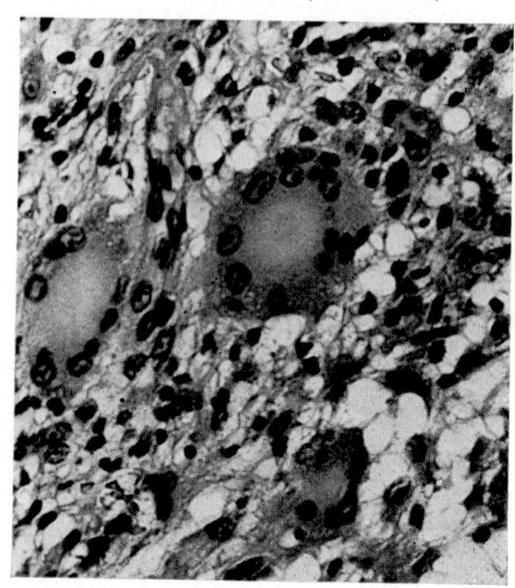

FIG. 305.—Touton giant cells in the granulomatous tissue.

the only known manifestation of the disease (12 out of 13 cases, Zimmerman, 1965; Daicker, 1966; and others) (Figs. 300–1), but occasionally there are multiple tumours widely scattered in great numbers involving the lids as well as extensive areas of the body, giving a picture somewhat resembling disseminated xanthomata; a striking example was published by Fleischmajer and Hyman (1960) (see Zimmerman, 1965) (Figs. 302–3).

The diagnosis of an isolated case can only be made after excision. Histopathologically the lesion on the lids differs in no way from those occurring elsewhere; under a normal epithelium the dermis contains a mass of xanthoma cells with occasional giant Touton cells (Figs. 304–5). The disease is benign apart from lesions in the uveal tract; after growing for a few months the nodules tend to undergo spontaneous resolution within a year or two leaving umbilicated scars that eventually disappear.

Adamson. *Brit. J. Derm.*, **17,** 222 (1905).
Altman and Winkelmann. *Arch. Derm.*, **85,** 633 (1962); **87,** 164 (1963).
Anderson. *Arch. Ophthal.*, **81,** 692 (1969).
Bacskulin, J. and E. *Acta ophthal.* (Kbh.), **43,** 610 (1965).
Barlow. *Brit. med. J.*, **1,** 998 (1884).
Bauer. *Klin. Mbl. Augenheilk.*, **133,** 478 (1958).
Bayer. *Klin. Mbl. Augenheilk.*, **154,** 728 (1969).
Birch-Hirschfeld. *v. Graefes Arch. Ophthal.*, **58,** 207 (1904).
Blodi. *Trans. Amer. Acad. Ophthal.*, **68,** 1012 (1964).
Cairns. *Brit. J. Derm.*, **75,** 441 (1963).
Chhetri, Chowdhury and De. *J. Ass. Phycns. India*, **15,** 405 (1967).
Christian. *Contrib. med. biol. Res.*, N.Y., **1,** 391 (1919).
Med. Clin. N. Amer., **3,** 849 (1920).
Codling, Soni, Barry and Martin-Walker. *Brit. J. Ophthal.*, **56,** 517 (1972).
Custović and Streiff. *Bull. Soc. franç. Ophtal.*, **84,** 145 (1971).
Daicker. *Ophthalmologica*, **152,** 267 (1966).
Davidson, Pilz and Zeller. *Amer. J. Ophthal.*, **34,** 233 (1951).
Dunphy. *Amer. J. Ophthal.*, **33,** 1579 (1950).
Edwards and Reed. *Survey Ophthal.*, **13,** 335 (1969).
Fagge. *Trans. path. Soc. Lond.*, **24,** 242 (1873).
Farber. *Amer. J. Dis. Child.*, **84,** 499 (1952).
Farber, Cohen and Uzman. *J. Mt. Sinai Hosp.*, **24,** 816 (1957).
Fleischmajer and Hyman. *The Dyslipidoses*, Springfield, 329 (1960).
François and Bacskulin. *Ophthalmologica*, **153,** 241 (1967).
Garb. *Ann. intern. Med.*, **19,** 241 (1943).
Geber and Simon. *Arch. Derm. Syph.* (Prague), **4,** 305 (1872).
Grönvall. *The Relative Frequencies in different Age Groups and the Extent of Palpebral Xanthelasma* (Festival paper to Malthe Ljungdahl, 215), Lund (1947).
Acta ophthal. (Kbh.), **31,** 241 (1953).
Hand. *Arch. Pediat.*, **10,** 673 (1893).
Amer. J. med. Sci., **162,** 509 (1921).
Hartmann. *Klin. Mbl. Augenheilk.*, **109,** 555 (1943); **113,** 271 (1948).
Heath. *Arch. Ophthal.*, **10,** 342 (1933).
Helwig and Hackney. *Amer. J. Path.*, **30,** 625 (1954).
Hutchinson. *Med.-chir. Trans.* (Lond.), **54,** 171 (1871).
Lancet, **1,** 409 (1871).
Iliff and Ossofsky. *Tumors of the Eye and Adnexa in Infancy and Childhood*, Springfield, 110 (1962).
Jackson. *J. cutan. vener. Dis.*, **8,** 241 (1890).
Jones and Robertson. *Brit. med. J.*, **1,** 1137 (1948).
Kahán, A. and I. L., and Timár. *Amer. J. Ophthal.*, **63,** 320 (1967).
Kornerup. *Familial Hypercholesterolaemia and Xanthomatosis* (Thesis), Kolding, Denmark (1948).
Letterer. *Frankf. Z. Path.*, **30,** 377 (1924).
Lewis. *Brit. J. Ophthal.*, **34,** 506 (1950); **36,** 325 (1952).
Lichtenstein. *Arch. Path.*, **56,** 84 (1953).
Liebman, Crocker and Geiser. *Arch. Ophthal.*, **76,** 221 (1966).
Lussier and Grenier. *Un. méd. Canad.*, **96,** 885 (1967).
McCraney and Falk. *J. Dis. Childh.*, **95,** 214 (1958).
McDonaugh. *Brit. J. Derm.*, **21,** 254 (1909); **24,** 85 (1912).
MacLeod. *Brit. J. Derm.*, **19,** 241 (1907).
Mawas. *Ann. Oculist.* (Paris), **151,** 437 (1914).
Mishpel and Freeman. *Med. J. Aust.*, **1,** 74 (1966).
Montgomery. *Proc. Mayo Clin.*, **12,** 641 (1937).
J. invest. Derm., **1,** 325 (1938).
Med. Clin. N. Amer., **24,** 1249 (1940).
Moro. *Bull. Soc. franç. Ophtal.*, **85,** 453 (1972).
Mumford. *Brit. J. Derm.*, **70,** 460 (1958).
Pajtás. *Cs. Oftal.*, **13,** 164 (1957).
Pedace and Winkelmann. *J. Amer. med. Ass.*, **193,** 893 (1965).
Pegum and Wallis. *Proc. roy. Soc. Med.*, **55,** 1071 (1962).
Pelfini, Sacchi and Cairo. *Boll. Soc. ital. Biol. sper.*, **46,** 192 (1970).
Poensgen. *Virchows Arch. path. Anat.*, **102,** 410 (1885).
Reese. *J. Amer. med. Ass.*, **107,** 937 (1936).
Roe. *Arch. Derm.*, **97,** 436 (1968).
Rogers. *Amer. J. Ophthal.*, **17,** 1141 (1934).

Rosai and Dorfman. *Arch. Path.*, **87,** 63 (1969).
Rydberg. *Acta ophthal.* (Kbh.), **42,** 541 (1964).
Samuelsson and Zetterström. *Scand. J. clin. Lab. Invest.*, **27,** 393 (1971).
Sanders. *Amer. J. Ophthal.*, **53,** 455 (1962).
Schaaf and Werner. *Arch. Derm. Syph.* (Berl.), **162,** 217 (1930).
Schindler. *Klin. Mbl. Augenheilk.*, **48** (2), 629 (1910).
Schüller. *Fortschr. Roentgenstr.*, **23,** 12 (1915).
Send and Cerboni. *Arch. Oftal. B. Aires*, **23,** 79 (1948).
Serpell. *Trans. ophthal. Soc. Aust.*, **17,** 143 (1957).
Silvers. *J. Amer. med. Ass.*, **105,** 796 (1935).
Siwe. *Z. Kinderheilk.*, **55,** 212 (1933).
Adv. Pediat., **4,** 117 (1949).
Sosman. *J. Amer. med. Ass.*, **98,** 110 (1932).
Stecher and Hersh. *J. clin. Invest.*, **23,** 699 (1944).
Sutton, R. L. and R. L. *Handbook of Diseases of the Skin*, London, **1,** 391 (1949).
Thannhauser. *Endocrinology*, **26,** 189 (1940).
Villard. *Arch. Ophtal.*, **23,** 364 (1903).
Walsh. *Clinical Neuro-ophthalmology*, Baltimore, 844 (1947).
Weber. *Brit. J. Derm.*, **36,** 335 (1924).
Wheeler. *Arch. Ophthal.*, **5,** 161 (1931).
Wile, Eckstein and Curtis. *Arch. Derm. Syph.* (Chic.), **19,** 35 (1929).
Williams. *Arch. Ophthal.*, **2,** 443 (1929).
Wolff. *Brit. J. Derm.*, **63,** 296 (1951).
Zetterström. *Acta paediat.* (Stockh.), **47,** 501 (1958).
Zimmerman. *Trans. Amer. Acad. Ophthal.*, **69,** 412 (1965).

Lipid Proteinosis

(*Urbach-Wiethe*)

This rare hereditary disease described by Siebenmann (1908) was fully reported as a clinical entity by the dermatologist Urbach and the otolaryngologist Wiethe in 1929, who called it *lipoidosis cutis et mucosæ*; Urbach (1934) re-named it *lipoid proteinosis*. Subsequent names have included *hyalinosis cutis et mucosæ* (Lundt, 1949) and *lipoglycoproteinosis* (McCusker and Caplan, 1962). More than 150 cases have been described (Grosfeld *et al.*, 1965). The ætiology of the condition is obscure. It is transmitted hereditarily through a recessive gene (Rosenthal and Duke, 1967; François *et al.*, 1968), and since Urbach and Wiethe (1929) found that the homogeneous hyaline deposits stained for fat they postulated a disturbance of the lipid metabolism, and in a few cases an increase in the lipoproteins and phospholipids has been found. In the vast majority of cases, however, the serum lipids and cholesterol are normal. Diabetes is an occasional association and the importance of the carbohydrate metabolism was stressed by Potter and Weinmann (1959), but a more constant association is an anomaly in the serum proteins, particularly the globulins; for this reason the condition has been classified as a dysproteinæmia (Laymon and Hill, 1957; Eberhartinger and Reinhardt, 1958; Blodi *et al.*, 1960). On the other hand, other authorities cast doubt upon the systemic nature of the condition and assume that it is due to a local breakdown of collagen and fatty tissue followed by hyalinization and is therefore a primary disorder of connective tissue (Borda and Abulafia, 1955; Ungar and Katzenellenbogen, 1957).

The disease starts in infancy or early childhood with hoarseness due to the deposition of the characteristic material in the larynx and this is followed by an extension of the deposition through the upper respiratory tract with a hardening and thickening of the tongue and lips so that after some years swallowing may be difficult and the voice is reduced to a whisper. After the

FIGS. 306 to 308.—LIPID PROTEINOSIS (Urbach-Wiethe).

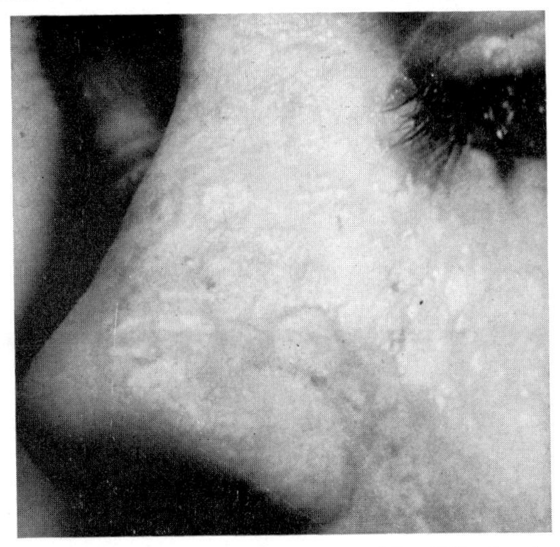

FIG. 306.—In a boy aged 14, the same case as Fig. 307, showing the typical depressed pale scars with light brown surrounds on the face giving a pock-marked appearance (G. E. Hewson).

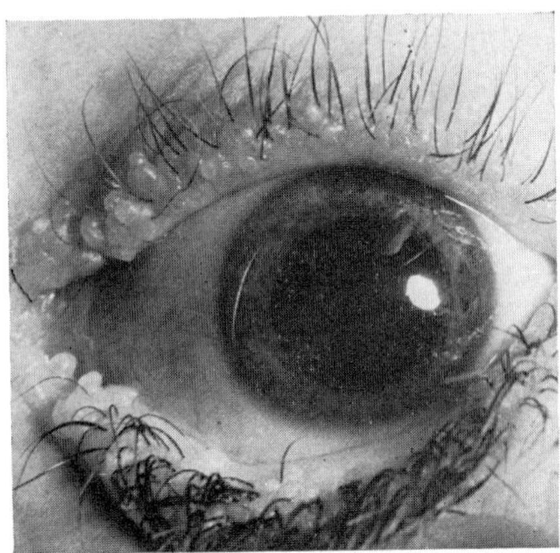

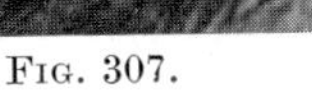

FIG. 307.

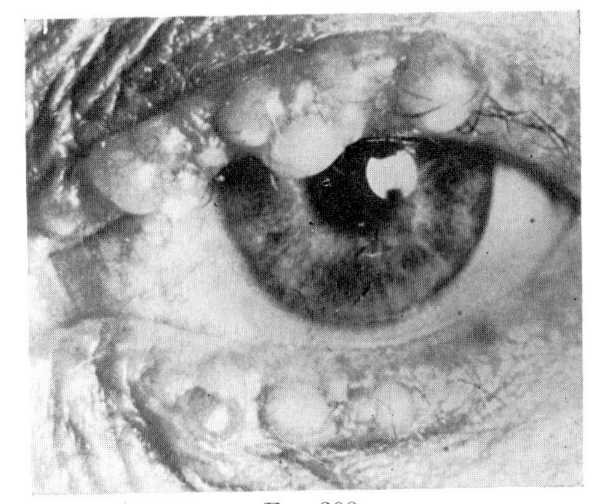

FIG. 308.

FIG. 307.—In a boy aged 14 years (G. E. Hewson).

FIG. 308.—In a 60-year-old patient (J. and E. Bacskulin).

The discrete nodules are seen on the lid-margins.

hoarseness a skin eruption appears over the face and scalp giving it a thickened, pale yellow-brown, pock-marked appearance (Fig. 306). Such discrete or confluent papular eruptions may occur anywhere on the body, while the extensor surfaces of the skin such as the elbows and knees show marked hyperkeratosis. A pathognomonic feature is the appearance of nodules on the margins of all four lids (Figs. 307–8). These nodules are discrete and resemble solid beads of yellow-brown wax among the lashes and can eventually exceed 5 mm. in diameter and approach the posterior border of the lid-margin without implicating the conjunctiva; the lashes are lost although trichiasis may occur (Hewson, 1963). They are characteristic and integral features of the disease and have been amply noted in ophthalmic as well as dermatological and otolaryngological literature.[1]

Other ocular involvement is rare. Deposits may occur in the choriocapillaris,[2] drusen or degeneration may be seen at the macula,[3] and in one instance a small nodule was found on the conjunctiva (Sanchez-Caballero *et al.*, 1954).

FIGS. 309 and 310.—LIPID PROTEINOSIS (F. C. Blodi *et al.*).

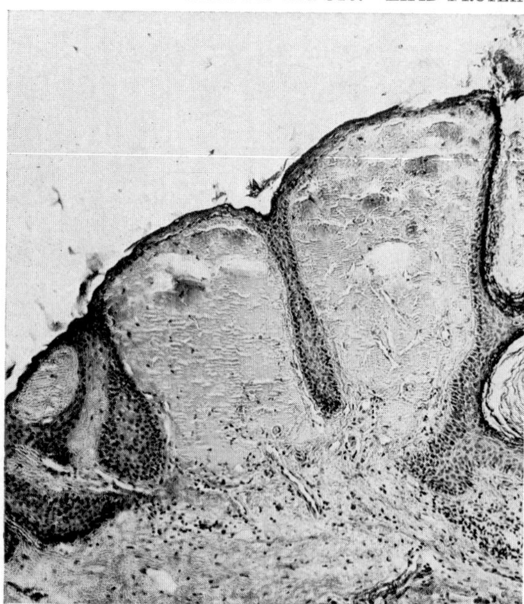

FIG. 309.—The histological appearance of the nodules on the lids.

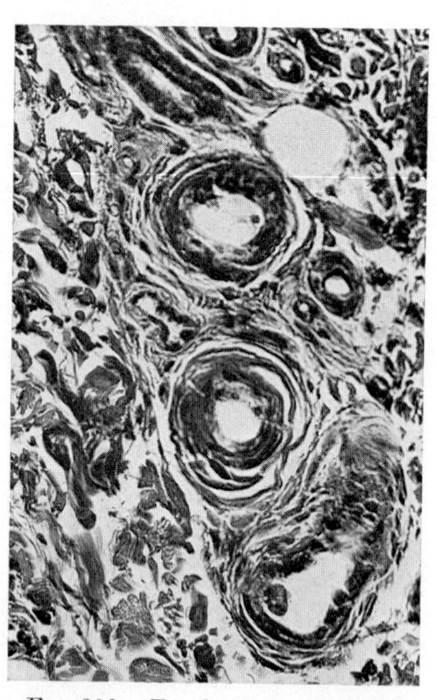

FIG. 310.—To show the diffuse infiltration of the walls of the capillaries in the cutis of the lid.

Histopathologically the material deposited is generally considered to be a lipid bound to a glycoprotein (McCusker and Caplan, 1962) (Figs. 309–10). It is first deposited around the capillaries and the ducts of the sweat glands and eventually in great masses in the upper layers of the dermis. The epidermis shows hyperkeratosis and some acanthosis and fine droplets of extracellular lipids are found in the hyalinized masses which also contain carbohydrates (Ungar and Katzenellenbogen, 1957).

[1] Urbach and Wiethe (1929), Montgomery and Havens (1939), Sulzberger (1942), Blodi *et al.* (1960), Scott and Findlay (1960), Muirhead and Jackson (1963), J. and E. Bacskulin (1965), Rosenthal and Duke (1967), MacKinnon (1968), François *et al.* (1968), Jensen *et al.* (1972), and many others.

[2] Vol. IX, p. 654.

[3] Vol. X, p. 493.

The disease is chronic and relatively benign although it is progressive until early adult life, the usual end-result being the formation of atrophic scars. The only danger lies in the upper respiratory passages for the infiltration of the larynx may necessitate a tracheostomy (Tripp, 1936; Wise and Rein, 1938). No effective treatment is known.

Bacskulin, J. and E. *Acta ophthal.* (Kbh.), **43,** 610 (1965).
Blodi, Whinery and Hendricks. *Trans. Amer. ophthal. Soc.*, **58,** 155 (1960).
Borda and Abulafia. *Arch. Argent. Derm.*, **5,** 311 (1955).
Eberhartinger and Reinhardt. *Hautarzt*, **9,** 503 (1958).
François, Bacskulin and Follmann. *Ophthalmologica*, **155,** 433 (1968).
Grosfeld, Spaas, van de Staak and Stadhouders. *Dermatologica*, **130,** 239 (1965).
Hewson. *Brit. J. Ophthal.*, **47,** 242 (1963).
Jensen, Khodadoust and Emery. *Arch. Ophthal.*, **88,** 273 (1972).
Laymon and Hill. *Arch. Derm.*, **75,** 55 (1957).
Lundt. *Arch. Derm. Syph.* (Berl.), **188,** 128 (1949).
McCusker and Caplan. *Amer. J. Path.*, **40,** 599 (1962).
MacKinnon. *Acta oto-laryng.* (Stockh.), **65,** 403 (1968).
Montgomery and Havens. *Arch. Otolaryng.*, **29,** 650 (1939).
Muirhead and Jackson. *Arch. Ophthal.*, **69,** 174 (1963).
Potter and Weinmann. *Arch. Derm.*, **80,** 149 (1959).
Rosenthal and Duke. *Amer. J. Ophthal.*, **64,** 1120 (1967).
Sanchez-Caballero, Ambrosetti and Lopez Lacarrere. *Sem. méd.* (B. Aires), **105,** 835 (1954).
Scott and Findlay. *S. Afr. med. J.*, **34,** 189 (1960).
Siebenmann. *Arch. Laryng. Rhin.* (Berl.), **20,** 101 (1908).
Sulzberger. *Laryngoscope*, **52,** 286 (1942).
Tripp. *N.Y. St. J. Med.*, **36,** 619 (1936).
Ungar and Katzenellenbogen. *Arch. Path.*, **63,** 65 (1957).
Urbach. *Klin. Wschr.*, **13,** 577 (1934).
Urbach and Wiethe. *Virchows Arch. path. Anat.*, **273,** 285 (1929).
Wise and Rein. *Arch. Derm. Syph.* (Chic.), **37,** 201 (1938).

Amyloidosis

In a previous Volume[1] the general pathology of hyalinosis and amyloidosis has been discussed, where it was pointed out that hyaline degeneration is the commonest degenerative change that occurs in the body. Hyalin is usually defined as embracing a number of substances which have the common features of being homogeneous, translucent, highly refractile and resistant to solvents such as acids and alkalis. Among these materials amyloid has been identified as a fairly stable chemical entity with characteristic staining reactions; it is a complex of glycoproteins and neutral mucopolysaccharides, staining mahogany-brown with iodine, pinkish red with basic aniline dyes such as methyl violet or gentian violet, red with Congo red, often demonstrating birefringence and dichroism, and fluorescing with thioflavine T with varying shades of blue and green (Pearse, 1960; Malak and Smith, 1962; Hashimoto *et al.*, 1965; Heyl, 1966; and others). The essential feature of its ultrastructure is the presence of fine fibrils composed of protein (Shirahama and Cohen, 1965; Cohen, 1966).

The mechanism of the formation of amyloid is far from clear; some workers considered that it was produced in endothelial cells (Cohen and Calkins, 1960; Cohen *et al.*, 1965), others by these and plasma cells (Kennedy, 1962), others by histiocytes (Gueft and Ghidoni, 1963), others that it is associated with fibroblasts (Hashimoto *et al.*, 1965), and others that it represents a perversion of the protein-synthesizing function of the reticulo-endothelial system, especially of its derivatives, the plasma cells (Teilum, 1952–56); possibly it can be formed by all these elements depending on the type of

[1] Vol. VII, p. 175.

amyloidosis and its site. The cause of the deposition is equally unknown; the condition was initially considered to be a simple degeneration (Virchow, 1858), and at a later date a derangement of protein metabolism. The most attractive theory, however, is that it represents a disorder of the reticulo-endothelial system which results in the deposition of antibody or antigen-antibody complexes as amyloid. This may occur as a result of chronic immunological stimulation, inflammation, infection, neoplasia, hereditary determination through enzyme defects or alterations in the ground-substance of the tissues wherein it is deposited (Teilum, 1956; Cohen, 1967). It is interesting that the disease can be produced in animals by a great number of stimuli, the most common of which has been casein (Cohen, 1967).

Two main types of amyloidosis occur—primary (which may be familial) and secondary, both of which may be systemic or local (Reimann *et al.*, 1935; Symmers, 1956). In all cases there is no inflammatory reaction associated with the deposits which are harmful only from their bulk or until they have almost or completely replaced the vital tissue of an organ.

PRIMARY SYSTEMIC AMYLOIDOSIS

If systemic amyloidosis (the *Lubarsch-Pick disease*) occurs in an otherwise healthy subject it is termed primary; it is characterized by the predominant involvement of muscles, nerves, the cardio-vascular system and the skin, as well as the viscera—liver, spleen, kidneys and adrenals. The symptoms are therefore protean. Sensory-motor neuropathy is common, macroglossia is a frequent feature, digestive disturbances occur and the commonest cause of death is cardiac failure. Cohen (1967) found that 8·4% of these patients had ocular involvement of some type. An early sign may be localized bilateral purpura of the lids (Goltz, 1952; Ramsdell and Winder, 1970). The skin of the eyelids is a frequent site for the deposition of amyloid[1] (Figs. 311–312). The typical eruption in the skin of the lids is small smooth papules, discrete or confluent, yellowish or waxy in colour and sometimes hæmorrhagic owing to the involvement of the blood vessels. These lesions are characteristic of the primary systemic type of the disease; conjunctival nodules are rare (Stansbury, 1965).

The relationship between primary systemic amyloidosis and multiple myelomata is interesting. The invariable presence of abnormal plasma cells in patients with the former (Kyle and Bayrd, 1961), the occurrence of multiple myeloma in some 10 or 15% of these patients, the invariable occurrence of Bence-Jones protein in their urine, an electrophoretic component typical of the myeloma protein, and the possibility that the plasma cells produce an antigen that stimulates the production of amyloid, have all given rise to the suggestion that the basic disease in every patient with primary systemic amyloidosis is multiple myelomatosis and that most patients die before it becomes apparent (Osserman, 1959; Kyle and Bayrd, 1961; Stobbe *et al.*, 1963).

[1] Michelson and Lynch (1934), Castelló and Pérez (1943), Miescher (1945), Wells (1952), Goltz (1952), Kennedy *et al.* (1957), Hellier (1959), Stansbury (1965), Hällén and Rudin (1966), Desai *et al.* (1966), Muller *et al.* (1969), Lange (1970), Brownstein *et al.* (1970).

A *heredo-familial type* has repeatedly been described, particularly from Portugal since Andrade's (1952) original observation. A dominant transmission is usual. Almost all the organs in the body may be implicated and the disease, starting in the second and third decades, is usually fatal within 10 or 12 years. The ocular complications involving the uveal tract and vitreous are prominent.[1] The lids are less commonly affected than in the non-hereditary type of the disease, but they may be thickened and hæmorrhages due to vascular fragility may occur (Kaufman, 1958; Paton and Duke, 1966).

FIGS. 311 and 312.—PRIMARY SYSTEMIC AMYLOIDOSIS (M. H. Brownstein *et al.*).

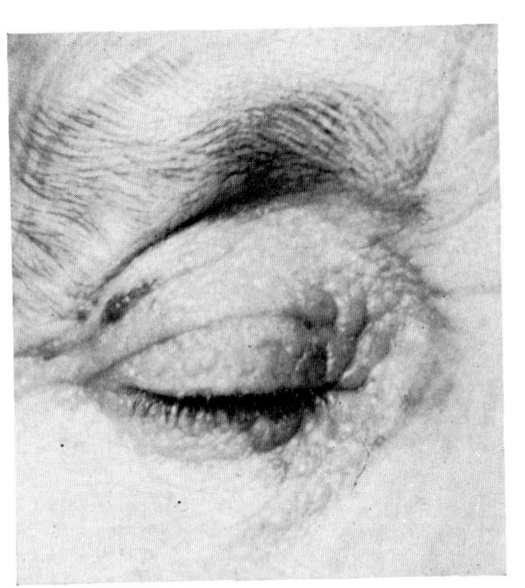

FIG. 311.—Waxy amyloid papules and purpura on the skin of the lids.

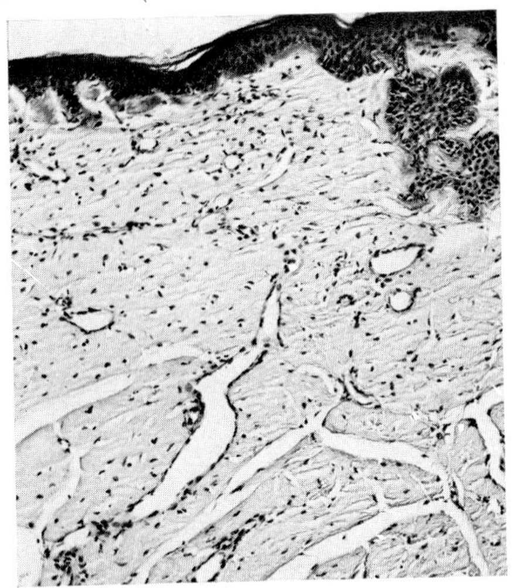

FIG. 312.—The histopathology of the lesion from a section of the lids, showing deposits of amyloid (H. & E.; ×100).

SECONDARY SYSTEMIC AMYLOIDOSIS

This is an amyloid infiltration of the tissues associated with a large number of diseases, particularly chronic infections such as tuberculosis, leprosy and syphilis, inflammatory diseases such as rheumatoid arthritis or Reiter's syndrome, neoplasms, metabolic diseases such as diabetes mellitus and the dysproteinæmias. The kidneys, liver, spleen and adrenals are the organs predominantly affected, but deposits in the eye and its adnexa are rare (Schmidt, 1905).

LOCALIZED AMYLOIDOSIS

Localized nodular amyloidosis may be primary or secondary. The PRIMARY AFFECTION shows a predilection for the skin, the larynx and tracheo-bronchial tree, heart, ureters, urinary bladder and the eyes and their

[1] Vol. IX, p. 738; Vol. XI, p. 330.

adnexa (Brownstein, 1968). It occurs in the lids and sometimes in the orbit in patients in whom these structures are and have been healthy. In the literature the condition has usually been described as conjunctival amyloidosis, under which heading it has already been described[1]; but the deposition is found not only in the conjunctiva itself and its substantia propria but also in the tarsal plate, the muscles of the lids, the blood vessels and sometimes Tenon's capsule. This may determine the development of ptosis associated with infiltrations of the levator muscle (Rogman, 1898; Guerry and Wiesinger, 1960; Richlin and Kuwabara, 1962; Smith and Zimmerman, 1966). The affection usually starts in the 3rd and 4th decades of life but sometimes earlier; it often appears as a swelling of the upper lid although all

Figs. 313 and 314.—Localized Amyloid Infiltration of the Lids.

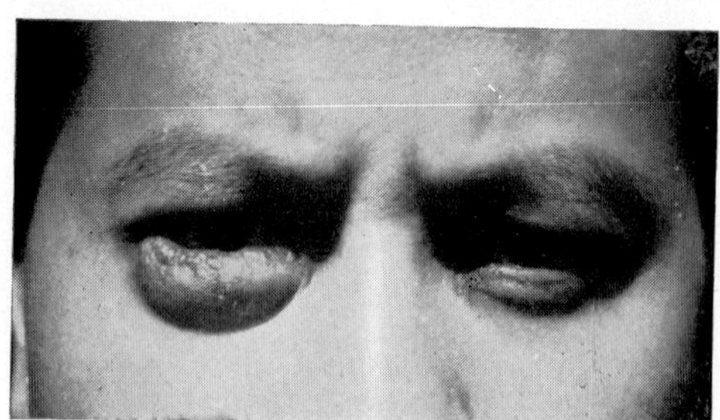

Fig. 313.

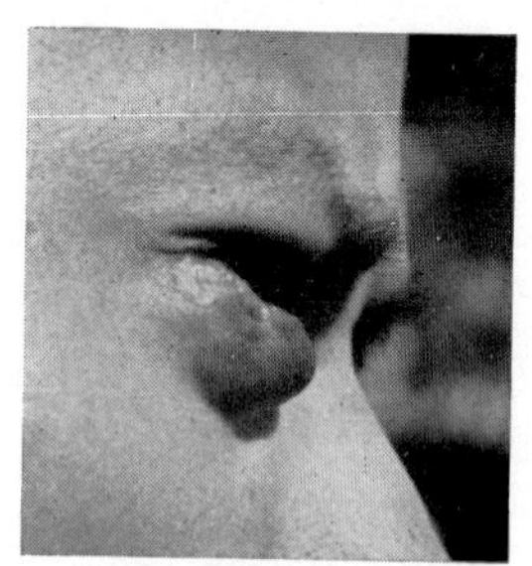

Fig. 314.

In a Siamese aged 22. Five years previously two small tumours in the lower lid were diagnosed as chalazia. The figures illustrate the swelling in the right lower lid (A. Rey).

four may become involved (Figs. 313–4). The swelling is discrete, non-ulcerative, waxy, firm and slowly growing. It is not tender and is asymptomatic so that many patients do not seek advice until it has become gross, sometimes so large that the lids cannot be opened. In this event the only treatment is by surgical excision which frequently cannot be complete; fortunately the material left behind has a habit of shrinking spontaneously although recurrences have been recorded (Kubik, 1924). The pathological appearances are the same as in amyloidosis generally (Figs. 315–6) (Ashton and Rey, 1951; Smith and Zimmerman, 1966; and many others).

A large number of cases has now been recorded from many countries throughout the world. The more important cases dating from Raehlmann (1881–82) to 1964 will be found in a previous Volume[2]; subsequent cases are appended.[3]

SECONDARY LOCALIZED AMYLOIDOSIS. Small deposits of amyloid have been found in the eyelids in several types of lesion; thus Brownstein and his colleagues (1970) noted them in lipid proteinosis. Large deposits, however,

[1] Vol. VIII, p. 586.
[2] Vol. VIII, p. 588, footnote 2.
[3] Behal (1964), Pisano (1964), Renard *et al.* (1965), Isbell (1966), Jain and Gupta (1966), Smith and Zimmerman (1966), Bisaria *et al.* (1967), Kojima *et al.* (1967).

FIGS. 315 and 316.—LOCALIZED AMYLOIDOSIS OF THE LIDS.

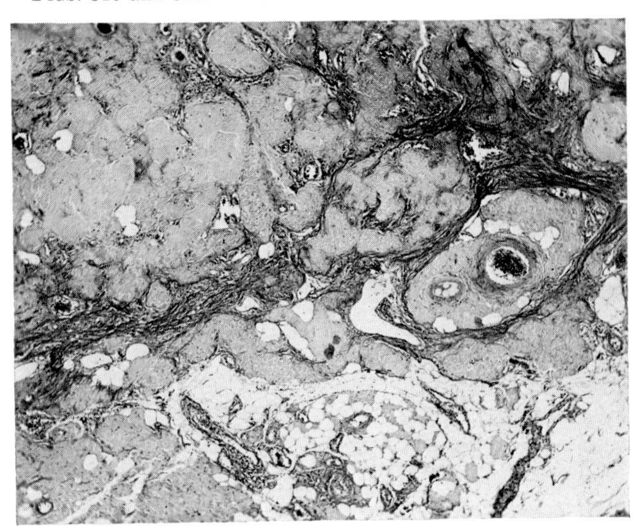

FIG. 315.—Amyloidosis of the muscular and adipose tissues of the lids (M. E. Smith and L. E. Zimmerman).

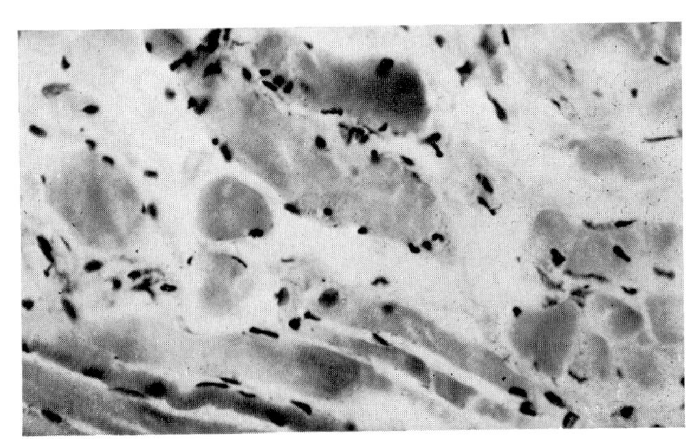

FIG. 316.—The levator muscle partly replaced by amyloid material (J. J. Richlin and T. Kuwabara).

may occur after chronic disease, usually conjunctivitis, such as trachoma,[1] after which the deposits in the lids may be massive (Cascio, 1960) or an old pneumococcal conjunctivitis (Stansbury, 1965). The pathology is the same as in other forms of amyloidosis; calcareous degeneration and bone formation may occur (Hameed and Nath, 1960).

LICHEN AMYLOIDOSUS is a form of amyloidosis which is confined to the skin usually affecting the legs but sometimes elsewhere. The deposits are limited to the subepidermal region of the dermis; they are smaller than those in primary or localized

[1] See Vol. VIII, p. 588, footnote 1; Halasa (1965).

amyloidosis and consist of discrete closely-set papules resembling lichen planus which usually itch severely.

Andrade. *Brain*, **75**, 408 (1952).
Ashton and Rey. *Brit. J. Ophthal.*, **35**, 125 (1951).
Behal. *Brit. J. Ophthal.*, **48**, 622 (1964).
Bisaria, Garg and Sud. *Brit. J. Ophthal.*, **51**, 136 (1967).
Brownstein. *Med. Times* (N.Y.), **96**, 232 (1968).
Brownstein, Elliott and Helwig. *Amer. J. Ophthal.*, **69**, 423 (1970).
Cascio. *G. ital. Oftal.*, **13**, 355 (1960).
Castelló and Pérez. *Vida nueva*, **52**, 215 (1943).
Cohen. *Lab. Invest.*, **15**, 66 (1966).
New Engl. J. Med., **277**, 522, 574, 628 (1967).
Cohen and Calkins. *J. exp. Med.*, **112**, 497 (1960).
Cohen, Gross and Shirahama. *Amer. J. Path.*, **47**, 1079 (1965).
Desai, Sethi and Mehta. *J. Ass. Phycns. India*, **14**, 1 (1966).
Goltz. *Medicine* (Balt.), **31**, 381 (1952).
Gueft and Ghidoni. *Amer. J. Path.*, **43**, 837 (1963).
Guerry and Wiesinger. *Amer. J. Ophthal.*, **49**, 1413 (1960).
Halasa. *Arch. Ophthal.*, **74**, 298 (1965).
Hällén and Rudin. *Acta med. scand.*, **179**, 483 (1966).
Hameed and Nath. *Amer. J. Ophthal.*, **49**, 814 (1960).
Hashimoto, Gross and Lever. *J. invest. Derm.*, **45**, 204 (1965).
Hellier. *Vida nueva*, **71**, 61 (1959).
Heyl. *Trans. St. John's Hosp. derm. Soc.*, **52**, 84 (1966).
Isbell. *J. Tenn. med. Ass.*, **59**, 995 (1966).
Jain and Gupta. *Brit. J. Ophthal.*, **50**, 102 (1966).
Kaufman. *Arch. Ophthal.*, **60**, 1036 (1958).
Kennedy. *J. Path. Bact.*, **83**, 165 (1962).
Kennedy, Henington and McAndrew. *J. Louisiana St. med. Soc.*, **109**, 365 (1957).
Kojima, Watanabe, Niimi and Yamada. *Acta Soc. ophthal. jap.*, **71**, 719 (1967).
Kubik. *v. Graefes Arch. Ophthal.*, **114**, 544 (1924).
Kyle and Bayrd. *Arch. intern. Med.*, **107**, 344 (1961).
Lange. *Sth. med. J.*, **63**, 321 (1970).
Malak and Smith. *Arch. Derm.*, **86**, 465 (1962).
Michelson and Lynch. *Arch. Derm. Syph.* (Chic.), **29**, 805 (1934).
Miescher. *Dermatologica*, **91**, 177 (1945).
Muller, Sams and Dobson. *Arch. Derm.*, **99**, 739 (1969).
Osserman. *New Engl. J. Med.*, **261**, 1006 (1959).
Paton and Duke. *Amer. J. Ophthal.*, **61**, 736 (1966).
Pearse. *Histochemistry, Theoretical and Applied*, 2nd ed., London, 281 (1960).
Pisano. *Ann. Ottal.*, **90**, 338 (1964).
Raehlmann. *Arch. Augenheilk.*, **10**, 129 (1881); **11**, 402 (1882).
Ramsdell and Winder. *Sth. med. J.*, **63**, 822 (1970).
Reimann, Koucky and Eklund. *Amer. J. Path.*, **11**, 977 (1935).
Renard, Dhermy and Nguyen. *Arch. Ophtal.*, **25**, 149 (1965).
Richlin and Kuwabara. *Arch. Ophthal.*, **67**, 138 (1962).
Rogman. *Ann. Oculist.* (Paris), **120**, 89 (1898).
Schmidt. *Zbl. allg. Path. path. Anat.*, **16**, 49 (1905).
Shirahama and Cohen. *Nature* (Lond.), **206**, 737 (1965).
Smith, M.E., and Zimmerman. *Arch. Ophthal.*, **75**, 42 (1966).
Stansbury. *Amer. J. Ophthal.*, **59**, 24 (1965).
Stobbe, Schoffke, Haase and Herrmann. *Dtsch. Gesundh.-Wes.*, **18**, 1295 (1963).
Symmers. *J. clin. Path.*, **9**, 187 (1956).
Teilum. *Ann. rheum. Dis.*, **11**, 119 (1952).
Amer. J. Path., **32**, 945 (1956).
Virchow. *Die Cellularpathologie u. ihrer Begründung auf physiologische u. pathologische Gewebelehre*, Berlin (1858).
Wells. *Brit. J. Derm.*, **64**, 196 (1952).

The Mucinoses

Changes in the ground-substance involving the deposition of mucopolysaccharides such as hyaluronic acid are relatively common in the skin due either to metabolic anomalies or secondary degenerative changes. The only conditions of this type which markedly affect the eyelids are myxœdema and sclerœdema.

MYXŒDEMA

MYXŒDEMA. The *solid œdema* associated with hypothyroidism, first accurately described by William Gull (1874) and named myxœdema by Ord

(1878), is well known, and the puffiness of the lids associated with the expressionless eyes with their narrowed palpebral fissures is one of the most characteristic features of the heavy mask-like face (Figs. 317–8). The baggy lower lids share particularly in this, the skin being dry and rough without any tendency to pit on pressure. The swelling and levelling out of the

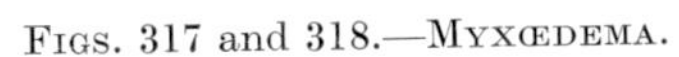

Figs. 317 and 318.—Myxœdema.

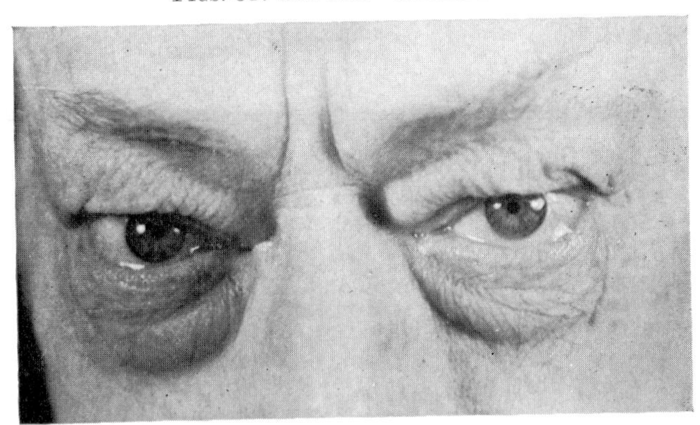

Fig. 317.—In a man aged 62, showing the puffiness of the lids and epiphora 5 years after deep x-ray therapy of the pituitary for acromegaly (W. Hamilton Smith *et al.*).

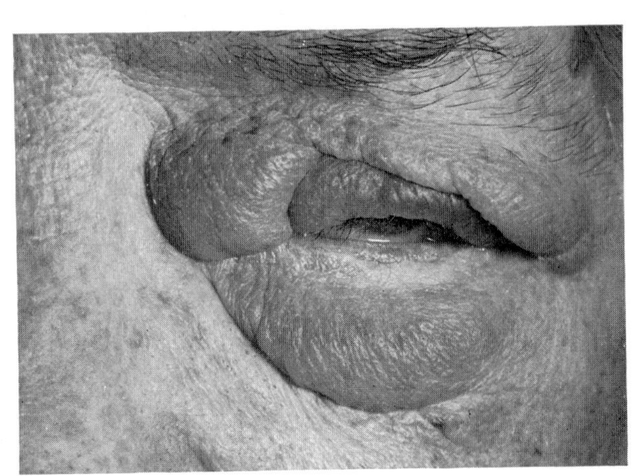

Fig. 318.—Uniocular solid œdema in a case of advanced myxœdema (Inst. Ophthal.).

normal contours are due essentially to infiltration of the cutis with a metachromatic intercellular substance of an acid polysaccharide nature (predominantly hyaluronic acid) associated with the appearance of immature mast cells (Asboe-Hansen, 1950; Gabrilove and Ludwig, 1957). The root-sheaths of the hairs show narrowings and irregular protrusions, possibly due to constriction, so that the follicles degenerate and the hairs fall out (Ord,

1878), a symptom frequently seen first in the lateral third of the eyebrows and lashes (Mahto, 1972) (Fig. 317). The sebaceous glands become swollen at first and later, like the hair-follicles, degenerate, a process well seen in the tarsal glands. In most cases the eyes are slightly exophthalmic (Galli-Mainini, 1942; Zondek, 1953), but occasionally enophthalmos has been noted (Lemoine, 1938; Schen, 1960).

Treatment is by the administration of L-thyroxine.

Asboe-Hansen. *J. invest. Derm.*, **15,** 25 (1950).
Gabrilove and Ludwig. *J. clin. Invest.*, **17,** 925 (1957).
Galli-Mainini. *Ann. intern. Med.*, **16,** 415 (1942).
Gull. *Trans. clin. Soc. Lond.*, **7,** 180 (1874).
Lemoine. *Arch. Ophthal.*, **19,** 184 (1938).
Mahto. *Brit. J. Ophthal.*, **56,** 546 (1972).
Ord. *Med.-chir. Trans.* (Lond.), **61,** 57 (1878).
Schen. *Brit. J. Ophthal.*, **44,** 567 (1960).
Smith, Howsam and Billings. *Brit. J. Ophthal.*, **43,** 622 (1959).
Zondek. *Die Krankheiten d. Endokrinen Drüsen*, Basel, 332 (1953).

SCLERŒDEMA

SCLERŒDEMA ADULTORUM of Buschke (1900–2) is a somewhat rare disease characterized by a diffuse brawny non-pitting œdema and induration of the skin. Despite its name it usually occurs in childhood or early adult life but may appear as late as the seventh decade. The ætiology is unknown but it frequently follows an acute infection (Rook, 1954), typically streptococcal (Greenberg *et al.*, 1963), and usually has a benign course, resolving in some three-quarters of the cases within a few months; relapses, however, may occur and in some cases patches of œdema persist as long as 30 years (in the lids, Adler, 1926). The œdema commences in the face and neck and extends rapidly to the upper part of the trunk, sparing the extremities. Œdema of the lids and conjunctiva is relatively common[1] (Fig. 319).

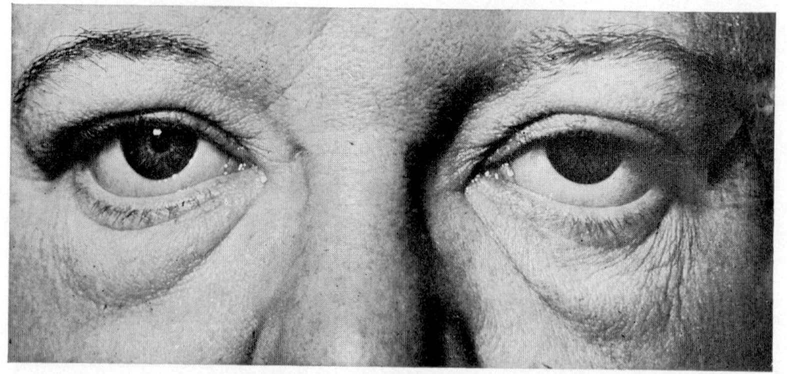

FIG. 319.—SCLERŒDEMA ADULTORUM.
In a woman aged 45 showing the puffy lids and chemosis (G. M. Breinin).

[1] Buschke (1900), Hoffmann (1927), Epstein (1932), Guy and Amshel (1934), Frank (1937), Oliver (1938), Vallee (1946), Breinin (1953), Pegum (1972), and others.

Ocular complications have been few: Stenbeck (1940) described a bilateral orbital infiltration with exophthalmos resulting in blindness, and Breinin (1953) noted the presence of numerous endothelial-lined channels in the conjunctiva and a peripheral punctate keratitis in a case of 8 years' duration. Effusions may occur in other organs such as the tongue, the pleura, the pericardium or the peritoneum; for this reason one case has terminated fatally (Leinwand, 1951).

Histologically a mild chronic non-inflammatory infiltrate is present with swelling and splitting of the collagen bundles; the fluid contains hyaluronic acid in various stages of polymerization.

The pathogenesis of the disease is unknown and several speculative suggestions have been made including endocrine disorders, a " collagen " disease, and others. It seems to be a systemic disturbance following an initial infection, consisting of changes in the mesenchymal ground-substance; Selye (1944) suggested that it represented a response to stress with an out-pouring of adrenocortical steroids related to the œstrogens, a view which has received some support (Duran-Reynals *et al.*, 1950; Breinin, 1953). Curtis and Shulak (1965), on the other hand, considered it to be dependent on an auto-immune mechanism.

The differential diagnosis may be difficult. Scleroderma may resemble it in the early stages, but in sclerœdema the hands and feet are not affected and pigmentation, telangiectasia and atrophy do not occur; the presence of hyaluronic acid is distinctive while the œdema is found more within than between the bundles of collagen.

Treatment is ineffective and the many remedies tried, including heat, vasodilators, antihistamines, hormones, have all been disappointing, as also has the administration of corticosteroids. Fortunately the affection is self-limiting and usually ends in spontaneous recovery.

Adler. *Zbl. Haut- u. Geschl.-Kr.*, **20**, 259 (1926).
Breinin. *Arch. Ophthal.*, **50**, 155 (1953).
Curtis and Shulak. *Arch. Derm.*, **92**, 526 (1965).
Buschke. *Arch. Derm. Syph.* (Wien), **53**, 383 (1900).
Klin. Wschr., **39**, 955 (1902).
Duran-Reynals, Bunting and van Wagenen. *Ann. N.Y. Acad. Sci.*, **52**, 1006 (1950).
Epstein. *J. Amer. med. Ass.*, **99**, 820 (1932).
Frank. *Arch. Derm. Syph.* (Chic.), **36**, 1052 (1937).
Greenberg, Geppert, Worthen and Good. *Pediatrics*, **32**, 1044 (1963).
Guy and Amshel. *Arch. Derm. Syph.* (Chic.), **29**, 777 (1934).
Hoffmann. *Med. Klin.*, **23**, 392 (1927).
Leinwand. *Ann. intern. Med.*, **34**, 226 (1951).
Oliver. *Arch. Derm. Syph.* (Chic.), **37**, 694 (1938).
Pegum. *Proc. roy. Soc. Med.*, **65**, 528 (1972).
Rook. *Postgrad. med. J.*, **30**, 30 (1954).
Selye. *Arch. Derm. Syph.* (Chic.), **50**, 261 (1944).
Stenbeck. *Acta ophthal.* (Kbh.), **18**, 76 (1940).
Vallee. *New Engl. J. Med.*, **235**, 207 (1946).

Connective-tissue Diseases

For some time it has been customary following the work of Klemperer and his colleagues (1942) to group together a number of diseases under the heading of " collagen diseases "—discoid and systemic lupus erythematosus, systemic scleroderma, dermatomyositis, polyarteritis nodosa, Wegener's granulomatosis, rheumatic fever and rheumatoid arthritis. The suggestion was that collagen was primarily at fault and the distinctive pathological appearance was fibrinoid necrosis. There is, however, no evidence that collagen is primarily at fault, nor that fibrinoid necrosis is a specific hypersensitive reaction resulting from the degeneration of collagen; on the other hand, this change is of multiple origin from the mucopolysaccharides of the ground-substance, fibrin and other plasma proteins.[1] The term is therefore best discarded and in the absence of a better, we are using the loose term "connective-tissue diseases" to denote this group in which conspicuous alterations occur in the intermediary substances of the connective tissue in a systemic manner (Klemperer, 1950). Their ætiology is unknown although such factors as hypersensitivity or auto-immunity have been suggested (Criep, 1959; Mackay and Burnet, 1963).

LUPUS ERYTHEMATOSUS

LUPUS ERYTHEMATOSUS is an inflammatory disorder with characteristic pathological changes, possibly an expression of an auto-immune reaction. It is usually divided into two types, discoid and systemic (or disseminated), the former confined to the skin and the latter with widespread systemic implications. It is true that the two have several features in common, the cutaneous lesions may be clinically and histologically indistinguishable and similar hæmatological, biochemical and immunological changes can be demonstrated in both, but the age and sex distributions are strikingly different and they are genetically distinct. It is probable that both are initiated by the occurrence of somatic mutations in lymphocytic stem-cells of predisposed individuals, the genotype in the discoid variety involving one X-linked allele, and in the systemic type three dominant X-linked alleles with three "forbidden clones" of lymphocytes synthesizing cellular auto-antibodies[2] (Burnet, 1959; Burch and Rowell, 1963–66). The two conditions are therefore better considered as two distinct entities.

DISCOID LUPUS ERYTHEMATOSUS

Discoid lupus erythematosus is a benign and very chronic disease localized to the skin, characterized by well-defined reddish or purple scaly patches with dilated capillaries which tend in the course of some years to heal with atrophy, scarring and pigmentary changes. It occurs twice as frequently in females as in males with the highest onset in the fourth decade, but in a series of 1,045 cases studied by Damm and Sönnichsen (1964) 3% began under the age of 15 and 2·5% over the age of 70 years. The disease is world-wide in distribution and is found in some 0·5% of dermatological patients. The essential histopathological characteristics of the disease are thinning of the epidermis with degeneration of its basal layer, hyperkeratosis with plugging of the mouths of the follicles, degenerative changes in the connective tissue consisting of œdema, hyalinization and fibrinoid necrosis, a patchy dermal infiltrate mainly of lymphocytes occasionally reaching granulomatous proportions and dilatation of the blood vessels. The lesions may itch and are usually made worse by sunlight.

The face is commonly attacked and the typical distribution is in the shape of a butterfly, one wing occupying each cheek and lower lid

[1] Vol. VII, p. 162. [2] Vol. VII, p. 220.

FIGS. 320 to 322.—DISCOID LUPUS ERYTHEMATOSUS.

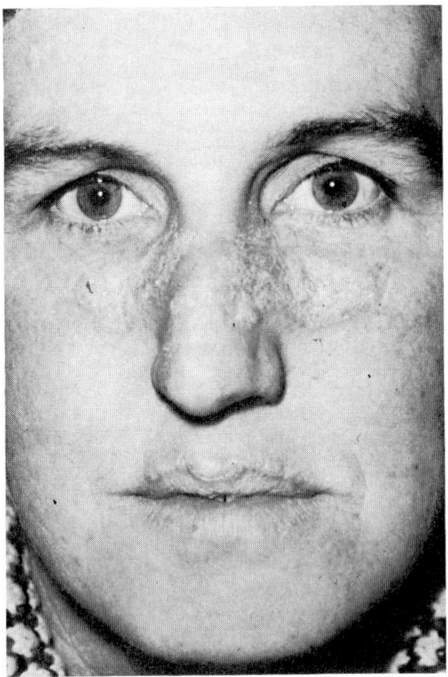

FIG. 320.—Showing the typical butterfly distribution on the face and lesions also on the lip (P. Watson, Addenbrooke's Hospital, Cambridge).

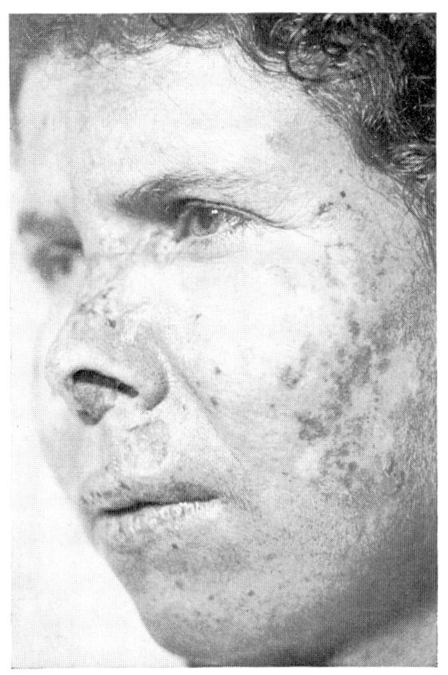

FIG. 321.—Showing sharply marginated, depressed, atrophic, bluish-white areas (J. V. Klauder and P. DeLong).

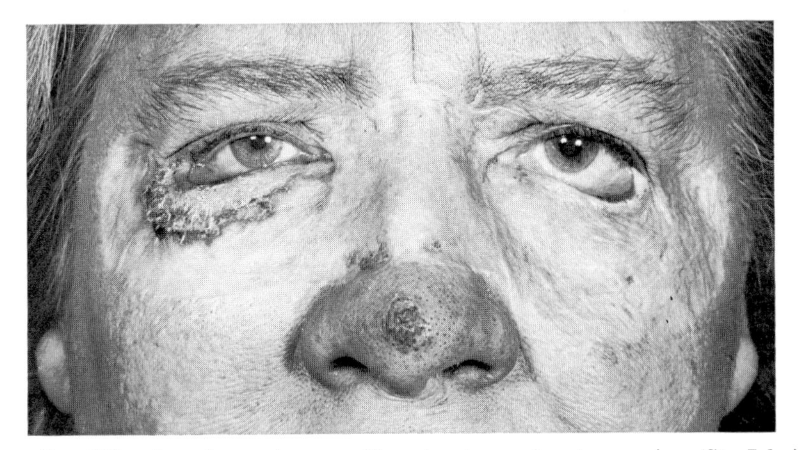

FIG. 322.—An advanced case with ectropion owing to scarring (St. John's Hospital for Diseases of the Skin).

symmetrically, the two being joined by a narrow strip over the bridge of the nose. The gradual development of such a patch can often be followed for several years, the lesions either starting in each cheek and slowly fusing in the midline, or commencing on the nose and spreading outwards. Less typical sites are the temple, the ears, the scalp and the hands and forearms. Considerable scaling may be present, with horny plugs penetrating into the epidermis leaving pits in the subsequent scars. Symptoms are usually absent, but deformities may arise from scarring (Fig. 322).

The lower lids are usually involved in this characteristic distribution, but exceptionally they are affected by isolated lesions.[1] Here these form typical plaques as occur elsewhere on the skin (Figs. 320–1). The lid-margins may also be affected when an appearance resembling blepharitis results[2] (Fig. 322). The margins are dry without matting of the lashes, they have a violet tinge and are covered with fine adherent scales; eventually the lashes fall out and atrophic changes set in, making the ciliary borders irregular. The conjunctiva may be involved wherein the localized lesion exhibits a vivid violaceous velvety appearance; a keratitis or a scleritis is much more rare.[3]

When untreated the lesions tend to be persistent over many years, but in the meantime the general health remains good; relapses are unfortunately common and may be precipitated by trauma, exposure to sunlight or cold, or mental stress, but they are generally less persistent than the initial lesions.

Treatment by salts of bismuth or gold given systemically which used to be the standard method of therapy has now been replaced by oral anti-malarial drugs. Of these, chloroquine sulphate (Nivaquine) is the most effective (200 mg. twice a day initially) but depending on the degree of improvement this should be reduced after six weeks. If this drug is not tolerated owing to nausea and vomiting, the less toxic hydroxychloroquine (Plaquenil) (400 to 600 mg. twice a day) may be substituted, but it is often less effective (Dubois, 1966). With this treatment recent lesions of tumid consistency and little scaling may resolve in a few months but more chronic lesions with much scarring may require prolonged administration; in this event a careful watch should be maintained to anticipate the considerable side-effects which may develop—bleaching of the hair, exfoliative dermatitis, lichenoid rashes, myasthenia, neuropathy, mental disturbances and ocular changes which include corneal deposits and, most important of all, retino-pathy which may involve the bilateral loss of central vision.[4] For this reason throughout the course of treatment frequent tests for visual symptoms are imperative, including an examination of the fundus and tests for colour vision at each visit. That antimalarial drugs are suppressive rather than curative is

[1] Chaillous and Polack (1907), Koenigstein (1911), Chaillous (1912), Thibierge (1923), Lutz (1932), von Grósz (1936), Palić-Szántó (1955), Rossi (1957), Sidi and Mawas (1963).

[2] Chaillous and Polack (1907), Ehrmann and Falkenstein (1922), Klauder and DeLong (1932–36), Örgen and Salma (1955).

[3] Vol. VIII, p. 1099.

[4] Vol. XIV, p. 1275.

suggested by the occurrence of relapses in about 6 months in some 75% of all patients; repeated courses of therapy are therefore frequently required.

Topical treatment with steroids can often control a small lesion or aid systemic therapy if it is not successful within a short period. The application of fluocinolone cream (0·025%) or triamcinolone acetonide cream (0·1%) or betamethasone three times a day frequently acts as a suppressive (Jansen *et al.*, 1965). Alternatively, the intralesional injection of triamcinolone acetonide (0·05 ml. of a suspension of 10 mg./ml.) may be useful in resistant cases. Corticosteroids in small doses given systemically may also be used as an adjunct to supplement treatment by chloroquine if the response to the latter in addition to local treatment is unable to induce resolution of the lesion.

General therapeutic measures should not be forgotten. Trauma and cold should be avoided and if mental stress seems to be an ætiological factor, a mild sedative such as chlordiazepoxide may be advisable. Care should also be taken to avoid sunlight; a broad-brimmed hat and a sun-screening cream (such as benzophenone or para-aminobenzoic acid) or a titanium dioxide powder is suitable.

SYSTEMIC LUPUS ERYTHEMATOSUS

Systemic (disseminated) lupus erythematosus is a rare disease which preferentially attacks young women. Its ætiology is obscure but the evidence suggests that it has a genetic basis and depends on somatic mutations in specific lymphoid stem-cells giving rise to cellular antibodies. Precipitating factors are a bacterial infection, exposure to sunlight or mental stress, or the administration of certain drugs such as hydrallazine, a hypotensive agent, cardio-vascular drugs such as procainamide (Fakhro *et al.*, 1967) and practolol (Raftery and Denman, 1973), and anticonvulsants. The essential pathological changes are fibrinoid necrosis, vascular endothelial thickening, the sclerosis and necrosis of collagen and the presence of hæmatoxylin bodies containing depolymerized nucleic acid (Klemperer *et al.*, 1942). In the skin the histological picture resembles that of discoid lupus erythematosus with thinning of the epithelium, hyperkeratosis with plugging of the hair follicles and glandular orifices, degeneration of the basal-cell layer, perivascular lymphocytic infiltration and dilatation of the superficial vessels (McCreight and Montgomery, 1950; Gardner, 1965).

The course of the disease may be very varied. In chronic cases the cutaneous eruption may be the presenting sign to be followed by systemic disturbances; fulminating cases are characterized by marked constitutional disturbances with fever and malaise and a subsequent involvement of the skin. In all cases the cutaneous erythema is present at some stage in 80% and is the presenting sign in 25%. The most common manifestation is the appearance of successive crops of purplish spots on the face, hands and arms (Fig. 323); occasionally an almost universal eruption appears. These spots coalesce to form large infiltrated purplish-red patches, the most typical site having a symmetrical distribution over the cheeks, lower lids and nose forming the characteristic butterfly pattern. In the most acute cases the patches may become bullous and hæmorrhagic and widespread systemic symptoms occur which include fever, leucopenia, splenomegaly, arthralgias, nephritis with characteristic "wire-loop" lesions of the glomerular vessels, and an atypical non-bacterial verrucous endocarditis with extensive fibrinoid swelling of the cardiac valves (*Libman-Sacks disease,*

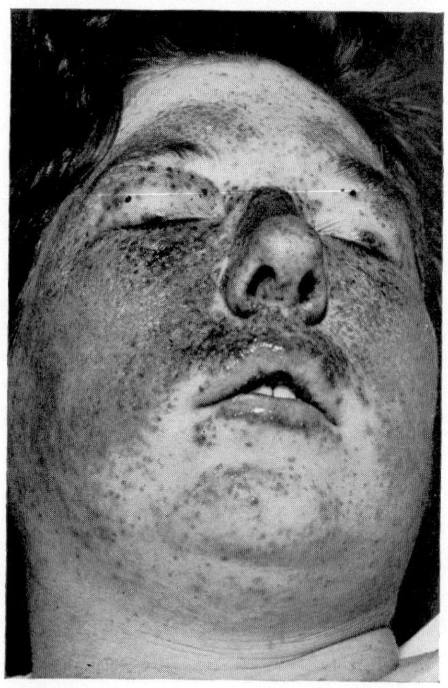

Fig. 323.—Systemic Lupus Erythematosus.
Erythematous lesions and purpura in a fatal case (N. R. Rowell).

1924). The complications which may occur in the uvea[1] and the retina where the typical appearance is of cotton-wool spots at the posterior pole, often as a terminal phenomenon,[2] have already been discussed; the ocular complications were summarized by Gold and his colleagues (1972).

The *diagnosis*, apart from the clinical appearances, depends on laboratory tests, the most valuable of which is the *LE cell test* which is positive in about 80% of cases of systemic lupus erythematosus. It was first established by Hargraves and his colleagues (1948) that LE cells are formed *in vitro* analogous to the hæmatoxylin bodies seen in the tissues. Their presence is due to a factor in the serum of patients (the *LE factor*), an antibody directed against nuclear protein, capable of reacting with polymorphonuclear leucocytes to produce a basophilic cytoplasmic inclusion body which when extruded is phagocytosed by a healthy neutrophil.

[1] Vol. IX, p. 551. [2] Vol. X, p. 502.

The prognosis of systemic lupus erythematosus is very varied; subacute cases may linger on for many years, but the outlook in fulminating cases was poor before steroids were introduced as a method of therapy; but multiple spontaneous remissions tend to occur. Death is most frequently due to renal failure or secondary infection leading to bronchopneumonia. Steroids such as prednisone should be given in high doses which should be reduced to a maintenance dose as quickly as the condition warrants. Cytotoxic drugs such as nitrogen mustard, 6-mercaptopurine and cyclophosphamide have also been used with a view to suppressing delayed-sensitivity reactions (Dubois, 1960; Hill and Scott, 1964; Hadidi, 1970; Cameron *et al.*, 1970; Feng *et al.*, 1970; and others). Antimalarial drugs are less effective than in discoid lupus erythematosus but may well allow the dose of steroids to be reduced, particularly in mild cases, but for continued treatment their side-effects may be dangerous.

Burch and Rowell. *Lancet*, **2,** 507 (1963); **1,** 977 (1966).
Amer. J. Med., **38,** 793 (1965).
Burnet. *The Clonal Selection Theory of Acquired Immunity*, Camb. (1959).
Cameron, Boulton-Jones, Robinson and Ogg. *Lancet*, **2,** 846 (1970).
Chaillous. *Bull. Soc. Ophtal. Paris*, 141 (1912).
Chaillous and Polack. *Bull. Soc. Ophtal. Paris*, 250 (1907).
Criep. *Int. Arch. Allergy*, **14,** 27 (1959).
Damm and Sönnichsen. *Derm. Wschr.*, **150,** 268 (1964).
Dubois. *J. Amer. med. Ass.*, **173,** 1633 (1960).
Lupus Erythematosus, N.Y., 278, 351 (1966).
Ehrmann and Falkenstein. *Arch. Derm. Syph.* (Berl.), **141,** 408 (1922).
Fakhro, Ritchie and Lown. *Amer. J. Cardiol.*, **20,** 367 (1967).
Feng, Jayaratnam, Tock and Seah. *Brit. med. J.*, **2,** 450 (1973).
Gardner. *Pathology of the Connective Tissues*, London (1965).
Gold, Morris and Henkind. *Brit. J. Ophthal.*, **56,** 800 (1972).
von Grósz. *Klin. Mbl. Augenheilk.*, **96,** 636 (1936).
Hadidi. *Ann. rheum. Dis.*, **29,** 673 (1970).
Hargraves, Richmond and Morton. *Proc. Mayo Clin.*, **23,** 25 (1948).
Hill and Scott. *Brit. med. J.*, **1,** 370 (1964).
Jansen, Dillaha and Honeycutt. *Arch. Derm.*, **92,** 283 (1965).
Klauder and DeLong. *Arch. Ophthal.*, **7,** 856 (1932); **16,** 321 (1936).
Klemperer. *Amer. J. Path.*, **26,** 505 (1950).
Klemperer, Pollack and Baehr. *J. Amer. med. Ass.*, **119,** 331 (1942).
Koenigstein. *Wien. klin. Wschr.*, **24,** 1143 (1911).
Libman and Sacks. *Arch. intern. Med.*, **33,** 701 (1924).
Lutz. *Kurzes Hb. d. Ophthal.*, Berlin, **7,** 330 (1932).
McCreight and Montgomery. *Arch. Derm. Syph.*, **61,** 1 (1950).
Mackay and Burnet. *Auto-immune Diseases*, Springfield (1963).
Örgen and Salma. *Birinci Türk. Oftal. Kong. Bült.*, 93 (1955).
Palić-Szántó. *Ophthalmologica*, **130,** 186 (1955).
Raftery and Denman. *Brit. med. J.*, **2,** 452 (1973).
Rossi. *Rass. ital. Ottal.*, **26,** 3 (1957).
Sidi and Mawas. *Ann. Oculist.* (Paris), **196,** 969 (1963).
Thibierge. *Bull. Soc. franç. Derm. Syph.*, **30,** 295 (1923).

DERMATOMYOSITIS

This rare disease, first described by Wagner (1863) and named by Unverricht (1887), is characterized by a specific œdema and eruption on the skin and inflammation and atrophy of the skeletal muscles. The ætiology is unknown; but it is now generally assumed to depend on an anomaly of auto-immunity. It occurs both in children and adults.

The clinical picture is variable; in some cases the dermal condition is the most prominent feature, in others the muscular disability which can occur without cutaneous involvement and is then called *polymyositis*. The rash is diagnostic and the eyelids, the upper parts of the cheeks, the forehead and the temple are frequently affected at an early stage. The œdema and puffiness may be marked, and a peculiar bluish erythema is a characteristic

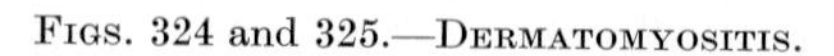

Figs. 324 and 325.—Dermatomyositis.

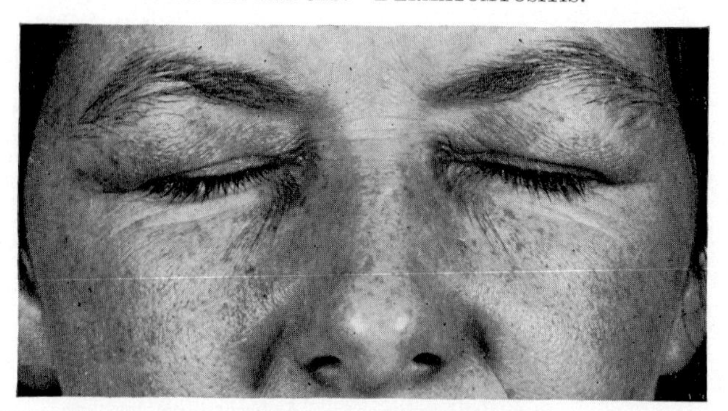

Fig. 324.—Showing œdema and puffiness of the lid and the typical heliotrope appearance (St. John's Hospital for Diseases of the Skin).

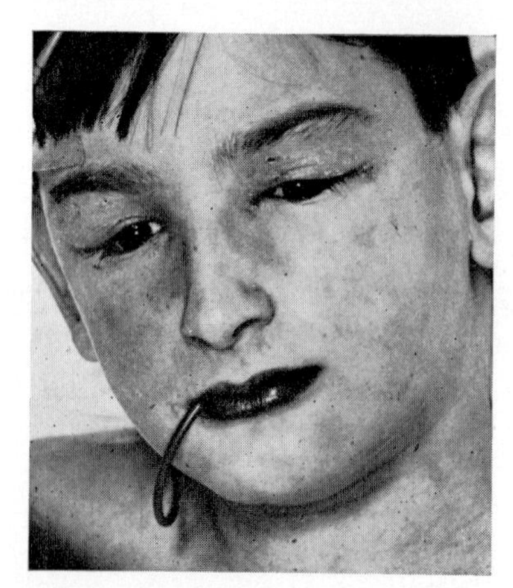

Fig. 325.—The facies of a 10-year-old boy, showing the swollen face and scaly erythematous condition of the skin with its butterfly distribution (S. Munro).

feature, usually described as heliotrope, due to numerous closely set telangiectasias, while on the dry surface a scaly desquamation is common. An involvement of the lids in this manner is frequent[1] (Figs. 324–5). Usually the œdematous areas pit on pressure (Munro, 1959; Manschot, 1961), but they may be firm (Lisman, 1947) and the swelling may be so gross that the

[1] 25 out of 40 cases, O'Leary and Waisman (1940); the most frequent lesion, Hollenhorst and Henderson (1951); 23 out of 26 children, Wedgwood *et al.* (1953).

eyes cannot be opened (Sament and Klugman, 1957). The œdema and the rash with its marked telangiectatic element may spread to the upper limbs and the body, and when resolution occurs the reticulate telangiectatic erythema combined with areas of atrophy and scarring with pigmentation may remain, giving rise to the picture of poikiloderma. Alopecia and calcinosis of the skin may be prominent features, but hypertrichosis may occur.

The myositis may vary greatly in degree and is usually most profound in the muscles of the shoulder and pelvis, but may result in complete prostration and may be associated with considerable pain and tenderness. The orbicularis oculi may be affected, making the lids tender on palpation, as well as the extrinsic ocular muscles.[1] Speech, swallowing and breathing may become difficult so that the usual cause of death is respiratory or cardiac failure.

Ocular complications include episcleritis, iritis and a marked retinopathy of which a characteristic feature is groups of soft cotton-wool patches.[2]

Histopathologically in the skin the œdema affects all the layers of the dermis with an infiltration of lymphocytes, histiocytes and plasma cells, while in the later stages the epidermis is atrophic and in the dermis the collagen shows thickening, homogenization and sclerosis. The affected muscles show a loss of cross-striation and hyalinization with a cellular infiltrate mainly of lymphocytes and in the later stages fibrosis and even calcification and eventually atrophy and sclerosis. The blood vessels undergo intimal proliferation and sometimes thrombosis (Freudenthal, 1940; Dowling, 1955).

The course of the disease is unpredictable. It may be fulminating and some 20% of patients die despite therapy within a year. On the other hand, an acute phase may slowly resolve leaving no muscular weakness and minimal cutaneous changes. Alternatively, particularly in children, a chronic course may result in gross contractures. It is interesting that in some 20% of adults over the age of 40 a carcinoma develops, successful treatment of which may lead to a resolution of the dermatomyositis.

Treatment must include rest, while corticosteroids are the most effective means of therapy given initially in high doses (50 to 80 mg. prednisone daily) to be reduced to a maintenance dose as rapidly as seems wise. Methandienone (10 mg. daily) is an alternative if these are not tolerated (Armstrong and Murdoch, 1960).

Armstrong and Murdoch. *Brit. med. J.*, **2,** 1929 (1960).
Dowling. *Brit. J. Derm.*, **67,** 275 (1955).
Freudenthal. *Brit. J. Derm.*, **52,** 289 (1940).
Hollenhorst and Henderson. *Amer. J. med. Sci.*, **221,** 211 (1951).
Lisman. *Arch. Ophthal.*, **37,** 155 (1947).
Manschot. *Adv. in Ophthal.*, **11,** 1 (1961).
Munro. *Brit. J. Ophthal.*, **43,** 548 (1959).
O'Leary and Waisman. *Arch. Derm. Syph.* (Chic.), **41,** 1001 (1940).
Sament and Klugman. *S. Afr. med. J.*, **31,** 430 (1957).
Unverricht. *Münch. med. Wschr.*, **34,** 488 (1887).
Wagner. *Arch. Heilk.*, **4,** 282 (1863).
Wedgwood, Cook and Cohen. *Pediatrics*, **12,** 447 (1953).

[1] Vol. VI, p. 762.
[2] Vol. X, p. 508.

SCLERODERMA

The two types of scleroderma, systemic and localized, are often classified together but, although there are several similarities between them and there is some discussion on their relationship, they are probably best considered as separate entities, one with systemic manifestations and the other without; the localized form which is confined to the skin is better designated as *morphœa*. Our knowledge, however, is not yet sufficiently advanced to state pragmatically that all forms of scleroderma can be clearly separated clinically into systemic and cutaneous forms.

SYSTEMIC SCLERODERMA (PROGRESSIVE SYSTEMIC SCLEROSIS) is a chronic disease of unknown ætiology mainly affecting females, usually about the fourth decade. It is generally initiated by the features of Raynaud's

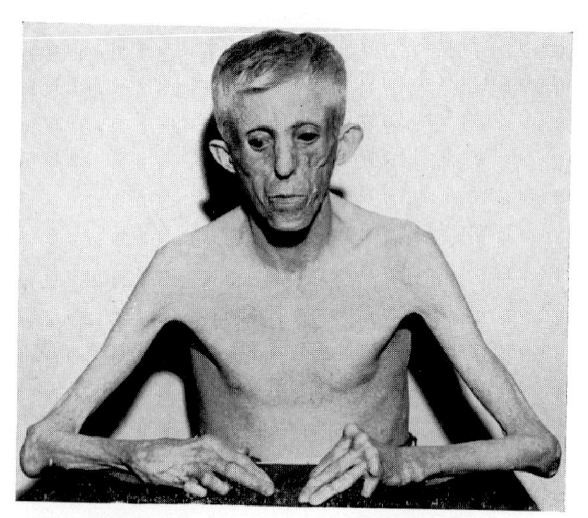

FIG. 326.—SYSTEMIC SCLERODERMA.

Generalized scleroderma in a man aged 61, showing the taut facial skin and severe atrophy of the skeletal muscles; there are arthritic deformities of the hands (W. A. Manschot).

phenomenon which are followed by sclerotic and atrophic changes with a widespread vasculitis affecting the skin and internal organs, adequately described by Jonathan Hutchinson (1896) as *acroscleroderma following Raynaud's phenomenon*, a term changed to *acrosclerosis* by workers at the Mayo Clinic (Sellei, 1934; O'Leary and Waisman, 1943; Brunsting, 1959). The disease is usually but not invariably ushered in by the signs of Raynaud's phenomenon, paræsthesias, painful and swollen joints, generalized myasthenia, malaise and loss of weight. At first the skin may be œdematous and then it gradually hardens, the stiffness appearing in the skin of the fingers and extending up the arms to the upper thorax, the neck and the face and sometimes down the legs to the feet. In the hands a characteristic flexion deformity develops often accompanied by necrosis or gangrene of the finger-tips, indolent paronychias, ulcers and distressing vasospastic attacks, while calcium deposits may break down to liberate chalky material. In the face the skin markings and folds become obliterated, and the expressionless " scleroderma mask " develops wherein the hardened atrophic integument is tightly bound to the underlying tissues and the forehead is smooth and shiny. Tightness of the lids with a woody hardness, blepharophimosis, lagophthalmos, difficulty in elevating the upper lid or depressing the lower so that the conjunctiva cannot be examined may all be features of the condition (Agatston, 1953; Pollack and Becker, 1962; Horan, 1969) (Fig. 326). Reduced lacrimal secretion (Shearn, 1960; Stucchi and Geiser, 1967), kerato-conjunctivitis sicca (Ramage and Kinnear, 1956; Stoltze *et al.*, 1960; Horan, 1969) or a fully developed Sjögren's syndrome may occur (Harrington and Dewar, 1951; Tuffanelli and Winkelmann, 1961; Rodnan, 1963; Kirkham, 1969; Horan, 1969). Finally, telangiectasias and pigmentation may be present.

These relentlessly progressive changes in the skin are usually associated with changes of a similar type in the internal organs. Parenchymatous and interstitial degenerative and inflammatory and fibrotic events occur affecting especially the alimentary tract, particularly the œsophagus leading to dysphagia, the lungs leading to fibrosis, the kidneys where a necrotizing arteritis may occur leading to progressive renal failure, the heart and vascular tree; in the retina cytoid bodies, œdema, cystoid and lipid degeneration have been noted[1] and have been well described by Manschot (1965). The occurrence of cataract is rare (Hartmann *et al.*, 1948; Leinwand *et al.*, 1954; Stucchi and Geiser, 1967).[2] The systemic changes vary in the time of their onset, in their distribution and intensity, for any organ in the body may be affected, and unexpected death may occur at any time from cardiac or renal failure. The prognosis is not good; reviewing 271 cases at the Mayo Clinic, Farmer and his colleagues (1960) found that half the patients had died 5 to 13 years after the diagnosis.

Histopathologically the dermis shows a hyalinization and thickening of the collagen and there is general atrophy. Hyalinization and intimal thickening of the blood vessels often leading to complete occlusion are prominent but fibrinoid changes are rare.

The ætiology is unknown and has been much disputed (Orabona and Albano, 1958; Rodnan, 1963). There is no evidence of an infective or endocrine cause; the calcinosis is probably secondary to the local dystrophy. It is disputed whether the changes in the collagenous tissue or the vascular system are primary, but the two are probably evidences of the same condition which has recently been ascribed to a disorder of auto-immunity (Burch and Rowell, 1963).

No specific *treatment* is known which will alter the course of the disease.

[1] Vol. X, p. 511. [2] Vol. XI, p. 214.

The corticosteroids in small dosage may relieve the articular symptoms but are otherwise ineffective, as also is the host of other drugs advocated from time to time. The only measures available are symptomatic, but it is important to keep the patient warm.

MORPHŒA is sometimes termed *localized scleroderma*, an unfortunate term since it is a condition confined to the skin without any systemic complications and probably a separate disease from systemic scleroderma. It is a disorder of unknown ætiology characterized by sclerosis and atrophy of the skin occurring three times more commonly in women than in men, with a peak incidence between 20 and 40 years of age. The whole thickness of the dermis is affected, first by œdema and a lymphocytic infiltrate and later by hypertrophy and sclerosis of the collagen fibres with atrophy of the glands and hair-follicles.

Clinically the disease may appear spontaneously as plaques or as linear bands, or be generalized. The *plaque* appears as an area, oval or irregular in shape, at first as an itchy patch of red or violet. After some months it loses its colour and is replaced by an indurated smooth area with an ivory centre and a lilac border; the lilac colour is due to dilatation of the blood vessels, and dilated capillaries may traverse the central area and irregular pigmentation may appear. The floor of the plaque may be depressed or elevated or maintain the level of the skin. The patches may disappear spontaneously leaving few traces or an atrophic area may persist resembling the scar of a wound. They may be multiple and bilateral but are asymmetrically distributed and develop anywhere on the body, the limbs and the face; involvement of the lids is comparatively rare (Despagnet, 1895; Mühsam, 1905; Schwab, 1949; and others) (Fig. 327).

The *linear* lesions are similar to the plaques but the lilac rim is absent or presents only at the advancing border. The lesion is usually single and may run down a limb or extend horizontally across the trunk as if following the posterior nerve roots or, following the direction of the supra-orbital nerve, runs from the eyebrow upwards on the forehead into the scalp "en coup de sabre" (Fig. 328); or includes the lid involving a localized loss of the lashes (Fig. 329); in severe cases the lesion may run downwards to involve the cheek, mouth, chin and neck. The underlying tissues may be affected so that a facial hemiatrophy appears with involvement of the bones of the skull, the teeth, gums and jaw.

A variety of ocular lesions has been noted: iritis (Dollfus, 1958; Segal *et al.*, 1960–61), retinal hæmorrhages (Dollfus, 1958; Josten, 1958; Meunier and Toussaint, 1958), phthisis bulbi with uveitis and cataract, an ipsilateral neurocytoma of the optic nerve (Curtis and Jansen, 1958), and pareses of ocular muscles (superior and medial rectus, Cords, 1928).

Generalized morphœa usually starts insidiously on the upper part of the body and the shoulders as large plaques and gradually spreads symmetrically over a period

FIGS. 327 to 329.—MORPHŒA.

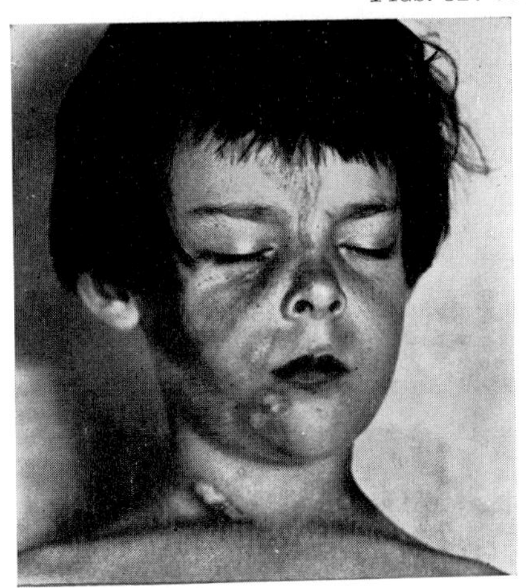

FIG. 327.—The patches affect the right side of the face and there was partial atrophy of the right eyeball (Vanderbilt Clinic, New York).

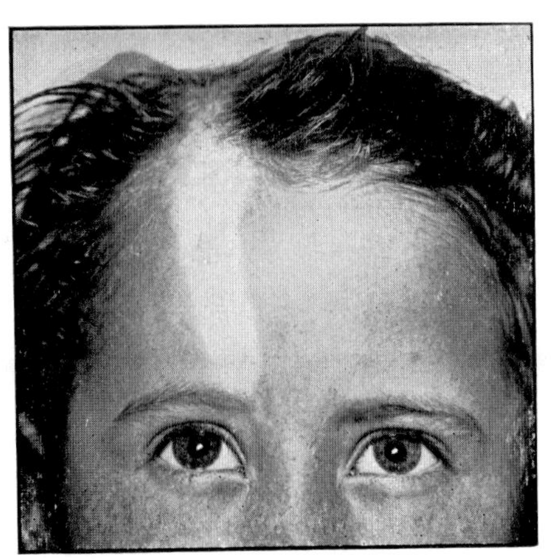

FIG. 328.—There is a white smooth strip *en coup de sabre* adherent to the skull in the area of distribution of the supra-orbital nerve (D. Paterson and A. Moncrieff, *Diseases of Children*, Arnold).

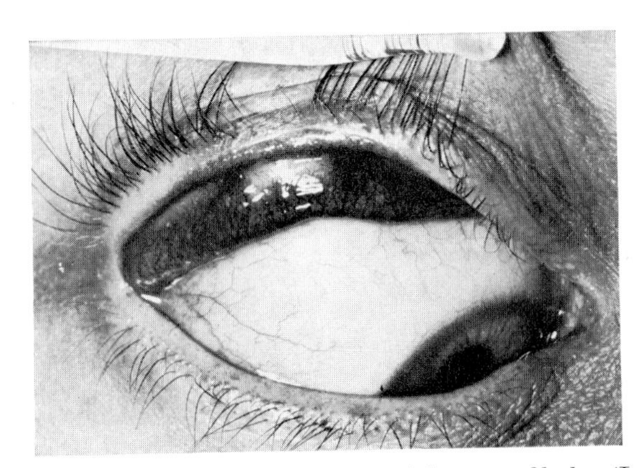

FIG. 329.—Affecting the upper lid with loss of the zone of lashes (Inst. Ophthal.).

of months or years until the whole surface of the body may be involved. The affected skin is bound down to the subcutaneous tissues, is smooth, hairless, glossy and thickened. It is readily injured and trauma is apt to result in ulceration, and pressure-atrophy of the muscles and bones may develop with deformities and contractures. The condition may resolve after some years, or remain stationary, but when it becomes widespread a pitiable hide-bound state results wherein the rigidity of the skin may interfere with swallowing and respiration leading to a fatal termination (Dowling, 1958).

Treatment is ineffective for no specific remedy is known, but symptomatic measures, such as physiotherapy in the generalized condition to prevent contractures, are important. Improvement has been reported with relaxin (Casten and Boucek, 1958), but the results have not yet been adequately assessed (Wells, 1963).

Agatston. *Amer. J. Ophthal.*, **36,** 120 (1953).
Brunsting. *Proc. Mayo Clin.*, **34,** 53 (1959).
Burch and Rowell. *Lancet*, **2,** 507 (1963).
Casten and Boucek. *J. Amer. med. Ass.*, **166,** 319 (1958).
Cords. *Ber. dtsch. ophthal. Ges.*, **47,** 53 (1928).
Curtis and Jansen. *Arch. Derm.*, **78,** 749 (1958).
Despagnet. *Ann. Oculist.* (Paris), **113,** 273 (1895).
Dollfus. *Bull. Soc. belge Ophtal.*, No. 118, 377 (1958).
Dowling. *Med. Press*, **239,** 92 (1958).
Farmer, Gifford and Hines. *Circulation*, **21,** 1088 (1960).
Harrington and Dewar. *Brit. med. J.*, **1,** 1302 (1951).
Hartmann, Collin and Vergne. *Ann. Oculist.* (Paris), **181,** 220 (1948).
Horan. *Brit. J. Ophthal.*, **53,** 388 (1969).
Hutchinson. *Clin. J.*, **7,** 240 (1896).
Josten. *Klin. Mbl. Augenheilk.*, **133,** 567 (1958).
Kirkham. *Brit. J. Ophthal.*, **53,** 131 (1969).
Leinwand, Duryee and Richter. *Ann. intern. Med.*, **41,** 1003 (1954).
Manschot. *Ophthalmologica*, **149,** 131 (1965).
Meunier and Toussaint. *Bull. Soc. belge Ophtal.*, No. 118, 369 (1958).
Mühsam. *Beitr. Augenheilk., Festschr. J. Hirschberg*, 192 (1905).
O'Leary and Waisman. *Arch. Derm. Syph.* (Chic.), **47,** 382 (1943).
Orabona and Albano. *Acta med. scand.*, **160,** Suppl. 333 (1958).
Pollack and Becker. *Amer. J. Ophthal.*, **54,** 655 (1962).
Ramage and Kinnear. *Brit. J. Ophthal.*, **40,** 416 (1956).
Rodnan. *Bull. rheum. Dis.*, **16,** 301 (1963).
J. chron. Dis., **16,** 929 (1963).
Schwab. *Wien. klin. Wschr.*, **61,** 271 (1949).
Segal, Jablonska and Mrzyglod. *Klin. oczna*, **30,** 381 (1960).
Amer. J. Ophthal., **51,** 807 (1961).
Sellei. *Brit. J. Derm.*, **46,** 523 (1934).
Shearn. *Ann. intern. Med.*, **52,** 1352 (1960).
Stoltze, Hanlon, Pease and Henderson. *Arch. intern. Med.*, **106,** 513 (1960).
Stucchi and Geiser. *Docum. ophthal.*, **22,** 72 (1967).
Tuffanelli and Winkelmann. *Arch. Derm.*, **84,** 359 (1961).
Wells. *Trans. St. John's Hosp. derm. Soc.*, **49,** 149 (1963).

IDIOPATHIC ATROPHODERMA

(*Pasini and Pierini*)

This rare condition described by Pasini (1923) and Pierini and Vivoli (1936) appears as sharply demarcated depressed patches of a slate-blue or violaceous colour. Many authors regard it as an idiopathic entity while others consider it a form of morphœa. It is true that a morphœa-like induration may precede (Rupec, 1962) or follow atrophoderma (Miller, 1965), and the two conditions may occur in different areas in the same patient (Kee *et al.*, 1960); on the other hand, two forms have been suggested, one associated with morphœa and the other a separate entity which may be congenital (Ramos e Silva, 1966).

The histopathological changes are slight. An initial homogenization of the collagen bundles with a scattered lymphocytic infiltrate and œdema is followed by induration and atrophy with dense packing of thickened bundles of collagen (Quiroga and Woscoff, 1961; Miller, 1965).

The patches extend slowly, increasing in size for 10 years or more and then persist unchanged. The back and front of the trunk are usually affected, but a case wherein the typical lesion occurred strictly limited to the eyelids was described by Tosti and Pintucci (1957).

No treatment is effective.

Kee, Brothers and New. *Arch. Derm.*, **82,** 100 (1960).
Miller. *Arch. Derm.*, **92,** 653 (1965).
Pasini. *G. ital. Mal. vener.*, **64,** 785 (1923).
Pierini and Vivoli. *G. ital. Derm. Sif.*, **77,** 403 (1936).
Quiroga and Woscoff. *Ann. Derm. Syph.* (Paris), **88,** 507 (1961).
Ramos e Silva. *G. ital. Derm. Sif.*, **107,** 1179 (1966).
Rupec. *Z. Haut- u. Geschl.-Kr.*, **33,** 114 (1962).
Tosti and Pintucci. *Boll. Oculist.*, **36,** 723 (1957).

POLYARTERITIS NODOSA, a widespread necrotizing panarteritis accompanied by a systemic illness, will be discussed more fully among diseases of the orbit.[1] When the

[1] p. 980.

orbital arteries are affected, œdema particularly of the upper lid may be an incidental symptom (Comberg, 1957).

WEGENER'S GRANULOMATOSIS (MALIGNANT GRANULOMA), an almost invariably fatal disease in which a progressive granulomatous destruction of the respiratory tract is associated with a necrotizing arteritis, may invade the orbit through the nasal sinuses.[1] As a result, œdema of the lids with proptosis may be a feature (Pecoldowa and Szmeja, 1970); alternatively the necrosis may spread to the lids (Norgaard and Pindborg, 1960) while the disease may take the form of a necrotizing chalazion (Verrey and Landolt, 1967).

PSEUDOXANTHOMA ELASTICUM, a rare disease affecting the elastic tissue throughout the body, first described by Balzer (1884), which we have already seen to be associated with angioid streaks in the fundus in the Grönblad-Strandberg syndrome,[2] seldom affects the lids; lesions have, however, been noted in the form of yellow flecks at both canthi of each eye (Darier, 1896). The cause of the disease is unknown; pathologically the flecks consist of swollen and degenerated masses of elastic tissue associated with giant cells. Treatment is unsatisfactory.

Balzer. *Arch. Physiol.* (Paris), **3**, 65 (1884).

Comberg. *Klin. Mbl. Augenheilk.*, **130**, 850 (1957).

Darier. *Mh. prakt. Deran.*, **23**, 609 (1896).

Norgaard and Pindborg. *Ugeskr. Laeg.*, **122**, 1533 (1960).

Pecoldowa and Szmeja. *Klin. oczna*, **40**, 107 (1970).

Verrey and Landolt. *Ophthalmologica*, **153**, 309 (1967).

NODULAR FASCIITIS

This is a benign nodular proliferation of connective tissue of unknown ætiology involving the superficial fascia, and from its alarming histological resemblance to a sarcoma and its rapid growth it has been variously called *subcutaneous pseudosarcomatous fibromatosis* or *fasciitis*. It was first described as a clinical entity by Konwaler and his colleagues (1955) and since that time several reports have appeared in the literature.[3] The lesion is usually encountered in the subcutaneous and fascial tissue of the trunk and upper extremities; lesions in the face are rare, but Font and Zimmerman (1966) described 10 cases affecting the neighbourhood of the eye, five of which involved the eyelids, one the eyebrow, and the remainder affected the fascial tissues of the orbit—Tenon's capsule, the muscular sheaths and check ligaments and the periorbital membrane; one situated initially at the limbus grew rapidly to overlie the cornea and invade the anterior chamber, the iris and the ciliary body.

On the lids the lesion appears at any age (from 3 to 81 years) as a subcutaneous lump often associated with some pain and tenderness. It grows rapidly to a size varying from 0·5 to 1·0 cm. in diameter, and in Font and Zimmerman's series excision was practised at periods varying from 10 days to 3 months after it was noticed.

The diagnosis can be made only from the histopathology on excision,

[1] p. 981. [2] Vol. IX, p. 725.

[3] Stout (1960–61) 123 cases; Price *et al.* (1961) 65 cases; Soule (1962) 56 cases; Hutter *et al.* (1962) 70 cases; and others.

and the prognosis is excellent. The lesion seems to be a non-neoplasic proliferation of connective-tissue elements due to an unknown stimulus. The mass consists of plump, stellate or spindle-shaped fibroblasts usually arranged in parallel bundles (Fig. 330) but sometimes haphazardly (Fig. 331). Among these cells there are abundant reticulin fibres and moderate amounts of collagen fibres intermingled with an intercellular myxoid ground-substance and a serous exudate which may be contained in cystic spaces.

FIGS. 330 and 331.—NODULAR FASCIITIS (R. L. Font and L. E. Zimmerman).

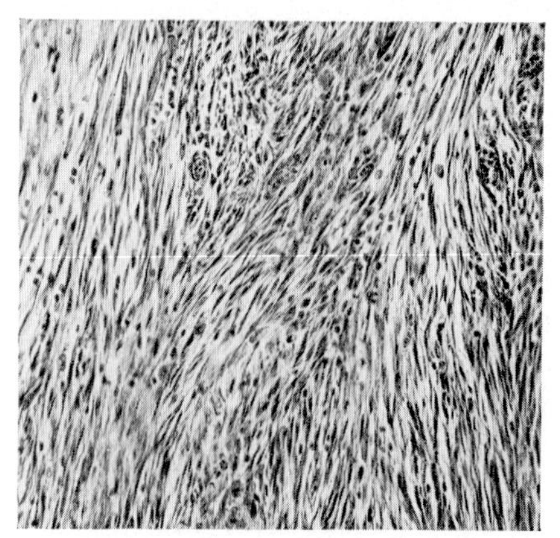

FIG. 330.—An area showing plump or spindle-shaped fibroblasts arranged in parallel bundles and fascicles (H. & E.).

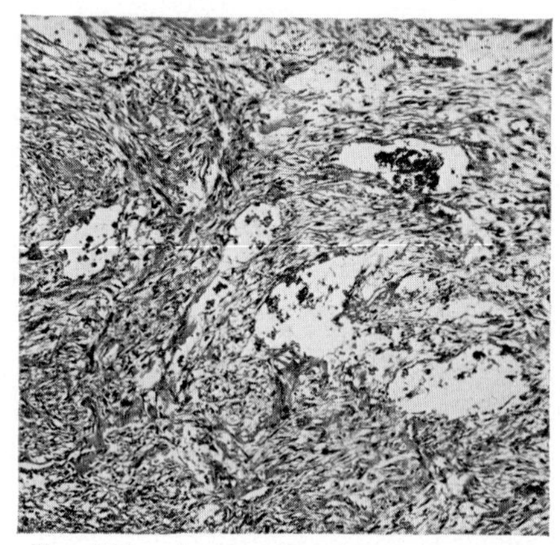

FIG. 331.—An area showing the haphazard arrangement of plump fibroblasts around irregular cyst-like spaces (H. & E.).

Font and Zimmerman. *Arch. Ophthal.*, **75,** 475 (1966).
Hutter, Stewart and Foote. *Cancer,* **15,** 992 (1962).
Konwaler, Keasby and Kaplan. *Amer. J. clin. Path.*, **25,** 241 (1955).
Price, Silliphant and Shuman. *Amer. J. Path.*, **35,** 122 (1961).
Soule. *Arch. Path.*, **73,** 437 (1962).
Stout. *Minnesota Med.*, **43,** 455 (1960).
Cancer, **14,** 1216 (1961).

MALIGNANT ATROPHIC PAPULOSIS (*Degos's Disease*)

This is a rare and usually fatal cutaneo-visceral disease first described by Degos and his colleagues (1942); only 27 cases were traced for analysis up to the time of writing by Howard and his co-workers (1968). It usually occurs in young adult males, and appears with an eruption of dermal papules which suffer necrosis owing to a thrombotic arteriolitis, while similar lesions develop in the intestine and central nervous system. Death usually results from intestinal perforation and subsequent peritonitis or progressive neurological deterioration.

The cutaneous lesions have been observed in the lids (Feuerman, 1966; Howard *et al.*, 1968; Howard and Nishida, 1969) (Fig. 332). They are asymptomatic, unique and diagnostic. They always occur scattered all over the body varying in number from 30 to over 300. They appear first as

FIGS. 332 and 333.—DEGOS'S DISEASE (R. O. Howard and S. Nishida).

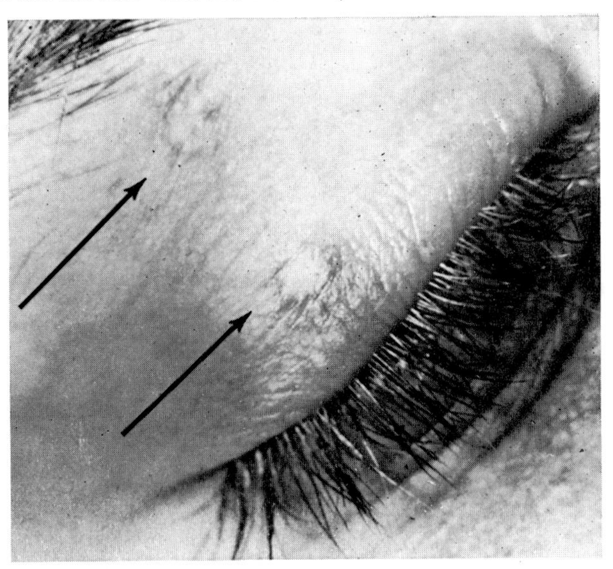

FIG. 332.—In a woman aged 21 showing the characteristic lesions in the upper lid; she also had a lesion on the bulbar conjunctiva.

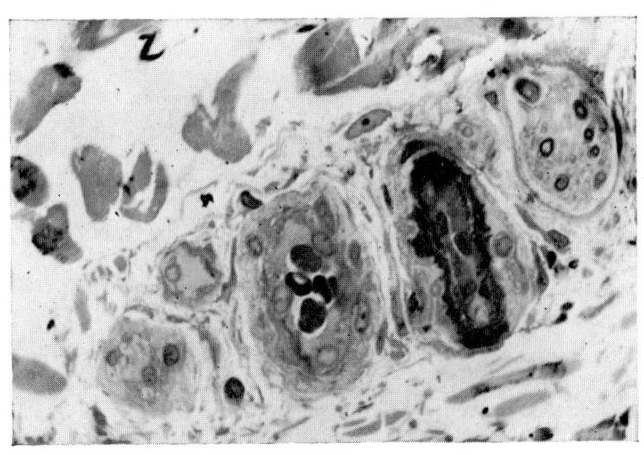

FIG. 333.—The typical histological picture of a lesion on the lid showing the lower third of the reticular dermis with acellular fibrous tissue incorporating a small artery, vein, nerve and lymphatic vessel.

pink or yellow papules measuring from 1 to 6 mm. in diameter; within some days they become umbilicated with porcelain-white centres crowned by scales and with a telangiectatic margin, eventually reaching a diameter of 1 cm. to remain for a period varying from days to several years when they may disappear leaving a scar; subsequent crops, however, frequently appear. Histopathologically the lesions are similar in all sites; the capillaries are narrowed by endothelial proliferation and fibrinoid degeneration affecting the intima; as a result of these changes thrombosis results in ischæmic infarcts (Fig. 333). Inflammatory changes are minimal. The appearance of cytoplasmic inclusions in the endothelial cells seen on electron microscopy suggested the presence of a virus to Howard and Nishida (1969).

Other ocular lesions have been observed: yellow plaques in the conjunctiva[1] and sometimes telangiectasias and micro-aneurysms (Feuerman, 1966), choroidal (Köhlmeier, 1941; Craps, 1959; Sidi *et al.*, 1960; Strole *et al.*, 1967; Howard *et al.*, 1968), and episcleral lesions (Winkelmann *et al.*, 1963; Henkind and Clark, 1968) which can be explained by the presence of multiple infarcts; papillœdema may occur with involvement of the central nervous system.

No effective treatment is known; all conceivable remedies have been proposed and tried with negative results; even the resection of the segment of gut which had perforated merely led to the recurrence of necrotic papules and a very temporary relief. Sidi and his colleagues (1960) found that 7 out of 10 cases were fatal and Howard and his team (1968) 18 out of the 27 reported in the literature.

Craps. *Arch. belges Derm.*, **15**, 188 (1959).
Degos, Delort and Tricot. *Bull. Soc. franç. Derm. Syph.*, **49**, 148 (1942).
Feuerman. *Arch. Derm.*, **94**, 440 (1966). *Harefuah*, **71**, 203 (1966).
Henkind and Clark. *Amer. J. Ophthal.*, **65**, 164 (1968).
Howard, Klaus, Savin and Fenton. *Arch. Ophthal.*, **79**, 262 (1968).
Howard and Nishida. *Trans. Amer. Acad. Ophthal.*, **73**, 1097 (1969).
Köhlmeier. *Arch. Derm. Syph.* (Berl.), **181**, 783 (1941).
Sidi, Reinberg, Spinasse and Hincky. *J. Amer. med. Ass.*, **174**, 1170 (1960).
Strole, Clark and Isselbacher. *New Engl. J. Med.*, **276**, 195 (1967).
Winkelmann, Howard, Perry and Miller. *Arch. Derm.*, **87**, 54 (1963).

Disorders of the Protein Metabolism

Several conditions associated with disturbances of the *metabolism of proteins and amino acids* have incidental manifestations in the eyelids.

GOUT, a hereditary disturbance of purine metabolism, has many systemic implications.[2] Complications in the lids are few, but tophi, aggregations of crystals of monosodium urate, may appear on the skin of the lids, especially near the lid-margins. These are, however, a great rarity and differ in no way from tophi elsewhere (Hirsch, 1899; Meller, 1900; Ebstein, 1912).

ACRODERMATITIS ENTEROPATHICA, a disorder described by Brandt (1936) and Danbolt and Closs (1943), probably transmitted by a recessive gene, is said to depend on a disturbance of the metabolism of tryptophan (Hansson, 1963) or other essential amino acids (Truckenbrodt *et al.*, 1966). It is an insidious disease occurring in early life usually before the end of the first year, characterized by diarrhœa, alopecia and a cutaneous eruption, vesiculo-bullous in form, appearing in crops which may involve the eyelids. The vesicles crust over and are replaced by erythematous patches and scaling, but secondary infection is common, particularly with *Candida*; indeed, some observers consider that the disorder is due to a genetic susceptibility to this organism. The disease tends to improve after puberty if the child survives. Diodoquin (400 to

[1] Vol. VIII, p. 1105. [2] Vol. VII, p. 137.

600 mg. daily) is the most effective treatment, but it is to be remembered that this drug may induce cataract (Berggren and Hansson, 1966).

ALKAPTONURIA is a rare metabolic disease inherited as a recessive character in which the failure of an enzyme leads to the accumulation of homogentisic acid causing a darkness of the urine on exposure to air, a melanin-like pigmentation of the skin, the connective tissue and cartilage (*ochronosis*) and arthritis. In addition to the pigmentation of the sclera and cornea[1] the skin of the lids and forehead may be markedly pigmented while the tarsal plates may appear blue on transillumination.

HARTNUP DISEASE, a metabolic disorder called after the name of the first affected family, transmitted as a recessive trait, is probably due to the malabsorption of tryptophan and other amino acids (Baron *et al.*, 1956; Milne *et al.*, 1960). It is characterized by the development of a pellagra-like rash,[2] a renal aminoaciduria and intermittent symptoms of cerebellar ataxia. The disease starts in childhood but the symptoms decrease with increasing age. The cutaneous eruption affects the areas exposed to light, notably the forehead, cheeks and eyelids where a dry scaly rash appears which on exposure turns erythematous and exudes, the appearance resembling atopic eczema. Sunlight should be avoided and treatment with high doses of nicotinamide usually controls the condition (K. and S. Halvorsen, 1963).

PORPHYRIA, an anomaly of porphyrin metabolism distinguished by the presence of porphyrinuria,[3] may be divided into two classes. (a) Erythropoietic types wherein the metabolic defect is in the bone marrow, in which two forms occur–*congenital porphyria* (*Günther's disease; erythropoietic porphyria*) appearing usually in infancy as a recessively inherited condition, and the dominantly inherited *protoporphyria* occurring in early childhood. (b) Hepatic types wherein the metabolic defect is in the liver, in which three forms occur: *acute intermittent porphyria*, occurring in the third or later decades of life, a frequently fatal disease without cutaneous (and therefore palpebral) complications; *porphyria cutanea tarda* occurring in persons over 40 induced by alcohol and other drugs; and the *combined variegated form* wherein cutaneous and systemic manifestations occur.

In congenital porphyria the outer eye[4] and lids are prominently affected; the usual evidences are the development of bullæ which heal with scarring and hyperpigmentation and result in gross deformities and ectropion with an exposure keratoconjunctivitis, corneal leucomata and scleral ulcers which may perforate (Douglas, 1972); there is hypertrichosis, with thick eyebrows and long cilia. In the late cutaneous form the lids participate in the scarring, hypertrichosis of the face and eyebrows occurs (Kingery, 1966), while elastotic degeneration of the sclera may develop which may lead to perforation (Barnes and Boshoff, 1952; Sevel and Burger, 1971); cataract may be a complication (Calmettes *et al.*, 1966). In the erythropoietic protoporphyria after exposure to sunlight an eruption resembling hydroa æstivale may appear with the formation of vesicles which may or may not heal with scarring. The ocular complications of porphyria have been summarized by Calmettes and his colleagues (1966), Aguade and his co-workers (1969) and Douglas (1972).

Aguade, Mascaro, Galy-Mascaro and Capdevila. *Ann. Derm. Syph.* (Paris), **96,** 265 (1969).
Barnes and Boshoff. *Arch. Ophthal.*, **48,** 567 (1952).
Baron, Dent, Harris *et al.* *Lancet*, **2,** 421 (1956).
Berggren and Hansson. *Lancet*, **1,** 52 (1966).
Brandt. *Acta derm. venereol.* (Stockh.), **17,** 513 (1936).

[1] Vol. VIII, p. 1064.
[2] p. 340.
[3] Vol. VII, p. 135.
[4] Vol. VIII, p. 1069.

Calmettes, Déodati, Bec and Delpech. *Bull. Soc. franç. Ophtal.*, **79**, 569 (1966).
Danbolt and Closs. *Acta derm. venereol.* (Stockh.), **23**, 127 (1943).
Douglas. *Trans. ophthal. Soc. U.K.*, **92**, 541 (1972).
Ebstein. *Dtsch. med. Wschr.*, **38**, 1236 (1912).
Halvorsen, K. and S. *Pediatrics*, **31**, 29 (1963).
Hansson. *Acta derm. venereol.* (Stockh.), **43**, 465 (1963).
Hirsch. Vossius's *Samml. zwangl. Abhdl. Geb. Augenheilk.*, **3** (2) (1899).
Kingery. *J. Amer. med. Ass.*, **195**, 571 (1966).
Meller. *v. Graefes Arch. Ophthal.*, **50**, 63 (1900).
Milne, Crawford, Girão and Loughridge. *Quart. J. Med.*, **29**, 407 (1960).
Sevel and Burger. *Arch. Ophthal.*, **85**, 580 (1971).
Truckenbrodt, Hövels, Sitzmann and Weber. *Ann. pœdiat.* (Basel), **207**, 99 (1966).

Deficiency Diseases

In VITAMIN A DEFICIENCY the skin of the lids may become dry with a heaping of keratinized cells between the lash follicles; in addition to the follicular keratosis the lashes frequently become fine, dry and straight or unusually long (Pillat, 1929; Oomen, 1955; Ten Doesschate, 1972).

PELLAGRA, a disease of widespread distribution probably associated with a deficiency of the vitamin B complex, particularly nicotinic acid (niacin) in the diet, although mainly characterized by gastro-intestinal, nervous and

FIGS. 334 and 335.—PELLAGRA (R. P. Wilson).

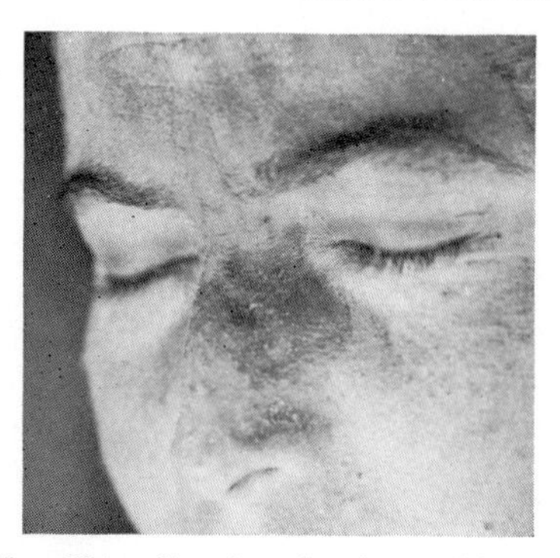

FIG. 334.—Showing the dermatitis on the face and the lids.

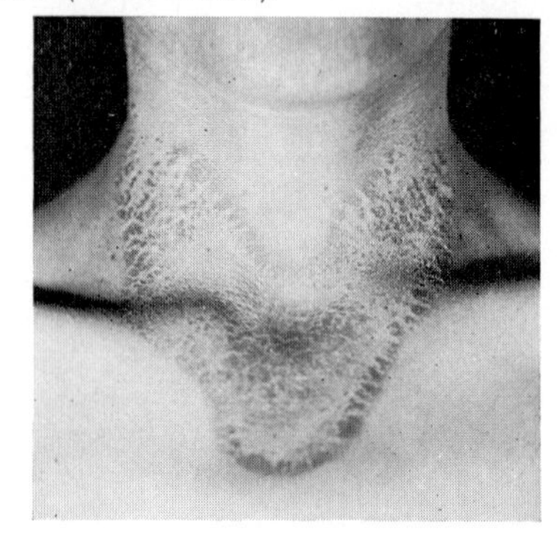

FIG. 335.—Showing Casal's necklace on the neck.

psychological disturbances, is associated also with changes in the skin, from which, indeed, the name of the disease is derived (*pelle*, skin; *agra*, rough). The skin changes, which are seasonal appearing in the spring and dying away in the winter, suggesting that solar energy acts as a traumatizing factor, affect the exposed parts of the body in a symmetrical distribution, spreading as a rule from the cheeks to the lids and the neck (Casal's necklace) (Figs. 334–5). At first an erythema resembling sunburn with redness, swelling and tension appears; later a deep dermatitis develops; eventually on further annual recurrences pigmentation of a magenta colour and thickening occur, and ultimately atrophy. Histopathologically there is initially an inflammatory infiltrate in the upper dermis to be followed by hyperkeratosis and parakeratosis, an increase of melanin and a chronic inflammatory exudate with hyalinization (Moore *et al.*, 1942). It will be remembered that an optic neuritis may also develop[1] as well as conjunctival symptoms[2] and dust-like opacities in the lens (Djacos, 1949). Treatment consists of varying the diet, adding especially meat, milk and eggs, with the specific daily administration of nicotinamide and multivitamins particularly of the B group (Smith *et al.*, 1937; Spies *et al.*, 1935–44; Djacos, 1949; Mathur *et al.*, 1966).

In RIBOFLAVINE (vitamin B_2) DEFICIENCY[3] the dryness and crusted eruptions around the mouth (cheilosis) may also be seen around the eyes

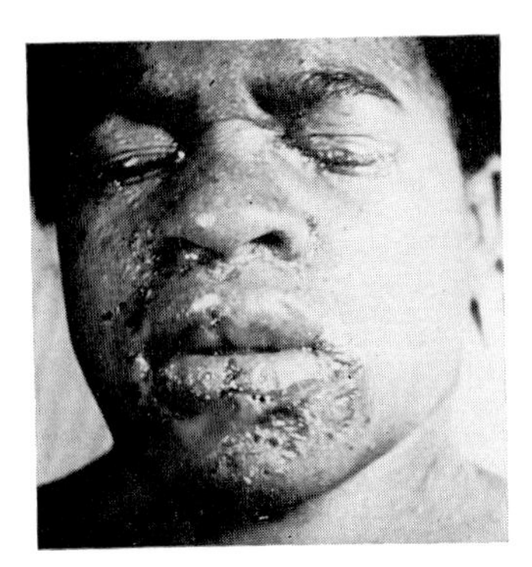

FIG. 336.—RIBOFLAVINE DEFICIENCY.

Showing the cheilosis of the lips and the lesions on the lids giving rise to photophobia, and blepharospasm due to keratitis (V. P. Sydenstricker).

and a scaly dermatitis resembling seborrhœic dermatitis may affect the lids, nose and ears (Sydenstricker *et al.*, 1940) (Fig. 336). Treatment should be by correction of the diet and the administration of riboflavine.

VITAMIN C DEFICIENCY (SCURVY) may be associated with hæmorrhages in the lids, a subject already noted.[4]

In KWASHIORKOR due to a general deficiency of protein[5] there may be a considerable œdema of the lids with chemosis and a " crazy-pavement skin " in the region of the forehead, the brows and the lids (Cockrum *et al.*, 1948; Ten Doesschate, 1972).

Cockrum, Lynch, Slaughter and Austin. *J. Indiana med. Ass.*, **41,** 489 (1948).

Djacos. *Ann. Oculist.* (Paris), **182,** 279 (1949).

Mathur, Shah and Makhija. *Proc. All-India ophthal. Soc.*, **23,** 195 (1966).

Moore, Spies and Cooper. *Arch. Derm. Syph.* (Chic.), **46,** 100 (1942).

Pillat. *Arch. Ophthal.*, **2,** 256 (1929)

Oomen. *Docum. med. geogr. trop.* (Amst.), **7,** 1 (1955).

Smith, D. T., Ruffin and Smith, S. G. *J. Amer. med. Ass.*, **109,** 2054 (1937).

Spies. *J. Amer. med. Ass.*, **104,** 1377; **105,** 1028 (1935).

Spies, Bean and Stone. *J. Amer. med. Ass.*, **111,** 584 (1938).

Spies, Chinn and McLester. *J. Amer. med. Ass.*, **108,** 853 (1937).

Spies, Cogwell and Vilter. *J. Amer. med. Ass.*, **126,** 752 (1944).

[1] Vol. XII, p. 137.
[2] Vol. VIII, p. 1130.
[3] Vol. VIII, p. 1128.
[4] p. 8.
[5] Vol. VIII, p. 1113.

Spies, Cooper and Blankenhorn. *J. Amer. med. Ass.*, **110**, 622 (1938).
Sydenstricker, Sebrell, Cleckley and Kruse. *J. Amer. med. Ass.*, **114**, 2437 (1940).
Ten Doesschate. *Causes and Prevention of Blindness* (ed. I. C. Michaelson and E. R. Berman), London, 164 (1972).

Sarcoidosis

SARCOIDOSIS as it affects the outer[1] and inner eye[2] has already been described where its pathology, its distribution in the body and the enigma of its ætiology have been discussed; in this Volume we shall confine ourselves to its manifestations in the eyelids. It will be remembered that it is a relatively common disease occurring anywhere in the world and may affect almost every organ of the body. The skin changes were first described by Jonathan Hutchinson (1878) who considered them to be a type of lupus,[3] but the condition was recognized as a dermatological entity by Caesar Boeck (1899) (*Boeck's sarcoid*) and was later related by Schaumann (1921–36) to a widely disseminated involvement of the reticulo-endothelial system implicating the skin, lymph nodes, tonsils, bones and internal organs including the lungs and kidneys (*benign lymphogranulomatosis*). It will also be remembered that it is a chronic granulomatous disease histopathologically consisting of non-caseating tubercles comprised of epithelioid cells with giant cells; it may be symptomless and usually pursues an indolent course for many years, often with exacerbations and remissions, sometimes to show spontaneous recovery but occasionally is progressive. Ætiologically it has been associated with various infections, particularly tuberculosis in patients with a high degree of resistance and a low allergy, but the same type of reaction may result from non-organismal factors. It is therefore possible that the condition may represent a peculiar cellular response to a non-specific stimulus depending to some extent on some constitutional immunological defect as yet undefined.

Cutaneous lesions occur in some 40% of all cases of sarcoidosis, sometimes being the only manifestation of the disease and sometimes in combination with lesions elsewhere. A common manifestation is erythema nodosum, a feature of early cases and often associated with hilar lymphadenopathy (Kerley, 1942; James *et al.*, 1956; Uehlinger, 1964). In the lids the most common expression of the disease is a nodular or papular lesion (Fig. 337).[4] It appears as sharply defined reddish or brownish nodules in the deeper parts of the dermis or the subcutaneous tissues, sometimes localized and asymmetrical and sometimes distributed widely over the skin of the face; the more deeply situated lesions are frequently associated with an overlying cyanosis. The condition is symptomless and the nodules may remain for

[1] Vol. VIII, p. 561. [2] Vol. IX, p. 517; Vol. X, pp. 221, 276.
[3] In 1898 (*Arch. Surg.*, **9**, 307) Hutchinson termed the disease *Mortimer's malady* after the name of one of his patients.
[4] Derby and Verhoeff (1917), Lehrfeld (1927), Wilmer (1933), Ernsting (1937), Levitt (1941), Benedict (1949), Attiah and Mortada (1956), Duperrat and Pringuet (1961, occurring 30 years after an injury with a pellet with no systemic signs), Fusco *et al.* (1968).

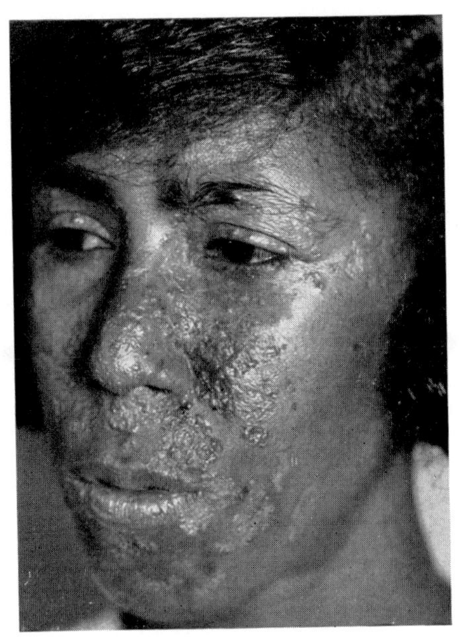

FIG. 337.—SARCOIDOSIS.

Involving the lids, skin of the face, lungs, bones and tendon sheaths (Dept. of Dermatology, Vanderbilt Clinic, N.Y.).

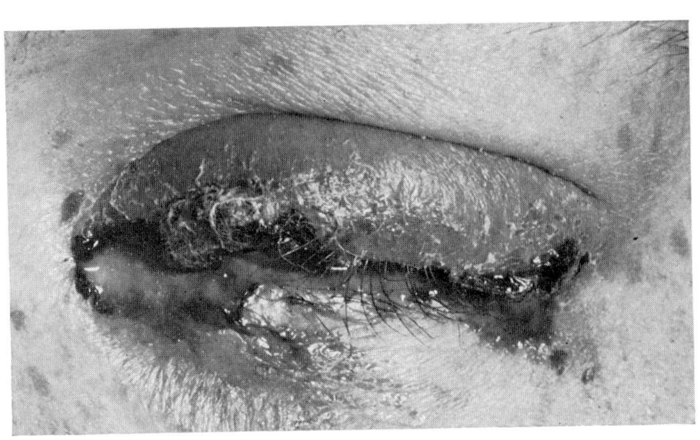

FIG. 338.—SARCOIDOSIS.

Affecting the lids (Inst. Ophthal.).

years without breaking down; if they resolve spontaneously they may leave atrophic scars. The pathological picture is typical of sarcoidosis (Fig. 340). Diffuse extensions of such lesions giving rise to a plaque-like form may affect the lids causing a thickened blepharitic involvement (Fig. 338). Finally, in the chronic stages a more extensive lesion consisting of a vividly red or

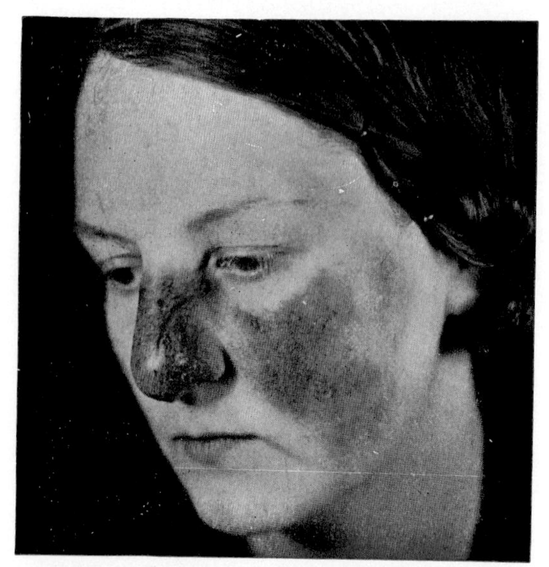

FIG. 339.—LUPUS PERNIO.

The hands, feet, small bones and glands were involved and the Mantoux reaction was negative (Sequeira's *Diseases of the Skin*, Churchill).

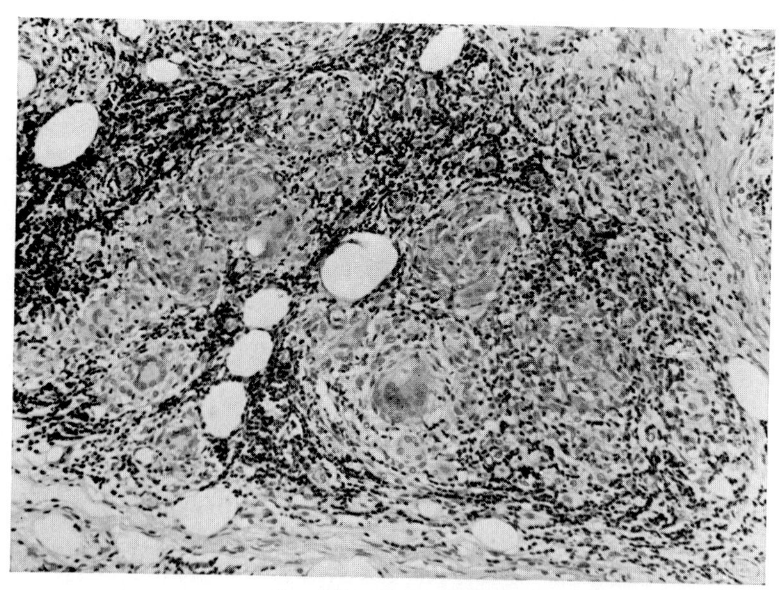

FIG. 340.—SARCOIDOSIS.

The typical appearance of a sarcoid nodule in the lid ($\times 120$) (N. Ashton).

purplish area frequently disposed symmetrically over the cheeks and nose and affecting the lower lid is known as *lupus pernio* (Fig. 339); it also appears on the fingers. The affected parts feel soft, doughy and indurated, showing dilated veins on the surface and often indurated nodules at the edges. Histologically, circumscribed collections of epithelioid cells are seen with occasional giant cells. Cysts of the bones, particularly the phalanges, are common giving rise to fusiform swellings, ocular complications such as uveitis are frequent, and pulmonary involvement almost invariable.

The diagnosis of cutaneous sarcoidosis depends on the clinical appearance, the presence of other typical lesions elsewhere such as in the phalanges and lungs, a frequent rise in the serum globulin to 3·0 or 5·0%, a frequent increase in the serum calcium, an anergy to tuberculin, the typical histological picture on biopsy, and a positive Kveim test—the development of a typical nodule 3 to 4 weeks after the intradermal injection of a sterile saline suspension of tissue from a lymph node or a cutaneous nodule taken from a patient with sarcoidosis; a biopsy of the nodule after six weeks has the characteristic histological picture in positive cases seen in 70% to 75% of active cases.

Since the natural course of sarcoidosis is towards spontaneous healing with fibrosis, the prognosis of cutaneous sarcoidosis is good and, in the absence of the involvement of other tissues, the main disadvantage is cosmetic.

No *treatment* affects the disease except systemic corticosteroids and these are often disappointing in the deep nodular or lupus pernio types. The use of these drugs over a long period, however, is unjustifiable unless the cosmetic disability is serious or systemic complications are grave, affecting the lungs, kidneys or eyes. If the dermal lesions are few and unsightly, the intralesional injection of triamcinolone is simple and often effective (Sullivan *et al.*, 1953), it may flatten nodular lesions and prevent scarring and may be combined with systemic chloroquine (Siltzbach and Teirstein, 1964); relapses, however, tend to occur within six months (James, 1971).

Attiah and Mortada. *Bull. ophthal. Soc. Egypt*, **49,** 95 (1956).
Benedict. *Arch. Ophthal.*, **42,** 546 (1949).
Boeck. *J. cutan. Dis.*, **17,** 543 (1899).
Derby and Verhoeff. *Arch. Ophthal.*, **46,** 312 (1917).
Duperrat and Pringuet. *Bull. Soc. franç. Derm. Syph.*, **68,** 816 (1961).
Ernsting. *Arch. Ophthal.*, **17,** 493 (1937).
Fusco, Vecchione and Romano. *Ann. Ottal.*, **94,** 675 (1968).
Hutchinson. *Illustrations of Clinical Surgery*, London, **1,** 42 (1878).
James. *Dermatology in General Practice* (ed. Fitzpatrick *et al.*), N.Y., 1549 (1971).
James, Thomson and Willcox. *Lancet*, **2,** 218 (1956).
Kerley. *Brit. J. Radiol.*, **15,** 155 (1942).
Lehrfeld. *Amer. J. Ophthal.*, **10,** 255 (1927).
Levitt. *Arch. Ophthal.*, **26,** 358 (1941).
Schaumann. *Acta derm.-venereol.* (Stockh.), **2,** 409 (1921).
Brit. J. Derm., **48,** 399 (1936).
Siltzbach and Teirstein. *Acta med. scand.*, Suppl. 425, 302 (1964).
Sullivan, Maycock, Jones and Beerman. *J. Amer. med. Ass.*, **152,** 308 (1953).
Uehlinger. *Acta med. scand.*, Suppl. 425, 7 (1964).
Wilmer. *Trans. Amer. ophthal. Soc.*, **31,** 59 (1933).

FIG. 341.—HERMANN SCHMIDT-RIMPLER [1838–1915].

CHAPTER V

ATROPHIES, HYPERTROPHIES, DEGENERATIONS, PIGMENTATIONS

IN VIEW of the fact that he was the first to describe accurately senile atrophy of the lids with the fatty bulge which he termed " fatty hernia ", HERMANN SCHMIDT-RIMPLER [1838–1915] (Fig. 341) is a suitable introduction to this Chapter. Born as Hermann Schmidt, he compounded his name when he married Fraulein Hedwig Rimpler. He studied at the Military Institute in Berlin and became chief of the clinic in 1863 under Albrecht von Graefe. In 1871 he became professor of ophthalmology in Marburg, in which city he played an important part in politics, eventually becoming its vice-mayor. In 1890 he was appointed professor in Göttingen and at a later date at Halle where he worked until his retirement in 1910. Schmidt-Rimpler was one of the most distinguished of the German ophthalmologists at the end of the nineteenth century and his writings were profuse. Among the most famous were his extensive contribution on *Glaucom und Ophthalmomalacia* in the first edition of the *Graefe-Saemisch Handbuch* (1873), his classical text-book, *Augenheilkunde und Ophthalmoskopie* (Braunschweig, 1884; 7th ed., 1901) which was translated into English, Italian and Russian, and his treatise on systemic ophthalmology, *Die Erkrankungen des Auges im Zusammenhang mit andren Krankheiten* (Vienna, 1898).

ATROPHIES

Atrophy of the skin may be the result of many conditions, physiological and pathological. Clinically it is characterized by thinning, a loss of elasticity and of the hair follicles; pathologically by a reduction or even a loss of the connective tissue in the dermis. In addition to congenital atrophy,[1] the lids may share in any generalized atrophy of the skin, and they may suffer local atrophy as a result of trauma or any inflammatory process such as tertiary syphilis or some deep mycoses. Finally, atrophic conditions may be confined to the lids themselves, such as blepharochalasis. *Facial hemiatrophy* which affects the lids, among other structures, has been considered in a previous Volume.[2]

SENILE ATROPHY

SENILE ATROPHY is particularly evident in the lax skin of the eyelids, generally developing slowly after the fiftieth year, particularly in thin subjects. The skin, losing its normal turgescence, becomes yellow, thin, harsh and inelastic, a change due chiefly to loss of the normal greasy protective coating of lipids following the aplasia of the sebaceous glands which accompanies the declining activities of the sex hormones. The hair-papillæ

[1] Vol. III, p. 882. [2] Vol. III, p. 1025.

become shrunken and the lashes and brows fall out owing to cornification of the outer root-sheaths; the sebaceous glands are enlarged, largely owing to failure to empty because of relaxation of the muscles round their ducts. In the upper lid the physiological furrows become deepened and prolonged and a pronounced orbital fold hangs loosely over the upper margin (Fig. 342). The lower lids, lacking support owing to relaxation of the tarso-orbital fascia and atony of the fibres of the orbicularis muscle, droop downwards in the condition of senile ectropion[1]; this is aggravated by the accompanying epiphora which, together with exposure, induces a chronic

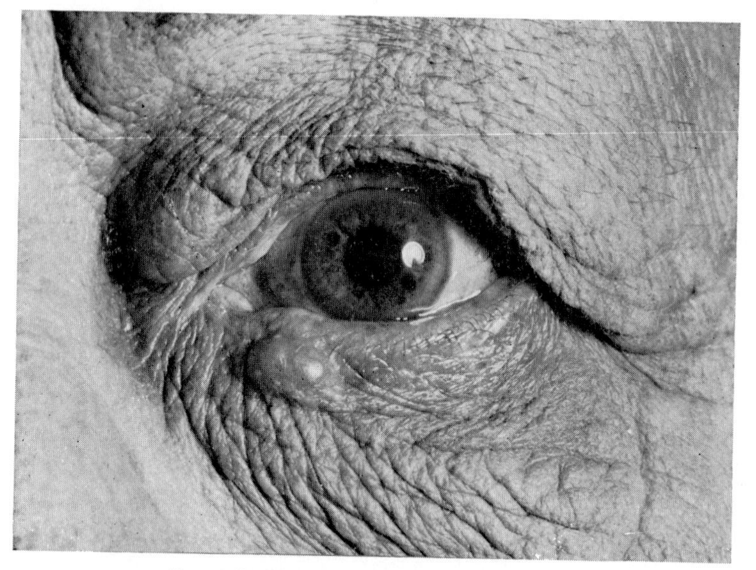

FIG. 342.—SENILE ATROPHY OF THE LIDS.
In a woman aged 82 with a hidrocystoma on the lower lid (Inst. Ophthal.).

hypertrophic conjunctivitis. In fat people a loss of elasticity of the tarso-orbital fascia may allow the orbital fat to herniate into the lids, producing a characteristic bulge in the inner half of the upper lid and a sac-like swelling of the lower (*fatty hernia*, Schmidt-Rimpler, 1899). Histopathologically the dermis is atrophic with thinning of the collagen fibres, but degenerative changes are absent (Hill and Montgomery, 1940).

Treatment of these conditions may be palliative since a considerable improvement in the dry, withering and wrinkling skin can be obtained by the lubricating effect of daily care with a bland lanolin cream.[2] Senile ectropion should be treated operatively, a matter which will be discussed later.

[1] p. 582.
[2] Adipis lanæ, liq. acidi borici 2%, aa. 10; petrol. albi, 5.

SENILE ELASTOSIS. While atrophic changes are the general rule in age, degenerative changes may occur in areas exposed to the sun, involving a disappearance of the collagen fibres and their replacement by granular amorphous material and the aggregation of thick interwoven fibres staining like elastic tissue (*elastosis*). The nature of this elastotic material has been a subject of controversy but it is probable that it is mainly a degenerative product of collagenous fibres and to a less extent of elastic tissue (Keech *et al.*, 1956; Niebauer and Stockinger, 1965). Ulcers have occurred on the lids and a case has been recorded wherein the elastotic degeneration involved the substantia propria of the conjunctiva, vascularization of the cornea and ulceration at the limbus (Offret and Haye, 1959). Owing to its dependence on sunlight the condition may be termed *solar elastosis*.

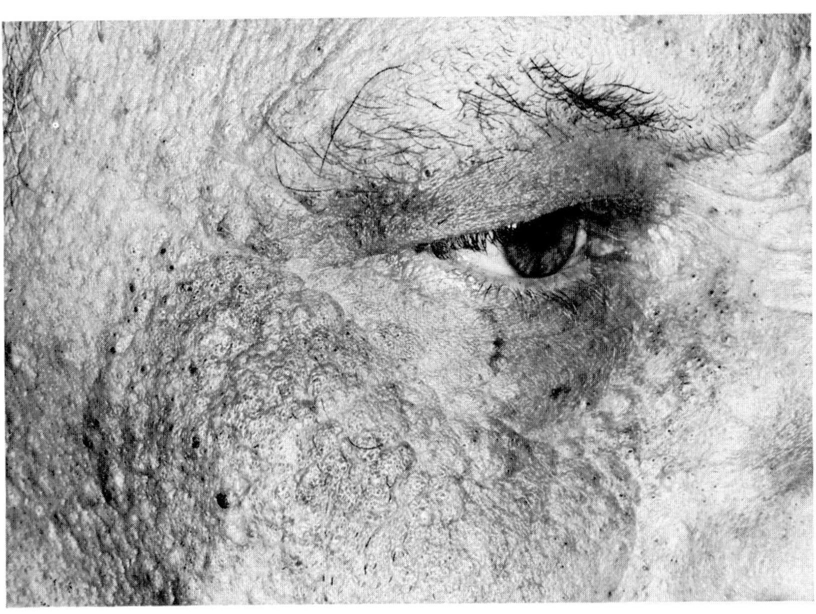

FIG. 343.—THE FAVRE–RACOUCHOT DISEASE.

Nodular cutaneous elastosis with cysts and comedones, of 6 years' duration in a woman aged 50 (Gerd Plewig).

NODULAR ELASTOSIS WITH CYSTS AND COMEDONES is a pronounced solar elastotic degeneration described by Favre and Racouchot (1951) which occurs preferentially in the periorbital region. Raised yellow patches and numbers of comedones appear; the pilosebaceous orifices are enlarged and both they and the cysts are lined by layers of horny material and a lipid substance (Fig. 343). In severe cases treatment by dermabrasion has been recommended (English *et al.*, 1971) and good results have been claimed after the repeated application of 0·1% vitamin A acid in an alcoholic solution twice daily (Plewig and Braun-Falco, 1971; Plewig, 1972). An alternative is plastic surgery with the removal of the diseased skin (Vakilzadeh and Goebel, 1968).

BLEPHAROCHALASIS

BLEPHAROCHALASIS (βλέφαρον, lid; χάλασις, slackening)[1] (DERMATOLYSIS PALPEBRARUM) is characterized by atrophy and relaxation of the tissues of the upper lids following chronic or recurrent œdema of the anterior structures of the orbit. The ætiology is unknown, but it is probably developmental in origin.

An atrophic condition of stretching and relaxation in the upper lid following repeated attacks of œdema was first described by Beer (1817); Sichel (1844) differentiated two types, PTOSIS ATONICA and PTOSIS ADIPOSA; and Fuchs (1896) in his classical description of the anomaly introduced the term BLEPHAROCHALASIS. Other names have been suggested—*ptosis atrophica* (Weinstein, 1909), *acrodermatitis chronica atrophicans* (Herxheimer and Hartmann, 1902; von Michel, 1906; Schreiber, 1924). The literature was collected by Loeser (1908), Weinstein (1909), Weidler (1913), Benedict (1926), Alvis (1935), Panneton (1936), Tapasztó *et al.* (1963), and others.

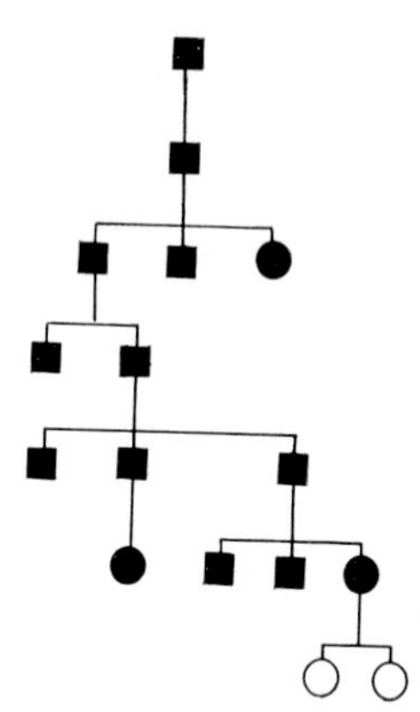

FIG. 344.—BLEPHAROCHALASIS.
Autosomal dominant inheritance (after F. Schulze).

The condition usually occurs in young people especially at puberty—one-third of the cases show evidence of it before the age of 10 and one-half between 11 and 20; it may be congenital (Urbach, 1927) or infantile (age 3, Löhr, 1949), and may be transmitted hereditarily as a dominant characteristic (five generations, Usher, 1932; four, Lafon and Villemonte, 1906; 51 persons in one family, Panneton, 1936; six generations, Schulze, 1965) (Fig. 344). The sexes are equally affected (Heckel, 1921; Michail, 1931).

The malady starts insidiously with intermittent œdematous swellings of the upper lid which are usually misinterpreted until permanent changes occur; the transient attacks last a day or two without pain but with some reddening of the skin resembling angio-œdema. As these succeeding attacks become more frequent, a permanent bagginess with thinning and wrinkling of the skin occurs so that it hangs down in loose folds over the lid-margin (Figs. 345–51). This stage of *intumescence* or *œdema* is followed by one of *atonic ptosis* when the skin, deeply wrinkled, discoloured reddish-brown, heavily venuled and flabby, hangs over the lashes and the power to lift the lid is diminished. Further progress follows relaxation of the tissues of the orbital septum when the orbital fat drops into the relaxed lid, weighing

[1] The term has nothing to do with chalazion: χάλαζα, hail-stone.

FIGS. 345 to 348.—BLEPHAROCHALASIS.

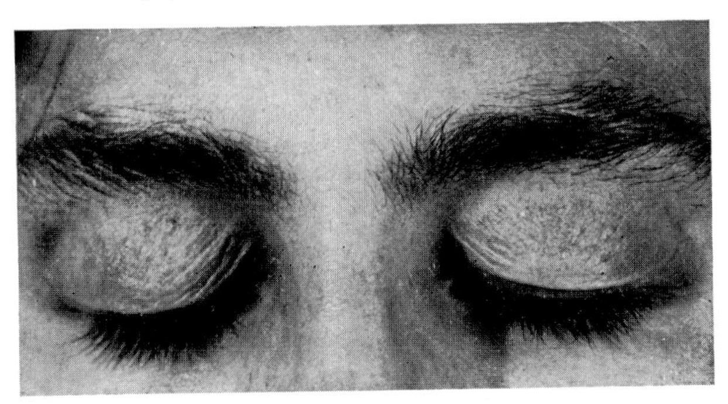

FIG. 345.—A typical early case (A. J. Rook, *Textbook of Dermatology*, Blackwell, Oxford).

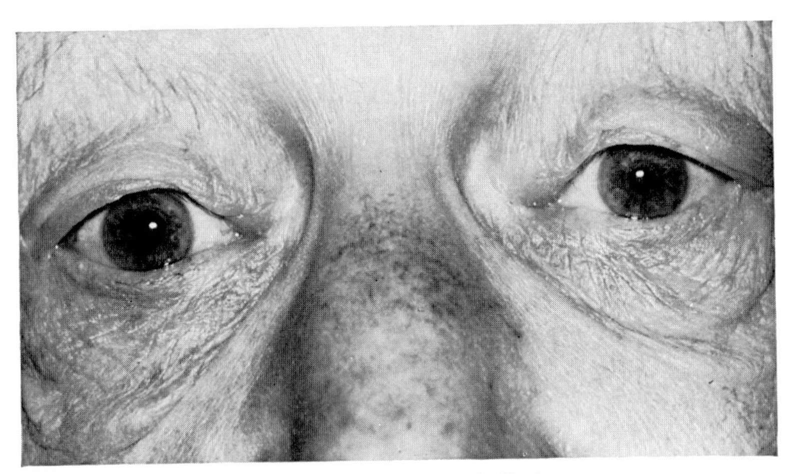

FIG. 346.—A later case (B. Jay).

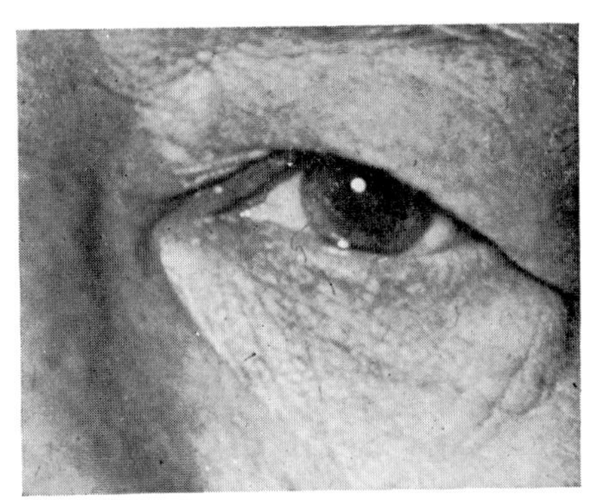

FIG. 347.—In a man aged 74, to show the configuration looking straight ahead.

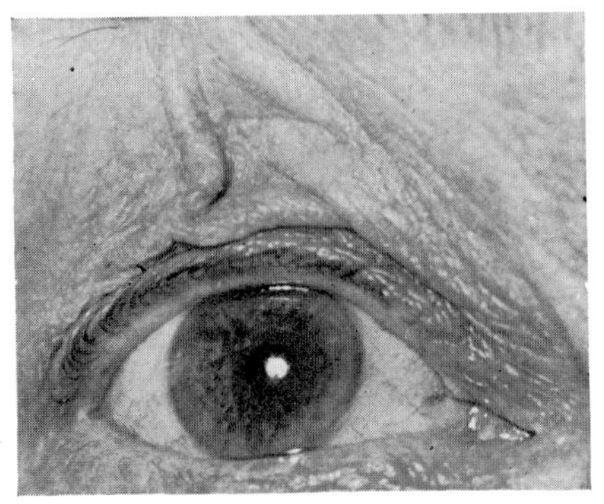

FIG. 348.—The configuration when the skin of the upper lid is raised (F. Schulze).

down the skin in full, heavy, transverse folds and narrowing the palpebral fissure in such a way as sometimes to interfere with vision and pull the lacrimal gland below the orbital margin, giving the face an appearance of tired debauchery; this is the stage of *ptosis adiposa* or *fat hernia*. Sometimes the overhanging lids turn the lashes into the globe (Sayoc, 1957). As a rarity the condition may occur in the lower lid as well as in the upper (Stein, 1930) (Fig. 347), or in the lower lid only (Pintucci and di Tizio, 1963).

Histopathologically there is general atrophy of all the layers of the skin, a loss and fragmentation of elastic tissue and degenerative changes in the

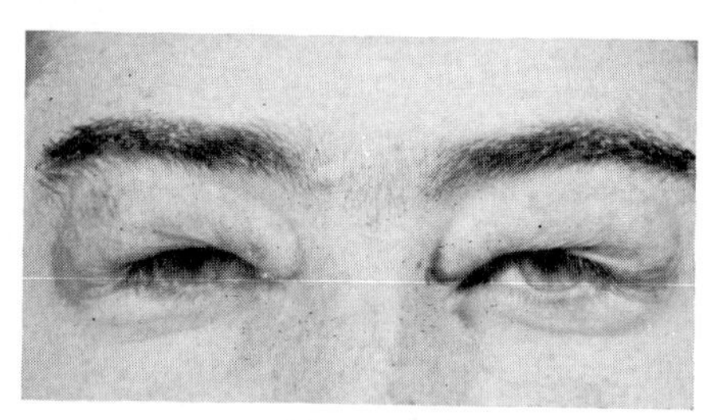

Fig. 349.

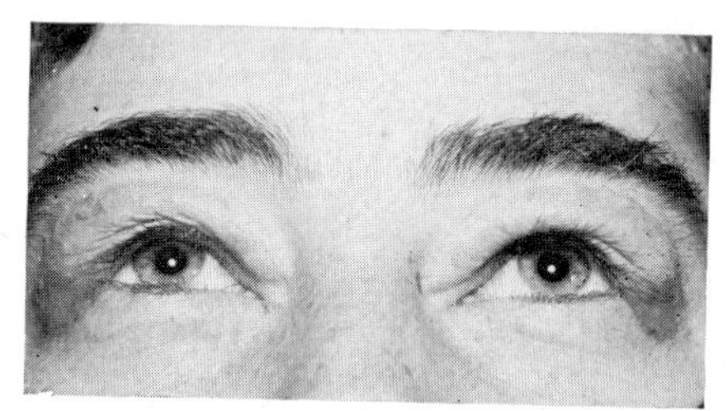

Fig. 350.

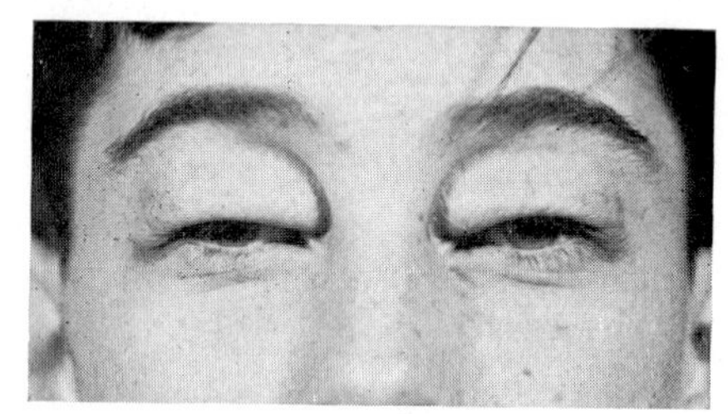

Fig. 351.

Figs. 349 to 351.—Blepharochalasis.

A typical attack in a boy aged 13 (Fig. 349) with a history of 7 years of periodic swelling of the lids. Fig. 350, 13 months after the operative removal of the redundant tissue when another attack of œdema occurred. Fig. 351, 16 months later during a further acute attack of œdema (L. W. Eisenstodt).

collagen, an increase in vascular supply and a proliferation of the capillary endothelium together with a round-celled infiltration somewhat suggestive of a chronic low-grade inflammation[1] (Figs. 352–3). The condition is to be differentiated from a redundant fold of normal skin which falls over the lashes of the upper lid occurring without previous attacks of œdema, a condition called *dermochalasis* by Beard (1969), and *superior epiblepharon* by Paterson and his colleagues (1969). An anomaly has been described in the levator palpebræ muscle, which may be primary or secondary owing to loosening of its anchorage in the skin because of the degeneration in the

[1] Fuchs (1896), Fehr (1898), Verhoeff and Friedenwald (1922), Friedenwald (1923), Stein (1930), Towbin and Adamyk (1931), Tapasztó *et al.* (1963), Gueli *et al.* (1968).

collagenous and elastic fibres (Kreiker, 1929; Schulze, 1965; Kettesy, 1970).

The only effective treatment is correction of the deformity by plastic surgery, excising redundant skin and prolapsed fat or lacrimal gland, attaching the lower margin of the incision to the tarsus, reinforcing the

FIGS. 352 and 353.—BLEPHAROCHALASIS (F. Schulze).

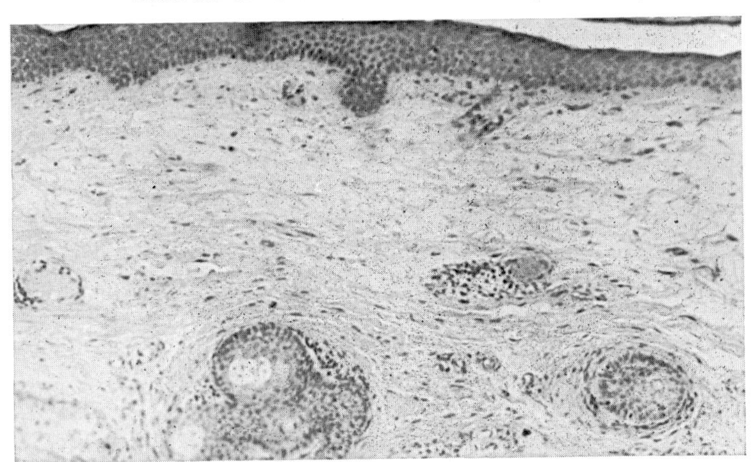

FIG. 352.—Section of the subcutaneous tissues from the upper lid.

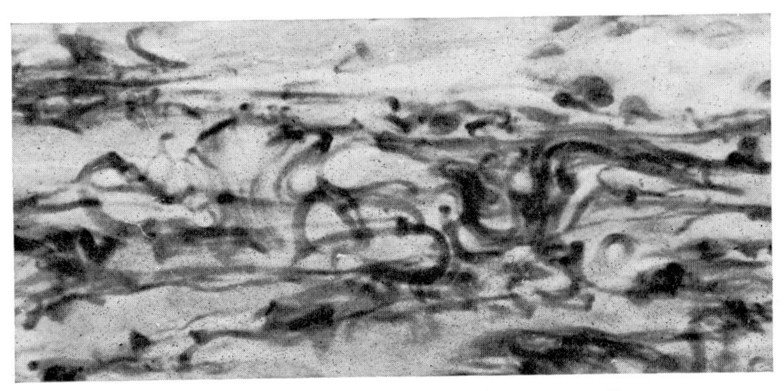

FIG. 353.—The elastic fibres in the subcutaneous tissues.

septum orbitale by suturing to the periorbita, while shortening and restoring the anchorage of the levator tendon may be necessary. If subsequently the attacks of œdema continue, the surgical attack may have to be repeated.

THE (LAFFER-) ASCHER SYNDROME

The occurrence of symmetrical blepharochalasis with a thickening of the upper lip was described by Laffer (1909) in a paper which did not attract international attention, and at a later date the complete syndrome of blepharochalasis, thickening of the upper lip and non-toxic hypertrophy of the thyroid gland was detailed by Ascher (1919–22). Since then several cases have been reported[1]; the presence of goitre is not invariable. The thickened upper lip shows a fold of mucosa hanging down adjoining the gums (*double lip*) (Figs. 354–5). The lower lid and the lower lip may also be slightly involved (Stein, 1930; Franceschetti, 1955).

A curious relationship is the occurrence of this syndrome with a unilateral paralysis of both elevator muscles of the globe[2] (Pirodda and Volpi, 1964; Malbrán, 1969); the presence of two rare syndromes in one patient is interesting.

FIGS. 354 and 355.—THE LAFFER–ASCHER SYNDROME (J. Malbrán).

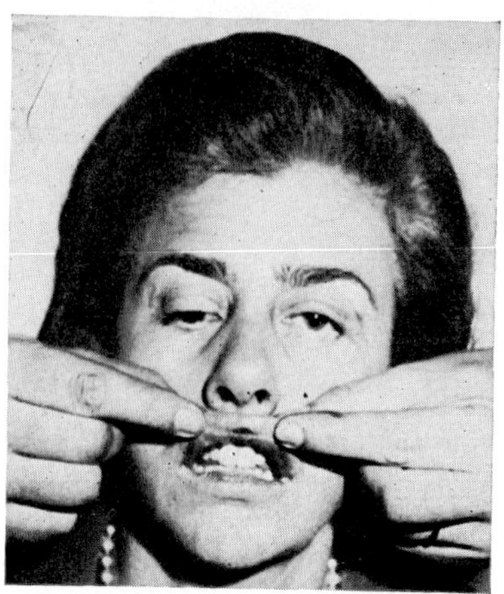

FIG. 354.—To show the blepharochalasis and the double upper lip.

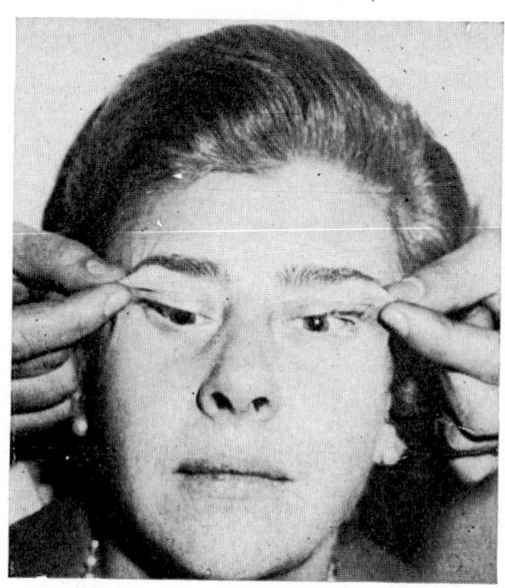

FIG. 355.—To demonstrate the blepharochalasis on raising the upper lids.

FAT HERNIA into the lids occurs with other conditions besides age and blepharochalasis, sometimes no doubt due to a disposition on the part of the fat to protrude, at other times due to laxity of the musculo-fascial structures (Parkes and Griffiths, 1967). It may occur in young people, particularly in the lower lid associated with a large orbital opening and a high globe (*tear sac of the lower lid*); it may also occur rarely in trophoneurosis, thyrotoxicosis, myxœdema, nephritis and uncompensated cardiac lesions (Elschnig, 1930). Pressure on the globe tends to increase the protrusion so that it can be grasped between the fingers. Frequently the herniated fat is separate from the main mass of the orbital fat and, if it is cosmetically desirable, it can usually be successfully excised under local anæsthesia together with the resection of a fold of skin and the restoration of the tonicity of the orbital septum by suturing (Gonzalez-Ulloa and Stevens, 1961; Rees and Dupuis, 1969; Beard, 1969; Obear, 1969; Tenzel, 1969).

Alvis. *Amer. J. Ophthal.*, **18,** 238 (1935).
Ascher. *Dtsch. med. Wschr.*, **45,** 1400 (1919).
Klin. Mbl. Augenheilk., **65,** 86 (1920).
Klin. Wschr., **1,** 2287 (1922).
Beard. *Trans. Amer. Acad. Ophthal.*, **73,** 1141 (1969).
Beer. *Lehre v. d. Augenkrankheiten*, Wien, **2,** 109 (1817).
Benedict. *J. Amer. med. Ass.*, **87,** 1735 (1926).
Bonaccorsi. *Boll. Oculist.*, **38,** 202 (1959).
Eigel. *Dtsch. med. Wschr.*, **51,** 1947 (1925).
Eisenstodt. *Amer. J. Ophthal.*, **32,** 128 (1949).

[1] Wirths (1920), Eigel (1925), Stein (1930), Hartmann (1932), Rosenstein (1932), Klemens (1940), Gallino and Cora Eliseht (1945), Eisenstodt (1949), Bonaccorsi (1959), Segal and Jablonska (1961–62), Orlowski *et al.* (1963), and others.
[2] Vol. VI, p. 722.

Elschnig. *Klin. Mbl. Augenheilk.*, **84,** 763 (1930).
English, Martin and Reisner. *Arch. Derm.*, **104,** 92 (1971).
Favre and Racouchot. *Ann. Derm. Syph.* (Paris), **78,** 681 (1951).
Fehr. *Zbl. prakt. Augenheilk.*, **22,** 74 (1898).
Franceschetti. *J. génét. hum.*, **4,** 181 (1955).
Friedenwald. *Arch. Ophthal.*, **52,** 367 (1923).
Fuchs. *Wien. klin. Wschr.*, **9,** 109 (1896).
Gallino and Cora Eliseht. *Prensa méd. argent.*, **32,** 2423 (1945).
Gonzalez-Ulloa and Stevens. *Plast. reconstr. Surg.*, **27,** 381 (1961).
Gueli, Garcovich and Testa. *G. ital. Derm.*, **109,** 271 (1968).
Hartmann. *Klin. Mbl. Augenheilk.*, **89,** 376 (1932).
Heckel. *Amer. J. Ophthal.*, **4,** 273 (1921).
Herxheimer and Hartmann. *Arch. Derm. Syph.* (Wien), **61,** 57, 255 (1902).
Hill and Montgomery. *J. invest. Derm.*, **3,** 231 (1940).
Keech, Reed and Wood. *J. Path. Bact.*, **71,** 477 (1956).
Kettesy. *Klin. Mbl. Augenheilk.*, **156,** 318 (1970).
Klemens. *Klin. Mbl. Augenheilk.*, **105,** 474 (1940).
Kreiker. *Klin. Mbl. Augenheilk.*, **83,** 302 (1929).
Laffer. *Cleveland med. J.*, **8,** 131 (1909).
Lafon and Villemonte. *Arch. Ophtal.*, **26,** 639 (1906).
Löhr. *Klin. Mbl. Augenheilk.*, **115,** 28 (1949).
Loeser. *Arch. Augenheilk.*, **61,** 252 (1908).
Malbrán. *Klin. Mbl. Augenheilk.*, **155,** 597 (1969).
Michail. *Z. Augenheilk.*, **73,** 337 (1931).
von Michel. *Zbl. prakt. Augenheilk.*, **30,** 75 (1906).
Niebauer and Stockinger. *Arch. klin. exp. Derm.*, **221,** 122 (1965).
Obear. *Trans. Amer. Acad. Ophthal.*, **73,** 1150 (1969).
Offret and Haye. *Bull. Soc. Ophtal. Fr.*, 263 (1959).
Orlowski, Stepniak and Zwierzchowski. *Ann. Oculist.* (Paris), **196,** 362 (1963).
Panneton. *Arch. Ophtal.*, **53,** 729 (1936).
Parkes and Griffiths. *Arch. Otolaryng.*, **86,** 201 (1967).
Paterson, McGavin and Williamson. *Brit. J. Ophthal.*, **53,** 134 (1969).
Pintucci and di Tizio. *Boll. Oculist.*, **42,** 120 (1963).
Pirodda and Volpi. *Ann. Ottal.*, **90,** 577 (1964).
Plewig. *Arch. Derm.*, **105,** 294 (1972).
Plewig and Braun-Falco. *Hautarzt*, **22,** 341 (1971).
Rees and Dupuis. *Plast. reconstr. Surg.*, **43,** 381 (1969).
Rosenstein. *Wien. klin. Wschr.*, **45,** 1017 (1932).
Sayoc. *Amer. J. Ophthal.*, **43,** 970 (1957).
Schmidt-Rimpler. *Zbl. prakt. Augenheilk.*, **23,** 297 (1899).
Schreiber. *Die Krankheiten d. Augenlider*, Berlin, 164 (1924).
Schulze. *Klin. Mbl. Augenheilk.*, **147,** 863 (1965).
Segal and Jablonska. *Ann. Oculist.* (Paris), **194,** 511 (1961).
Klin. oczna, **32,** 31 (1962).
Sichel. *Ann. Oculist.* (Paris), **12,** 187 (1844).
Stein. *Klin. Mbl. Augenheilk.*, **84,** 553, 846 (1930).
Tapasztó, Liszkay and Vass. *Acta ophthal.* (Kbh.), **41,** 167 (1963).
Tenzel. *Trans. Amer. Acad. Ophthal.*, **73,** 1154 (1969).
Towbin and Adamyk. *v. Graefes Arch. Ophthal.*, **126,** 367 (1931).
Urbach. *Zbl. Haut- u. Geschl.-Kr.*, **24,** 743 (1927).
Usher. *Biometrika*, **24,** 1 (1932).
Vakilzadeh and Goebel. *Hautarzt*, **19,** 527 (1968).
Verhoeff and Friedenwald. *Arch. Ophthal.*, **51,** 554 (1922).
Weidler. *J. Amer. med. Ass.*, **61,** 1128 (1913).
Weinstein. *Klin. Mbl. Augenheilk.*, **47** (2), 190 (1909).
Wirths. *Z. Augenheilk.*, **44,** 176 (1920).

HYPERTROPHIES

Hypertrophies of the lids may be due to DIFFUSE NEOPLASIC CONDITIONS such as neurofibromatosis (*elephantiasis neuromatodes*),[1] hæmangiomata (*elephantiasis telangiectodes*)[2] or lymphangiomata (*elephantiasis mollis* or *lymphangiectatica*)[3]; these will be dealt with more conveniently in the section on tumours.

CHRONIC OR RECURRENT INFLAMMATORY CONDITIONS cause the only common form of generalized hypertrophy encountered in the lids. In this category four types are important, three of which are of local and one of general geographical distribution.

[1] p. 508. [2] p. 488. [3] p. 507.

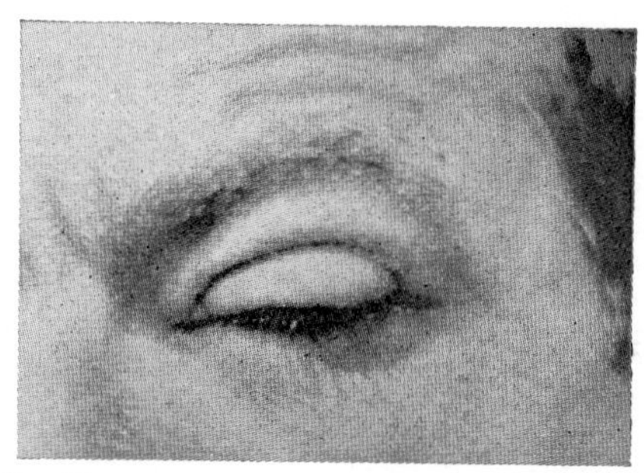

FIG. 356.—ELEPHANTIASIS NOSTRAS OF THE UPPER LID.

Occurring in a man aged 35. The right eye was normal. The left had been affected for a considerable time and for 8 years the upper lid had been gradually and progressively becoming thicker (see Figs. 359 to 361) (M. N. Beigelman).

FIGS. 357 and 358.—ELEPHANTIASIS NOSTRAS OF THE LIDS (M. A. Heintz).

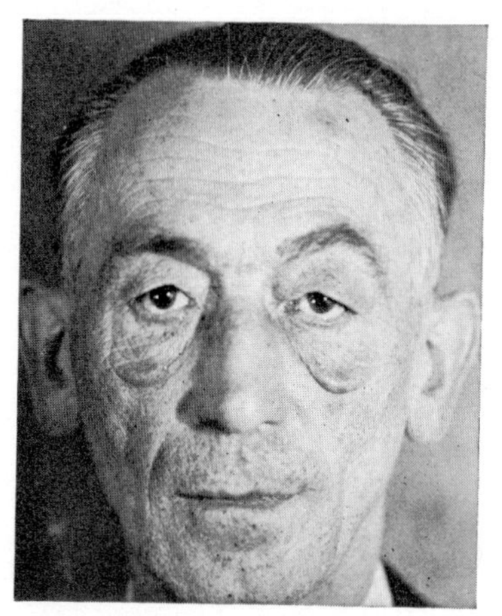

FIG. 357.—In a man aged 52. The hypertrophy of the lower lids had been progressive for 4 or 5 years.

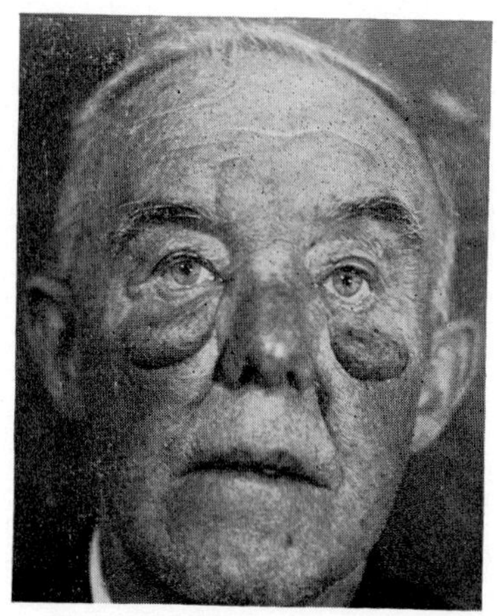

FIG. 358.—A more advanced case in a man aged 50. The swellings of the lower lid were intermittent for 20 years and eventually formed irreducible pouches.

The hideous leonine overgrowths characteristic of nodular leprosy (*elephantiasis græcorum*), very common in leprotic countries, have already been described.[1] As we have seen when discussing filarial infections, the hypertrophy caused by the lymphatic stasis due to the parasite—*elephantiasis lymphangioides* (*arabicum* or *filariosa*)—is rarely encountered in the lids (Becker, 1895). Leishmaniasis of the nose also causes a hypertrophic solid œdema of the lids with the formation of much new connective tissue[2] (*elephantiasis leishmaniana*).

ELEPHANTIASIS NOSTRAS OR SOLID ŒDEMA is a generalized hypertrophy of the cutaneous and subcutaneous tissues of the lids following a chronic or recurrent inflammation and œdema. The condition has not been commonly reported, and Carletti (1924) was able to trace only 24 cases in the literature

[1] p. 112. [2] p. 184.

FIGS. 359 to 361.—ELEPHANTIASIS NOSTRAS (M. N. Beigelman).

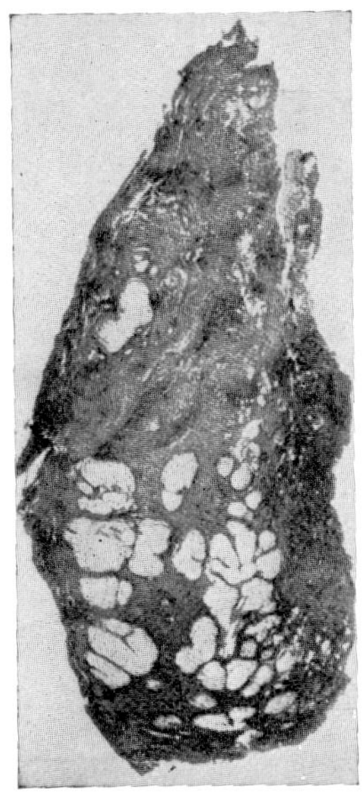

FIG. 359.—Section through the tarsus showing much general thickening.

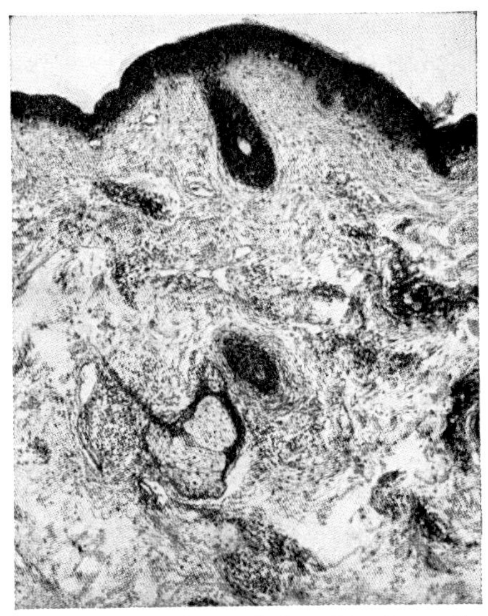

FIG. 360.—Section of the skin from the affected eyelid (Fig. 356) showing diffuse inflammatory infiltration, some œdema and newly formed connective tissue.

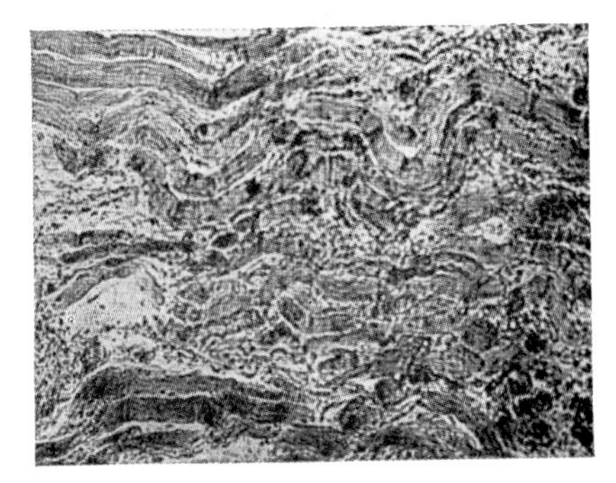

FIG. 361.—Longitudinal section of the orbicularis muscle showing interstitial infiltration and degenerative changes.

up to his time. The most usual cause is a chronic eczema or erysipelas, particularly of the recurrent type, an association which has long been known.[1] It has also been noted after granulomatous tumours of the orbit (Heinc *et al.*, 1966; Suda *et al.*, 1968). The lids appear swollen and œdematous, without, however, inflammatory symptoms and on palpation the skin has a characteristic, solid consistency into which the fingers can sink; it cannot,

[1] Doutrelepont (1879), Pedraglia and Deutschmann (1888), Anderson-Smith (1889), Fage (1892), Critchett (1899), Ormsby (1948), Andersen and Asboe-Hansen (1951), Keidan (1954), Bobb (1955), Georgiadès (1963), Emüler (1965), and others.

however, be pitted or folded. Subjectively the lids feel large and heavy and in severe cases it may be difficult to open the eyes, but the swelling ends abruptly at the line where the fascia is bound down to the deep tissues along the orbital margin. Pain and tenderness are absent (Figs. 15, 356–8).

Pathologically the main gross changes are a dense fibrosis of newly formed connective tissue and an infiltration of the chronic inflammatory type extending throughout the dermis, the subcutaneous tissue and the orbicularis muscle (Figs. 359–61). The enormous thickening is essentially due to an accumulation in the reticular layer of the skin of masses of connective-tissue fibrils arranged in dense partly hyalinized bundles and irregular œdematous networks. Of elastic fibres there is comparatively little increase. Changes in the blood vessels vary; endothelial proliferation and obliteration with thrombosis are sometimes prominent; usually they are increased in number and size. Some histologists have found lymphatic dilatation in the subcutaneous structures; others have seen little change in the lymphatics. The most prominent and invariable feature, however, is a diffuse infiltration by lymphocytes and plasma cells which extends throughout the cutis and the subcutaneous tissues and permeates the orbicularis muscle. Partly because of this infiltration but largely owing to the vastly increased amount of fibrous tissue, there is considerable atrophy of the glandular and follicular elements of the skin and a degeneration of the fibres of the orbicularis and even of the tarsus.[1]

Classically, the condition has been assumed to be due to stagnation of the lymphatic circulation following its blockage by chronic inflammatory processes. There is evidence, however, that the elephantoid state cannot arise from lymphœdema alone, but is dependent rather upon the infection itself, a sequence emphasized by the lack of lymphatic derangement in some cases and the presence of cellular infiltration in them all (Matas, 1913; Alberca, 1924; Bertwistle and Gregg, 1928; Kuntzen, 1930).

Elephantiasis has been noted as a sequel to long-standing trachoma wherein both inflammatory reaction and scarring are very evident (Beigelman, 1932).

Kalt (1909) described a case as fibromatous hypertrophy of the palpebral border wherein a keloid-like thickening of the upper ciliary margin occurred in a child without apparent cause. Rössler (1912) described a case of palpebral elephantiasis following a suppurating infection of the parotid and submaxillary glands; histological examination of excised and redundant lid-tissue showed microscopic changes resembling pachydermia lymphangiectatica.

The *differential diagnosis* must include fibroma, lipoma, angioma, neurofibroma and leucæmic infiltrations. A soft-tissue radiogram, the film being inserted between the lid and the globe, may be of diagnostic value by demonstrating the extensive network of trabecula formed by the subcutaneous fibromatosis (Reichert, 1930). In doubtful cases a biopsy should be made of some excised tissue.

[1] Teillais (1882), Rombolotti (1898), Alberca (1924), Beigelman (1932), Heintz (1947), Grom (1948), Bobb (1955).

Treatment should always include an investigation for the cause of a continued low-grade infection by the elimination of local or focal sepsis and attention to constitutional disorders. Local treatment may be necessary for cosmetic reasons or even to allow the eyes to open easily; in this case surgical removal of the superabundant tissue is indicated.

PACHYDERMO-PERIOSTOSIS

This condition, the *osteodermopathic syndrome of Touraine-Solente-Golé* (1935), initially described by Friedreich (1868), has now been fully annotated.[1] It is a rare mesenchymal disease of unknown ætiology occurring predominantly in males with a familial incidence probably transmitted by an autosomal dominant gene of variable expressivity (Collier, 1962; Rimoin, 1965). The condition has an insidious onset about puberty, progressively increases in severity for 5 or 10 years, and thereafter remains stationary throughout life. It is painless and the main symptom is of excessive sweating due to hyperhidrosis of the hands and feet.

The evidences of the syndrome are gross clubbing of the fingers and toes owing to persistent thickening of the bones (Fig. 362) and a similar enlargement of the extremities of the long bones. The skin is thickened, particularly on the face, the forehead and the scalp which is grossly hypertrophied and thrown into thick folds with a marked seborrhœic hyperplasia resulting in a greasy surface with widely open sebaceous pores giving a most unsightly acromegalic appearance (*cutis verticis gyrata*). The lids are massively enlarged, projecting over the globe like the roof of a penthouse, causing ptosis so marked because of their weight that surgical correction by the excision of an ellipse of the lid may be necessary. It is interesting that there is an occasional association with presenile macular dystrophy (Collier, 1962; Hervouët, 1968; Pietruschka *et al.*, 1969) and lacunar dystrophy of Bowman's membrane (Collier, 1971) or a corneal leucoma (Davidson and Smith, 1970).

Histopathologically the tarsal plate is enlarged partly by the hypertrophy of the meibomian glands and partly by diffuse thickening of the connective tissue of the plate itself. The general structure of the glands is maintained but their ducts are dilated and granulomatous foci with occasional giant cells are numerous associated with localized destruction of glandular tissue. In the dermis the loose areolar tissue is replaced by a compact deposition of collagen fibres forming a thick layer. The normally rudimentary sebaceous glands of the skin are greatly enlarged. The conjunctiva is richly coated with hypertrophic papillæ with numerous crypts in which goblet cells abound (Roy, 1936; Huriez *et al.*, 1962; Davidson and Smith, 1970) (Figs. 363–4). The histological picture in the lids suggests a hyperplasia of the mesenchymal tissue together with a progressive enlarge-

[1] Roy (1936), Rintelen (1937), Lievre *et al.* (1948), François *et al.* (1960), Collier (1962–71), Huriez *et al.* (1962), Hervouët (1968), Pietruschka *et al.* (1969), Davidson and Smith (1970).

FIGS. 362 to 364.—PACHYDERMO-PERIOSTOSIS (J. N. Roy).

FIG. 362.—A patient aged 37 showing bilateral symmetrical hypertrophy of the tarsi making the lids stand out of contact with the eyeball; the pachydermic skin is very obvious in the face and generalized osteo-periostosis is seen in the enlarged wrists, hands and fingers.

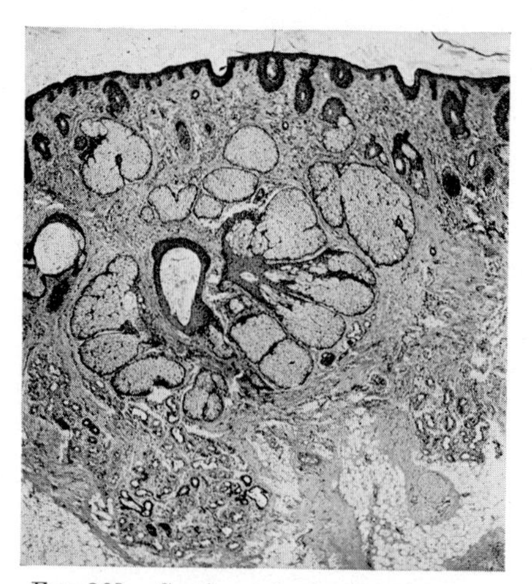

FIG. 363.—Section of the hypertrophied skin with many hair follicles and gigantic multilobulated sebaceous glands. Deep in the dermis there are groups of enlarged sweat glands.

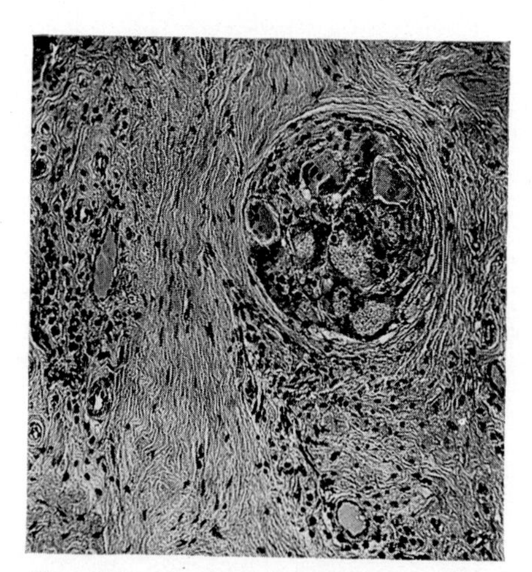

FIG. 364.—Section through the upper lid. A group of giant cells contains the remains of a degenerated sebaceous gland and to the left of this there is a mass of fibrous tissue on either side of which are columns of macrophages.

ment and degeneration of the tarsal glands which excite a granulomatous foreign-body reaction.

Rosenthal-Kloepfer syndrome. It is interesting that in acromegaly a somewhat similar appearance may occur with cutis verticis gyrata, a hypertrophy of the nose and supra-orbital ridges with clubbing of the fingers. A syndrome, transmitted by an autosomal dominant gene, wherein this acromegaloid appearance, without ptosis, was combined with bilateral developmental corneal leucomata was described by Rosenthal and Kloepfer (1962).

Alberca. *C.R. Soc. Biol.* (Paris), **91,** 813 (1924).
Andersen and Asboe-Hansen. *Acta ophthal.* (Kbh.), **29,** 487 (1951).
Anderson-Smith. *Illustr. med. News,* 197 (1889).
Becker. *v. Graefes Arch. Ophthal.,* **41** (3), 169 (1895).
Beigelman. *Arch. Ophthal.,* **8,** 218 (1932).
Bertwistle and Gregg. *Brit. J. Surg.,* **16,** 267 (1928).
Bobb. *Amer. J. Ophthal.,* **40,** 730 (1955).
Carletti. *Elefantiasi delle palpebre,* Roma (1924).
Collier. *Ann. Oculist.* (Paris), **195,** 721 (1962). *Boll. Oculist.,* **50,** 3 (1971).
Critchett. *Trans. ophthal. Soc. U.K.,* **19,** 7 (1899).
Davidson and Smith. *Trans. ophthal. Soc. U.K.,* **90,** 539 (1970).
Doutrelepont. *S.B. niederrhein. Ges. Natur- u. Heilk.,* Bonn, 166 (1879).
Emüler. *Oto Nöro Oftal.,* **20,** 65 (1965).
Fage. *Ann. Oculist.* (Paris), **107,** 276 (1892).
François, P., Clay, Asseman and LeRoyer. *Bull. Soc. belge Ophtal.,* No. 124, 924 (1960).
Friedreich. *Virchows Arch. path. Anat.,* **43,** 83 (1868).
Georgiadès. *Arch. ophthal. Hetair. borei hellad.,* **12,** 25 (1963).
Grom. *Trans. ophthal. Soc. U.K.,* **68,** 255 (1948).
Heinc, Svec and Cerný. *Cs. Oftal.,* **22,** 455 (1966).
Heintz. *Bull. Soc. belge Ophtal.,* No. 87, 147 (1947).
Hervouët. *Travaux d'anatomie pathologique oculaire* (4th series), Paris, 138 (1968).
Huriez, François, P. and Agache. *Ann. Derm. Syph.* (Paris), **89,** 372 (1962).
Kalt. *Ann. Oculist.* (Paris), **141,** 189 (1909).
Keidan. *Lancet,* **1,** 79 (1954).
Kuntzen. *Arch. klin. Chir.,* **158,** 543 (1930).
Lievre, Breton, Bloch-Michel and Betourne. *Bull. Soc. méd. Hôp. Paris,* **64,** 954 (1948).
Matas. *Amer. J. trop. Dis.,* **1,** 60 (1913).
Ormsby. *Arch. Derm. Syph.* (Chic.), **57,** 463 (1948).
Pedraglia and Deutschmann. *v. Graefes Arch. Ophthal.,* **34** (1), 161 (1888).
Pietruschka, Bibergeil, Schill *et al. Klin. Mbl. Augenheilk.,* **154,** 525 (1969).
Reichert. *Arch. Surg.,* **20,** 543 (1930).
Rimoin. *New Engl. J. Med.,* **272,** 923 (1965).
Rintelen. *Z. Augenheilk.,* **92,** 1 (1937).
Rombolotti. *Arch. Augenheilk.,* **36,** 301 (1898).
Rosenthal and Kloepfer. *Arch. Ophthal.,* **68,** 722 (1962).
Rössler. *Klin. Mbl. Augenheilk.,* **50** (2), 325 (1912).
Roy. *Ann. Oculist.* (Paris), **173,** 637 (1936).
Suda, Okamura and Otsubo. *Folia ophthal. jap.,* **19,** 1000 (1968).
Teillais. *Arch. Ophtal.,* **2,** 42 (1882).
Touraine, Solente and Golé. *Presse méd.,* **43,** 1820 (1935).

DEGENERATIONS

DEGENERATIONS OF THE SKIN

COLLOID MILIA

This condition, first described and named by Wagner (1866), is characterized by numerous round yellow nodules on the skin, usually 1 to 2 mm. in size. They occur on areas exposed to the sun—the face, nose, forehead and the dorsa of the hands; the lids may be especially affected (Philippson, 1891), the eyebrows (Zaun, 1966) or the lower lid and the nose (Prakken, 1951) (Fig. 365). Two forms have been differentiated, a juvenile form which starts before puberty and is dominantly inherited and an adult form associated with senile elastosis (Percival and Duthie, 1948). The nodules may be closely packed but they do not usually coalesce; occasionally large plaque-like deposits occur to which the term *colloid degeneration in plaques* has been applied (Reuter and Becker, 1942).

The nodules have a translucent appearance and on expression may discharge a gelatinous material. In colloid milia the masses of colloid accumulate in individual papillæ (Prakken, 1951); in the plaque-type it is spread diffusely throughout the dermis (Sullivan and Ellis, 1961). The colloid material belongs to the family of hyalin and is probably a form of collagenous degeneration (Pirreda, 1960). The diagnosis depends on biopsy. Treatment has been suggested by destruction of the nodules by electrolysis, carbon dioxide snow or the curette. Vitamin C has been said to have a favourable effect (Robinson and Tasker, 1945) but others have found that this or any other vitamin is equally useless (Percival and Duthie, 1948; Agius, 1963).

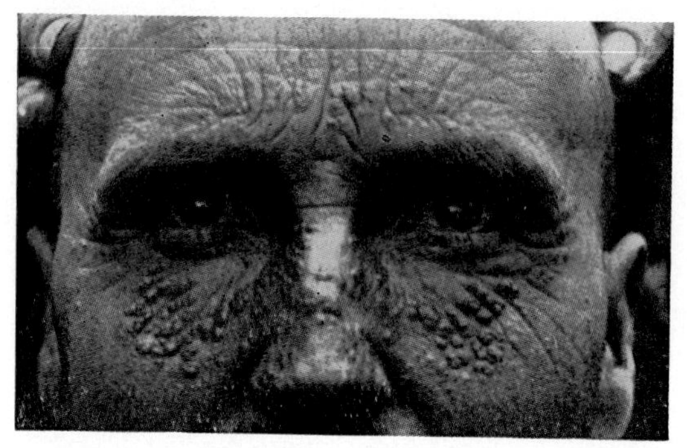

Fig. 365.—Colloid Milia (R. L. and R. L. Sutton).

DEGENERATIONS OF THE TARSUS

FATTY DEGENERATION OF THE TARSUS occurs as a rarity after such chronic inflammations as trachoma; in post-trachomatous scarred conditions the tarsus may be studded with groups of fatty cells representing probably a fatty degeneration of the meibomian glands (Ginsberg, 1903; Birch-Hirschfeld, 1925).

The primary and secondary types of *amyloid infiltration* have already been discussed.[1]

BONE FORMATION in the tarsus is also rare. True bone-formation may occur as an end-result in cases of amyloidosis particularly following trachoma (Vossius, 1889; Wilson, 1934; Hameed and Nath, 1960); it has also been noted following syphilitic tarsitis (Stanka, 1923) and in a chalazion (Franklin and Cordes, 1924) or in the tarsus after recurrent chalazia (Nath *et al.*, 1966) (Figs. 366–7). A curious case was reported by Herbert (1901) in a Hindu wherein one tarsus was hypertrophied but not greatly thickened and almost completely osseous; on the inner surface a small piece of dead bone had ulcerated through the conjunctiva and was exposed causing a conjunctivitis.

[1] p. 313.

FIGS. 366 and 367.—HETEROTOPIC OSSIFICATION OF THE TARSUS (K. Nath, A. H. and S. L. Rahi).

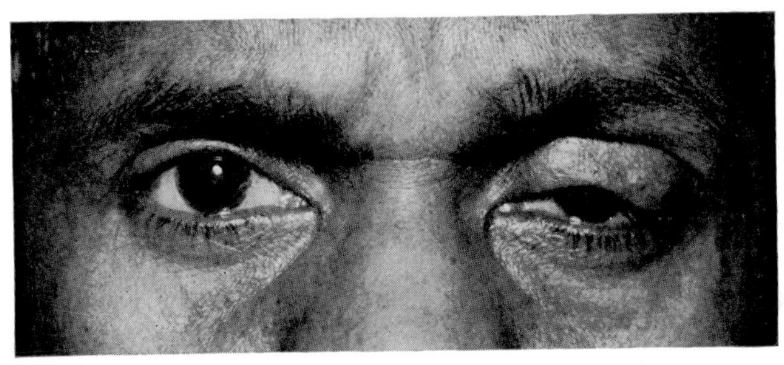

FIG. 366.—In a Hindu male, affecting the left upper lid, after recurrent chalazia.

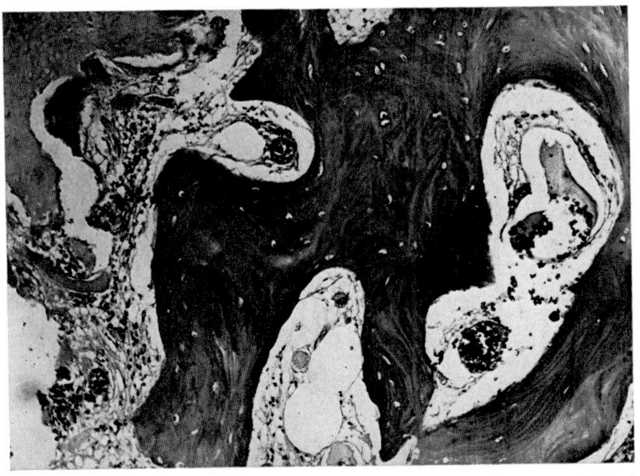

FIG. 367.—Metastatic bone formation in the tarsus surrounded by dilated blood spaces and a zone of fibrosis.

Agius. *Brit. J. Derm.*, **75,** 55 (1963).

Birch-Hirschfeld. *Zur Pathologie d. Granulose (Schriften d. Königsberger gelehrten Ges., naturwiss. Kl.,* **2** (1), 1), Berlin (1925).

Franklin and Cordes. *J. Amer. med. Ass.*, **82,** 519 (1924).

Ginsberg. *Grundriss d. pathologischen Histologie d. Auges*, Berlin (1903).

Hameed and Nath. *Amer. J. Ophthal.*, **49,** 814 (1960).

Herbert. *Trans. ophthal. Soc. U.K.*, **21,** 202 (1901).

Nath, Rahi, A. and S. *Arch. Ophthal.*, **75,** 642 (1966).

Percival and Duthie. *Brit. J. Derm.*, **60,** 399 (1948).

Phillippson. *Brit. J. Derm.*, **3,** 35 (1891).

Pirreda. *Ital. gen. Rev. Derm.*, **1** (5), 35 (1960).

Prakken. *Acta derm.-venereol.* (Stockh.), **31,** 713 (1951).

Reuter and Becker. *Arch. Derm. Syph.* (Chic.), **46,** 695 (1942).

Robinson and Tasker. *Arch. Derm. Syph.* (Chic.), **52,** 180 (1945).

Stanka. *Klin. Mbl. Augenheilk.*, **71,** 348 (1923).

Sullivan and Ellis. *Arch. Derm.*, **84,** 816 (1961).

Vossius. *Beitr. path. Anat.*, **4,** 335; **5,** 291 (1889).

Wagner. *Arch. Heilk.*, **7,** 463 (1866).

Wilson. *Rep. Mem. ophthal. Lab. Giza*, **9,** 44 (1934).

Zaun. *Arch. klin. exp. Derm.*, **224,** 408 (1966).

ANOMALIES OF PIGMENTATION

Congenital anomalies of pigmentation have already been discussed[1]; acquired anomalies are more common. These may be either an increase (HYPERCHROMIA) or lack (ACHROMIA) of the natural pigmentation of the skin by melanin, or coloration by another pigment (DYSCHROMIA). The conditions of melanosis associated with neoplasia will be discussed subsequently.[2]

One peculiar and very common type of coloration should be mentioned in passing—" dark rings round the eyes ". This may be due to hyperpigmentation, as will be noted presently, but sometimes the appearance results from a dilatation of the deep venules of the skin in a ring-shaped distribution round the lower orbital margin and is prone to occur, particularly in women, in states of fatigue, during menstruation or at the menopause. The reason for the localization of the vasomotor disturbance to this region is unknown, but it may be partly determined by the sudden lack of fascial support to the subcutaneous tissues on the transition from the cheek to the lids, the laxity of the lids permitting vasodilatation in the same way as it favours the development of œdema.[3]

In conditions of hyperpigmentation the pigment involved may be endogenous derived from melanin or hæmoglobin, or may be exogenous due to some foreign substance either absorbed systemically or applied locally.

MELANOTIC HYPERPIGMENTATION

It will be remembered that melanin is formed by the progressive oxidation of tyrosine by the melanocytes, dendritic cells found in the skin mainly in the epidermis, dermis and hair follicles; this action is to some extent controlled by pituitary hormones and its main function in man is to act as a protection from short-waved light. Solar energy and hormonal disturbances as well as genetic influences may therefore modify the degree of pigmentation.

Melanotic hyperpigmentation can be differentiated into two types—local (lentigo) and diffuse (melanoderma).

Of the local type, *ephelides* (freckles) are the most common; these have already been described[4] (Fig. 64). They are also present in considerable numbers in the *Peutz-Touraine* (*Peutz-Jeghers*) *syndrome*[5] on the lips, nostrils and lids associated with intestinal polyposis. In this dominantly inherited disorder the histological appearance is that of ephelides with no elongation of the rete ridges and no increase in the number of melanocytes which, however, are longer and contain more pigment (Jeghers *et al.*, 1949; Bologa *et al.*, 1965).

LENTIGO is a larger area with brown or brownish-black macules usually circular in shape caused by an increase in the number of melanocytes along the dermo-epidermal junction which may fade or disappear over the years. They may be present at birth or soon thereafter in large numbers, being transmitted as an autosomal dominant character (A. and S. Pipkin, 1950), or appear in old age (*senile lentigo*). In lentigenes the rete ridges are elongated and the number of melanocytes and their content of pigment are greatly increased (Hodgson, 1963).

[1] Vol. III, p. 885. [2] p. 518. [3] p. 11.
[4] p. 77. [5] Vol. III, p. 1131.

The malignant lentigo of Hutchinson (*intra-epithelial melanoma*), xeroderma pigmentosum, as well as nævi and melanotic tumours will be discussed at a later stage (Chapter VII).

MELANODERMA

MELANODERMA is characterized by the occurrence of yellowish-brown macular lesions or of a diffuse pigmentation due to the deposition of melanin throughout the layers of the rete. The skin shows no other changes. The localized areas are usually round or oval but, by the coalescence of several, large irregular patches may be formed. Normally a certain amount of pigment is found among and under the deeper epidermal cells, and with this the lids are supplied more richly than skin elsewhere with the exception of the regions of the nipples, the axillæ and the genitalia. The dusky hue may

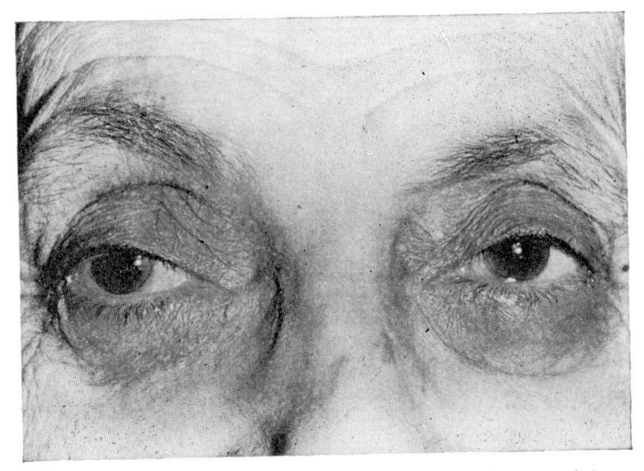

FIG. 368.—SENILE MELANODERMA (A. L. Kornzweig).

be physiological and start sharply as a ring round the orbital margin, and in pathological states of general hyperpigmentation the change tends to be particularly evident in this locality. A hyperpigmentation of this type affecting the periorbital area has been observed with a genetic character inherited as an autosomal dominant trait (20 members in 5 generations, Peters, 1918; 10 members, Hunziker, 1962; 22 in 6 generations, Goodman and Belcher, 1969; 20 in 3 generations, Maruri and Diaz, 1969). A diffuse pigmentation of this type involving the skin of both the upper and lower lids may occur as a senile phenomenon (Kornzweig, 1964) (Fig. 368).

CHLOASMA (MELASMA) is the term usually applied to the occurrence of symmetrical brown patches around the eyes in a mask-like distribution, as well as on the nipples, axillæ and genitalia. It occurs typically in women during pregnancy, at the menopause or when taking contraceptive pills (Esoda, 1963), and has been attributed to a variety of ovarian disorders; it also occurs occasionally in men (Newcomer *et al.*, 1961). The ætiology is not clearly understood.

RIEHL'S MELANOSIS is a distinctive pigmentation described by Riehl (1917); brownish-grey pigmentation develops over the face, eyelids, forehead and temples. There is degeneration of the basal cells of the epidermis and an infiltration of the dermis together with a large amount of melanin in macrophages and lying free in the dermis (Storck, 1946). It has been associated with war in Europe and also occurs in the Argentine and in African Bantu (Pierini, 1952; Findlay, 1952) so that nutritional factors, among others, may be involved, suggesting a relation to pellagra[1] wherein a somewhat similar pigmentation may occur (Poehlmann, 1947); exposure to sunlight may be involved.

Hypermelanosis is common in *endocrine disorders*. This is seen most markedly in *Addison's disease* wherein the quantity of melanin is increased in the basal layer of the epidermis and upper dermis (Montgomery and O'Leary, 1930). The diffuse hyperpigmentation is probably due to an increased output of pituitary hormones to compensate for the failing adrenal glands (Lerner *et al.*, 1953). A somewhat similar pigmentation may occur with pituitary tumours, acromegaly (Lerner, 1961), Cushing's syndrome (Ross *et al.*, 1966), in some patients with thyrotoxicosis (Readett, 1964) and after the administration of ACTH (Cass, 1964).

Hyperpigmentation may occur in *systemic diseases*; the cause of the condition is obscure and in some may be due to an increase in the activity of the pituitary hormones. Genetic predisposition may sometimes be present and it is not always certain if the pigmentation is due to melanin or some other related substance. Such diseases include chronic infections such as syphilis and tuberculosis, neoplasic conditions, cirrhosis of the liver, rheumatoid arthritis, chronic renal disease, and some chronic diseases of the central nervous system such as Schilder's disease or post-encephalitic parkinsonism (Cottini, 1947; Whitlock, 1951; Meerloo, 1957; Liddle *et al.*, 1965; and many others). In Gaucher's disease and the Niemann-Pick disease pigmentation of the lids of the Addisonian type may occur (Bloem *et al.*, 1936).

Localized areas of hyperpigmentation may be associated with *discoid lupus erythematosus*, *scleroderma* and *morphœa*. In *endogenous ochronosis* the violet-black pigmentation of the skin is due to a melanin-like substance.[2]

Local conditions may also induce hyperpigmentation, particularly acanthosis nigricans, urticaria pigmentosa, lichen planus and, less commonly and particularly in brunettes or Negroes, many chronic inflammatory processes such as pityriasis rosea, furuncles, erythema multiforme, acne, pediculosis, and so on (Rothman, 1952; Papa and Kligman, 1965; Ippen, 1966).

Pathological hyperpigmentation may be due to *physical agents*: *erythema ab igne* is an example; sunlight, ultra-violet rays, x-rays and friction may also be causal factors. The tanning effect of ultra-violet light results from the biosynthesis of new melanin (Snell, 1963).

Melanoderma may follow exposure to or the absorption of certain *chemicals*, whether acquired industrially by contact or inhalation or by drugs. Several of them are examples of photosensitization. These have all been discussed in another Volume with their clinical features and pathology so that a short mention only is required here. Among them are *arsenic*[3] which favours the oxidation of tyrosine to produce a

[1] p. 340.
[2] Vol. VIII, p. 1064.
[3] Vol. XIV, pp. 1096, 1248.

bronzy stain; the phenothiazine derivatives, particularly *chlorpromazine*, may cause a violet or black pigmentation in skin exposed to sunlight producing the typical *visage mauve* due to melanin, probably owing to the fact that these drugs localize in the melanocytes; the pigmentation fades spontaneously but very slowly when the drug is stopped[1] (Blois, 1965; Rives and Pellerat, 1965). A pigmentation resembling chloasma may follow the administration of *hydantoin* which also exerts a direct action on melanocytes; it fades slowly when the drug is discontinued (Kuske and Krebs, 1964). Among the *antimalarial drugs*, chloroquine and hydroxychloroquine, which have an affinity for dermal melanin, may cause a dusky pigmentation on exposed skin (Zachariae, 1963; Sams and Epstein, 1965).

Workers with *tars* and hydrocarbon oils are apt to acquire an occupational melanosis owing to the photodynamic action of such substances as anthracene and phenanthrene (Foerster and Schwartz, 1939). The phenol compounds, particularly

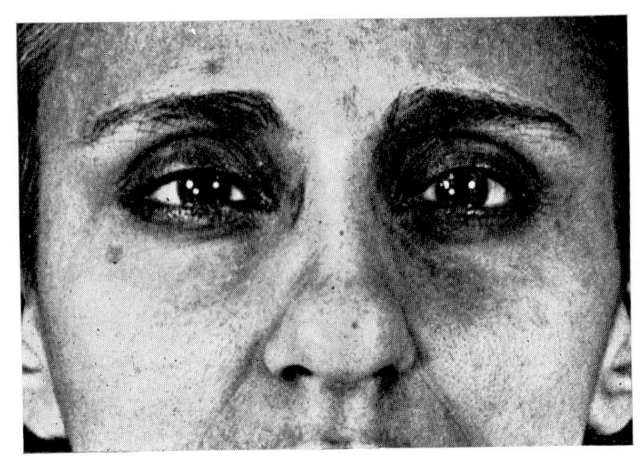

FIG. 369.—MELANODERMA.

Arising after the cosmetic use of petroleum jelly (Dept. of Dermatology, Vanderbilt Clinic, N.Y.).

hydroquinone, may similarly cause a pigmentation in workers as in aniline plants (*exogenous ochronosis*) due to melanin-like substances.[2] A pigmentation of this type may follow the use of cosmetics such as petroleum jelly (Fig. 369) and is dramatically seen in the *Berloque dermatitis* caused by toilet waters, notably eau-de-Cologne, containing bergamot oil which owing to its content of furocoumarin stimulates melanogenesis by light (Harber *et al.*, 1964; Bundick, 1966).

HÆMOCHROMATOSIS

Hæmochromatosis, whether primary due to an inborn error of metabolism causing an increased absorption of iron from the intestinal tract or secondary as a result of hepatic cirrhosis, has been described in a previous Volume.[3] It entails a widespread deposition of iron in the skin and viscera, leading in the first to pigmentation, in the liver to cirrhosis and in the pancreas to diabetes, the combination known as *bronzed diabetes*; certain mucous membranes may also be affected, particularly in the mouth, and in

[1] Vol. XIV, p. 1294. [2] Vol. XIV, p. 1119. [3] Vol. VII, p. 155.

the retina micro-aneurysms resembling those of diabetes have been described as well as the typical hæmochromic discoloration.[1] In the conjunctiva the same type of micro-aneurysms may be present (Hudson, 1953) and also the typical pigmentation (Davies *et al.*, 1972). In 9 out of 44 cases studied by Davies and his colleagues (1972) a slaty-blue coloration was seen in the lids, especially at their margins extending throughout the length of the lids, particularly around the follicles of the lashes (Fig. 370). Histologically an increased formation of melanin was seen in the basal layers of the epidermis; no free ferric iron was found in the lids, as is usual in dermal lesions (Perdrup and Poulsen, 1964) although it was present in minute amounts in

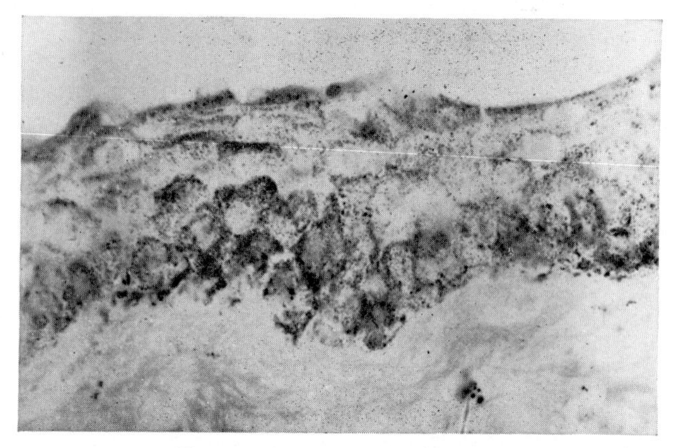

Fig. 370.—Hæmochromatosis.

Section of the lid to show melanin pigment in the epithelium of the skin (G. Davies *et al.*).

the conjunctiva and within the eye. The usual treatment is by repeated venesection which tends to diminish the cutaneous pigmentation (Williams *et al.*, 1969).

Hæmatogenous pigmentation is more rare than that due to melanin if exception be made of the sometimes vivid sequelæ of a bruise. The yellow pigmentation of jaundice is well known; an interesting case of this was reported by Taylor (1886) wherein the lower lid retained permanently a deep blue-green shade. Pigmentation may also be associated with malaria, the anæmias and purpura.

DYSCHROMIA

A pigmentation of the skin including that of the lids may result from certain metals by their application as ointments or their absorption industrially or as medicaments. These are all discussed in another Volume and therefore can be rapidly dismissed here. The most common and dramatic are mercury which is deposited in the skin,[2] silver, causing argyrosis through the deposition of silver albuminate resulting sometimes in a massive black pigmentation,[3] and gold causing chrysiasis[4] which rarely causes discoloration of the lids.

[1] Vol. X, p. 493.
[2] Vol. XIV, pp. 1096, 1248.
[3] Vol. XIV, pp. 1093, 1247.
[4] Vol. XIV, p. 1246.

The vivid green discoloration of the skin occurring with the administration of atebrin[1] and the yellow discoloration after the absorption of picric acid (trinitrophenol) or following chronic industrial poisoning with nitrogen mustard[2] are noted in another Volume.

Bloem, Groen and Postma. *Quart. J. Med.*, **5**, 517 (1936).
Blois. *J. invest. Derm.*, **45**, 475 (1965).
Bologa, Bene and Pasztor. *Ann. Derm. Syph.* (Paris), **92**, 277 (1965).
Bundick. *Arch. Derm.*, **93**, 424 (1966).
Cass. *Curr. ther. Res.*, **6**, 601 (1964).
Cottini. *G. ital. Derm. Sif.*, **88**, 131 (1947).
Davies, Dymock, Harry and Williams. *Brit. J. Ophthal.*, **56**, 338 (1972).
Esoda. *Arch. Derm.*, **87**, 486 (1963).
Findlay. *S. Afr. med. J.*, **26**, 273 (1952).
Foerster and Schwartz. *Arch. Derm. Syph.* (Chic.), **39**, 55 (1939).
Goodman and Belcher. *Arch. Derm.*, **100**, 169 (1969).
Harber, Harris, Leider and Baer. *Arch. Derm.*, **90**, 572 (1964).
Hodgson. *Arch. Derm.*, **87**, 197 (1963).
Hudson. *Brit. J. Ophthal.*, **37**, 242 (1953).
Hunziker. *J. Génét. hum.*, **11**, 16 (1962).
Ippen. *Derm. Wschr.*, **152**, 281 (1966).
Jeghers, McKusick and Katz. *New Engl. J. Med.*, **241**, 993 (1949).
Kornzweig. *Int. Ophthal. Clin.*, **4**, 55 (1964).
Kuske and Krebs. *Dermatologica*, **129**, 121 (1964).
Lerner. *Arch. Derm.*, **83**, 97 (1961).
Lerner, Shizume and Fitzpatrick. *J. invest. Derm.*, **21**, 337 (1953).
Liddle, Givens, Nicholson and Island. *Cancer Res.*, **25**, 1057 (1965).
Maruri and Diaz. *Cutis*, **5**, 979 (1969).
Meerloo. *Psychosom. Med.*, **19**, 89 (1957).
Montgomery and O'Leary. *Arch. Derm. Syph.* (Chic.), **21**, 970 (1930).
Newcomer, Lindberg and Sternberg. *Arch. Derm.*, **83**, 284 (1961).
Papa and Kligman. *J. invest. Derm.*, **45**, 465 (1965).
Perdrup and Poulsen. *Arch. Derm.*, **90**, 34 (1964).
Peters. *Zbl. prakt. Augenheilk.*, **42**, 8 (1918).
Pierini. *Arch. argent. Derm.*, **2**, 315 (1952).
Pipkin, A. and S. *J. Hered.*, **41**, 79 (1950).
Poehlmann. *Derm. Wschr.*, **119**, 454 (1947).
Readett. *Brit. J. Derm.*, **76**, 126 (1964).
Rives and Pellerat. *Ann. Derm. Syph.* (Paris), **92**, 517 (1965).
Riehl. *Wien. klin. Wschr.*, **30**, 280 (1917).
Ross, Marshall-Jones and Friedman. *Quart. J. Med.*, **35**, 149 (1966).
Rothman. *Acta derm.-venereol.* (Stockh.), **32**, Suppl. 29, 316 (1952).
Sams and Epstein. *J. invest. Derm.*, **45**, 482 (1965).
Snell. *J. invest. Derm.*, **40**, 127 (1963).
Storck. *Dermatologica*, **92**, 246 (1946).
Taylor. *Trans. ophthal. Soc. U.K.*, **6**, 145 (1886).
Whitlock. *Arch. Derm. Syph.* (Chic.), **64**, 23 (1951).
Williams, Smith, Spicer *et al.* *Quart. J. Med.*, **38**, 1 (1969).
Zachariae. *Acta derm.-venereol.* (Stockh.), **43**, 149 (1963).

MELANOTIC HYPOPIGMENTATION

Congenital ALBINISM and PARTIAL ALBINISM (LEUCISM), genetically determined disorders due to an absence of the enzyme, tyrosinase, have been described in another Volume.[3]

VITILIGO (LEUCODERMA)

An acquired hypopigmentation of the skin was initially called *vitiligo* by Celsus. It affects both sexes equally and in many cases occurs familially suggesting transmission by an autosomal dominant gene with varying expressivity (Lerner, 1959). It usually starts about the age of 20 years when patches of perfectly white but otherwise normal skin appear surrounded by a hyperpigmented areola on any area of the skin, especially the face, neck, hands, lower abdomen and thighs. These are usually symmetrical and increase slowly in size until large areas of the skin are un-

[1] Vol. XIV, p. 1128. [2] Vol. XIV, p. 1259.
[3] Vol. III, pp. 803, 810.

FIGS. 371 to 373.—VITILIGO WITH UVEITIS.

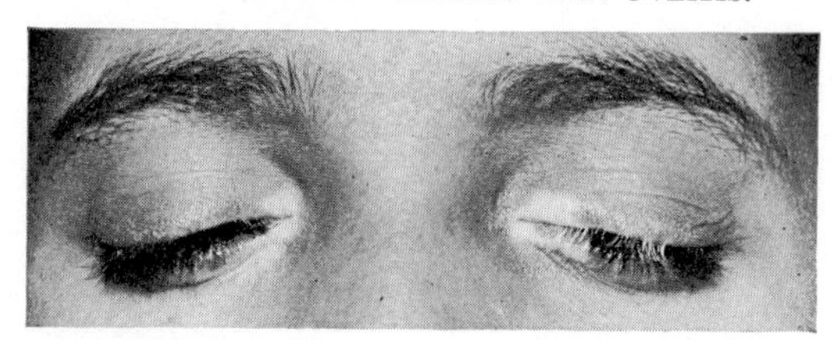

FIG. 371.—In a Caucasian boy aged 12. Note the whitening of the lashes (Inst. Ophthal.).

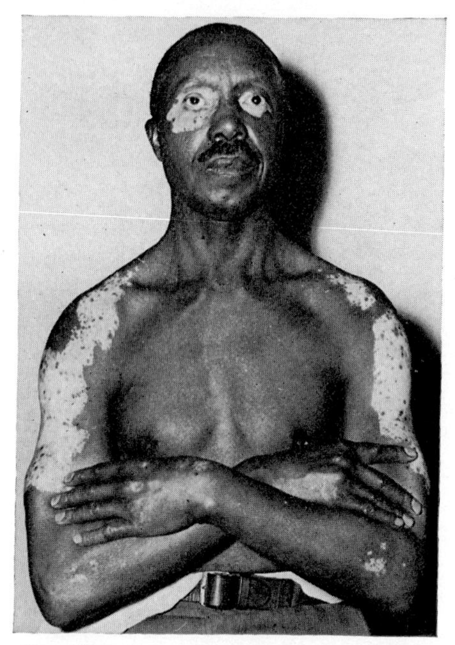

FIG. 372.—In a Negro with the Vogt-Koyanagi syndrome (Mary Bruno and S. D. MacPherson).

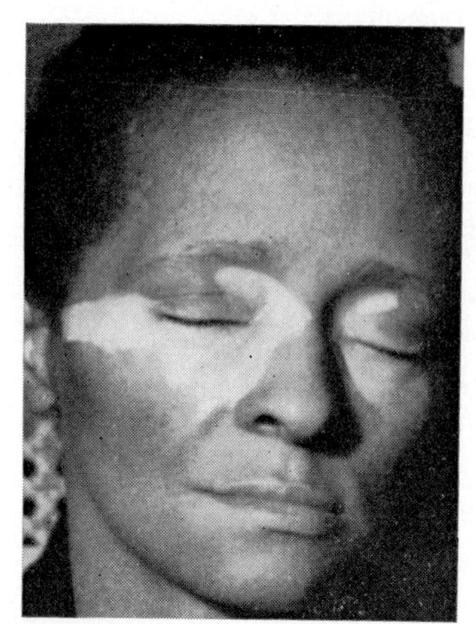

FIG. 373.—In a Negress with sympathetic ophthalmitis (Mary Bruno and S. D. MacPherson).

FIGS. 374 and 375.—VITILIGO WITH DRUGS.

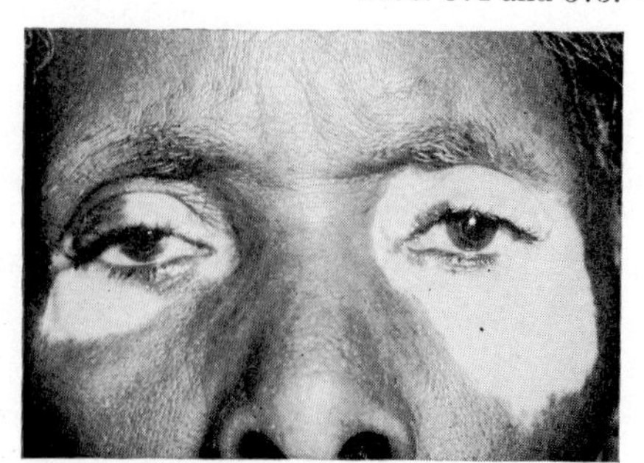

FIG. 374.—Two years after the initial instillation of thiotepa drops (J. W. Berkow *et al.*).

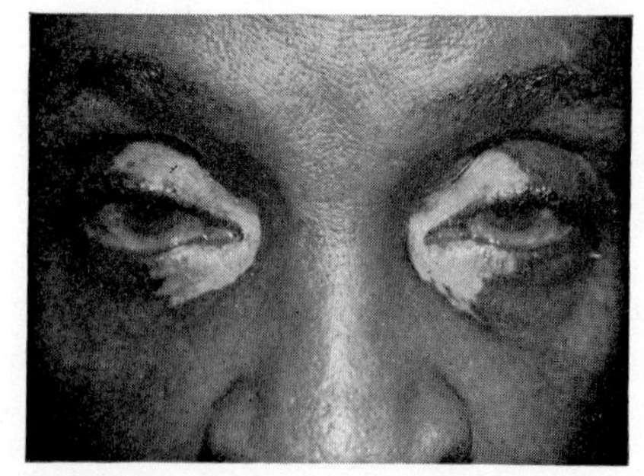

FIG. 375.—Five years after the initial application of eserine ointment (H. N. Jacklin).

pigmented; any associated hairs in the affected areas, such as brows and lashes, are similarly depigmented (Fig. 371). Apart from the cosmetic appearance the only symptom is the development of a painful erythema on exposure to sunlight. In the depigmented area there is a complete absence of melanin, and the melanocytes are replaced by cells of Langerhans; in the surrounding areola, melanocytes are abundant (Birbeck *et al.*, 1961; Niebauer, 1965).

Vitiligo may be associated with other syndromes, the most interesting of which are the *Vogt-Koyanagi syndrome* in which uveitis occurs with deafness and encephalitic or meningeal symptoms[1] (Fig. 372), *Harada's disease* wherein the posterior uvea is principally involved accompanied by a retinal detachment, and less commonly in *sympathetic ophthalmitis*[2] (Fig. 373). It has also been noted in association with benign mucosal pemphigoid (Tripodi, 1971).

Depigmentation may follow the topical use of drugs. This has been reported following the experimental and clinical application of hydroquinone (Arndt and Fitzpatrick, 1965; Spencer, 1965) and its monobenzyl and monomethyl ether analogues (Brun, 1959; Sidi *et al.*, 1961; Snell, 1964), the mercapto-ethyl amines (Frenk *et al.*, 1968), guanonitrofurazone (Iijima and Tokunaga, 1953; Stegmaier, 1960), and thiotepa, a radiomimetic drug of the nitrogen mustard family used to inhibit corneal vascularization or the regrowth of pterygia (Berkow *et al.*, 1969; Howitt and Karp, 1969) (Fig. 374). The same type of vitiligo has been seen after the use of eserine in allergic subjects (Jacklin, 1965) (Fig. 375).

A *consecutive leucoderma* may follow local diseases of the skin such as syphilis, pinta, psoriasis, leprosy, neuro-dermatitis, connective-tissue diseases such as lupus erythematosus or scleroderma, parasitic affections or ionizing radiation (Fig. 387). Such a loss of pigment is a characteristic feature of eczema in the Negro.

The *treatment* of vitiligo is difficult and so far as the lids are concerned reliance is best placed on cosmetic applications and cover-creams, or by painting with dihydroxyacetone; protection from the sun by a benzophenone barrier cream may be advisable. The administration of methoxypsoralen with chloroquine in cases wherein all tyrosinase activity is not lost has been advocated but the results are not reliable and any re-pigmentation which may develop is not permanent (Chanco-Turner and Lerner, 1965).

Arndt and Fitzpatrick. *J. Amer. med. Ass.*, **194,** 965 (1965).
Berkow, Gills and Wise. *Arch. Ophthal.*, **82,** 415 (1969).
Birbeck, Breathnach and Everall. *J. invest. Derm.*, **37,** 51 (1961).
Brun. *Dermatologica*, **118,** 202 (1959).
Chanco-Turner and Lerner. *Arch. Derm.*, **91,** 390 (1965).
Frenk, Pathak, Szabó and Fitzpatrick. *Arch. Derm.*, **97,** 465 (1968).
Howitt and Karp. *Amer. J. Ophthal.*, **68,** 473 (1969).
Iijima and Tokunaga. *Jap. J. Derm.*, **63,** 490 (1953).
Jacklin. *Amer. J. Ophthal.*, **59,** 89 (1965).
Lerner. *J. invest. Derm.*, **32,** 285 (1959).
Niebauer. *Dermatologica*, **130,** 317 (1965).
Sidi, Bourgeois-Spinasse and Planat. *Presse méd.*, **69,** 2369 (1961).
Snell. *Arch. Derm.*, **90,** 63 (1964).
Spencer. *J. Amer. med. Ass.*, **194,** 962 (1965).
Stegmaier. *J. Amer. med. Ass.*, **172,** 559 (1960).
Tripodi. *Boll. Oculist.*, **50,** 26 (1971).

[1] Vol. IX, p. 373.
[2] Vol. IX, p. 573.

FIG. 376.—HAROLD GLENDON SCHEIE
[1909——].

CHAPTER VI

DISORDERS OF THE EYEBROWS AND LASHES

IN MANY sections of this *System* we have encountered the name of HAROLD GLENDON SCHEIE [1909 ——] (Fig. 376); we are inserting his photograph here in view of the fact that, with D. M. Albert, he indicated that aberrant lashes could arise as an acquired condition from the meibomian glands after prolonged inflammations of the lids. His contributions to our specialty, published in more than 185 scientific and clinical papers, have been varied and prolific. Elsewhere in this Volume we have already noted his authoritative studies on the ophthalmological complications and treatment of zoster, one of the most complete to appear since the original classical paper of Jonathan Hutchinson, and in a later Chapter we shall discuss his several contributions to our knowledge of malignant tumours of the orbit; but his reputation depends essentially on his surgical innovations, the most important of which is the operation now universally known as Scheie's operation which is widely (and advisedly) practised for simple glaucoma. Scheie, a mid-western American, was educated at the University of Minnesota and became the first resident in ophthalmology at the University of Pennsylvania (1937); here he has remained for the rest of his working life, being Professor of Ophthalmology and eventually building the Scheie Eye Institute (1972), a magnificent multi-million building financed entirely by his patients and friends. During the Second World War he served in the U.S. Army for four years, achieving the rank of Brigadier General and obtaining the Legion of Merit and the Order of the British Empire.

Congenital anomalies of the eyelashes have been discussed in a previous Volume.[1] They include congenital hypotrichosis and hypertrichosis, either an increase in length (trichomegaly) or in number (polytrichia), reduplication of the ciliary follicles, ectopic and inverse cilia, pili torti and monolethrix.

Similar congenital anomalies of the eyebrows have been discussed,[2] including hypoplasia, hypertrichosis and duplicated brows. The Typus Amstelodamensis (Cornelia de Lange syndrome) wherein a diffuse overgrowth and synophrys of the eyebrows and hypertrophy of the lashes occur has also been described[3] (Milot and Demay, 1972). Here we are concerned essentially with acquired conditions.

Disorders of the lashes include exuberance or deficiency of growth, disturbances in the direction of growth, and pathological changes in structure or pigmentation. In many of these diseases of the lashes the eyebrows participate. Inflammatory diseases affecting the lash follicles have already been discussed with blepharitis.[4]

Disturbances of Growth

HYPERTRICHOSIS

ACQUIRED HYPERTRICHOSIS (ὑπέρ, over; θρίξ, τριχός, a hair), or overgrowth of the hairs (*trichomegaly*), may affect the cilia, the brows and the

[1] Vol. III, p. 872. [2] Vol. III, p. 878.
[3] Vol. III, p. 1116. [4] p. 205.

fine hairs normally present on the surface of the lids. All these may be increased in number and length, a phenomenon sometimes affecting all the lids in a general hirsutism, and at others affecting particular regions, especially the upper lid in its outer half near the lateral canthus. In women such a condition may be associated with other evidences of masculinity including an exuberance of hair on the upper lip and chin. The ætiology is obscure, but it would seem that the main influences are the adrenals, thyroid, pituitary and gonads. The overgrowth usually starts about puberty or after mid-life and may become apparent in old age in association with tumours of the adrenals, pituitary or ovaries (Broster, 1950). Once developed it is usually permanent unless the hairs are destroyed by electrolysis, but repeated treatments at intervals of 6 to 9 months are usually necessary. This is the only safe method in the region of the eye, using a direct current of 1 to 3 milliamperes—a long and painstaking business (see Cipollaro, 1938). Rubbing with pumice-stone has been advocated (Stelwagon, 1914), disguising by bleaching with hydrogen peroxide (Unna, 1914), or the local use of an epilating wax or lotion. Thallium acetate (Sabouraud, 1936) is toxic and must be used with great care and in small quantity, and even then is not free from risk.

Acquired hypertrichosis lanuginosa, wherein fine lanugo hair develops to unusual lengths, is a rare condition, often particularly affecting the lids and nose but liable to spread over the whole body. The hair may grow very rapidly, sometimes up to 2·5 cm. weekly to reach a length of more than 10 cm. (Lyell and Whittle, 1951). This dramatic syndrome is rare; it tends to occur in adults in mid-life and is also associated with serious illness. It is probably related to endocrine disorders but its ætiology has not been fully explored and is obscure (Le Marquand and Bohn, 1951).

A *nævoid hypertrichosis* wherein a circumscribed bundle of stout hairs are associated with nævi may be found on the lids; the hair may be present from infancy but may appear about puberty or later (Figs. 563, 566–9).

A *symptomatic hypertrichosis* may occur in a wide variety of pathological conditions. The ætiology of many cases is obscure but little understood endocrine factors may be operative or some anomaly of the connective tissue of the dermis. These include epidermolysis bullosa (Cofano, 1955), porphyria (Kingery, 1966), Hurler's syndrome (Korting and Corin-Thenberg, 1964), and endocrine disorders such as hypo- or hyper-thyroidism (Perloff, 1955) or pituitary and diencephalic disturbances (Stegagno and Vignetti, 1955). It has also been reported with acrodynia (Holzel, 1951), dermatomyositis (Reich and Reinhart, 1948), in states of gross malnutrition (Holzel, 1951) and anorexia nervosa (Ryle, 1936).

Hypertrichosis has also been reported following the administration of *drugs* such as corticosteroids, streptomycin (Fono, 1950), diphenylhydantoin given to epileptic children (Livingston *et al.*, 1955) and, in women, anabolic steroids with androgenic activity (Wilkins, 1962).

CILIARY POLYTRICHOSIS. An increase in the number of the lashes may come on in adult or late life, in which case the extra lashes are arranged irregularly and not in well-defined rows as is seen in congenital polytrichia (*distichiasis*), due to a heterotypical anomaly of the meibomian glands which develop into hair follicles. Perhaps because of an inflammatory stimulus, from the follicles of the cilia two or three side-buds sprout, each of which grows into a separate lash (Unna, 1876; Herzog, 1904). As would be expected, these are frequently stunted and, growing obliquely, lose the parallelism of normal lashes. In other cases, after prolonged inflammations such as the Stevens-Johnson syndrome (Scheie and Albert, 1966), similarly stunted and non-pigmented cilia may arise from the meibomian glands at

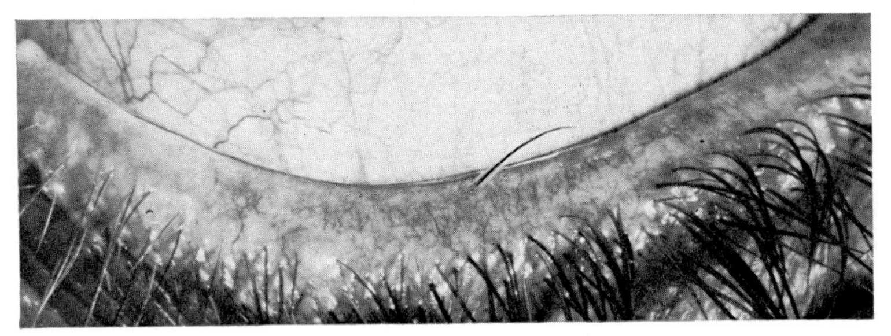

FIG. 377.—AN EYELASH GROWING FROM A MEIBOMIAN GLAND IN THE LOWER LID (Inst. Ophthal.).

various sites along the inner aspect of the margins of the lids, constituting *aberrant lashes* (Fig. 377); they may be difficult to see and electrolysis is most easily done with the help of the operating microscope.

The occurrence of congenital distichiasis with lymphœdema and other anomalies (the Nonne-Milroy-Meige syndrome) has already been noted.[1]

CILIARY TRICHOMEGALY, abnormally long lashes, has been reported by several observers (Figs. 378–9); as an exception a single lash may be involved (Fig. 380). The length of the lashes has varied between 20 and 40 mm. Most of the cases have been congenital and some familial[2]; some have been associated with various anomalies such as dwarfism, pigmentary degeneration of the retina, and mental retardation (Oliver and McFarlane, 1965), ecto-dermal dysplasia (Cant, 1967), or cataract and spherocytosis (Goldstein and Hutt, 1972). It is obvious that all these conditions cannot be referred to a common basis. An acquired condition developing in an adult is rare (Reitter, 1926–31, presumably of endocrine origin).

A SUPERCILIARY TRICHOMEGALY occasionally occurs. Thus Fuchs (1956), measuring the eyebrows of a Serbian, found them to be 14·5 cm. long; he pointed out their immense length in certain Holy Men in China, grown as a distinguishing feature, up to 16 cm. and falling down onto the shoulders (Fig. 381). This anomaly may occur with ciliary trichomegaly (Marquez, 1928).

[1] p. 21.
[2] Marquez (1928), Bab (1931), Gray (1944), Majewski (1958), Goldstein and Hutt (1972).

FIGS. 378 and 379.—TRICHOMEGALY (J. H. Goldstein and A. E. Hutt).

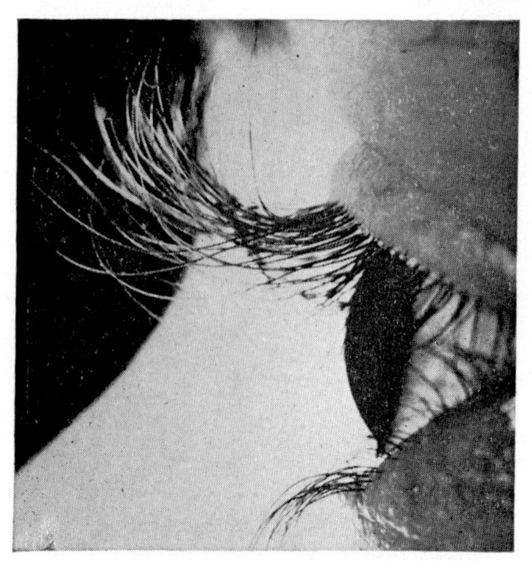

FIG. 378.—In an 18-year-old girl.

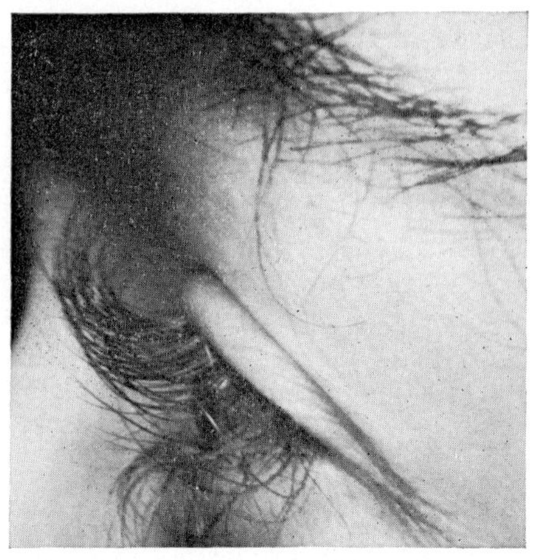

FIG. 379.—In a 14-year-old boy, the brother of the girl seen in Fig. 378.

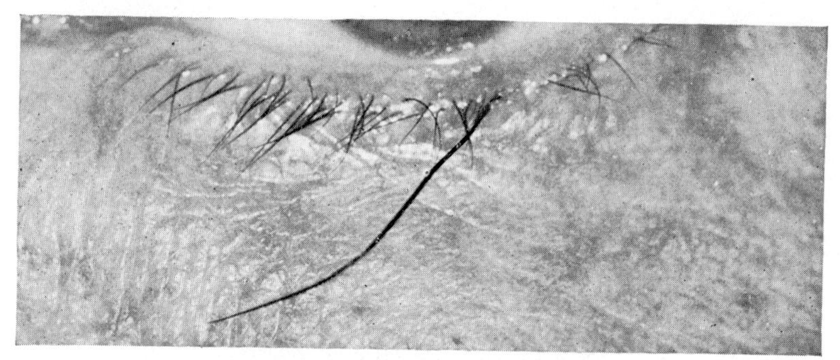

FIG. 380.—A SINGLE LONG EYELASH.
Growing from the lower lid-margin (Inst. Ophthal.).

FIG. 381.—SUPERCILIARY TRICHOMEGALY.

A Holy Man in the Ming Hing Temple in Hangchow, showing exuberant growth of the eyebrows (13 to 16 cm. long) (A. Fuchs).

Bab. *Klin. Mbl. Augenheilk.*, **87**, 804 (1931).
Broster. *Brit. med. J.*, **1**, 1171 (1950).
Cant. *J. pediat. Ophthal.*, **4**, 13 (1967).
Cipollaro. *J. Amer. med. Ass.*, **111**, 2488 (1938).
Cofano. *Ann. ital. Derm. Sif.*, **10**, 195 (1955).
Fono. *Ann. paediat.* (Basel), **174**, 389 (1950).
Fuchs. *Klin. Mbl. Augenheilk.*, **129**, 121 (1956).
Goldstein and Hutt. *Amer. J. Ophthal.*, **73**, 333 (1972).
Gray. *Stanford med. Bull.*, **2**, 157 (1944).
Herzog. *Z. Augenheilk.*, **12**, 180 (1904).
Holzel. *Acta paediat.* (Stockh.), **40**, 59 (1951).
Kingery. *J. Amer. med. Ass.*, **195**, 571 (1966).
Korting and Corin-Thenberg. *Z. Haut- u. Geschl.-Krankh.*, **37**, 65 (1964).
Le Marquand and Bohn. *Proc. roy. Soc. Med.*, **44**, 155 (1951).
Livingston, Petersen and Boks. *J. Pediat.*, **47**, 351 (1955).
Lyell and Whittle. *Proc. roy. Soc. Med.*, **44**, 576 (1951).
Majewski. *Klin. oczna*, **28**, 121 (1958).
Marquez. *Siglo méd.*, **82**, 279 (1928).
Milot and Demay. *Amer. J. Ophthal.*, **74**, 394 (1972).
Oliver and McFarlane. *Arch. Ophthal.*, **74**, 169 (1965).
Perloff. *J. Amer. med. Ass.*, **157**, 651 (1955).
Reich and Reinhart. *Arch. Derm. Syph.* (Chic.), **57**, 725 (1948).
Reitter. *Z. Augenheilk.*, **59**, 354 (1926); **73**, 387 (1931).
Ryle. *Lancet*, **2**, 893 (1936).
Sabouraud. *Nouvelle pratique dermatologique*, (ed. Darier *et al.*), Paris, **7**, 92 (1936).
Scheie and Albert. *Amer. J. Ophthal.*, **61**, 718 (1966).
Stegagno and Vignetti. *Arch. ital. Pediat. Pueric.*, **17**, 421 (1955).
Stelwagon. *Treatise on Diseases of the Skin*, 7th ed., Phila., (1914).
Unna. *Arch. mikr. Anat.*, **12**, 665 (1876).
Wilkins. *Amer. J. Dis. Child.*, **104**, 449 (1962).

HYPOTRICHOSIS: ALOPECIA AND MADAROSIS

In ALOPECIA[1]—a falling out of the hair—the brows and lashes may be involved along with the hair of the scalp and other hairy regions; when the loss of hair is confined to the lashes and is due to destructive processes, the term MADAROSIS (μαδαρός, bald) is used.

Generalized alopecia may be total occurring as a congenital defect[2]

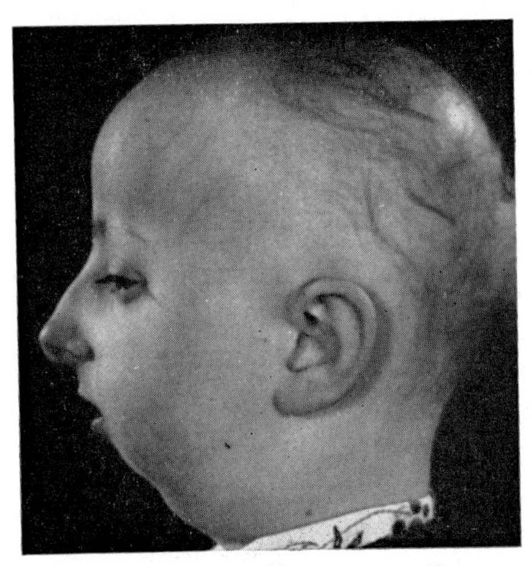

FIG. 382.—CONGENITAL ECTODERMAL DYSPLASIA.

Showing the typical facies, with the depression of the bridge of the nose and the under-development of the jaw, the scarcity of hair on the scalp and absence of eyebrows. There was bilateral cataract (Irene Gregory).

[1] ἀλώπηξ, a fox; ἀλωπηκία, fox-mange, baldness. Foxes were allegedly supposed to suffer from patchy loss of hair, and the term was first applied to alopecia areata; it is now applied to loss of hair of any type, diffuse or patchy.

[2] Vol. III, p. 872.

transmitted usually as a recessive but sometimes as an irregular dominant trait. The hair is usually normal at birth but is shed between the second and sixth months of life all over the body including the eyebrows and lashes, not to be replaced; no other anomalies occur. A total or almost total alopecia may occur in progeria or anhidrotic ectodermal dysplasia wherein the loss of hair follicles is associated with absence of the sebaceous and sweat glands[1] (Fig. 382). A generalized hypotrichosis, wherein the hairs are sparse, brittle and poorly pigmented is more common and may be associated with several hereditary syndromes such as the mandibulo-oculo-facial dyscephaly of Hallermann and Streiff,[2] Werner's syndrome[3] and the Marinesco-Sjögren syndrome.[4]

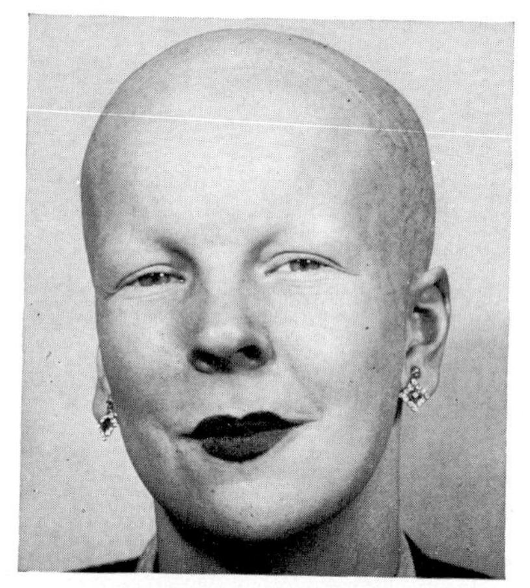

Fig. 383.—Alopecia Totalis.
Showing loss of hair, eyebrows and lashes (P. Borrie).

ALOPECIA AREATA is characterized by the sudden appearance of bald patches particularly on the scalp often in an apparently normal person without obvious local cause; relapses tend to occur but, in young persons at any rate, the prognosis is frequently good, the new crop of hair generally starting as a thin, white lanugo-like growth. Sometimes the lashes fall out or the eyebrows (Fig. 384) and occasionally the lashes in a segment of one lid (Kile, 1960). It may occur at any age between 5 and 30, but is rare over the age of 45. The follicles are formed but no hair develops, only keratinous debris. The course of the disease is unpredictable; the hair in a bald patch may re-grow in a few months, fresh patches may appear and coalesce, all the hair of the head may be lost including the brows and lashes (ALOPECIA TOTALIS) (Fig. 383) or the whole body may be implicated (ALOPECIA UNIVERSALIS).

[1] Vol. III, p. 1128.
[2] Vol. III, p. 1022.
[3] Vol. XI, p. 200.
[4] Vol. III, p. 1147.

Symptomatic alopecia is a relatively common condition and has a complex ætiology embracing both constitutional and local conditions, but in many cases the cause is obscure. It may be divided into two types, non-cicatricial and cicatricial. In the former an interference with the physiology may result in temporary loss of the hair so that re-growth is possible; in the latter scarring leads to the permanent destruction of the hair follicles.

It occurs in general infections, of which the commonest used to be early syphilis (van der Ploeg and Stagnone, 1964); thus in 136 cases Wilbrand and Staelin (1897) found a loss of the lashes in 5·1%, of the brows in 8·8% and of both in 33·3%. A loss of the outer half of the two eyebrows is a common syphilitic manifestation—significantly called *la signe d'omnibus* by the French; it is, however, not diagnostic. Other infections which may act similarly are leprosy,[1] severe cachectic tuberculosis, and serious acute infections such as typhoid, scarlet or puerperal fever, influenza and so on, as well as viral infections such as infective hepatitis (Amalric and Bessou, 1960).

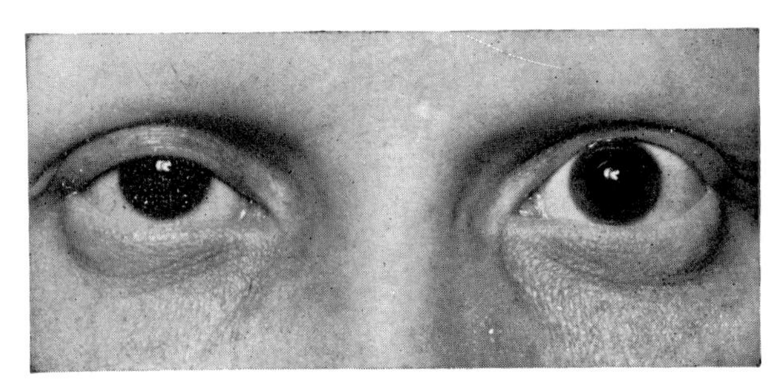

FIG. 384.—ACQUIRED ALOPECIA.

Showing total absence of brows and lashes ascribed to shock (there is also retraction of the left upper lid) (S. P. Meadows).

Endocrine disorders also cause alopecia, particularly disturbances of the thyroid, both myxœdema (Church, 1965; Kingery, 1966; Mahto, 1972) and thyrotoxicosis (Hoffmann, 1908; Rook, 1965), and of the pituitary, as in Simmond's disease or hypopituitarism wherein the brows and lashes are lost along with the hair of the scalp, axillæ and pubic regions (Wahlberg, 1936).

Dermatoses such as exfoliative dermatitis, seborrhœic dermatitis or psoriasis are also sometimes accompanied by alopecia, and particularly the follicular dermatoses wherein hyperkeratosis occurs in the hair follicles accompanied by a loss of the brows and lashes. Thus in keratosis pilaris decalvans the whole body may be affected, but a typical effect is a loss of the outer portion of the eyebrows and alopecia of the scalp (Franceschetti *et al.*, 1956; Adler and Nyhan, 1969) (Fig. 385). A similar loss of the lateral portions of the eyebrows may occur in keratosis pilaris atrophicans faciei (Davenport, 1964) (Fig. 249), and a loss of all the brows and lashes may characterize acanthosis nigricans.[2]

Other factors causing alopecia include:

Chronic diseases and neoplasms such as the reticuloses in the cachectic state, and conditions of gross malnutrition (Godwin, 1962).

[1] Loutfy *et al.* (1937), Guerrero Santos *et al.* (1961), Kingery (1966), Emiru (1970), Weerekoon (1972). See p. 111.

[2] p. 263.

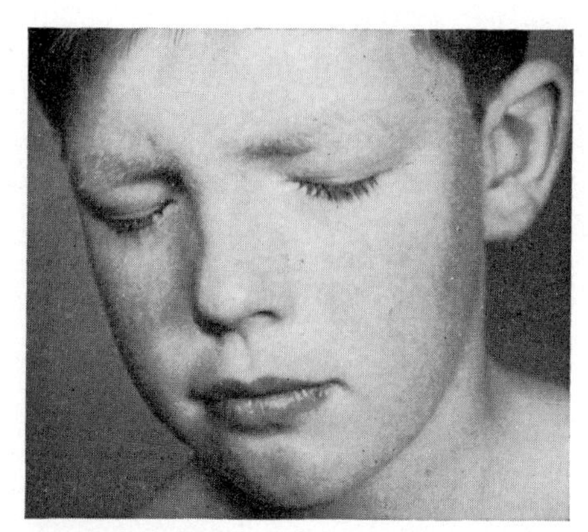

FIG. 385.—KERATOSIS PILARIS DECALVANS.
Showing loss of the outer half of the eyebrow (W. Jadassohn and R. Paillard).

FIGS. 386 and 387.—LOSS OF EYELASHES DUE TO IRRADIATION
(P. A. MacFaul and M. A. Bedford).

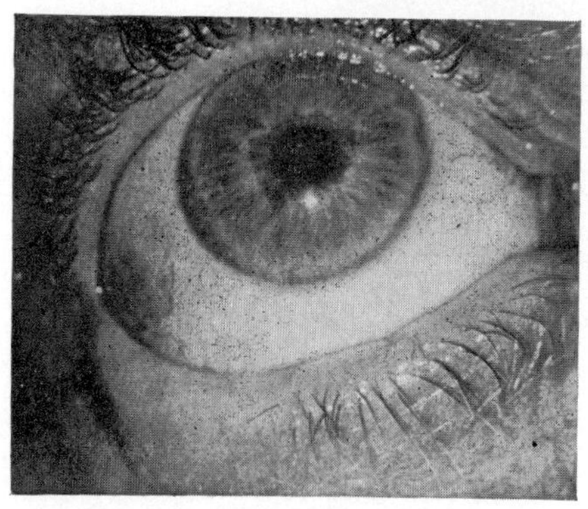

FIG. 386.—Loss of the eyelashes after radiotherapy for a basal-cell carcinoma of the eyelid.

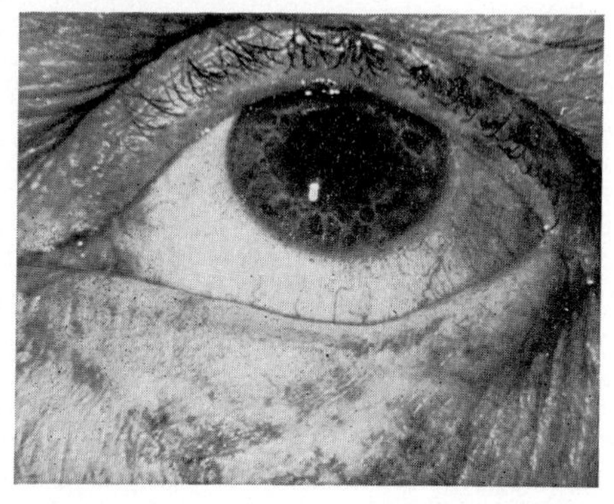

FIG. 387.—Changes in the skin, showing depigmentation, telangiectasia and atrophy as well as loss of the eyelashes following irradiation for a basal-cell carcinoma of the eyelid.

Psychological factors such as an emotional shock are often obvious in alopecia areata (Greenberg, 1955; Kile, 1960) (Fig. 384).

A disturbance of the sympathetic nerve in alopecia areata was suggested by Langhof and Lemke (1962) who found Horner's syndrome in 52 out of 63 patients.

Chemical agents include thallium,[1] cytotoxic drugs and arsenic.

Physical agents may have the same effect, particularly x-rays[2] (Figs. 386–7).

An extremely interesting and suggestive group of cases wherein alopecia is associated with uveitis, vitiligo, poliosis and dysacousia (the Vogt-Koyanagi syndrome) has already been described,[3] as has its occurrence with perforating wounds of the eye followed by traumatic iridocyclitis.[4]

[1] Vol. XIV, p. 1250. [2] Vol. XIV, p. 957.
[3] Vol. IX, p. 375. [4] Vol. XIV, p. 390.

The cicatricial type of alopecia may follow such diseases as lupus erythematosus, lichen planus, scleroderma or any condition wherein the hair-follicles are destroyed. This includes loss of the eyebrows after a carbuncle or zoster, or of the lashes after ulcerative blepharitis or trachoma, or of either after acid or alkaline burns or mycotic infections. Traumatic alopecia may be present in the scars left by trauma.

ALOPECIA ARTEFACTA is finally to be noted, practised on the eyebrows from time immemorial by women of fashion to produce the required fine, curved arch which seems to be a perennial desire, and sometimes in addition on the lashes by neurotes and hysterics—mainly young girls (*trichomania*)—for reasons more difficult to understand but frequently seen in those suffering from a severe emotional disturbance (Rohrbach, 1963; Greenberg and Sarner, 1965).

The *treatment* of alopecia is difficult apart from those symptomatic types wherein an ætiological factor can be attacked. In this connection the question of seborrhœa should always be remembered. Once the follicles are destroyed, restoration of the hair is, of course, impossible. Any measures to stimulate their growth have a very problematical effectivity, and the value of repeated shaving or oiling (Trotter, 1923) as well as the application of rubefacients, high-frequency currents or other physical measures with a supposed stimulatory effect, despite popular belief, is largely illusory—a clinical fact which applies to the loss of hairs, brows or lashes whatever the ætiology. In alopecia areata systemic corticosteroids may result in a re-growth within a short period in those conditions in which recovery would normally occur, but when the prognosis is poor the condition recurs on cessation of the drug (Kile, 1960; Alverdes and Schmidt, 1961); the topical infiltration of the bald area may stimulate re-growth, as on the eyebrows (Berger, 1961).

To remedy a permanent alopecia of either the lashes or the brows, many expedients have been tried to simulate the cilia, including the application of cosmetics or tattooing a line of indian ink (Kuhnt, 1922; El Bakry, 1937). The most effective procedure, however, is the grafting of new eyebrows from the scalp or the post-auricular region with a free graft or by a vascularized pedicle-graft from the temporal region, which is a relatively simple plastic operation[1]; but the grafting of eyelashes is a more tricky matter. Dzondi (1818) was the first to practise this by meticulously inserting hair-follicles into the lid-margins; Krusius (1914) used a special cylindrical needle for this, accomplishing 20 transplants in each session, a procedure used by Strampelli (1947). The simpler technique of transplanting a strip from the eyebrows to the ciliary margin by a flap-graft was advocated by Beck (1838), while a free graft of hair-bearing skin from the brow was suggested by Hirschberg (1892) and Knapp (1908–17), a procedure employed with various

[1] Clarkson (1948), Guerrero Santos *et al.* (1961), Longacre *et al.* (1962), González Fontana (1964), Mukhin (1965), Bracciolini and Travia (1968), Thomas *et al.* (1969), and others.

modifications by several surgeons.[1] Mir y Mir (1966) performed a tarsorrhaphy of the inner half of the upper and lower lids, sutured a graft from the eyebrow into this prepared layer and two months later opened the tarsorrhaphy thus dividing the graft into two parts, one on each lid. Other expedients include transplants from the scalp (Schuessler and Filmer, 1947; Tóth, 1950; Mutou and Khoo, 1962) or from the nape of the neck (Lexer, 1919; Wick, 1925). It is important that the three or four rows of hairs in the grafted strip lie in the same direction as the lashes; the outer rows atrophy and fall out in a few weeks leaving the inner rows to obtain the desired cosmetic result and any irregular lashes may be destroyed by electrolysis. Variations in these techniques are described in books on plastic surgery of the lids.[2]

Adler and Nyhan. *J. Pediat.*, **75,** 436 (1969).

Alverdes and Schmidt. *Derm. Wschr.*, **144,** 1342 (1961).

Amalric and Bessou. *Bull. Soc. belge Ophtal.*, No. 124, 881 (1960).

El Bakry. *Bull. ophthal. Soc. Egypt*, **30,** 99 (1937).

Beck. von Ammon's *Mschr. Med. Augenheilk. Chir.*, **1,** 24 (1838).

Berger. *Arch. Derm.*, **83,** 151 (1961).

Beyer and Smith. *Arch. Ophthal.*, **85,** 445 (1971).

Bracciolini and Travia. *Riv. ital. Tracoma*, **20,** 429 (1968).

Callahan. *Reconstructive Surgery of the Eyelids and Ocular Adnexa*, Bgham., Ala. (1966).

Church. *Brit. J. Derm.*, **77,** 661 (1965).

Clarkson. *Guy's Hosp. Gaz.*, **67,** 256 (1948).

Converse. *Reconstructive Plastic Surgery*, Phila. (1964).

Davenport. *Arch. Derm.*, **89,** 74 (1964).

Dzondi. *J. prakt. Heilk.*, **47,** 99 (1818).

Emiru. *Brit. J. Ophthal.*, **54,** 740 (1970).

Esser. *Klin. Mbl. Augenheilk.*, **62,** 202 (1919).

Fox. *Ophthalmic Plastic Surgery*, 4th ed., N.Y. (1970).

Franceschetti, Rossano, Jadassohn and Paillard. *Dermatologica*, **112,** 512 (1956).

Godwin. *Wld. Rev. Nutr. Diet*, **3,** 107 (1962).

González Fontana. *Rev. esp. Oto-neuro-oftal.*, **23,** 283 (1964).

Greenberg. *Arch. Derm.*, **72,** 454 (1955).

Greenberg and Sarner. *Arch. gen. Psychiat.*, **12,** 482 (1965).

Guerrero Santos, Matus and Vera. *Plast. reconstr. Surg.*, **27,** 316 (1961).

Hirschberg (1892). *Graefe-Saemisch Hb. d. ges. Augenheilk.*, 2nd ed., Leipzig, **14** (2), 105 (1911).

Hirshovitz. *Acta med. orient.*, **14,** 200 (1955).

Hoffmann. *Arch. Derm. Syph.* (Wien), **89,** 381 (1908).

Hughes. *Reconstruction of the Lids*, St. Louis (1958).

Kile. *Arch. Derm.*, **81,** 979 (1960).

Kingery. *J. Amer. med. Ass.*, **195,** 571 (1966).

Knapp. *Klin. Mbl. Augenheilk.*, **46** (2), 317 (1908); **59,** 447 (1917).

Krusius. *Dtsch. med. Wschr.*, **40,** 958 (1914).

Kuhnt. *Hb. d. ärztlichen Erfahrungen im Weltkriege* 1914/1918 (ed. von Schjerning), Leipzig, **5,** 449 (1922).

Langhof and Lemke. *Derm. Wschr.*, **146,** 585 (1962).

Lexer. *Klin. Mbl. Augenheilk.*, **62,** 486 (1919).

Longacre, de Stefano and Homstrand. *Plast. reconstr. Surg.*, **30,** 638 (1962).

Loutfy, Fahmy and Ismail. *Bull. ophthal. Soc. Egypt*, **30,** 181 (1937).

Mahto. *Brit. J. Ophthal.*, **56,** 546 (1972).

Mir y Mir. *Ann. Chir. plast.*, **11,** 102 (1966).

Mukhin. *Acta chir. plast.* (Praha), **7,** 15 (1965).

Mustardé. *Repair and Reconstruction in the Orbital Region*, Edinb. (1966).

Mutou and Khoo. *Plast. reconstr. Surg.*, **29,** 573 (1962).

Pérel. *Presse méd.*, **74,** 2161 (1966).

vander Ploeg and Stagnone. *Arch. Derm.*, **90,** 172 (1964).

Rohrbach. *Hautarzt*, **14,** 122 (1963).

Rook. *Brit. med. J.*, **1,** 609 (1965).

Schuessler and Filmer. *Plast. reconstr. Surg.*, **2,** 345 (1947).

Strampelli. *Atti Cong. Soc. oftal. ital.*, **36,** 728 (1947).

Thomas, Stricker and Reny. *Bull. Soc. Ophtal. Fr.*, **69,** 719 (1969).

Tóth. *Klin. Mbl. Augenheilk.*, **116,** 209 (1950).

Trotter. *Arch. Derm. Syph.* (Chic.), **7,** 93 (1923).

Wahlberg. *J. Amer. med. Ass.*, **106,** 1968 (1936).

Weerekoon. *Brit. J. Ophthal.*, **56,** 106 (1972).

[1] Esser (1919), Wheeler (1920–22), Hirshovitz (1955), Pérel (1966), Beyer and Smith (1971), and others.

[2] Hughes (1958), Converse (1964), Callahan (1966), Mustardé (1966), Fox (1970), and others.

Wheeler. *J. Amer. med. Ass.*, **75**, 1055 (1920). *Amer. J. Ophthal.*, **5**, 828 (1922).
Wick. *Z. Augenheilk.*, **55**, 322 (1925).
Wilbrand and Staelin. *Mitt. a. d. Poliklin. d. allg. Krankenhauses Hamburg*, **1**, 416 (1897).

Anomalies in Direction

TRICHIASIS

TRICHIASIS (θρίξ, τριχός, a hair) is a misalignment of the lashes so that instead of being directed outwards they turn inwards to come in contact with the cornea with all the unfortunate effects which this may have.

Owing largely to its association with trachoma and the constant and acute distress which it causes, trichiasis and its relief have excited attention from the earliest times. The Ebers Papyrus, which dates back to the 16th century B.C., is full of complaints about it; epilation is recommended and a host of prescriptions advanced to prevent re-growth of the lashes. Hippocrates (5th century B.C.) was well acquainted with it, and Celsus[1] burned out the follicles with a red-hot needle. Paul of Ægina, who flourished in Alexandria in the 7th century, tried to induce a cicatricial ectropion by including a fold of the skin of the lids between two pieces of reed or stick tied so tightly together that the skin underwent pressure-necrosis and fell off with the sticks; sometimes the lids became everted and lagophthalmos frequently resulted. The Chinese practised a somewhat similar operation, clamping the lids in bamboo forceps.[2] Suśruta, of ancient Hindu culture,[3] practised an incision of the skin of the lids, the sutures being tied to a band round the forehead to evert the margins, followed if necessary by cauterization by fire or alkali. So distressing is the complaint that such remedies were acceptable.

An inturning of the lashes is met with in distichiasis[4] and in entropion of the lower lid, senile, spastic or cicatricial, wherein the fault lies not in the cilia but in the lid-margin itself.[5] Trichiasis, however, is usually cicatricial in nature, the follicles being distorted and turned inwards in post-traumatic or post-inflammatory scarring. Thus, a localized trichiasis involving a few lashes may follow a stye, or a generalized inturning may follow ulcerative blepharitis. The most common cause, however, is trachoma, wherein the local scarring is aided by softening, thickening and inturning of the tarsal plate and therefore of the whole lid. In most of these conditions the associated cicatricial atrophy of the hair follicles leads to the formation of fine, downy or stunted and frequently unpigmented cilia which may be difficult to see except under the magnification of the loupe or the slit-lamp. To a certain extent, also, the mal-direction and hypoplasia are due to the presence of oblique and dwarfed accessory hairs arising from side-buds to the follicles, a condition thus related to polytrichia (Herzog, 1904) (Figs. 388–9).

The seriousness of the condition lies in the damage it causes to the cornea with its clamant symptoms ultimately leading in marked cases to opacity and gross visual damage. Initially the corneal epithelium may respond with a protective thickening of its outer layers which may undergo

[1] *De re medicina*, A.D. 29.
[2] Chen Hou-Hsi, *The Six Essences of Ophthalmology*, 1821.
[3] Vol. II, p. 18. [4] Vol. III, p. 873. [5] p. 573.

vascularization as a pannus; eventually ulcers are formed and ultimately a permanent and considerable opacity may result. Recurrent corneal irritation or ulceration the ætiology of which is difficult to explain should always suggest the possibility of trichiasis, for the presence of one or two minute hairs on the lid-margin is very readily and frequently missed.

The *treatment* of trichiasis varies with the number of lashes involved. If these are few, epilation should only be practised as a measure of temporary relief until more effective methods are available, for the cilia re-grow persistently and rapidly; it is most easily carried out with an epilation forceps, an operation sometimes performed with extraordinary efficiency after a little practice by a deft patient himself. If the trichiasis is local and involves a few lashes only, destruction of the offending follicles by electrolysis is sufficient; the negative pole is formed by a slender needle which is

FIGS. 388 and 389.—TRICHIASIS.

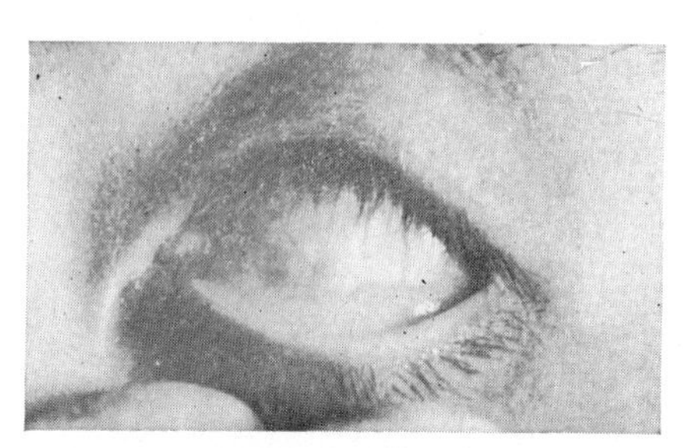

FIG. 388.—Involving the upper lid in a trachomatous Chinese (H. G. Scheie).

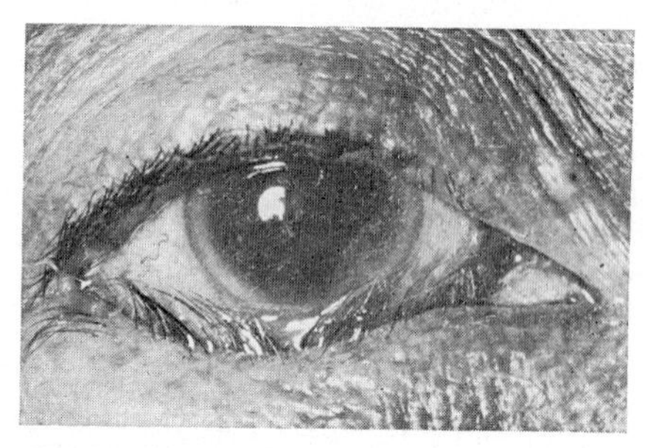

FIG. 389.—Involving the lower lid in a woman aged 48 (Inst. Ophthal.).

introduced into the follicle after anæsthetization of the lid-margin and retained until a light foam of hydrogen bubbles exudes around the root of the cilium, whereupon the lash is readily lifted out with forceps.

If, however, the trichiasis is general, surgical procedures are necessary. In cases due to entropion the offending lid is everted,[1] but when the lashes themselves are misaligned a large number of techniques has been employed.

(1) Excision of the ciliary border with the skin and part of the tarsal plate bearing the lashes (Flarer, 1824), has been abandoned. In Stellwag's (1883) modification of this the excised strip is placed in the reversed position, thus covering the raw area left by Flarer's " scalping " procedure.

(2) The ciliary border is displaced upwards after an incision in the intermarginal line towards the gap produced by excising a crescentic area of skin (Jaesche, 1844; von Arlt, 1874). The technique was improved by Waldhauer (1883–98) who placed the excised skin on the raw border below the out-turned lashes. A modification of this was made by Spencer Watson (1894) who, after splitting the entire lid in front of the tarsus from the intermarginal line, made two flaps and interchanged them so that the lower flap bearing the lashes is moved up to the anterior surface

[1] p. 578.

of the lid and the skin-flap lies below it. All these techniques have been modified in subsequent procedures.

(3) The grafting of flaps of mucous membrane derived from the fornix if it is healthy, or from the lower lip if it is not, into a split incision in the lid-border was a procedure introduced by van Millingen (1888–89) in order to eliminate the irritating effect of a flap of skin on the globe. This procedure has undergone several modifications and is still widely in use today (Forough *et al.*, 1972; and others).

(4) In cases wherein the trichiasis is due to a thickening and incurving of the tarsus (as in trachoma), attempts have been made to bend it outwards. For this purpose a deep groove was made on the anterior surface of the tarsus by Streatfeild (1858). The first to suggest the use of tension-sutures to draw the ciliary border towards the upper border of the tarsus after excising the orbicularis muscle anterior to it was Anagnostakis (1857). A somewhat similar procedure was advised by Hotz (1880) who later improved it by excising a wedge from the anterior surface of the tarsus to aid its eversion (1888); again, this type of operation has undergone many modifications.

(5) A longitudinal incision through the tarsus from the conjunctival surface so that the ciliary border can fall away from the globe to be maintained in eversion by deep sutures tied over the skin was practised by several surgeons; if necessary a strip of skin may be excised (Burow, 1873; Green, 1880; Ewing, 1907; and others). This procedure with modifications is still performed (Hadija, 1960). Panas (1882) used an anterior approach through the skin for the incision.

(6) An excision of the tarsus and conjunctiva has been advocated in severe inturning of the lid by several surgeons. The tarsus may be excised completely (Saunders, 1811; Heisrath, 1882; Kuhnt, 1897). In cases of marked deformity due to trachoma this procedure has been widely employed (MacCallan, 1913; Boase, 1952; Raïs *et al.*, 1970; and many others); alternatively a partial excision may be made of the lower half of the tarsal plate (Motais, 1934).

All these techniques have undergone innumerable modifications for which text-books of plastic surgery should be consulted. In cases of simple trichiasis the grafting of mucous membrane is very suitable; where the tarsus is not markedly deformed the excision of a wedge-shaped gutter from its anterior surface may be suitable; but in advanced cases of trachoma the combined excision of the tarsus and the conjunctiva is the most effective measure.

Anagnostakis. *Ann. Oculist.* (Paris), **38,** 5 (1857).
von Arlt. *Graefe-Saemisch Hb. d. ges. Augenheilk.*, 1st ed., Leipzig, **3,** 447 (1874).
Boase. *Brit. J. Ophthal.*, **36,** 645 (1952).
Burow. *Berl. klin. Wschr.*, **10,** 295 (1873).
Ewing. *Ophthal. Rec.*, **16,** 490 (1907).
Flarer. *Riflessioni sulla trichiasi*, Milano (1824).
Forough, Mohsen-Zadeh and Emami. *Ann. Oculist.* (Paris), **205,** 1215 (1972).
Green. *Trans. Amer. ophthal. Soc.*, **3,** 167 (1880).
Hadija. *Brit. J. Ophthal.*, **44,** 436 (1960).
Heisrath. *Berl. klin. Wschr.*, **19,** 428, 450, 461 (1882).
Herzog. *Z. Augenheilk.*, **12,** 180 (1904).
Hotz. *Klin. Mbl. Augenheilk.*, **18,** 149 (1880); **26,** 98 (1888).
Jaesche. *Med. Times* (Russia), No. 9 (1844). (Quoted by von Arlt, 1874).
Kuhnt. *Klin. Jb.*, **6,** Suppl. (1897).
MacCallan. *Ophthalmoscope*, **11,** 538 (1913).
van Millingen. *Arch. Ophtal.*, **8,** 60 (1888). *Zbl. prakt. Augenheilk.*, **13,** 193 (1889).
Motais. *Rev. int. Trachome*, **11,** 65 (1934).
Panas. *Arch. Ophtal.*, **2,** 208 (1882).
Raïs, Bessaïs and Triki. *Ann. Oculist.* (Paris), **203,** 845 (1970).
Saunders. *A Treatise on some Practical Points relating to the Eye*, London (1811).
Stellwag. *Arch. Augenheilk.*, **13,** 495 (1883).
Streatfeild. *Roy. Lond. ophthal. Hosp. Rep.*, **1,** 121 (1858).
Waldhauer. *Klin. Mbl. Augenheilk.*, **21,** 432 (1883); **36,** 47 (1898).
Watson, Spencer. *Trans. ophthal. Soc. U.K.*, **14,** 17 (1894).

CILIA INCARNATA

CILIA INCARNATA correspond to the phenomenon of *pili incarnati* (ingrowing hairs) which is more common in the beard region or the legs than on the lids and may occasionally be hereditary (Weninger, 1928). In this anomaly the hair grows out of the orifice of the follicle but so obliquely that after reaching the epidermis it grows into and under it; indeed, it may not penetrate the outer epidermal layer but grow beneath it for the whole of its extent. In the eyelid the phenomenon was first noted by Makrocki (1883) who termed it a *perverse subcutaneous growth of the cilium.* One lash or a clump of them may be affected and they may grow outwards under the surface of the skin of the lid (*cilium incarnatum externum*) or in the intermarginal area (*cilium intermarginalis*, Agarwala, 1963); in either event they are clearly visible under the epidermis. The cilium may become encased in an epithelial tunnel or surrounded by granulation tissue or a wart-like neoplasia which shows inflammatory changes and may develop into a papule.[1] Treatment is simple: the lash should be cut down upon and epilated.

CILIUM INCARNATUM INTERNUM is a more rare occurrence wherein the lash turns inwards under the conjunctiva tarsi which it usually pierces so that its termination moves freely in the cul-de-sac. Such behaviour gives rise to no harm unless its aberrant growth carries it onto the cornea. In this event an adult who has hitherto been without discomfort may suddenly complain of symptoms suggestive of a foreign body in the eye, the trouble being due to a perfectly formed lash curving directly inwards to impinge upon the eye. Epilation puts the matter right, but recurrences may have to be similarly treated.[2]

CILIUM INVERSUM is a congenital condition wherein the follicle is misplaced so that the lash grows directly into the tissues of the lid.[3]

Agarwala. *Amer. J. Ophthal.*, **55,** 648 (1963).

Agarwala and Munshi. *J. All-India ophthal. Soc.*, **14,** 253 (1966).

Bloch. *Arch. Ophthal.*, **37,** 772 (1947).

Casanovas. *Arch. Soc. oftal. hisp.-amer.*, **15,** 235 (1955).

Cherno. *Z. Augenheilk.*, **18,** 1 (1907).

Eisner. *Klin. Mbl. Augenheilk.*, **85,** 810 (1930).

Herzog. *Z. Augenheilk.*, **12,** 180 (1904).

Hirose and Hirayama. *Acta Soc. ophthal. jap.*, **36,** 1423 (1932).

Jess. *Klin. Mbl. Augenheilk.*, **83,** 47 (1929).

Kreiker. *Klin. Mbl. Augenheilk.*, **91,** 193 (1933).

Makrocki. *Zbl. prakt. Augenheilk.*, **7,** 129 (1883).

Mathur. *Int. Surg.*, **50,** 14 (1968).

Palich-Szántó. *Klin. Mbl. Augenheilk.*, **131,** 107 (1957).

Schreiber. *Graefe-Saemisch Hb. d. ges. Augenheilk.*, 3rd ed., *Die Krankheiten d. Augenlider*, Berlin, 380 (1924).

Sen, Mohan and Gupta. *Brit. J. Ophthal.*, **53,** 207 (1969).

Vershinin. *Vestn. Oftal.*, **80,** 85 (1967).

Weninger. *Ann. Derm. Syph.* (Paris), **9,** 687 (1928).

[1] Makrocki (1883), Herzog (1904), Cherno (1907), Schreiber (1924), Jess (1929), Bloch (1947), Casanovas (1955), Palich-Szántó (1957), Sen *et al.* (1969).

[2] Eisner (1930), Kreiker (1933), Hirose and Hirayama (1932), Bloch (1947), Agarwala and Munshi (1966), Vershinin (1967), Mathur (1968).

[3] Vol. III, p. 876.

Ciliary Trichoses

No diseases affect the lashes or the hairs of the brows alone since they themselves are inert structures, their state being determined by the condition of the parent follicle. Several atrophic changes may occur which are sometimes determined by severe constitutional disturbances, sometimes by local diseases such as seborrhœa, and sometimes by no apparent cause.

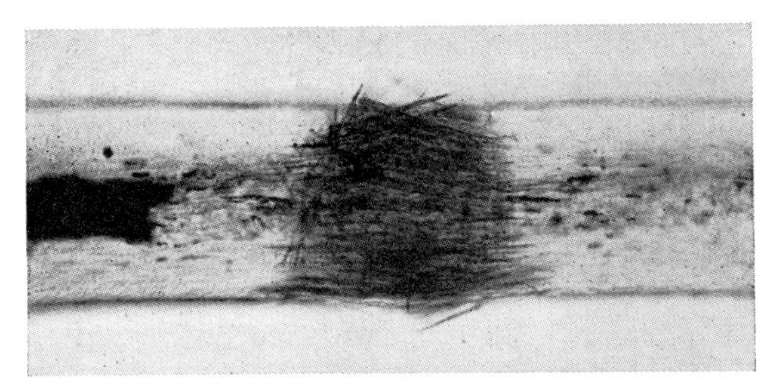

FIG. 390.—TRICHORRHEXIS NODOSA (R. L. and R. L. Sutton).

TRICHORRHEXIS NODOSA (NODOSITAS CRINIUM) is characterized by the formation of minute grey nodes on the shafts of the hair. These are due to a localized longitudinal splitting of the shaft so that the appearance is presented of two small brushes telescoped end-to-end (Fig. 390). The condition was first described by Beigel (1855). Occurring physiologically, it is the normal method of arresting an otherwise unlimited growth of hair generally (Heidingsfeld, 1905); occurring pathologically, multiple nodes appear in the shafts resulting in stunted growth, frequently as a result of trauma; this may be due to damage from chemicals used cosmetically or in permanent waving (Chernosky and Owens, 1966). The condition may affect the brows and lashes as part of a general disturbance or be confined to them (Coudert and Etienne, 1956).

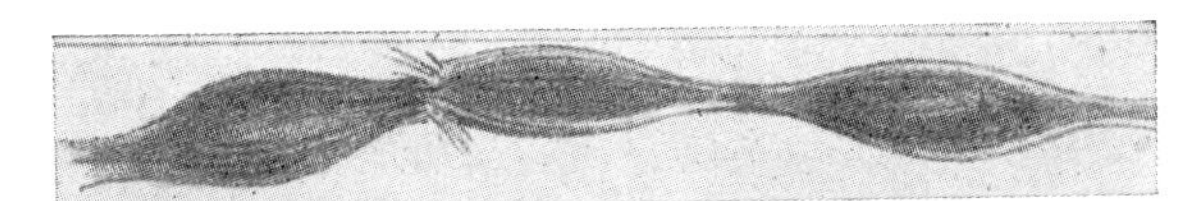

FIG. 391.—MONILETHRIX (R. L. and R. L. Sutton).

MONILETHRIX (L. *monile*, a necklace; θρίξ, a hair) (APLASIA PILORUM INTERMITTENS) is a condition first described by Walter Smith (1879–80) wherein the hair assumes a beaded appearance, nodular swellings alternating with annular atrophic constrictions (*moniliform hair*) (Fig. 391). It may be transmitted, usually as a dominant but occasionally as a recessive trait (Sabouraud, 1892; Tobias, 1923; Larsen, 1936). The hair is usually normal at birth but is eventually lost, frequently before the second month but sometimes not until adolescence or later life. The shafts are fragile and frequently break at the constrictions leading to alopecia, and there are corresponding constrictions in the parent follicle (*follicular hyperkeratosis*); eventually the hairs of the brows and the lashes, like those of the head and elsewhere on the body, consist mainly of broken stumps which are themselves extremely brittle, while the ciliary margins are usually irritated and inflamed. At first glance, indeed, the condition may resemble an ordinary blepharitis (Beatty and Scott, 1892; Galloway, 1896;

Collins, 1899; and others). Some cases remain permanently but others improve or completely recover in later life; systemic corticosteroids have been recommended.

PILI TORTI is a condition wherein the hair, including the lashes, emerge at unusual angles, are fragile, short and scanty, and are twisted at 180° on their axes at varying intervals along their length (Beare, 1952) (Fig. 392). Cases may appear sporadically, but transmission as a dominant character may occur. Sometimes the child is born bald[1]; in other cases the hair is normal at birth and the condition becomes evident between the second and third year or is delayed until adolescence or adult life. Some cases improve after puberty; others remain stationary throughout life. The anomaly may be confined to the lashes (three generations, Jaeger, 1956) but the eyebrows and scalp may be affected as well (two generations, Touraine *et al.*, 1938; three generations, Scott, 1953) or the axillary and pubic hair may also be involved (Friederich and Seitz, 1955). It has been noted to form part of the clinical picture of an ectodermal dysplasia (Friederich and Seitz, 1955; Gregory, 1955).

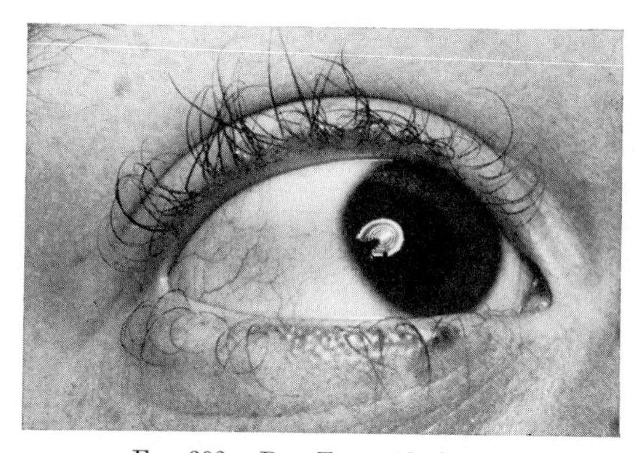

FIG. 392.—PILI TORTI (O. Scott).

FRAGILITAS CRINIUM. Brittleness of the hairs is shown by a tendency to split at either end or in the middle of the shaft. It may be associated with follicular atrophy.

None of these atrophies is amenable to treatment. In a condition such as pili torti wherein the eyebrows are lost they can be reconstituted by grafting[2] (Tanzer, 1956).

Beare. *Brit. J. Derm.*, **64,** 366 (1952).
Beatty and Scott. *Brit. J. Derm.*, **4,** 171 (1892).
Beigel. *Denkschr. wien. kais. Acad. Wiss.*, **17,** 612 (1855).
Chernosky and Owens. *Arch. Derm.*, **94,** 577, 586 (1966).
Collins. *Trans. ophthal. Soc. U.K.*, **19,** 1 (1899).
Coudert and Etienne. *Bull. Soc. Ophtal. Fr.*, 327 (1956).
Friederich and Seitz. *Derm. Wschr.*, **131,** 277 (1955).
Galloway. *Brit. J. Derm.*, **8,** 41 (1896).
Gregory. *Brit. J. Ophthal.*, **39,** 44 (1955).
Heidingsfeld. *J. cutan. Dis.*, **23,** 246 (1905).
Jaeger. *1st int. Cong. hum. Genet.*, Copenhagen, 142 (1956).
Larsen. *Hospitalstidende*, **79,** 129 (1936).
Sabouraud. *Bull. Soc. franç. Derm. Syph.*, **3,** 362 (1892).
Scott. *Clinical Genetics* (ed. Sorsby), London, 220 (1953).
Smith. *Brit. med. J.*, **2,** 291 (1879); **1,** 654 (1880).
Tanzer. *Plast. reconstr. Surg.*, **17,** 406 (1956).
Tobias. *Arch. Derm. Syph.* (Chic.), **8,** 655 (1923).
Touraine, Huber, Weissenbach *et al.* *Bull. Soc. franç. Derm. Syph.*, **45,** 441 (1938).

[1] Vol. III, p. 877. [2] p. 381.

Anomalies of Pigmentation

CANITIES : POLIOSIS

Acquired decolorization of the cilia and brows, in contradistinction to the congenital condition of leucotrichia[1] is usually termed CANITIES (*canus*, white) when the condition is generalized, POLIOSIS (πολίος, grey) when localized.

SENILE CANITIES is a physiological manifestation of the ageing process wherein the hair, preferentially of the head beginning at the temples and sometimes involving all the hairs on the body, turns grey and white. The loss of pigment is due to a progressive deficiency in the tyrosinase activity

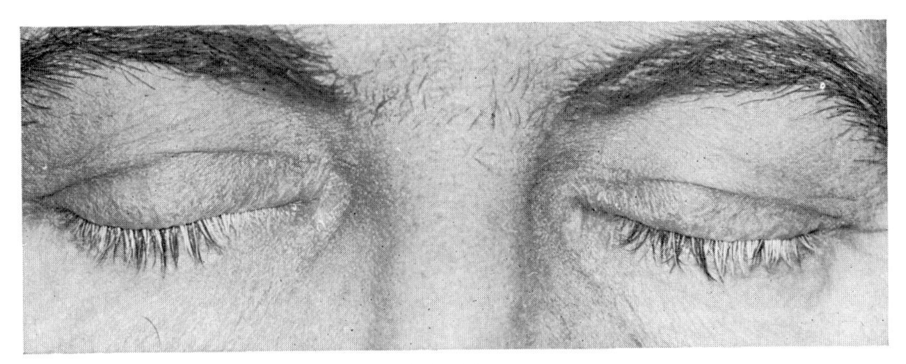

FIG. 393.—POLIOSIS.
In a woman aged 44 (Inst. Ophthal.).

FIGS. 394 and 395.—POLIOSIS AND VITILIGO.

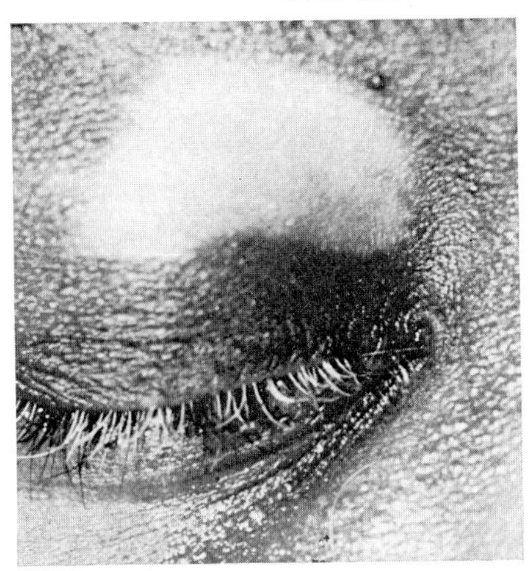

FIG. 394.—In a Negro girl aged 19 suffering from Harada's disease (M. Bruno and S. D. McPherson).

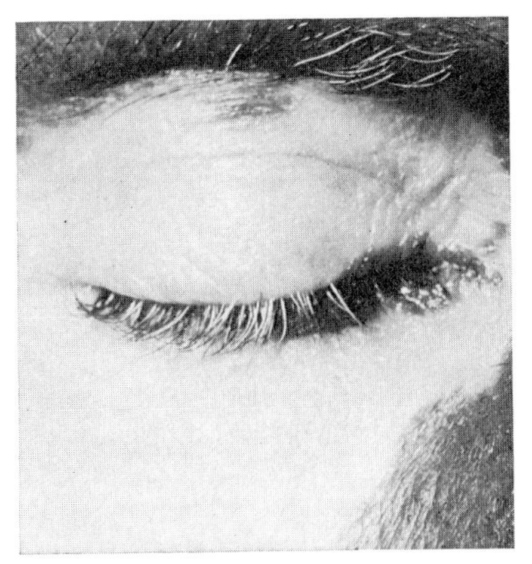

FIG. 395.—In a Negro aged 64 after topically administered thiotepa (J. W. Berkow *et al.*).

[1] Vol. III, p. 886.

of the melanocytes in the hair bulbs. The age of onset is determined to a considerable extent by heredity; a generalized premature canities of this nature may affect the eyebrows and lashes (Cockayne, 1933) (Fig. 393).

SYMPTOMATIC POLIOSIS of the lashes and sometimes of the brows may occur in several conditions. It is usually combined with vitiligo affecting the lids.[1] The most interesting is that associated with chronic uveitis (Hutchinson, 1892), particularly sympathetic ophthalmitis,[2] chorioretinitis (Erdmann, 1911), Harada's disease (Bruno and McPherson, 1949)[3] (Figs. 371, 394) or the Vogt-Koyanagi syndrome comprised of a diffuse bilateral uveitis of self-limiting tendency, alopecia, vitiligo and deafness.[4] A depigmentation of the hair may also occur in leprosy (Beretti and Cahuzac, 1970) or after the administration of drugs notable among which are chloroquine (Saunders *et al.*, 1959; Lazar *et al.*, 1963; Fitzpatrick *et al.*, 1965) and thiotepa used topically (Fig. 395) (Berkow *et al.*, 1969; Howitt and Karp, 1969)[5]. Sobhy Bey and Tobgy (1932) reported from Egypt a case of parasitic poliosis which appeared to be due to the presence of a rare hair-mite. *Poliosis neurotica* has, as a rarity, been associated with nervous disturbances or with severe emotional stress (Touraine, 1945; Ephraim, 1959). The hairs may become white in the course of a few days, a change of appearance due to the falling out of the pigmented hair in an alopecia of acute onset. Accounts of sudden blanching, such as that recorded by Roose (1899) of a girl whose lashes turned white overnight after a fright, should be viewed with suspicion; most of such incidents are probably the result of the removal of cosmetics. A rapid poliosis of the lashes has been reported after trigeminal neuralgia (Bock, 1890), migraine (Vogt, 1906) and in hysteria (Hutchinson, 1892). A cataract operation has had such a sequel on the operated side (Reich, 1881), as has severe trachoma (Herzog, 1904; Oesterreicher, 1913) or an injury (Bach, 1905; Steindorff, 1914).

In other cases of apparently spontaneous poliosis no cause has been found; most of them occurred in young people, frequently the blanching is patchy and sometimes it is associated with whitening of the brows (Ponti, 1859; Hirschberg, 1888; Rindfleisch, 1902; and others).

Bach. *Z. Augenheilk.*, **14**, 246 (1905).
Beretti and Cahuzac. *Arch. Ophtal.*, **30**, 313 (1970).
Berkow, Gills and Wise. *Arch. Ophthal.*, **82**, 415 (1969).
Bock. *Klin. Mbl. Augenheilk.*, **28**, 484 (1890).
Bruno and McPherson. *Amer. J. Ophthal.*, **32**, 513 (1949).
Cockayne. *Inherited Abnormalities of the Skin*, London, 63 (1933).
Ephraim. *Arch. Derm.*, **79**, 228 (1959).
Erdmann. *Klin. Mbl. Augenheilk.*, **49** (1), 129 (1911).
Fitzpatrick, Szabó and Mitchell. *Advances in Biology of the Skin*, N.Y., **6**, 35 (1965).
Herzog. *Z. Augenheilk.*, **12**, 180 (1904).
Hirschberg. *Zbl. prakt. Augenheilk.*, **12**, 15 (1888).
Howitt and Karp. *Amer. J. Ophthal.*, **68**, 472 (1969).
Hutchinson. *Arch. Surg.*, **4**, 357 (1892–93).
Jacobi. *Klin. Mbl. Augenheilk.*, **12**, 153 (1874).
Lazar, Regenbogen and Sepin. *Ophthalmologica*, **146**, 411 (1963).
Nettleship. *Trans. ophthal. Soc. U.K.*, **4**, 83 (1884).
Oesterreicher. *Prag. med. Wschr.*, **38**, 485 (1913).
Ponti. *G. Oftal. ital.*, **2**, 105 (1859).
Reich. *Arch. Ophtal.*, **1**, 307 (1881).
Rindfleisch. *Klin. Mbl. Augenheilk.*, **40** (2), 53 (1902).
Roose. *Ann. Oculist.* (Paris), **122**, 314 (1899).
Saunders, Fitzpatrick, Seiji *et al.* *J. invest. Derm.*, **33**, 87 (1959).
Schenkl. *Arch. Derm. Syph.* (Wien), **4**, 137 (1873).
Sobhy Bey and Tobgy. *Bull. ophthal. Soc. Egypt*, **25**, 155 (1932).
Steindorff. *Klin. Mbl. Augenheilk.*, **53**, 188 (1914).
Tay. *Trans. ophthal. Soc. U.K.*, **12**, 29 (1892).
Touraine. *Progr. méd.* (Paris), **73**, 47 (1945).
Vogt. *Klin. Mbl. Augenheilk.*, **44** (1), 228 (1906).

[1] p. 369.
[2] Schenkl (1873), Jacobi (1874), Nettleship (1884), Bock (1890), Tay (1892), Vogt (1906) and others. Vol. IX, p. 573.
[3] Vol. IX, p. 377. [4] Vol. IX, p.375. [5] p. 371.

CHAPTER VII

CYSTS AND TUMOURS

IT IS only fitting that a chapter should be introduced by the portrait of an Irishman; and ARTHUR JACOB [1790–1874] (Fig. 396), the first Irish ocular pathologist, certainly deserves this mention. For 41 years professor of anatomy and physiology to the Royal College of Surgeons in Ireland and three times its president, he practised ophthalmology as an able clinician and surgeon during a long and prolific professional life. He was the first to describe the nervous layer of the retina (1819) (*Jacob's membrane*), but is most remembered for the first accurate description of a rodent ulcer ("Observation respecting an Ulcer of Peculiar Character which attacks the Eyelid and other Parts of the Face", *Dubl. Hosp. Rep.*, **4,** 232, 1827) (Fig. 441). In addition to his scientific and clinical work, inspired by his early studies at Moorfields Eye Hospital in London, he founded two ophthalmic hospitals in Dublin, the first when he was but 27 years of age; moreover, he was joint founder and sole editor for 21 years of the *Dublin Medical Press*, from the pages of which he issued, with robust and exuberant Irish enthusiasm, a constant stream of medico-political articles wherein he showed himself a dauntless champion of doctors' rights and a fearless opponent of all forms of professional impropriety and quackery. Socially, politically and professionally, he was one of the great figures of his age in Ireland.

CYSTS OF THE LIDS

Cysts of the lids are relatively unimportant. They comprise:

(1) Cysts of the sweat glands, either on the skin (hidrocystoma) or on the lid-margin (Moll's cysts).

(2) Cysts of the sebaceous glands, either on the skin (milium, atheroma) or on the lid-margin (Zeis's glands) or of the meibomian glands.

(3) Cysts of Krause's glands.[1]

CYSTS OF THE SWEAT GLANDS

HIDROCYSTOMATA (ἱδρώς, sweat), a rare cystic condition of the sweat glands first described by Robinson (1884–85), are deep-seated, non-inflammatory, translucent vesicles of the size of a pin-head which appear on the face occasionally over large areas, and sometimes in great numbers on the lids, particularly the lower (Fig. 397). They give rise to no symptoms and never rupture spontaneously, but after some weeks or months may undergo desiccation. Exposure to moist heat in persons inclined to hyperhidrosis is frequently an antecedent factor. Histopathologically two distinct types exist. Those arising from an eccrine gland are comprised of a large cyst extending throughout most of the dermis, lined by two rows of flattened epithelial cells; leading into it is a dilated sweat duct but no duct leads out of

[1] p. 642.

Fig. 396.—Arthur Jacob
[1790–1874].

Jacob as a young man, from a painting in the offices of *The Medical News and Review* (London), a paper that he founded.

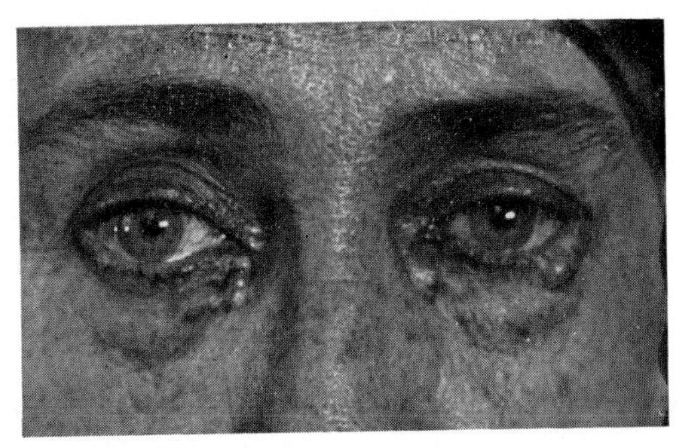

FIG. 397.—HIDROCYSTOMATA OF BOTH LOWER LIDS.
(Dept. of Dermatology, Vanderbilt Clinic, N.Y.)

FIGS. 398 and 399.—HIDROCYSTOMATA OF THE EYELIDS.

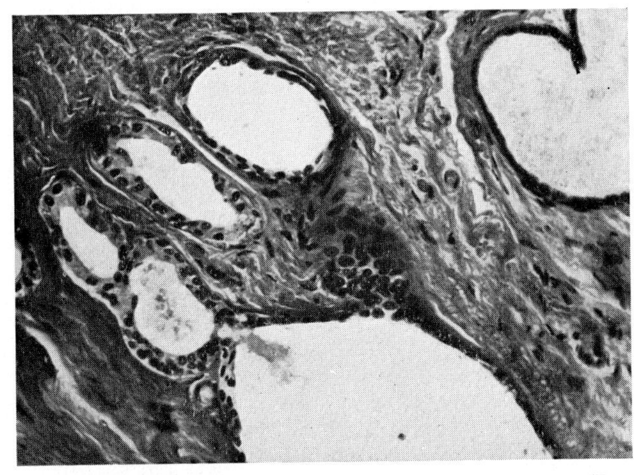

FIG. 398.—Showing sweat gland tubules and sweat gland cysts (A. Hagedoorn).

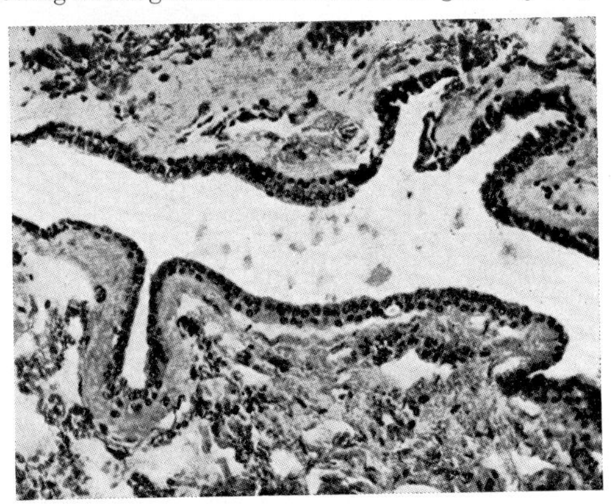

FIG. 399.—A portion of one of the cystic spaces, showing the double lining of epithelium the inner of which is cubical and regular, the outer smaller and less regular ($\times 150$) (G. Coats).

it; an apocrine cyst which is usually single (Fig. 342) contains several cystic spaces lined by a layer of secretory cells peripheral to which is a row of elongated myo-epithelial cells (Figs. 398–9).[1] A congenital case in the upper lid was reported by Saunders (1973). A suitable method of treatment is by electrolysis or excision; mild doses of X-rays cause lasting disappearance of the lesions in recurrent cases.

Cavara (1919) described a large retention cyst of a palpebral sweat gland due to hyperkeratosis.

CYSTS OF MOLL'S GLANDS differ in several respects from the retention cysts in the skin of the lids. Instead of developing rapidly and tending to disappear without rupturing, they grow slowly and continuously and may reach a considerable size without any tendency to disappear or shrink; thus

FIGS. 400 and 401.—RETENTION CYSTS OF MOLL'S GLANDS.

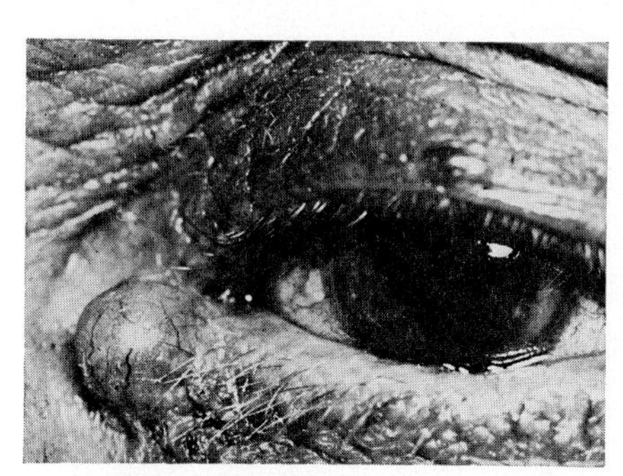

FIG. 400.—A typical sudoriferous cyst of Moll's gland (B. Daicker).

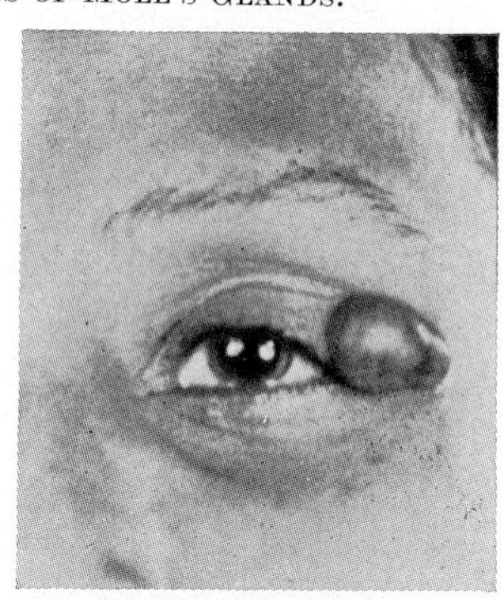

FIG. 401.—A cyst of unusual size recurring after removal (A. Kamel).

Orzalesi (1937) described one of dimensions of 21 × 14 × 19 mm. They form pinkish transparent vesicles filled with clear fluid and one or more may be found at any point in the ciliary margin, although they have a predilection for the lower lid near the lacrimal punctum (Figs. 400–1).

They are generally said to be retention cysts due to inflammatory processes or hyperkeratosis of the ciliary margin, but pathological evidence shows that a proliferative element may be present. The wall consists of fibrous tissue in which is embedded striped muscle fibres and the cavity is lined with epithelium. If the tubule of the gland is involved this epithelium is composed of tubular cells arranged in a single layer, and the cyst may be multilocular; if the duct is implicated there is a double or treble layer of flattened epithelium and the cavity is unilocular. The fluid contents may contain crystals of cholesterol and calcium sulphate.[2]

[1] von Michel (1901), Coats (1912), Arzt (1922), Herzberg (1962), Gross (1965), Castellazzo and Ciurlo (1967).

[2] Desfosses (1880), Wedl and Bock (1886), Wintersteiner (1896–1900), von Michel (1901), Ahlström (1903), Hagedoorn (1938), Vancea *et al.* (1959).

Occasionally proliferation of the walls occurs to form papillomatous projections into the cavity of the cyst, a connecting link thus being formed with adenomata[1] and, exceptionally, malignant changes have occurred (Zeeman, 1923; Weizenblatt, 1957). If the cyst is small, treatment may be by electrolysis; if it is large, by excision of the cyst wall.

CYSTS OF THE SEBACEOUS GLANDS

MILIUM (L. *millet*). Milia are frequently seen on the lids. They appear as rounded, white, sharply circumscribed, pearly tumours about the size of a pin-head, elevated above the surface of the skin, without inflammatory reaction and unassociated with symptoms (Fig. 402). There is no reliable evidence regarding their mode of origin but they may spring from a pilosebaceous follicle, constituting a retention cyst (Love and Montgomery, 1943); on the other hand, many have been regarded as keratinizing benign tumours consisting of accumulations of horn cells (Epstein and Kligman, 1956). They may arise spontaneously but may develop after trauma, zoster or during the course of bullous diseases such as epidermolysis bullosa. Treatment is electrolysis, diathermy or excision.

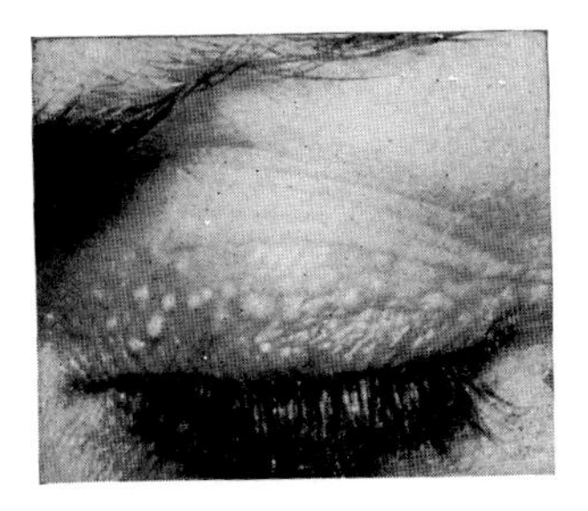

FIG. 402.—MILIA (R. L. and R. L. Sutton).

PILAR (SEBACEOUS) and EPIDERMOID CYSTS occur in the lids as elsewhere, particularly in the regions of the brows and inner canthus where hair follicles are numerous (Socin, 1871; Daicker, 1966; and others) (Figs. 403–4). They form smooth, globular subcutaneous tumours closely attached to the skin and over the summit the glandular orifice may sometimes persist in the centre as a waxy, comedo-like, blackened plug. Their contents are epithelial cells

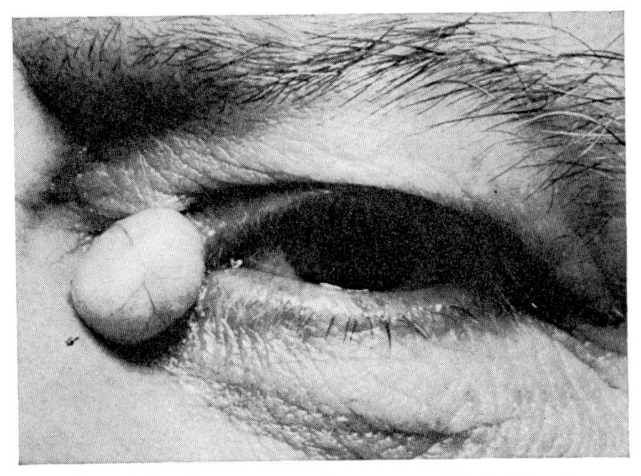

FIG. 403.—RETENTION ATHEROMA (B. Daicker).

[1] p. 459.

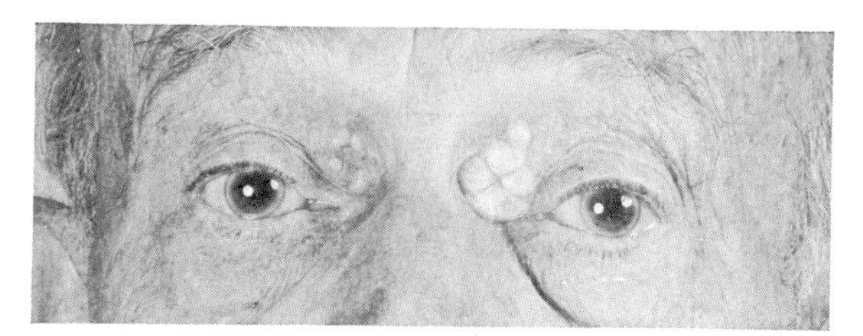

Fig. 404.—Sebaceous Cysts.
Affecting the upper lids (B. Jay).

Figs. 405 and 406.—Cyst of Zeis's Gland.

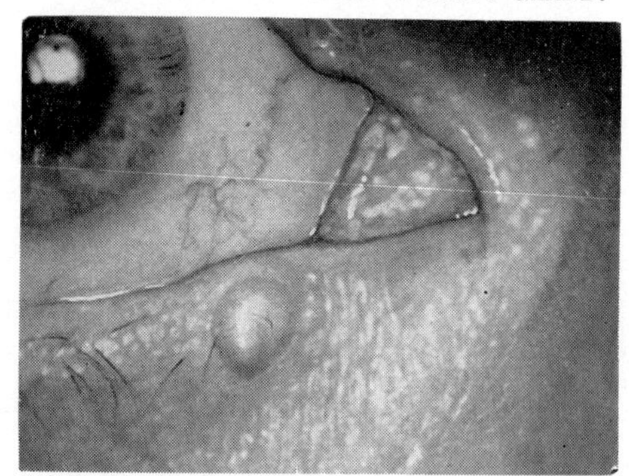

Fig. 405.—Near the inner canthus (M. J. Reeh and K. C. Swan).

Fig. 406.—Section showing the general configuration. It is lined by layers of squamous epithelium and contains keratin and sebaceous material (E. Wolff).

in various stages of disintegration, keratin, fatty granules and cholesterol crystals, and the cyst itself is lined with a thick layer of stratified epithelium showing fatty and atheromatous changes surrounded by a capsule of fibrous tissue. They were generally considered to be retention cysts of sebaceous glands, but the distension of the follicle of the associated hair probably figures largely in their formation (Shattock, 1897; Grönvall, 1953; Lever, 1967); most of them are epidermoid cysts filled with laminated keratin (Broders and Wilson, 1930; Cramer, 1966).

Such cysts may remain quiescent for a long time or may slowly grow to reach a considerable size. Sometimes inflammatory changes develop

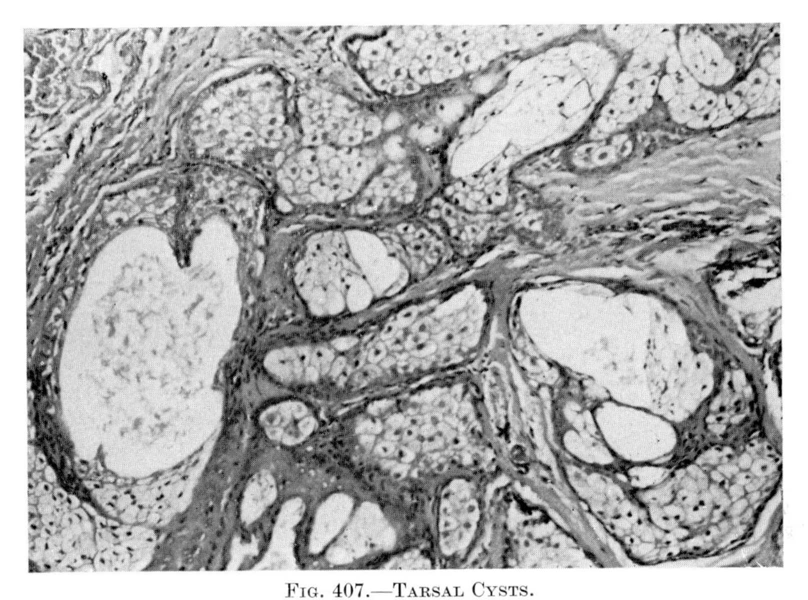

FIG. 407.—TARSAL CYSTS.
Early tarsal cysts forming in a meibomian gland (× 120) (N. Ashton).

particularly after trauma. Calcification sometimes occurs (Sternberg, 1904) and malignant changes (Bishop, 1931; Collins, 1936).

The ideal *treatment* is complete excision together with a small elliptical piece of skin containing the orifice of the gland, care being taken to avoid rupture during the operative manipulations. If infection has occurred systemic antibiotics should be given and excision postponed until the inflammation has subsided. The method of incision, squeezing out the contents and applying caustics to the walls is not satisfactory, for recurrences follow unless the lining epithelium is completely destroyed.

CYSTS OF THE GLANDS OF ZEIS have a similar ætiology, springing from the follicle of a cilium and its associated sebaceous gland at the lid-margin as small rounded white tumours (Fig. 405). They have a similar pathology, being lined by layers of squamous epithelium and contain keratin and sebaceous material (Wolff, 1949) (Fig. 406).

MEIBOMIAN CYSTS in the true sense of retention cysts of the tarsal glands are rare; they have been produced artificially by obliteration of the ducts by cicatrization and are usually met with in long-standing trachoma when deformity and cicatrization are advanced. The cysts are lined by flattened epithelium lying upon a dense fibrous basis representing the thin and distended tarsus and contain a fibrinous coagulum and granular debris (de Vincentiis, 1875; Wintersteiner, 1896) (Fig. 407). Clinically imperceptible cystic dilatations of the meibomian gland have been observed during histological examination of the tarsal plate adjacent to neoplasms and surgical incisions (Straatsma, 1959) and in a case described by Leibiger (1961) recurrent cysts in the lid were said to be due to over-secretion from the meibomian glands. An epithelial implantation cyst following incision and curettage of a chalazion is a rare occurrence (Sen, 1967).

Ahlström. *Ann. Oculist.* (Paris), **129,** 107 (1903).
Arzt. *Frankf. Z. Path.*, **28,** 507 (1922).
Bishop. *Ann. Surg.*, **93,** 109 (1931).
Broders and Wilson. *Surg. Clin. N. Amer.*, **10,** 127 (1930).
Castellazzo and Ciurlo. *Ann. Ottal.*, **93,** 967 (1967).
Cavara. *Arch. Ottal.*, **26,** 93 (1919).
Coats. *Roy. Lond. ophthal. Hosp. Rep.*, **18,** 266 (1912).
Collins. *Canad. med. Ass. J.*, **35,** 370 (1936).
Cramer. *Arch. klin. exp. Derm.*, **224,** 168 (1966).
Daicker. *Ophthalmologica*, **152,** 267 (1966).
Desfosses. *Arch. Ophtal.*, **1,** 82 (1880).
Epstein and Kligman. *J. invest. Derm.*, **26,** 1 (1956).
Grönvall. *Acta ophthal.* (Kbh.), **31,** 235 (1953).
Gross. *Arch. Derm.*, **92,** 706 (1965).
Hagedoorn. *Amer. J. Ophthal.*, **21,** 487 (1938).
Herzberg. *Arch. klin. exp. Derm.*, **214,** 600 (1962).
Hogan and Zimmerman. *Ophthalmic Pathology*, 2nd ed., Phila., 194 (1962).
Leibiger. *Klin. Mbl. Augenheilk.*, **138,** 876 (1961).
Lever. *Histopathology of the Skin*, 4th ed., Phila. (1967).
Love and Montgomery. *Arch. Derm. Syph.* (Chic.), **47,** 185 (1943).
von Michel. *Arch. Augenheilk.*, **42,** 1 (1901).
Orzalesi. *Boll. Oculist.*, **16,** 167 (1937).
Robinson. *J. cutan. Dis.*, **2,** 362 (1884).
Manual of Dermatology, N.Y., 84 (1885).
Saunders. *Arch. Ophthal.*, **89,** 205 (1973).
Sen. *Orient. Arch. Ophthal.*, **5,** 123 (1967).
Shattock. *Trans. path. Soc. Lond.*, **48,** 224 (1897).
Socin. *Virchows Arch. path. Anat.*, **52,** 550 (1871).
Sternberg. *Zbl. allg. Path. path. Anat.*, **15,** 988 (1904).
Straatsma. *Arch. Ophthal.*, **61,** 918 (1959).
Vancea, Balan, Lazarescu and Vaighel. *Ophthalmologica*, **137,** 233 (1959).
de Vincentiis. *Ann. Ottal.*, **4,** 208 (1875).
Wedl and Bock. *Pathologische Anatomie d. Auges*, Wien (1886).
Weizenblatt. *Amer. J. Ophthal.*, **43,** 968 (1957).
Wintersteiner. *Arch. Augenheilk.*, **33,** Erg., 114 (1896); **40,** 291 (1900).
Wolff. *Ann. roy. Coll. Surg.*, **4,** 58 (1959).
Zeeman. *Ned. T. Geneesk.*, **67,** 1194 (1923).

TUMOURS OF THE LIDS

As would be expected in a region with so complex a structure, tumours of the lids form a heterogeneous collection. Most of them are skin-tumours and many of them are benign, but sufficient of them are malignant to necessitate the greatest care in the diagnosis of each. They may be classified as follows:

I. EPITHELIAL TUMOURS (A) CUTANEOUS

(a) *Benign*
1. Papilloma.
2. Senile keratosis.
3. Seborrhœic keratosis.

4. Kerato-acanthoma.
5. Inverted follicular keratosis.
6. Tricho-epithelioma.
7. Benign calcified epithelioma of Malherbe.
8. Cornu cutaneum.

(b) *Malignant.* 1. Carcinoma
(i) Squamous-cell epithelioma.
(ii) Basal-cell epithelioma (rodent ulcer).
(iii) Intra-epithelial carcinoma.

2. Xeroderma pigmentosum.

(B) GLANDULAR

(a) *Tumours of the Sebaceous Glands*

1. Adenoma
Sebaceous adenoma of the skin; adenoma of the meibomian glands; of Zeis's glands.

2. Adenocarcinoma
Adenocarcinoma of the skin; of the meibomian glands; of Zeis's glands.

(b) *Tumours of the Sweat Glands*

1. Hidradenoma
Hidradenoma of the skin; syringoma; pleomorphic adenoma; adenoma of Moll's glands.
2. Hidradenocarcinoma
Hidradenocarcinoma of the skin; of Moll's glands.

(c) Papillary cystadenoma lymphomatosum

(d) Oncocytoma

II. MESENCHYMAL TUMOURS

(a) *Benign.* Fibroma; tuberous sclerosis; lipoma; rhabdomyoma; leiomyoma; myxoma; chondroma
(b) *Malignant.* Sarcoma.

III. TUMOURS OF LYMPHO-RETICULAR TISSUE

(a) Benign lymphoma
(b) Lymphosarcoma
(c) Reticulum-cell sarcoma
(d) Giant follicular lymphoma
(e) Burkitt's lymphoma
(f) Hodgkin's disease
(g) Mycosis fungoides
(h) Plasmocytoma.

IV. VASCULAR TUMOURS

(a) Hæmangioma: capillary, cavernous, plexiform; hæmangio-endothelioma, hæmangio-pericytoma; spider angioma; senile angioma; angioma serpiginosum.
(b) Telangiectatic granuloma.
(c) Angiokeratoma of Mibelli.

(d) Multiple hæmorrhagic sarcoma of Kaposi.
(e) Glomus tumour.
(f) Lymphangioma; lymphangio-endothelioma.

V. NERVOUS TISSUE TUMOURS

(a) Neurofibromatosis: plexiform neuroma, diffuse neurofibromatosis, molluscum fibrosum, multiple mucosal neuroma syndrome.
(b) Neurilemmoma.
(c) Granular-cell schwannoma of Abrikossoff.
(d) Ganglioneuroma.
(e) Amputation neuroma.

VI. PIGMENTED TUMOURS

(a) Nævus.
(b) Malignant melanoma.

VII. METASTATIC TUMOURS

VIII. DEVELOPMENTAL TUMOURS: dermoids, teratoma, phakomatous choristoma.

The incidence of these lesions is interesting. Statistics have been published by O'Brien and Braley (1936) on 100 consecutive cases seen at Iowa; Welch and Duke (1958) on 617 cases seen at the Wilmer Institute, Baltimore; Kwitko and his colleagues (1963) on 1,176 cases examined at the Armed Forces Institute of Pathology in Washington; and H. and J. Allington (1968), a dermatologist and an ophthalmologist in California, on 1,687 seen in practice.

Welch and Duke (1958) found the following incidence: papilloma 125 (20·3 %), nævus 89 (14·4 %), chalazion 80 (13 %), basal-cell carcinoma 68 (11 %), epidermoid cyst 45 (7·3 %), dermoid cyst 26 (4·2 %), epithelial cyst 23 (3·7 %), xanthelasma 19 (3·1 %), inflammatory 18 (2·9 %), hæmangioma 17 (2·8 %), basal-cell papilloma 16 (2·6 %), sebaceous cyst 13 (2·1 %), baso-squamous carcinoma 9 (1·5 %), squamous-cell carcinoma 8 (1·3 %), verruca vulgaris 7 (1·1 %), hyperkeratosis 7 (1·1 %), miscellaneous 7·6 %.

H. and J. Allington (1968) found the following incidence: benign keratosis and papilloma 546 (20·6 %), milium 407 (15·4 %), nævus 350 (13·2 %), hæmangioma 144 (5·4 %), syringoma 62 (2·3 %), keratinous cyst 58 (2·2 %), dermatitis papulosa nigra 29 (1·1 %), sudoriferous cyst 25 (0·9 %), actinic keratosis 16, lentigo 15, xanthelasma 13, basal-cell carcinoma 7, chalazion 6, cyst of Krause's or Wolfring's gland 4, senile sebaceous nævus 2, Bowen's disease 1, verrucose epidermal nævus 1, neurofibroma 1.

The Clinical Approach to Tumours of the Eyelids

Many solid tumours of the eyelids are often diagnosed clinically as papillomata or rodent ulcers, while cystic lesions are frequently regarded as derived from sebaceous glands, but as the foregoing list indicates the wide variety of structures comprising the eyelid can give rise to a multiplicity of tumours so that accuracy in the clinical diagnosis requires an awareness of the many different lesions which may occur in the palpebral tissues. Some of these lesions have characteristic features which permit a definite clinical assessment, while others may be non-specific and, although benign, may easily be confused with malignant conditions. The difficulties in diagnosis are seen in that of the 617 conditions reported by Welch and Duke (1958),

50% were misdiagnosed clinically before histological examination. Advances in our knowledge of the pathology of the skin together with the long-term follow-up and histological reappraisal of treated lesions have shown that many of them formerly regarded as malignant are, in fact, benign. There is a tendency to over-diagnose squamous carcinoma when a more simple lesion is present; thus Kwitko and his collaborators (1963) found that of 115 lesions clinically diagnosed as this, only 9 were histologically confirmed. On the other hand, a condition which appears to be benign may be malignant, as when a chalazion-like swelling represents a basal-cell carcinoma or a metastasis. The conditions which have most frequently given rise to confusion with carcinoma of the eyelid include senile keratosis, inverted follicular keratosis, kerato-acanthoma, benign calcifying epithelioma of Malherbe, and a number of pseudo-epitheliomatous lesions induced by inflammatory processes. Recognition of the relatively harmless nature of most of these has led to a more conservative approach to treatment often by simple excision without the need for extensive reconstructive surgery or radiotherapy (Boniuk, 1962; Zimmerman, 1967). In certain countries where fungal diseases are common their presence in the eyelid may give rise to confusion with malignant tumours.

The accurate assessment of a tumour of the eyelid requires a consideration of the history of its development and a family history of similar occurrences, a careful clinical examination of the lesion and the associated lymph nodes and a search for other similar or associated lesions, visceral or cutaneous, elsewhere in the body. The recurrence of a previously treated lesion should raise a suspicion of malignancy particularly in the case of a chalazion which not infrequently mimics a carcinoma of the meibomian gland or even a rodent ulcer. Finally, the presence of systemic diseases such as leucæmia, lymphosarcoma or other reticuloses or of a carcinoma or neurofibromatosis elsewhere may explain the nature of an otherwise obscure palpebral tumour. The examination of the lesion should include a note of the exact site and size together with its shape, colour and consistency, the degree of elevation above the surface or of fixation to neighbouring structures, the presence or absence of inflammatory signs, vascularization or pigmentation. Photographic documentation is helpful for follow-up purposes.

Ultimately the diagnosis will be made by histological examination and in all cases in which there is doubt or a suspicion of malignancy a biopsy is essential. If this can be a complete excision-biopsy, so much the better, but certainly when radiotherapy is contemplated a histological diagnosis should be regarded as an essential preliminary step.

Allington, H. and J. *Arch. Derm.*, **97**, 50 (1968).
Boniuk. *Int. Ophthal. Clin.*, **2**, 237 (1962).
Kwitko, Boniuk and Zimmerman. *Arch. Ophthal.*, **69**, 693 (1963).
O'Brien and Braley. *J. Amer. med. Ass.*, **107**, 993 (1936).
Welch and Duke. *Amer. J. Ophthal.*, **45**, 415 (1958).
Zimmerman. *Arch. Ophthal.*, **78**, 166 (1967).

Descriptions of palpebral tumours are given in the following books:

Boniuk (editor). Ocular and Adnexal Tumors, St. Louis (1964).
François (editor). The Tumors of the Eye and its Adnexa, Basel (1966).
Hogan and Zimmerman. Ophthalmic Pathology, 2nd ed., Phila. (1962).

Iliff and Ossofsky. Tumors of the Eye and Adnexa in Infancy and Childhood, Springfield (1962).
Offret and Haye. Tumeurs de l'oeil et des annexes oculaires, Paris (1971).
Reeh. Treatment of Lid and Epibulbar Tumors, Springfield (1963).
Reese. Tumors of the Eye, 2nd ed., N.Y. (1963).

Cutaneous Epithelial Tumours

The lids share in the common tumours which arise from the surface epithelium, both simple hyperplasias and malignant neoplasms—carcinoma in its various forms including the virulently malignant reaction to light seen in xeroderma pigmentosum. Between these two extremes there lies a group of cutaneous diseases which may give rise to malignant infiltrating neoplasms; the precancerous dermatoses which involve the eyelids include senile keratosis, Bowen's disease and radiation dermatitis.[1]

PAPILLOMA

True PAPILLOMATA are to be sharply distinguished from the various types of infective warts which we have already discussed.[2] Pathologically the latter are not true blastomata but are inflammatory papillary hypertrophies. A true papilloma, on the other hand, shows proliferation of the epithelium with thickening and hyperkeratosis around a central fibrous core, and blunt

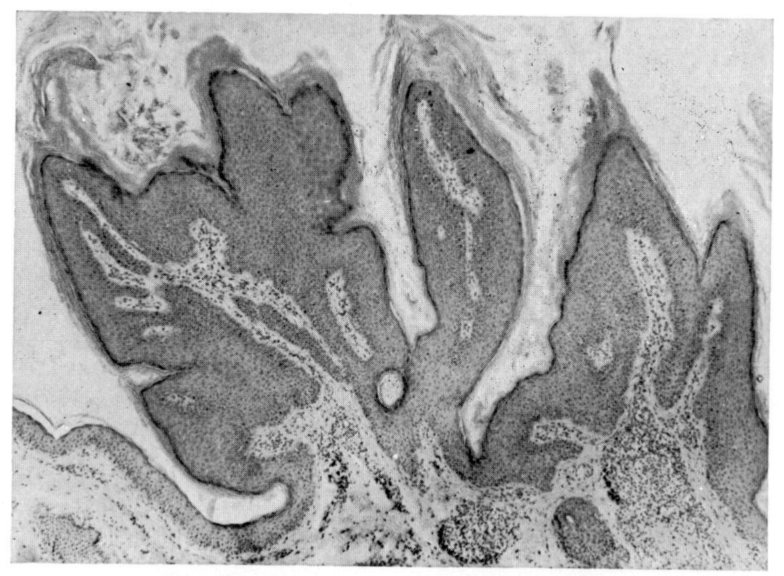

FIG. 408.—SIMPLE PAPILLOMA OF THE EYELID.

Showing the irregular proliferation of the squamous epithelium, thickening of the epidermis and hyperkeratosis with a chronic inflammatory reaction in the dermis (H. & E.; ×48) (N. Ashton).

[1] Vol. XIV, p. 956. [2] p. 150.

FIGS. 409 to 412.—PAPILLOMATA (Inst. Ophthal.).

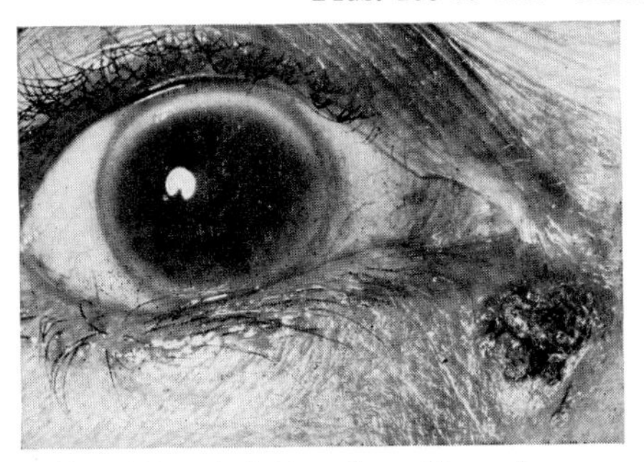

FIG. 409.—A flattened papilloma in a man aged 65.

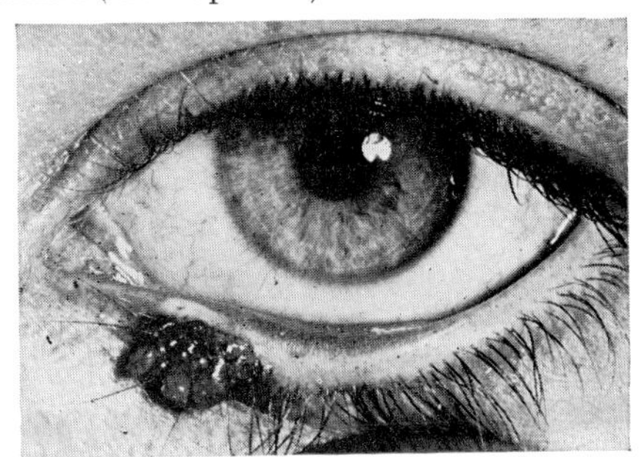

FIG. 410.—In a boy aged 14.

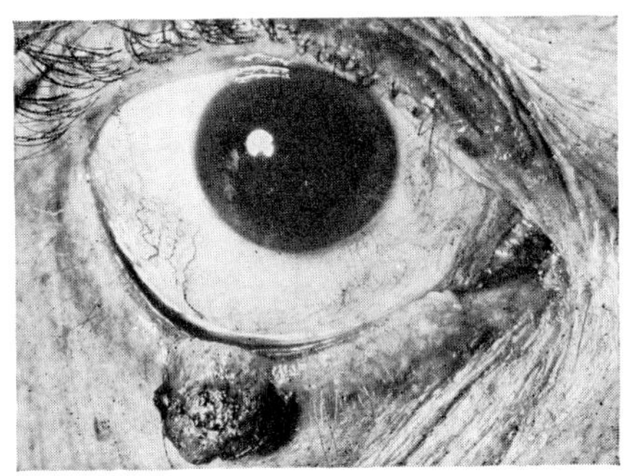

FIG. 411.—In a man aged 52.

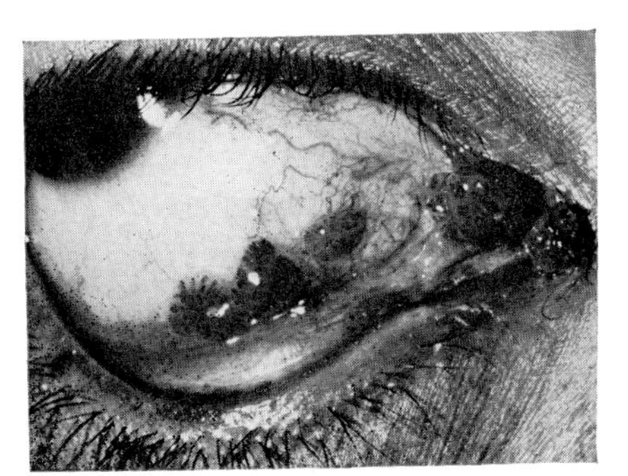

FIG. 412.—Recurrent papilloma, initially in the caruncle. Eight months after removal, multiple recurrences are seen.

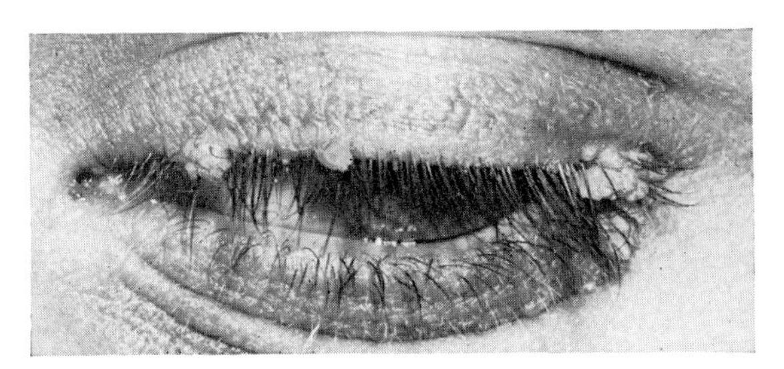

FIG. 413.—MULTIPLE PAPILLOMATA ON THE EYELIDS (M. Lederman).

processes of epithelial cells reaching down into the corium (Fig. 408). It is neither infectious nor inoculable, occurs usually in adults or old people, appears slowly and not suddenly in crops, and is stationary or slowly progressive with no tendency to spontaneous cure.

True papillomata are the commonest tumours of the lids (34% of palpebral neoplasms, O'Brien and Braley, 1936), where one or more may be evident (Figs. 409-413). Their site of predilection is a region where the epithelium undergoes transition. They may occur *on the lid-margin*, particularly near the inner canthus, where they may be seen in young children (8 months, Walker, 1945); in this site they usually resemble papillomata of the conjunctiva,[1] appearing as sessile, raspberry-like growths with squat processes (Fig. 410); here they may remain for a considerable time (10 years, Bose, 1965). They sometimes appear in considerable numbers along the margins of both lids (Fig. 413); as in a case described by Schrire (1949) wherein the tumours were noted soon after birth, were removed at the age of 8 and again successfully at the age of 70 when they had attained a size such that for visual purposes the lids had to be manually separated. Occasionally attempts at removal seem to stimulate multiple recurrences not only on the lid but on the caruncle and conjunctiva (Doherty, 1932) (Fig. 412). In the case reported by Walker (1945) wherein the tumour originated in the upper punctum of a baby, recurrences of this type which had disseminated over most of the palpebral conjunctiva and still retained their benign character had to be removed 14 times.

Bose. *Calcutta med. J.*, **62**, 334 (1965).
Doherty. *Amer. J. Ophthal.*, **15**, 1016 (1932).
O'Brien and Braley. *J. Amer. med. Ass.*, **107**, 933 (1936).
Schrire. *S. Afr. med. J.*, **23**, 307 (1949).
Walker. *Amer. J. Ophthal.*, **28**, 751 (1945).

SENILE KERATOSIS

SENILE (SOLAR) KERATOSIS is a common condition characterized by the appearance of flat or verrucose, brownish, circumscribed growths which appear on the skin of adults in regions exposed to sunlight so that the face, including the eyelids, is frequently involved (Fig. 414). It resembles the lesions caused by carcinogenic substances or those of xeroderma pigmentosum, tending to develop on an inherited or acquired type of skin, harsh, dry and prone to freckles, particularly in blond people (Sutton, 1938). The keratoses which sometimes appear as a cutaneous horn, may drop off, remain static, or take on malignant characteristics (Montgomery, 1935–39; Sutton, 1942; Coman, 1944–47); indeed, senile keratosis can be regarded as a precancerous lesion which may give rise to a squamous-cell carcinoma but with little or no tendency to metastasize (Graham and Helwig, 1963; Hirata, 1971). Moreover, patients with this condition frequently have other cutaneous neoplasms such as a basal or squamous carcinoma (Boniuk, 1962–64) or adeno-acanthoma. Histologically there is hyperkeratosis, parakeratosis and irregular acanthosis with bizarre epithelial cells and numerous mitoses, while the dermal papillæ and underlying tissues are usually infiltrated by large numbers of chronic inflammatory cells (Hogan and Zimmerman, 1962) (Fig. 415). The involvement of the deeper layers of the epidermis distinguishes this condition from Bowen's disease in which all layers of the epidermis are involved. In the early stages treatment by the removal of the

[1] Vol. VIII, p. 1159.

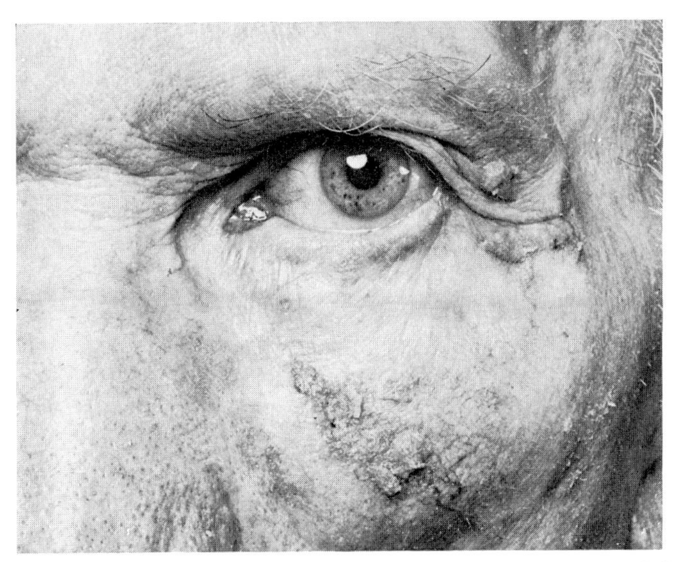

FIG. 414.—SOLAR (SENILE) KERATOSIS (St. John's Hospital for Diseases of the Skin).

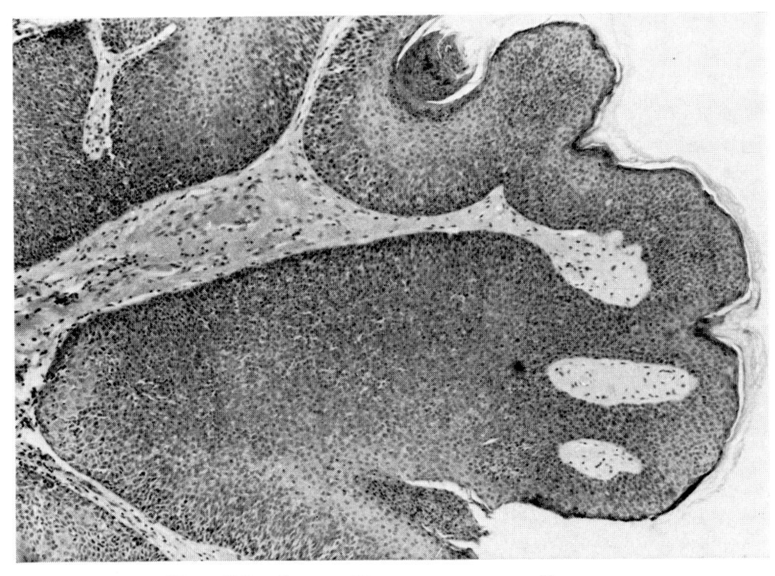

FIG. 415.—SENILE KERATOSIS OF THE EYELID.

There is hyperkeratosis and papillomatosis with an irregular downward extension of the proliferating cells of the stratum malpighii. Numerous mitoses are present (H. & E.; ×60) (N. Ashton).

outer horny layers by grease or an ointment of salicylic acid and sulphur may be sufficient but if the warty growth persists local destruction by freezing or curettage is advisable.

Boniuk. *Int. Ophthal. Clin.*, **2**, 239 (1962).
Ocular and Adnexal Tumors, St. Louis (1964).
Coman. *Cancer Res.*, **4**, 625 (1944).
Science, **105**, 347 (1947).
Graham and Helwig. *Nat. Cancer Inst. Monog.*, **10**, 323 (1963).
Hirata. *Folia ophthal. jap.*, **22**, 457 (1971).
Hogan and Zimmerman. *Ophthalmic Pathology*, 2nd ed., Phila., 198 (1962).
Montgomery. *Arch. Derm. Syph.* (Chic.), **32**, 218 (1935); **39**, 387 (1939).
Sutton. *Arch. Derm. Syph.* (Chic.), **37**, 737 (1938); **46**, 1 (1942).

SEBORRHŒIC KERATOSIS

SEBORRHŒIC KERATOSIS (*seborrhœic wart, senile wart, basaloid-cell papilloma*, Pinkus, 1953–65) is a common benign tumour occurring usually in the elderly. It is a sharply circumscribed wart-like lesion frequently pigmented, of a soft friable consistency and appears as if it is stuck on the surface of the skin. It may occur in large numbers in any area where there are pilosebaceous follicles, particularly on the face and lids, the trunk and the arms, and usually has a diameter of a few millimetres but occasionally reaches a size of over a centimetre (15 × 13 × 13 mm., Miki, 1972); seborrhœa plays no part in its genesis. The commonest appearance is that of a verrucose plaque varying from dirty yellow to black in colour; on the eyelids, however, and on the major flexures they may be pedunculated and not keratotic.

Histopathologically three types have been recognized (Braun-Falco and Kint, 1963)—hyperkeratotic, acanthotic and adenoid. In the first the wart is produced by an accumulation of immature epidermal cells, hyperkeratosis and papillomatosis being prominent. In the second type acanthosis is prominent and most of the cells are basaloid in character; cystic inclusions of horny material are frequent (*pseudo-horn cysts*). In the adenoid type there is minimal acanthosis and papillomatosis and fine fissures may break the contour; in this type of tumour thin tracts composed of a double row of epidermal cells form the main bulk of the lesion and there are no cystic inclusions. In these lesions irritation is common, associated with dermal inflammation causing a swelling of the wart sometimes with oozing and crusting. In this event there is a proliferation of squamous cells that may reach the proportions of a pseudo-carcinomatous hyperplasia which is sometimes difficult to differentiate from a true squamous-cell carcinoma (Rowe, 1957). In this phase the pathology resembles that of an inverted follicular keratosis, and between the two the differential diagnosis may be very difficult (Boniuk and Zimmerman, 1963).

Treatment for a smaller lesion is generally easy. Curettage usually leaves a flat surface which becomes covered with normal epidermis within a week, while satisfactory results may be obtained from refrigeration with the ethyl chloride spray which can be repeated if necessary. Pedunculated lesions of any size should be excised.

DERMATITIS PAPULOSA NIGRA is a variant of seborrhœic keratosis; it usually starts in adolescents and is most frequently seen in adult Negroes. The lesions appear as small pigmented papules occurring predominantly on the face including the lids

(29 cases, H. and J. Allington, 1968). A related pigmented lesion, termed MELANO-ACANTHOMA by Mishima and Pinkus (1960), has been described causing multiple, minute, shiny papules on the ciliary border of the lid by David and his colleagues (1972).

Allington, H. and J. *Arch. Derm.*, **97**, 50 (1968).
Boniuk and Zimmerman. *Arch. Ophthal.*, **69**, 693, 698 (1963).
Braun-Falco and Kint. *Arch. klin. exp. Derm.*, **216**, 615 (1963).
David, Wood and Heaton. *Arch. Derm.*, **105**, 898 (1972).
Miki. *Folia ophthal. jap.*, **23**, 421 (1972).
Mishima and Pinkus. *Arch. Derm.*, **81**, 539 (1960).
Pinkus. *Arch. Derm. Syph.* (Chic.), **67**, 598 (1953).
Hautarzt, **16**, 184 (1965).
Rowe. *J. invest. Derm.*, **29**, 165 (1957).

KERATO-ACANTHOMA

This tumour was first described as a " crateriform ulcer " by Jonathan Hutchinson (1889) (Fig. 416); the term KERATO-ACANTHOMA was introduced in 1950 (Rook and Whimster, 1950; Musso and Gordon, 1950) to replace the name *molluscum sebaceum*, applied by MacCormac and Scarff (1936). It is

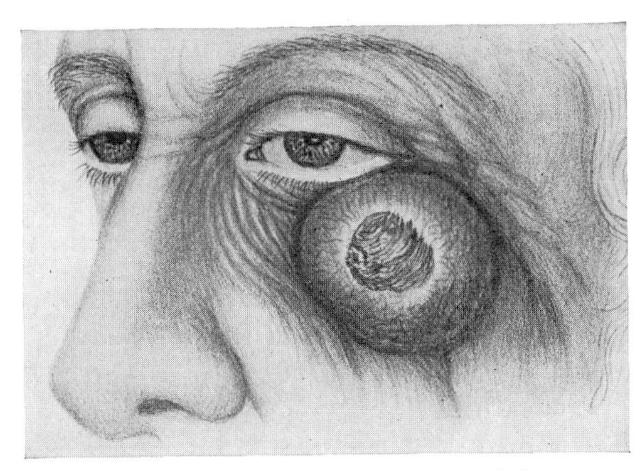

FIG. 416.—JONATHAN HUTCHINSON'S ILLUSTRATION OF A " CRATERIFORM ULCER " (1889).

In a man aged 62. The tumour had been present for 4 to 6 months. It was freely excised and a flap of skin transplanted into the wound; there was no recurrence.

another benign epithelial lesion which has many times been confused with a squamous-cell carcinoma (Kwitko *et al.*, 1963; Boniuk and Zimmerman, 1967) (Figs. 417–8). The condition occurs in older people and may affect the skin of the hands or face including the eyelids, usually the lower lid where a single lesion may be present; although a solitary lesion is most common, a generalized form of the disease may occur with multiple self-healing lesions which may be familial and arise in early life (Smith, 1934; Epstein *et al.*, 1957; Lever, 1967). Three types were described by Grinspan and Abulafia (1955): a verrucose type covered with a scale that is difficult to detach, a molluscum sebaceum type with a crater-like central opening from which soft greasy keratinous material may be extruded, and a nodulo-vegetating type with a lobulated surface. Cases with involvement of the periorbital skin have frequently been recorded.[1] Baer and Kopf (1962) collected 592 cases from the literature of which 420 affected the skin of the face, 33 the eyelids, 2 the eyebrows, 161 the cheek, 100 the nose and 13 the temple.

The lesion is characterized by its rapid growth, reaching its maximum size of between 1 and 2 cm. within 6 to 8 weeks, after which spontaneous regression may occur over 2 to 3 months. In most cases the presence of a rapidly growing lump has been the main symptom, but occasionally complaints have been made of pain or discomfort on pressure or of irritation; lymphadenopathy does not occur (Offret and Haye, 1971). In the typical case there is a sharply elevated hemispherical lesion with a crusted surface

FIGS. 417 and 418.—KERATO-ACANTHOMA.

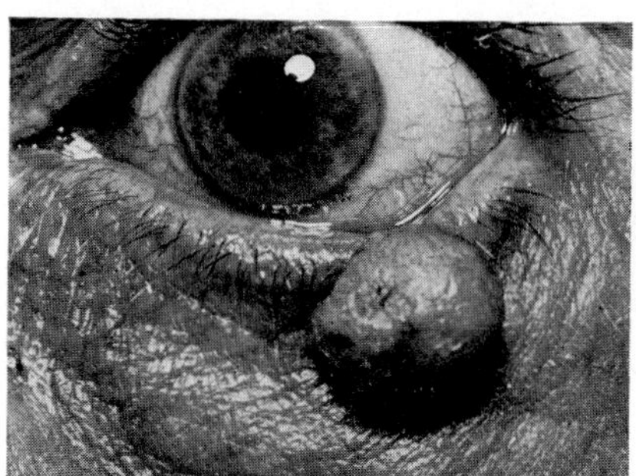

FIG. 417.—On the lower lid (B. Daicker).

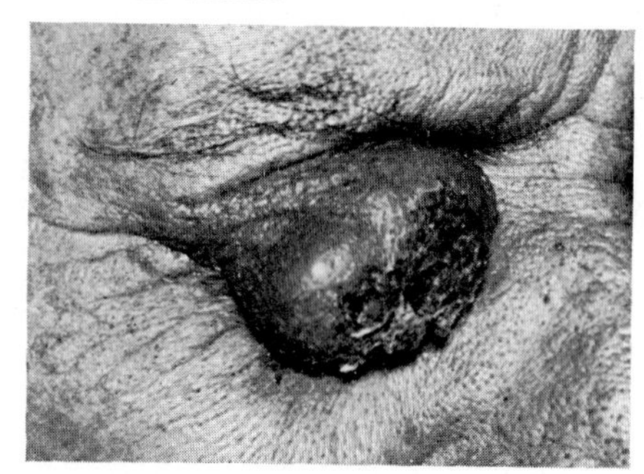

FIG. 418.—On the upper lid, showing ulceration (Inst. Ophthal.).

and keratin-filled central crater (Fig. 418); sometimes the appearance is warty and occasionally the patient presents with a cutaneous horn. Histologically the typical kerato-acanthoma has a cup-shaped configuration with a central keratin-filled crater into which protrude papillary epithelial projections, and from which keratinous debris accumulates on the surface (Figs. 419–20). There is marked acanthosis and an inflammatory infiltrate with large numbers of plasma cells and eosinophils in the corium. The surrounding epidermis, which shows no evidence of reactive proliferation, extends along the margin into the elevated lesion and partially covers its surface.

The main problem in diagnosis lies in its differentiation from a squamous-cell carcinoma; in the latter the growth is much less rapid and the lesion does not usually have the typical cup-shaped appearance of a kerato-acanthoma while inflammatory cells do not generally invade the proliferating epithelium in a true carcinoma.

[1] Ruedemann and Hoak (1955), Christensen and Fitzpatrick (1955), Hager (1957), Rodenhäuser (1957–66), Barry (1962), Renard *et al.* (1963), Mortada (1964), Freeman and Rossman (1965), Boniuk and Zimmerman (1967), Lalla and Thomas (1968), Várhegyi *et al.* (1968), Cardia *et al.* (1968), Slem *et al.* (1970).

FIGS. 419 and 420.—KERATO-ACANTHOMA (D. R. Barry).

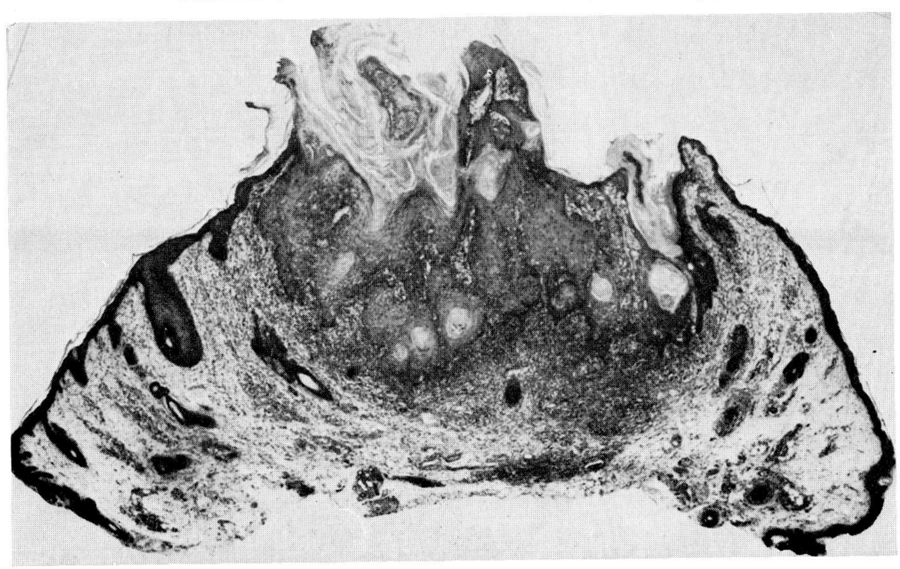

FIG. 419.—Showing the general appearance of the lesion with an intense underlying inflammatory reaction (H. & E.; ×24).

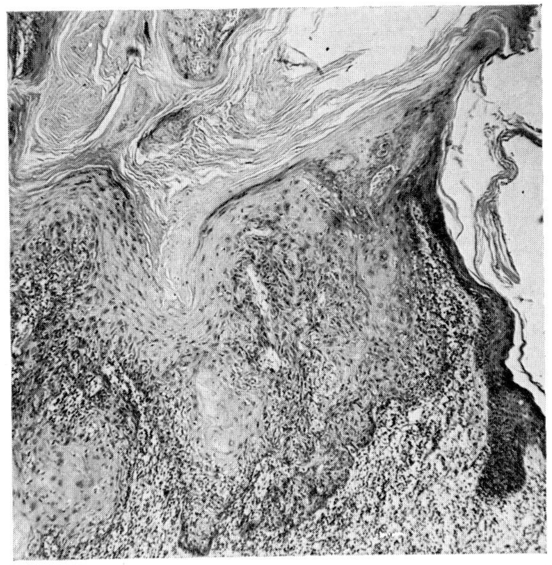

FIG. 420.—Acanthosis is very marked in this section suggesting a pseudo-epitheliomatous hyperplasia (H. & E.; ×56).

The pathogenesis of this condition is not established although numerous possible agents have been incriminated including trauma, sunlight or exposure to carcinogenic substances; the rapidity of its appearance and development has suggested a viral ætiology and, indeed, virus-like particles have been identified in some cases by electron microscopy (Ereaux *et al.*, 1955; Zelickson and Lynch, 1961), but attempts to culture the virus have so far been unsuccessful. Boniuk and Zimmerman (1967) considered that it was a specialized form of pseudo-epitheliomatous hyperplasia.

Since kerato-acanthoma is usually a self-healing tumour in the majority of cases, conservative *treatment* is justified. To reduce the extent of scarring and improve the cosmetic appearance several methods have been used such as the application of podophyllin, curettage, cauterization, electro-coagulation and freezing, while even radiotherapy and the intralesional injection of steroids have had their advocates. A histological differentiation from carcinoma may be difficult unless the whole lesion is examined and if there is any doubt complete excision should be undertaken, preferably at an early stage when the lesion is still small before the tumour reaches a size which may make surgical removal difficult. Recurrences are uncommon but have been reported after incomplete excision or more conservative methods of treatment (Baer and Kopf, 1962; Belisario, 1965); although the lesion is considered to be essentially benign, the remote possibility of malignant transformation has been held to justify complete surgical excision in every case (Iverson and Vistnes, 1973). Indeed, an instance has occurred in which a basal-cell carcinoma was found beneath a kerato-acanthoma (Einaugler *et al.*, 1968).

Baer and Kopf. *Year-Book of Dermatology*, Chicago, 7 (1962).
Barry. *Brit. J. Ophthal.*, **46**, 528 (1962).
Belisario. *Aust. J. Derm.*, **8**, 65 (1965).
Boniuk and Zimmerman. *Arch. Ophthal.*, **77**, 29 (1967).
Cardia, Balestrazzi and Sborgia. *Riv. ital. Tracoma*, **20**, 279 (1968).
Christensen and Fitzpatrick. *Arch. Ophthal.*, **53**, 857 (1955).
Einaugler, Henkind, de Oliveira and Bart. *Amer. J. Ophthal.*, **65**, 922 (1968).
Epstein, Biskind and Pollack. *Arch. Derm.*, **75**, 210 (1957).
Ereaux, Schopflocher and Fournier. *Arch. Derm.*, **71**, 73 (1955).
Freeman and Rossman. *E.E.N.T. Digest*, **27**, 93 (1965).
Grinspan and Abulafia. *Cancer*, **8**, 1047 (1955).
Hager. *v. Graefes Arch. Ophthal.*, **158**, 393 (1957).
Hutchinson. *Trans. path. Soc. Lond.*, **40**, 275 (1889).
Iverson and Vistnes. *Amer. J. Surg.*, **126**, 359 (1973).
Kwitko, Boniuk and Zimmerman. *Arch. Ophthal.*, **69**, 693 (1963).
Lalla and Thomas. *Brit. J. Ophthal.*, **52**, 876 (1968).
Lever. *Histopathology of the Skin*, 4th ed., Phila., 514 (1967).
MacCormac and Scarff. *Brit. J. Derm.*, **48**, 624 (1936).
Mortada. *Amer. J. Ophthal.*, **58**, 813 (1964).
Musso and Gordon. *Proc. roy. Soc. Med.*, **43**, 838 (1950).
Offret and Haye. *Tumeurs de l'oeil et des annexes oculaires*, Paris (1971).
Renard, Duperrat, Dhermy and Mascaro. *Arch. Ophtal.*, **23**, 5 (1963).
Rodenhäuser. *v. Graefes Arch. Ophthal.*, **158**, 468 (1957).
2nd Cong. europ. ophthal. Soc. (Wien, 1964) (ed. François), Basel, 473 (1966).
Rook and Whimster. *Arch. belges Derm.*, **6**, 137 (1950).
Ruedemann and Hoak. *Amer. J. Ophthal.*, **40**, 199 (1955).
Slem, Ilcayto and Turan. *Haseki Tip. Bült.*, **8**, 178 (1970).
Smith, J. F. *Brit. J. Derm.*, **46**, 267 (1934).
Várhegyi, Meszléri and Fehér. *Klin. Mbl. Augenheilk.*, **153**, 56 (1968).
Zelickson and Lynch. *J. invest. Derm.*, **37**, 79 (1961).

INVERTED FOLLICULAR KERATOSIS

This term was used by Helwig (1955) to describe the histological appearance of a lesion occurring mainly on the face and particularly on the eyelids as a solitary nodule or papule projecting from the surface of the skin (Duperrat and Mascaro, 1963; Boniuk and Zimmerman, 1963; Mehregan, 1964). The lesion is nearly always solitary, occurring predominantly in men, and is a non-malignant keratotic tumour involving the epithelial surface of the skin or lid-margin presenting as a nodular, papillomatous, verrucose or occasionally cystic lesion or even as a cutaneous horn; it may rarely be heavily pigmented (Fig. 421). Although the clinical appearance may thus vary, the condition has certain distinctive histological features

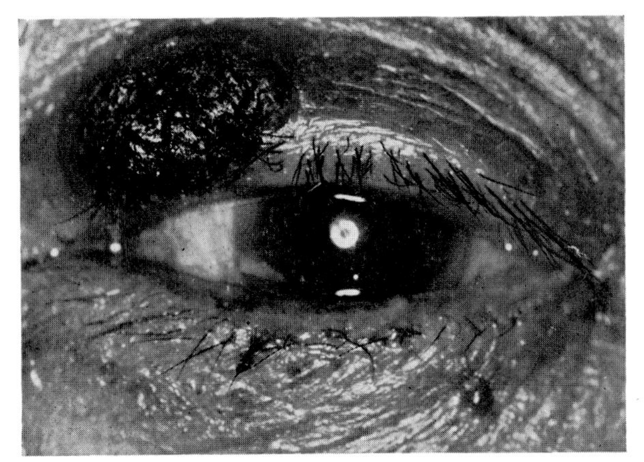

Fig. 421.—Inverted Follicular Keratosis.
Affecting the upper lid (M. Boniuk and L. E. Zimmerman).

(Kwitko *et al.*, 1963). Microscopically it often has an inverted cup-shaped configuration with the surrounding epidermis extending up over its margin thus simulating a kerato-acanthoma, an appearance that suggested to Helwig an analogy to the opening of a pilosebaceous complex (Fig. 422). There is hyperkeratosis and parakeratosis and discrete epithelial folds extending into the corium. In these folds squamoid cells may be seen in the superficial layers and basaloid cells in the deeper, and at the junction between these two types there is considerable inter- and intra-cellular œdema and clusters of squamoid cells arranged concentrically in small nests—the " squamous eddies " of Helwig (Fig. 423). Desquamation of the abnormal surface epithelium may give rise to a scab and probably accounts for the symptoms of burning, bleeding, itching and oozing frequently present in these cases. Follow-up studies have shown that they are benign, although in the past many of them were regarded as carcinomatous. Treatment should be by simple excision. The cause is not definitely known, but some have

FIGS. 422 and 423.—INVERTED FOLLICULAR KERATOSIS (M. Boniuk and L. E. Zimmerman).

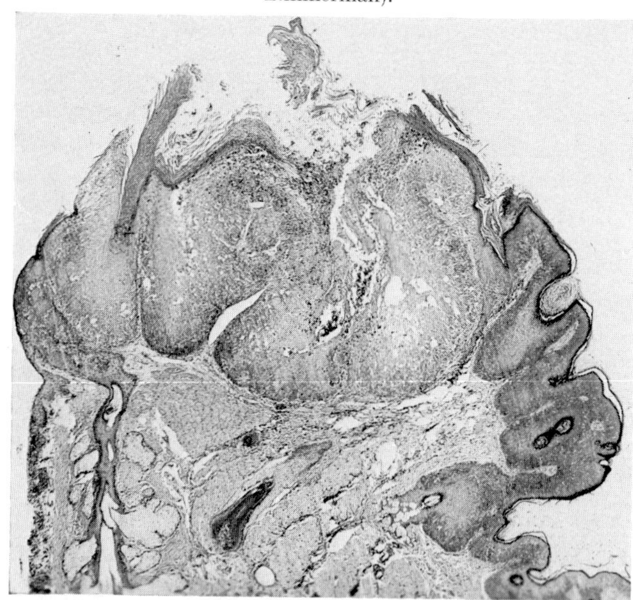

FIG. 422.—Affecting the margin of the lid; the conjunctival surface is on the left, the cutaneous surface on the right. There is marked hyperplasia of the adjacent epidermis (H. & E.).

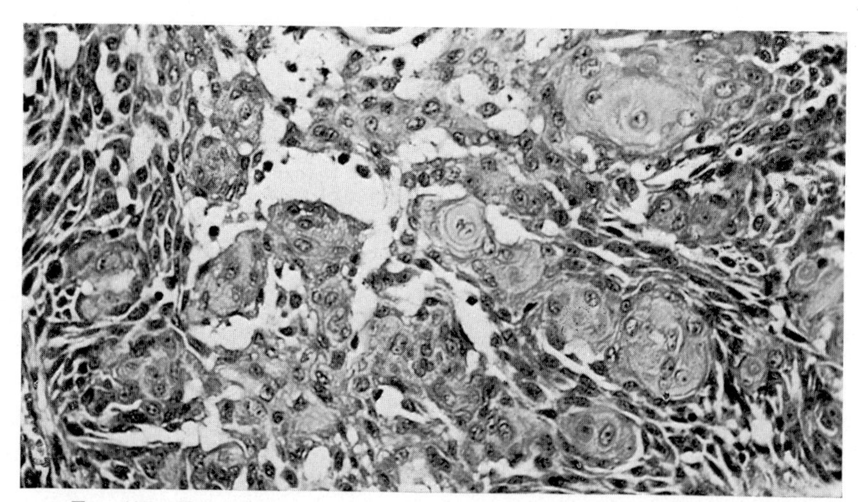

FIG. 423.—From the same case, showing nests of squamoid cells (" squamous eddies ") at the junction between the zones of basaloid and squamoid cells. Around the " squamous eddies " there is marked inter- and intra-cellular œdema.

thought that the condition is of viral origin (Boniuk, 1962; Boniuk and Zimmerman, 1963), while Lever (1967) considered it a form of irritated seborrhœic keratosis.

Boniuk. *Int. Ophthal. Clin.*, **2**, 239 (1962).

Boniuk and Zimmerman. *Arch. Ophthal.*, **69**, 698 (1963).

Duperrat and Mascaro. *Dermatologica*, **126**, 291 (1963).

Helwig. *Proc. XX Sem. Amer. Soc. clin. Path.*, Wash. (1955).

Kwitko, Boniuk and Zimmerman. *Arch. Ophthal.*, **69**, 693 (1963).

Lever. *Histopathology of the Skin*, 4th ed., Phila., 490 (1967).

Mehregan. *Arch. Derm.*, **89**, 229 (1964).

TRICHO-EPITHELIOMA

TRICHO-EPITHELIOMA, so named by Jarisch (1894), called MULTIPLE BENIGN CYSTIC EPITHELIOMA by Fordyce (1892) and EPITHELIOMA ADENOIDES CYSTICUM by Brooke (1892) (*Brooke's tumour*), is a rare neoplasm

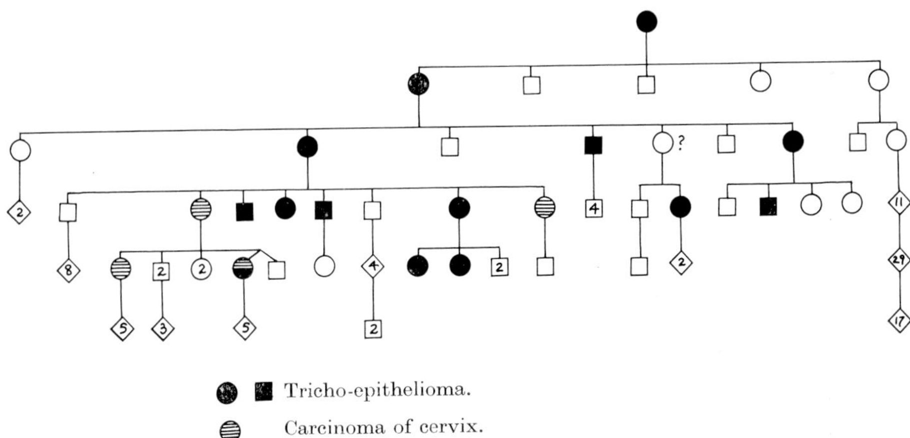

FIG. 424.—TRICHO-EPITHELIOMA.

Autosomal dominant inheritance, with carcinoma of the cervix (S. H. Wolken *et al.*).

showing a comparative lack of malignancy with the characteristics of a hyperplasia, not of the epidermis but of the epithelial cells of the hair follicles.[1] The solitary type occurs mainly in adults and is not inherited; this is the most usual form seen in the lids. Keyes and Queen (1945) and S. P. and B. P. Mathur (1959) reported its occurrence on the lid-margin, Focosi (1948) on the eyebrows, and Giacomelli (1959) and Yamada and Fujimura (1966) on the skin of the lids. Multiple tricho-epitheliomata, on the other hand, begin to appear at adolescence as a dominantly inherited condition (Fliegelman and Kruse, 1948; Gaul, 1953; Lever, 1967; Wolken

[1] For summaries of the general literature, see Summerill and Hutton (1932), Warvi and Gates (1943), Gray and Helwig (1963), Lever (1967).

FIGS. 425 and 426.—TRICHO-EPITHELIOMA.

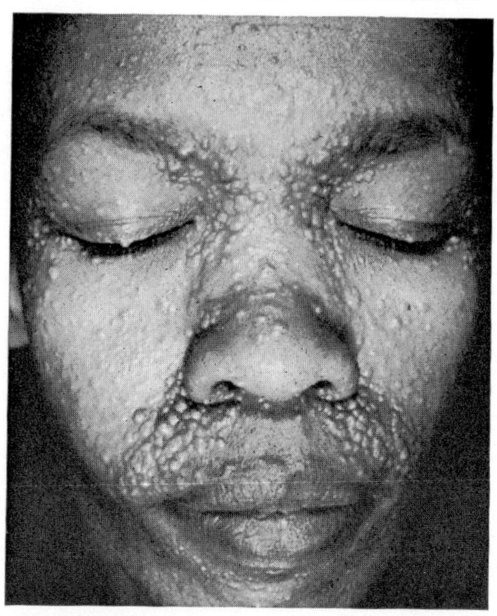

FIG. 425.—With a wide distribution over the face and lids with a predilection for the naso-labial folds (M. Boniuk).

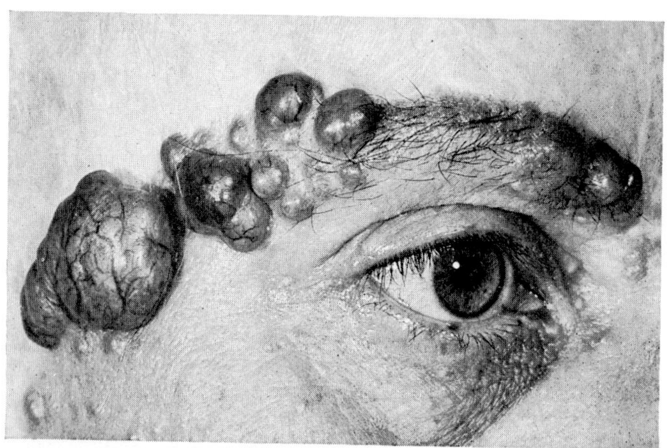

FIG. 426.—Large tumours growing on the eyebrows (A. Nover and G. W. Korting).

et al., 1969) (Fig. 424). They occur mainly on the face with a predilection for the naso-labial folds, the forehead, nose and lower eyelids, sometimes in considerable numbers and frequently symmetrically (Fig. 425). A most dramatic case was described by Bishop (1965) in a woman aged 67 in whom the tumours were first noted at the age of 12 and were present over most of the face and the lids; they reached such a size that the eyes could be opened only with difficulty. Among 57 adults in a family with this disease reported by Wolken and his colleagues (1969), 14 had multiple small papules appearing at puberty, and in one case in which the lesions of the face and eyelids had reached a large size and had become ulcerated the histological changes of a basal-celled carcinoma were found; it is of interest that in 4 of the 26 women evaluated in this pedigree a carcinoma of the cervix was found. The association of multiple lesions of tricho-epithelioma with multiple cylindromata is said to be common (Lever, 1967).

The lesions commence as subcutaneous, well-defined tiny papules or wart-like growths, oval and lobulated in shape, yellowish or bluish in colour,

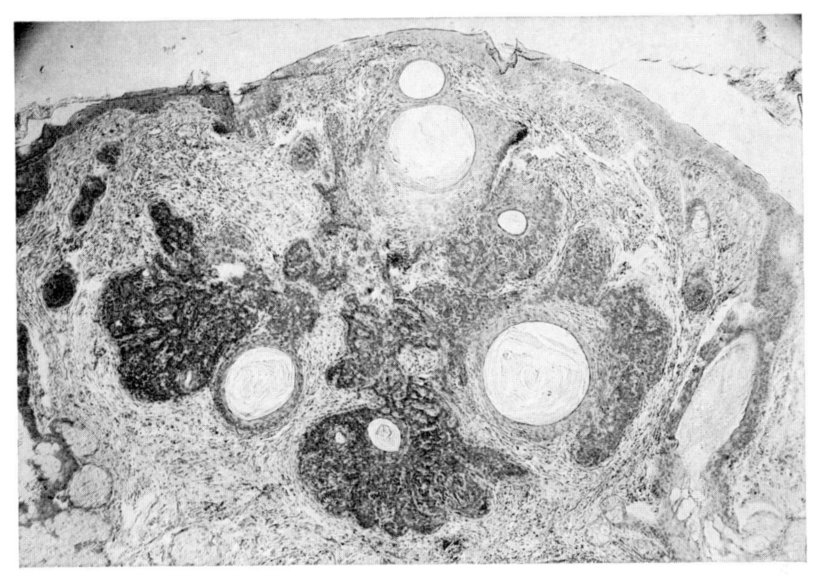

Fig. 427.—Tricho-epithelioma.

There are many keratin-filled cysts and in the centre is the beginning of a basal-cell epithelioma (H. & E.; ×48) (S. H. Wolken *et al.*).

firm in consistency and sometimes associated with telangiectasia. Usually they remain stationary although they may grow in size but they may ulcerate and very occasionally they turn malignant showing the characteristics of basal-celled epitheliomata (5 out of 116 reported cases, Warvi and Gates, 1943). Histologically the growth shows a circumscribed cellular arrangement with the characteristic "horn-cysts" and a network of basaloid cells which may, in some areas, be reminiscent of a basal-celled epithelioma, the cellular downgrowths being usually in continuity with the hair follicles[1] (Fig. 427).

Treatment has consisted of surgical excision, irradiation, curettage, electrolysis, cryotherapy and also dermabrasion; with multiple tumours

[1] Sutton and Dennie (1912), Savatard (1922), Ingels (1935), Schwab (1950), Knoth and Ehlers (1960), Bishop (1965), Lever (1967), Wolken *et al.* (1969).

the excised area may have to be repaired by a skin-graft (Bishop, 1965).

Bishop. *Arch. Ophthal.*, **74,** 4 (1965).
Boniuk. *Int. Ophthal. Clin.*, **2,** 239 (1962).
Brooke. *Brit. J. Derm.*, **4,** 269 (1892).
Fliegelman and Kruse. *J. invest. Derm.*, **11,** 189 (1948).
Focosi. *Boll. Oculist.*, **27,** 417 (1948).
Fordyce. *J. cutan. Dis.*, **10,** 459 (1892).
Gaul. *Arch. Derm. Syph.* (Chic.), **68,** 517 (1953).
Giacomelli. *Ann. Ottal.*, **85,** 551 (1959).
Gray and Helwig. *Arch. Derm.*, **87,** 102 (1963).
Ingels. *Arch. Derm. Syph.* (Chic.), **32,** 75 (1935).
Jarisch. *Arch. Derm. Syph.* (Wien), **28,** 163 (1894).
Keyes and Queen. *Amer. J. Ophthal.*, **28,** 189 (1945).
Knoth and Ehlers. *Hautarzt*, **11,** 535 (1960).
Lever. *Histopathology of the Skin*, 4th ed., Phila., 553 (1967).
Mathur, S. P. and B. P. *Brit. J. Ophthal.*, **43,** 762 (1959).
Nover and Korting. *Klin. Mbl. Augenheilk.*, **156,** 621 (1970).
Savatard. *Brit. J. Derm.*, **34,** 381 (1922).
Schwab. *v. Graefes Arch. Ophthal.*, **150,** 388 (1950).
Summerill and Hutton. *Arch. Derm. Syph.* (Chic.), **26,** 854 (1932).
Sutton and Dennie. *J. Amer. med. Ass.*, **58,** 333 (1912).
Warvi and Gates. *Amer. J. Path.*, **19,** 765 (1943).
Wolken, Spivey and Blodi. *Amer. J. Ophthal.*, **68,** 26 (1969).
Yamada and Fujimura. *Rinsho Ganka*, **20,** 59 (1966).

BENIGN CALCIFIED EPITHELIOMA OF MALHERBE (PILOMATRICOMA)

This is a relatively rare benign tumour occurring in the skin, the origin of which has excited some speculation. Originally it was thought to represent a neoplasic entity of low malignancy, possibly derived from the anlage of a sebaceous gland exhibiting structural features intermediate in type between an epidermoid cyst and a basal-celled carcinoma (Malherbe and Chenantais, 1880; Malherbe, 1905). More recent work on the histogenesis of the lesion has suggested that it is derived from the matrix cells of a hair or primitive cells that are capable of differentiating into these (Bingul *et al.*, 1962); the term *pilomatrixoma*[1] was therefore introduced by Forbis and Helwig (1961) to indicate this origin rather than from a sebaceous gland and to avoid the use of the word " epithelioma " with its sinister implication.

The earlier published cases were analysed by Fink (1933), Ch'in (1933; 126 cases) and Côté (1936); Castigliano and Rominger (1954) noted 243 cases in the literature and Forbis and Helwig (1961) 300. In his survey of the literature Ch'in (1933) found 12 cases affecting the lids and eyebrows; subsequent tumours on the skin of the eyelids have been reported by several authors.[2]

Clinically the tumour appears at any time of life but more commonly in children and young adults under twenty years of age as a slowly growing hard or occasionally a cystic subcutaneous lump, which may be mobile or fixed to the overlying skin and may have a bluish or reddish appearance (Figs. 428–9). Occasionally it is tender, a feature which has been correlated histologically with hæmorrhage into the area. As a rule the lesion is solitary, most frequently on the upper lid, but more than one have been recorded

[1] *Pilomatricoma* is etymologically the more correct term.

[2] Ashton (1951), Roberts (1957), Mitchell and Newell (1957), Nikolowski and Seitz (1960), Boniuk and Zimmerman (1963), Carbone *et al.* (1965), Dhermy and Mascaro (1966), Lohse and Tost (1967), Durán and de Buen (1968), Howard (1968), Spelsberg (1969), Sobieska-Clarowa and Dudek (1970), Nover *et al.* (1971), Offret and Haye (1971).

FIGS. 428 and 429.—MALHERBE'S BENIGN CALCIFIED EPITHELIOMA.

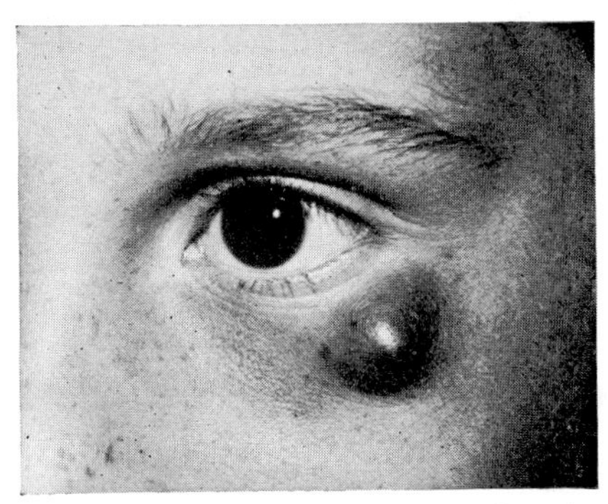

FIG. 428.—On the lower lid (A. Nover).

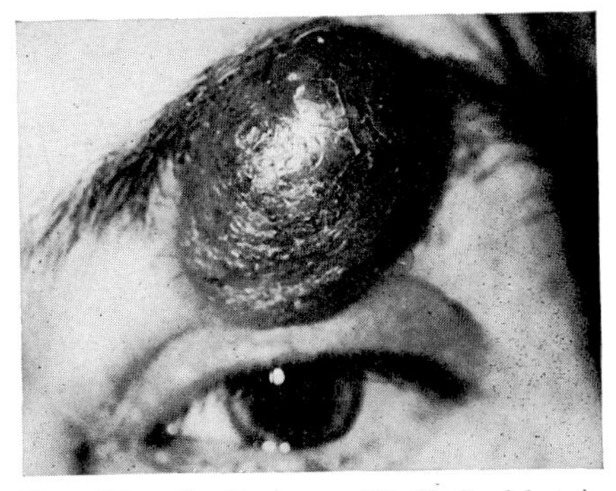

FIG. 429.—On the upper lid (E. Spelsberg).

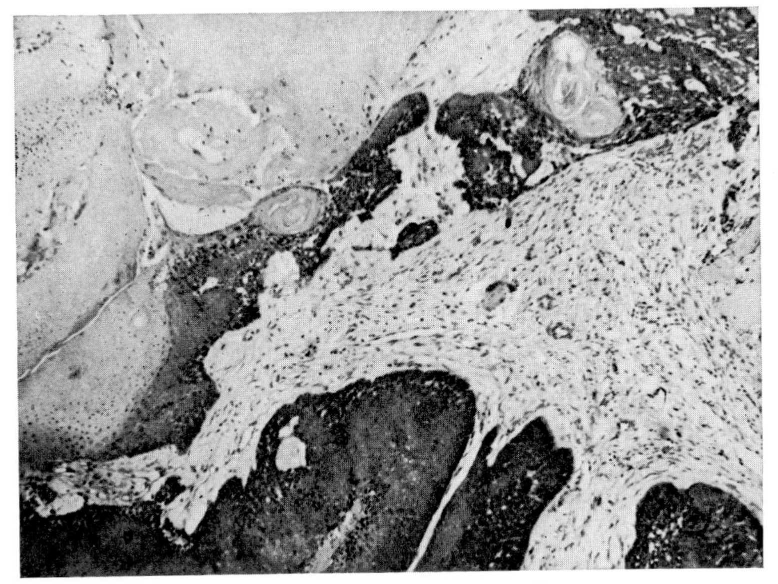

FIG. 430.—MALHERBE'S BENIGN CALCIFIED EPITHELIOMA.

The squamous epithelium is necrotic (top left) and viable proliferating basophilic cells (bottom left) lie in a fibrous stroma (H. & E.; ×74) (N. Ashton).

(both eyebrows, Carbone *et al.*, 1965). Histopathologically the tumour is circumscribed and encapsulated with cysts containing yellowish-grey granular or cheesy material; the typical lesion shows one or more areas of deeply staining basophilic cells and masses of pale pink "shadow cells" with transitional stages between the two (Fig. 430). Calcification and foreign-body giant cells are present in most cases and in a few ossification may be evident, particularly when the lesion has been present for several years; occasionally hæmorrhage may be present where there has been a recent history of trauma to the area. It is essentially benign and is usually cured by simple excision, but if this is incomplete recurrence may ensue (Bartkowska, 1966).

Ashton. *Trans. ophthal. Soc. U.K.*, **71**, 301 (1951).
Bartkowska. *Klin. oczna*, **36**, 601 (1966).
Bingul, Graham and Helwig. *Pediatrics*, **30**, 233 (1962).
Boniuk and Zimmerman. *Arch. Ophthal.*, **70**, 399 (1963).
Carbone, Zingirian and Ciurlo. *Ann. Ottal.*, **91**, 401 (1965).
Castigliano and Rominger. *Arch. Derm. Syph.* (Chic.), **70**, 590 (1954).
Ch'in. *Amer. J. Path.*, **9**, 497 (1933).
Côté. *J. Path. Bact.*, **43**, 575 (1936).
Dhermy and Mascaro. *Arch. Ophtal.*, **26**, 33 (1966).
Durán and de Buen. *An. Soc. mex. Oftal.*, **41**, 109 (1968).
Fink. *Virchows Arch. path. Anat.*, **289**, 527 (1933).
Forbis and Helwig. *Arch. Derm.*, **83**, 606 (1961).
Howard. *Ind. J. ocular Path.*, **2**, 33 (1968).
Lohse and Tost. *Acta ophthal.* (Kbh.), **45**, 876 (1967).
Malherbe. *Rev. Chir.*, **32**, 651 (1905).
Malherbe and Chenantais. *Bull. Soc. Anat. Paris*, **15**, 169 (1880).
Progr. Méd. (Paris), **8**, 826 (1880).
Mitchell and Newell. *Amer. J. Ophthal.*, **44**, 629 (1957).
Nikolowski and Seitz. *Klin. Mbl. Augenheilk.*, **136**, 825 (1960).
Nover, Heinrich and Rockert. *Klin. Mbl. Augenheilk.*, **159**, 501 (1971).
Offret and Haye. *Tumeurs de l'oeil et des annexes oculaires*, Paris (1971).
Roberts. *Brit. J. Ophthal.*, **41**, 492 (1957).
Sobieska-Clarowa and Dudek. *Klin. oczna*, **40**, 241 (1970).
Spelsberg. *Klin. Mbl. Augenheilk.*, **154**, 807 (1969).

CORNU CUTANEUM

A CUTANEOUS HORN is not a specific pathological lesion but a clinical term to describe a hyperkeratotic condition which as we have seen may arise from papillomata, senile or seborrhœic keratosis, inverted follicular keratosis, Bowen's disease, or verruca vulgaris[1] (Figs. 431–3). It constitutes a rare and striking lesion which may appear on the skin of the lids particularly near the outer canthus, or on their margins; other sites of predilection in the body are the scalp, the forehead, the nose and the glans penis. Such epidermal growths consist of corneous material which accumulates in concentric laminæ, the resulting appearance resembling the horn of an animal, being yellowish or brownish in colour and conical or cylindrical, straight, twisted or angular in form (Fig. 431). They occur usually in old people of both sexes, but have occasionally been seen in young children (one on the right and two on the left upper lid in a child of 3, Shaw, 1869; Pecoraro, 1927) or adolescents (Natanson, 1899). As a rule they are small and cylindrical, a few millimetres long and broad, but sometimes they reach considerable dimensions: thus Carl Theodor (1892) described one on the upper lid of a 78-year-old woman, 4·5 cm. in length, and Schöbl (1892) at the root of the nose and inner canthus in an 82-year-old woman, 5 cm. long and 3 cm. in diameter at the base. More than one may occur in the same person (Ascher, 1914); Spietschka (1898) described a patient with two horns near the outer canthus, one on each lid.

As a general rule the horn remains without disturbance or signs of inflammation but occasionally, particularly after trifling trauma, irritation

[1] Unna (1879), Mitvalsky (1894), Baas (1897), Bland-Sutton (1902), Pasini (1902), Boniuk (1964), Lennox and Sayed (1964), Cramer and Kahlert (1964), Mishra and Agrawal (1967).

FIGS. 431 and 432.—CUTANEOUS HORNS (Inst. Ophthal.).

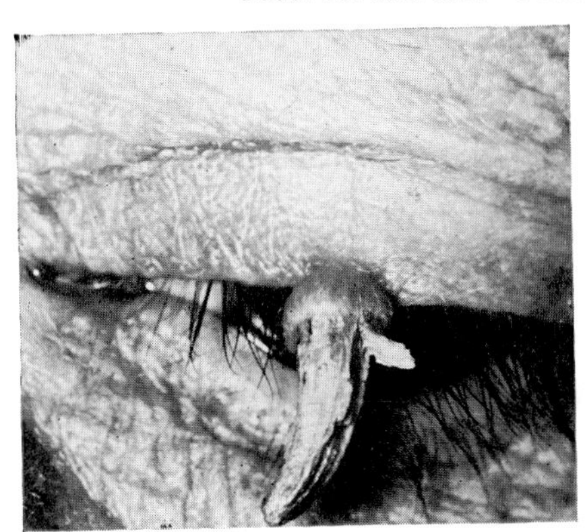

FIG. 431.—On the upper lid-margin.

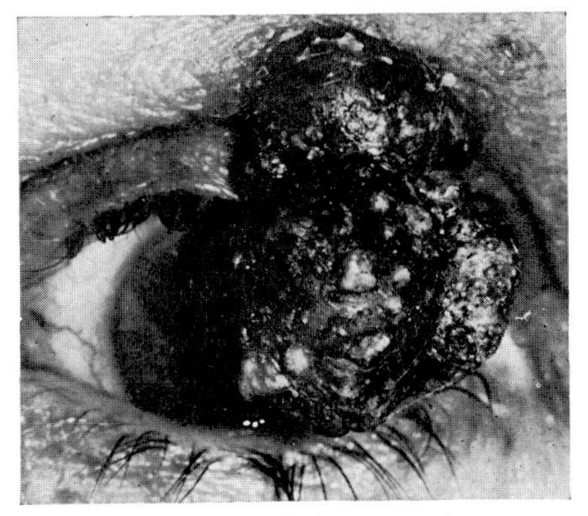

FIG. 432.—At the inner canthus.

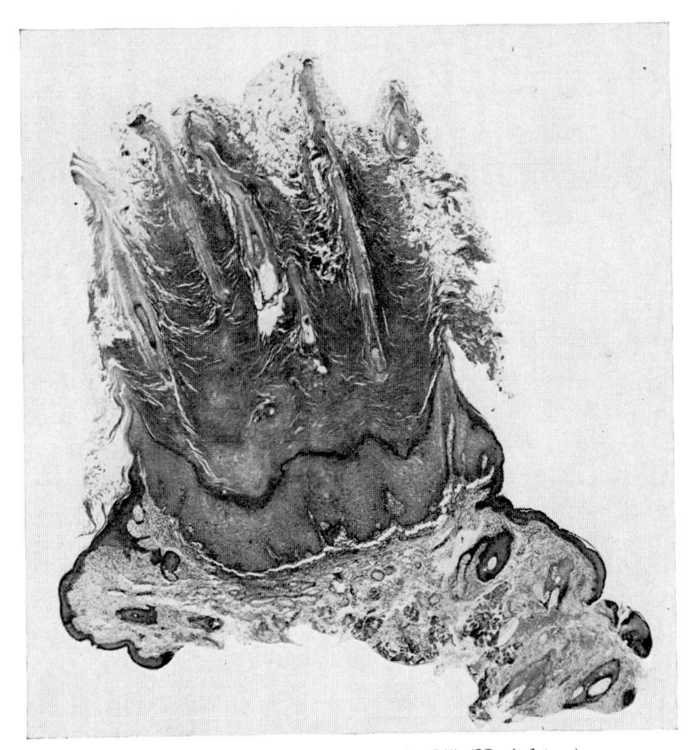

FIG. 433.—CUTANEOUS HORN ($\times$15) (N. Ashton).

of the base occurs. It may be shed spontaneously not to appear again; recurrences, however, are the rule and precarcinomatous changes may occasionally develop (Hine, 1920; Gualdi, 1931; Dorello, 1952). Total excision including the base is the treatment of election.

Ascher. *Klin. Mbl. Augenheilk.*, **52,** 144 (1914).
Baas. *Zbl. allg. Path. path. Anat.*, **8,** 295 (1897).
Bland-Sutton. *Tumours*, London (1902).
Boniuk. *Ocular and Adnexal Tumors*, St. Louis (1964).
Cramer and Kahlert. *Derm. Wschr.*, **150,** 521 (1964).
Dorello. *G. ital. Oftal.*, **5,** 30 (1952).
Gualdi. *Ann. Ottal.*, **59,** 79 (1931).
Hine. *Proc. roy. Soc. Med.*, **13,** Sect. Ophthal., 86 (1920).
Lennox and Sayed. *J. Path. Bact.*, **88,** 575 (1964).
Mishra and Agrawal. *J. All-India ophthal. Soc.*, **15,** 123 (1967).
Mitvalsky. *Arch. Derm. Syph.* (Wien), **27,** 47 (1894).
Natanson. *Arch. Derm. Syph.* (Wien), **50,** 203 (1899).
Pasini. *G. ital. Mal. vener.*, **37,** 475 (1902).
Pecoraro. *Ann. Ottal.*, **55,** 284 (1927).
Schöbl. *Ueber einige seltene Keratome d. Auges*, Prague (1892).
Shaw. *Boston med. surg. J.*, **3,** 17 (1869).
Spietschka. *Arch. Derm. Syph.* (Wien), **42,** 35 (1898).
Theodor. *Klin. Mbl. Augenheilk.*, **30,** 310 (1892).
Unna. *Dtsch. Z. Chir.*, **12,** 267 (1879).

CARCINOMA (EPITHELIOMA)

The *incidence* of carcinoma in its common cutaneous forms—basal-celled and squamous-celled—is high in the lids.

Morax (1926) found palpebral cancer to comprise 0·8% of all ophthalmic patients; it constitutes some 3 to 5% of all epitheliomata over the body generally. Laborde (1933) found it to comprise 12% of all cutaneous cancers; Papolczy (1935) 12%, Shulman (1962) 40%, and Dollfus (1966) 30% of all facial cancers, while in a total of 2,556 skin lesions at all sites Rank (1973) found 63 on the eyelids. They are the most common malignant tumours in the ocular region. Several long series of cases have been reported in the literature since that of Weber (1866) of 211 cases; the more recent publications are appended.[1]

The *age of the patient* affected is usually from 50 to 70 years; before 40 and after 80 an epithelioma is rare, but children of 3½ years (Birge, 1938), 7 (de Castro, 1948), 12 (Safar, 1927), 16 (Rinaldi *et al.*, 1970), and adults of 85 (Hirschfelder and Frost, 1948) and 88 years (Payne *et al.*, 1969) have been affected. Reese (1963) and Payne and his colleagues (1969) noted the highest incidence in the sixth decade, while del Regato (1949), Barron (1962) and Aurora and Blodi (1970) found more cases in the seventh and eighth decades. One thousand cases taken from the literature show the following approximate age distribution: before 40 years 5·5%, 41 to 50 years 15%, 51 to 60 years 25%, 61 to 70 years 30%, 71 to 80 years 20%, over 81 4·5%.

The *sex incidence* shows a slight predominance of males which in some series has been marked (Cobb *et al.*, 1964; Holland and Bellmann, 1965; Wynn-Williams, 1967). In Dollfus's series (1966) there were slightly more females. Some authors, however, have found no difference between the sexes (Breed, 1964; Payne *et al.*, 1969), while Aurora and Blodi (1970) found that this was so below the age of 50 but that after this age males were more frequently affected than females.

[1] Moldenhauer (1958), Wellauer and Alesch (1960), Mitasov (1961), Barron (1962), Fayos and Wildermuth (1962), Shulman (1962), Kwitko *et al.* (1963), Breed (1964), Cobb *et al.* (1964), Lederman (1964), Domonkos (1965), Holland and Bellmann (1965), Dollfus (1966), Wollensak and Meythaler (1967), Wynn-Williams (1967), Halnan and Britten (1968), Quiñones and Risco (1968), Petersen *et al.* (1968), Payne *et al.* (1969), Aurora and Blodi (1970).

With regard to *race*, it is generally a disease of Caucasians; in white races the basal-cell carcinoma is the commonest malignant tumour of the skin.

The *duration* of the lesion before treatment is of interest. In the majority of cases it has been present for a considerable time, up to twelve months in the case of a squamous carcinoma and up to five years or even longer with basal-cell carcinomata.[1] Because of their slow growth and relative lack of symptoms many of them are not treated until the lesions are far advanced; indeed, many small carcinomata are discovered only during the course of a routine examination for other reasons. The course of the disease after the initial treatment may be very prolonged: thus Vit (1960) reported a case which after several recurrences lasted over 18 years, and in another patient described by Baxter and Pirozynski (1967) the disease smouldered relentlessly for 26 years until the development of metastases resulted in a fatal termination.

The *size* of the lesion is, of course, related to its duration, most commonly 5 to 10 mm. in diameter when diagnosed, but lesions much larger than 2 cm. can be seen even on exposed sites such as the eyelids (Wynn-Williams, 1967; Payne *et al.*, 1969; and others).

The *site* on the lids is characteristic, the most common being in the skin near the margin of the lower lid. Of 1,000 cases from the literature the following approximate distribution was found for basal-cell carcinoma—lower lid 50%, inner canthus 30%, upper lid 14%, outer canthus 6%. The much less common squamous-cell carcinoma is found particularly in the upper lid and at the outer canthus; Baclesse (1966) gave the following figures—in the upper lid 13 squamous-cell carcinomata and 54 basal-celled; lower lid 40 and 477 respectively.

As a general rule carcinomata of the eyelids occur as solitary lesions with the rare exception of those cases associated with chronic arsenical dermatosis. Multiple independent tumours involving the lids are very rare but have been reported; thus Cochran and Robinson (1931) found a squamous-cell epithelioma of the right lower lid which was apparently adequately treated by radium; 26 months later similar tumours appeared on the left eyeball and right conjunctiva. Multiple epitheliomata have been seen to develop in the eyelid of a young woman with chronic seborrhœic blepharitis (Lister and Dalley, 1960). Wollensak and Meythaler (1967) noted the case of a woman aged 26 who had had multiple basal-cell carcinomata of both lower eyelids at the age of 20. Six of 270 cases reported by Payne and his colleagues (1969) had two lesions. In 12 cases reported by Aurora and Blodi (1970), basal-cell carcinomata were present also on the face and in 4 of these multiple small tumours were found elsewhere in the skin. Other cases with multiple superficial carcinomata involving the periorbital skin have been annotated

[1] Nearly 40% over 2 years, McCallum and Kinmont (1966); 10% for 10 years, Payne *et al.* (1969); 15 to 17 years, Cobb *et al.* (1964); 20 years, Wynn-Williams (1967).

by Williams and Klein (1970); they are a feature of the basal-cell nævus syndrome.[1]

The relative frequency of the different *histological types* of carcinoma of the eyelid has been studied by many workers and it is clear that basal-cell carcinoma occurs more often than the squamous-cell form.

Regaud and his colleagues (1926) found a ratio of basal-cell to squamous-cell carcinomata of 11·3 to 1, Driver and Cole (1939) 4·5 to 1, del Regato (1949) (excluding baso-squamous carcinoma) 17·1 to 1, Welch and Duke (1958) 9·6 to 1, Shulman (1962) 12·6 to 1, Lederman (1964) 7 to 1, Dollfus (1966) 7 to 1, Halnan and Britten (1968) 5·2 to 1, and Aurora and Blodi (1970) 11·5 to 1. In the series of 550 malignant neoplasms of the eyelid reported by Kwitko and his colleagues (1963), 502 were of the basal-cell and only 13 were of the squamous-cell type, giving a ratio of almost 39 to 1; this unusually high figure, however, reflects the selected nature of their cases and also the improved histological diagnosis of these tumours; out of 28 lesions in their series originally regarded as squamous-cell carcinoma only 13 were subsequently accorded this diagnosis. These workers also reported their findings in a reappraisal of 115 cases diagnosed as squamous-cell carcinoma before 1955, and in only nine instances did this histological diagnosis stand up to scrutiny.

The *œtiology* of carcinomata of the lids is unknown, but it is significant that some 30% of all varieties are said to be associated with trauma or irritation such as chronic blepharitis, eczema, injuries, the pressure of spectacles (Healey *et al.*, 1967) or the squeezing of blackheads or pimples (Broders, 1919; Birge, 1938). Such an association may in many cases be spurious; and it is interesting that, although examples have been reported (Milian and Garnier, 1928; Goulden and Stallard, 1933), it has been found that carcinoma of the lids occurs less frequently in workers with carcinogenic agents, such as tar and pitch, than in the ordinary population (Kennaway, 1925; Lane, 1937). Malignant changes may develop on an old cicatrix; lupus vulgaris may also show malignant degeneration, as also may senile papillomata, senile keratosis, moles or cutaneous horns. Squamous carcinomata may develop from other precancerous dermatoses such as Bowen's disease and xeroderma pigmentosum. Basal-cell carcinomata seem to be more frequent in areas of skin with a high density of sebaceous glands (P. G. Graham and McGavran, 1964). Radiotherapy may also have a predisposing influence in certain cases, as in the reports of basal-cell epitheliomata following the radiation of tuberculous lesions (Archambault and Marin, 1930; Fazekas, 1933), angiomata of the eyelids, intra-ocular tumours such as retinoblastoma (Forrest, 1962) and other conditions (Haye *et al.*, 1965).[2] Actinic radiation, particularly the combination of bright sunlight and low humidity over a long period of time, seems to be a causal factor in non-pigmented peoples, particularly those with outdoor occupations such as farmers and sailors (Blumenthal, 1936; Büngeler, 1937; Mathews, 1937); thus such tumours are very common in Australia. Rank (1973) reported that in the State of Victoria the incidence of skin cancer was 1 in 100 for the

[1] p. 433. [2] Vol. XIV, p. 949.

age-group 20–40, 1 in 20 for those between 40 and 60, and 1 in 4 for those over 70, an extraordinarily high incidence which suggests that white immigrants from Northern Europe who today spend most of the time bare down to the waists and almost up to it are a genetically susceptible population in an actinic environment admirably suited for a black population. The same high incidence is seen in cattle in Australia. Findlay (1930) recorded that of 190 cases of skin cancer in 900,000 cattle in New South Wales, all but three began in white areas of the skin and these three developed upon branding scars. Heredity, also, may be of some importance.

SQUAMOUS-CELL CARCINOMA

SQUAMOUS-CELL CARCINOMA (EPITHELIOMA) (EPIDERMOID CARCINOMA; EPITHELIOMA SPINO-CELLULARE of Krompecher, 1902) is relatively rare in the eyelids (Figs. 434–8). It starts slowly, insidiously and painlessly and hence may have been present for some considerable time before the patient takes serious note of it. Commencing as a small, hard, indurated nodule or as a roughened warty keratotic patch, after some months it begins to show erosions and fissures which tend to crust, until eventually an ulcer develops at first hidden by the overlying scales and crusts. In this common ulcerative type the base is always sharply defined, indurated and hyperæmic and the edges hard and undermined (Figs. 437–8). Less frequently it starts as a papillomatous growth (Fig. 434) or even resembles a cyst on the lid-margin (Fig. 435); a cutaneous horn has been simulated (Lamba *et al.*, 1971). Local extension occurs gradually and relentlessly, skin, connective tissue, cartilage, periosteum and bone being attacked and eaten away until the lids, the conjunctiva, the lacrimal passages, the eyeball, the orbit, the nasal sinuses and large areas of the nose, face and temple are destroyed in a fœtid-smelling, ulcerating, fungating crater which may eventually reach the cranial cavity. The sclera resists invasion for a long time, penetration occurring as a rule through the perivascular and perineural lymphatics of the limbus.[1] Indirect destruction of the eye, however, may occur long before this owing to exposure of the cornea and the development of a hypopyon ulcer; alternatively, the eye is, as it were, dissected away by destruction of the soft tissues of the orbit and may, indeed, fall out spontaneously (Rollet and Bussy, 1923; Tallei, 1925). Sooner or later the regional lymphatic nodes are involved, in most cases owing to secondary infection, and only in a proportion, probably some 20%, due to cancerous extension[2] (Fig. 439). In 3 out of 14 cases the cervical lymph nodes were involved and death occurred within 2 years (Shulman, 1962), and of 9 out of 55 cases with lymphatic involvement treated by Baclesse and Dollfus (1960) only 2 survived 5 years. Only very occasionally do distant metastases occur. Pain, at first absent, eventually becomes severe and constant and when the infra-orbital or supra-orbital

[1] Vol. VIII, p. 1168.
[2] Morax (1926), Regaud *et al.* (1926), Nicolini (1935), Lacassagne (1936).

FIGS. 434 to 438.—SQUAMOUS-CELL CARCINOMATA.

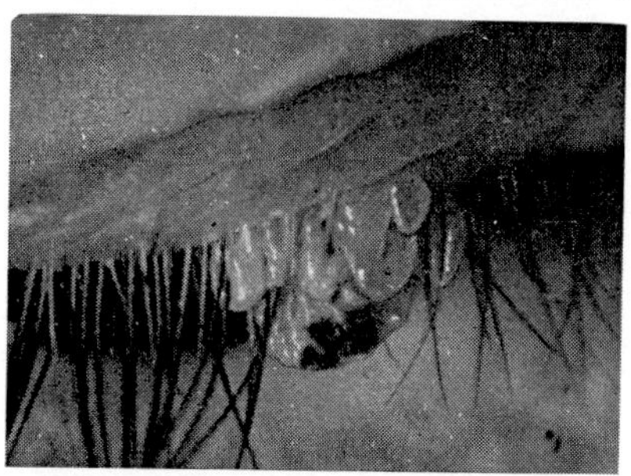

FIG. 434.—Very early squamous-cell carcinoma at the lid-margin resembling a pedunculated papilloma (M. J. Reeh and K. C. Swan).

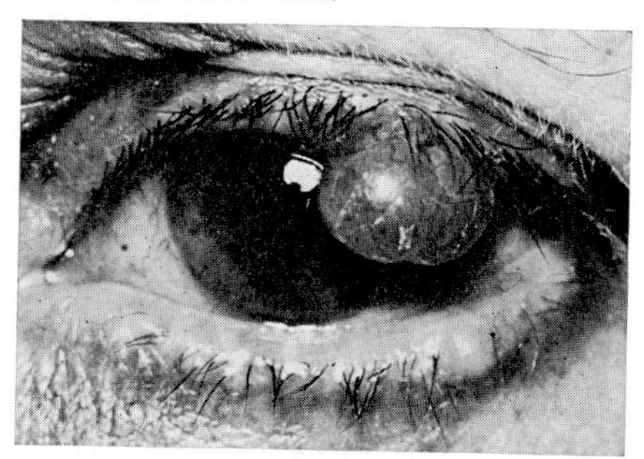

FIG. 435.—An early tumour at the lid-margin resembling a cyst (B. Daicker).

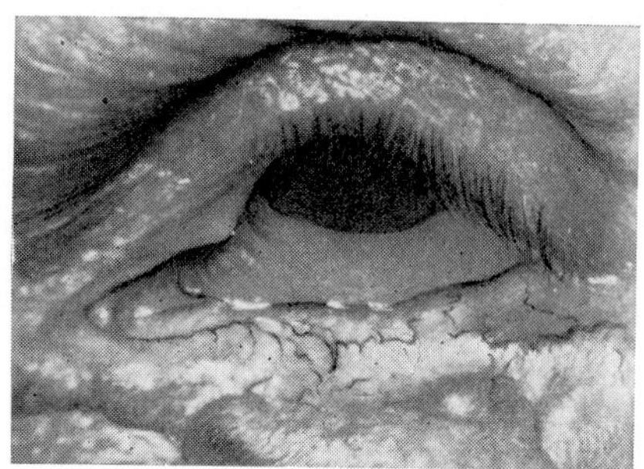

FIG. 436.—Destroying the lower lid and invading the orbit (M. J. Reeh and K. C. Swan).

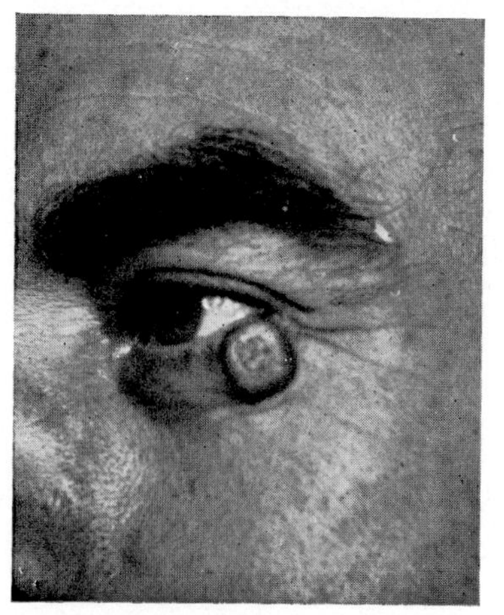

FIG. 437.—On the lower lid at the pre-ulcerative stage (D. W. Smithers).

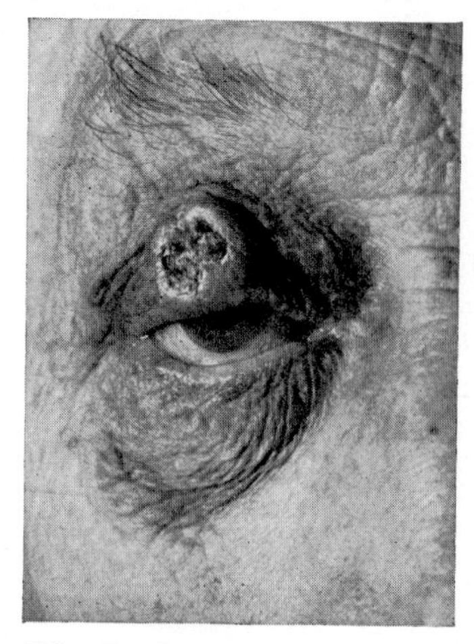

FIG. 438.—On the upper lid at the ulcerative stage (D. W. Smithers).

nerves and bony orbit are involved, it may become excruciating; as time goes on it is associated with progressive cachexia and emaciation. Eventually after a period varying from some months to 10 or 20 years, the patient dies of hæmorrhage, meningitis, general exhaustion, septic absorption or cachexia.

Less commonly a *vegetative form* of carcinoma may be met with when the tumour or a part of an ulcerating growth, becoming sessile or pedunculated, appears as a red, eroded, strawberry-like mass projecting from an indurated base on the lid (Villard and Dejean, 1929). At first it grows rapidly, but usually develops into the common ulcerative type or shows cicatricial tendencies.

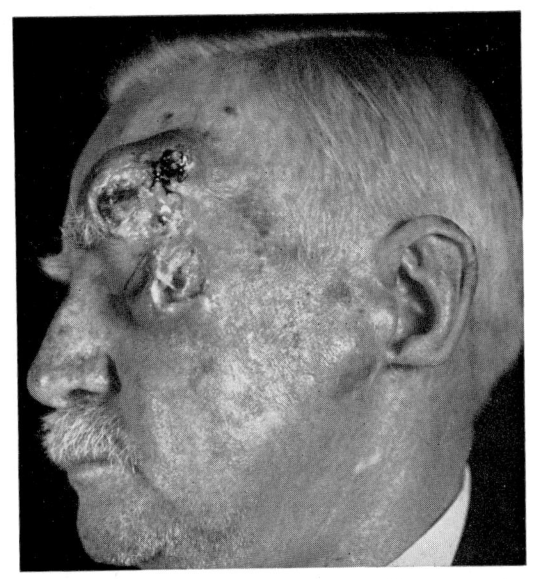

Fig. 439.—Squamous-cell Carcinoma of the Brow and Lower Lid.

At an advanced stage. Note the lymphatic involvement in the pre-auricular lymph nodes (L. Hollander).

The *histopathology* of these tumours corresponds to that met with in other cutaneous areas. All types of carcinoma show an epidermal invasion of the corium with a typical epithelial arrangement of cells in groups or alveoli penetrating the lymph spaces of the underlying connective tissue, with fibrous tissue between the groups but not between the cells of the groups. In sections the downgrowth of cells may appear to be made up of large numbers of separate masses, but these are really finger-like claws[1] extending from the main mass and in continuity with it, the appearance of separation being merely fictitious. The downgrowth of columns of epithelial cells is therefore the reverse of that seen in papillomata. From these downgrowths secondary processes bud off laterally as well as terminally

[1] Lat. *cancrum*, a crab; Gk. *καρκίνωμα*, from *καρκίνος*, a crab.

and anastomose to form a network. There is no basement membrane and the stroma shows a marked irritative inflammatory reaction with lymphocytic and plasma-cell infiltration. As a rule the tumour eventually disintegrates and sloughs in the central area owing to obliteration of the vessels to form a malignant ulcer.

A squamous-cell carcinoma is characterized by cellular processes which in general retain the structure of the epidermis—peripherally there are cylindrical cells, internal to which are particularly large and well-developed prickle cells, while in the centre are squamous cells undergoing the same

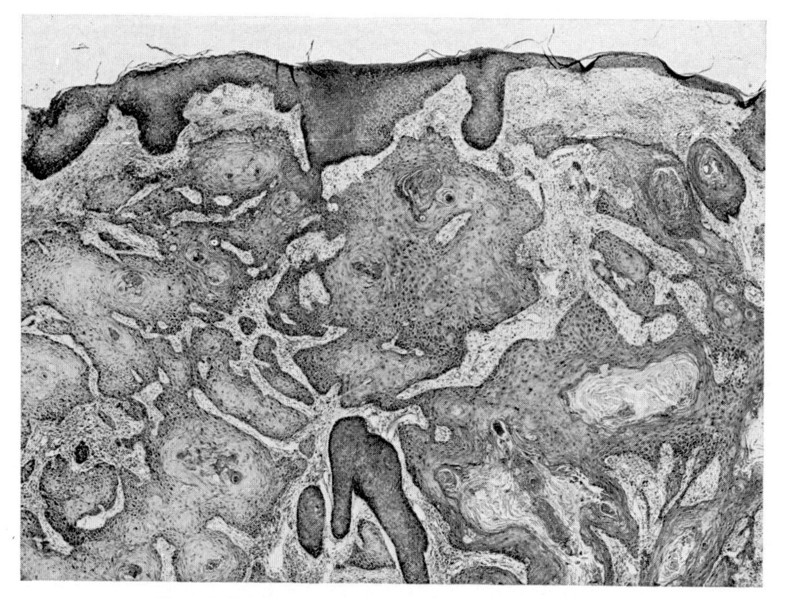

FIG. 440.—EPIDERMOID CARCINOMA OF THE LID.

A squamous-cell carcinoma arises from the surface epithelium and extends downwards in irregular masses to invade the dermis; epithelial pearls are present (H. & E.; ×24) (N. Ashton).

process of cornification as occurs on the surface of the skin, arranged in compressed, laminated masses staining strongly eosinophilic with acid dyes (CELL-NESTS or EPITHELIAL PEARLS) (Fig. 440). If these cell-types are clearly differentiated the tumour has less malignancy than if no differentiation is obvious and the whole mass is anaplasic showing many mitotic figures, an absence of pearls and deficient keratinization. The degree of malignancy may vary in different parts of the tumour so that *several sections should be examined* (Broders, 1932; Edmundson, 1948; and others). The lattice-fibre network around the cellular masses is not dense, and in the more malignant types cancer cells may permeate into it (Way, 1927) but the proliferating cells are not invaded by the inflammatory infiltrate as occurs in pseudo-epitheliomatous hyperplasia (Boniuk, 1964). Spread is by the perineural lymphatics so that the regional lymph nodes are first affected; metastatic spread is unusual and late when a carcinoma has developed in sun-damaged skin, but much more frequent in other predisposing situations such as radiation dermatitis or burns (Lund, 1965). Tumours of the upper lid drain into the pre-auricular nodes, of the lower lid first into the sub-maxillary nodes, and metastatic growths are found near the angle of the jaw.

BASAL-CELL CARCINOMA

BASAL-CELL CARCINOMA (RODENT ULCER: BASALIOMA: EPITHELIOMA BASO-CELLULARE of Krompecher, 1902) is the most common malignant neoplasm of the eyelids. It was first adequately described by Arthur Jacob of Dublin (1827) (Fig. 396) and his original description of *Jacob's ulcer* can only with difficulty be bettered (Fig. 441):

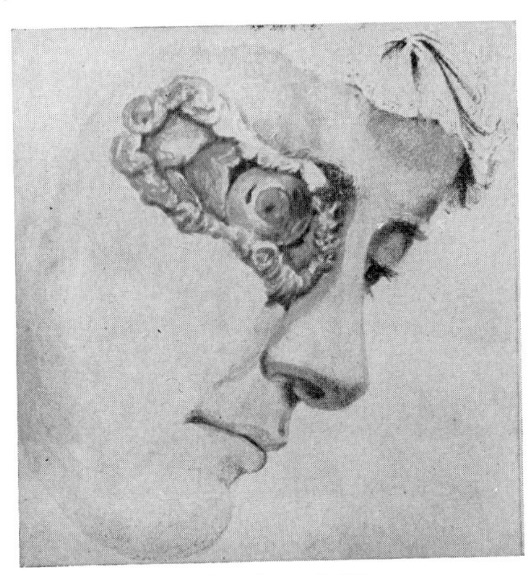

FIG. 441.—JACOB'S ULCER.

An advanced stage of a rodent ulcer as illustrated in the original article in the *Dubl. Hosp. Rep.*, **4,** 232 (1827).

"The characteristic features of this disease are the slowness of its progress, the peculiar condition of the edges and surface of the ulcer, the comparatively inconsiderable suffering produced by it, its incurable nature unless by extirpation, and its not contaminating the neighbouring lymphatic glands."

These tumours are comparatively benign, of very slow growth and although involvement of the regional lymph nodes and metastases are extremely rare, in their growth they are locally invasive and may extend deeply into the orbit and nasal bones.

The *solid type* of lesion which occurs almost exclusively on hair-bearing skin begins as a small, shiny, translucent nodule or a scaly patch; sometimes it appears as a slowly growing sessile wart (Fig. 442). No symptoms are

FIGS. 442 to 446.—RODENT ULCERS OF THE LIDS (Inst. Ophthal.).

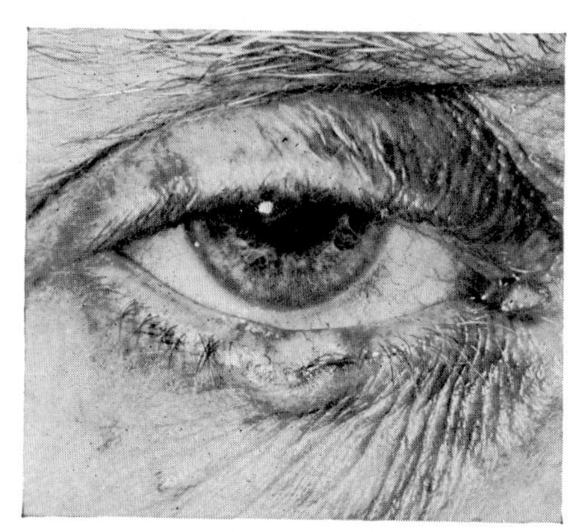

FIG. 442.

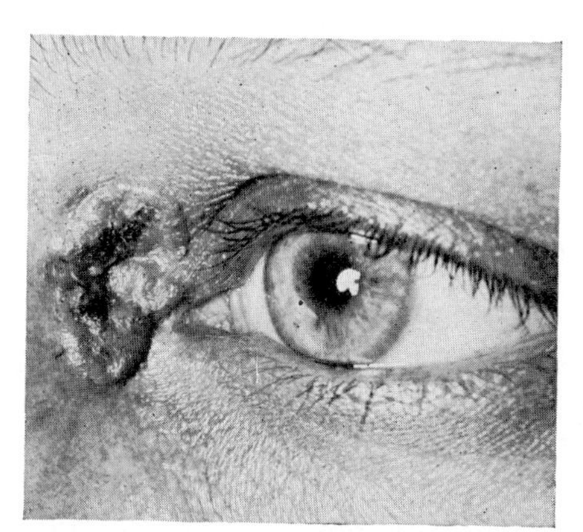

FIG. 443.

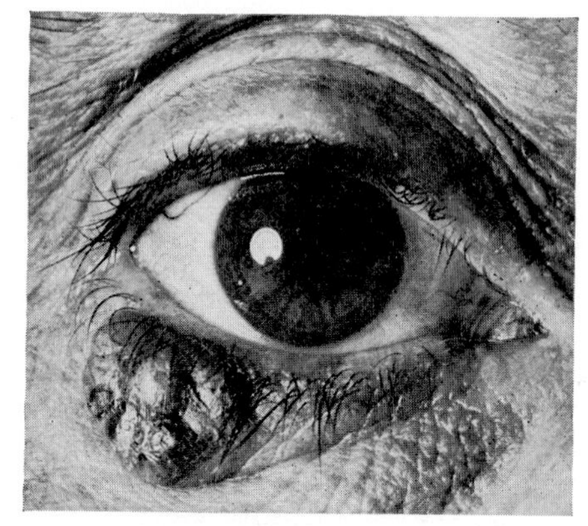

FIG. 444.

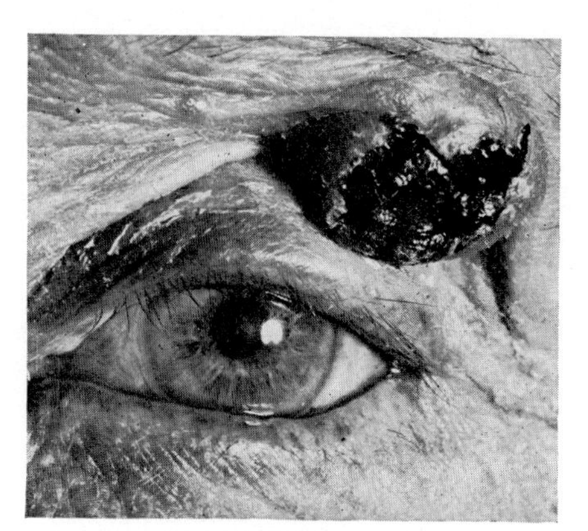

FIG. 445.

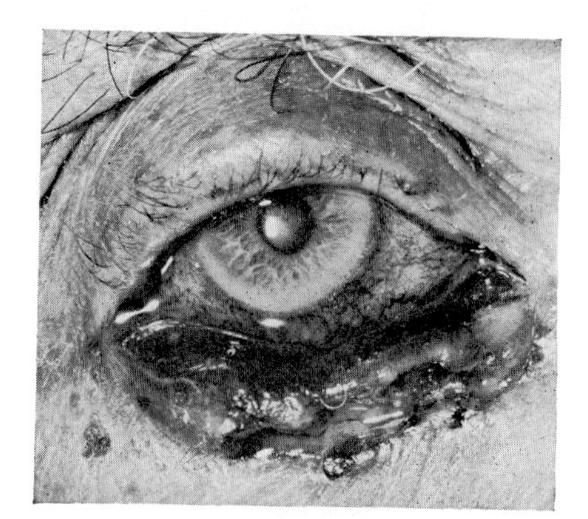

FIG. 446.

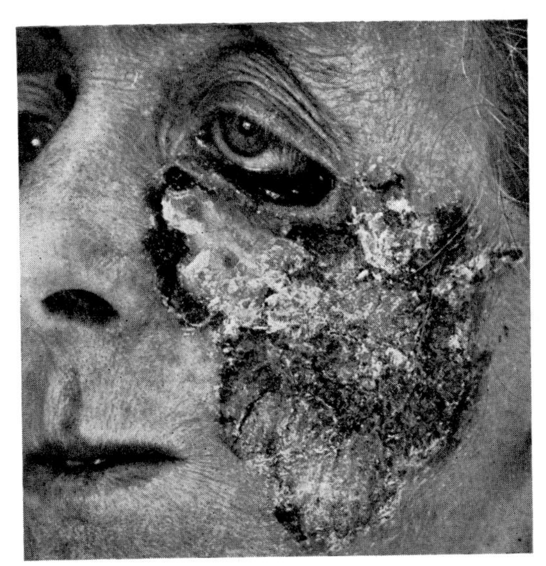

FIG. 447.—BASAL-CELL CARCINOMA.
The superficial sclerosing type affecting the cheek (D. W. Smithers).

FIGS. 448 and 449.—ADVANCED RODENT ULCERS.

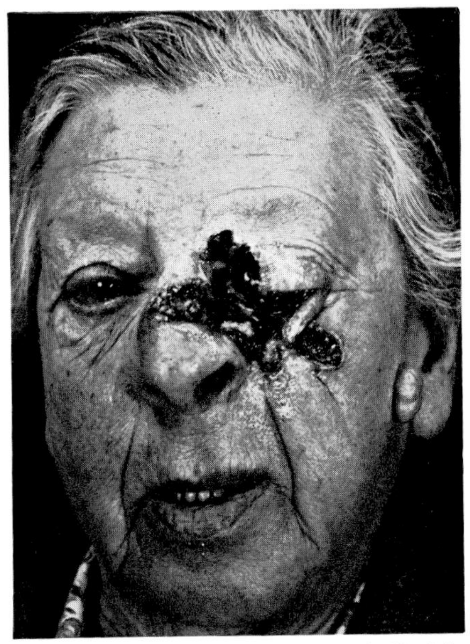

FIG. 448.—Destroying the inner half of the orbit and nose (D. W. Smithers).

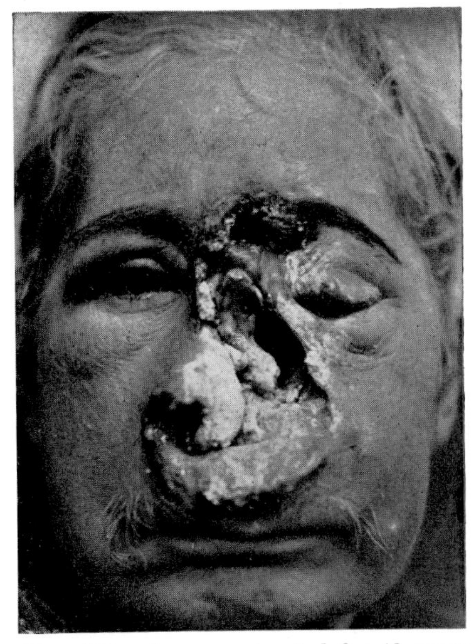

FIG. 449.—A case untreated for 42 years (C. P. G. Wakeley).

noted until after some weeks or months when ulceration commences in the central area, while towards the periphery extension occurs by the appearance of small pearly satellite nodules. Eventually the typical rodent ulcer is formed with its raised nodular border, best observed by stretching the skin, and its indurated base (Figs. 443–5). Telangiectasia may be present and if the lid-margin is involved the eyelashes may be absent. Slowly, and for long painlessly, it undergoes progressive evolution until eventually, extending superficially and then deeply, it eats away all the surrounding structures creating a crateriform deformity sometimes involving large areas of the face and nose (*ulcus* or *basalioma terebrans*) after the manner of a squamous epithelioma (Fig. 449). Occasionally, the tumour grows outwards producing a large polypoid mass, particularly at the inner canthus where the nasal bones provide a natural barrier and prevent inward extension for some time while the tumour is spreading along the surface (Boniuk, 1962). The tendency to destruction, however, is neither so great nor so rapid as in the squamous-cell type, although quite hideous deformities can result from an untreated tumour in the mid-facial region (Hayes, 1962; Garzón, 1964; Hirshowitz and Mahler, 1971) (Figs. 448–9). The lymphatic nodes are rarely involved in the absence of superadded infection, and metastases almost never occur. It is to be noted that the nodular type of rodent ulcer may appear so like a chalazion that it is frequently treated in the usual way by incision and curettage; failure to heal or recurrence at the same site should always lead to biopsy (Wolkowicz, 1962; Freiwald and Kravitz, 1970).

In the *cystic type* of basal-cell carcinoma ulceration occurs very much later, if at all, so that the correct diagnosis may be long delayed, many of them being regarded clinically as soft-tissue cysts or other benign conditions which may have been left untreated: Petersen and his colleagues (1968) examined histologically 364 cases of basal-cell carcinoma, of which 179 were solid tumours and 90% were correctly diagnosed clinically before excision; in 185 cases in which the lesion was cystic the correct diagnosis was made in only 60% of cases before excision.

A *cicatricial* or *sclerosing form* also occasionally occurs. The tumour extends superficially, cicatrizing at one point and ulcerating at another, leaving the skin atrophic. Although appearing as if a partial cure had been spontaneously attained, slow invasion of the surrounding tissue gradually proceeds (Fig. 447).

Metastases from basal-cell carcinoma are extremely rare, about 70 cases having been reported in the literature (Cotran, 1961; Binkley and Rauschkolb, 1962; Hirshowitz and Mahler, 1968). In the majority of these cases the lymph nodes were affected but visceral metastases have been reported (Meadows, 1964); the primary lesion, in most cases, has generally been present for some considerable time and been subjected to repeated attempts at cure, particularly by irradiation (Conway and Hugo, 1965; Hirshowitz and Mahler, 1968; Almeyda and Mantell, 1971).

Histopathology. The microscopic appearances of a basal-cell carcinoma differ from those of the squamous-cell type, although the general structure of downward-growing columns of cells is evident. The processes grow down to a uniform level, their ends have an expanded club-shaped form, and the cells,

FIGS. 450 and 451.—BASAL-CELL CARCINOMA OF THE LID (N. Ashton).

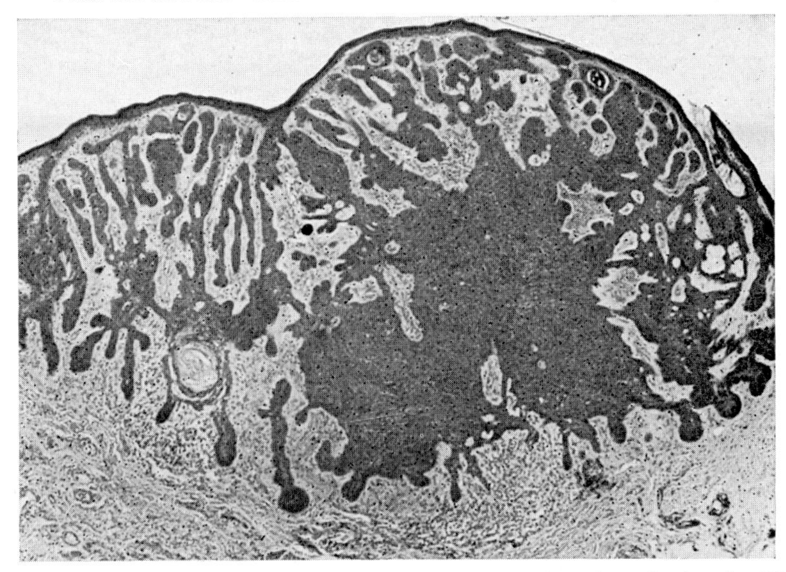

FIG. 450.—Solid masses of darkly staining cells grow down into the dermis. The columns have a characteristically club-shaped appearance. Note the absence of cell nests (H. & E.; ×114).

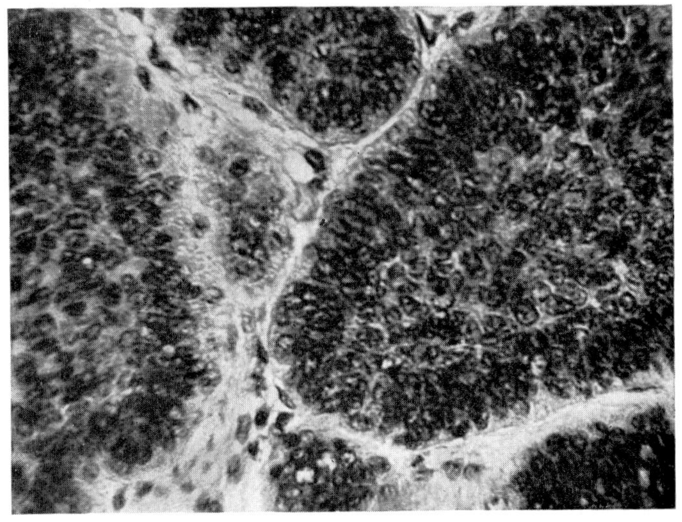

FIG. 451.—The cells at the periphery of the mass are seen to be columnar in shape (H. & E.; ×405).

which are basophilic with scanty cytoplasm, are all of one type corresponding to the basal cells of the epidermis (Figs. 450–1). There are no prickle cells, no cell-nests, no cornification and no retention of the eosin stain so characteristic of the squamous type; mucoid degeneration may occur in the centre of the cell-masses sometimes giving rise to cystic formations and vesicles near the surface, mitoses are rare, and cellular infiltration in the surrounding tissue scanty. Some tumours are associated with an excessive production of melanin (Boniuk, 1962). The uniform appearance and deep basophilia are characteristic and help to differentiate the basal-celled from the squamous-celled type.

The general arrangement of the cells of the tumour, however, varies considerably. In the more frequent or *solid type* the cells grow in cords or sheets with a haphazard disposition of the nuclei centrally and with polarization or palisading at the periphery so that the oval nuclei point towards the centre of the mass; in the rarer *scirrhous form* the connective-tissue proliferation is extreme so that long thin strands of neoplasic cells are seen embedded in a dense fibrous stroma (Hogan and Zimmerman, 1962). From the nature of the cells the tumour can be classified into two types, those with undifferentiated basal cells (the solid type) and those with cells differentiated towards hair structures (keratotic), sebaceous glands (cystic), and towards apocrine or eccrine glands (adenoid basal-cell epithelioma). In some cases differentiation may be in the direction of more than one of these cutaneous structures so that the distinction may be difficult, and the clinical appearance cannot always be relied upon as a guide to the cellular nature of the tumour (Lever, 1967).

BASO-SQUAMOUS CELLED, TRANSITIONAL OF MIXED CELLED EPITHELIOMA (EPITHELIOMA BASO-SPINO-CELLULARE) is claimed by some authors to represent a mixed picture between the squamous and basal-celled types. The cells are cylindrical at the periphery, oval, spindle-shaped or irregular at the centre and there may be large nests of cells but without keratinization. These tumours with areas of squamous proliferation have a low malignancy and a slow course and their potentialities for harm are probably no greater than those of the pure basal type (Darier and Ferrand, 1922; Montgomery, 1928); they are invasive but do not have the metastasizing capacity of squamous-cell carcinoma (Boniuk, 1964). It has been suggested that there is one primary form of carcinoma of the skin which assumes the basal-cell appearance when growing slowly, the squamous type when growing rapidly (MacCormac, 1924); the intermediary type may be a transitional stage between the two or part of the tumour may be basal-celled and the other squamous (Way, 1927). The existence of baso-squamous carcinoma, however, has been questioned by many, and according to Lever (1967) and Sanderson (1968) the entirely different histogenesis of squamous- and basal-cell carcinomata makes the occurrence of true transitional forms impossible. It is likely that the mixed type represents a keratotic basal-cell carcinoma and the intermediate type one in which the cells with large, round, pale nuclei represents an attempted differentiation either towards the secretory or ductal cells of apocrine or eccrine glands.

A MORPHŒA-LIKE OF FIBROSING BASAL-CELL EPITHELIOMA is a rare type which manifests itself as a slightly elevated plaque with an ill-defined border over which the skin remains intact for a considerable time. At this stage it has been mistaken for a chalazion (Wolkowicz, 1962). Eventually, however, the central area breaks down into an ulcer and extensive destruction may occur wherein the typical method of growth is lateral and superficial, its edge having a rolled wax-like border. Histopathologically the proliferation of connective tissue is much greater than in other types of basal-cell epithelioma but numerous groups of closely packed neoplasic cells arranged in elongated strands are buried in the dense fibrous stroma; the cells show considerable pleomorphism as well as hyperchromatic nuclei (Ketron, 1919; Caro and Howell, 1951; Wolkowicz, 1962) (Fig. 452).

A PIGMENTED BASAL-CELL CARCINOMA similar to the nodulo-ulcerative type is a rare occurrence and is easily confused with a malignant melanoma or seborrhœic keratosis, but the pigmented flare characteristic of a melanoma is absent and the lesion has a long duration (Bodenham and Lloyd, 1963; Haye *et al.*, 1972).

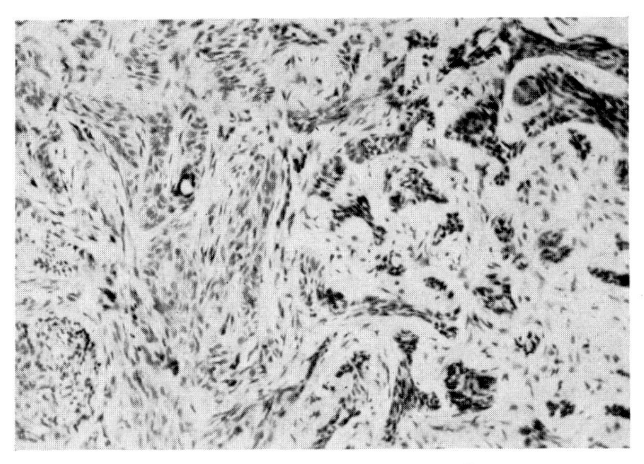

FIG. 452.—CARCINOMA OF THE LID.

In a woman aged 70. Morphœa-like basal-cell carcinoma showing small cords of darkly staining epithelial cells, pleomorphism and hyperchromatic nuclei in a fibrous stroma (M. I. Wolkowicz).

THE BASAL-CELL NÆVUS SYNDROME (*Gorlin's syndrome*) is a condition which was first reported by Nomland (1932) and Binkley and Johnson (1951) but the full implications were first adequately assessed by Gorlin and Goltz (1960). It is relatively common and in an extensive review Gorlin and his colleagues (1965) analysed some 150 cases. Multiple tumours in the skin arise in childhood and continue to appear in adult life; they are frequently associated with widespread malformations of the skeleton, with cysts of the jaws, ectopic calcification and defective teeth, hypertelorism giving a typical cranio-facial appearance (Fig. 453) and abnormalities of the spine and the central nervous system such as agenesis of the corpus callosum and medulloblastoma.[1] The radiological findings in the many skeletal anomalies were analysed by McEvoy and Gatzek (1969); they are of prime importance in the diagnosis of the syndrome. Various associated ocular abnormalities have been noted such as squint, epicanthal folds (Nover and Korting, 1970), coloboma of the choroid and optic nerve (Gorlin *et al.*, 1963), pigmentary retinal dystrophy and vitreous hæmorrhages (Happle *et al.*, 1971), and in one case a malignant melanoma of the iris

[1] Howell *et al.* (1964), Gorlin *et al.* (1965), Mason *et al.* (1965), Hermans *et al.* (1965), Rater *et al.* (1968), McEvoy and Gatzek (1969).

developed (Kedem *et al.*, 1970). Because of the ocular changes Hermans and his colleagues (1965) suggested that the condition should be regarded as a fifth type of phakomatosis. It is inherited as an autosomal dominant condition. The multiple small tumours found on the skin vary from a minute size to 1 or 2 cm. in diameter sometimes numbering hundreds, occurring anywhere in the body and frequently on

FIGS. 453 to 455.—THE BASAL-CELL NÆVUS SYNDROME.

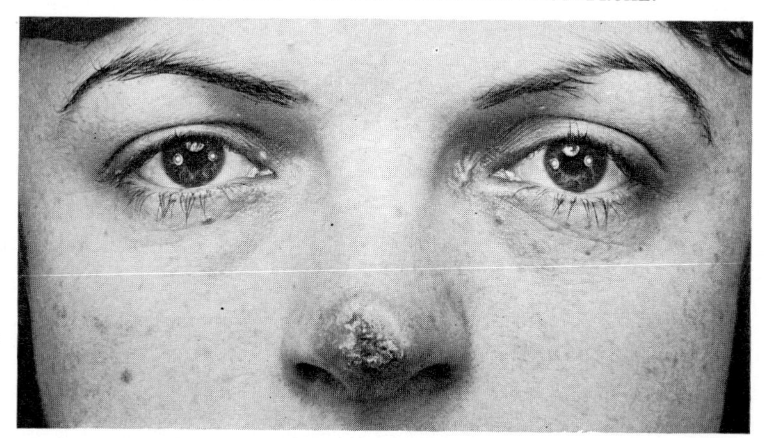

FIG. 453.—Basal-cell carcinoma on the inner canthus and nose in a girl with slight hypertelorism (St. John's Hospital for Diseases of the Skin).

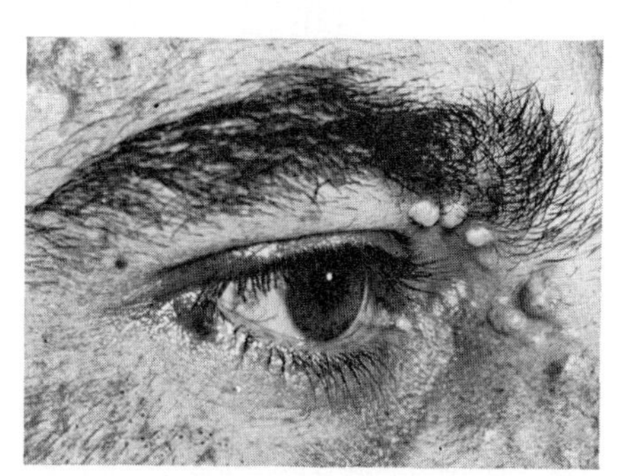

FIG. 454.

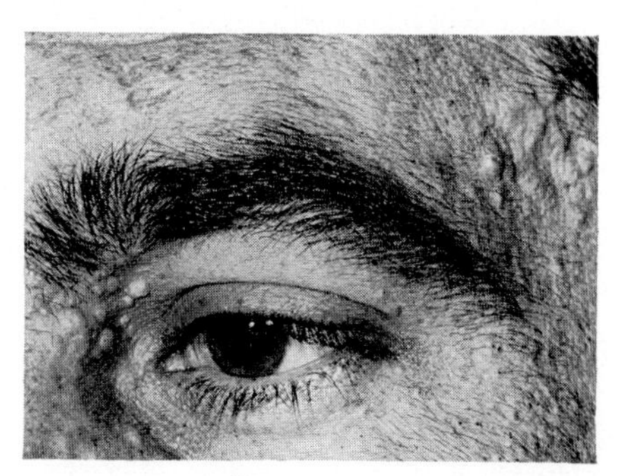

FIG. 455.

FIGS. 454 and 455.—Occurring in a man aged 27 who showed widespread lesions over the body (A. Nover and G. W. Korting).

the lids where they may be pedunculated (Figs. 453–5); in appearance they resemble basal-cell carcinomata and sometimes tricho-epitheliomata. Histopathologically they are indistinguishable from basal-cell carcinoma except that deep invasion is absent (Fig. 456); despite the histological appearance most behave in a benign manner although the periocular lesions are said to be more prone to become invasive and eventually malignant (Sanderson, 1968). Treatment by excision or local chemotherapy by the application of cytotoxic agents such as 5-fluoro-uracil has produced satisfactory results (Williams and Klein, 1970).

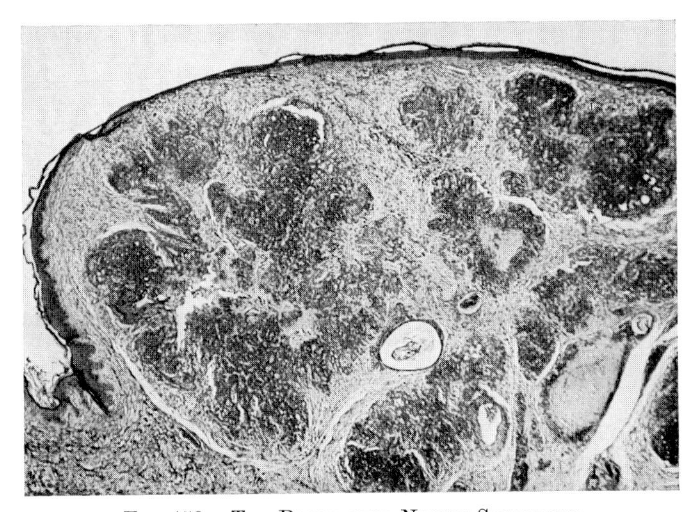

FIG. 456.—THE BASAL-CELL NÆVUS SYNDROME.

The typical histological appearance resembling a basal-cell carcinoma without penetration into the subjacent tissues (×45) (A. Nover and G. W. Korting).

The *diagnosis* of these tumours is important. *Any small eroded or crusted nodule or persistent ulceration in the region of the lids occurring in a person over 40 should always excite suspicion, as should a chalazion if it has not healed.* If to this is added an indurated base, peripheral pearly nodules and the absence of regional adenopathy, suspicion should give immediate place to action, and if the entire lesion is not quickly and completely eradicated, at the very least a biopsy through the full thickness of the lesion should be made, the dangers or discomforts of which are inconsequential.

Treatment. The treatment of carcinoma of the lids should aim at the complete destruction of the growth throughout its entire extent; at the same time due regard should be taken where possible for the preservation of the function of the eye and the cosmetic appearance. In early tumours this is usually possible, the best means to employ being excision or radiotherapy, both of which are probably equally effective in controlling the disease when efficiently carried out. Each case should be considered individually, taking into account the age and general health of the patient and his occupation and accessibility to specialized centres of treatment. Excision, if it is easily accomplished, is frequently preferable owing to the occasional uncertainty of radiation and its possible complications, the more serious of which may not become apparent for some time.[1] These have included ectropion, leucoplakia of the conjunctiva, keratitis, necrosis of the cornea, iridocyclitis and secondary glaucoma, cataract, epiphora, radiodermatitis and bony necrosis, but the incidence of these complications has been very much

[1] Vol. XIV, p. 956.

reduced with modern radio-therapeutic techniques. On the other hand, surgical treatment which usually produces a better cosmetic result is not always entirely free from complications which, apart from continued growth of the tumour due to incomplete excision, may include epiphora from damage to the lacrimal passages, ectropion and irregularity of the lid-margin. On the whole, basal-celled tumours are the most radio-sensitive and large numbers of patients are now being treated in this way. Cauterization or diathermic coagulation are inappropriate and other methods such as scraping, the application of caustic substances and freezing are probably best avoided for carcinomata involving the margins of the eyelids.

In advanced cases, however, treatment may be very much more difficult; wide surgical excision may require sacrifice of the eye and radiotherapy on its own may be indicated; indeed, large tumours can be effectively cured by irradiation, although the eye may not survive. Difficult problems are presented in the management of a recurrent tumour as a wider excision must be undertaken to include the whole of the previously treated area; there is no doubt that recurrences are even more frequent after the second than after the first attempt at treatment, which in early cases should lead to a cure.

Whatever method of treatment is chosen an adequate biopsy must be done to provide histological confirmation of the clinical diagnosis. After treatment a careful watch should always be kept for the first signs of recurrence; so long as the patient lives, this surveillance should never cease.

It is not intended here to enter into the details of each method of treatment, for which the appropriate specialized books should be consulted; a summary of the principles of the various methods may, however, be helpful.

(1) SURGICAL TREATMENT. Excision should always be wide and free extending well beyond the apparent superficial and deep limits of the tumour so as to be in healthy tissue only. The defect is made good by a reparative plastic operation for which numerous procedures have been devised.[1] When the margins of the lids are involved the operation should include the entire thickness of the lids; fortunately, most carcinomata occur on the lower eyelid which is more easily repaired than the upper. The following procedures may be required.

(a) *Excision and Direct Suture.* Small lesions in loose skin away from the ciliary margin and not deeply infiltrating may be removed by excision and simple closure of the undermined skin if this can be done without tension, but if the margin or full thickness of the lid is involved a wedge resection and closure by sliding flaps, if necessary facilitated by canthoplasty, are indicated.

(b) *Free Skin Grafting.* Larger lesions not involving the lid-margin may be treated by an adequately wide excision of the neoplasm and underlying muscle with closure of the defect by a full-thickness graft of skin, preferably taken from the other lid or, if this is not available in sufficient quantity, post-auricular skin is functionally and cosmetically effective.

[1] See Stallard (1959–69), Reese (1963), Mustardé (1966–70), Fox (1970), King and Wadsworth (1970), Mustardé *et al.* (1970).

(c) *Local skin flaps* can be of value for the treatment of lesions at the canthi, either the glabella-flap using skin rotated from the dorsum of the nose and the glabellar area for lesions at the inner canthus, or for tumours of the outer canthus the skin may be rotated from the temporal area and cheek. A flap of skin may be transposed from the upper to the lower eyelid and a rotation or pedicle flap is necessary when the defect extends down to the periosteum or into bone or is the site of radio-necrosis. On the other hand, cover by a flap which is necessarily thicker than a graft may be undesirable since a recurrence may be concealed for some time to reappear at the edge of the flap. When up to one half of either lid is involved, various procedures are available for the advancement of a lid-flap to fill the defect, a procedure which requires mobilization of the opposite canthal region; alternatively, a free graft may meet the case (Fox, 1969).

(d) *Total removal of the eyelid* presents difficult problems of reconstruction involving the provision of skin flaps supported by a cartilage graft and lined preferably by conjunctiva or mucosa from the nose or mouth. In the reconstruction of less extensive defects the tarsal plate of the upper lid should be preserved but it is permissible to utilize tarsal plate from the lower lid to fill a defect in the upper. In all those cases which require several operations over a period of time, it is, of course essential to take measures to protect the eye from exposure.

(e) *Repair by granulation* may be advisable when neoplasic invasion at the inner canthus is extensive. The lesion is widely excised down to the bone, the medial ends of the lids are secured to the periosteum and the wound is allowed to heal slowly without a formal plastic repair; a skin graft can be carried out later if necessary (Fox and Beard, 1964; Fox, 1970).

(f) *Exenteration* is indicated when the growth in the lids is so far advanced, as to involve the eye or is extending deeply into the orbit so that there is no hope of preserving the function of the eye. When the nasal sinuses and orbital bones are involved a combined intra- and extra-cranial approach may be required (Reese, 1963; Ketcham *et al.*, 1966; Beare, 1969; Littlewood and Maisels, 1970; Spaeth, 1971; Hirshowitz and Mahler, 1971).

The ideal surgical excision removes the whole of the tumour and at the same time the least possible amount of normal tissue. At operation the surgeon may have some difficulty in determining the exact dimensions of the tumour; this problem of apparent incomplete removal has been well shown by Einaugler and Henkind (1969) who found that in 20 of 40 consecutive cases neoplasic cells were present at the margin of the excised tissue, three of them from recurrent lesions. There is no doubt that excision offers an almost certain cure in the early cases provided it is sufficiently wide; great care should therefore be attached to determining the full extent of the growth.

Reese's method (1963) of clinical assessment is useful; the lesion is grasped between the fingers with the index finger in the fornix and the thumb on the outer surface. In this way it should be possible to discover where the tumour ends and the normal tissue begins and the excision should be so placed as to be several millimetres beyond this site. A modification of this procedure has been described by Cole (1970) in which the tumour is outlined with ink and a second line is drawn 3 mm. outside the first while a sketch of the area is prepared for the pathologist, numbering each part of the border, the tumour and the base. The whole area is then excised and the numbered pieces sent to the pathologist; the wound is left unclosed. If the pathologist reports that the tumour has been completely excised the area is then covered with a skin graft. In his series of 293 cases with up to 10 years follow-up, in only one patient did the tumour recur. The chemosurgical treatment introduced by Mohs (1948–49)[1] utilized somewhat similar principles with the application of a tissue-fixative and the systematic excision of specimens of fixed tissue, the microscopic examination of which served as a guide to further treatment.

The only safe assessment of complete excision is microscopic examination of the excised tissue. If this shows active tumour cells at the margin, a policy of careful observation rather than immediate further intervention is justified. In a study of 1,197 patients, Gooding and his colleagues (1965) found extension of the tumour to the margin of the excised lesion in 66 cases (5·5%), but in only 23 of these did recurrence develop within 4 years. *Prolonged follow-up* is, however, essential as these tumours grow slowly and recurrence may be long delayed. According to Stallard (1959–64) recurrence after adequate surgery is rare; Rintala (1971) reported a recurrence-rate of 4% in 204 early cases treated surgically but others have found a higher figure (14% Bedford, 1969; 12% Payne *et al.*, 1969) and in a small proportion of cases these are persistent and in spite of numerous methods of treatment are eventually incurable (Hirshowitz and Mahler, 1971).

(2) RADIOTHERAPY is very effective in destroying carcinomata of the eyelid, particularly the basal-cell type which is perhaps more sensitive than the squamous variety. The cosmetic results are generally good but loss of the lashes and some cutaneous and mucosal changes are to be expected in the treated area although these do not usually cause significant disability (MacFaul and Bedford, 1970) (Figs. 457–8). The eye itself must be protected from the harmful effects of irradiation and with modern methods damage to normal structures can be kept to a minimum.[2] In advanced cases, however, such protective measures may be impossible and loss of the eye may inevitably follow. It should be noted that the morphœa-like carcinoma responds poorly to radiation.

Three principal methods are available, the choice being determined by the site, surface extent and depth of the lesion.

(a) *Contact therapy* for very small and superficial lesions using low-voltage X-rays (up to 60 kV) produces excellent results particularly on the lid-margins or canthi (see Haye *et al.*, 1965; Domonkos, 1965; Offret and Haye, 1971).

(b) *External irradiation* is suitable for small lesions of the lid and lid-margin. The method of choice is external irradiation with X-rays using a beam of relatively low energy—100–140 kV (Lederman, 1962–66) or 140–180 kV (Dollfus, 1966; Offret and Haye, 1971; and others). With more deeply extending lesions doses up to 250 kV may be required (Newall, 1970). Although a single dose of sufficient intensity may cure the lesion, the immediate reaction may be severe; more satisfactory cosmetic

[1] p. 440. [2] Vol. VII, p. 790.

results are generally obtained if treatment is given in small doses over a more extended period. The total dose and the number of fractions in which it is given will depend on the dimensions of the growth and the number of fields required. For lesions up to 3 cm. a dose of 3,000 to 4,500 r is usually adequate although higher doses are sometimes recommended (Lederman, 1966, 4,800 to 5,400 r over a period of 18 to 21 days; Offret and Haye, 1971, 4,600 to 6,000 r in fractions of 300 r over 6 weeks). For extensive lesions the dose may go up to 6,000 r using more than one field, which should always include a margin of apparently normal tissue around the growth. For the details of these techniques the reader is referred to textbooks of radiotherapy or to a previous Volume.[1]

FIGS. 457 and 458.—RODENT ULCER (P. MacFaul).

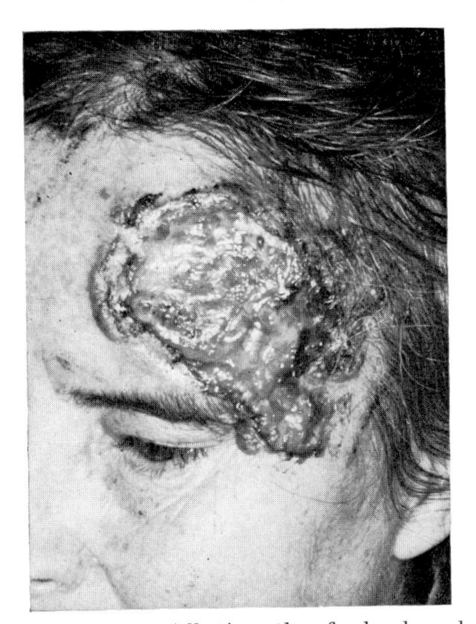

FIG. 457.—Affecting the forehead and temporal area.

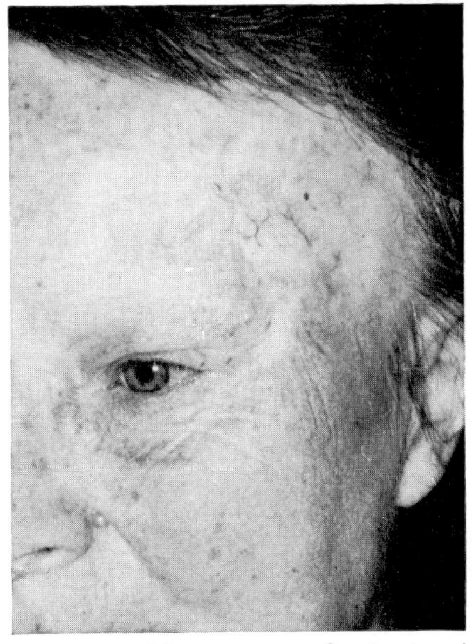

FIG. 458.—The same case after treatment by irradiation.

The changes that occur in the irradiated tissues, both those which are transient and of no significance and those which are more serious and may lead to loss of sight or of the eye, are fully described elsewhere in this *System*.[2]

(c) Techniques based on the use of radiation from *radium* in the form of needles or radon-gold seeds or tantalum wire inserted into and around the lesion are now rarely indicated although they have in the past been widely used (see Haye *et al.*, 1965). Because of the difficulty in shielding the eye, the incidence of serious ocular complications such as cataract and corneal damage is much higher with these methods than with external irradiation (McAuley, 1963; Lederman, 1964; Levitt *et al.*, 1966; Halnan and Britten, 1968).

Other methods of treatment are available.

(3) *Curettage* with *electro-cauterization* and *diathermic coagulation* have had their advocates, particularly in small growths near the inner canthus; if efficiently carried

[1] Vol. VII, p. 764. [2] Vol. XIV, p. 956 *et seq.*

out the operation destroys the tumour and seals the blood vessels by thromboses, thus lowering the risk of spread. Neither suturing nor grafting is necessary since the wound granulates under a simple dressing leaving a white supple scar.[1] There is, however, little place for this type of treatment since a complete excision-biopsy for small tumours is generally easy and in any case desirable.

(4) *Chemosurgery* was advocated by Mohs (1948–56); the area is painted with dichloracetic acid to coagulate the epidermal proteins and subsequently fixed for 24 hours with a paste of zinc chloride (a 40% solution of zinc chloride in stibnite paste); thereafter sections of the tissue approximately 1 cm. square in area and 2 mm. in thickness are surgically excised and a microscopic examination made of the frozen section of the tissue at the edge; if tumour cells are present the process is repeated and the final layer of fixed tissue separates after a few days with the formation of healthy granulation tissue. Carcinomata of the lid and periorbital skin can thus be completely removed with few complications (Robins, 1970). A five year cure-rate of 98·5% in 61 cases of basal-cell carcinoma was reported by Mohs (1956–68); in other cases when other methods had been used previously a cure-rate of 83·3% was obtained and even better figures were claimed with squamous carcinomata. This method is not ideal for lesions involving the lid-margin and is probably best suited to deep infiltrating lesions at the inner canthus which lack a line of demarcation and often have to be removed piecemeal (Fox, 1970), or for recurrences following radiotherapy (Robins *et al.*, 1971).

(5) *Cytotoxic agents* have been applied in considerable variety either as ointments or by injection into the lesion; thus E.39, a derivative of ethyleniminoquinone, was found effective in cases of basal-cell carcinoma which healed without scarring (Pillat, 1960; Heinrich, 1961). Podophyllin (Nicolov, 1953; Belisario, 1959) is also effective but there is a risk of chemosis and conjunctivitis if this substance comes into contact with the eye[2] and it is therefore contra-indicated for lesions involving the lid-margin. Another cytotoxic agent, 5-fluoro-uracil (5% to 20%) in ointment applied to the lesion, has produced satisfactory results (Ebner, 1969; Belisario, 1969) in cases of cutaneous carcinomata by introducing degeneration in the neoplasic cells but apparently not in normal skin; this method is not advised for a solitary nodular lesion but is indicated for multiple tumours where the usual methods of surgery or radiation are inappropriate because of the number and extent of the lesions and the propensity for repeated appearances of new lesions.

(6) *Cryosurgery*, a technique introduced by Arnott as early as 1855 but revived in recent years, has been effective for the destruction of small tumours although repeated applications may be necessary (Imachi *et al.*, 1969; Zacarian, 1970–72). *Local freezing* by a spray of ethyl chloride repeatedly applied to the tumour was suggested by Cavka (1965) and was found to give good results in a series of 22 cases (Cavka *et al.*, 1972).

(7) *Systemic chemotherapy* has been advanced as a method of treatment, such as the antibiotic, Bleomycin, which was found by Hijikata and his team (1970) to produce central necrosis of the lesion; such compounds may have a place as a palliative measure in advanced cases.

Immunotherapy, by which cutaneous carcinoma can be destroyed as a result of a hypersensitivity reaction induced by certain substances applied to the lesion, is another technique at present under investigation (Williams and Klein, 1970).

[1] Hazen (1917), Lehmann (1934), Sadler (1934), Ducuing *et al.* (1936), Hollander and Krugh (1944), Sweet (1963), Whelan and Deckers (1973).

[2] Vol. XIV, p. 1192.

Of all these techniques, surgery and radiotherapy are the most popular. The choice between the two continues to be a subject of debate between surgeons and radiotherapists, each being an ardent advocate of his own technique and a critic of the methods of others. Modern techniques of irradiation used with meticulous care for the protection of the eye are probably as effective as adequate surgical excision when dealing with small tumours. It must, however, be stated that several workers have found that recurrences occur more frequently after radiotherapy than after surgery (Hayes, 1962; Cobbett, 1965; Hirshowitz and Mahler, 1971; Rank, 1973; and others).

While each case must be considered individually on its merits, there are certain circumstances when one or other method is to be preferred, bearing in mind that radiotherapy is certainly contra-indicated if there is any doubt that the condition is malignant. A biopsy is therefore essential, which in small lesions may of itself amount to a complete excision. Radiotherapy may be preferred in elderly people with a short life-expectancy and in neglected cases when life is at stake and surgery must be extensive and mutilating; in simpler cases a single admission to hospital for excision and repair may be more reasonable than repeated visits for treatment if the patient is required to travel long distances. Tumours at the inner canthus are amenable to both methods of treatment and epiphora may well result from both, but irradiated skin does not easily tolerate the weight of spectacles. Tumours of the upper lid in the centre and medially should be excised because the keratinization of the conjunctiva which frequently follows their treatment by radiation may damage the cornea and cause considerable discomfort. Recurrent tumours following radiotherapy should, if possible, be treated surgically although this may well be difficult when dealing with devitalized tissue. There is no doubt that large numbers of cases of carcinoma of the eyelid are treated satisfactorily by irradiation, but the criterion of a satisfactory result should not be just whether or not the growth is eradicated but also whether the eye remains functionally intact and cosmetically acceptable. Clearly the best results will be obtained through collaboration between ophthalmic and plastic surgeons and radiotherapists interested in this disease, ideally working in a combined clinic.

Prognosis

Given early diagnosis and adequate treatment it is probable that the vast majority of skin cancers may be permanently cured. Unfortunately, neither of these conditions is met with so frequently as might be expected since, as we have seen, even although the lids are constantly in view the average length of time before a patient seeks advice is considerably over a year. It is also true that the recurrences of cutaneous cancer are too often due to a failure to follow the drastic but essential principles of treatment. In cases wherein malignancy is high and the disease is at an advanced stage, no treatment will suffice to prevent blindness or death; but fortunately most cases develop so slowly that many opportunities are offered for relief.

The prognosis of squamous epitheliomata is worse than that of the basal-cell type; in the former the mortality-rate after surgical excision is about 12%. Birge (1938) found in a series of 42 adequately traced patients treated by excision sometimes followed by radiation, that 83% were alive 3 years after, 62% 5 years, 29% 10 years, and 20% 15 years after operation. The prognosis, however, both for vision and for life, depends on the malignancy as shown histologically and on the site of the lesion: the upper lid and the inner canthus carry the worst prognosis and the outer canthus the best (Birge, 1938).

In basal-cell carcinomata the prognosis is better. Payne and his colleagues (1969) reported a mortality of 2% in 273 cases treated surgically, 8 of which required exenteration; on the other hand, Cobb and his team (1964) and Lederman (1964) reported no fatalities in their series treated by radiation although recurrences were observed. The region of the upper lid and inner canthus are again associated with the greatest incidence of blindness and the highest mortality.

Almeyda and Mantell. *Proc. roy. Soc. Med.*, **64**, 611 (1971).

Archambault and Marin. *Un. méd. Canad.*, **59**, 340 (1930).

Aurora and Blodi. *Amer. J. Ophthal.*, **70**, 329 (1970).

Surv. Ophthal., **15**, 94 (1970).

Baclesse. *2nd Cong. europ. ophthal. Soc.* (Wien, 1964) (ed. François), Basel, lv (1966).

Baclesse and Dollfus. *Arch. Ophtal.*, **20**, 473 (1960).

Barron. *Acta derm.-venereol.* (Stockh.), **42**, Suppl. 51 (1962).

Baxter and Pirozynski. *Amer. J. clin. Path.*, **48**, 53 (1967).

Beare. *Proc. roy. Soc. Med.*, **62**, 1087 (1969).

Bedford. *Proc. centen. Symp. Manhattan Eye, Ear, Thr. Hosp.* (ed. Turtz), St. Louis, **1**, 165 (1969).

Belisario. *Cancer of the Skin*, London (1959).

Med. J. Aust., **2**, 1136 (1969).

Binkley and Johnson. *Arch. Derm. Syph.* (Chic.), **63**, 73 (1951).

Binkley and Rauschkolb. *Arch. Derm.*, **86**, 332 (1962).

Birge. *Arch. Ophthal.*, **19**, 700; **20**, 254 (1938).

Blumenthal. *Arch. Derm. Syph.* (Chic.), **33**, 1042 (1936).

Bodenham and Lloyd. *Postgrad. med. J.*, **39**, 278 (1963).

Boniuk. *Int. Ophthal. Clin.*, **2**, 239 (1962).

Ocular and Adnexal Tumors, St. Louis (1964).

Breed. *Illinois med. J.*, **125**, 237 (1964).

Broders. *J. Amer. med. Ass.*, **72**, 856 (1919).

N.Y. St. J. Med., **32**, 667 (1932).

Treatment of Cancer and Allied Diseases (ed. Pack and Livingston), N.Y. (1958).

Büngeler. *Z. Krebsforsch.*, **46**, 130 (1937).

Caro and Howell. *Arch. Derm. Syph.* (Chic.), **63**, 53 (1951).

de Castro. *Arch. bras. Oftal.*, **11**, 90 (1948).

Cavka. *Bull. Soc. franç. Ophtal.*, **78**, 133 (1965).

Cavka, Curković and Janev. *Bull. Soc. franç. Ophtal.*, **85**, 365 (1972).

Cobb, Thompson and Allt. *Canad. med. Ass. J.*, **91**, 743 (1964).

Cobbett. *Brit. J. Surg.*, **52**, 347 (1965).

Cochran and Robinson. *Arch. Ophthal.*, **5**, 936 (1931).

Cole. *Amer. J. Ophthal.*, **70**, 240 (1970).

Conway and Hugo. *Amer. J. Surg.*, **110**, 620 (1965).

Cotran. *Cancer*, **14**, 1036 (1961).

Darier and Ferrand. *Ann. Derm. Syph.* (Paris), **3**, 385 (1922).

Dollfus. *Ophthalmologica*, **151**, 23 (1966).

Domonkos. *Arch. Derm.*, **91**, 364 (1965).

Driver and Cole. *Amer. J. Roentgenol.*, **41**, 617 (1939).

Ducuing, Couadau and Lu-Van Xuong. *Arch. Ophtal.*, **53**, 800 (1936).

Ebner. *Wien. klin. Wschr.*, **81**, 838 (1969).

Edmundson. *Arch. Derm. Syph.* (Chic.), **57**, 141 (1948).

Einaugler and Henkind. *Amer. J. Ophthal.*, **67**, 413 (1969).

Fayos and Wildermuth. *Arch. Ophthal.*, **67**, 298 (1962).

Fazekas. *Klin. Mbl. Augenheilk.*, **91**, 413 (1933).

Findlay. *Lancet*, **1**, 1229 (1930).

Forrest. *Int. Ophthal. Clin.*, **2**, 543 (1962).

Fox. *Amer. J. Ophthal.*, **67**, 941 (1969).
Ophthalmic Plastic Surgery, 4th ed., N.Y. (1970).
Fox and Beard. *Amer. J. Ophthal.*, **58**, 947 (1964).
Freiwald and Kravitz. *E.E.N.T. Mthly.*, **49**, 31 (1970).
Garzón. *Hospital* (Rio de J.), **65**, 1205 (1964).
Gooding, White and Yatsuhashi. *New Engl. J. Med.*, **273**, 923 (1965).
Gorlin and Goltz. *New Engl. J. Med.*, **262**, 908 (1960).
Gorlin, Vicker, Kelln and Williamson. *Cancer*, **18**, 89 (1965).
Gorlin, Yunis and Tuna. *Acta derm.-venereol.* (Stockh.), **43**, 39 (1963).
Gougerot and Patte. *Bull. Soc. franç. Derm. Syph.*, **46**, 288 (1939).
Goulden and Stallard. *Proc. roy. Soc. Med.*, **26**, 1477 (1933).
Graham, J. H. and Helwig. *Arch. Derm.*, **80**, 133 (1959); **83**, 738 (1961).
Graham, P. G. and McGavran. *Cancer*, **17**, 803 (1964).
Halnan and Britten. *Brit. J. Ophthal.*, **52**, 43 (1968).
Happle, Mehrle, Sander and Höhn. *Arch. derm. Forsch.*, **241**, 96 (1971).
Haye, Calle and Mazabraud. *Bull. Soc. franç. Ophtal.*, **85**, 448 (1972).
Haye, Jammet and Dollfus. *L'oeil et les radiations ionisantes*, Paris (1965).
Hayes. *Plast. reconstr. Surg.*, **30**, 273 (1962).
Hazen. *J. cutan. Dis.*, **35**, 590 (1917).
Healey, Wilske and Sagebiel. *New Engl. J. Med.*, **277**, 7 (1967).
Heinrich. *Ber. dtsch. ophthal. Ges.*, **64**, 436 (1961).
Hermans, Grosfeld and Spaans. *Dermatologica*, **130**, 446 (1965).
Hijikata, Kudo, Koseki *et al.* *Rinsho Ganka*, **24**, 733 (1970).
Hirschfelder and Frost. *Amer. J. Ophthal.*, **31**, 999 (1948).
Hirshowitz and Mahler. *Cancer*, **22**, 654 (1968).
Brit. J. plast. Surg., **24**, 205 (1971).
Hogan and Zimmerman. *Ophthalmic Pathology*, 2nd ed., Phila., 212 (1962).
Holland and Bellmann. *Ophthalmologica*, **150**, 138 (1965).
Hollander and Krugh. *Amer. J. Ophthal.*, **27**, 244 (1944).
Howell, Anderson and McClendon. *J. Amer. med. Ass.*, **190**, 274 (1964).
Imachi, Isayama, Mizusawa *et al.* *Folia ophthal. jap.*, **20**, 409 (1969).
Jacob. *Dubl. Hosp. Rep.*, **4**, 232 (1827).
Kedem, Even-Paz and Freund. *Dermatologica*, **140**, 99 (1970).
Kennaway. *J. industr. Hyg.*, **7**, 69 (1925).
Ketcham, Hoye, van Buren *et al.* *Amer. J. Surg.*, **112**, 591 (1966).
Ketron. *J. cutan. Dis.*, **38**, 22 (1919).
King and Wadsworth. *An Atlas of Ophthalmic Surgery*, 2nd ed., Phila. (1970).
Krompecher. *Der Basalzellenkrebs*, Jena (1902).
Kwitko, Boniuk and Zimmerman. *Arch. Ophthal.*, **69**, 693 (1963).
Laborde. *Presse méd.*, **2**, 1548 (1933).
Lacassagne. *Nouvelle pratique dermatologie* (ed. Darier *et al.*), Paris, **6**, 937 (1936).
Lamba, Shukla, Madhavan and Ratnakar. *Orient. Arch. Ophthal.*, **9**, 270 (1971).
Lane. *Surg. Gyn. Obstet.*, **64**, 458 (1937).
Lederman. *System of Ophthalmology* (ed. Duke-Elder), London, **7**, 772, 796 (1962).
Ocular and Adnexal Tumors (ed. Boniuk), St. Louis, 104 (1964).
2nd Cong. europ. ophthal. Soc. (Wien, 1964) (ed. François), Basel, xliii (1966).
Lehmann. *Arch. Derm. Syph.* (Chic.), **29**, 270 (1934).
Lever. *Histopathology of the Skin*, 4th ed., Phila., 505, 576 (1967).
Levitt, Bogardus and Brandt. *Radiology*, **87**, 340 (1966).
Lister and Dalley. *Brit. J. Ophthal.*, **44**, 638 (1960).
Littlewood and Maisels. *Proc. roy. Soc. Med.*, **63**, 681 (1970).
Lund. *Arch. Derm.*, **92**, 635 (1965).
McAuley. *Brit. J. Ophthal.*, **47**, 257 (1963).
McCallum and Kinmont. *Brit. J. Derm.*, **78**, 141 (1966).
MacCormac. *Brit. med. J.*, **2**, 457 (1924).
McEvoy and Gatzek. *Brit. J. Radiol.*, **42**, 24 (1969).
MacFaul and Bedford. *Brit. J. Ophthal.*, **54**, 237 (1970).
Mason, Helwig and Graham. *Arch. Path.*, **79**, 401 (1965).
Mathews. *Arch. Path.*, **23**, 399 (1937).
Meadows. *Aust. J. Derm.*, **7** (4), 254 (1964).
Milian and Garnier. *Bull. Soc. franç. Derm. Syph.*, **35**, 793 (1928).
Mitasov. *Med. Radiol.* (Mosk.), **6**, 5 (1961).
Mohs. *Arch. Ophthal.*, **39**, 43 (1948).
Calif. Med., **71**, 173 (1949).
Chemosurgery in Cancer, Gangrene and Infections, Springfield (1956).
J. Ark. med. Soc., **65**, 203 (1968).
Moldenhauer. *Klin. Mbl. Augenheilk.*, **132**, 335 (1958).
Montgomery. *Arch. Derm. Syph.* (Chic.), **18**, 50 (1928).
Morax. *Cancer de l'appareil visuel*, Paris (1926).
Mustardé. *Repair and Reconstruction in the Orbital Region*, Edinb. (1966).
Plast. reconstr. Surg., **45**, 146 (1970).
Trans. ophthal. Soc. U.K., **90**, 1 (1970).
Mustardé, Jones and Callahan. *Ophthalmic Plastic Surgery—up to date*, B'ham, Ala. (1970).
Newall. *Ophthalmic Plastic Surgery* (ed. Fox), 4th ed., N.Y. (1970).
Nicolini. *Sem. méd.* (B. Aires), **1**, 1630 (1935).
Nicolov. *Krebsarzt* (Wien), **8**, 334 (1953).
Nomland. *Arch. Derm. Syph.* (Chic.), **25**, 1002 (1932).

Nover and Korting. *Klin. Mbl. Augenheilk.*, **156,** 621 (1970).
Offret and Haye. *Tumeurs de l'oeil et des annexes oculaires*, Paris (1971).
Papolczy. *Szemészet*, **70,** 186 (1935).
Payne, Duke, Butner and Eifrig. *Arch. Ophthal.*, **81,** 553 (1969).
Petersen, Aaberg and Smith. *Arch. Ophthal.*, **79,** 31 (1968).
Pillat. *Wien. med. Wschr.*, **110,** 975 (1960).
Quiñones and Risco. *Arch. Soc. oftal. hisp.-amer.*, **28,** 883 (1968).
Rank. *Ann. roy. Coll. Surg.*, **52,** 148 (1973).
Rater, Selke and van Epps. *Amer. J. Roentgenol.*, **103,** 589 (1968).
Reese. Tumors *of the Eye*, 2nd ed., N.Y., 1 (1963).
del Regato. *Radiology*, **52,** 564 (1949).
Regaud, Coutard, Monod and Richard. *Ann. Oculist.* (Paris), **163,** 1 (1926).
Rinaldi, Ferlito and Campos. *Ann. Ottal.*, **96,** 171 (1970).
Rintala. *Scand. J. plast. Surg.*, **5,** 87 (1971).
Robins. *Ophthalmic Plastic Surgery* (ed. Fox), 4th ed., N.Y., 569 (1970).
Robins, Henkind and Menn. *Trans. Amer. Acad. Ophthal.*, **75,** 1228 (1971).
Rollet and Bussy. *Lyon med.*, **132,** 1129 (1923).
Sadler. *Bull. Soc. Ophtal. Paris*, 509 (1934).
Safar. *Z. Augenheilk.*, **61,** 389; **62,** 261 (1927).
Sanderson. *Textbook of Dermatology* (ed. Rook *et al.*), Oxford, **2,** 1658, 1698, 1706 (1968).
Shulman. *Brit. J. plast. Surg.*, **15,** 37 (1962).
Spaeth. *Trans. ophthal. Soc. U.K.*, **91,** 611 (1971).
Stallard. *Brit. J. Ophthal.*, **43,** 159 (1959).
Ocular and Adnexal Tumors (ed. Boniuk), St. Louis (1964).
Eye Surgery, 4th ed., Bristol, 239 (1965).
Proc. roy. Soc. Med., **62,** 1083 (1969).
Sweet. *Brit. J. Derm.*, **75,** 137 (1963).
Tallei. *Boll. Oculist.*, **4,** 475 (1925).
Vancea. *v. Graefes Arch. Ophthal.*, **129,** 191 (1932).
Villard and Dejean. *Arch. Soc. Sci. méd. biol.* (Montpellier), **10,** 167 (1929).
Vit. *Klin. Mbl. Augenheilk.*, **136,** 682 (1960).
Way. *Arch. Derm. Syph.* (Chic.), **16,** 25 (1927).
Weber. *Hb. d. allg. u. spec. Chirurgie* (ed. Pitha and Billroth), Erlangen, **3** (1), 118 (1866).
Welch and Duke. *Amer. J. Ophthal.*, **45,** 415 (1958).
Wellauer and Alesch. *Strahlentherapie*, **113,** 485 (1960).
Whelan and Deckers. *Cancer*, **31,** 159 (1973).
Williams and Klein. *Cancer*, **25,** 450 (1970).
Wolkowicz. *Amer. J. Ophthal.*, **54,** 249 (1962).
Wollensak and Meythaler. *Klin. Mbl. Augenheilk.*, **150,** 388 (1967).
Wynn-Williams. *Brit. J. plast. Surg.*, **20,** 315 (1967).
Zacarian. *Ann. Ophthal.* (Chic.), **2,** 706 (1970); **4,** 473 (1972).

INTRA-EPITHELIAL CARCINOMA

EXTRA-MAMMARY PAGET'S DISEASE. Paget's disease of the skin, first described by Sir James Paget (1874) (Fig. 998) as an eczematoid condition of the skin on the breast, is occasionally found as an extra-mammary lesion, usually in the ano-genital area or the axillæ. Histologically there is an infiltration of large round Paget cells, singly or in groups throughout the lower part of the epidermis with a pale cytoplasm containing mucopolysaccharide and a large vesicular nucleus. Frequently there is an underlying carcinoma although sometimes it is separate and deep-seated. It has usually been associated with an adenocarcinoma arising from apocrine glands, but in view of the fact that the underlying tumour may be some distance away or may not exist, there is a growing consensus of opinion that the Paget cells arise from keratinocytes in the epithelium and that the disease itself is a type of intra-epithelial carcinoma, the lesions representing a multifocal primary cancer arising in an extensive field of neoplasia (Fenn *et al.*, 1971; Medenica and Sahihi, 1972). Its occurrence in the lids is very rare, when it has usually been claimed to arise from a carcinoma of Moll's glands (Hagedoorn, 1937; Whorton and Patterson, 1955; Knauer and Whorton, 1963). In the first two cases an underlying carcinoma necessitated eventual exenteration of the orbit; the third case was successfully treated by a block section of the lid.

BOWEN'S DISEASE, another type of intra-epithelial carcinoma or precancerous dermatosis wherein the basal cells proliferate and lose their palisaded arrangement to form elongated pegs and multinucleated giant cells may occur in the epidermis, is relatively common in the conjunctiva and cornea.[1] Its occurrence in the lid is

[1] Vol. VIII, p. 1154.

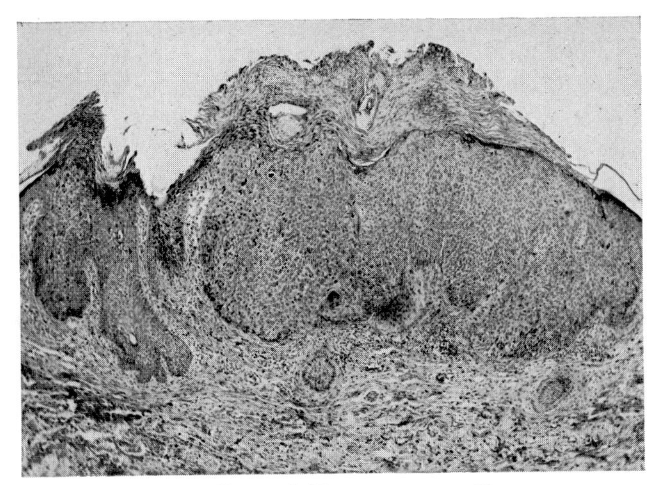

FIG. 459.—BOWEN'S DISEASE OF THE EYELID.

There is acanthosis, hyperkeratosis and parakeratosis with disorientation of the epidermal layers and mitotic figures occurring at all levels of the epidermis (M. Boniuk).

exceptional (Boniuk, 1962; Krassai, 1969). Cases affecting the conjunctiva and also involving the margin of the lid were reported by Willard (1953) and Trevor-Roper (1957). Clinically on the lids the lesion appears characteristically as an erythematous, pigmentary, scaling plaque; histopathologically this area shows marked acanthosis, hyperkeratosis and parakeratosis (Fig. 459). It is as yet unknown how often the lesion on the lids progresses to an invasive carcinoma or if there is an association with carcinomata of the viscera.

Boniuk. *Int. Ophthal. Clin.*, 2 (2), 239 (1962).
Fenn, Morley and Abell. *Obstet. and Gynec.*, **38,** 660 (1971).
Hagedoorn. *Brit. J. Ophthal.*, **21,** 234 (1937).
Knauer and Whorton. *Trans. Amer. Acad. Ophthal.*, **67,** 829 (1963).
Krassai. *Klin. Mbl. Augenheilk.*, **154,** 411 (1969).
Medenica and Sahihi. *Arch. Derm.*, **105,** 236 (1972).
Paget. *St. Bart's Hosp. Rep.*, **10,** 87 (1874).
Trevor-Roper. *Brit. J. Ophthal.*, **41,** 167 (1957).
Whorton and Patterson. *Cancer*, **8,** 1009 (1955).
Willard. *Amer. J. Ophthal.*, **36,** 1750 (1953).

XERODERMA PIGMENTOSUM

This disease, called XERODERMA PIGMENTOSUM by Moriz Kaposi (1879) and ATROPHODERMA PIGMENTOSUM by Crocker (1884), has already been described[1] where it was pointed out that, starting in the first few years of life, it makes itself evident by the appearance of freckles in the areas exposed to the sun, small telangiectasias, and eventually atrophic patches; these are followed by the development of warty growths, superficial ulceration, and eventually carcinomatous degeneration (Figs. 462–4). The malady is an actinic dermatosis of an unusually severe type, presumably due to a lack of the capacity to repair damage by ultraviolet radiation to deoxyribonucleic acid thus producing cumulative effects (Cleaver, 1968–70).

[1] Vol. VIII, p. 551.

The deficiency is transmitted as a recessive character so that the disease tends to occur in several members of a family, frequently appearing after a consanguineous marriage (Macklin, 1936–44; Rocha, 1942; Saebø, 1948; Sood, 1969) (Fig. 460); a partially sex-linked dominant heredity is rare (Anderson and Begg, 1950) (Fig. 461). There is thus an inherited tendency

FIGS. 460 and 461.—PEDIGREES OF XERODERMA PIGMENTOSUM.

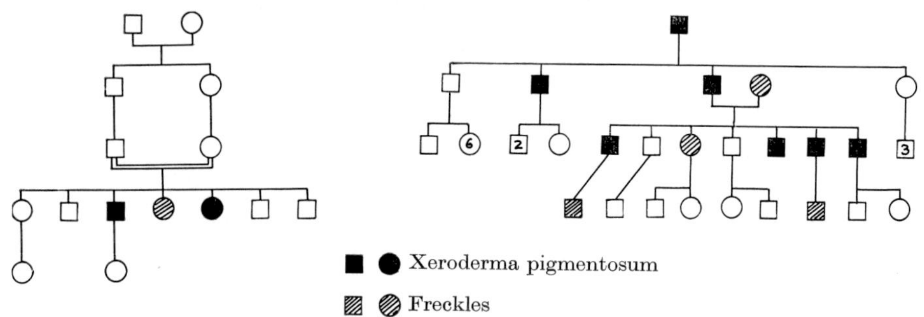

FIG. 460.—Recessive heredity (after J. Saebø).

FIG. 461.—Partially sex-linked dominant heredity (after T. Anderson and M. Begg).

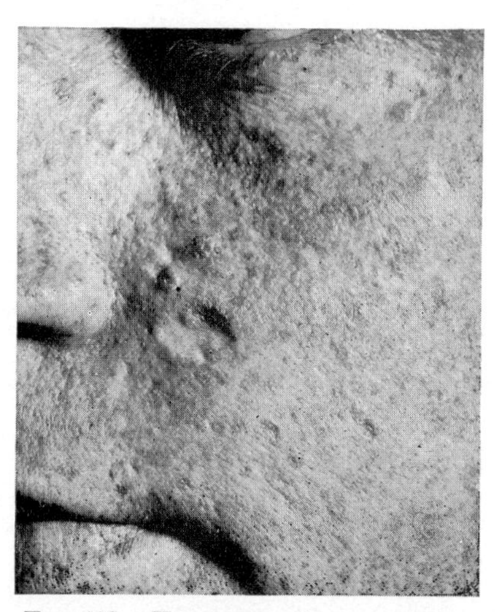

FIG. 462.—XERODERMA PIGMENTOSUM.

An early case in an adolescent girl showing freckling, dryness and telangiectasia; keratoses and scars are conspicuous (A. J. Rook, *Textbook of Dermatology*, Blackwell, Oxford).

of the epidermis to neoplasia, and in their unstable precancerous state the cells in exposed areas are liable to undergo malignant transformation at multiple foci. It was the general belief that the Negro, protected by his pigmentation, was immune to the disorder, but this is not so (Loewenthal and Trowell, 1938; Targowsky and Loewenthal, 1956; Sevel, 1963).

FIGS. 463 to 465.—XERODERMA PIGMENTOSUM.

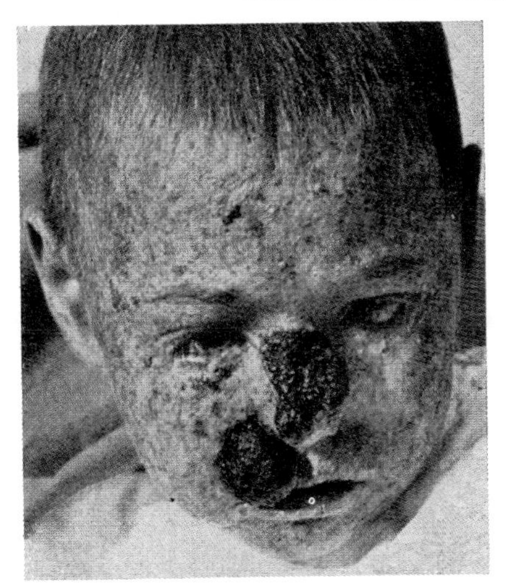

FIG. 463.

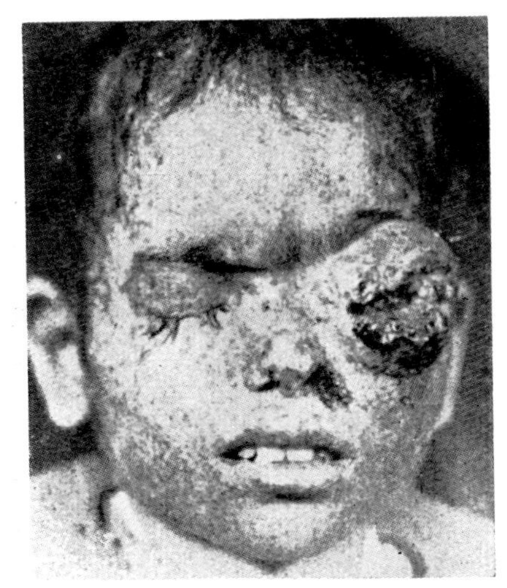

FIG. 464.

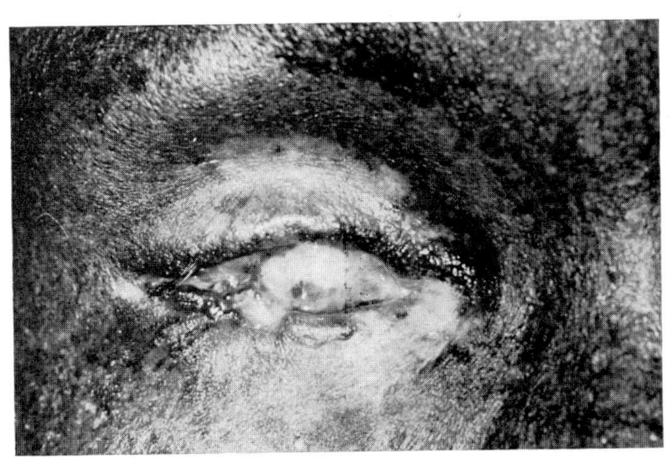

FIG. 465.

FIG. 463.—An advanced case in a girl aged 5, the second of 6 siblings the 5th and 6th of whom developed the same disease. Large scab nodules are seen on the nose and upper lip. Both lower lids, particularly at the inner half, showed a disappearance of their margins. The child died 2 years after the photograph was taken (A. B. Reese and I. E. Wilber).

FIG. 464.—An advanced case in a boy aged 10. The history began 5 years previously with the removal of a wart on the right lower lid. There are several ulcers on the right lids and on the left side both lids are enormously swollen and infiltrated, while between their ulcerating margins protrudes a huge necrotic, fungating tumour. The eye was destroyed. The skin of the face, neck, chest, back and extremities was covered with pigmented, scaly, atrophic patches (R. P. Wilson).

FIG. 465.—In a boy aged 10, to show disappearance of the lid-margin and exposure keratitis (Inst. Ophthal.).

The eyelids are frequently affected and are often the first area of skin to show changes. In the milder cases, apart from an unusual tendency to freckles and sunburn, no untoward event may occur until adolescence or early adult life when multiple tumours, papillomatous and eventually carcinomatous and frequently pigmented, begin to appear; the first tumour may become evident on the eye itself (Saebø, 1948). In the typical severe case, on the first exposure to sunlight in early life, an acute erythema develops followed by a diffuse pigmentation resembling numerous gigantic

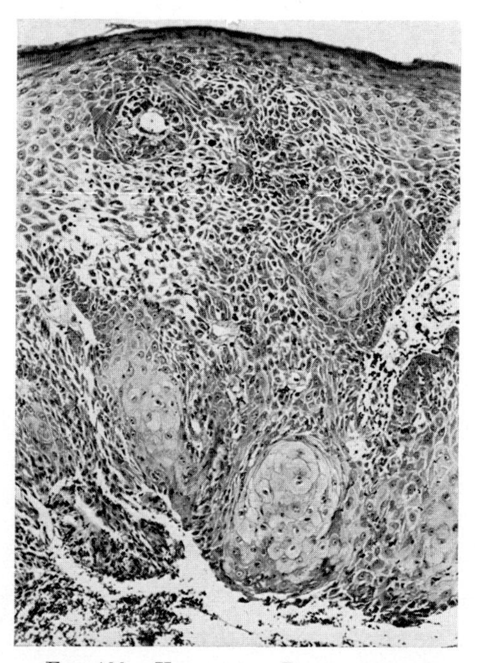

Fig. 466.—Xeroderma Pigmentosum.

Section through the skin of the right upper lid showing a squamous-cell carcinoma (D. Sevel).

freckles. This stage is followed by atrophy affecting particularly the lower lid which becomes denuded of cilia, suffers entropion (Blackmar, 1931) or ectropion (Viallefont, 1935; Martins, 1965) and eventually may entirely disappear resulting in the development of an exposure keratitis and secondary corneal ulcers[1] (Fig. 465). In the diseased area Wilson (1937) reported the occurrence of a typical capillary angioma appearing as a red sessile tumour on the lid-margin, but the more usual development is the appearance of telangiectasias and warty growths on the pigmented areas which eventually, taking on malignant characteristics, fungate and ulcerate

[1] Greeff (1901), Velhagen (1903), Heine (1906), Max (1912), Sulzer (1913), Cross (1915), Lederer (1919), Reese and Wilber (1943), Hadida *et al.* (1963).

as multiple cancers either of the squamous- or basal-cell type (Figs. 463–6).[1] In some cases malignant melanomata have occurred either as solitary (Gulati and Ahluwalia, 1967) or multiple lesions, which may or may not give rise to metastases (Ronchese, 1953; McGovern, 1962).

Malignant tumours developed in 21 (38%) of 55 patients ranging from one to fifteen years of age reported by Mortada (1968) thus: intra-epithelial 3 and squamous-cell carcinoma 7 at the limbus; basal-cell 5 and squamous-cell 5 carcinomata of the lid, and in one case a malignant melanoma of the conjunctiva.

The disease usually appears before the second year of life and is almost invariably fatal, usually in a few years, although the termination may be delayed (40 years, Darier, 1928; Saebø, 1948); the orbit is frequently invaded and eventually cachexia, meningitis or metastases end the scene. Life may be prolonged by assiduously protecting the skin from the sun by all possible means including the use of creams containing such substances as titanium dioxide (25%), removing warty growths, keeping ulcers clean and excising carcinomatous tumours, a method preferable to radiotherapy owing to the atrophic and degenerative condition of the skin; extensive plastic repair may be necessary, but the condition of the patient becomes progressively more pitiful. Cytotoxic drugs such as 5-fluoro-uracil may be helpful.

Abdin. *Bull. ophthal. Soc. Egypt*, **47**, 223 (1954).
Anderson and Begg. *Brit. J. Derm. Syph.*, **62**, 402 (1950).
Bengisu. *Oto Nöro Oftal.* (Istanbul), **2**, 33 (1947).
Blackmar. *Amer. J. Ophthal.*, **14**, 884 (1931).
Chaves. *Gaz. med. Port.*, **2**, 134 (1949).
Cleaver. *Nature* (Lond.), **218**, 652 (1968). *J. invest. Derm.*, **54**, 181 (1970).
Crocker. *Med.-chir. Trans.*, **67**, 169 (1884).
Cross. *Trans. ophthal. Soc. U.K.*, **35**, 202 (1915).
Darier. *Précis de dermatologie*, 4th ed., Paris (1928).
El-Hefnawi and Mortada. *Brit. J. Derm.*, **77**, 261 (1965).
Greeff. *Arch. Augenheilk.*, **42**, 99 (1901).
Gulati and Ahluwalia. *J. All-India ophthal. Soc.*, **15**, 233 (1967).
Hadida, Marill and Sayag. *Ann. Derm. Syph.* (Paris), **90**, 467 (1963).
Heine. *Klin. Mbl. Augenheilk.*, **44** (2), 460 (1906).
Ibrahim. *Bull. ophthal. Soc. Egypt*, **26**, 145 (1933).
Kaposi. *Pathologie u. Therapie d. Hautkrankheiten*, Wien (1879).
Katzenellenbogen. *Acta med. orient.*, **6**, 117 (1947).
Lamba, Shukla and Madhavan. *Canad. J. Ophthal.*, **4**, 148 (1969).
Larmande and Timsit. *Acta XVII int. Cong. Ophthal.*, Montreal-N.Y., **3**, 1643 (1954).
Lederer. *v. Graefes Arch. Ophthal.*, **100**, 32 (1919).
Loewenthal and Trowell. *Brit. J. Derm.*, **50**, 66 (1938).
Lynch. *Arch. Derm. Syph.* (Chic.), **29**, 858 (1934).
McGovern. *Aust. J. Derm.*, **6**, 190 (1962).
Macklin. *Arch. Derm. Syph.* (Chic.), **34**, 656 (1936); **49**, 157 (1944).
Martins. *Brit. J. Surg.*, **52**, 526 (1965).
Mathur. *Ophthalmologica*, **140**, 33 (1960).
Max. *Klin. Mbl. Augenheilk.*, **50** (1), 750 (1912).
Mortada. *Bull. ophthal. Soc. Egypt*, **61**, 231 (1968).
Reese and Wilber. *Amer. J. Ophthal.*, **26**, 901 (1943).
Rocha. *Soc. Oftal. Minas Geraes*, **25**, 1 (1942).
Ronchese. *Arch. Derm. Syph.* (Chic.), **68**, 355 (1953).

[1] Blackmar (1931), Ibrahim (1933), Wilson (1935), Bengisu (1947), Katzenellenbogen (1947), Sundareson (1948), Chaves (1949), Abdin (1954), Larmande and Timsit (1954), Mathur (1960), Sevel (1963), Martins (1965), El-Hefnawi and Mortada (1965), Mortada (1968), Lamba *et al.* (1969).

Saebø. *Brit. J. Ophthal.*, **32,** 398 (1948).
Sevel. *Brit. J. Ophthal.*, **47,** 687 (1963).
Sood. *J. pediat. Ophthal.*, **6,** 36 (1969).
Sulzer. *Ann. Oculist.* (Paris), **150,** 20 (1913).
Sundareson. *J. Indian med. Ass.*, **17,** 353 (1948).
Targowsky and Loewenthal. *S. Afr. med. J.*, **30,** 984 (1956).
Velhagen. *Arch. Augenheilk.*, **46,** 232 (1903).
Viallefont. *Arch. Ophtal.*, **52,** 32 (1935).
Wilson. *Rep. Mem. ophthal. Lab. Giza*, **10,** 37 (1935); **12,** 47 (1937).

Glandular Tumours

As would be expected in a small area rich in specialized glands, the number of types of adenoma which may appear on the lids is large, much larger than their importance in practice. Each of the glands may be involved both in a simple and a malignant hyperplasia. On the skin of the lids a solitary adenoma of a sebaceous or a sweat gland may occur, while each of the specialized glands of the lid-margin may similarly suffer a benign or malign hyperplasia—the meibomian glands and those of Zeis and Moll. Finally, the sweat glands of the skin may show the pleomorphic changes resembling those seen in the lacrimal gland.

Pathologically an adenoma is a benign overgrowth of glandular tissue which retains its essential structure and secretory function; the tumour cells remain resting upon their basement membrane arranged in giant acini, differing little in structure from that of the normal gland. An encapsulated lobulated tumour results which grows slowly, does not ulcerate and is unassociated with lymphatic involvement, metastases or recurrences when it is removed. All the glands in the lids may show such a neoplasic tendency. An adenocarcinoma, on the other hand, shows considerable cytological variations, while the cells may be arranged in solid cords, clusters or sheets; they are locally invasive and metastases may occur.

TUMOURS OF THE SEBACEOUS GLANDS

ADENOMA

A true SOLITARY ADENOMA OF A SEBACEOUS GLAND of the skin is rare, but has been noted on the face (Woolhandler and Becker, 1942), on the eyebrow (Lever, 1948) and on the skin of the lid (Rumschewitsch, 1890); such tumours have been noted on the caruncle (Coats, 1912; Mohamed, 1961). Histologically such a tumour is composed of circumscribed sebaceous lobules sheathed with connective tissue; there are two main types of cell, undifferentiated germinal cells and mature sebaceous cells as well as transitional types, within the last two of which lipid material is present (Fig. 467). It is interesting that while this is true in so far as the sebaceous glands of the skin are concerned, the modifications of these glands (meibomian) not uncommonly show a tendency to blastomatous growth both simple and malignant (adenoma and carcinoma).

The multiple lesions seen in tuberous sclerosis (the adenoma sebaceum of Pringle) are angiofibromata containing only atrophic remnants of sebaceous glands.[1]

[1] p. 469.

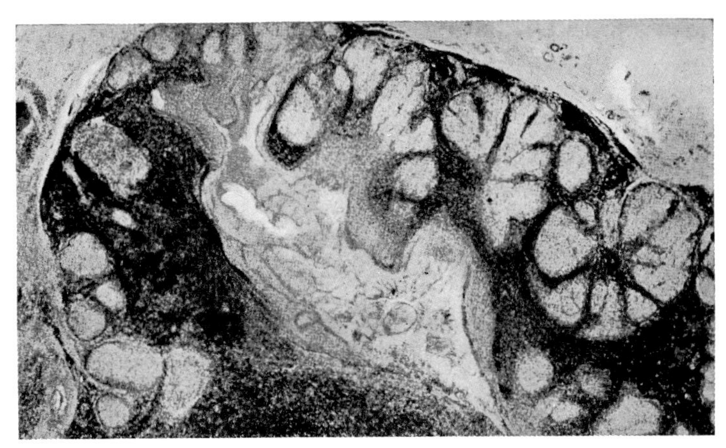

FIG. 467.—SEBACEOUS ADENOMA OF THE CARUNCLE.

Glandular masses of typical sebaceous structure are seen surrounding a central cavity filled with sebaceous material and debris. There is much lymphocytic infiltration among the acini (G. Coats).

ADENOMATA OF THE MEIBOMIAN GLANDS, although rare, are probably more common than is supposed. Clinically they are indistinguishable from chalazia, and it is only when the tumour is incised and is found to be uniformly hard and probably recurs after incision and scraping that suspicions are aroused and a histological diagnosis is made (Fig. 468). They may grow to a considerable size (Baldauf, 1870, 3·8 × 2·4 cm; Bock, 1888, 2 × 3 × 2 cm.) and ultimately ulcerate through the skin surface (Wilson, 1934) or become pedunculated from the conjunctival surface (Wilson, 1938) and even hang out over the cheek (Wright, 1938). Histologically these tumours consist of cellular acini of various sizes, separated by vascularized

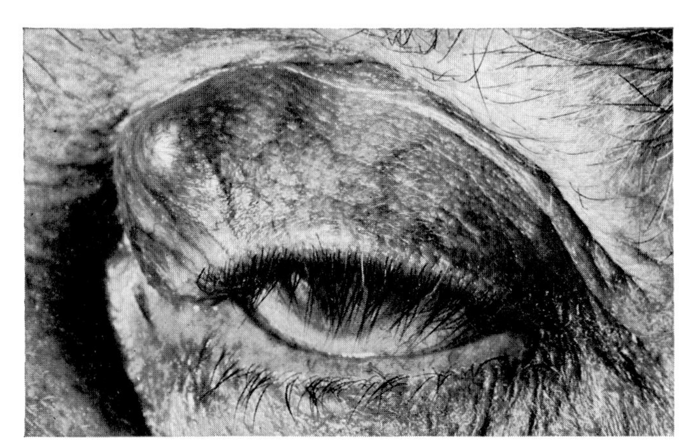

FIG. 468.—ADENOMA OF A MEIBOMIAN GLAND.

Affecting the upper lid of a woman aged 68 (Inst. Ophthal.).

connective tissue; the peripheral cells resting on the basement membrane are cylindrical and show no fatty vacuoles, the central cells are polygonal, gradually losing their nuclei and becoming fatty, the entire conformation exhibiting a regular evolution from typical glandular cells to sebaceous maturation.[1] Treatment is by total excision.

ADENOMATA OF ZEIS'S GLANDS are very rare (Balzer and Ménétrier, 1885; Rumschewitsch, 1890; Marri, 1909). They appear clinically as hard grey nodules the size of a pea on the lid-margin. Histologically they show the usual adenomatous structure with columnar cells peripherally and sebaceous cells in the central area of the acini. Again, the treatment is by excision.

SEBACEOUS NÆVUS

The SEBACEOUS NÆVUS of Jadassohn (1895) is usually located on the scalp or the face as a single lesion; it is a verrucose hamartoma with papillomatous hyperplasia of the epidermis and changes in the sebaceous and sweat glands. At birth it is a circumscribed raised plaque characterized by the presence of underdeveloped small sebaceous glands and hair follicles that do not possess a hair. At puberty the lesion usually becomes verrucose and nodular owing to the development of large numbers of mature sebaceous glands and frequently apocrine glands located deep in the dermis beneath the masses of sebaceous lobules (Mehregan and Pinkus, 1965). In 20% of cases a basal-cell epithelioma is said to develop (Nikolowski, 1951; Michalowski, 1962) and rarely a squamous-cell carcinoma (Parkin, 1950). Baquis (1969) described two cases occurring on the lids, one of which became malignant. The lesion may be associated with many developmental anomalies including several ocular abnormalities which have been summarized by Haslam and Wirtschafter (1972). Similar cases have been described wherein in addition to these multiple anomalies, painful epileptiform seizures and other neurological symptoms have occurred (the *oculo-neuro-cutaneous syndrome* of Schimmelpenning, 1957); these have been summarized by Denk (1971).

Baldauf. *Ein Fall von adenom der Meibomschen Drüsen* (Diss.), München (1870).

Balzer and Ménétrier. *Arch. Physiol.* (Paris), **6,** 564 (1885).

Baquis. *Ann. Ottal.*, **95,** 35 (1969).

Bertoldi. *Rass. ital. Ottal.*, **10,** 211 (1941).

Bock. *Wien. klin. Wschr.*, **1,** 799 (1888).

Cabannes and Lafon. *Arch. Ophtal.*, **26,** 422 (1906).

Coats. *Roy. Lond. ophthal. Hosp. Rep.*, **18,** 280 (1912).

Contino. *Clin. Oculist.*, **11,** 353 (1910).

Denk. *Med. Welt* (Berl.), **22,** 666 (1971).

Fuchs. *v. Graefes Arch. Ophthal.*, **24** (2), 121, 158 (1878).

Haslam and Wirtschafter. *Arch. Ophthal.*, **87,** 293 (1972).

Hesse. *Klin. Mbl. Augenheilk.*, **48** (2), 145 (1910).

Jadassohn. *Arch. Derm. Syph.* (Wien), **33,** 355 (1895).

Knapp. *Trans. Amer. ophthal. Soc.*, **9,** 328 (1901).

Lever. *Arch. Derm. Syph.* (Chic.), **57,** 102 (1948).

Marri. *Ophthalmologica* (Torino), **1,** 126 (1909).

[1] Baldauf (1870), Nettleship (1871), Fuchs (1878), Rumschewitsch (1890), Wadsworth (1895), Knapp (1901), Pause (1905), Cabannes and Lafon (1906), Contino (1910), Hesse (1910), Scheerer (1914), Wilson (1931–38), Soetojo (1933), Bertoldi (1941), Purtscher (1950), Miki (1972).

Mehregan and Pinkus. *Arch. Derm.*, **91,** 574 (1965).
Michalowski. *Dermatologica*, **124,** 326 (1962).
Miki. *Folia ophthal. jap.*, **23,** 421 (1972).
Mohamed. *Bull. ophthal. Soc. Egypt*, **54,** 19 (1961).
Nettleship. *Roy. Lond. ophthal. Hosp. Rep.*, **7,** 220 (1871).
Nikolowski. *Arch. Derm. Syph.* (Berl.), **193,** 340 (1951).
Parkin. *Brit. J. Derm.*, **62,** 167 (1950).
Pause. *Klin. Mbl. Augenheilk.*, **43** (1), 88 (1905).
Purtscher. *Wien. klin. Wschr.*, **62,** 54 (1950).
Rumschewitsch. *Klin. Mbl. Augenheilk.*, **28,** 387 (1890).
Scheerer. *Klin. Mbl. Augenheilk.*, **52,** 86 (1914).
Schimmelpenning. *Fortschr. Geb. Röntgenstr.*, **87,** 716 (1957).
Soetojo. *Geneesk. T. Ned.-Ind.*, **73,** 401 (1933).
Wadsworth. *Trans. Amer. ophthal. Soc.*, **7,** 383 (1895).
Wilson. *Rep. mem. ophthal. Lab. Giza*, **6,** 45 (1931); **9,** 38 (1934); **13,** 46 (1938).
Woolhandler and Becker. *Arch. Derm. Syph.* (Chic.), **45,** 734 (1942).
Wright. *Selected Pictures of Extraocular Affections*, Madras (1938).

ADENOCARCINOMA

It has been contended that carcinomata of the sebaceous glands of the skin do not exist but cases suggestive of such a condition have been reported (Beach and Severance, 1942; 29 cases in 4,000 cutaneous carcinomata, Warren and Warvi, 1943); malignant tumours of the modified sebaceous glands on the lids do, however, occur. Eighty-eight cases of carcinoma arising in the sebaceous glands of the ocular adnexa were studied by Boniuk and Zimmerman (1968): their distribution was caruncle 8, eyebrow 2, and eyelid 78, 24 of which arose in the meibomian glands, 9 in Zeis's glands, 9 in both and in 36 the exact site of origin was uncertain. Many of these presented with a chalazion-like lesion and in a few cases the appearance had suggested a basal-cell carcinoma.

ADENOCARCINOMA OF THE MEIBOMIAN GLANDS has had a somewhat curious appreciation in the literature. Although cases were reported by Allaire (1891), Panas (1894), Sourdille (1894), Snell (1896), Scott and Griffith (1900) and others and were noted by Parsons (1904) and Lagrange (1904), some authors such as von Michel (1908) in the second edition of the *Graefe-Saemisch Handbuch*, and Schreiber (1924) in the third, expressed doubts as to the existence of such tumours. Their occurrence, however, is undoubted although rare; the view of Morax (1926) that all meibomian neoplasms were carcinomatous is certainly wrong. It is probable, indeed, that adenomata and carcinomata are equally infrequent, the former predominating under 40 years of age, the latter being more common in older persons.

Thus Scheerer (1914), analysing 31 cases of meibomian tumours in the literature, concluded that 14 were adenomata and 17 carcinomata; Cavara (1920) found a proportion of 15 to 21, Riva (1922) 24 to 16; Lazarescu and his colleagues (1930) collected 48 cases of carcinoma, Lebensohn (1935) 52 cases, and Toulant and Morard (1936) 80 cases; Hagedoorn (1937) found 10 further reports after his first paper (1934), Rice and Lindeke (1950) brought the total to 97 and Ginsberg (1965) to 142. Several recent cases of interest have been described.[1]

[1] Kennedy and King (1954), Hartz (1955), Straatsma (1956), Sweebe and Cogan (1959), Gubareva (1963), Scheie *et al.* (1964), Subramaniam *et al.* (1965), Meythaler and Fleck (1965), Gowey and Kern (1965), Radnót and Lovas (1967), Ward (1967), Rao *et al.* (1967), Ide *et al.* (1968), Taylor and Lehman (1969), Mathur and Bhasin (1969), Freiwald and Kravitz (1970), Sood *et al.* (1972), Reddy *et al.* (1972), Portney (1973).

Meibomian carcinoma usually occurs in persons, particularly women, of middle age and over, although it has occasionally been reported in younger patients (11 years, Allaire, 1891; Lagrange, 1904); it is met with on either lid but, in contradistinction to carcinoma of the skin, with a predilection for the upper. At the commencement its progress is insidious and painless (Radnót, 1938; Pagès *et al.*, 1938); it usually appears as a tough

FIGS. 469 and 470.—CARCINOMA OF A MEIBOMIAN GLAND.

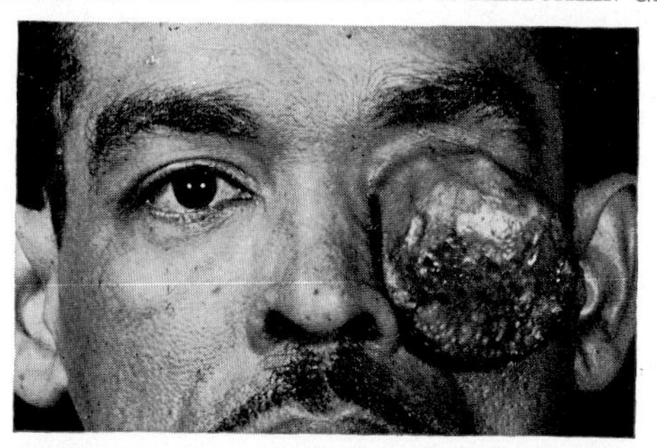

FIG. 469.—Affecting the upper lid of a Negro aged 35. It had been present for 20 months and grew rapidly in the last 6 months (D. M. Ward).

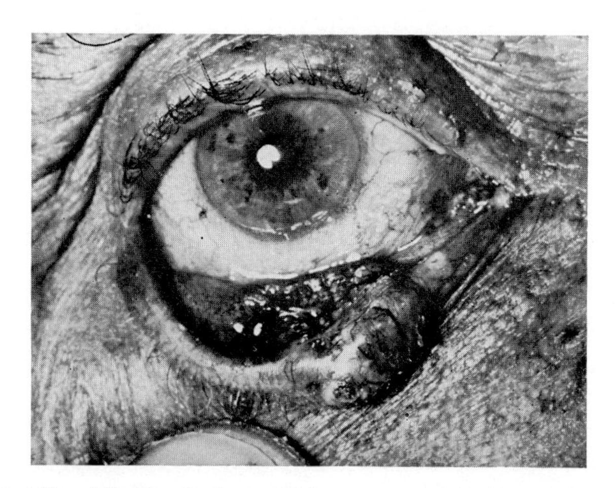

FIG. 470.—Affecting the lower lid in a man aged 73 (Inst. Ophthal.).

elastic tumour over which the skin is freely movable, without the ulceration characteristic of epitheliomata, in no way distinguishable from an ordinary chalazion. In the majority of cases it is mistaken for this common condition and incised; and as a rule it is only after persistent recurrences and continued growth that suspicions are aroused and a biopsy is made (Hughes, 1932; Wolkowicz, 1962; Boniuk and Zimmerman, 1968) (Figs. 469–70).

The case reported by Scheie and his colleagues (1964) was unique since it appeared as a diffuse eczematoid thickening involving both lids adjacent to the outer canthus. Unfortunately, after curetting or incomplete operation the slow growth characteristic of the tumour in its early stages may suddenly be replaced by rapid malignancy, quick recurrences and progressive extension involving the lid-margin and the tarsus, while the tarsal conjunctiva becomes infiltrated and ulcerated. Sometimes at a relatively early stage the orbital tissues become involved and the eye immobilized,[1] and the pre-auricular and cervical lymph nodes may be involved (Cavara, 1920; Hagedoorn, 1934; Radnót, 1949; Gowey and Kern, 1965; 17% of cases, Ginsberg, 1965; 28%, Boniuk and Zimmerman, 1968). Death may occur from multiple metastases (Magnus, 1947; Scheie *et al.*, 1964); 13·5% of the patients followed-up by Boniuk and Zimmerman (1968) died from this cause and in Ginsberg's (1965) review of 142 cases 6% were fatal.

Knapp's (1936) case is typical. A tarsal swelling resembling a large chalazion was removed by tarsectomy; 4 months later a local recurrence was removed and 12 months later a pre-auricular node, an operation which had to be repeated in 9 months. Despite radium treatment, nodes appeared in the neck, and 7 years after the original operation on the lid the patient died from abdominal metastases.

Histopathologically, alongside acini which are of normal size, are others which become of giant size surrounded by connective-tissue lamellæ and separated into lobules by septa rich in capillaries. In the acini the cells are of two types. The great mass is composed of sebaceous mother-cells peripherally, internal to which is a successive metamorphosis into sebaceous cells filled with fatty granules. The arrangement, however, is irregular; pavement-cells of an epitheliomatous nature are interposed, sometimes arranged in concentric nests and containing kerato-hyalin, while areas of mucoid degeneration and necrosis are common (Fig. 471); cystic formation may also occur (Pereyra, 1922). Considerable cytological variations may be seen with the cells arranged in sheets, solid cords or in clusters; some of them may be unusually large with abundant vacuolated cytoplasm in which large quantities of fat may be demonstrated by the appropriate staining of frozen sections (Straatsma, 1956; Sweebe and Cogan, 1959; Hogan and Zimmerman, 1962). Mitotic figures may be conspicuous and are an indication of the malignancy of the lesion. Poorly differentiated tumours may simulate a squamous carcinoma. The metastases in the lymph nodes show the same preponderance of fatty cells, but pavement epithelial cells are usually more numerous. The conjunctival epithelium or ciliary margin frequently shows changes of a hyperplasic or pagetoid nature[2]; whether this is a simple reaction or a more active participation in the neoplasic process is not always clear.

[1] Shoji (1929), Dupuy-Dutemps (1938), Sie-Boen-Lian (1941), Poddany and Bartos (1967); 17 % of cases, Boniuk and Zimmerman (1968).

[2] Gabriélidès (1910), van Duyse (1914), Létulle and de Lapersonne (1923), Hagedoorn (1934), Hogan and Zimmerman (1962), Scheie *et al.* (1964), Ward (1967), and others.

Exceptionally the tumour is composed of pavement cells arising perhaps from the duct, or a basal-celled configuration may occur; such a tumour was described by Dollfus and his colleagues (1939) in a very slow-growing tumour which persisted for 13 years with ulceration and cicatrization without adenopathy. In some cases the sebaceous differentiation may be lacking while in others it may be preponderant. Finally, transitional atypical types are sometimes seen.

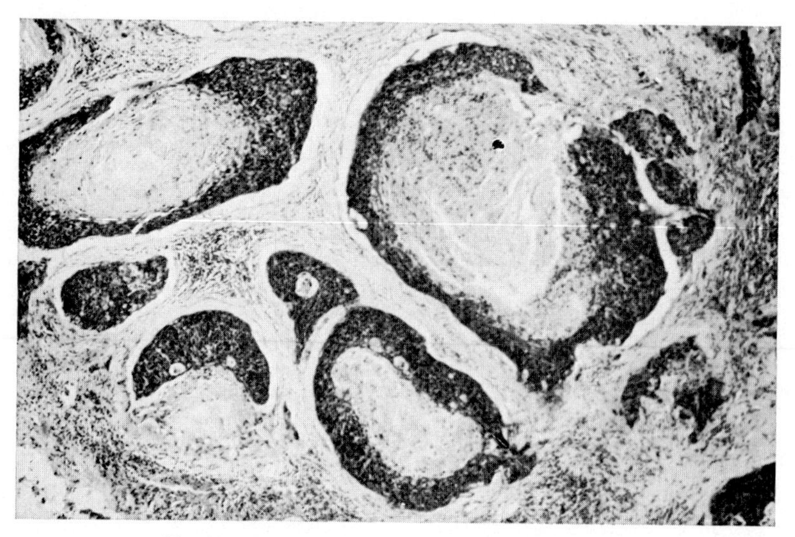

Fig. 471.—Carcinoma of a Meibomian Gland.

Lobular masses of malignant sebaceous cells showing necrosis in their central areas (H. & E.; ×60) (N. Ashton).

Diagnosis is by no means easy, a chalazion or tarsitis often being simulated. Occasionally an inflammatory reaction induced by the growth may conceal its neoplasic nature, and Reese (1963) considered that the initial lesion usually resembled a chalazion with or without suppuration. Suspicion should be excited in older people by a very hard chalazion-like tumour without inflammatory signs, having a protuberant appearance on the conjunctival surface. Atrophy of the skin and protrusion of the lid-margin, and adenopathy are even more characteristic signs; ulceration of the lid-margin suggests the simultaneous occurrence of a carcinoma of Zeis's gland. In all such cases the tumour should be completely excised and examined, and if a biopsy shows carcinomatous changes more radical measures should be undertaken. Similarly, every chalazion which appears atypical or which recurs should be examined histologically.

Treatment should be by amply wide excision followed by plastic reconstruction of the lid or radiational therapy, the same measures being adopted for lymphatic metastases if they are present. Exenteration of the orbit has been necessary on several occasions.[1] The tendency to metastases,

[1] Cabannes and Lafon (1906), Grignolo (1909), Scheerer (1914), Cavara (1920), Shoji (1929), and others.

coupled with the difficulty of early diagnosis, probably makes the prognosis more serious than that of carcinoma of the skin. Modern methods of radiotherapy, however, even when previous treatment has been inadequate or where resection is impracticable, have considerably improved the chances of a successful result (Ide *et al.*, 1968).

It is interesting that a case of melanocarcinoma associated with a tarsal gland has been reported in a dog (Bernoulli, 1948).

ADENOCARCINOMA OF ZEIS'S GLANDS is very rare indeed and can only be differentiated histologically. The tumour appears on the ciliary margin as a small ulcerating growth, the structure of which is similar to the corresponding growth in a meibomian gland (Michaïl, 1924; Lugli, 1932; Scheie *et al.*, 1964; Boniuk and Zimmerman, 1968; Best *et al.*, 1970).

Allaire. *Contribution à l'étude du polyadénome ou épithéliome intra-glandulaire* (Thèse), Paris (1891).

Beach and Severance. *Ann. Surg.*, **115**, 258 (1942).

Bernoulli. *Ophthalmologica*, **116**, 101 (1948).

Best, de Chabon, Park and Galin. *N.Y. St. J. Med.*, **70**, 433 (1970).

Boniuk and Zimmerman. *Trans. Amer. Acad. Ophthal.*, **72**, 619 (1968).

Cabannes and Lafon. *Arch. Ophtal.*, **26**, 422 (1906).

Cavara. *Arch. Sci. med.*, **43**, 1 (1920).

Dollfus, Joseph, Courtial and Desvignes. *Bull. Soc. Ophtal. Paris*, 23 (1939).

Dupuy-Dutemps. *Bull. Soc. Ophtal. Paris*, 70 (1938).

van Duyse. *Arch. Ophtal.*, **34**, 355 (1914).

Freiwald and Kravitz. *E.E.N.T. Monthly*, **49**, 31 (1970).

Gabriélidès. *Arch. Ophtal.*, **30**, 178 (1910).

Ginsberg. *Arch. Ophthal.*, **73**, 271 (1965).

Gowey and Kern. *Calif. Med.*, **103**, 126 (1965).

Grignolo. *Ophthalmologica* (Torino), **1**, 250 (1909).

Gubareva. *Arkh. Pat.*, **25**, 50 (1963).

Hagedoorn. *Arch. Ophthal.*, **12**, 850 (1934); **18**, 50 (1937).

Hartz. *Amer. J. clin. Path.*, **25**, 636 (1955).

Hogan and Zimmerman. *Ophthalmic Pathology*, 2nd ed., Phila., 214 (1962).

Hughes. *Trans. ophthal. Soc. U.K.*, **52**, 557 (1932).

Ide, Ridings, Yamashita and Buesseler. *Arch. Ophthal.*, **79**, 540 (1968).

Kennedy and King. *Amer. J. Ophthal.*, **37**, 259 (1954).

Knapp. *J. Amer. med. Ass.*, **107**, 937 (1936).

Lagrange. *Tumeurs de l'oeil*, Paris, **2**, 733 (1904).

Lazarescu, D. and E., and Ionescu. *Brit. J. Ophthal.*, **14**, 588 (1930).

Lebensohn. *Amer. J. Ophthal.*, **18**, 552 (1935).

Létulle and de Lapersonne. *Arch. Ophtal.*, **40**, 641 (1923).

Lugli. *Boll. Oculist.*, **11**, 49 (1932).

Magnus. *Trans. ophthal. Soc. U.K.*, **67**, 432 (1947).

Mathur and Bhasin. *Ind. J. ocular Path.*, **3**, 10 (1969).

Meythaler and Fleck. *Klin. Mbl. Augenheilk.*, **147**, 745 (1965).

Michaïl. *Cluj. med.*, **5**, 18 (1924).
Ann. Oculist. (Paris), **161**, 817 (1924).

von Michel. *Graefe-Saemisch Hb. d. ges. Augenheilk.*, 2nd ed., Leipzig, **5** (3), 323 (1908).

Morax. *Cancer de l'appareil visuel*, Paris, 101 (1926).

Pagès, Stora and Duguet. *Bull. Soc. Ophtal. Paris*, 621 (1938).

Panas. *Traité des maladies des yeux*, Paris, **2** (1894).

Parsons. *Pathology of the Eye*, London, **1**, 28 (1904).

Pereyra. *Arch. Ottal.*, **29**, 271, 320 (1922).

Poddany and Bartos. *Cs. Pat.*, **3**, 55 (1967).

Portney. *Ann. Ophthal.* (Chic.), **5**, 193 (1973).

Radnót. *Ophthalmologica*, **96**, 22 (1938).
G. ital. Oftal., **2**, 100 (1949).

Radnót and Lovas. *Klin. Mbl. Augenheilk.*, **151**, 508 (1967).

Rao, S. and K., and Reddy. *Brit. J. Ophthal.*, **51**, 492 (1967).

Reese. *Tumors of the Eye*, 2nd ed., N.Y., 6 (1963).

Reddy, Rao and Satyendra. *Orient. Arch. Ophthal.*, **10**, 1 (1972).

Rice and Lindeke. *Amer. J. Ophthal.*, **33**, 1434 (1950).

Riva. *Arch. Ottal.*, **29**, 261 (1922).

Scheerer. *Klin. Mbl. Augenheilk.*, **52**, 86 (1914).

Scheie, Yanoff and Frayer. *Arch. Ophthal.*, **72**, 800 (1964).

Schreiber. *Graefe-Saemisch Hb. d. ges. Augenheilk.*, 3rd ed., *Die Krankheiten d. Augenlider*, Berlin, 335 (1924).

Scott and Griffith. *Trans. ophthal. Soc. U.K.*, **20**, 44 (1900).

Shoji. *Arch. Ophtal.*, **46**, 144 (1929).
Sie-Boen-Lian. *Ophthalmologica*, **101**, 205 (1941).
Snell. *Trans. ophthal. Soc. U.K.*, **16**, 1 (1896).
Sood, Sofat, Chandel and Longani. *Orient. Arch. Ophthal.*, **10**, 31 (1972).
Sourdille. *Arch. Ophtal.*, **14**, 179 (1894).
Straatsma. *Arch. Ophthal.*, **56**, 71 (1956).
Subramaniam, Sreedharan and Kutty. *Brit. J. Ophthal.*, **49**, 93 (1965).
Sweebe and Cogan. *Arch. Ophthal.*, **61**, 282 (1959).
Taylor and Lehman. *Arch. Ophthal.*, **82**, 66 (1969).
Toulant and Morard. *Bull. Soc. franç. Ophtal.*, **49**, 27 (1936).
Ward. *Brit. J. Ophthal.*, **51**, 193 (1967).
Warren and Warvi. *Amer. J. Path.*, **19**, 441 (1943).
Wolkowicz. *Amer. J. Ophthal.*, **54**, 249 (1962).

TUMOURS OF THE SWEAT GLANDS

Tumours arising in the sweat glands of the eyelid are uncommon and the differences in their structure have given rise to much confusion and a wide variety of descriptive terms. Thus in the literature any of the following may be found—syringoma, porosyringoma, eccrine acrospiroma, clear-cell hidradenoma, sweat gland adenoma, mixed tumours, dermal sweat duct tumour and others (Lever, 1967; Ferry and Haddad, 1970). They may develop from either the eccrine or apocrine glands but opinions have differed on which part of the structures gives rise to the lesion. These tumours may occasionally be multiple but as a rule are solitary and almost invariably benign; malignant forms are rare and may be associated with local invasion and metastases.

BENIGN TUMOURS

ADENOMATA (CLEAR-CELL HIDRADENOMA—ἱδρώς, sweat; SPIRADENOMA—σπεῖρα, a coil) may appear on the lids as solitary or multiple small white tumours (Krompecher, 1919); occasionally they may grow slowly over a period of years to a considerable size (Fig. 472) (Clifton and Gordon, 1947; Tulman and Jack, 1965; Greer, 1968). These tumours are essentially benign but recurrences may appear after excision, presumably due to imperfect removal (20% of cases, Boniuk and Halpert, 1964); malignant types with metastases are exceptional (Kersting, 1963; Johnson and Helwig, 1969).

Histopathologically the clear-cell hidradenoma is composed mainly of polygonal epithelial cells with clear unstaining cytoplasm and deeply staining basophilic nuclei (Fig. 473); fusiform cells with a basophilic cytoplasm may also be present; the presence of myo-epithelial cells, as suggested by Lever and Castleman (1952) and Boniuk and Halpert (1964), was denied by Allen (1954) who considered that they represented cells from the outer layer of the glandular epithelium. It is probable that the tumour arises from either eccrine or apocrine glands, usually the former. Opinions are divided, however, on the actual origin of the cells, whether from the ducts (Kersting, 1963; Greer, 1968; Johnson and Helwig, 1969; Ferry and Haddad, 1970) or the secretory cells (O'Hara *et al.*, 1966).

These adenomata have a strong tendency to become cystic (HIDROCYSTOMA) in this way linking up with the cysts of Moll's glands (Salzmann,

FIGS. 472 and 473.—CLEAR-CELL HIDRADENOMA (C. H. Greer).

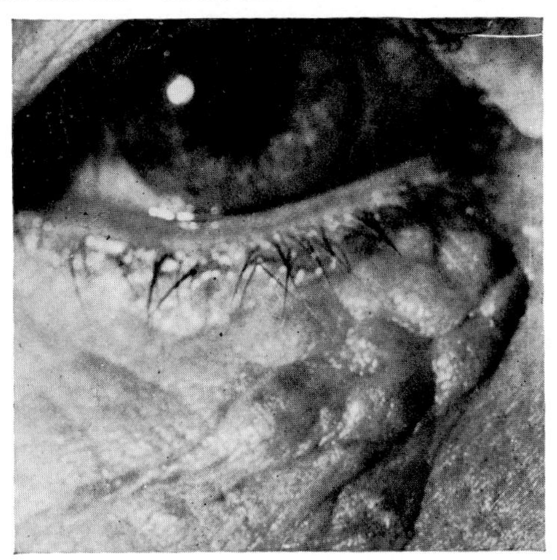

FIG. 472.—In a woman aged 70.

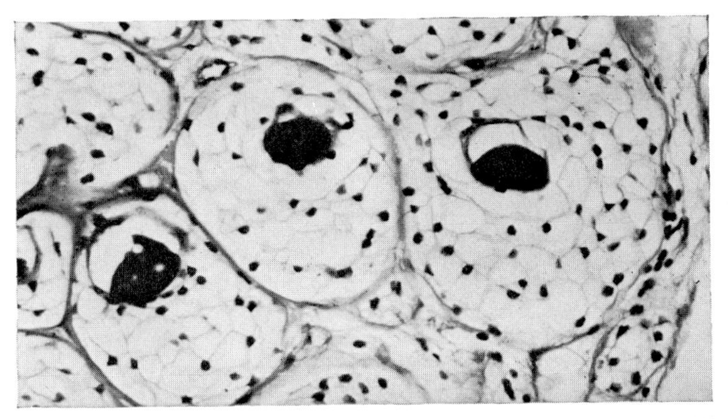

FIG. 473.—The acini of the tumour made up of polygonal clear cells with deeply-staining basophilic nuclei. The cytoplasm contains a PAS-positive diastase-resistant substance.

1891; Krompecher, 1919).[1] Others show papillomatous proliferations into the cavity of the cyst (HIDRADENOMA PAPILLIFERUM CYSTICUM). These are formed by two layers of epithelium based on a scaffolding of delicate fibrous strands folded in a most complex manner (von Michel, 1906; Wollensak and Meythaler, 1967).

SYRINGOMA is a relatively common type of tumour arising from the eccrine glands on the skin of the lids. In clinical appearance these multiple lesions resemble milia or tricho-epitheliomata, from which the diagnosis can only be made on histological examination (Fig. 474). Such neoplasms, indeed,

[1] p. 395.

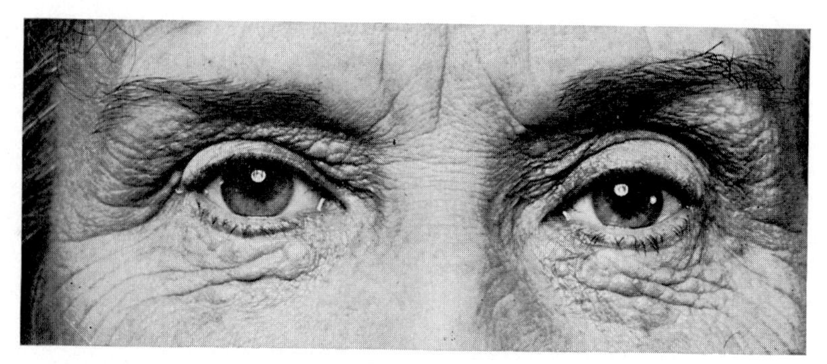

FIG. 474.—SYRINGOMA.
Affecting both eyelids (St. John's Hospital for Diseases of the Skin).

may be more common than is supposed owing to the difficulty in obtaining material for histological section. The lesions on the skin of the lids consist of collections of dilated and convoluted sweat ducts usually lined by a double layer of cells and the lumina of the ducts are filled with colloid material.[1] They arise from elements of eccrine glands, mainly ductal but sometimes secretory in nature (Winkelmann and Muller, 1964; Lever, 1967). Treatment is by diathermy or electrolysis.

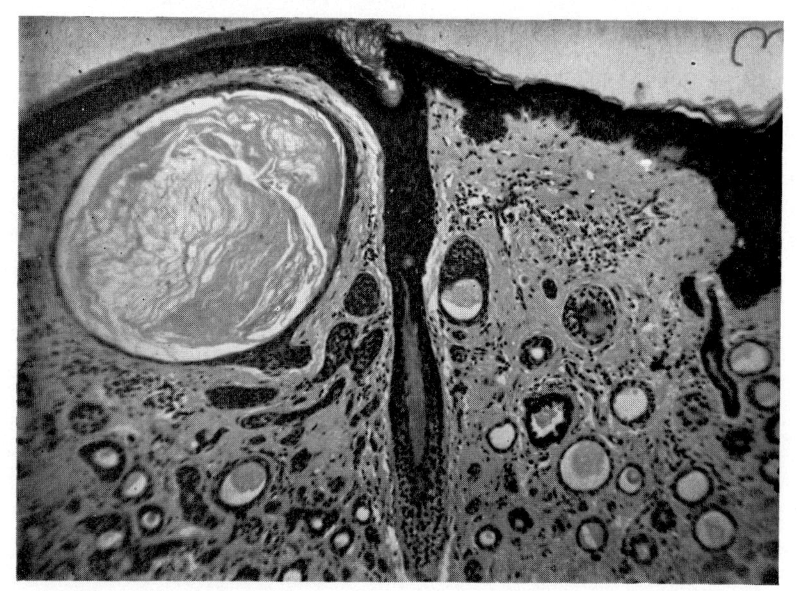

FIG. 475.—SYRINGOCYSTADENOMA PAPILLIFERUM OF THE EYELID.
There is cystic dilatation of the sweat gland ducts which are more numerous than normal. Some of the ducts show epithelial hyperplasia with villous projections (H. & E.; ×90) (N. Ashton).

[1] Jarisch (1900), Greenbaum (1928), Hagedoorn (1938), Franceschetti and Stadlin (1950) Gaul (1955), Daicker (1963), H. and J. Allington (1968).

Somewhat similar tumours of the ducts of the apocrine sweat glands occur with a congenital basis, becoming raised and nodular at puberty and are thus in a sense nævi or hamartomata (SYRINGOCYSTADENOMA PAPILLIFERUM); the ducts are numerous and they show epithelial hyperplasia with villous formation (Fig. 475).

PLEOMORPHIC ADENOMATA of the sweat glands, frequently known as *mixed tumours* of the skin, have a histological appearance similar to the pleomorphic adenomata of the lacrimal and major salivary glands[1]; for

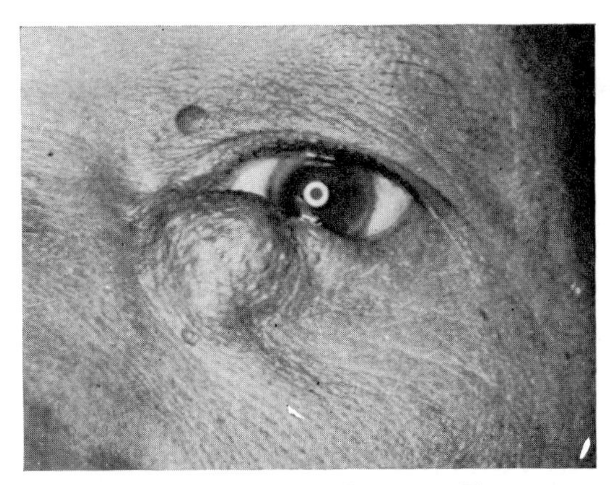

FIG. 476.—PLEOMORPHIC ADENOMA OF LACRIMAL TISSUE IN THE EYELID.
Occurring in a female African aged 50. The tumour on the left lower lid had been present for 2 years without any symptoms (P. A. S. Evans).

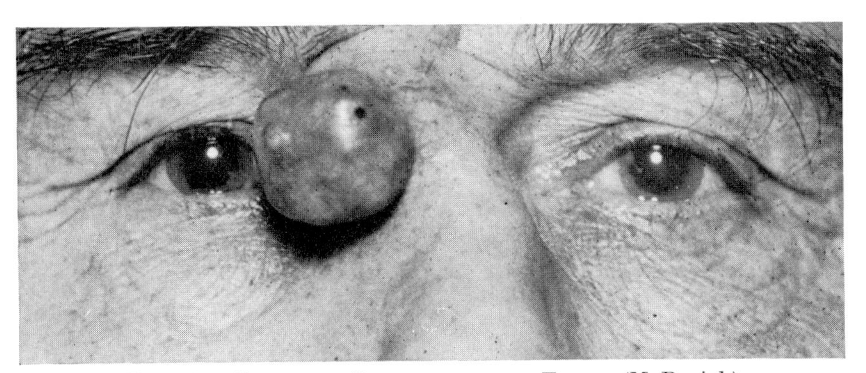

FIG. 477.—CHONDROID SYRINGOMA OF THE EYELID (M. Boniuk).

this reason they have frequently been thought to arise from accessory islands of one or other of these glands, and in some cases this may be so[2] (Fig. 476). They may arise from eccrine or apocrine glands and appear as hard circumscribed nodules in the skin, slowly growing and asymptomatic, but may assume a considerable size with a polypoid form. They are frequently

[1] p. 652.
[2] p. 661.

encapsulated, and the cellular content exhibits a glandular structure; the lumina are lined with two layers of cells in apocrine tumours and a single layer in eccrine tumours, and epithelial cells are arranged singly or in groups throughout the abundant stroma. The stroma itself is very varied; in most cases it is mucoid consisting largely of mucopolysaccharides, and may have a hyaline or cartilaginous appearance (the *chondroid syringomata* of Hirsch and Helwig, 1961; Boniuk, 1962) (Fig. 477). It would seem probable, however, that these appearances are the result of metaplasia of the epithelium and its associated stroma so that the term " pleomorphic adenomata " is probably more appropriate than " mixed tumours ". These tumours are benign but if the capsule is ruptured they may become locally invasive. Treatment is by excision which must be wide; enucleation may lead to recurrences.

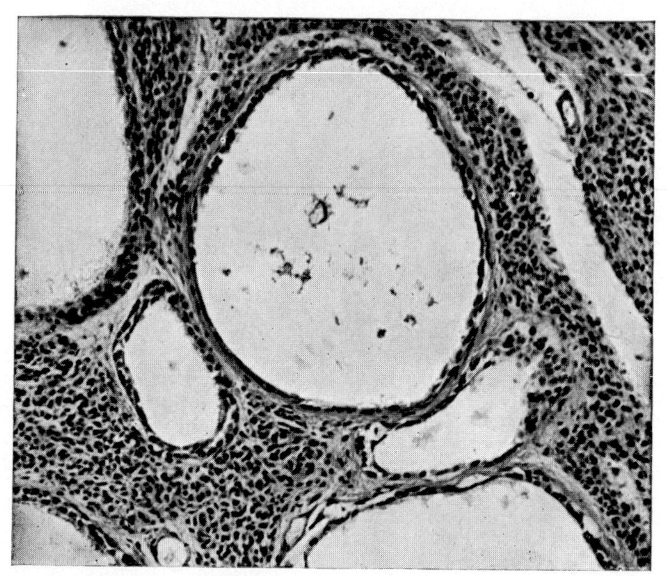

Fig. 478.—Cystadenomatous Tumour of Moll's Gland.
Showing the cystic cavities lined by a single layer of cells (G. Baquis).

ADENOMATA OF MOLL'S GLANDS are also rare. They form small nodular growths the size of a pea on the ciliary margin, growing slowly with an irregular and sometimes ulcerated surface.[1] The adenomatous structure is in conformity with the configuration of these glands with hypertrophied alveoli or cysts lined with irregularly arranged cubical, cylindrical or flattened epithelium, sometimes with finger-like processes projecting into the lumen. The stroma is of loose fibrous tissue poor in vessels. Baquis (1968) described a cystadenomatous tumour which he categorized as a syringoma (Fig. 478).

[1] Rumschewitsch (1890), Wintersteiner (1900), Letulle and Duclos (1911), Pereyra (1924), Nichellati (1931), Huber and Picena (1933), Gault and Legait (1935), O'Brien and Braley (1936), Orbán (1955), Renard *et al.* (1963).

H.-J. Meyer (1968) reported a case wherein a cystadenoma (originally treated as a chalazion) recurred in association with a similar tumour in the parotid gland.

Allen. *The Skin, a Clinico-pathologic Treatise*, St. Louis (1954).
Allington, H. and J. *Arch. Derm.*, **97**, 50 (1968).
Baquis. *Ann. Ottal.*, **94**, 1204 (1968).
Boniuk. *Int. Ophthal. Clin.*, **2** (2), 239 (1962).
Boniuk and Halpert. *Arch. Ophthal.*, **72**, 59 (1964).
Clifton and Gordon. *Brit. J. Ophthal.*, **31**, 697 (1947).
Daicker. *Ophthalmologica*, **145**, 281 (1963).
Evans. *Brit. J. Ophthal.*, **48**, 234 (1964).
Ferry and Haddad. *Arch. Ophthal.*, **83**, 591 (1970).
Franceschetti and Stadlin. *Arch. Oftal. B. Aires*, **25**, 366 (1950).
Gaul. *Arch. Ophthal.*, **53**, 371 (1955).
Gault and Legait. *Bull. Soc. Ophtal. Paris*, 263 (1935).
Golay. *Acta derm. venereol.* (Stockh.), **4**, 479 (1923).
Greenbaum. *Amer. J. Ophthal.*, **11**, 275 (1928).
Greer. *Arch. Ophthal.*, **80**, 220 (1968).
Hagedoorn. *Amer. J. Ophthal.*, **21**, 487 (1938).
Hirsch and Helwig. *Arch. Derm.*, **84**, 835 (1961).
Huber and Picena. *Rev. méd. Rosario*, **23**, 13 (1933).
Jarisch. *Die Hautkrankheiten*, Wien, 902 (1900).
Johnson and Helwig. *Cancer*, **23**, 641 (1969).
Kersting. *Arch. Derm.*, **87**, 323 (1963).
Krompecher. *Arch. Derm. Syph.* (Berl.), **126**, 765 (1919).
Letulle and Duclos. *Ann. Oculist.* (Paris), **145**, 203 (1911).
Lever. *Histopathology of the Skin*, 4th ed., Phila., 552, 566 (1967).
Lever and Castleman. *Amer. J. Path.*, **28**, 691 (1952).
Meyer. *Klin. Mbl. Augenheilk.*, **153**, 662 (1968).
von Michel. *Festschr. f. J. Rosenthal.*, Leipzig, **2**, 213 (1906).
Nichellati. *Boll. Oculist.*, **10**, 882 (1931).
O'Brien and Braley. *J. Amer. med. Ass.*, **107**, 933 (1936).
O'Hara, Bensch, Ioannides and Klaus. *Cancer*, **19**, 1438 (1966).
Orbán. *Klin. Mbl. Augenheilk.*, **127**, 474 (1955).
Pereyra. *Boll. Oculist.*, **3**, 338 (1924).
Renard, Duperrat, Mascaro and Dhermy. *Bull. Soc. Ophtal. Fr.*, **63**, 823 (1963).
Rumschewitsch. *Klin. Mbl. Augenheilk.*, **28**, 387 (1890).
Salzmann. *Arch. Augenheilk.*, **22**, 302 (1891).
Tulman and Jack. *Amer. J. Ophthal.*, **60**, 1116 (1965).
Winkelmann and Muller. *Arch. Derm.*, **89**, 827 (1964).
Wintersteiner. *Arch. Augenheilk.*, **40**, 291 (1900).
Wollensak and Meythaler. *Klin. Mbl. Augenheilk.*, **150**, 388 (1967).

ADENOCARCINOMATA

Carcinomata of the sweat glands (HIDRADENOCARCINOMATA) of the skin are rare; there are a few references in general dermatological literature (Gates *et al.*, 1943). Certain rodent ulcers, it will be remembered, are claimed by some histologists to arise in part from sweat glands. Cases of carcinomata apparently originating in the apocrine sweat glands of the skin of the lids have been reported by Stout and Cooley (1951) and Pandya (1959). In the first of these cases a tumour of the lower lid near the inner canthus recurred after excision and eventually invaded the orbit and surrounding bones with a fatal termination due to metastases.

A MUCINOUS ADENOCYSTIC CARCINOMA occurring on the skin is rare and on the eyelid rarer still, but two cases have been reported in the lower lid (Mendoza and Helwig, 1971; Rodrigues *et al.*, 1973). The lesion appears as a papillomatous growth; histologically the tumour is multilobulated divided by connective-tissue septa and mucinous pools are prominent surrounded by epithelial cells arranged in cords, columns or irregular nests. It is not clear whether it arises from the apocrine or eccrine glands.

Authentic cases of ADENOCARCINOMATA OF MOLL'S GLANDS are also rare[1] (Fig. 479). A small, hard tumour appears on the lid-margin which may ulcerate and bleed and grows very slowly. Histologically it is made up of irregular glandular tubules maintaining the coiled arrangement of sweat glands; in some places the epithelium has the normal double layer of cells

FIGS. 479 and 480.—ADENACARCINOMA OF A GLAND OF MOLL (A. L. Aurora and M. N. Luxenberg).

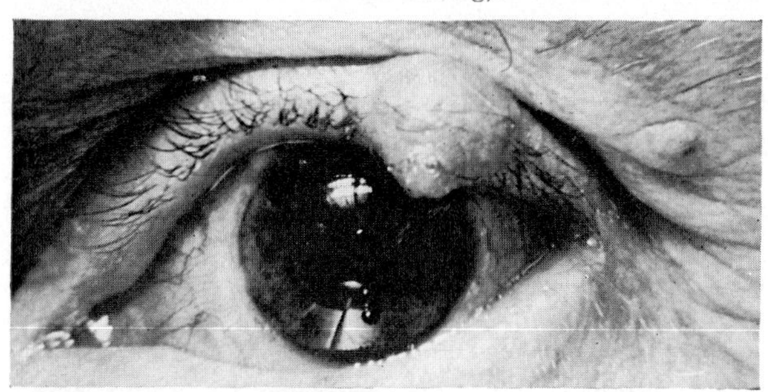

FIG. 479.—The lesion in the upper lid of a man aged 58.

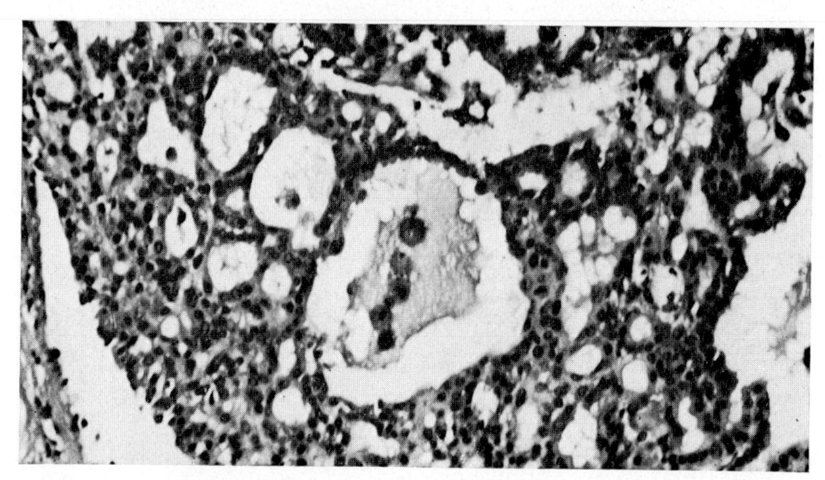

FIG. 480.—Section of the tumour seen in Fig. 479 made up of acini of different sizes with variations in the size and shape of the nuclei (H. & E.).

but in many tubules there is irregular cellular proliferation sometimes entirely filling up the lumen. In the deeper tissues there is widespread and characteristically malignant infiltration where the glandular structure is less faithfully reproduced (Fig. 480). Apart from histological section the tumour cannot be recognized. Treatment is by wide excision.

Tumours such as those described by von Graefe (1864) and Alt (1910) were probably rodent ulcers of the lid-margin. Cases of extra-mammary Paget's disease

[1] Coats (1912), Zeeman (1923), Dusseldorp (1929), Weizenblatt (1957), Aurora and Luxenberg (1970), Futrell *et al.* (1971).

affecting the margin of the lid have sometimes been referred to as arising from a carcinoma of the glands of Moll; these are discussed elsewhere.[1]

Alt. *Amer. J. Ophthal.*, **27,** 263 (1910).
Aurora and Luxenberg. *Amer. J. Ophthal.*, **70,** 984 (1970).
Coats. *Roy. Lond. ophthal. Hosp. Rep.*, **18,** 266 (1912).
Dusseldorp. *Arch. Oftal. B. Aires*, **4,** 695 (1929).
Futrell, Krueger, Chretien and Ketcham. *Cancer*, **28,** 686 (1971).
Gates, Warren and Warvi. *Amer. J. Path.*, **19,** 591 (1943).
von Graefe. *v. Graefes Arch. Ophthal.*, **10,** 176, 206 (1864).
Mendoza and Helwig. *Arch. Derm.*, **103,** 68 (1971).
Pandya. *Ind. J. med. Sci.*, **13,** 632 (1959).
Rodrigues, Lubowitz and Shannon. *Arch. Ophthal.*, **89,** 493 (1973).
Stout and Cooley. *Cancer*, **4,** 521 (1951).
Weizenblatt. *Amer. J. Ophthal.*, **43,** 968 (1957).
Zeeman. *Ned. T. Geneesk.*, **67,** 1194 (1923).

Two tumours which may occur in the caruncle should be noted here.

A PAPILLARY CYSTADENOMA LYMPHOMATOSUM (WARTHIN'S TUMOUR) consisting of an epithelial parenchyma with a lymphoid stroma and irregular cystoid spaces occurs

FIGS. 481 and 482.—PAPILLARY CYSTADENOMA (C. H. Greer).

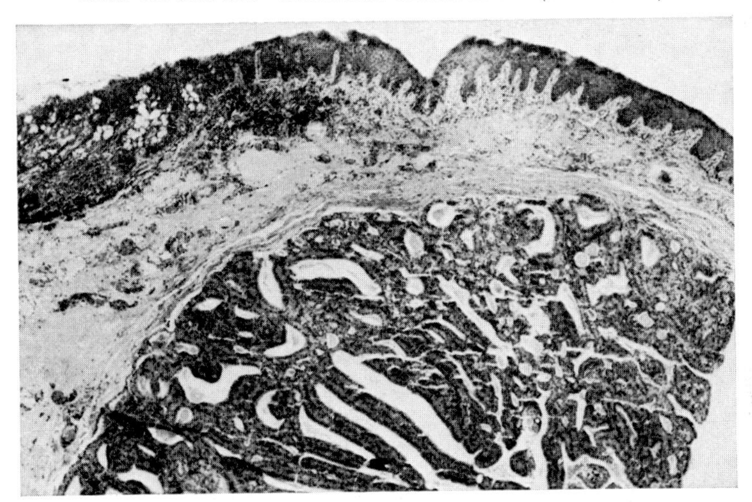

FIG. 481.—Affecting the caruncle in a woman aged 68 (H. & E.; ×13).

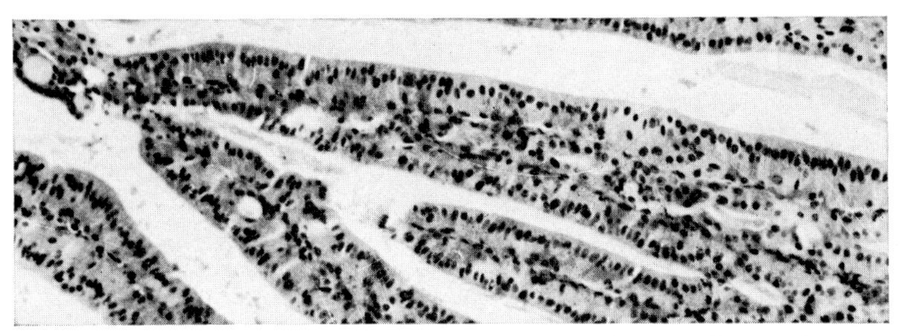

FIG. 482.—The papillary processes with inconspicuous fibrous tissue cores (H. & E.; ×56).

[1] p. 444.

characteristically in the parotid gland. Such a tumour may occur as a rarity in the caruncle where it is said to represent accessory salivary tissue (Oaks and Jenson, 1963). Papillary cystadenomata of the caruncle without the lymphoid elements arising from accessory lacrimal glands have been reported (Coats, 1912; Mackenzie and Patience, 1959; Forbes and Crawford, 1963; Lennox *et al.*, 1968; Greer, 1969) (Figs. 481–2).

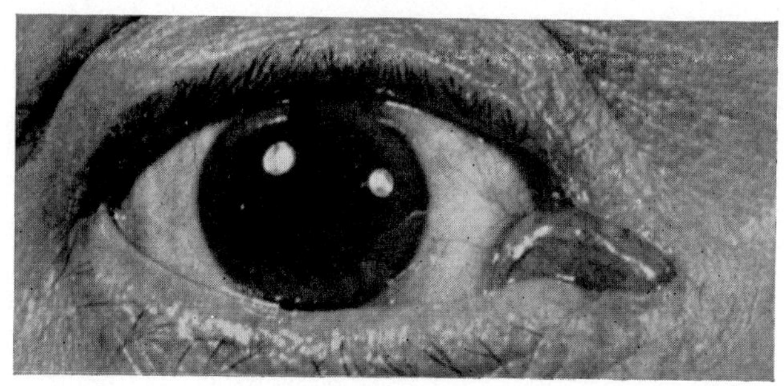

FIG. 483.—ONCOCYTOMA.

An oxyphil-cell adenoma of the right caruncle in a Mexican woman aged 63. After excision of the caruncle with the plica there was no recurrence after 2 years (T. H. Noguchi and E. R. Lonser).

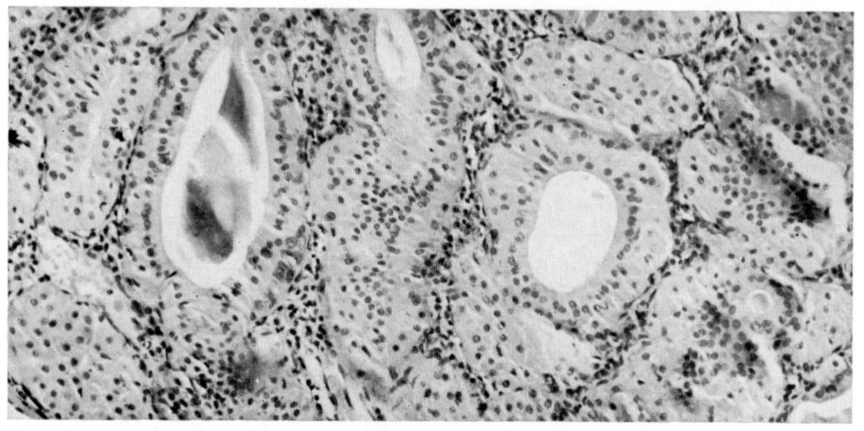

FIG. 484.—OXYPHIL-CELL ADENOMA.

Occurring in the caruncle of a woman aged 80. The appearance is typical. The tumour is composed of distinctive eosinophilic granular cells arranged in cords, tubules and acini. The stroma is infiltrated by lymphocytes (H. & E.; ×100) (C. H. Greer).

An ONCOCYTOMA (OXYPHIL ADENOMA), a benign tumour composed of granular acidophilic cells, may also occur in the caruncle, presumably arising from accessory lacrimal glands (Radnót, 1947; Noguchi and Lonser, 1960; Kloucek, 1963; Klein, 1965; Deutsch and Duckworth, 1967; Radnót and Lapis, 1970) (Fig. 483). Greer (1969) considered that these cells, the significance of which is unknown, could originate from the surface epithelium of the caruncle rather than from accessory lacrimal glands (Fig. 484). The *treatment* of all these growths is by excision.

ADENOMATA OF KRAUSE'S GLANDS are noted elsewhere.[1]

[1] p. 660.

Coats. *Roy. Lond. ophthal. Hosp. Rep.*, **18**, 266 (1912).
Deutsch and Duckworth. *Amer. J. Ophthal.*, **64**, 458 (1967).
Forbes and Crawford. *Brit. J. Ophthal.*, **47**, 177 (1963).
Greer. *Brit. J. Ophthal.*, **53**, 34, 198 (1969).
Klein. *Klin. Mbl. Augenheilk.*, **146**, 343 (1965).
Kloucek. *Cs. Oftal.*, **19**, 117 (1963).
Lennox, Timperley, Murray and Kellett. *J. Path. Bact.*, **96**, 321 (1968).
Mackenzie and Patience. *J. Path. Bact.*, **78**, 288 (1959).
Noguchi and Lonser. *Arch. Path.* (Chic.), **69**, 516 (1960).
Oaks and Jenson. *Amer. J. Ophthal.*, **56**, 459 (1963).
Radnót. *Ophthalmologica*, **113**, 270 (1947).
Radnót and Lapis. *Ophthalmologica*, **161**, 63 (1970).

Mesenchymal Tumours

Tumours of the mesenchymal tissues of the lids may be divided into two groups: those derived from definitive tissue cells and retaining their characteristics—fibromata, lipomata, myxomata, myomata, chondromata and osteomata—with their anaplasic transformation into sarcomata, and those associated with the more primitive reticulo-endothelial system—lymphomata, and their malignant expressions as lymphosarcomata and reticulosarcomata. Plasmocytomata are rare and unimportant occurrences in the lids.

FIBROMA

Fibromata are rare in the lids, forming some 5% of palpebral tumours (O'Brien and Braley, 1936). Two types are met with, the hard and the soft varieties.

HARD FIBROMA (FIBROMA DURUM) may be divided into two types, dermatofibroma composed predominantly of fibroblasts, and histiocytoma containing an admixture of histiocytes; no sharp line of differentiation, however, can be drawn between them, and intermediate lesions exist. A DERMATOFIBROMA is a small, firm, pearly white, sharply defined benign tumour of slow growth, which cannot be diagnosed except by histological section. Its site may be dermal or subdermal or it may be associated with the tarsus or the lid-margin (Henkind and Schultz, 1968) (Figs. 485–6). Histopathologically the neoplasm is dry and hard on section, made up of bundles of dense collagen often together with fibroblasts with spindle-shaped nuclei arranged in whorls and cords (Fig. 487). KELOIDS, developing on scars, are of a similar structure.

A HISTIOCYTOMA contains less collagen but a varying number of histiocytes containing lipids, sometimes assuming the appearance of foam cells and sometimes multinucleated giant cells; such tumours are sometimes called XANTHOFIBROMATA. Massimeo (1959) described two enormous pedunculated tumours of this type (weighing 62 g. and 260 g.) growing from each lower lid in a woman aged 65 (Fig. 488); histologically they had the appearance of a fibroma with considerable peripheral œdema and circumscribed foci of xanthomatous cells. Where blood vessels are prominent as well as cells containing lipid material the term SCLEROSING HÆMANGIOMA is frequently

FIGS. 485 to 487.—DERMATOFIBROMATA OF THE LIDS.

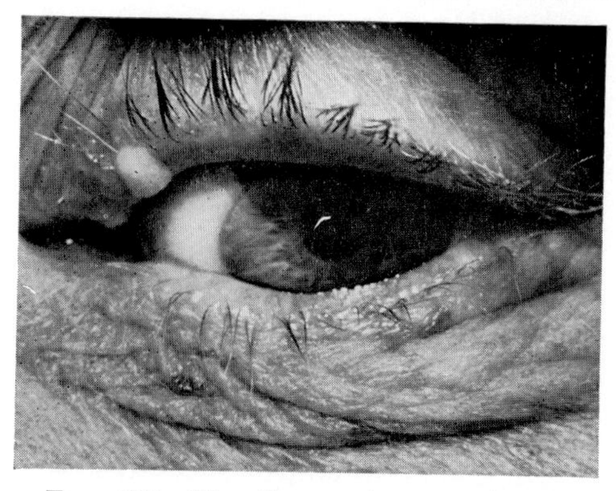

FIG. 485.—The fibroma is a pearly white elevated mass on the inner third of the margin of the upper lid (P. Henkind and G. Schultz).

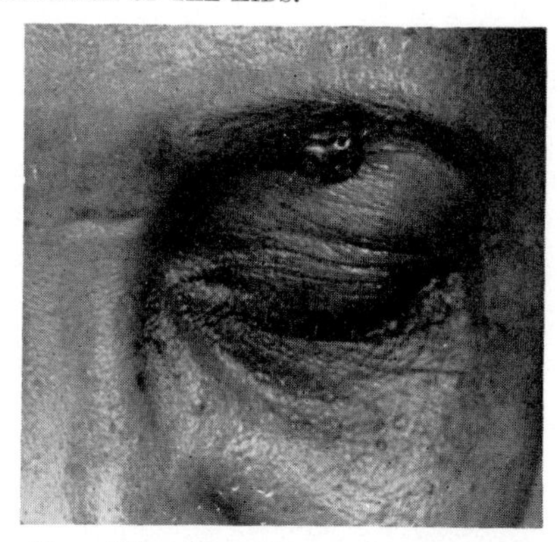

FIG. 486.—The fibroma is on the left upper lid in the region of the eyebrow (Dept. of Dermatology, Vanderbilt Clinic, N.Y.).

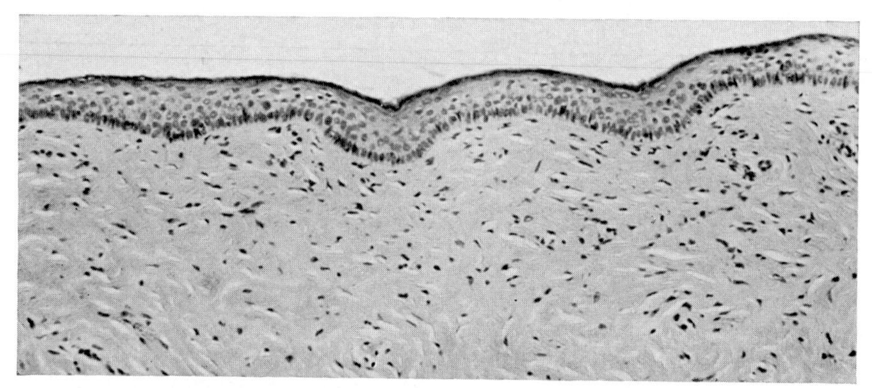

FIG. 487.—Histological section from the case seen in Fig. 485. There is hyperplasia of the collagen tissue with fibroblasts interspersed between collagen bundles, an absence of adnexal structures and a normal epithelium except for flattened rete pegs (P. Henkind and G. Schultz).

applied (Gross and Wolbach, 1943; Reeh, 1963). Excision of these tumours is the treatment of choice.

SOFT FIBROMA (MOLLUSCUM PENDULUM) are polypoid, succulent, reddish tumours showing marked vascular congestion and, being frequently exposed to trauma, are liable to repeated hæmorrhages. Histopathologically the fibrous tissue is usually sparse, hyaline infiltration is the rule, the blood supply is abundant and there is frequently an infiltration of plasma and mast cells (Kalt, 1909; Bakry, 1929; Wilson, 1937). Such tumours may be subcutaneous attached to the deeper tissues, or cutaneous; most of the latter are in reality neurofibromata (*molluscum fibrosum*).[1]

The treatment of fibromata is total excision.

[1] p. 511.

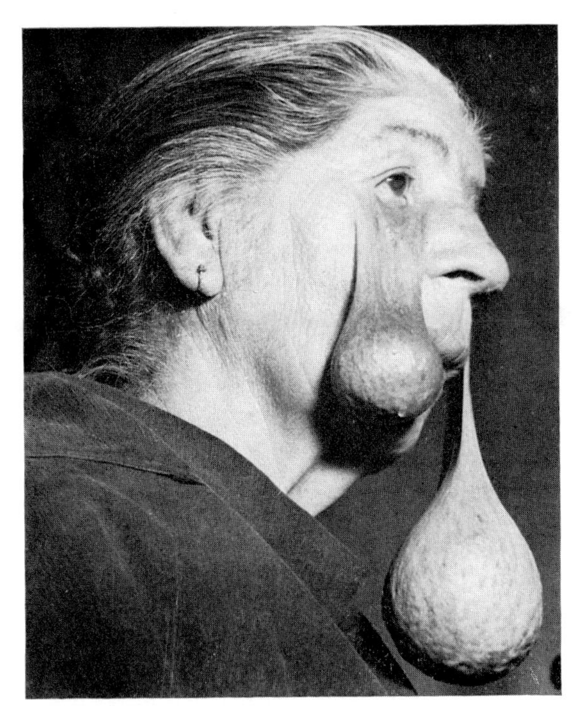

FIG. 488.—PENDULOUS FIBROMATA OF THE EYELIDS.

Fibromata of enormous size growing from the lower lids in a woman aged 65. The histology showed the general fibromatous structure with much œdema and circumscribed areas of xanthomatous cells (A. Massimeo).

TUBEROUS SCLEROSIS

TUBEROUS SCLEROSIS (EPILOIA: BOURNEVILLE'S DISEASE, 1880) is one of the systemic syndromes classified as phakomatoses by van der Hoeve (1921–23)[1] transmitted as an irregular autosomal dominant character (Fig. 489); it is considered to be a disturbance of organogenesis involving a primary defect of connective tissue. It is characterized by a triad of mental deficiency, epilepsy and multiple angiofibromatous lesions on the skin; these may also occasionally be associated with mulberry gliomata in the brain, the retina and the optic nerve[2] and tumours elsewhere. The cutaneous lesions take the form of many small pink or brownish pin-head growths which appear typically in early childhood or at puberty, usually distributed in considerable numbers symmetrically in butterfly-fashion over the nose, cheeks and lower lids and persist indefinitely[3] (Fig. 490); occasionally isolated tumours appear on the upper lid (Purtscher and Wendlberger, 1938). The dermal manifestations were at one time mistakenly considered to be adenomata (the *sebaceous adenoma of Pringle*, 1890), but it has been conclusively shown by Nickel and Reed (1962–69) that the essential component is fibrous tissue with capillary dilatations, the sebaceous glands being atrophic; in the older lesions there is a considerable proliferation of collagen, a change seen to a marked degree in the larger asymmetrical lesions with a diminution or absence of the

[1] Vol. X, p, 736. [2] Vol. X, p. 754.

[3] Bock (1880), Balzer and Grandhomme (1886), Pringle (1890), Vogt (1908), Sutton (1911), Shelmire (1918), van der Hoeve (1921–23), von Herrenschwand (1929), James (1937), Koch and Walsh (1939), Constantine (1943), Yamamoto *et al.* (1972), and others.

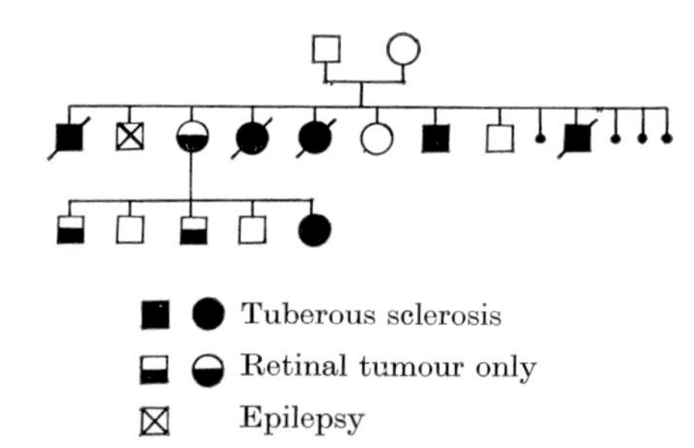

FIG. 489.—TUBEROUS SCLEROSIS.

The father in generation I had a leiomyosarcoma of the stomach (after F. S. van Bouwdijk-Bastiaanse and K. Landsteiner, 1922).

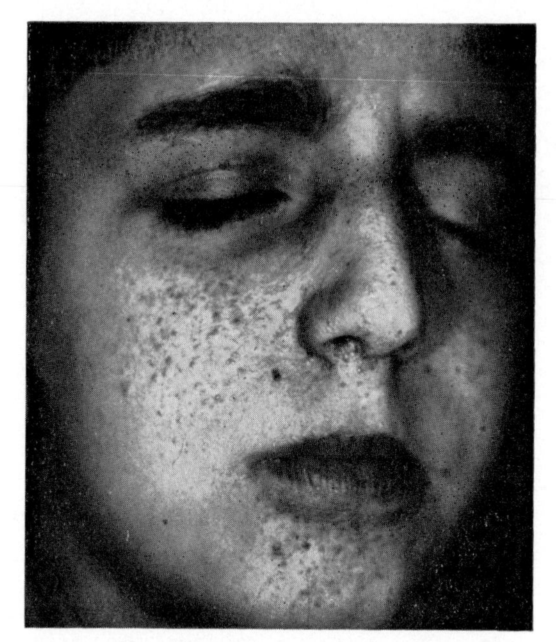

FIG. 490.—TUBEROUS SCLEROSIS.

A 10-year-old boy with the typical " adenoma sebaceum " of the face with a butterfly distribution across the nose and cheeks (E. F. Constantine).

dilated capillaries. Treatment of the nodules on the skin is unnecessary except for cosmetic reasons; the most effective methods are dermabrasion and electrolysis. The prognosis of the disease, however, is poor; if it is fully developed in infancy, some 75% of patients die before the age of 25.

LIPOMA

LIPOMATA of the lids are very rare, and since subcutaneous adipose tissue is absent in this region these neoplasms are probably secondary extensions from the orbit, usually of a congenital nature, which may send processes between the muscles and the tarsus. A true lipoma has a thin fibrous capsule. They may be symmetrical or even occur in all four lids (Wingenroth, 1900). The tumours are round or lobulated, of soft and pseudo-fluctuant consistency, and usually appear on the upper lid immediately under the skin or in front of the tarsus. Histologically they are made up of lobules of fatty cells separated by vascularized connective-tissue strands (Knapp, 1877; Vossius, 1895–1910; Bach, 1906; Wirths, 1920). Myxomatous degeneration may be marked, and other neoplasic processes may be associated with the lipoma (angiolipoma, Velhagen, 1899; Laspeyres, 1906). Such cases show a strong affinity with blepharochalasis, ptosis adiposa or fat hernia. Treatment is by total excision.

MYOMA

RHABDOMYOMA of the orbicularis muscle is extremely rare. A small, hard nodular, subcutaneous tumour, it is composed of fusiform cells and new-formed striated muscle fibres in a dense connective-tissue stroma rich in fibroblasts and round cells. Amyloid degeneration and necrosis may occur, and the possibility of sarcomatous degeneration necessitates complete removal (Alt, 1896; Schnaudigel, 1910–13; Mayer, 1929; López Enriquez, 1933; Cárdenas Pupo, 1935; Cristini, 1946).[1]

A LEIOMYOMA of the lid has been described by Wassenaar (1931).

MYXOMA. An unencapsulated tumour with a cystic appearance occurring in a woman of 39 near the inner angle of the lower lid was found to have the histological structure of a myxoma and seemed to be derived from the intermuscular connective tissue of the orbicularis (Town and Reese, 1945).

CHONDROMA

CHONDROMA is an extremely rare tumour growing as a small, hard mass in association with the tarsus (Fuchs, 1878; Drake-Brockman, 1891; Keyser, 1895); an osteochondroma in the lower lid of a child was studied by Wildi (1925) who suggested that it was a congenital inclusion.

SARCOMA

The frequency of SARCOMATA of the lids is difficult to assess because of the lack of histological differentiation in the early literature, but a primary tumour of this nature is rare.

Histopathologically sarcomata of the lids resemble in structure similar tumours elsewhere, the neoplasm being composed of embryonic cells bordering on blood-spaces frequently without definite vascular walls and separated by little reticular tissue. Several varieties have been described according to the predominant type of cell; the neoplasm which used to be termed a " round-cell sarcoma " is now considered to be lymphomatous in nature. The sarcomata may be classified according to the type of tissue on which they are based.

FIBRO- (SPINDLE-CELL) SARCOMATA of the lids have seldom been described (Shannon, 1923; Clapp, 1925; Roy, 1927; Vele, 1933; Morgan, 1936) (Fig. 491). They are firmer, more circumscribed and less infiltrating than other varieties and show all gradations in their cellular content until the differentiation from a fibroma becomes difficult, in which case the prognosis after removal is relatively good (Fig. 1047). MYXOSARCOMATA have also been reported (van Duyse and Cruyl, 1887; Flack, 1892; Moehle, 1949) as well as a FIBROMYXOSARCOMA (Shukla and Agrawal, 1967) (Figs. 492–3).

[1]See further, p. 1049.

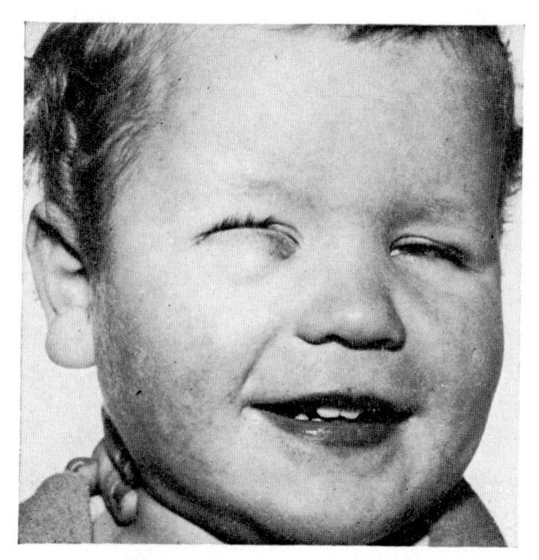

FIG. 491.—FIBROSARCOMA.
In a boy aged 2½ years (Inst. Ophthal.).

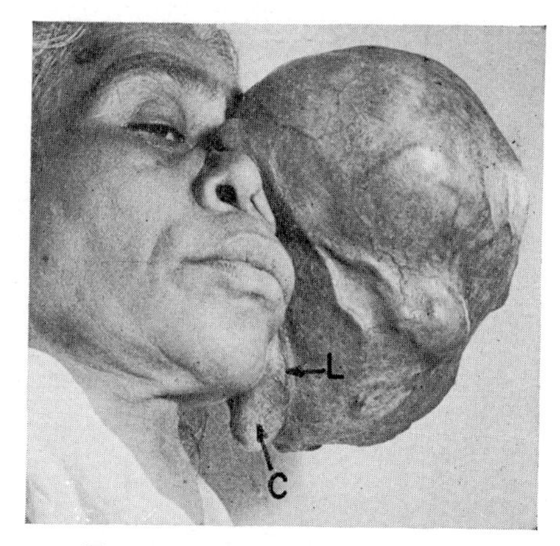

FIG. 492.—FIBROMYXOSARCOMA.
In a Hindu woman aged 54. The tumour had been growing for 30 years. L, lid-margin; C, palpebral conjunctiva (I. M. Shukla and S. Agrawal).

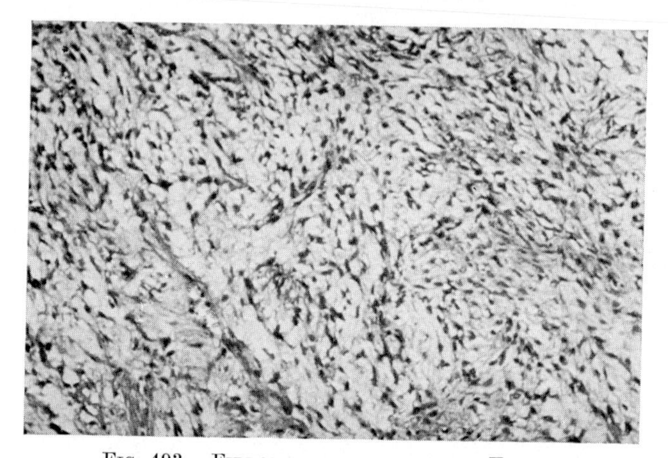

FIG. 493.—FIBROMYXOSARCOMA OF THE EYELID.
The tumour in Fig. 492, showing spindle cells separated by a mucoid matrix (I. M. Shukla and S. Agrawal).

Terry (1934) reported a case which was rapidly fatal despite exenteration of the orbit and radium treatment; it seemed to have been a LEIOMYOSARCOMA derived from Müller's muscle.

A number of RHABDOMYOSARCOMATA presents clinically in the lids; indeed, analysing the records of 191 cases reported in the literature, Schuster and his colleagues (1972) found that 21% of these tumours occurring in the region of the eye appear initially at this site, either the upper or the lower lid. The clinical features, pathology and prognosis of these tumours will be dealt with in the section dealing with neoplasms of the orbit.[1] It would seem that the site of presentation does not greatly influence the prognosis (Jones *et al.*, 1965).

The growth of a sarcoma is rapid until it may extend from the root of the nose to the zygoma and may eventually attain the size of the patient's head (van Duyse, 1889; Shukla and Agrawal, 1967), all the neighbouring tissues being invaded—the lid, the conjunctiva and the orbit. Recurrences are common after removal and, although adenopathy is exceptional, general metastases and death from cachexia is the usual termination. The prognosis, however, in the case of a myxosarcoma (Moehle, 1949) or a fibrosarcoma may be relatively good after extirpation (Shukla and Agrawal, 1967).

Diagnosis is frequently difficult in the early stages, particularly in the fixed deep varieties which resemble a chalazion or tarsitis. In many cases a biopsy is the only recourse, but preparations should be made in suspicious cases to follow it up by adequate treatment without delay.

Treatment must be radical. The safest method is excision either by the knife or preferably by diathermy followed by irradiation. As a rule the lid, the eye and the orbital contents should be sacrificed.

Alt. *Amer. J. Ophthal.*, **13,** 109 (1896).
Bach. *Arch. Augenheilk.*, **54,** 73 (1906).
Bakry. *Bull. ophthal. Soc. Egypt*, **22,** 78 (1929).
Balzer and Grandhomme. *Arch. Physiol.* (Paris), **8,** 93 (1886).
Bock. *Virchows Arch. path. Anat.*, **81,** 503 (1880).
Bourneville. *Arch. Neurol.* (Paris), **1,** 81 (1880).
Cárdenas Pupo. *Bol. Liga Cáncer* (Habana), **10,** 255 (1935).
Clapp. *Trans. Amer. ophthal. Soc.*, **23,** 228 (1925).
Constantine. *Arch. Ophthal.*, **30,** 494 (1943).
Cristini. *Rass. ital. Ottal.*, **15,** 207 (1946).
Drake-Brockman. *Trans. S. Ind. Branch, Brit. med. Ass.* (1889–90), 226 (1891).
van Duyse. *Ann. Oculist.* (Paris), **101,** 227 (1889).
van Duyse and Cruyl. *Ann. Oculist.* (Paris), **98,** 112 (1887).
Flack. *Ueber Sarkome d. Augenlider*, Königsberg (1892).
Fuchs. *v. Graefes Arch. Ophthal.*, **24** (2), 121 (1878).
Gross and Wolbach. *Amer. J. Path.*, **19,** 533 (1943).
Henkind and Schultz. *Amer. J. Ophthal.*, **65,** 420 (1968).
von Herrenschwand. *Klin. Mbl. Augenheilk.*, **83,** 732 (1929).
van der Hoeve. *v. Graefes Arch. Ophthal.*, **105,** 880 (1921); **111,** 1 (1923).
James. *Lancet*, **1,** 1223 (1937).
Jones, Reese and Krout. *Trans. Amer. ophthal. Soc.*, **63,** 224 (1965).
Kalt. *Ann. Oculist.* (Paris), **141,** 189 (1909).
Keyser. *J. Amer. med. Ass.*, **30,** 798 (1895).
Knapp. *Arch. Augenheilk.*, **6,** 38 (1877).
Koch and Walsh. *Arch. Ophthal.*, **21,** 465 (1939).
Laspeyres. *Z. Augenheilk.*, **15,** 527 (1906).
López Enriquez. *Arch. Oftal. hisp.-amer.*, **33,** 419 (1933).
Massimeo. *Boll. Oculist.*, **38,** 693 (1959).
Mayer. *Amer. J. Ophthal.*, **12,** 293 (1929).
Moehle. *Arch. Ophthal.*, **41,** 317 (1949).
Morgan. *Arch. Ophthal.*, **16,** 472 (1936).
Nickel and Reed. *Arch. Derm.*, **85,** 209 (1962).
Cutis, **5,** 73 (1969).
O'Brien and Braley. *J. Amer. med. Ass.*, **107,** 933 (1936).
Pringle. *Brit. J. Derm.*, **2,** 1 (1890).
Purtscher and Wendlberger. *v. Graefes Arch. Ophthal.*, **138,** 388 (1938).
Reeh. *Treatment of Lid and Epibulbar Tumors*, Springfield, 221 (1963).
Roy. *Sth. med. J.*, **20,** 436 (1927).
Schnaudigel. *v. Graefes Arch. Ophthal.*, **74,** 372 (1910); **85,** 252 (1913).
Schuster, Ferguson and Marshall. *Arch. Ophthal.*, **87,** 646 (1972).
Shannon. *Trans. Amer. ophthal. Soc.*, **21,** 119 (1923).
Shelmire. *J. Amer. med. Ass.*, **71,** 963 (1918).
Shukla and Agrawal. *Brit. J. Ophthal.*, **51,** 403 (1967).
Sutton. *J. cutan. Dis.*, **29,** 480 (1911).
Terry. *Arch. Ophthal.*, **12,** 689 (1934).
Town and Reese. *Amer. J. Ophthal.*, **28,** 68 (1945).
Vele. *Boll. Oculist.*, **12,** 788 (1933).

[1] p. 1051.

Velhagen. *Klin. Mbl. Augenheilk.*, **37**, 253 (1899).
Vogt. *Zbl. Nervenheilk. Psychiat.*, **19**, 653 (1908).
Vossius. *Ber. dtsch. ophthal. Ges.*, **24**, 55 (1895).
v. Graefes Arch. Ophthal., **74**, 526 (1910).
Wassenaar. *J. med. Ass. S. Africa*, **5**, 818 (1931).
Wildi. *Klin. Mbl. Augenheilk.*, **74**, 473 (1925).
Wilson. *Rep. mem. ophthal. Lab. Giza*, **12**, 40 (1937).
Wingenroth. *v. Graefes Arch. Ophthal.*, **51**, 380 (1900).
Wirths. *Z. Augenheilk.*, **44**, 176 (1920).
Yamamoto, Tabuchi, Hanada *et al.* *Folia ophthal. jap.*, **23**, 130 (1972).

Tumours of Lympho-reticular Tissue

The classification of tumours of lymphoid tissue has long been a matter of difficulty and dispute because of the remarkable variability of their clinical and histological manifestations as well as the frequent difficulty in recognizing precisely their cytological character but they are all derived by divergent differentiation from the same source.[1] Two different terms have been widely used: the *reticuloses*, which include benign and malignant tumours, and *malignant lymphomata* describing diseases of the lymphoid system characterized clinically by progressive tumour-like enlargements of these tissues with an eventual fatal outcome, and histologically by a multiplication of one or more of the elements normally present in the lymph nodes with the loss of their normal structure. The two principal types of cell which have a common mesenchymal origin from the primitive undifferentiated pluripotent stem-cells from which all the hæmopoietic elements are derived are the reticulum cells, and the more mature lymphocytes. Proliferation of the former gives rise to reticulum-celled sarcomata, of the latter to benign and malignant lymphomata, while in Hodgkin's disease both types are involved. In each case the cells may be well or poorly differentiated and there may be a tendency for a transformation of one type of cell to another. Starting usually with enlargement of a single node the disease may spread to involve other lymph nodes until the whole lympho-reticular system is implicated including the spleen and the liver; a proportion shows involvement of the bone marrow with the resulting appearance of an abnormal number of cells in the circulating blood giving rise to a leucæmia. Young persons are frequently affected and the degree of malignancy varies from the relatively benign lymphoma to the highly malignant reticulum-cell sarcoma (Whitby and Britton, 1969; Boyd, 1970; de Gruchy, 1970).

In all these changes the tissues of the eyelids may share, although less frequently than those of the orbit[2]; they present a series of clinical pictures which vary from a localized hyperplasia with no systemic but only local significance to a highly malignant and rapidly fatal riot of growth of the reticular system throughout the body.

From the clinical point of view three types of lymphomatous mass may

[1] Gall and Mallory (1942), Robb-Smith (1947), Israels (1953), Jelliffe and Thomson (1955), Hilton and Sutton (1962), Lukes and Butler (1966), Ultmann (1966), Willis (1967), Thiel (1970), Zimmerman (1970), Lukes (1971), Kaplan (1972), and others.

[2] p. 1061.

occur (Zimmerman, 1970): (1) non-neoplasic inflammatory pseudo-tumours originating as a reactive hyperplasia with a polymorphic cytological picture, (2) a group usually of pure lymphocytic proliferations, and (3) malignant lymphomata with definite anaplasic cytological features.

A REACTIVE LYMPHOCYTIC HYPERPLASIA may form a tumefaction of the lids alone or in common with a similar involvement of the conjunctiva, the lacrimal gland and orbit; frequently the condition is bilateral and symmetrical.[1] Histologically the mass is composed of fully-formed lymphocytes, with or without follicular aggregations, although there is frequently considerable polymorphism with an admixture of reticular and plasma cells, eosinophils and neutrophils. The condition is a benign "pseudo-lymphoma" and is presumably a response to chronic inflammation and not a true neoplasm; treatment by corticosteroids is indicated.

LYMPHOMA

LYMPHOMATA may be considered to embrace the large group of lymphoid tumours which lack the polymorphic appearance of a reactive inflammatory hyperplasia and the anaplasic types of cell and infiltrative tendencies characteristic of the frankly malignant neoplasms. The cytological picture may present difficulties but usually lymphocytic cells predominate and the tumour is typically radio-sensitive.

The occurrence of localized autonomous lymphomata in the lids without simultaneous evidence of constitutional blood disease is relatively rare (Scuderi, 1971) (Figs. 494–6). Somewhat more frequently but by no means commonly, a lymphomatous hyperplasia of the same clinical and histological appearance occurs in the course of generalized hæmopoietic disease, particularly acute or chronic lymphocytic or monocytic leucæmia (Figs. 497–9). In some cases the swelling of the lids is associated with an infiltration of the orbit,[2] in others with an infiltration of the conjunctiva[3] and in others with a diffuse involvement of the skin (*leucæmia cutis*).[4] Of 21 cases of lymphomatous disease in the region of the eye, McGavic (1943) found that in 17 the peri-ocular tumour was primary, while in 4 it appeared after the generalized disease had become apparent: of these 21 tumours, 10 involved the lids. Of 76 similar cases of ocular lymphomatous disease Heath (1949) found that 20% involved the lids and 20% the orbit.

Lymphomata as a rule form ill-defined nodular masses in the lids lying underneath the skin which is freely movable over them. Such tumours may have a sudden onset and run an acute course (Neugebauer, 1907), but more usually grow slowly. Eventually they may reach a considerable size, inducing ptosis or sometimes completely closing the eyes with their bulk

[1] Hogan and Zimmerman (1962), Reese (1963), Mortada (1963–66), Zimmerman (1970), Thiel (1970), and others.

[2] Leber (1878), Delens (1886), Axenfeld (1891), Glinski (1903), Reese and Guy (1933), Leinfelder and O'Brien (1935), Dacie *et al.* (1950), Massa *et al.* (1966), Haye and Haut (1966), Stärk and Thiel (1970).

[3] Vol. VIII, p. 1190.

[4] Fröhlich (1893), Hazen (1911), Epstein and MacEachern (1937), Gates (1938).

FIGS. 494 to 496.—LYMPHOMA OF THE LID-MARGIN
(G. Scuderi and E. Balestrazzi).

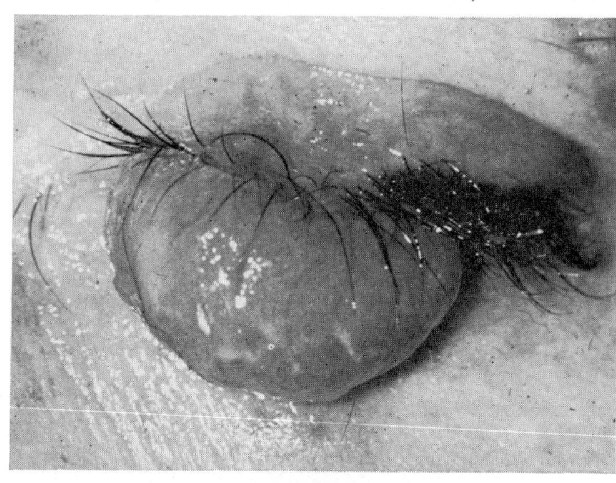

FIG. 494.—In a woman aged 35 with a history of 6 months. The large tumour was attached to the margin of the upper lid. There was no generalized disease. Excision followed by radiotherapy induced a cure (follow-up period of 11 months).

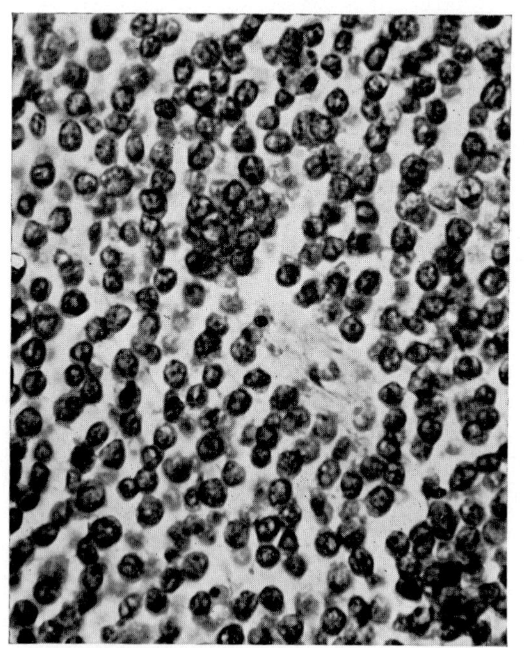

FIG. 495.—Section of the tumour showing the lymphocytes of equal size and shape (H. & E.; ×300).

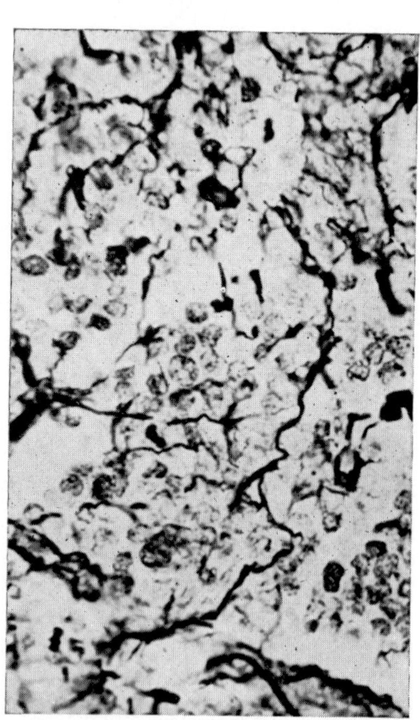

FIG. 496.—Reticulum delimiting alveolar spaces in an irregular mesh (Gomori; ×300).

(Fig. 499). Most frequently they are bilateral and symmetrical and may involve all four lids, but one eye or one lid of one eye may alone be affected (Leinfelder and O'Brien, 1935; Okutani and Tamura, 1971): the involvement of one lid has been associated with an episcleral nodule on the other eye (Bégué, 1932). The lacrimal glands and Krause's glands may become

FIGS. 497 to 499.—INVOLVEMENT OF THE LIDS IN LEUCÆMIA.

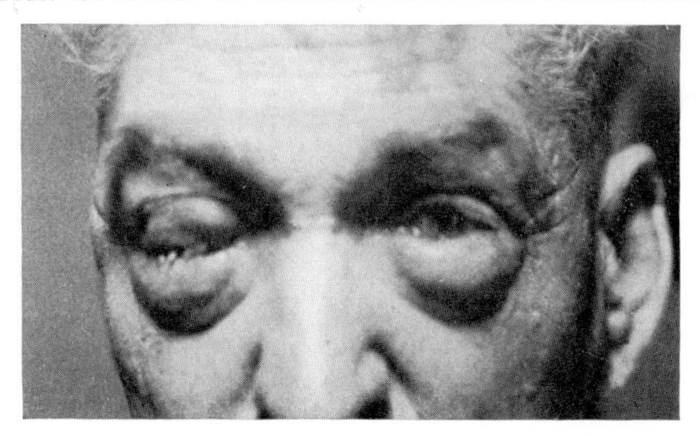

FIG. 497.—In lymphatic leucæmia, showing massive nodular thickening of all the lids (P. J. Leinfelder and C. S. O'Brien).

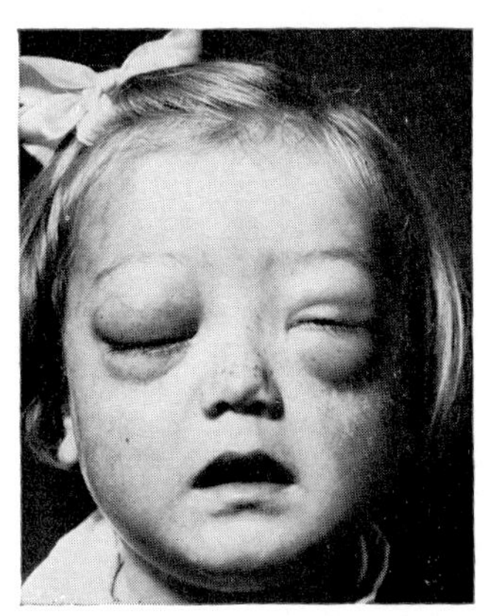

FIG. 498.—In acute myeloblastic leucæmia, in a child aged 3½ years with proptosis. The lids became normal after treatment with aminopterin and blood transfusion, but two months later the child died (J. V. Dacie *et al.*).

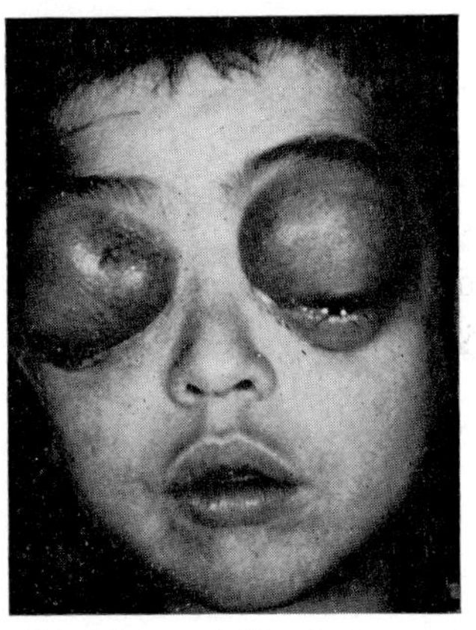

FIG. 499.—In myeloblastic leucæmia, in a boy aged 6. Massive involvement of the lids with gross limitation of ocular movements (K. Wybar, Royal Marsden Hospital).

infiltrated, the conjunctiva may show a follicular hypertrophy or a nodular infiltration and orbital invasion may produce proptosis.

Histologically these tumours are made up of a dense mass of lymphocytes (Figs. 495–6) and the diagnosis must usually rest on the appearance at biopsy unless the blood gives a characteristic picture. As will be explained more fully when we are dealing with lymphocytic tumours of the orbit,[1]

[1] p. 1062.

however, the histological differentiation between a benign and a malignant lymphocytic tumour is often impossible; a certain diagnosis can be made only after a long period of follow-up to decide whether the lesion disseminates or not. Benign growths are amenable to treatment by x-rays or radium although recurrences must frequently be expected[1]: *they are radio-sensitive, not necessarily radio-curable.* Excision does not give good results, being usually followed by recurrences, but subsequent radiational treatment may ensure a satisfactory outcome (Scuderi, 1971). Remedies for the constitutional disease are, of course, indicated, for immediately successful local treatment does not mean that the patient is cured.

MALIGNANT LYMPHOMA

LYMPHOSARCOMA

This intensely malignant tumour of lymphocytes or lymphoblasts which used to be termed *round-celled sarcoma* occurs relatively rarely in the lids, for less than 1% of all lymphosarcomatous growths originate in the region of the eye; in 196 cases, Sugarbaker and Craver (1940) found only one primary lesion in the eyelid, as also did Gall and Mallory (1942) in a series of 618. When it does occur in this situation it would seem to be somewhat less malignant than in most other localities for generalization of the disease may be delayed for months or years, and it prefers to extend locally in an invasive and destructive manner. Multiple metastases or pluricentric dissemination of the disease may, however, occur at any time, and all such tumours should be regarded potentially as forming a stage in a malady which some time or other may involve all the lymphoid tissues of the body. The tumours of the lids may be associated with universal lymphomatous changes throughout all the lymphatic nodes without leucæmic changes in the blood or bone marrow (Godtfredsen, 1947); or the blood itself may similarly be involved. It is to be remembered that the appearance of a leucæmic blood-picture is incidental, depending on whether the malignant cells enter the blood stream in demonstrable numbers.

Sometimes the neoplasm forms large, hard nodular masses (Fig. 500). Scalinci (1902) reported a case affecting one lower lid, Taubmann (1899) cases involving both lids of one eye, Croci (1934) and Erbakan and Aksu (1962) symmetrical tumours on both upper lids, and Leinfelder and O'Brien (1935) recorded three cases wherein both lids of both eyes were involved, and a fourth in which a tumour of the lower lid of one eye was followed by proptosis. Numerous similar cases have been recorded.[2] At other times the neoplasm may develop so rapidly that the clinical picture resembles an acute facial cellulitis for which it has initially been treated (Stansbury, 1948; Carpendale, 1949); this type is usually rapidly fatal (Fig. 501). Diagnosis is made by

[1] Tiscornia (1923), Speciale-Picciché (1927), Shannon and McAndrews (1932), Tooke (1939), Calle *et al.* (1971), and many others.

[2] Métivier (1937), Robolotti (1957), Jain (1964), Bascarán Collante and López Nieto (1966), Fusco *et al.* (1968), Nolan (1968), Lissia and Borellini (1969), Offret and Haye (1971).

FIGS. 500 and 501.—MALIGNANT LYMPHOMATA.

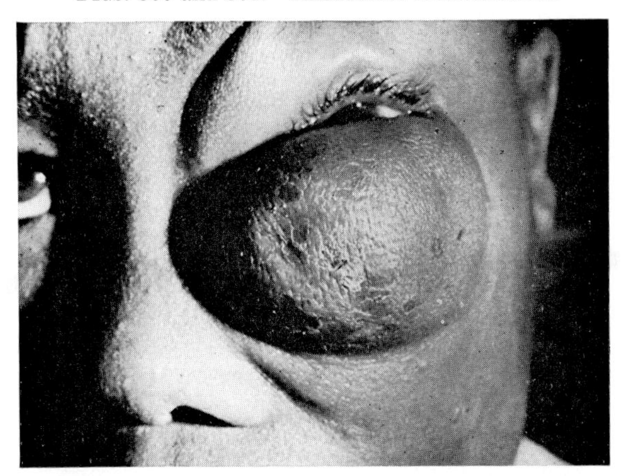

FIG. 500.—Affecting the lower lid in a Kenyan (G. G. Bisley).

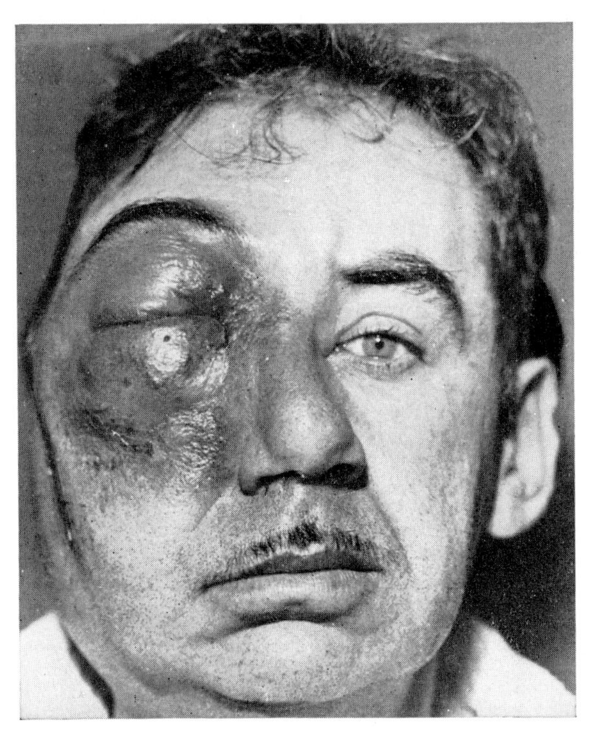

FIG. 501.—Lymphosarcoma occurring in a man aged 49. The clinical appearance suggested the diagnosis of orbital cellulitis and abscess of the right lower lid. Incision led to no result. Subsequently the swelling of the lids increased as did the mass in the neck while enlarged lymph nodes became palpable in the right axilla and elsewhere. A course of nitrogen mustard was given but after initial improvement the patient died (see Fig. 502) (F. C. Stansbury).

histological section which shows that the mesenchymal cells have differentiated essentially into lymphocytes or lymphoblasts (Fig. 502), but when the mass is composed of mature lymphocytes it may be difficult to distinguish from a benign reactive lymphocytic hyperplasia (Mortada, 1963).

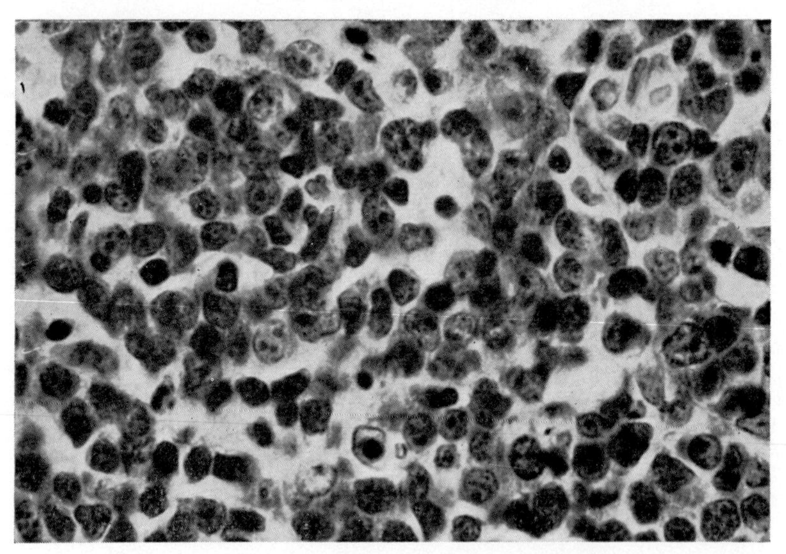

FIG. 502.—LYMPHOSARCOMA.

Histological section of the tumour seen in Fig. 501, showing large pale cells without grouping or structure, with ovoid or rounded nuclei and many mitoses. No reticular network is visible (F. C. Stansbury).

RETICULUM-CELL SARCOMA

In a reticulosarcoma any variant of the reticular cell may be involved but there is usually a general uniformity in the type of cell present in a particular tumour, while a fine reticulum is present throughout. It may arise primarily in the dermis from a single site or several points simultaneously; the hyperplasia is malignant in character, invading and destroying the surrounding tissues and spreading by lymphatics and the blood stream.[1] These growths occur at any age, and typically form infiltrating, nodular tumours with a tendency to ulceration, sometimes spreading superficially from one lid to the other or over the root of the nose to the other side (Figs. 503–6). They occur rarely on the caruncle (Molnár and Keresztury, 1963). Lymphatic involvement is not prominent, but metastases and recurrences are common. Occasionally they may be rapidly fatal, as the case of a child of 18 months reported by Nebeská (1948) who had a small tumour excised from an upper lid: it recurred within a week and

[1] Reese (1963), Reeh (1963), Gokhale and Desai (1966), Alezzandrini *et al.* (1967), Kamel (1967), and many others.

FIGS. 503 to 506.—RETICULUM-CELL SARCOMATA.

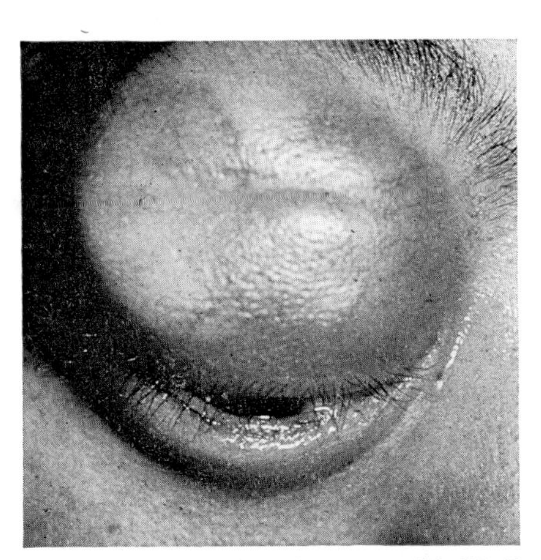

FIG. 503.—Affecting the upper lid (M. J. Reeh, *Treatment of Lid and Epibulbar Tumors*, Thomas, Springfield).

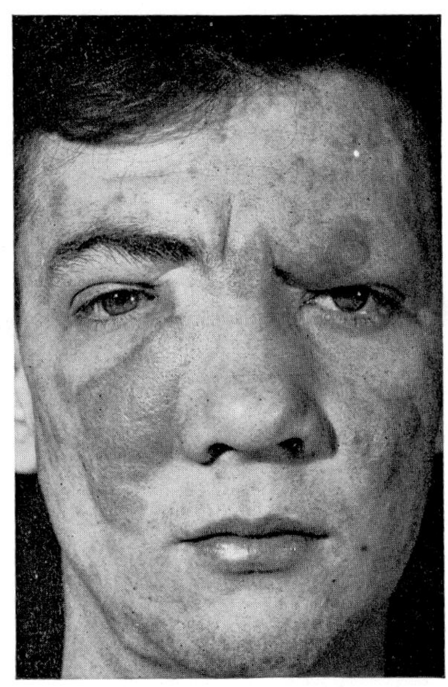

FIG. 504.—Spreading in a diffuse manner (C. S. Wright and R. Friedman).

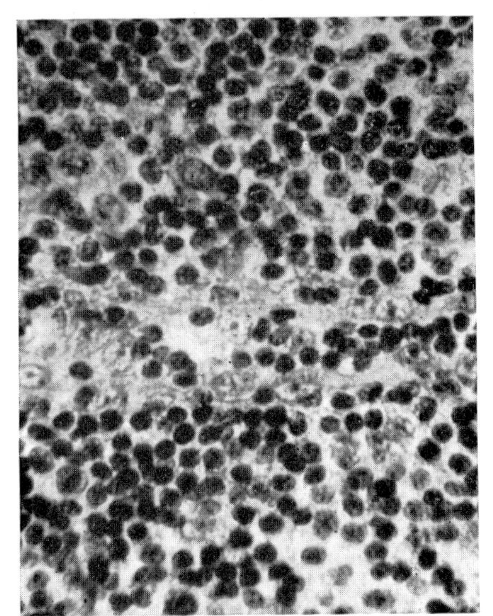

FIG. 505.—A tumour originating in the lacrimal gland. It consisted of a diffuse mass of mature lymphocytes and larger immature lymphoblasts (H. & E.; ×570) (N. Ashton).

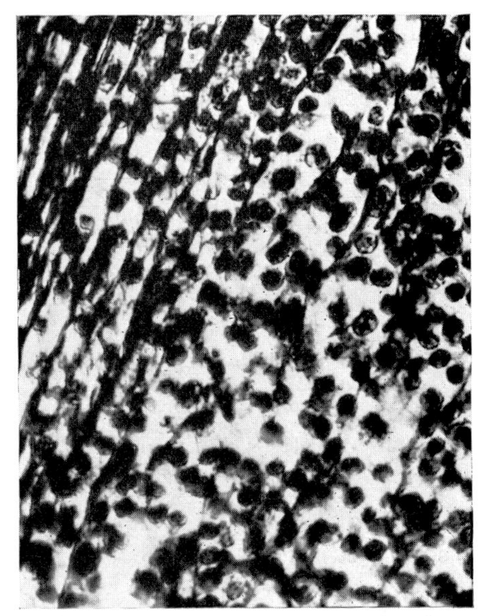

FIG. 506.—Another section of the tissue shown in Fig. 505 stained for reticulum fibres. It shows many reticulum fibres extending between the tumour cells (Wilder's stain; ×690) (N. Ashton).

despite repeated excision combined with irradiation, death from widespread metastases occurred within 2 months.

GIANT FOLLICULAR LYMPHOMA (FOLLICULAR LYMPHOBLASTOMA; MACROFOLLICULAR LYMPHOMA: *Brill-Symmers's disease*, Brill *et al.*, 1925; Symmers, 1927–38) is related to lymphosarcoma but of less malignancy; it is the least common and least malignant of this group of diseases, and is very radio-sensitive. It has a multicentric origin in the lymph follicles throughout the body, particularly of the lymph nodes of the neck and the malpighian bodies of the spleen. The tonsils and lymphoid tissue of the gastro-intestinal tract are not affected. Pathologically the lymph nodes show large follicles of highly differentiated neoplasic lymphocytes or reticulum cells, being thus distinguished from a lymphocytic hyperplasia (see Fig. 709). It is slowly progressive with gradual extension to other lymph nodes, the average duration of life being about 4 or 5 years after the appearance of the malady (50% showed a 5-year survival, Gall and Mallory, 1942). Rarely neoplasic masses may appear subcutaneously in the lids associated with more or less widespread lymphatic involvement in the neck, mediastinum and elsewhere (Goedbloed and Wyers, 1941; Chambers, 1950; Törnquist, 1950; Gemolotto, 1955). The lacrimal glands may also be affected with resulting proptosis.

BURKITT'S TUMOUR (*African lymphoma*) is a disease characterized by a proliferation of lympho-reticular cells in certain lymph nodes, particularly those in the head and neck; the lesions frequently arise in the jaws and occur predominantly in children with a curious geographical distribution mainly in the equatorial regions of Africa, although similar cases have also been reported from America. Since the orbit is frequently heavily involved, a description of the disease and its treatment by chemotherapy will be found in that section.[1]

HODGKIN'S DISEASE

THOMAS HODGKIN [1798–1866] (Fig. 507), a pathologist who consistently wore the dress peculiar to a Quaker, was a contemporary of Richard Bright and Thomas Addison at Guy's Hospital in London, and was never elected to the staff because of his eccentricities. He first described a disease characterized by a simultaneous enlargement of the spleen and lymph nodes in 1832[2]; the condition was called Hodgkin's disease in 1865 by Sir Samuel Wilks who introduced the eponymous terms Bright's disease and Addison's disease in honour of his colleagues at Guy's Hospital.

Hodgkin's disease in the lids is rarely associated with tumours of the typical lymphomatous and reticular structure; here, however, the first clinical sign of the disease may occur (Kravitz, 1939; Okuda *et al.*, 1967) or the lids and brow may be affected at a later stage (McGavic, 1943). Subconjunctival masses may also occur (Leinfelder and O'Brien, 1935; Avery and

[1] p. 1070.
[2] *Med.-chir. Trans.*, **17**, 68 (1832).

FIG. 507.—THOMAS HODGKIN [1798–1866].
(Courtesy of the Wellcome Institute of the History of Medicine.)

Warren, 1941). Clinically, the affected lymph nodes are large, firm, lobulated and rubbery. Histologically three sub-groups have been recognized: *Hodgkin's granuloma*, in which the tissue is replaced by reticulum cells, giant Sternberg-Reed cells (1898/1902) which may have a single or two mirror-image type nuclei or multiple nuclei, lymphocytes, eosinophils and fibrous tissue; the *paragranuloma*, in which lymphocytes and a few giant cells predominate with some fibrous stroma; and the *sarcoma*, which consists of a very cellular mass of stem cells, reticulum cells and Sternberg-Reed cells with numerous mitoses and little fibrous tissue. These are respectively characterized by mixed cellularity or nodular sclerosis, lymphocytic proliferation and lymphocytic depletion, the second and third types having best prognosis (Lukes *et al.*, 1966). Investigation of these cases now includes lymphography, laparotomy and splenectomy (Smithers, 1972). Among these, lymphography is particularly useful, whereby a contrast medium (Lipiodol Ultra-fluid) is injected into a lymphatic vessel; thus to assess the systemic implications of the disease if the vessel on the dorsum of the foot is used the whole of the lymphatic system in the thorax and abdomen is outlined when the nodes affected by Hodgkin's disease show a typical reticular appearance (Macdonald, 1973).

MYCOSIS FUNGOIDES (GRANULOMA FUNGOIDES OF SARCOMATODES) is characterized by a somewhat indefinite histological picture of histiocytes, lymphocytes, myelocytes and plasmocytes supported by a reticular scaffolding while micro-abscesses may occur; the cellular picture may closely resemble Hodgkin's disease, lymphoma or leucæmia, but systemic involvement of the lymphoid tissues is rare (Fig. 510). It commences typically as a chronic eczematous dermatitis, the numerous circinate patches of which may last for years; this is followed by the plaque stage wherein the patches become infiltrated and thickened, a period of tumour formation, and finally by a phase of ulceration (Figs. 508–9). The tumours vary in size from a pea to an orange and occur all over the body in large numbers; deep ulceration develops at their summits, the raw surface being covered with a sanguineous, purulent exudate; they occasionally regress (Gall, 1955). A rare form occurs (" *à tumeurs d'emblée* ") wherein the tumours appear suddenly without a precedent basis of dermatitis (Duboys de Lavigerie and Onfray, 1913). The usual form of the disease has been reported on the brows and lids

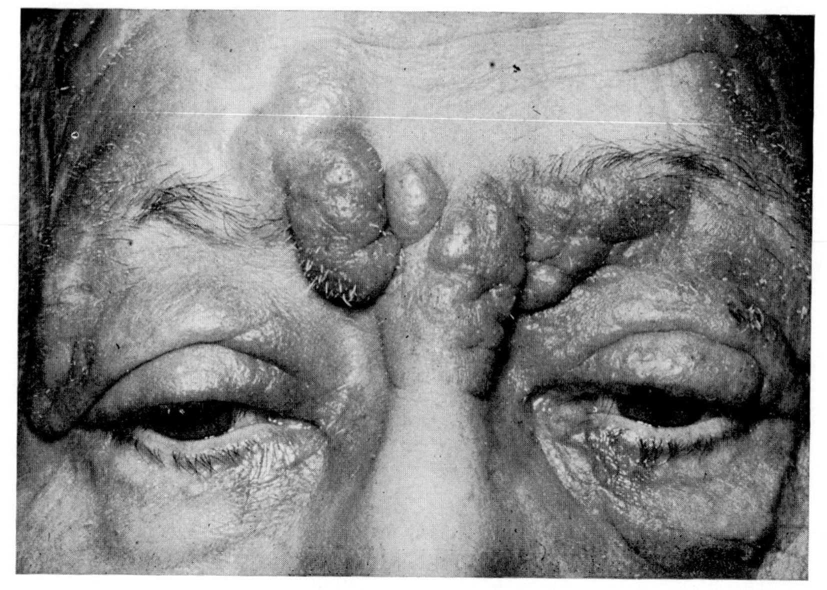

FIG. 508.—MYCOSIS FUNGOIDES.

In a woman aged 63. Pale raised circinate or papular lesions are seen, some of which are ulcerated and crusted (M. J. Reeh).

in which case ectropion may develop (Crull, 1897; Velhagen, 1903; Werther, 1906; Hallopeau and Dainville, 1910; Durand *et al.*, 1971; Domonkos, 1971); Deutsch and Duckworth (1968) described a large solitary tumour of the upper lid in a case of the generalized disease, Garrie and his colleagues (1972) an ulcerating tumour on the lower lid, and Fradkin and his colleagues (1969) one on the caruncle. The growths may disappear to reappear in another site and life-expectancy is limited although survival may extend to several decades (39 years, Sandbank and Katzenellenbogen, 1968). Irradiation and cytotoxic chemotherapy may induce temporary relief.

A malignant reticulosis was described by Sézary (1949) in elderly women, characterized by the development of erythroderma and œdema in the skin of the extremities and eyelids, followed by a dense infiltration of the skin leading, in the lids, to ectropion. The characteristic Sézary cells are giant histiomonocytes presumably derived from the reticulo-endothelial system, which are also sometimes found in mycosis fungoides.

FIGS. 509 and 510.—MYCOSIS FUNGOIDES (S. A. Garrie *et al.*).

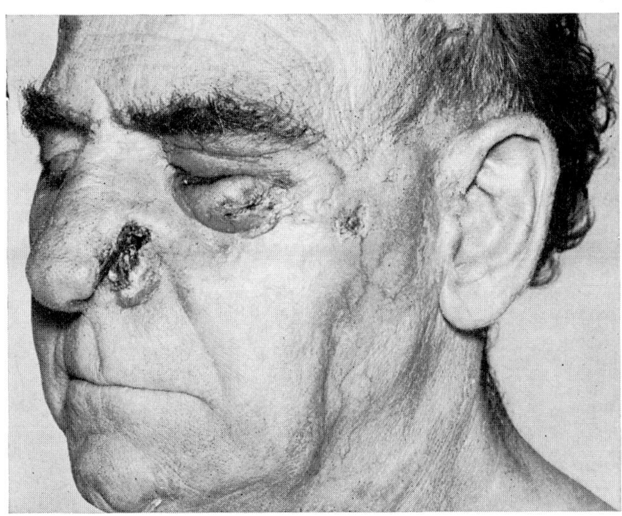

FIG. 509.—In a man aged 54, showing the œdematous tumour of the left lower eyelid, an ulcerating tumour on the left nostril and an ulcerated atrophic area on the left cheek which had been treated with superficial X-rays.

FIG. 510.—Section of a biopsy from the same case showing necrotic areas of acellularity surrounded by a palisading type of infiltrate including histiocyte-type cells, multinucleated giant cells and small mononuclear cells. The histological pattern resembled granuloma annulare.

X-ray therapy was ineffective and a fatal termination occurred between 18 months and 2 years.

PLASMOCYTOMA

A rare benign lesion affecting all four lids and composed of plasmocytes was described by Schwarzkopf (1921); it was apparently non-inflammatory, benign and associated with a normal blood-picture; surgical excision gave a satisfactory result. Speciale-Picciché (1924) described a mass composed partly of lymphocytes, partly of plasmocytes, the presence of the latter being interpreted by him as indicating an infective origin. Most such lesions occur in the conjunctiva and are of inflammatory origin[1]; those not of inflammatory origin are most commonly associated with multiple myeloma.[2]

The *treatment* of the malignant types of lymphoid tumour is usually distressing. Most are radio-sensitive and the local lesion may respond partially or completely to such therapy but subsequent recurrences or generalization of the disease is not unusual. Surgery and radiotherapy remain the treatment of choice for localized tumours but once spread has occurred beyond the point at which these two methods can be applied, reliance has to be placed on medical treatment. Initially this took the form of various drugs which would attack all dividing cells, but additional methods are now available involving the use of hormones, enzymes and immunotherapy. This subject is advancing so rapidly that for details of the drugs and their possible side-effects the reader is referred to more specialized works.[3] In many cases one or other of these different approaches may be used in combination. Radiotherapy is the definitive treatment for localized tumours as in Hodgkin's disease, lymphosarcoma and reticulosarcoma when it may be curative; in some circumstances it may be valuable as a palliative measure to reduce the size of the tumour and relieve pain. Treatment is usually directed not only to the involved tissue but also to the adjacent non-involved lymph nodes (Calle *et al.*, 1971). Most of the chemotherapeutic substances used in the treatment of malignant disease interfere by one mechanism or another with the synthesis of DNA, RNA or protein and act against both malignant and normal cells. They include the alkylating agents such as chlorambucil, melphalan, cyclophosphamide, busulphan and nitrosurea derivatives, the antimetabolites such as methotrexate, 6-mercaptopurine, azathioprine, cytosine arabinoside, and substances derived from bacteria or plants including antibiotics and the Vinca alkaloids. Corticosteroid therapy may also be useful in addition to these drugs.

Alezzandrini, Juegens and Kaufer. *Arch. Oftal. B. Aires*, **42**, 307 (1967).

Avery and Warren. *Arch. Ophthal.*, **26**, 1019 (1941).

Axenfeld. *v. Graefes Arch. Ophthal.*, **37** (4), 102 (1891).

Bascarán Collante and López Nieto. *Arch. Soc. oftal. hisp.-amer.*, **26**, 157 (1966).

Bégué. *Bull. Soc. Ophtal. Paris*, 205 (1932).

Boyd. *Textbook of Pathology*, 8th ed., London (1970).

[1] Vol. VIII, p. 1183.

[2] p. 1083; see Vol. VIII, p. 1093.

[3] See Jelliffe (1966), Whitby and Britton (1969), Hope-Stone (1969), de Gruchy (1970), Jelliffe *et al.* (1970), Todd (1970), Fairley (1971).

Brill, Baehr and Rosenthal. *J. Amer. med. Ass.*, **84**, 668 (1925).
Calle, Zajdela, Haye and Schlienger. *Bull. Cancer*, **58**, 329 (1971).
Carpendale. *Lancet*, **1**, 305 (1949).
Chambers. *Arch. Ophthal.*, **43**, 520 (1950).
Croci. *Lettura oftal.*, **11**, 3 (1934).
Crull. *Münch. med. Wschr.*, **44**, 627 (1897).
Dacie, Dresner, Mollin and White. *Brit. med. J.*, **1**, 1447 (1950).
Delens. *Arch. Ophtal.*, **6**, 154 (1886).
Deutsch and Duckworth. *Amer. J. Ophthal.*, **65**, 884 (1968).
Domonkos. Andrews' *Diseases of the Skin*, 6th ed., Phila., 823 (1971).
Duboys de Lavigerie and Onfray. *Ann. Oculist.* (Paris), **149**, 281 (1913).
Durand, Magnard and Trepsat. *Bull. Soc. Ophtal. Fr.*, **71**, 723 (1971).
Epstein and MacEachern. *Arch. intern. Med.*, **60**, 867 (1937).
Erbakan and Aksu. *Amer. J. Ophthal.*, **53**, 527 (1962).
Fairley. In *Recent Advances in Hœmatology*, (ed. Goldberg and Brain), London, 219 (1971).
Fradkin, Ruiz and Sloane. *Amer. J. Ophthal.*, **68**, 719 (1969).
Fröhlich. *Wien. med. Wschr.*, **43**, 285, 331, 384, 422 (1893).
Fusco, Vecchione and Romano. *Ann. Ottal.*, **94**, 675 (1968).
Gall. *Minn. Med.*, **38**, 674, 705 (1955).
Gall and Mallory. *Amer. J. Path.*, **18**, 381 (1942).
Garrie, Hirsch and Levan. *Arch. Derm.*, **105**, 717 (1972).
Gates. *Arch. Derm. Syph.* (Chic.), **37**, 1015 (1938).
Gemolotto. *Boll. Oculist.*, **34**, 705 (1955).
Glinski. *Virchows Arch. path. Anat.*, **121**, 101 (1903).
Godtfredsen. *Nord. Med.*, **34**, 1431 (1947).
Goedbloed and Wyers. *Acta ophthal.* (Kbh.), **19**, 28 (1941).
Gokhale and Desai. *J. All-India ophthal. Soc.*, **14**, 124 (1966).
de Gruchy. *Clinical Hœmatology in Medical Practice*, Oxford (1970).
Hallopeau and Dainville. *Bull. Soc. franç. Derm. Syph.*, **21**, 32 (1910).
Haye and Haut. *Bull. Soc. belge Ophtal.*, No. 142, 435 (1966).
Hazen. *J. cutan. Dis.*, **29**, 521 (1911).
Heath. *Amer. J. Ophthal.*, **32**, 1213 (1949).
Hilton and Sutton. *Lancet*, **1**, 283 (1962).
Hogan and Zimmerman. *Ophthalmic Pathology*, 2nd ed., Phila., 218 (1962).
Hope-Stone. *Brit. J. Radiol.*, **42**, 770 (1969).
Israels. *Lancet*, **2**, 525 (1953).
Jain. *J. All-India ophthal. Soc.*, **12**, 176 (1964).
Jelliffe. *Proc. roy. Soc. Med.*, **59**, 1261 (1966).
Jelliffe, Millett, Marston *et al. Clin. Radiol.*, **21**, 439 (1970).
Jelliffe and Thomson. *Brit. J. Cancer*, **9**, 21 (1955).
Kamel. *Indian J. ocular Path.*, **1**, 19 (1967).
Kaplan. *Proc. roy. Soc. Med.*, **65**, 62 (1972).
Kravitz. *Arch. Ophthal.*, **21**, 844 (1939).
Leber. *v. Graefes Arch. Ophthal.*, **24** (1), 295 (1878).
Leinfelder and O'Brien. *Arch. Ophthal.*, **14**, 183 (1935).
Lissia and Borellini. *Ann. Ottal.*, **95**, 893 (1969).
Lukes. *Recent Results in Cancer Research*, Berlin, No. 36, 6 (1971).
Lukes and Butler. *Cancer Res.*, **26**, 1063 (1966).
Lukes, Butler and Hicks. *Cancer*, **19**, 317 (1966).
Macdonald. *Brit. J. hosp. Med.*, **9**, 437 (1973).
McGavic. *Arch. Ophthal.*, **30**, 179 (1943).
Massa, Devloo and Jamotton. *Bull. Soc. belge Ophtal.*, No. 142, 1, 415 (1966).
Métivier. *Brit. J. Ophthal.*, **21**, 202 (1937).
Molnár and Keresztury. *v. Graefes Arch. Ophthal.*, **165**, 370 (1963).
Mortada. *Amer. J. Ophthal.*, **56**, 649 (1963). *2nd Cong. Europ. ophthal. Soc.* (Wien, 1964) (ed. François), Basel, 834 (1966).
Nebeská. *Cs. Oftal.*, **4**, 293 (1948).
Neugebauer. *Z. Augenheilk.*, **17**, 393 (1907).
Nolan. *Brit. J. Ophthal.*, **52**, 532 (1968).
O'Brien and Leinfelder. *Amer. J. Ophthal.*, **18**, 123 (1935).
Offret and Haye. *Tumeurs de l'oeil et des annexes oculaires*, Paris (1971).
Okuda, Matsuo, N. and H., and Nakano. *Folia ophthal. jap.*, **18**, 504 (1967).
Okutani and Tamura. *Folia ophthal. jap.*, **22**, 462 (1971).
Reed. *Johns Hopk. Hosp. Rep.*, **10**, 133 (1902).
Reeh. *Treatment of Lid and Epibulbar Tumors*, Springfield, 248 (1963).
Reese. *Tumors of the Eye*, 2nd ed., N.Y., 465 (1963).
Reese and Guy. *Amer. J. Ophthal.*, **16**, 718 (1933).
Robb-Smith. In *Recent Advances in Clinical Pathology* (ed. Dyke), London (1947).
Robolotti. *Atti Soc. oftal. Lombarda*, **12**, 236 (1957).
Sandbank and Katzenellenbogen. *Arch. Derm.*, **98**, 620 (1968).
Scalinci. *Ann. Ottal.*, **31**, 360 (1902).
Schwarzkopf. *Z. Augenheilk.*, **45**, 142 (1921).
Scuderi. *Ophthalmologica*, **162**, 178 (1971).
Sézary. *Ann. Derm. Syph.* (Paris), **9**, 5 (1949).
Shannon and McAndrews. *Amer. J. Ophthal.*, **15**, 821 (1932).
Smithers. *Proc. roy. Soc. Med.*, **65**, 61 (1972).
Speciale-Picciché. *Ann. Ottal.*, **52**, 774 (1924); **55**, 412 (1927).
Stärk and Thiel. *Klin. Mbl. Augenheilk.*, **157**, 308 (1970).
Stansbury. *Arch. Ophthal.*, **40**, 518 (1948).
Sternberg. *Z. Heilk.*, **19**, 21 (1898).
Sugarbaker and Craver. *J. Amer. med. Ass.*, **115**, 17 (1940).

Symmers. *Amer. J. med. Sci.*, **174,** 9 (1927).
Arch. Path., **3,** 816 (1927); **26,** 603 (1938).
Taubmann. *Ein Fall von Lymphosarkom d. Lider mit epidermidaler Metaplasie d. Conjunctivalepithels* (Diss.), Königsberg (1899).
Thiel. *Klin. Mbl. Augenheilk.*, **157,** 308 (1970).
Tiscornia. *Rev. Ass. med. Argent.*, **36,** Sect. Soc. Oftal., 92 (1923).
Todd. *Proc. roy. Soc. Med.*, **63,** 86 (1970).
Törnquist. *Arch. Ophthal.*, **44,** 842 (1950).
Tooke. *Brit. J. Ophthal.*, **23,** 444 (1939).
Ultmann. *Cancer*, **19,** 297 (1966).
Velhagen. *v. Graefes Arch. Ophthal.*, **55,** 175 (1903).
Werther. *Münch. med. Wschr.*, **53,** 1546 (1906).
Whitby and Britton. *Diseases of the Blood*, 10th ed., London (1969).
Willis. *Pathology of Tumours*, 4th ed., London, 772 (1967).
Zimmerman. *Trans. Aust. Coll. Ophthal.*, **2,** 83 (1970).

Vascular Tumours

HÆMANGIOMA

ANGIOMATA, as we have seen,[1] are nearly always developmental in origin and if they are not obviously present at birth the predisposing conditions are present and full development occurs during the first few months or years of life. These tumours usually have a period of rapid growth whereafter most of them spontaneously cease growing and many regress. On the other hand, there are some vascular lesions which are definitely neoplasic in that they show progressive growth and the more malignant types may give rise to metastases and cause death. They have been classified into two principal groups: *polymorphous angiomata*, developmental anomalies which include capillary, cavernous, racemose, telangiectatic and angioblastic angiomata; and *monomorphous angiomata* which appear to be derived from one cellular component of the vessel wall and are true neoplasms including benign and malignant varieties of hæmangio-endothelioma, hæmangio-pericytoma and tumours arising from the smooth muscle of the blood vessels. We shall discuss here those most commonly occurring in the lids where they are seen with considerable frequency (2·7% of 617 tumours, Welch and Duke, 1958; 5·4% of 2,651, H. and J. Allington, 1968); they have also been described on the caruncle (Göngül, 1949; Ahmad and Hafeez, 1968).

CAPILLARY ANGIOMA (ANGIOMA SIMPLEX; NÆVUS VINOSUS, VASCULOSUS or FLAMMEUS; *port-wine stain*) is the commonest form, appearing as a dark red, usually slightly elevated area of very variable extent. It may affect a small part of one lid only, or cover a large area of the face or involve the entire distribution of one or more branches of the Vth nerve (Figs. 511–13). At times the subcutaneous tissues are much thickened (ELEPHANTIASIS HÆMANGIECTATICA or TELANGIECTODES), while small nodular raspberry-like tumours may be scattered over the surface. Pathologically the tumour is made up of very superficial dilated capillaries providing a picture of a multitude of endothelial-lined vascular channels with little connective-tissue stroma (Figs. 514–15). These tumours are stationary,

[1] Vol. VIII, p. 1198.

FIGS. 511 to 513.—CAPILLARY HÆMANGIOMATA.

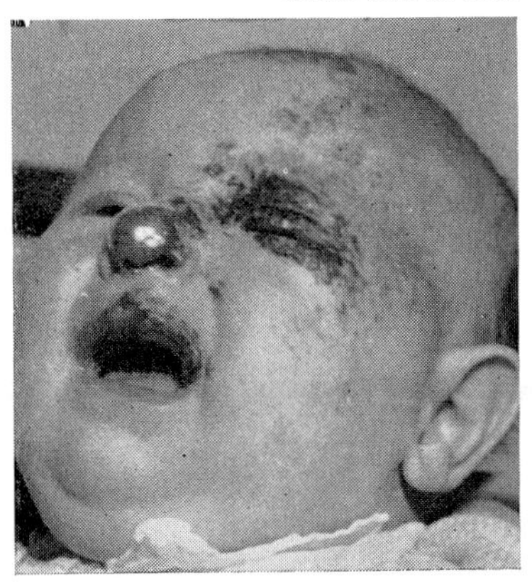

FIG. 511.—Multiple angiomata with widespread distribution over the face (Dept. of Dermatology, Vanderbilt Clinic, N.Y.).

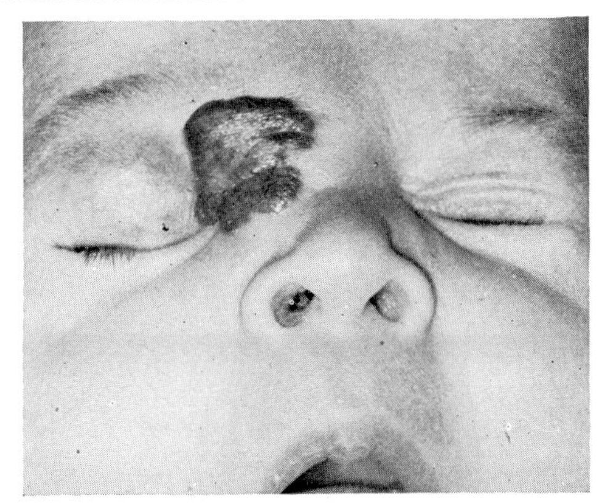

FIG. 512.—A strawberry mark on the inner part of the upper lid (B. Jay).

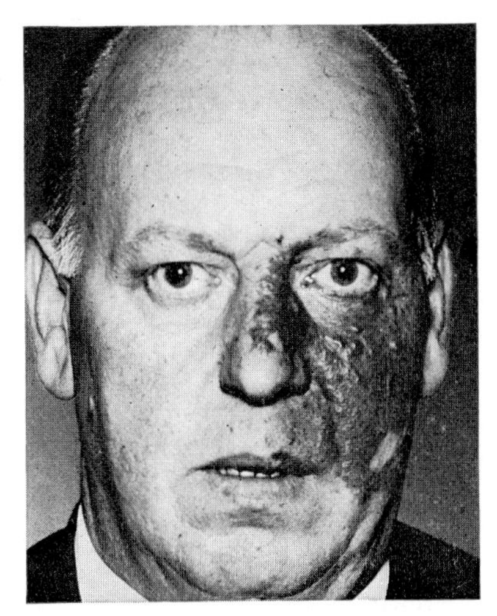

FIG. 513.—An extensive hæmangioma of the port-wine stain variety (M. F. Stranc).

their only effect being a cosmetic disability, although sometimes traumatic hæmorrhages are annoying.

The *strawberry mark* is a type of capillary hæmangioma consisting of one or more bright red, soft, often globulated tumours which arise shortly after birth, increase in size for a few months and then, in contrast to other types of capillary hæmangioma, regress spontaneously within a few years (Fig. 512). Occasionally a cavernous hæmangioma may be present beneath the superficial capillary lesion. Histologically the lesion consists of a proliferation of endothelial cells with narrow blood-spaces; with time the capillary lumina widen and the endothelial lining flattens and still later

FIGS. 514 and 515.—CAPILLARY HÆMANGIOMA OF THE LID (N. Ashton).

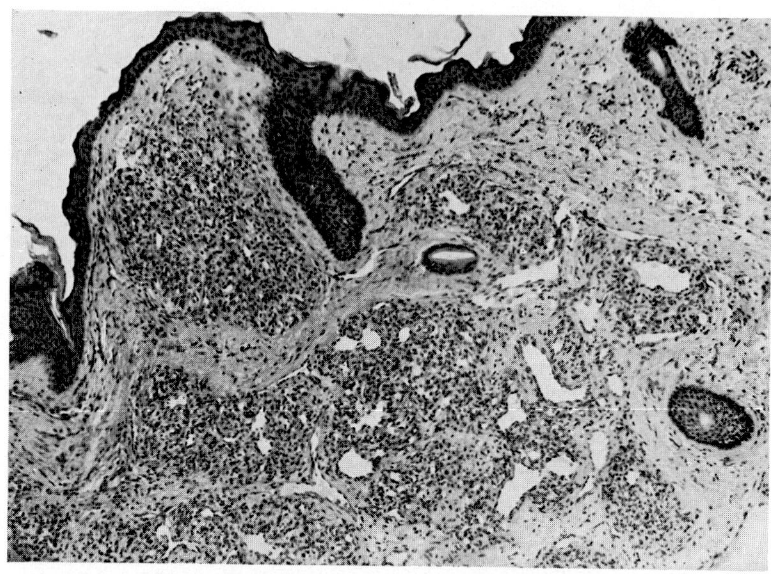

FIG. 514.—The lesion consists of masses of endothelial vasoformative tissue in which a few developed capillary vessels may be seen (H. & E.; ×75).

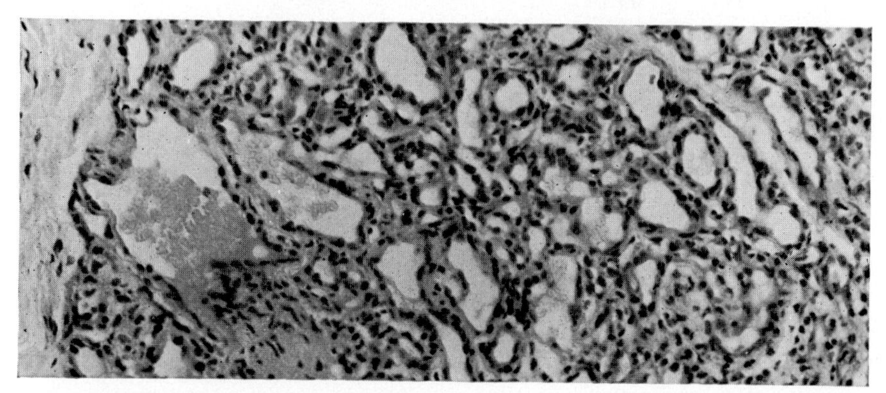

FIG. 515.—The lesion in Fig. 514, showing a mass of capillary channels some of which contain blood (×210).

increasing fibrosis obliterates the capillaries with eventual shrinkage of the lesion (Lever, 1967).

In this connection the STURGE-WEBER SYNDROME of ENCEPHALO-FACIAL ANGIOMATOSIS should be remembered, which has been fully described elsewhere.[1]

CAVERNOUS ANGIOMATA, consisting of encapsulated masses of endothelium-lined spaces resembling erectile tissue (Figs. 1126–7), occur more

[1] Vol. III, p. 1120.

rarely. As a rule they are superficial and form purplish elevated tumours, soft in consistency, red or violet in colour, sessile or pedunculated in form, and compressible by the finger (Figs. 516–19). More rarely they occupy the deeper subcutaneous tissues in which case the overlying epidermis is normal and the tumour may increase on bending downwards owing to the increased turgidity (Figs. 520–1); on the other hand, a hæmorrhage into the angioma may suddenly increase its size (Fig. 522). In the infant such a tumour may

Figs. 516 to 519.—Cavernous Hæmangiomata.

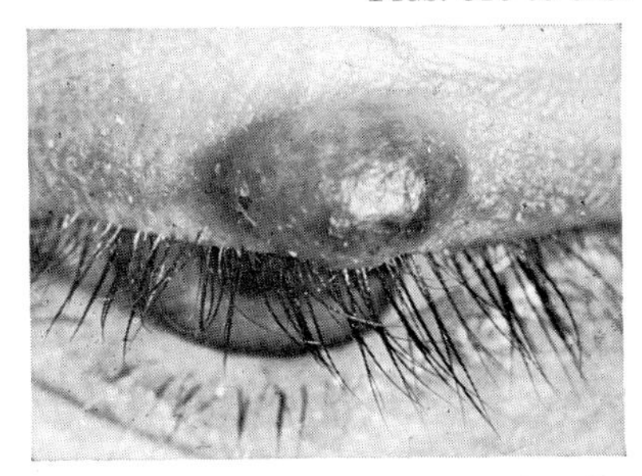

Fig. 516.—In the upper lid (P. MacFaul).

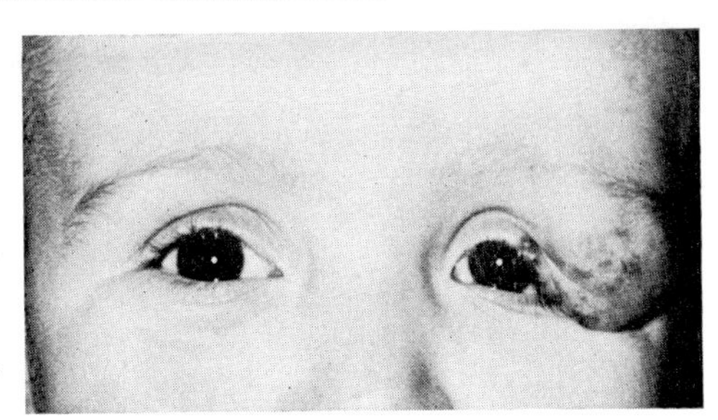

Fig. 517.—Extension to the forehead and temple (M. F. Stranc).

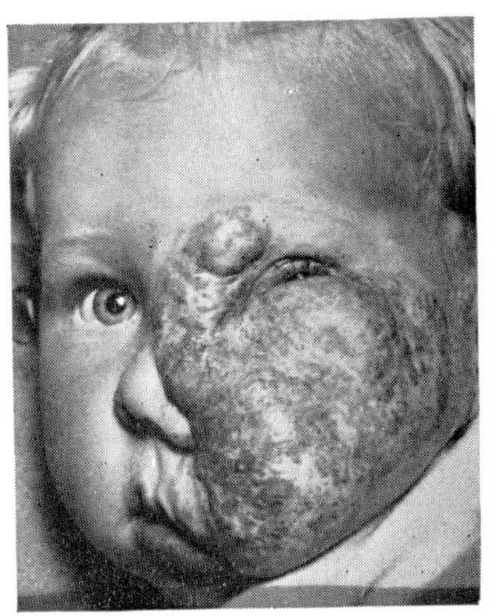

Fig. 518.—Covering most of the side of the face (W. H. Brown).

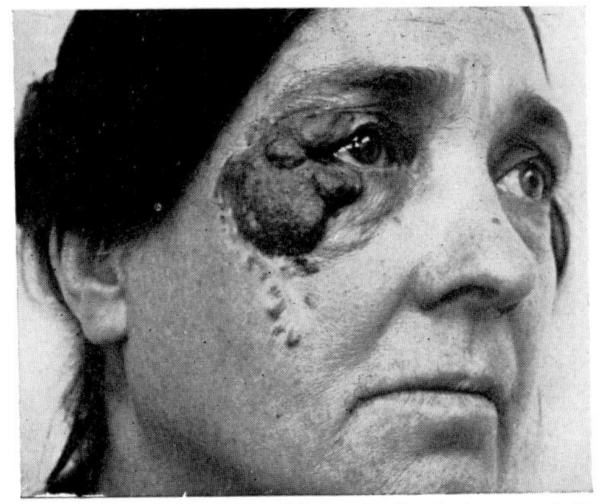

Fig. 519.—In the lids and cheek in an adult (W. H. Brown).

grow rapidly for a period to cover a large area including the lids, cheeks and temple and, invading the orbit, may cause exophthalmos; having thus attained very considerable dimensions it may remain stationary indefinitely without treatment (Fig. 518) (Rumschewitsch, 1897; Risley, 1906; Natale, 1935).

A plexiform angioma occurs very rarely in infants, growing rapidly into a red nodular tumour which ulcerates and bleeds readily (Bardanzellu, 1931; O'Brien and Braley, 1936).

Figs. 520 and 521.—Cavernous Angioma.
Occurring in a woman aged 32 (Inst. Ophthal.).

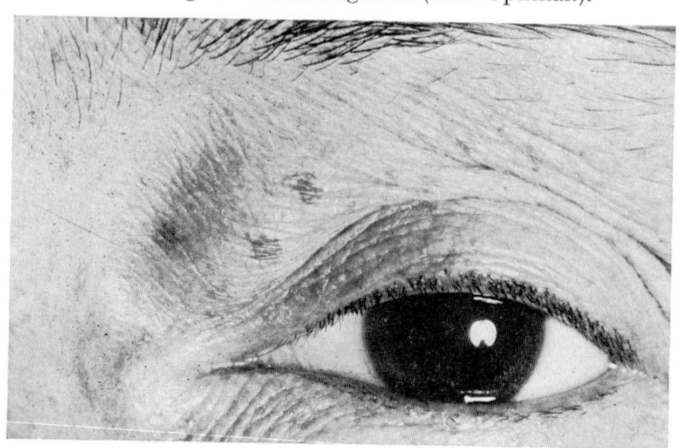

Fig. 520.—The patient's head is erect.

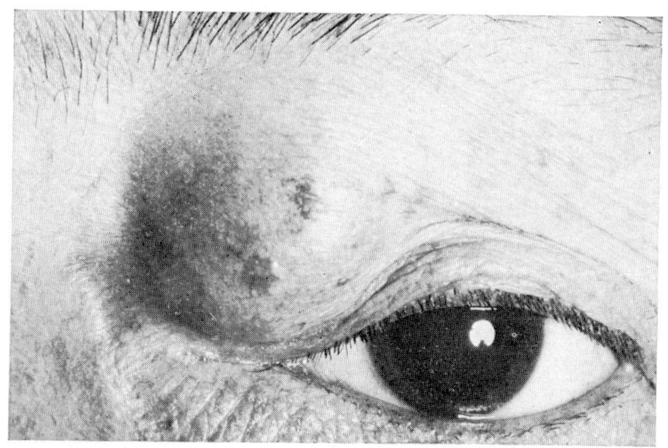

Fig. 521.—The patient is bending forwards; note the increase in size of the tumour.

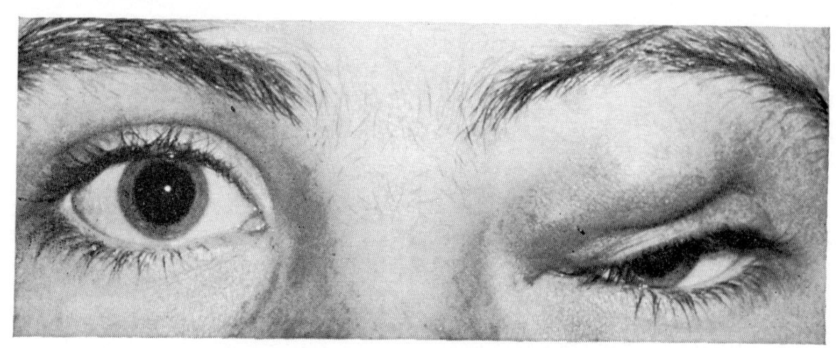

Fig. 522.—A Hæmorrhage into a Cavernous Hæmangioma.
In the left upper lid (Inst. Ophthal.).

HÆMANGIO-ENDOTHELIOMA, wherein the blood-spaces are reduced considerably by the proliferation of endothelial cells, is a rare occurrence in the lids (Wilson, 1936; Reeh, 1963; Barca *et al.*, 1971); such a tumour has been known to assume malignant characteristics (ANGIOSARCOMA; ANGIOBLASTIC SARCOMA, Rintelen, 1935; Wilson Jones, 1964; Roschin, 1964;

FIGS. 523 to 526.—HÆMANGIO-ENDOTHELIOMATA OF THE LIDS.

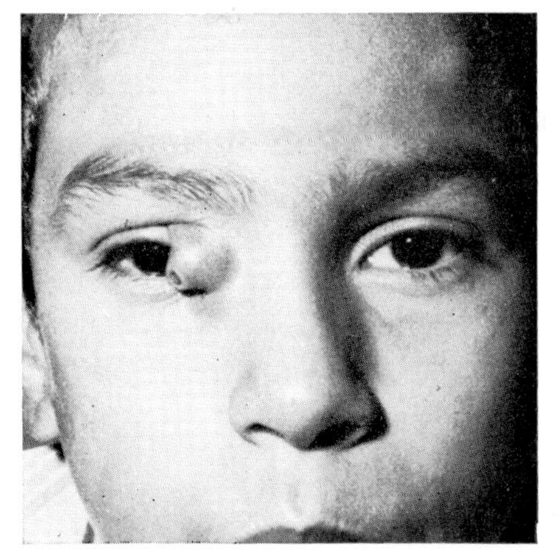

FIG. 523.—In a boy aged 8 years; the swelling on the upper lid was of 2 months' duration and was excised (A. Mortada).

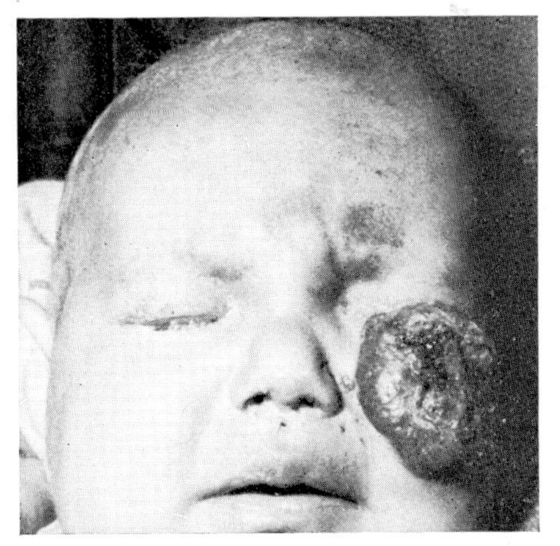

FIG. 524.—In a baby aged one month who had a gross swelling in the lower left lid from birth; it disappeared with radiational treatment (A. Mortada).

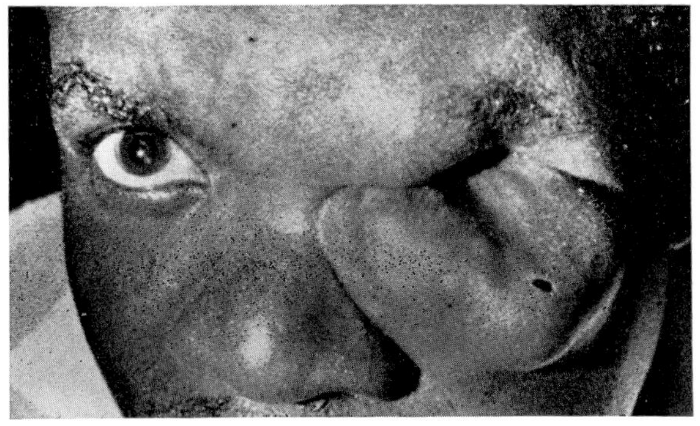

FIG. 525.—A large tumour in an East African (G. G. Bisley).

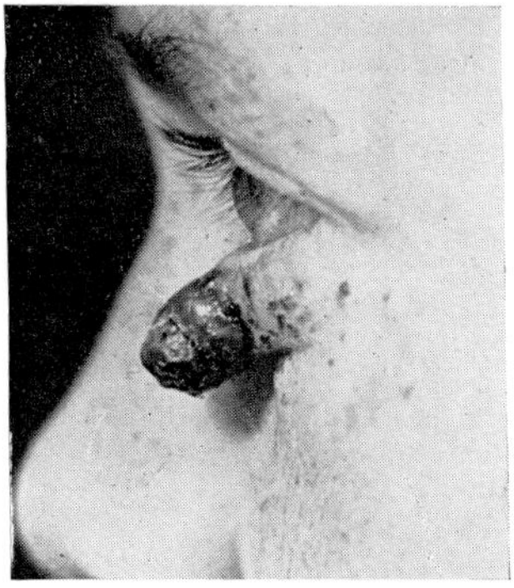

FIG. 526.—Occurring in a child aged 14 (F. Rintelen).

Okuda *et al.*, 1970) (Figs. 523–6). Stout (1943) stressed the potentially malignant character of vascular tumours composed of embryonic endothelial cells; among his 18 cases 2 occurred in the orbito-palpebral region and both recurred repeatedly after excision. The malignant type which is locally invasive occurs usually in elderly individuals; in children it may be associated with a typical port-wine stain (Girard *et al.*, 1970). The majority of hæmangio-endotheliomata, however, shows a more benign course and may present in one of two forms (Mortada, 1966), either small localized tumours of the superficial tissues of the lid seen usually in children or young adults which should be excised, or large non-encapsulated tumours affecting the lid and orbit and occurring in infants under the age of 12 months which usually spontaneously regress. Histologically the vascular channels show proliferation of the endothelial cells within the reticulin sheath of the vessels in an abundance greater than necessary to line the anastomosing vascular channels (Figs. 527–8); the younger the patient the more cellular and invasive the tumours may appear yet clinically they generally behave as

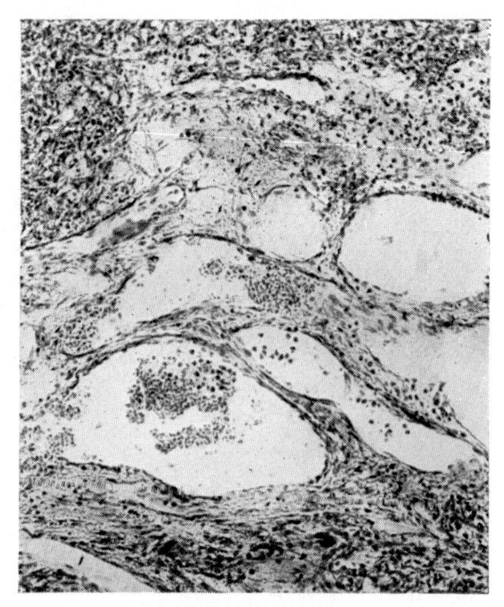

Fig. 527.

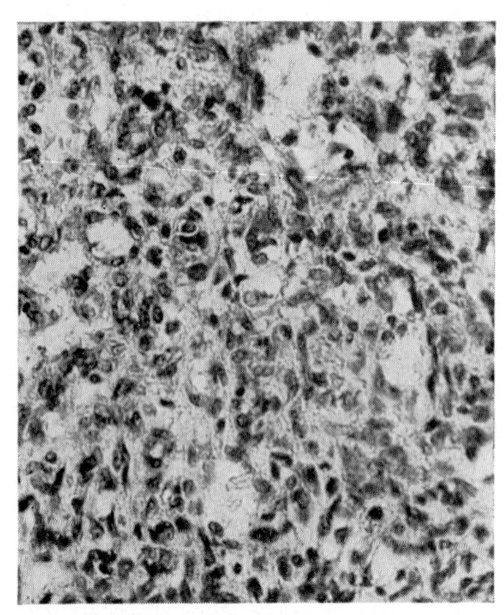

Fig. 528.

Figs. 527 and 528.—Hæmangio-endothelioma.

Histological section of the tumour seen in Fig. 526, showing areas with a cavernous structure (Fig. 527) and other areas where an exuberant and irregular growth of the endothelial cells occurs with attempts at the formation of capillaries (Fig. 528) (F. Rintelen).

benign lesions (Hogan and Zimmerman, 1962; François and Hanssens, 1962; Reese, 1963). The malignant type should be widely excised and recurrences may be treated by radiation.

Hæmangio-pericytoma is another variety of vascular tumour characterized histologically by a proliferation of the contractile pericytes situated outside the reticulin sheath of the capillary wall, the distinction from the hæmangio-endothelioma being made by the use of silver reticulin stains (Stout and Murray, 1942; Stout, 1949; Kauffman and Stout, 1961) (Figs. 529–30). Although endowed with considerable numbers of blood vessels, subcutaneous tumours may show no redness or discoloration to suggest their vascular nature. These tumours may be benign or malignant and may occur in any of the soft tissues where capillaries are found. They present as a slowly growing nodular mass which is usually painless (Macoul, 1968; Oshida *et al.*, 1970); they are slowly infiltrative, may grow to a considerable size and if not completely excised may recur; they are usually orbital in origin under which heading they will be discussed,[1] but invasion of the orbit may occur from a neoplasm starting in the lid with the result that exenteration is necessary (Reich, 1972). A dramatic case was reported by Macoul (1968) wherein a tumour starting in the lid spread to both orbits,

FIGS. 529 and 530.—HÆMANGIO-PERICYTOMA (K. L. Macoul).

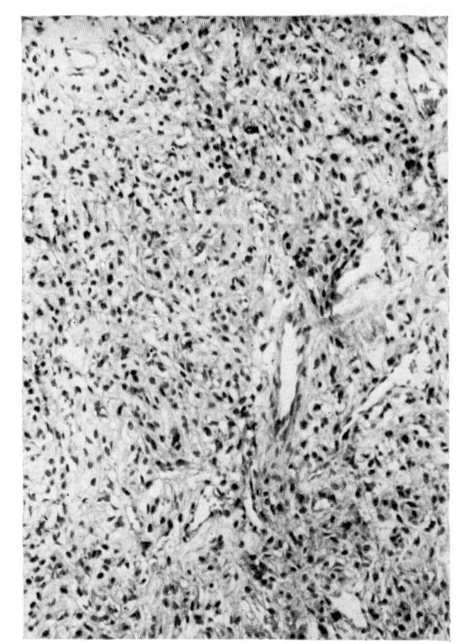

FIG. 529.—Broad sheets of oval and spindle-shaped cells, central nuclei and scanty cytoplasm (×250).

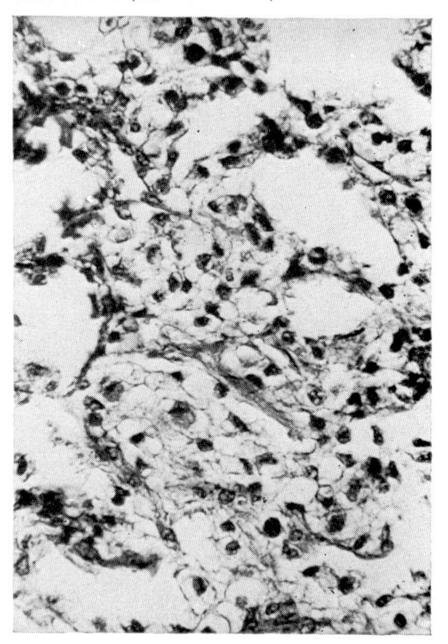

FIG. 530.—Capillaries with a basement membrane of reticulum fibres surrounded by pericytes lying outside the capillary lumen (×375).

the nose, the paranasal sinuses and the anterior cranial fossa. Most are radio-resistant and treatment should be by wide excision (Reese, 1963; Backwinkel and Diddams, 1970).

Acquired angiomatous lesions which may appear in the eyelids include the spider angioma and the senile angioma.

SPIDER ANGIOMATA (NÆVUS ARANEUS) occur usually on the upper half of the face often in considerable numbers arising spontaneously particularly in patients with hepatic disease and during pregnancy. The typical lesion has fine superficial blood vessels radiating from a central dot which usually pulsates. If necessary, for cosmetic reasons, they may be eliminated by electrolysis.

[1] p. 1099.

SENILE ANGIOMATA are small reddish papular or nodular lesions that may involve the skin of the face and lids in older persons. The typical histological appearance is of a focal collection of thin-walled blood vessels in the superficial layers rather like a true capillary hæmangioma although some cases represent merely telangiectasias of pre-existing vessels or venous lakes. Treatment if necessary is by excision.

Sclerosing hæmangiomata are discussed with fibromata.[1]

Treatment. The natural history of most congenital angiomata is towards spontaneous regression, the majority becoming obvious and growing rapidly during the first 6 months of life and tending to disappear more or less completely by the age of 5 years, particularly when active growth has stopped before the age of 12 months. Involution usually starts in the centre with softening of the mass and a decrease in its volume and thickness; the

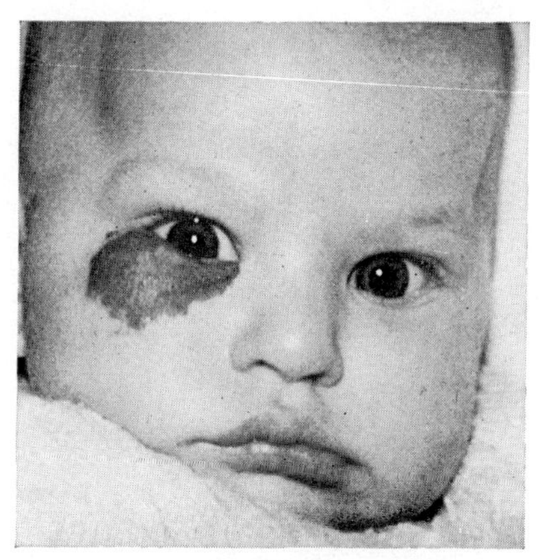

FIG. 531.

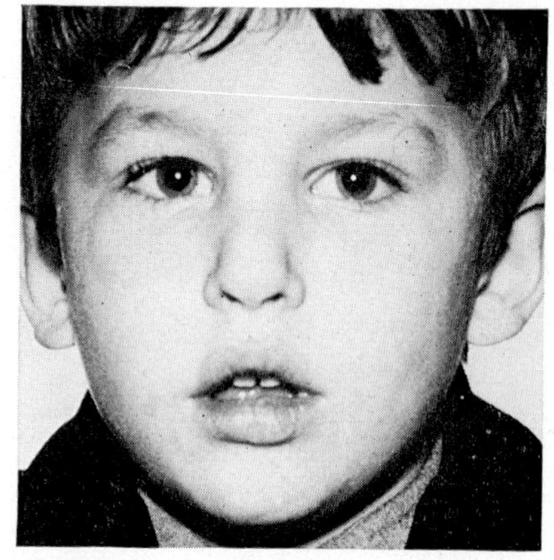

FIG. 532.

FIGS. 531 and 532.—A CONGENITAL HÆMANGIOMA.
To show complete spontaneous regression at the age of 8 years (M. F. Stranc).

colour fades to a mottled or dust grey as fibrosis develops and redundant skin usually shrinks until the normal colour is attained well before puberty. In most cases, therefore, no active measures should be undertaken apart from periodic observation with measurement and photography of the lesion, but it is important to reassure the parents of the benign nature of the disease and that spontaneous cure is the usual outcome (Figs. 531–2).[2] There are, however, cases wherein the location and size of the lesion around the eyelids cause great disfigurement, the obscuration of the visual axis entails the risk of amblyopia, or an associated orbital angioma gives rise to proptosis; moreover, appeals from the parents may necessitate some interference. On the

[1] p. 467.

[2] Lister (1938), Walter (1953), Kapuściński (1954), Modlin (1955), Walsh and Tompkins (1956), Andrews *et al.* (1957), Desvignes and Amar (1958), Simpson (1959), Lampe and Latourette (1959), Bowers *et al.* (1960), Reese (1963), Margileth and Museles (1965), Holland (1968), Barkhash (1970), Hood (1970), Offret and Haye (1971), Hiles and Pilchard (1971).

other hand, a few cases are so extensive as to defy any reasonable method of treatment (Fig. 533).

Many methods of treatment have been suggested; large superficial capillary angiomata are usually best left alone, but from the cosmetic point of view excision and skin-grafting may sometimes be satisfactory; circumscribed tumours may be easily excised (Wallace, 1953; Lekieffre *et al.*, 1972); so also should those with a suspicion of malignancy. It is to be remembered, however, that the lesion may be more extensive in the deeper tissues than its superficial appearance would suggest, and it may be prudent to arrange for angiography to determine the depth and extent of the lesion

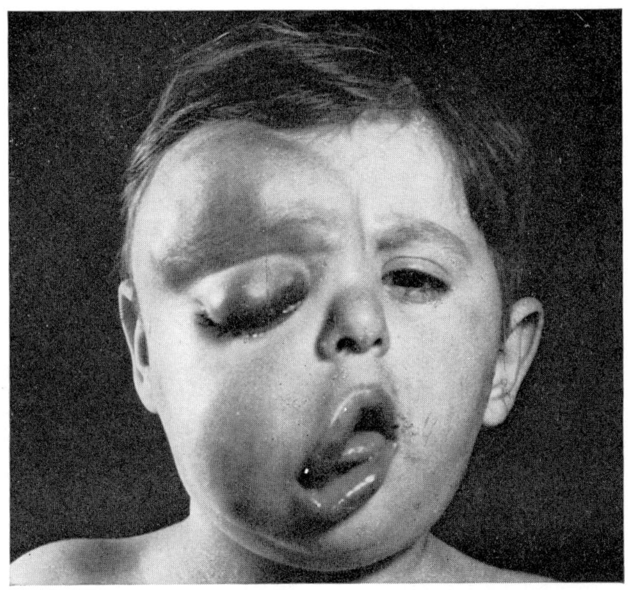

Fig. 533.—Cavernous Hæmangioma.

A congenital lesion affecting the entire right side of the face; syndactyly was also present. Such a case is beyond surgical relief (D. Silva).

as a guide in the planning of any surgical procedure (Offret and Aron Rosa, 1963; Kiehn *et al.*, 1964). For small superficial angiomata cryotherapy may give satisfactory results when the lesion is destroyed by freezing with carbon dioxide snow or other substances, the effects of which may penetrate deeply (Caspar, 1920; Carlotti and Spinetta, 1952; and others); repeated electrolysis has also caused regression but is a painful process which may leave some scarring (Castroviejo, 1925; Geronimi, 1931; Marin Amat, 1932). Stationary port-wine stains, when small, may conveniently be disguised by the use of cosmetics; larger areas can sometimes be excised or may be satisfactorily disguised by tattooing with the intradermal injection of pigments derived from titanium dioxide and cobalt, a procedure usually slow and painful

Figs. 534 to 537.—Radiational Treatment in Cavernous Hæmangiomata.

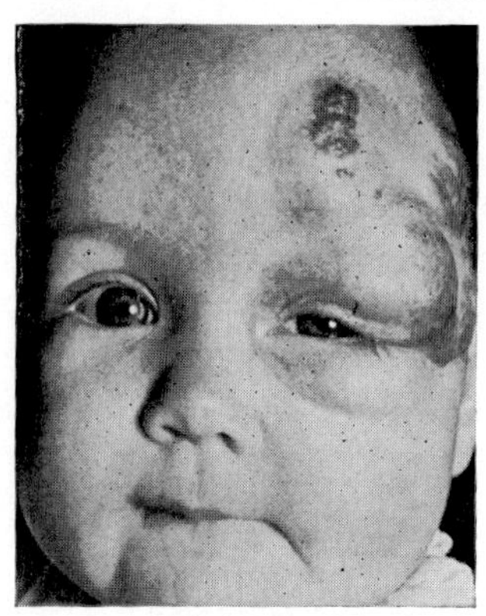

Fig. 534.

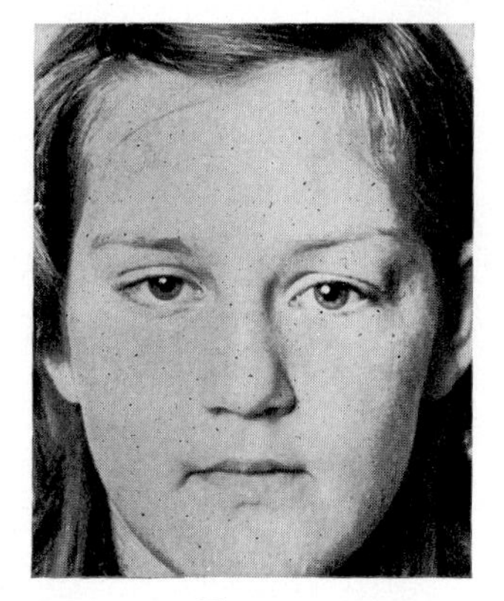

Fig. 535.

Figs. 534 and 535.—Treatment by x-rays, showing complete disappearance at the age of 6 (H. J. Peeters).

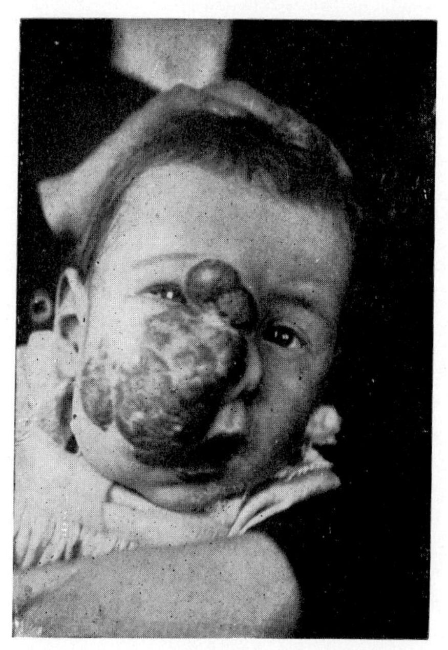

Fig. 536.

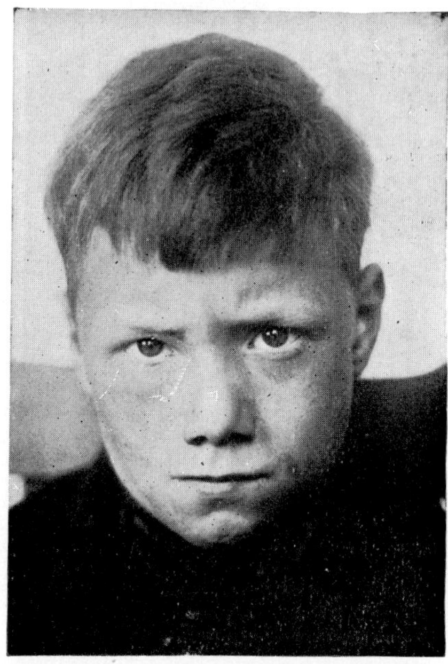

Fig. 537.

Figs. 536 and 537.—The appearance of the tumour in a child aged 21 months (Fig. 536). The same case after 7 applications of radium (radium bromide needles) over a period of 10 years and a plastic operation to remove some scar tissue on the cheek (H. B. Stallard).

(Conway and Montroy, 1965; Thomson and Wright, 1971). A cavernous hæmangioma on the eyelids can be satisfactorily excised.

Sclerosing solutions have been used in the treatment of deeper lesions involving the subcutaneous structures; the substances employed have included sodium morrhuate (Reese, 1945) and even boiling water (Cole and Hunt, 1949); repeated injections are made into the tumour or sometimes into the vessels feeding it and induce a gradual hardening and shrinkage of the lesion which, if it does not completely disappear, can usually be removed thereafter by surgical excision. Good results may be obtained by the use of sclerosing therapy (Prokoffieva, 1963) although some workers have found the technique unsatisfactory and it is always painful (Reese, 1963).

Irradiation using X-rays and radium was at one time a popular method of treatment in producing a fairly rapid regression of the growth (Stallard, 1933) but the high incidence of serious side-effects—not only cataract but also damage to the growth-centres of the bone around the orbit and the rare occurrence of malignant change in the irradiated tissues—brought this form of treatment into some disfavour (Haye *et al.*, 1965), although radium is still advocated by some workers for large tumours (Dana and Beyer, 1966; Offret and Haye, 1971) (Figs. 534–7). Low-voltage contact X-ray therapy has been found to be satisfactory in cavernous hæmangiomata of the eyelids (Domonkos, 1971). Occasionally, however, small doses of X-rays are used to initiate the process of regression.[1] Irradiation with radio-active isotopes such as phosphorus has been advocated to induce fibrosis and progressive obliteration of the vascular spaces.[2]

Dramatically successful results have followed the use of systemic cortico-steroids in cases of orbital and adnexal hæmangiomata acting possibly through an inhibitory effect on immature vascular tissue or by a non-specific anti-inflammatory reaction; the administration of systemic steroids in young infants, however, is not without its hazards, and if treatment is stopped too soon a rebound phenomenon may occur in which the angioma progresses with renewed vigour, requiring more intensive steroid therapy with its attendant hazards.[3]

Ahmad and Hafeez. *J. All-India ophthal. Soc.*, **16,** 103 (1968).
Allington, H. and J. *Arch. Derm.*, **97,** 50 (1968).
Andrews, Domonkos, Rodriguez and Bembenista. *J. Amer. med. Ass.*, **165,** 1114 (1957).
Backwinkel and Diddams. *Cancer*, **25,** 896 (1970).
Barca, Mancini and Venturi. *Ann. Ottal.*, **97,** 195 (1971).
Bardanzellu. *Arch. Ottal.*, **38,** 68 (1931).
Barkhash. *Oftal. Zh.*, **25,** 102 (1970).
Bek and Zahn. *Acta radiol.* (Stockh.), **54,** 443 (1960).
Bowers, Graham and Tomlinson. *Arch. Derm.*, **82,** 667 (1960).
Carlotti and Spinetta. *Bull. Soc. Ophtal. Fr.*, 297 (1952).
Caspar. *Klin. Mbl. Augenheilk.*, **65,** 584 (1920).
Castroviejo. *Arch. Oftal. hisp.-amer.*, **25,** 552 (1925).
Cole and Hunt. *Brit. J. Surg.*, **36,** 346 (1949).

[1] Lewis (1957), Lampe and Latourette (1959), Bek and Zahn (1960), Reese (1963), Katrakis (1963), Nordberg and Sundberg (1963), Offret and Haye (1971).
[2] Dutreix and Tubiana (1958), Kritter (1964).
[3] Zarem and Edgerton (1967), Fost and Esterly (1968), de Venecia and Lobeck (1970), Hiles and Pilchard (1971).

Conway and Montroy. *N.Y. St. J. Med.*, **65,** 876 (1965).
Dana and Beyer. *J. Radiol. Électrol.*, **47,** 325 (1966).
Desvignes and Amar. *Arch. Ophtal.*, **18,** 414 (1958).
Domonkos. Andrews' *Diseases of the Skin*, 6th ed., Phila., 699, 959 (1971).
Dutreix and Tubiana. *Ann. Radiol.*, **1,** 103 (1958).
Fost and Esterly. *J. Pediat.*, **72,** 351 (1968).
François and Hanssens. *Bull. Soc. belge Ophtal.*, No. 132, 513 (1962).
Geronimi. *Schweiz. med. Wschr.*, **61,** 589 (1931).
Girard, Johnson and Graham. *Cancer*, **26,** 868 (1970).
Göngül. *Göz Klin.* (Istanbul), **7,** 33 (1949).
Haye, Jammet and Dollfus. *L'oeil et les radiations ionisantes*, Paris (1965).
Hiles and Pilchard. *Amer. J. Ophthal.*, **71,** 1003 (1971).
Hogan and Zimmerman. *Ophthalmic Pathology*, 2nd ed., Phila., 741 (1962).
Holland. *Klin. Mbl. Augenheilk.*, **152,** 365 (1968).
Hood. *Arch. Ophthal.*, **83,** 49 (1970).
Kapuściński. *Klin. oczna*, **24,** 99 (1954).
Katrakis. *Bull. Soc. hellén. Ophtal.*, **31,** 30 (1963).
Kauffman and Stout. *Cancer*, **14,** 1186 (1961).
Kiehn, Desprez and Kaufman. *Plast. reconstr. Surg.*, **33,** 338 (1964).
Kritter. *Ann. Chir. plast.*, **9,** 280 (1964).
Lampe and Latourette. *Pediat. Clin. N. Amer.*, **6,** 511 (1959).
Lekieffre, Woillez, Asseman *et al.* *Bull. Soc. franç. Ophtal.*, **85,** 433 (1972).
Lever. *Histopathology of the Skin*, 4th ed., Phila., 642 (1967).
Lewis. *Plast. reconstr. Surg.*, **19,** 201 (1957).
Lister. *Lancet*, **1,** 1429 (1938).
Macoul. *Amer. J. Ophthal.*, **66,** 731 (1968).
Margileth and Museles. *J. Amer. med. Ass.*, **194,** 523 (1965).
Marin Amat. *Arch. Oftal. hisp.-amer.*, **32,** 604 (1932).
Modlin. *Surgery*, **38,** 169 (1955).
Mortada. *Bull. ophthal. Soc. Egypt*, **59,** 247 (1966).
2nd Cong. europ. ophthal. Soc. (Wien, 1964) (ed. François), Basel, 482 (1966).
Natale. *Arch. Oftal. B. Aires*, **10,** 587 (1935).
Nordberg and Sundberg. *Acta radiol. Ther. Phys. Biol.*, **1,** 257 (1963).
O'Brien and Braley. *J. Amer. med. Ass.*, **107,** 933 (1936).
Offret and Aron Rosa. *Bull. Soc. franç. Ophtal.*, **76,** 418 (1963).
Offret and Haye. *Tumeurs de l'oeil et des annexes oculaires*, Paris (1971).
Okuda, Doi and Hioka. *Folia ophthal. jap.*, **21,** 341 (1970).
Oshida, Hisatomi, Takemura and Kobayashi. *Folia ophthal. jap.*, **21,** 269 (1970).
Prokoffieva. *Oftal. Zh.*, **18** (1), 7 (1963).
Reeh. *Treatment of Lid and Epibulbar Tumors*, Springfield, 224 (1963).
Reese. *Amer. J. Ophthal.*, **28,** 209 (1945).
Tumors of the Eye, 2nd ed., N.Y., 364 (1963).
Reich. *Klin. Mbl. Augenheilk.*, **160,** 184 (1972).
Rintelen. *Klin. Mbl. Augenheilk.*, **94,** 463 (1935).
Risley. *Ophthal. Rec.*, **15,** 106 (1906).
Roschin. *Vestn. Oftal.*, **77,** 3 (1964).
Rumschewitsch. *Klin. Mbl. Augenheilk.*, **35,** 294 (1897).
Simpson. *Lancet*, **2,** 1057 (1959).
Stallard. *Brit. J. Ophthal.*, Monog. Suppl. VI (1933).
Stout. *Ann. Surg.*, **118,** 445 (1943).
Cancer, **2,** 1027 (1949).
Stout and Murray. *Ann. Surg.*, **116,** 26 (1942).
Thomson and Wright. *Plast. reconstr. Surg.*, **48,** 113 (1971).
de Venecia and Lobeck. *Arch. Ophthal.*, **84,** 98 (1970).
Wallace. *Brit. J. plast. Surg.*, **6,** 78 (1953).
Walsh and Tompkins. *Cancer*, **9,** 869 (1956).
Walter. *J. Fac. Radiol.*, **5,** 134 (1953).
Welch and Duke. *Amer. J. Ophthal.*, **45,** 415 (1958).
Wilson. *Rep. mem. ophthal. Lab. Giza*, **11,** 67 (1936).
Wilson Jones. *Brit. J. Derm.*, **76,** 21 (1964).
Zarem and Edgerton. *Plast. reconstr. Surg.*, **39,** 76 (1967).

ANGIOMA SERPIGINOSUM

ANGIOMA SERPIGINOSUM is a grouping of vascular puncta consisting essentially of dilated and tortuous capillaries in the papillary and subpapillary region of the dermis with micro-aneurysms (Frain-Bell, 1957; Barker and Sachs, 1965). It is an asymptomatic disorder usually beginning in early adult life in the lower limbs and extending slowly, composed of deeply red puncta on a cyanotic background arranged in mottled or net-like patterns the extending borders of which give them a serpiginous outline. Associated vascular anomalies of the retina and central nervous system may occur. The dermal lesion may become widespread; the cheek, forehead and eyelid are rarely affected (Gauthier-Smith *et al.*, 1971). No decision has yet been made whether the lesion is of an angiomatous nature or a telangiectatic dilatation of existing vessels

(Barker and Sachs, 1965); it is possibly due to a congenital defect in the mesodermal vasoformative tissues.

Barker and Sachs. *Arch. Derm.*, **82,** 613 (1965).
Frain-Bell. *Brit. J. Derm.*, **69,** 251 (1957).
Gauthier-Smith, Sanders and Sanderson. *Brit. J. Ophthal.*, **55,** 433 (1971).

TELANGIECTATIC GRANULOMA

This is a small raspberry-like vascular tumour which occurs as a rarity on the lids.

Originally described by Poncet and Dor (1897–1900) as identical with botryomycosis of horses, the disorder was for long considered to be of fungoid nature, although Bodin (1902) showed that the conglomerates of globules thought to be botryomycoses were staphylococci. Despite the fact that re-inoculation of the organism does not reproduce the lesion, the view was held for many years that it was caused by the infection of a small wound (*granuloma pyogenicum*); nevertheless, the histological picture of early lesions makes it clear that it is a capillary angioma (Nödl, 1955; Oehlschlaegel and Müller, 1964), and older lesions which show an infection in their eroded surfaces retain this appearance of an angioma in their deeper layers (Freund, 1932).

The lesion may arise spontaneously but may follow a trauma often so unimportant as hardly to excite attention. It varies in size from a pin-head to a cherry (3·25 × 3·75 cm., Malik *et al.*, 1964), is red, pedunculated or mushroom-shaped with a smooth, moist surface occasionally showing crusting, and is painless but usually tender (Figs. 538–9). It bleeds profusely, sometimes apparently spontaneously through the night; it grows rapidly to its full size in a few days and remains unchanged indefinitely; sometimes it may fall off, in which case it rapidly reappears, a tendency also seen if it is clipped off. Such lesions do not occur commonly in the lids; among 63 cases Aksu and his colleagues (1963) found 2 on the lids and 8 on the conjunctiva.[1] They appear preferentially on the lower lid, sometimes at the margin or at the canthus.[2]

Histopathologically the tumour is so vascularized as to constitute a hæmangioma; it is essentially made up of vessels in varying degrees of dilatation lined by a prominent endothelium (Figs. 540–1). It is covered by epithelium continuous with that of the skin; at the base this layer grows inwards to form an " epidermal collarette " which causes a slight pedunculation of the tumour. In early lesions there is no inflammatory reaction but at a later stage the thin epidermis erodes allowing infection to enter with the appearance of a secondary inflammatory reaction in the superficial layers (Müller, 1962). When this is marked the resemblance to granulation tissue is close, but there is no fibroblastic proliferation nor any penetration into the deeper tissues (Peterson *et al.*, 1964).

[1] Vol. VIII, p. 1202; see Friedman and Henkind (1971).
[2] Wescott (1916), Wohl (1917), Schreiber (1924), Tisserand (1924), Bivona (1925), Michelson (1925), Schmelzer (1933), Hagedoorn (1934), Malik *et al.* (1964), Mohan and Gupta (1968), Slem *et al.* (1970), Huck and Meythaler (1972).

FIGS. 538 to 541.—TELANGIECTATIC GRANULOMA (S. R. K. Malik *et al.*).

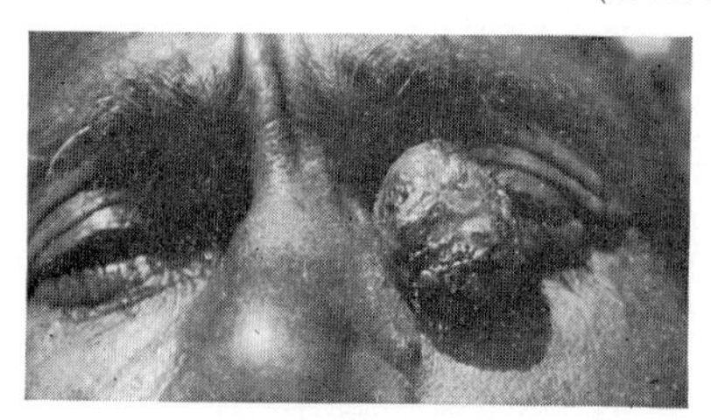

FIG. 538.—Growing from the left lower lid in a Hindu male aged 60.

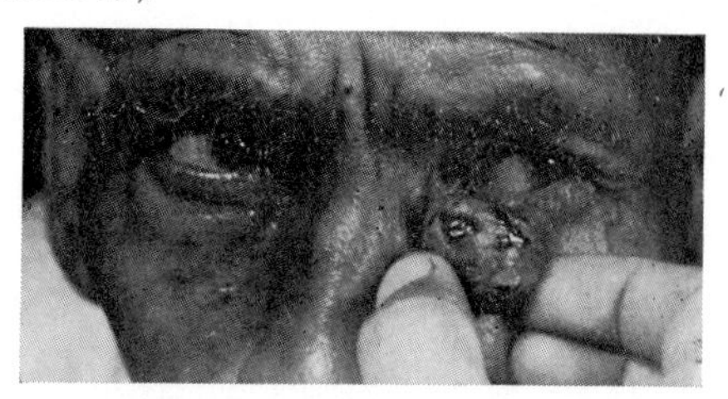

FIG. 539.—Showing the attachment to the skin of the lower lid when the tumour is moved.

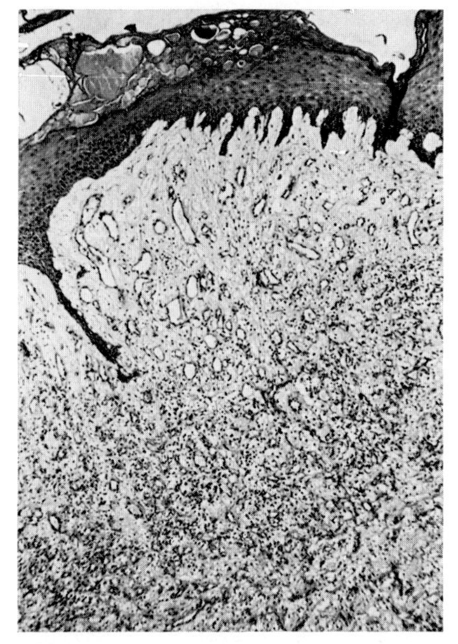

FIG. 540.—Section showing the attenuated epidermis and a maze of capillaries of varying degrees of maturity (H. & E.; ×90).

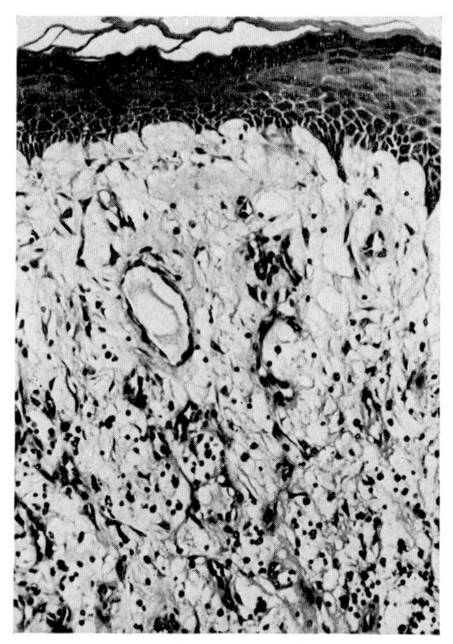

FIG. 541.—High power view showing the maze of capillaries (H. & E.; ×322).

The *diagnosis* is difficult, perhaps because of its rarity, and must usually wait for histological section. Of 26 cases seen in Freiburg, Luchs (1920) found that the diagnosis was made clinically in only one case; a malignant growth is usually suspected. A similar appearance in the palpebral conjunctiva may be caused by ligneous conjunctivitis, the papillary hyperplasia of vernal conjunctivitis, or more rarely the embryonal rhabdomyosarcoma of children; these conditions can, however, usually be excluded on the basis of the clinical course and the histological appearance.

Treatment has generally been by a radical excision followed by cauterization of the base as with the silver nitrate stick or fulguration by diathermy; clipping off merely invites a prompt return and is invariably accompanied by severe hæmorrhage. The tumour is not very radio-sensitive, but irradiation is sometimes effective.

Aksu, Abaci and Ugur. *Ege Univ. Tip. Fak. Mecmuasi*, **11**, 21 (1963).
Bivona. *Ann. Ottal.*, **53**, 258 (1925).
Bodin. *Ann. Derm. Syph.* (Paris), **3**, 289 (1902).
Freund. *Arch. Derm. Syph.* (Berl.), **166**, 669 (1932).
Friedman and Henkind. *Amer. J. Ophthal.*, **71**, 868 (1971).
Hagedoorn. *Brit. J. Ophthal.*, **18**, 561 (1934).
Huck and Meythaler. *Klin. Mbl. Augenheilk.*, **160**, 461 (1972).
Luchs. *Münch. med. Wschr.*, **67**, 1470 (1920).
Malik, Sood and Aurora. *Brit. J. Ophthal.*, **48**, 502 (1964).
Michelson. *Arch. Derm. Syph.* (Chic.), **12**, 492 (1925).
Mohan and Gupta. *Amer. J. Ophthal.*, **65**, 619 (1968).
Müller. *Hautarzt*, **13**, 120 (1962).
Nödl. *Z. Haut- u. Geschl.-Kr.*, **19**, 163 (1955).
Oehlschlaegel and Müller. *Arch. klin. exp. Derm.*, **218**, 126 (1964).
Peterson, Fusaro and Goltz. *Arch. Derm.*, **90**, 197 (1964).
Poncet and Dor. *Rev. Chir.* (Paris), **18**, 996 (1897).
Arch. gén. Méd., **3**, 129 (1900).
Schmelzer. *Klin. Mbl. Augenheilk.*, **91**, 470 (1933).
Schreiber. *Graefe-Saemisch Hb. d. ges. Augenheilk.*, 3rd ed., *Die Krankheiten d. Augenlider*, Berlin, 206 (1924).
Slem, Ilcayto and Turan. *Haseki Tip. Bült.*, **8**, 300 (1970).
Tisserand. *Bull. Soc. Anat. Paris*, **94**, 608 (1924).
Wescott. *J. Amer. med. Ass.*, **66**, 2067 (1916).
Wohl. *N.Y. med. J.*, **105**, 106 (1917).

ANGIOKERATOMA OF MIBELLI

This rare lesion, a combination of angiomatosis and hyperkeratosis wherein ectatic vessels and irregular blood-spaces occur in the dermis, has been noted in the eyelids (Offret and Dhermy, 1966). Treatment by diathermic destruction or cryotherapy may be undertaken for cosmetic reasons.

It should be noted that this lesion is not related to the *angiokeratoma corporis diffusum* of Fabry, a disturbance of glycolipid metabolism more properly called *hereditary dystopic lipidosis*[1]; in this condition, however, swellings and telangiectasia may appear on the lids (de Groot, 1964; Spaeth and Frost, 1965; Velzeboer and de Groot, 1971).

de Groot. *Dermatologica*, **128**, 321 (1964).
Offret and Dhermy. *Bull. Soc. Ophtal. Fr.*, **66**, 30 (1966).
Spaeth and Frost. *Arch. Ophthal.*, **74**, 760 (1965).
Velzeboer and de Groot. *Brit. J. Ophthal.*, **55**, 683 (1971).

MULTIPLE HÆMORRHAGIC SARCOMA: SARCOMATOSIS OF KAPOSI

This disease of obscure nature, originally described by the Hungarian dermatologist Moriz Kaposi (1872)[2] under the name of *idiopathic pigmented sarcoma*, is characterized by the appearance of multiple well-defined purplish areas of nodules in the skin which may persist unchanged, disappear or ulcerate (Fig. 542); hæmorrhages within the tumour are usual. New growths constantly appear (for example, on the upper and lower lids, the nose, neck, body and legs, Sezer and Erçikan, 1964), and the disease is slowly progressive until almost every organ in the body may be affected, and a fatal termination ensues in some 10 to 20% of cases (Cox and Helwig, 1959; Reynolds *et al.*, 1965). The great majority shows lesions first on the limbs, and as the

[1] Vol. VIII, p. 1080.
[2] Vol. VIII, Fig. 466.

FIGS. 542 to 544.—KAPOSI'S SARCOMA (N. Sezer and C. Erçikan).

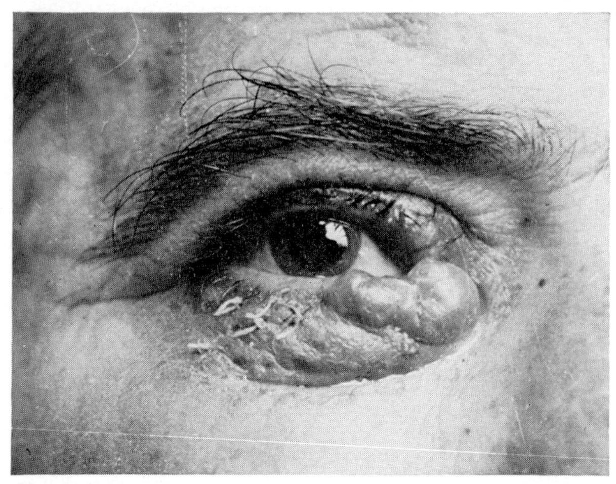

FIG. 542.—The lesion on the right lower lid of a man aged 53. There were multiple tumours on the face, trunk and legs.

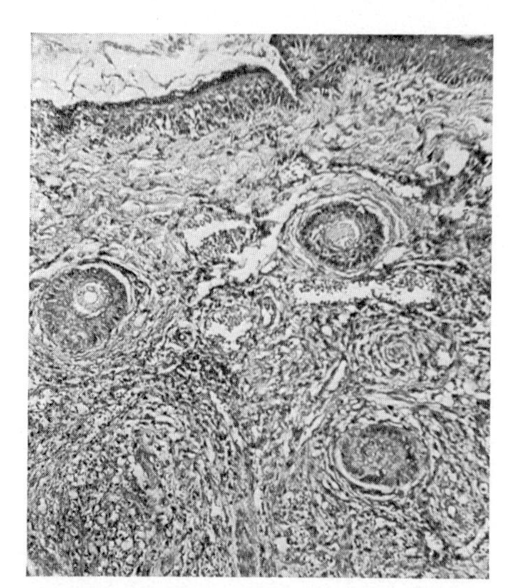

FIG. 543.—From the same case. There is hyperpigmentation of the cells of the basal layer. Sections of hair follicles are present in the dermis and around them areas of cellular infiltration giving a general angiomatous and fibroblastic appearance.

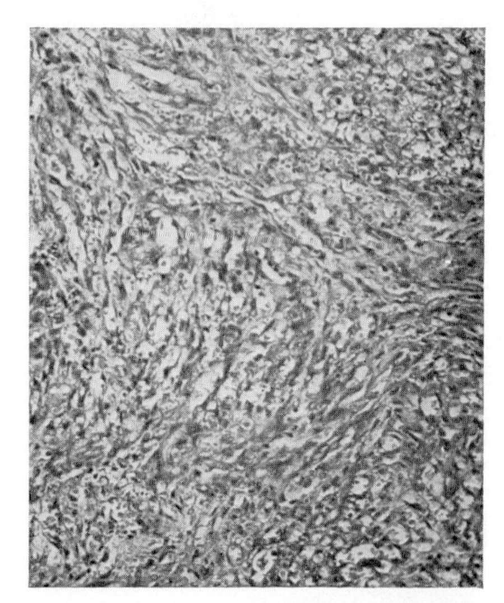

FIG. 544.—Reticulin stain reveals the presence of numerous reticulum fibres forming bands and networks in the areas of cellular infiltration.

condition becomes generalized, palpebral involvement occurs,[1] but a few cases have been reported in which the disease started on the face, nose and

[1] Tiscornia (1930), Kocsard (1949), Chilaris (1953), Appelmans *et al.* (1957), Lerman and Pinsky (1959), Cox and Helwig (1959), McLaren (1960), Boniuk (1962), Alexander (1963), Quéré *et al.* (1963), Cernea *et al.* (1968).

ears, and an initial lesion may appear on the lid (Graham, 1942). The condition is common on the lids of Bantu children. There has been some doubt as to the pathological nature of the condition. In a considerable number of cases it is associated with tumours of the lymphoid tissues such as Hodgkin's disease or mycosis fungoides and, indeed, some authors regard it as a disease of the reticulo-endothelial system. The lesion, however, is exceedingly vascular, often angiomatous (Dutz and Stout, 1960), and is now generally regarded as a neoplasic disease of the vascular system with multiple foci of origin arising from immature pluripotential vascular cells (Cox and Helwig, 1959). In the early stages the histological appearance may resemble granulation tissue, but later it may be either angiomatous if endothelial-cell proliferation predominates or fibroblastic if perithelial proliferation is profuse (Lever, 1967) (Figs. 543–4). *Treatment* is difficult although irradiation or cytotoxic drugs may sometimes prove helpful.

Alexander. *Amer. J. Ophthal.*, **55,** 625 (1963).
Appelmans, Michiels, Dehoux and van Hoonacker. *Bull. Soc. belge Ophtal.*, No. 117, 619 (1957).
Boniuk. *Int. Ophthal. Clin.*, **2,** 239 (1962).
Cernea, Balan and Brodicico. *Klin. Mbl. Augenheilk.*, **153,** 52 (1968).
Chilaris. *Arch. ophthal. Soc. N. Greece*, **2,** 145 (1953).
Cox and Helwig. *Cancer*, **12,** 289 (1959).
Dutz and Stout. *Cancer*, **13,** 684 (1960).
Graham. *Arch. Ophthal.*, **27,** 1188 (1942).
Kaposi. *Arch. Derm. Syph.* (Wien), **4,** 265 (1872).
Kocsard. *Dermatologica*, **99,** 43 (1949).
Lerman and Pinsky. *Arch. Ophthal.*, **62,** 320 (1959).
Lever. *Histopathology of the Skin*, 4th ed., Phila., 655 (1967).
McLaren. *Arch. Ophthal.*, **63,** 859 (1960).
Quéré, Basset and Camain. *Ophthalmologica*, **146,** 23 (1963).
Reynolds, Winkelmann and Soule. *Medicine* (Balt.), **44,** 419 (1965).
Sezer and Erçikan. *Brit. J. Ophthal.*, **48,** 223 (1964).
Tiscornia. *Arch. Oftal. B. Aires*, **5,** 392 (1930).

GLOMUS TUMOUR (GLOMANGIOMA)

A GLOMUS is a peculiar direct anastomosis between arterioles and venules thickened by the presence of " glomus epithelioid cells " and supplied with muscular and nervous elements, which probably shares in the control of the circulation of the blood and the peripheral regulation of temperature. These structures were first described as normal anatomical entities by Masson (1924–35) and they may form benign neuro-myo-arterial tumours, particularly in the fingers and toes. No glomus bodies have been described in the lids; but glomangiomata have been reported by Kirby (1941), Mortada (1963), Jensen (1965) and Robin and his colleagues (1968) (Fig. 545). The lesion is soft and bluish, either intracutaneous or subcutaneous, and usually tender or painful; clinically it may appear as a small nævus. Histologically a solitary glomus tumour is composed of vascular lumina lined by a single layer of flattened endothelial cells and by several layers of glomus cells with their eosinophilic cytoplasm and large oval nuclei resembling epithelioid cells, usually considered to be derived from pericytes, together with numerous nerve fibres, the whole being surrounded by a fibrous capsule (Figs. 546–8). Treatment is by complete and wide excision; in Kirby's (1941) case the tumour recurred twice after excision.

Jensen. *Arch. Ophthal.*, **73,** 511 (1965).
Kirby. *Arch. Ophthal.*, **25,** 228 (1941).
Masson. *Lyon Chir.*, **21,** 257 (1924).
Bull. Soc. franç. Derm. Syph., **42,** 1174 (1935).
Mortada. *Brit. J. Ophthal.*, **47,** 697 (1963).
Robin, Loubet and Chaput. *Bull. Soc. Ophtal. Fr.*, **68,** 885 (1968).

FIGS. 545 to 548.—GLOMANGIOMA OF THE LIDS.

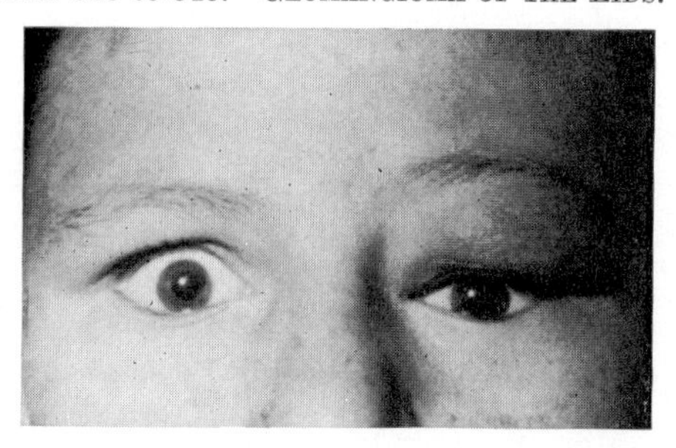

FIG. 545.—There is a swelling of the left upper lid resembling solid œdema in a girl aged 18 (A. Mortada).

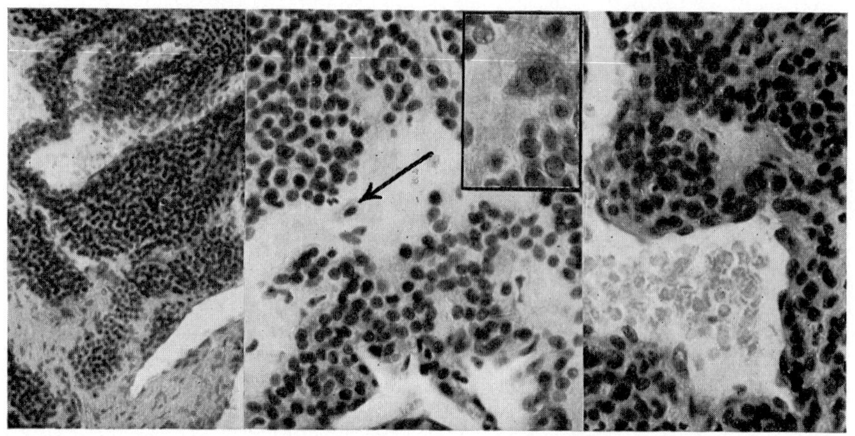

FIG. 546. FIG. 547. FIG. 548.

FIGS. 546 to 548.—Different areas of a glomus tumour showing the vascular lumina lined with endothelium and surrounded by uniform glomus cells with round nuclei. Fig. 546 (×150). Fig. 547. The arrow points to a mast cell (×350); the inset shows a mast cell (×600). Fig. 548 (×350) (O. E. Jensen).

LYMPHANGIOMA

LYMPHANGIOMATA, like hæmangiomata, are slowly progressive congenital tumours, and appear on the lids in two main forms.

SIMPLE CAPILLARY LYMPHANGIOMATA are rare superficial cutaneous neoformations, flat or slightly nodular on the surface, reddish in colour, and usually presenting several superficial vesicles. Histologically they are formed of masses of dilated lymphatics in the dermis. Occasionally they break down and ulcerate (Bishay, 1946).

CAVERNOUS LYMPHANGIOMATA, although also rare, are the more common variety.[1] They are subcutaneous tumours, usually occurring on the upper lid, often nodular, soft and translucent, and can be reduced by pressure.

[1] van Duyse (1899), Meyerhof (1902), Uhthoff (1916), von Hippel (1921), Bonnet and Colrat (1935), Jones (1959) 11 cases.

They tend to progress slowly during the growing period of life both superficially and deeply until the entire tissue of the lid may be invaded along with the conjunctiva, the orbit and the fronto-temporal region, sometimes producing a considerable deformity (*elephantiasis lymphangiectatica* or *mollis*) (van Duyse, 1899; Meyerhof, 1902; Hirschberg, 1906) (Fig. 549); a diffuse lymphangioma may also extend over the face and neck (Speciale-Picciché, 1924). Occasionally they do not become apparent until adult life (Jones, 1959). Histologically they show a spongy erectile structure, being made up of intercommunicating lymphatic spaces lined by flat endothelium bound together by a fibrous reticulum. The tumour may resemble a hæmangioma if a hæmorrhage occurs.

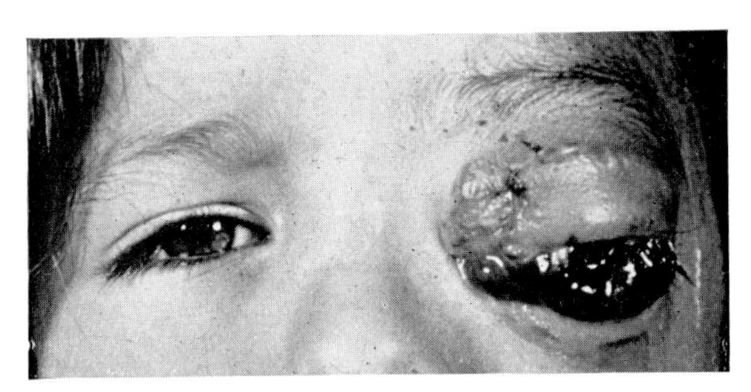

FIG. 549.—LYMPHANGIOMA.

In a child aged 5 years; the tumour involves the conjunctiva, the orbit and the lids and had been present since birth (I. S. Jones).

Lymphangioma must be differentiated from lymphangiectasia resulting from obstruction of the lymphatic vessels.[1]

Treatment is the same as for hæmangiomata.

A LYMPHANGIO-ENDOTHELIOMA is a rare occurrence in the lids (Bertoncini, 1961).

Bertoncini. *Arch. Ottal.*, **65**, 391 (1961).
Bishay. *Bull. ophthal. Soc. Egypt*, **39**, 47 (1946).
Bonnet and Colrat. *Arch. Ophtal.*, **52**, 683 (1935).
van Duyse. *Arch. Ophtal.*, **19**, 273 (1899). *Ann. Oculist.* (Paris), **102**, 157 (1899).
von Hippel. *Klin. Mbl. Augenheilk.*, **66**, 297 (1921).
Hirschberg. *Zbl. prakt. Augenheilk.*, **30**, 2 (1906).
Jones. *Trans. Amer. ophthal. Soc.*, **57**, 602 (1959).
Meyerhof. *Klin. Mbl. Augenheilk.*, **40** (1), 300 (1902).
Speciale-Picciché. *Ann. Ottal.*, **52**, 774 (1924).
Uhthoff. *Klin. Mbl. Augenheilk.*, **57**, 8 (1916)

Nervous Tissue Tumours

NEUROFIBROMATOSIS

The disease generally known as neurofibromatosis has given rise to a considerable amount of controversy and has been designated by several names. The first case of a plexiform neuroma involving the periorbital area was described and illustrated by Mott (1854) (Fig. 550), a condition he called

[1] p. 19.

pachydermatocele, while a typical case was presented by Billroth and Czerny (1869). von Recklinghausen, however, made the first elaborate study of the disease in 1882 and his name is usually associated with it. Since then the literature has grown enormously.

VON RECKLINGHAUSEN'S DISEASE, together with Bourneville's disease, von Hippel-Lindau's disease and the Sturge-Weber syndrome, forms one of the group of *phakomatoses* (φακός, a birth-mark) classed together by van der Hoeve (1923–32)[1] because of their congenital origin, familial incidence, their cutaneous manifestations and the widespread distribution of the

FIG. 550.—NEUROFIBROMATOSIS.

The classical illustration of V. Mott (1854) who called the condition pachydermatocele. It was noticed shortly after birth as a pimple which gradually grew to involve the scalp, the forehead, one side of the nose, the eyelids from which the tumour extended down to the lower part of the face, and the upper and lower lips. When the tumour was raised the eye appeared to be sound at the bottom of a canal 4 inches in depth. Several operations were performed but the condition recurred.

neoplasic formations by which they are characterized. von Recklinghausen's disease[2] (*neurofibromatosis*), originally described (1882) as a combination of neuromata and multiple fibromata arising from cutaneous nerve filaments, exhibits neurofibromatous lesions in the peripheral and sympathetic nerves and the central nervous system, with which are related anomalies of pigmentation (café-au-lait or coffee-coloured patches), hypertrophy of the skin and subcutaneous tissues (*elephantiasis neuromatosa, neuromatodes* or *nervorum*) and sometimes endocrine or epileptic disorders. In the lids, manifestations of the disease are relatively rare, but those that do occur may be dramatic.

[1] Vol. III, p. 1120; Vol. X, p. 736. [2] Vol. X, p. 761.

Three types of change are classically described—plexiform neurofibroma, diffuse neurofibromatosis in any degree up to facial hemi-hypertrophy, and molluscum fibrosum—but despite the great diversity of their clinical appearances, whether they be multiple, diffuse or solitary, the changes are essentially similar. It is now generally accepted that neurofibromata represent a developmental defect of the neuro-ectodermal tissues which is dominantly inherited (five generations, Frank, 1947). The peripheral nerve trunks are involved, sometimes in great numbers in the body, the main changes occurring in the nerve sheaths, although the development of other parts of the nervous system or its envelopes may also be disturbed. As a rule, the only evidences of the disease seen in childhood are the pigmented spots; the characteristic tumours may be present then but usually develop later, although a massive tumour involving the orbit may be present at birth (Moore, 1962); they may grow slowly and progressively, particularly during puberty or pregnancy, or may remain stationary. Active growth may resume at any time in adult life, and a fatal outcome may rarely result from extension of a lesion into the spinal cord or the development of malignant transformation with metastases (Reese, 1963).

The tumours, which are usually multiple, are not circumscribed or encapsulated but involve the nerves for considerable distances and merge into the surrounding dermal tissues. Their origin has been disputed; some authors held the classical view of von Recklinghausen (1882) that the neurofibromatous growths are mesodermal in origin, fibroblastic derivatives of the perineurium and endoneurium (Penfield, 1927; Bailey and Herrmann, 1938); others suggested that they are derived from the cells of the sheath of Schwann, an ectodermal derivative of the neural crest (Verocay, 1910; Stewart and Copeland, 1931; Masson, 1932), while others again believed that both elements play a part (del Rio Hortega, 1934). Pathologically these tumours are characterized by an irregular overgrowth of schwannian cells associated with an increase of reticulin and collagen intimately penetrated by nerve fibres either singly or in bundles; the predominant role of the schwannian cell in neurofibromata with fibroblasts playing a secondary role has been amply confirmed by electron microscopy (Girard *et al.*, 1962; Russell and Rubinstein, 1971).

PLEXIFORM NEUROFIBROMA, representing multiple involvements of the peripheral nerves, is the most characteristic palpebral manifestation of the disease, the lesion being present at birth and growing slowly thereafter, sometimes increasing rapidly at puberty.

The site is most commonly within the distribution of the trigeminal and cervical nerves. In 58 cases the distribution was—upper lid, temple, forehead, 18 cases; posterior part of the neck and behind the ear, 14 cases; nose and cheek, 4; lower jaw and front of the neck, 5; breast and back, 8; extremities, 9 cases. The nerves most commonly involved are thus the supra-orbital branches of the fifth; but the three divisions of this nerve may be implicated (Marin Amat, 1931). Many cases have shown typical involvement of the lid, the more recent of which are appended.[1]

The tumour on the upper lid may be relatively small at first (Figs. 551–2), or may be of considerable size so that the upper lid becomes thickened, flabby and pendulous, and sometimes so large that the eye can be opened only with difficulty and with the aid of the finger (Fig. 553).

Figs. 551 to 553.—Plexiform Neurofibromata.

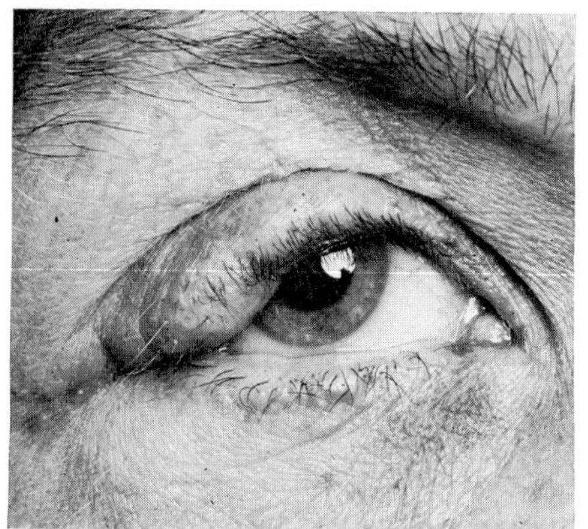
Fig. 551.—Confined to the lid-margin (S. P. Meadows).

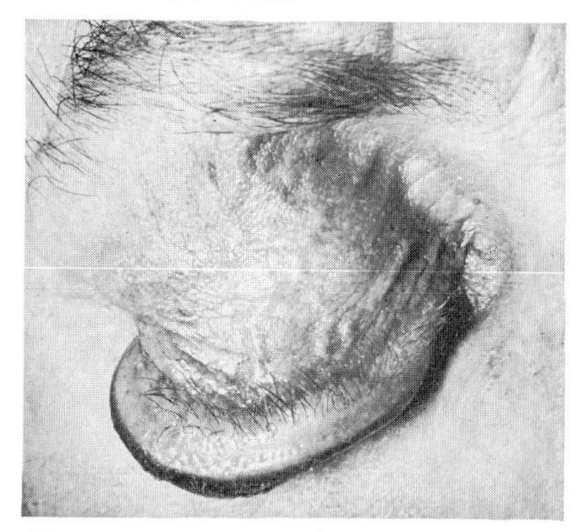
Fig. 552.—Affecting the upper lid which has been turned outwards (Inst. Ophthal.).

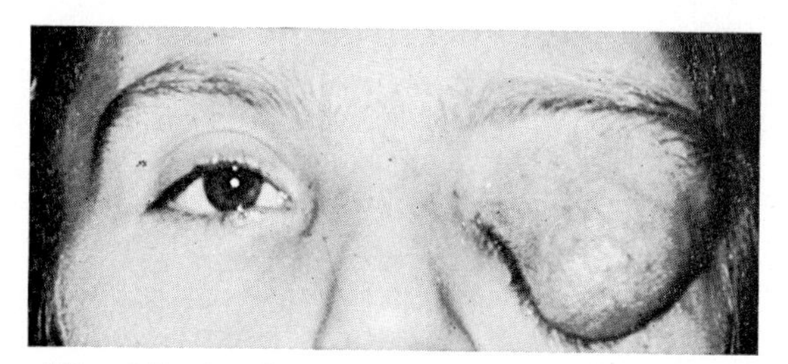
Fig. 553.—Affecting the upper lid which is pendulous (B. Jay).

The overlying skin is usually normal but may be pigmented, and is movable over the mass which itself is deeply fixed; sometimes the skin is hyper-elastic or thickened and coarsened. The mass is soft, and on palpation feels as if it were made up of knotted cords inextricably twisted together; occasionally such manipulation is tender. As a rule the swelling is not limited to the upper lid itself, but spreads over the fronto-temporal region, while it frequently penetrates into the orbit causing proptosis, enlargement of the bony orbit and hyperostoses or rarefactions of its wall including the

[1] Leo (1950), Charleux (1957), Larmande *et al.* (1957–61), Betetto (1959), Triandaf (1960), Moore (1962), Shoukry (1962), Girard *et al.* (1962), Reese (1963), Ehlers (1966), Rodriguez and Berthrong (1966), Walsh and Hoyt (1969), Smith and English (1970), Majima and Shirai (1972).

sphenoid and sella (Burrows, 1963; Turpin *et al.*, 1963). Increased vascularization may be present and lead to confusion with a vascular tumour (Walsh and Hoyt, 1969).

A DIFFUSE NEUROFIBROMATOSIS of this type may affect in varying degree the entire side of the face producing a condition of partial or complete *facial hemi-hypertrophy*, with its typical picture of congenital *elephantiasis neuromatodes* wherein both the upper and lower lids are involved as well as all the soft tissues of the affected side (nose, mouth, cheek, etc.)[1] (Fig. 550).

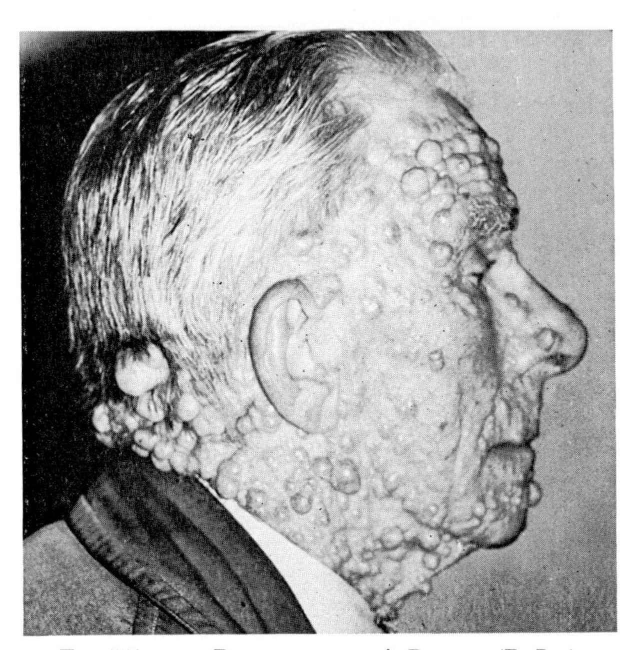

FIG. 554.—VON RECKLINGHAUSEN'S DISEASE (B. Jay).

MOLLUSCUM FIBROSUM (FIBROMA MOLLUSCUM) may appear in association with these manifestations, occasionally in great numbers as soft cutaneous tumours (Figs. 554). These may develop all over the body including the lids; in an unusual case wherein they occurred in multitudes universally (for example, between 25 and 40 per sq. inch on the back, varying in diameter from 1 to 15 mm.), they appeared on the ciliary margins of the lids causing a recurrent trichiasis (Wallach, 1948). Occasionally, as in the other types, they become pedunculated and may eventually form enormous pendulous folds which may on occasion hang far over the cheek; over them the skin is usually hypertrophied and corrugated. Gross deformities may be encoun-

[1] Sachsalber (1898), Goldzieher (1898), Snell and Collins (1903), Collins and Batten (1905), Sutherland and Mayou (1907), Schreiber (1924), Charleux (1957), Larmande *et al.* (1957–61), François and Katz (1961), Darabos *et al.* (1963), Blatt (1966).

tered wherein the pendulous upper lid may fall below the level of the jaw.[1] A lipoma may also occur with the neural lesions (Reese, 1963).

Associated with the condition of the lid, the eyeball may be affected: the uveal tract[2] frequently with buphthalmos or glaucoma,[3] the retina[4], the optic nerve,[5] the orbit[6] and the conjunctiva.[7]

Histopathologically such tumours are fundamentally of the same nature. A plexiform neuroma consists essentially of thickened, nodular vermiform cords held together in a compact mass by loose connective tissue which can readily be teased out and is highly vascularized; it is mainly subcutaneous but the lobulated mass penetrates the orbicularis muscle and the deeper structures of the lid. Early tumours contain numerous myelinated and non-myelinated nerve fibres of which some appear normal but the majority are irregularly swollen. Older tumours show a proliferation of the sheath of Schwann and of the endoneurium concomitant with the degeneration and disappearance of the nerve fibres (Lever, 1967). Electron microscopy has shown (in plexiform neuromata), apart from an increase in collagen fibres, a preponderance of schwannian cells and fibroblasts (Girard *et al.*, 1962). The inclusion of bundles of neurofibromatous nerves frequently determines well-defined patterns of whorls and fasciculi in the mass of growth (Figs. 555–6). In diffuse neurofibromatosis there is in addition a generalized massive hypertrophy of the skin and subcutaneous tissue, and throughout the latter course innumerable enlarged nerves, the nerve trunks being encased in thickened endoneurium. Molluscum fibrosum superficially resembles a soft fibroma in its histological picture, but careful examination demonstrates the presence of nerve fibrillæ; in this case the growth is probably derived from the perineurium without the participation of the nerve elements.

No known *treatment* is of any value in this disease, but if the neoplasic or hyperplasic masses become too large or unsightly, surgical removal can be undertaken which may necessitate considerable plastic reconstruction. Dissection may be very difficult if the neuromatous tissue penetrates deeply into the orbit, and if removal is not complete, recurrences are to be expected. The vascularity of the tumours may add to the technical difficulty: in infants it has proved fatal. Sarcomatous degeneration has been reported in such tumours, but it is exceptional.

In the *multiple mucosal neuroma syndrome* numerous small neuromatous nodules like grains of rice occur on the eyelids, lips and tongue associated with carcinoma of the thyroid and phæochromocytoma (Williams and Pollock, 1966; Offret and Haye, 1971). A variant of this syndrome, which consisted of highly visible corneal nerves,

[1] Horner (1871), de Vincentiis (1897), Seifert (1901), Siegrist (1905), Marx (1908), Wagenmann (1922), Schreiber (1924), and others.

[2] Vol. IX, p. 823.

[3] Vol. XI, p. 636.

[4] Vol. X, p. 761.

[5] Vol. XII, p. 248.

[6] p. 1108.

[7] Vol. VIII, p. 1239.

FIGS. 555 and 556.—NEUROFIBROMA OF THE EYELID.

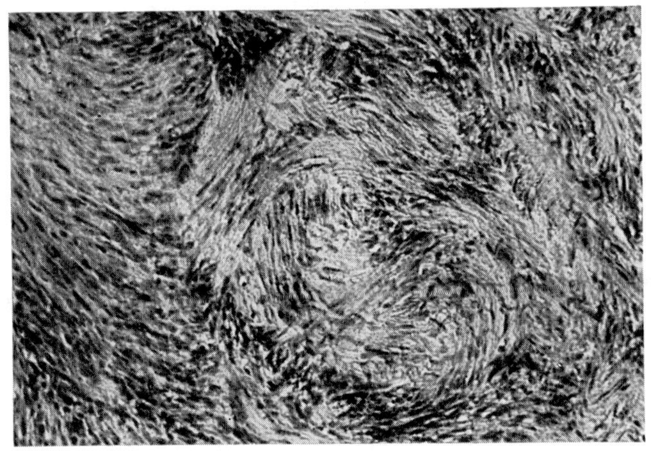

FIG. 555.—Showing interlacing bundles and whorls of spindle cells (N. Ashton).

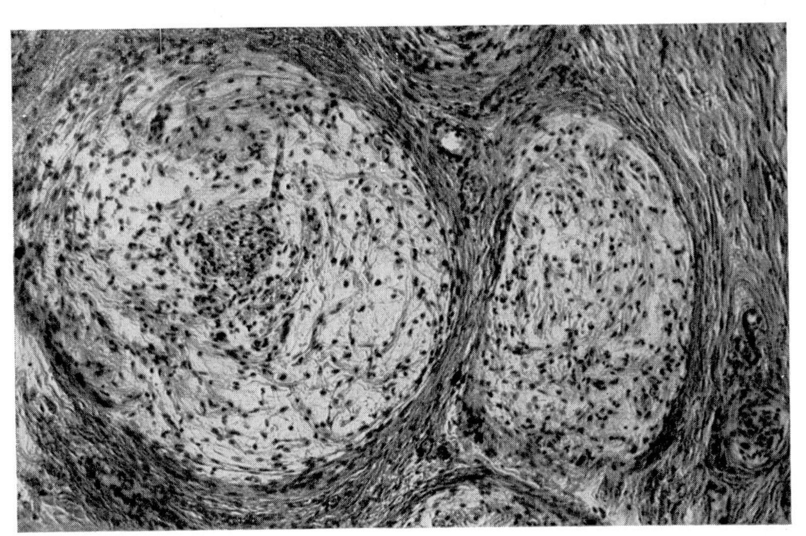

FIG. 556.—Showing great thickening of the nerves due to a hyperplasia of the endo- and peri-neurium (H. & E.; ×100) (N. Ashton).

thickened eyelids, an ectopic lacrimal punctum, lacrimal hyposecretion, phæochromocytoma, medullary carcinoma of the thyroid, multiple mucosal neuroma, intestinal ganglioneuromatosis and a marfanoid habitus, was described by Baum and Adler (1972) who cited 15 other cases from the literature with somewhat similar clinical manifestations.

NEURILEMMOMA

Isolated tumours of the nerve sheaths (NEURILEMMOMA, SCHWANNOMA) may also occur in the lids as a rarity (Mishra and Sharan, 1960; Reeh, 1963; Row *et al.*, 1967; Slem *et al.*, 1971) (Fig. 557); they are circumscribed, en-

FIGS. 557 and 558.—BENIGN NEURILEMMOMA
(M. J. Reeh, *Treatment of Lid and Epibulbar Tumors*, C. Thomas, Springfield).

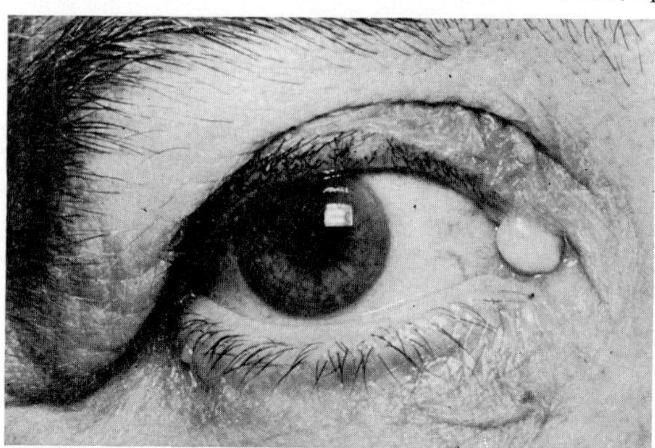

FIG. 557.—A firm white sessile nodule arises from the lid-margin near the inner canthus.

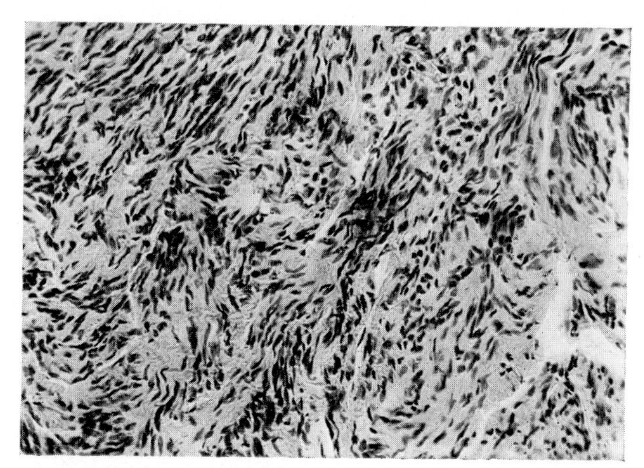

FIG. 558.—The tumour is composed of elongated cells suggesting an origin from schwannian cells.

capsulated usually benign tumours in the dermis or the subcutaneous tissue. Histologically two patterns may be seen : in some, nuclei are characteristically arranged in a palisading fashion with a fibrillar matrix and hyaline foci (Verocay bodies) (Fig. 558); in others a haphazard arrangement of the cells occurs with mucoid degeneration of the connective tissue and small cystic spaces. Few neurons are present and the growth represents primarily a proliferation of schwannian cells (Hogan and Zimmerman, 1962). A malignant schwannoma of the lids is very rare (Ahluwalia and Prem, 1972). Multiple tumours of this type may also occur in von Recklinghausen's disease. This lesion will be discussed more fully with tumours of the orbit.[1]

[1] p. 1105.

GRANULAR-CELL SCHWANNOMA (" MYOBLASTOMA ") OF ABRIKOSSOFF

This tumour, originally described by Abrikossoff (1926), was initially thought to be composed of immature striated muscle cells and was therefore named *granular-cell myoblastoma*. The cells are large and polyhedral with a pale cytoplasm filled with slightly eosinophilic granules and a small central nucleus and there is a slender reticulin and fibrous framework (Figs. 561–2). From histological and electron-microscopic studies, however, Fisher and Wechsler (1962) decided that it was of neural origin and called it a *schwannoma*. This view has subsequently been supported and it has been claimed that the cytoplasmic granules and other structures of the cells resemble alterations in schwannian cells seen with wallerian degeneration (Haisken and Langer, 1962; Sobel and Churg, 1964; Moscovic and Azar, 1967; Garancis *et al.*, 1970; Sobel *et al.*, 1971); moreover, the tumour may appear in tissues devoid of striated muscle such as the pituitary gland (Symon *et al.*, 1971; Waller *et al.*, 1972). On the other hand, the opinion has been expressed that the original theory is correct and that the tumour arises from undifferentiated mesenchymal cells with muscular affinities (Aparicio and Lumsden, 1969). The subject is still controversial but a neurogenic origin seems most likely on the present evidence.

The tumour is circumscribed, firm and not tender, usually solitary, and occurs most frequently in the tongue, skin and subcutaneous tissues, generally between the ages of 30 and 50 years, but has been found in virtually every tissue in the body (Strong *et al.*, 1970). In the region of the ocular adnexa it is rare (Figs. 559–60); a case occurring near the lacrimal sac was described by von Bahr (1938), and cases in the lids by Cristini (1946) and Friedman and his colleagues (1973) in an instance in which the margin of the upper lid and tarsus were implicated. Involvement of the upper lid and eyebrow was reported by Blodi (1956), Timm and Timmel (1966), and Dhermy and his colleagues (1966), and the lids and orbit by Dunnington (1948); it may also be localized in the orbit.[1] A unique case was reported by Cunha and Lobo (1966) wherein the tumour developed in the ciliary body. It is typically benign; a malignant form, however, which is exceptionally rare, may occur giving rise to extensive metastases (Dunnington, 1947–48; Ross *et al.*, 1952) (Fig. 560). A case was described by Powell (1946) of a primary tumour in the ovary involving the retroperitoneal space which gave rise to metastases including the upper lid. Treatment is by wide excision.

GANGLIONEUROMA. This is a very rare benign tumour which has occasionally originated in association with peripheral nerves. It forms a firm, encapsulated mass containing glial tissue and well-formed ganglion cells probably of sympathetic origin surrounded by a delicate reticulum. The tumours are usually found in the brain of children and young adults, but they may appear in the eye and have been found as

[1] p. 1121.

Figs. 559 to 562.—Granular-cell Schwannoma.

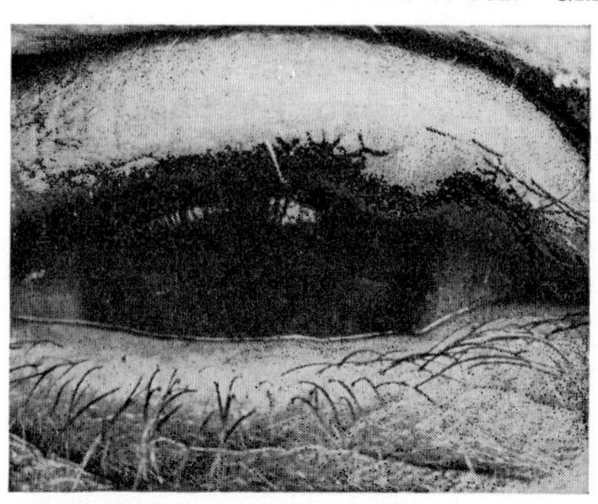

Fig. 559.—A recurrent tumour of the upper lid of the benign type in a man aged 45. The tumour was removed by simple excision but recurred after one month; after a wide wedge-shaped resection there had been no recurrence for a follow-up period of over one year (Z. Friedman *et al.*).

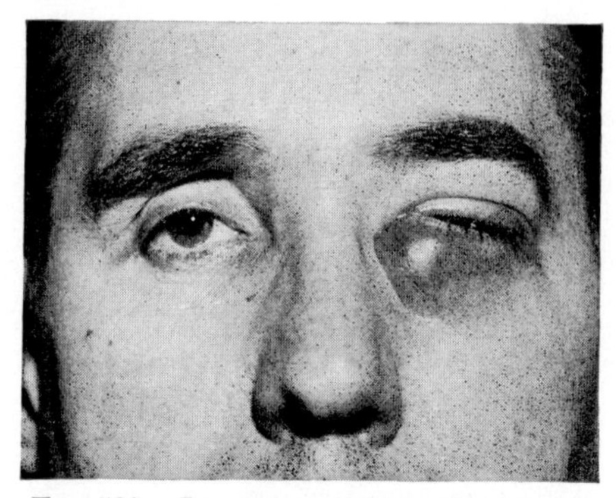

Fig. 560.—In a man aged 40. The tumour in the left lower lid was of one month's duration. It was of the malignant type and despite exenteration of the orbit and subsequent radiational treatment, general metastases ensued (J. H. Dunnington).

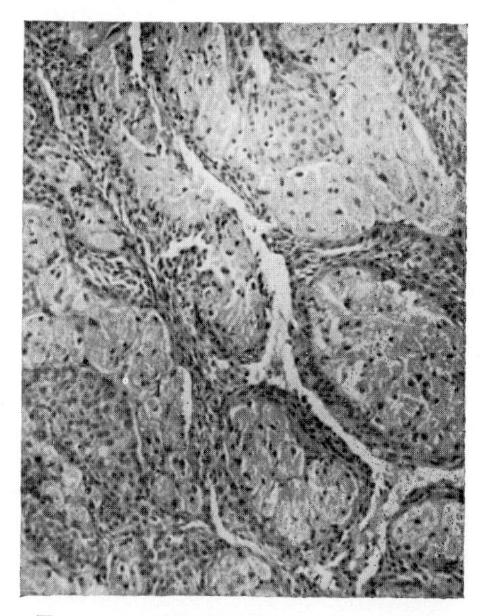

Fig. 561.—The histology of the tumour seen in Fig. 559, showing groups of large polyhedral cells with small hyperchromatic nuclei and eosinophilic granules in the cytoplasm (H. & E.; ×120) (Z. Friedman *et al.*).

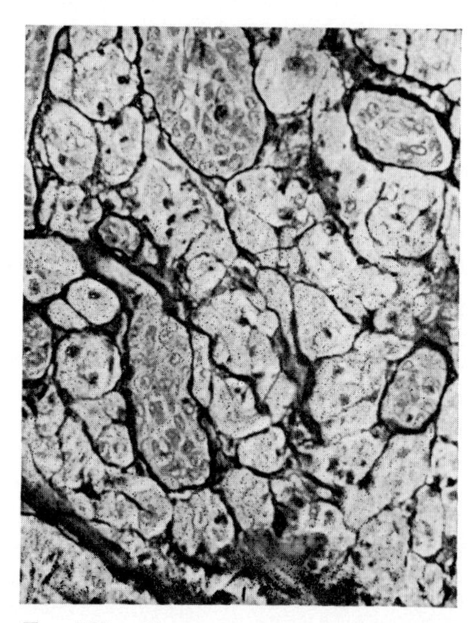

Fig. 562.—The same case, stained to show the reticulin and collagen fibrous network within the tumour (×225) (Z. Friedman *et al.*).

solitary lesions on the face (Zak, 1950) and an example was described in the upper lid by Krauss (1912).

amputation neuroma. This is not a true neoplasm but represents a regenerative overgrowth of axons at the proximal end of a severed nerve into a mass of connective tissue formed by the nerve sheaths; it is occasionally seen as a late complication following surgical procedures or injuries on the eyelids (Boniuk, 1962).

Abrikossoff. *Virchows Arch. path. Anat.*, **260,** 215 (1926).
Ahluwalia and Prem. *Orient. Arch. Ophthal.*, **10,** 35 (1972).
Aparicio and Lumsden. *J. Path.*, **97,** 339 (1969).
von Bahr. *Acta ophthal.* (Kbh.), **16,** 109 (1938).
Bailey and Hermann. *Amer. J. Path.*, **14,** 1 (1938).
Baum and Adler. *Arch. Ophthal.*, **87,** 574 (1972).
Betetto. *Riv. Oto-neuro-oftal.*, **34,** 201 (1959).
Billroth and Czerny. *Arch. klin. Chir.*, **11,** 230 (1869).
Blatt. *2nd Cong. europ. ophthal. Soc.* (Wien, 1964) (ed. François), Basel, 921 (1966).
Blodi. *Arch. Ophthal.*, **56,** 698 (1956).
Boniuk. *Int. Ophthal. Clin.*, **2,** 239 (1962).
Burrows. *Brit. J. Radiol.*, **36,** 549 (1963).
Charleux. *Les manifestations palpébrales et orbitaires de la neurofibromatose de Recklinghausen* (Thèse), Lyon (1957).
Collins and Batten. *Trans. ophthal. Soc. U.K.*, **25,** 248 (1905).
Cristini. *Rass. ital. Ottal.*, **15,** 207 (1946).
Cunha and Lobo. *Brit. J. Ophthal.*, **50,** 99 (1966).
Darabos, Valu and Zajácz. *Klin. Mbl. Augenheilk.*, **142,** 303 (1963).
Dhermy, Morax and Jolivet. *Ann. Oculist.* (Paris), **199,** 1025 (1966).
Dunnington. *Trans. Amer. ophthal. Soc.*, **45,** 93 (1947).
Arch. Ophthal., **40,** 14 (1948).
Ehlers. *Ophthalmologica*, **151,** 284 (1966).
Fisher and Wechsler. *Cancer*, **15,** 936 (1962).
François and Katz. *Ophthalmologica*, **142,** 549 (1961).
Frank. *Arch. Derm. Syph.* (Chic.), **55,** 109 (1947).
Friedman, Eden and Neumann. *Brit. J. Ophthal.*, **57,** 757 (1973).
Garancis, Komorowski and Kuzma. *Cancer*, **25,** 542 (1970).
Girard, Freeman and Makk. *Trans. Amer. Acad. Ophthal.*, **66,** 242 (1962).
Goldzieher. *Zbl. prakt. Augenheilk.*, **22,** 174 (1898).
Haisken and Langer. *Frankfurt. Z. Path.*, **71,** 600 (1962).
van der Hoeve. *v. Graefes Arch. Ophthal.*, **111,** 1 (1923).
Trans. ophthal. Soc. U.K., **43,** 534 (1923); **52,** 380 (1932).
Hogan and Zimmerman. *Ophthalmic Pathology*, 2nd ed., Phila., 207 (1962).
Horner. *Klin. Mbl. Augenheilk.*, **9,** 1 (1871).
Krauss. *Z. Augenheilk.*, **28,** 110 (1912).
Larmande, Margaillan and Giudici. *Bull. Soc. franç. Ophtal.*, **74,** 734 (1961).
Larmande, Timsit and Thomas. *Pédiatrie*, **12,** 682 (1957).
Leo. *Atti Soc. oftal. ital.*, **12,** 218 (1950).
Lever. *Histopathology of the Skin*, 4th ed., Phila., 685 (1967).
Majima and Shirai. *Folia ophthal. jap.*, **23,** 426 (1972).
Marin Amat. *Arch. Ophtal.*, **48,** 509 (1931).
Marx. *Z. Augenheilk.*, **19,** 528 (1908).
Masson. *Amer. J. Path.*, **8,** 367, 389 (1932).
Mishra and Sharan. *Brit. J. Ophthal.*, **44,** 252 (1960).
Moore. *Brit. J. Ophthal.*, **46,** 682 (1962).
Moscovic and Azar. *Cancer*, **20,** 2032 (1967).
Mott. *Med.-chir. Trans.* (Lond.), **37,** 155 (1854).
Offret and Haye. *Tumeurs de l'oeil et des annexes oculaires*, Paris (1971).
Penfield. *Surg. Gyn. Obstet.*, **45,** 178 (1927).
Powell. *Arch. Path.*, **42,** 517 (1946).
von Recklinghausen. *Ueber die multiplen Fibrome der Haut u. ihre Beziehung zu den multiplen Neuromen*, Berlin (1882).
Reeh. *Treatment of Lid and Epibulbar Tumors*, Springfield, 255 (1963).
Reese. *Tumors of the Eye*, 2nd ed., N.Y., 190 (1963).
del Rio Hortega. *Anat. micros. d. los tumores d. sistema nerviosa* (*Int. Cancer Cong., 1933*), Madrid (1934).
Rodriguez and Berthrong. *Arch. Neurol.*, **14,** 467 (1966).
Ross, Miller and Foote. *Cancer*, **5,** 112 (1952).
Row, Satyendran, Singha and Sharma. *Orient. Arch. Ophthal.*, **5,** 61 (1967).
Russell and Rubinstein. *Pathology of Tumours of the Nervous System*, 3rd ed., London, 294 (1971).
Sachsalber. *Beitr. Augenheilk.*, **3** (27), 523 (1898).
Schreiber. *Graefe-Saemisch Hb. d. ges. Augenheilk.*, 3rd ed.: *Die Krankheiten d. Augenlider*, Berlin, 211 (1924).
Seifert. *Münch. med. Wschr.*, **48,** 1197 (1901).
Shoukry. *Bull. ophthal. Soc. Egypt*, **55,** 259 (1962).
Siegrist. *Ber. dtsch. ophthal. Ges.*, **32,** 360 (1905).
Slem, Ayan and Ercan. *Türk. oftal. Gaz.*, **1,** 215 (1971).
Smith and English. *Brit. J. Ophthal.*, **54,** 134 (1970).
Snell and Collins. *Trans. ophthal. Soc. U.K.*, **23,** 157 (1903).
Sobel and Churg. *Arch. Path.*, **77,** 132 (1964).
Sobel, Marquet, Avrin and Schwarz. *Amer. J. Path.*, **65,** 59 (1971).
Stewart and Copeland. *Amer. J. Cancer*, **15,** 1235 (1931).
Strong, McDivitt and Brasfield. *Cancer*, **25,** 415 (1970).

Sutherland and Mayou. *Trans. ophthal. Soc. U.K.*, **27,** 179 (1907).
Symon, Ganz and Burston. *J. Neurosurg.*, **35,** 82 (1971).
Timm and Timmel. *Klin. Mbl. Augenheilk.*, **148,** 665 (1966).
Triandaf. *Oftalmologia* (Buc.), **4,** 141 (1960).
Turpin, Saraux, Caille *et al. Ann. Oculist.* (Paris), **196,** 776 (1963).
Verocay. *Beitr. path. Anat.*, **48,** 1 (1910).
de Vincentiis. *Lav. clin. ocul. Univ. Napoli,* **5,** 41, 65 (1897).
Wagenmann. *Ber. dtsch. ophthal. Ges.*, **43,** 282 (1922).
Wallach. *Amer. J. Ophthal.*, **31,** 1487 (1948).
Waller, Riley and Sundt. *Arch. Ophthal.*, **88,** 269 (1972).
Walsh and Hoyt. *Clinical Neuro-ophthalmology*, 3rd ed., Baltimore (1969).
Williams and Pollock. *J. Path. Bact.*, **91,** 71 (1966).
Zak. *Brit. J. Derm.*, **62,** 351 (1950).

Pigmented Tumours

Elsewhere in this *System* the nature and origin of melanin and pigmented cells have already been discussed,[1] and the pathological changes arising in these cells have been described in the uveal tract[2] and the conjunctiva.[3] There are no basic differences in the histological nature of pigmented lesions of the skin and conjunctiva. In previous Chapters we have also described the benign forms of epithelial melanosis, both localized (ephelides and lentigo) and diffuse (chloasma); the congenital subepithelial benign pigmentation of the periorbital region associated with melanosis of the eye, known as the nævus of Ota, has been discussed in a previous Volume.[4] Here we shall confine ourselves to the melanotic neoplasms.

It will be remembered that melanotic tumours contain cells termed *melanocytes* which are capable of forming melanin and are constant components of the basal layer of the epidermis. It is now generally agreed that these cells arise from the schwannian sheaths of dermal nerves and are neuro-ectodermal in origin, derived ultimately from the neural crests.[5] It is established that the junctional zone of cellular unrest in the basal layer of the epidermis at the dermo-epidermal junction gives rise to pigmented lesions. It must be remembered that the amount or even the presence of pigment is without clinical or prognostic significance; its presence is merely an expression of metabolic activity and although it adds drama to the clinical picture and greatly aids the problems of diagnosis, it is without fundamental importance.

NÆVUS

PIGMENTED NÆVI (*moles*; *birth-marks*) are small pigmented spots in the skin; they are to be distinguished from vascular nævi (angiomata) in that although probably present at birth they are not usually apparent until childhood or adolescence, while those that are first noticed in adult life usually show histological evidence of having been present for some considerable time (Jay, 1964–65). They have a characteristic pattern of growth with a period of active development in infancy followed by a stage of slow

[1] Vol. III, p. 273. [2] Vol. IX, p. 830.
[3] Vol. VIII, p. 1210. [4] Vol. III, p. 798.
[5] Vol. III, p. 276.

FIGS. 563 to 565.—NÆVI OF THE LIDS.

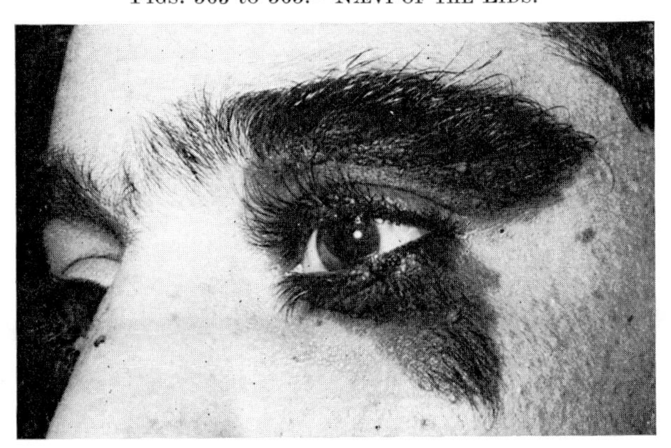

FIG. 563.—A boy aged 15 with a deeply pigmented divided hairy nævus on the upper and lower lids. It was removed surgically for cosmetic reasons, followed by a split-skin graft (B. Hirshowitz and D. Mahler).

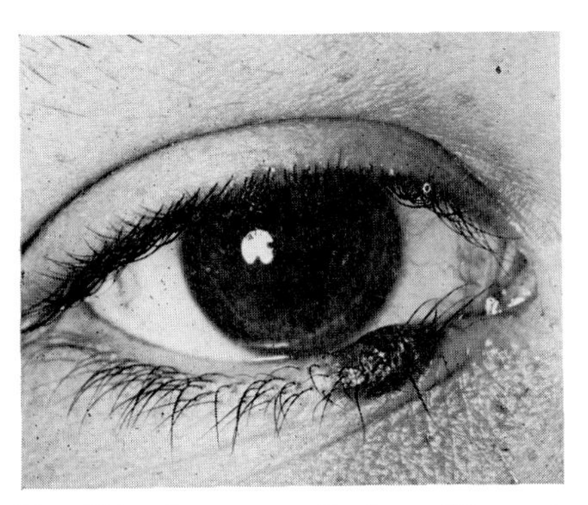

FIG. 564.—A nævus on the lower lid-margin in a boy aged 14 (Inst. Ophthal.).

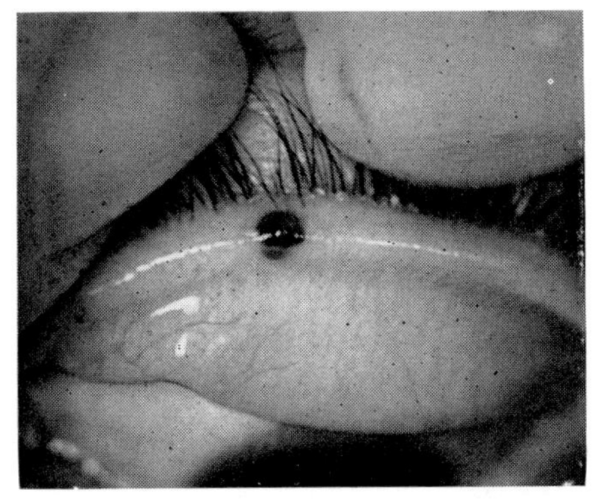

FIG. 565.—A small nævus on the under surface of the upper lid near the margin, previously unnoticed; there was an increase in the pigmentation during the menopause (M. J. Reeh and K. C. Swan).

growth and then a period of quiescence in early adult life when they remain stationary (Cook, 1951; Cade, 1961; Reese, 1963; Levene, 1972). This is usually followed by a final stage of atrophy but occasionally a sudden stimulus to rapid growth occurs in which the tumour acquires the capacity to become malignant (Fig. 579). Although the assumption of malignant characteristics is rare, it is clinically impossible to predict so that every nævus should be looked upon as potentially dangerous. Such a transition from the benign to the malignant form is marked clinically by an increase in the extent of the lesion, its elevation and deepening pigmentation, and the increasing vascularity of the surrounding tissues. At the same time to preserve perspective it must be remembered that nævi are common and malignant melanomata are the least common of palpebral tumours.

As a rule nævi vary in size from a pinhead to a diameter of 2 cm., but occasionally they may cover wide areas. They vary in colour from light brown to black depending on the amount of pigment they contain and the depth at which they lie. They may also vary in form, being sometimes flat and flush with the skin, sometimes slightly raised, dome-shaped or tuberous,

FIGS. 566 to 569.—DIVIDED NÆVI.

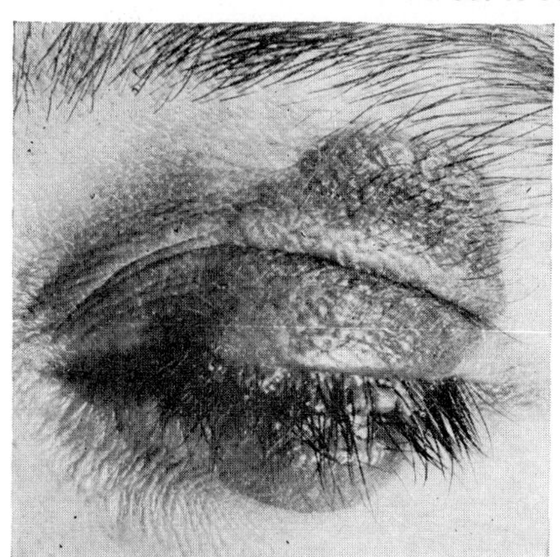

FIG. 566.

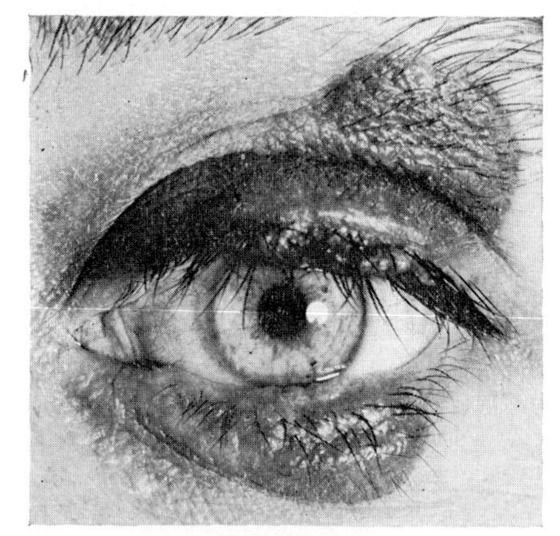

FIG. 567.

FIGS. 566 and 567.—Occurring in a woman aged 27 (Inst. Ophthal.).

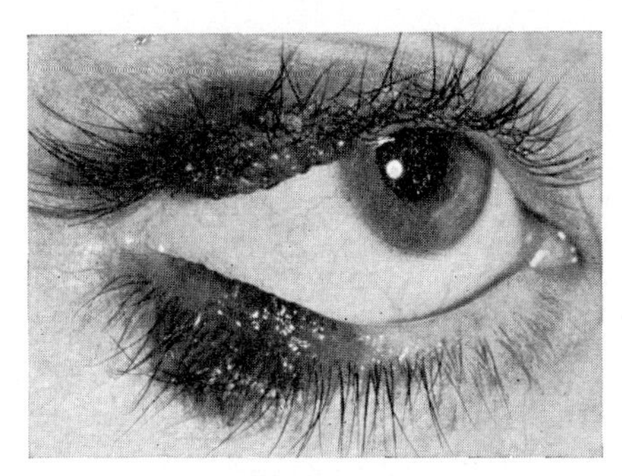

FIG. 568.

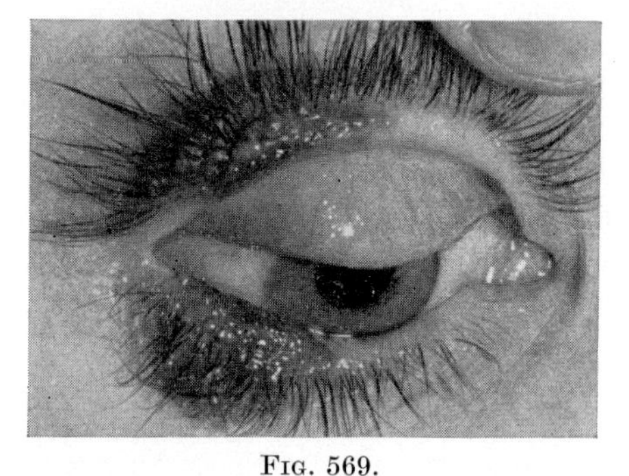

FIG. 569.

FIGS. 568 and 569.—Occurring on both upper and lower lids. Note the exuberant masses on the margin of the upper lid (A. Callahan).

sometimes pedunculated and warty with long coarse hairs issuing from their surface (Kaposi, 1895) (*nævus spilus, verrucosus, pilosus, mollusciformis* or *lipomatodes; hairy mole*) (Fig. 563). Nævi are extremely common, for few individuals do not have at least one small lesion of this type somewhere on their body; on the eyelids they are fairly frequent,[1] particularly near the ciliary margin and the outer canthus (Figs. 564–5). On the lid-margin the

[1] 12·7 % of all tumours of the lids, Welch and Duke (1958); 13·2 %, H. and J. Allington (1968); 17 %, O'Brien and Braley (1936).

Figs. 570 and 571.—Pigmented Nævus of the Lid.

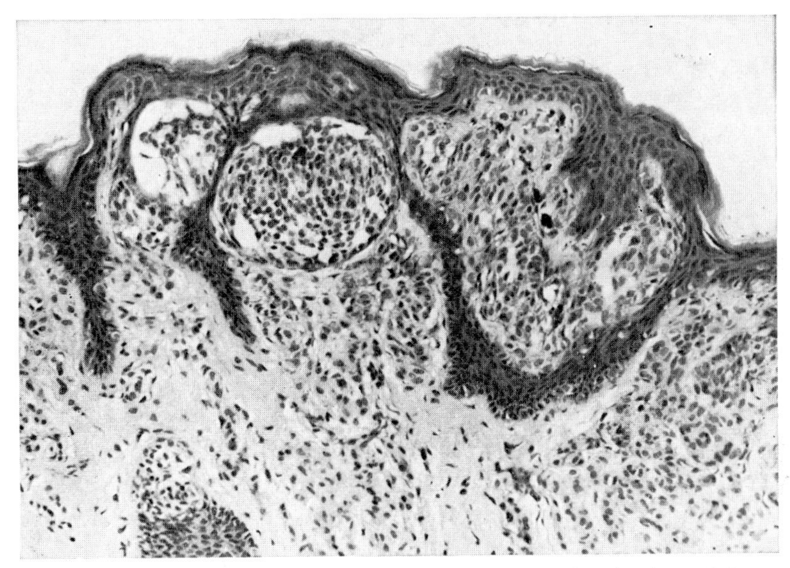

Fig. 570.—There are clusters of nævus cells in the dermis situated between downgrowths of the epidermis. Fusiform cells containing pigment are in close relation to the groups of nævus cells (H. & E.; ×150) (N. Ashton).

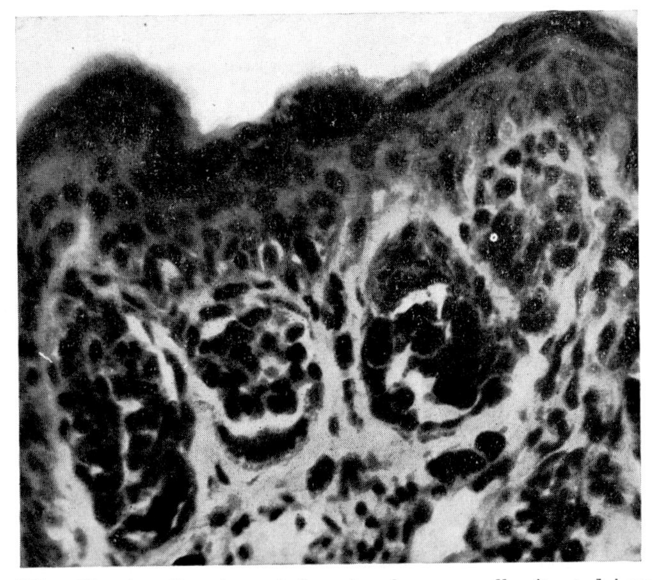

Fig. 571.—Showing the pigmented nests of nævus cells situated immediately beneath the epithelium from which they appear to arise (H. & E.; ×500) (N. Ashton).

tuberous variety may disturb the configuration of the lashes and cause trichiasis. An interesting type is that in which a nævus is situated partly on the upper lid and partly on the lower, both parts forming a symmetrical whole when the eye is closed (a *divided nævus*)[1] (Figs. 566–9); presumably such nævi are formed in fœtal life at the stage before the lids have separated.

Histologically the characteristic features of these tumours are *nævus cells*—small cells with a deeply staining nucleus and a scanty cytoplasm arranged in clusters, nests, sheets or strands (Figs. 570–1). They may be divided into two main groups composed either of characteristic nævus cells arising from melanocytes in the basal layer of the epidermis or of dermal melanocytes; the former group are much the more common and may be subdivided into three types, the junctional nævus, the compound nævus and the intradermal nævus. Each of these three types commences by the proliferation of cells at the dermo-epidermal junction and while the nests of nævus cells remain within or attached to the epidermis the nævus is *junctional* in type. In some cases of simple lentigo a few nests of nævus cells can be seen forming at the junctional zone where they resemble normal melanocytes; usually, however, particularly in children, an early nævus is composed of rather large epithelioid cells which evolve into typical nævus cells not unlike lymphocytes in size and shape; as the nævus grows the nests of cells pass into the dermis forming a *compound nævus*. In older nævi, occurring particularly in adults, the deepest cells become elongated and compressed to resemble fibroblasts or schwannian cells. In the majority of nævi junctional activity becomes less apparent as adult life is reached and when this activity ceases the nævus is called *intradermal* (Lund and Stobbe, 1949). Most nævi mature to the intradermal form and it is only from the junctional nævus and the compound nævus with its junctional component that a malignant melanoma is likely to arise. Although it may not always be possible to distinguish clinically between them, four types of nævi can be differentiated histologically depending on the location of these cells. Lesions derived from dermal melanocytes include the nævus of Ota and blue nævi.

(*a*) *The junctional nævus* when quiescent is entirely within the epidermis and confined for the most part to the basal layer. Numbers of nævus cells may be seen arranged in strips or clusters with clear areas between. Clinically such a lesion tends to be flat or only very slightly raised and, while it usually remains stationary or even regresses, it may spread into the subepidermal tissues or become malignant. It should be regarded as a potentially malignant condition and its removal before puberty is justified. Histologically the development of malignancy is preceded by cellular pleomorphism, anaplasia, hyperchromatism, increased mitosis and an inflammatory reaction in the adjacent dermis (Hogan and Zimmerman, 1962).

[1] A. Fuchs (1919–60), Collenza (1937), Callahan (1946), Lo (1951), Chu *et al.* (1960), Boniuk (1962), Ehlers (1965–69).

FIGS. 572 and 573.—JUVENILE MELANOMA OF SPITZ
(J. Wollensak and H. Meythaler).

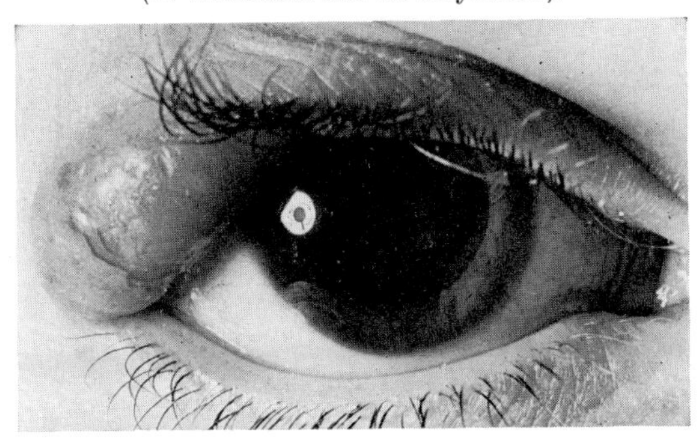

FIG. 572.—In a girl aged 9 years showing the melanoma on the right upper lid.

FIG. 573.—Histological section showing dilated capillaries and subepidermal œdema.

(*b*) *The compound nævus* has both junctional and intradermal elements; it tends to be elevated and because of its junctional element is capable of malignant change, but this is relatively rare. A particular type of compound nævus, the *juvenile melanoma* (*of Spitz*), occurs predominantly in children and young adults and contains giant, spindle and epithelioid cells, often associated with a considerable inflammatory infiltrate; surface ulceration may occur. It is frequently vascular and may not be heavily pigmented, appearing clinically as a firm, rounded, reddish-brown nodule (Figs. 572–3). Histologically it has many of the appearances of malignancy but does not infiltrate or metastasize. Before puberty it is clinically benign and thereafter is no more likely to become malignant than any other type, but excision is desirable.[1]

[1] Spitz (1948), Allen (1949–63), Hendrix (1954), Haber (1952–62), Kernen and Ackerman (1960), Jones and Dukes (1963), Lever (1967), Wollensak and Meythaler (1967), Sanderson (1968), Lerman *et al.* (1970).

(*c*) The *intradermal nævus* is composed of collections of nævus cells all lying within the dermis; it is the commonest type and is mature and stable, remaining benign. The appearance varies from a smooth and flat to a papillary or warty type of lesion.

(*d*) A *blue nævus* is a slightly elevated pigmented tumour arising from melanocytes in the dermis, the colour varying from blue-grey to black depending on its depth; over it the epidermis is unchanged. An interesting case was reported by Jay (1965) wherein several blue nævi occurred in a

FIGS. 574 and 575.—BLUE NÆVUS.

The tumour spread slowly over a period of 7 years and became invasive after an injury at the age of 12 (F. Braithwaite).

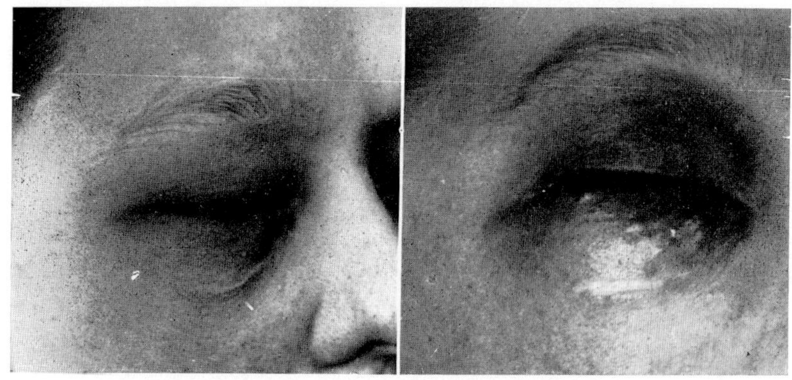

FIG. 574. FIG. 575.

FIG. 574.—The extent of the pigmentation in 1945 (aged 17) involving both lids.

FIG. 575.—The extent of the pigmentation in 1948.

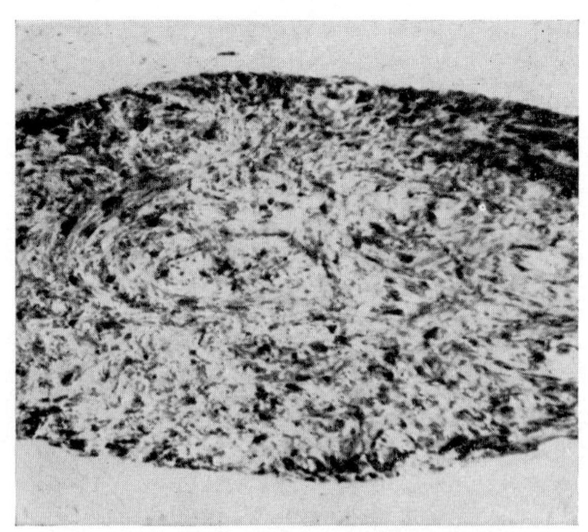

FIG. 576.—BLUE NÆVUS OF THE EYELID.

Showing the pigmented fusiform melanocytes (H. & E.) (B. Jay).

patient with oculodermal melanosis (nævus of Ota)[1] in association with a malignant melanoma of the orbit. It is composed of heavily pigmented fusiform cells usually lying deep in the dermis but sometimes in the superficial layers and is occasionally associated with a junctional nævus (Montgomery and Kahler, 1939). These lesions are thought to represent the arrested migration of melanocytes from the epidermal junction (Sanderson, 1968). The common type of blue nævus has a characteristic histological appearance but may occasionally be mistaken for a spindle-cell malignant melanoma (Fig. 576); as a rule, however, the cells show no significant anaplasia and the lesions which are present at birth or soon after usually remain unchanged throughout life suffering a gradual fibrosis with age; malignant change is rare (Allen and Spitz, 1953; Fisher, 1956). Nevertheless, after some time the lesion may become invasive; thus in a case described by Braithwaite (1948) affecting both lids the underlying orbicularis and the orbital tissues became invaded (Figs. 574–5).

A case of great interest was reported by Rank (1963). An Australian girl was born with an immense melanin-pigmented lesion of a bluish tinge apparently in the deeper layers of the skin and subcutaneous tissues, involving both lids and the whole left side of her face; the pigmentation was more dense about the eyelids which showed a considerable amount of puffy swelling. Owing to the somewhat hideous appearance, at the age of 7 a pigmented area was removed which showed the characteristics of a deeply seated nævus composed largely of spindle-shaped cells extending to the deeper layers including the subcutaneous fatty tissue with a normal epidermis; a molar tooth had to be removed and this showed a dense pigmentation about its roots. At the age of 11 when the girl reached puberty the pigmentation became darker and the eye became more proptosed; since the eye was amblyopic the orbit was exenterated and seven operations were undertaken wherein large abdominal skin flaps were implanted over the forehead and cheek. At the age of 17, wearing a camouflage prosthesis, her appearance became relatively normal and she commenced training as a nurse, eventually to be married (Figs. 577–8).

Treatment is usually unnecessary, for most nævi are best left alone although they should be kept under observation. It is true that many malignant melanomata develop from or in association with benign nævi and, therefore, it would appear reasonable to remove all such benign tumours as a prophylactic measure, an unrealistic approach since many adults have at least fifteen nævi scattered over the body (Pack *et al.*, 1952). Because of the high incidence of junctional activity in nævi of the eyelids, Reese (1963) recommended that they should all be excised preferably before puberty. If removal is desired either for cosmetic reasons, since some may cause psychological distress (Hirshowitz and Mahler, 1969), or owing to signs of activity such as an increase in size, in pigmentation or in vascularity or evidences of inflammation, or because of repeated exposure to irritation or trauma, local non-surgical attempts at removal as by electrolysis or freezing should be avoided. Radiotherapy is contra-indicated since nævi are generally radio-

[1] Vol. III, p. 798.

resistant; wherever possible they should be completely excised rather than subjected to incisional biopsy. Small flat lesions can sometimes be disguised by the use of cosmetics.

FIGS. 577 and 578.—BLUE NÆVUS (B. K. Rank).

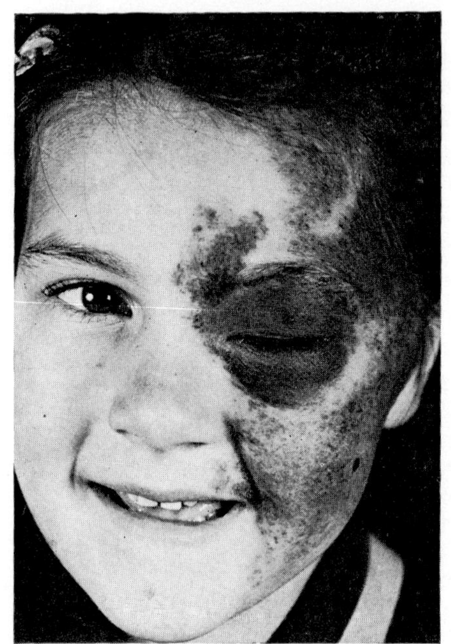

FIG. 577.—At $6\frac{1}{2}$ years of age the girl showed an extensive pigmented lesion affecting the lids, the forehead, the temple and the cheek. Seven operations were performed including an exenteration of the left orbit and extensive skin grafting.

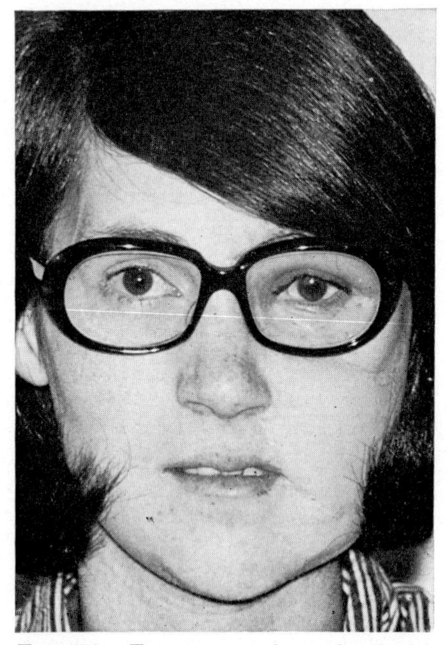

FIG. 578.—Twenty years later, fitted with an effective prosthesis, she had an almost normal appearance; she had trained as a nurse, become a theatre nurse, and subsequently married and had a family.

MALIGNANT MELANOMA

The literature on pigmented tumours of the skin contains a great diversity of nomenclature to describe apparently similar or related neoplasms. Moreover, variations occur in structure, the significance of which may be difficult to assess, while different types of cell may enter into the composition of one tumour. Thus, we find the descriptions nævo-carcinoma, malignant nævus, melano-carcinoma and other more indefinite terms. In the older literature these tumours have been considered variously as epitheliomata, endotheliomata or sarcomata, a circumstance which makes any selection of cases confusing. It is true that many of them are derived from a nævus (*nœvo-carcinoma*), a sequence which may be observed clinically or demonstrated histologically (Fig. 579), but in other cases the origin is apparently spontaneous; while an aggressively invasive tumour may make any recognition of the origin of the neoplasic cells impossible since, even although it may have arisen from the junctional elements of a small or unnoticed nævus, all traces of its true origin may have been destroyed. The indefinite generic name MALIGNANT MELANOMA is therefore the most convenient.

In contrast to the large numbers of benign pigmented lesions, malignant melanomata of the skin are rare, much less common than the basal- or squamous-celled type of epithelioma, and account for only 3% of all skin cancers (Cade, 1961; Ackerman and del Regato, 1970). Large series of cases have shown a predominance in blond individuals with red hair and pale complexion, many of whom have numerous cutaneous lesions. A geographical variation has been noted, the disease, for example, being common in Australia and New Zealand. Malignant melanomata are rare in Negroes: Helwig (1963) found in 392 cases of cutaneous melanoma that 96% were Caucasians, only 1% Negroes, and 3% were of other ethnic origins; and in a study of 3,852 cases of uveal melanomata, only 24 occurred in Negroes (Paul *et al.*, 1962). Familial cases are rare (Salamon *et al.*, 1963).

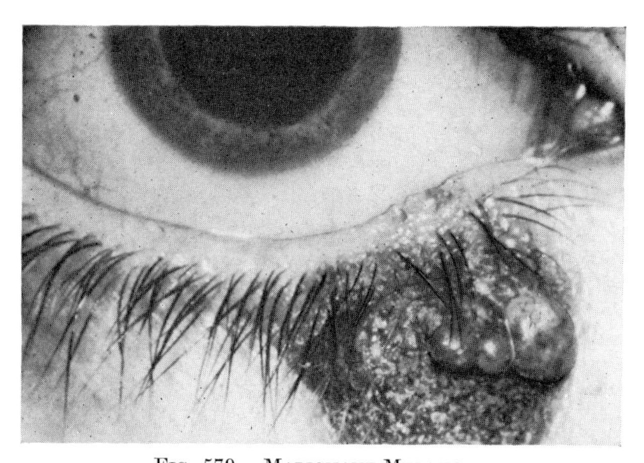

FIG. 579.—MALIGNANT MELANOMA.

Malignant nodules arising from the surface of a benign junctional nævus during pregnancy (M. J. Reeh and K. C. Swan).

The great majority of malignant melanomata appears to arise in association with a pre-existing nævus which has remained quiescent for many years: out of 317 cases occurring all over the body, Affleck (1936) found that 266 (84%) developed from pre-existing moles, but the literature on this varies from almost 100% (Pack and Davis, 1960) to 48% (Daland, 1959) and 25% (Becker, 1954) (Fig. 579). In some cases, of course, the presence of a small pigmented nævus may have been overlooked; it is said that an origin from a hairy nævus is unusual although such cases have occurred (Ebert, 1942; Oliver, 1942; Dobson, 1955; Lever, 1967). A malignant melanoma, however, may develop independently from pigmented nævi; since the typical nævus cells which constitute benign nævi tend to persist indefinitely, their absence in many malignant melanomata is evidence against the origin of these tumours from nævi (Allen and Spitz, 1953; Lund and Kraus, 1962; Jay, 1964–65).

Malignant melanomata rarely occur before puberty (Skov-Jensen *et al.*, 1966; Lever, 1967). The rare *malignant melanomata of childhood* are of three clinical types—those with a fatal outcome, those with metastases to the lymph nodes but indefinitely prolonged survival, and those due to trans-placental metastases from mother to fœtus. Lerman and his colleagues (1970) collected 48 cases from the literature and added 12 more; the ages at the time of diagnosis varied from 11 months to 14 years; 8 died within 18 months from generalized metastases. Whatever the type, the histological appearances are similar to those occurring in adults although many cells may appear more anaplasic (Spitz, 1948; McWhorter and Woolner, 1954; Allen, 1963).

Malignant melanomata of dermal origin are uncommon tumours of which there appear to be two types; one is the cellular blue nævus that metastasizes to lymph nodes but does not cause death, the other a truly malignant type which produces widespread metastases and is usually fatal (Allen and Spitz, 1953; Jay, 1964–65; Kwittken and Negri, 1966), although cases have been described in which despite widespread metastases the patient has survived for many years (Lund and Kraus, 1962). They have the usual histological features of malignancy including cellular pleomorphism, hyperchromatic nuclei and mitoses and generally show areas of necrosis.

INTRA-EPITHELIAL MELANOMA (ACQUIRED MELANOSIS). This is an acquired pigmentation affecting the skin and mucous membranes appearing in middle age usually between 40 and 50 years. First described by Jonathan Hutchinson (1891–1904) as originating in senile freckles (*lentigo maligna*), it has been given many names including *precancerous melanosis* by Dubreuilh (1894–1912), *intra-epithelial melanosis* by Ashton (1957) and, in order to avoid the element of prophecy in the common term, precancerous melanosis, to designate a lesion of very slow evolution remaining benign in the majority of cases, Zimmerman (1966) suggested that it should be termed *benign* and *malignant acquired melanosis*. In the region of the eye it involves the conjunctiva, occasionally the cornea, and the lids, and since the condition is fully described and illustrated in a previous Volume, together with the literature[1] when dealing with epibulbar lesions, it is unnecessary to repeat it here (Fig. 580) (Reese, 1938–43).

It will be remembered that clinically the lesion may arise from apparently normal skin or as a flat brown slowly progressive area. In its early stages it may be difficult to distinguish from a simple lentigo, but as it develops there is intra-epidermal proliferation of melanocytes which appear benign in all layers of the epidermis. At this stage the lesion may remain stationary for long periods (20 years, Reese, 1963) or may regress. This stage may pass gradually into the next when there is increased invasion of the epidermis by cells which now show distinct pleomorphism. Finally, in

[1] Vol. VIII, p. 1233.

17% of cases (Reese, 1966) malignant cells, all of which are probably derived from melanocytes in the basal layer of the epidermis, invade the dermis and a frankly malignant melanoma results (Fig. 581). In those lesions which arise from apparently normal skin this sequence of events may not be seen but frequently their edges show changes characteristic of an intra-epidermal melanoma. A superficial malignant melanoma with only slight dermal invasion has a less aggressive character and a better prognosis than a melanoma originating from a junctional nævus (Allen, 1949; Petersen *et al.*, 1962).

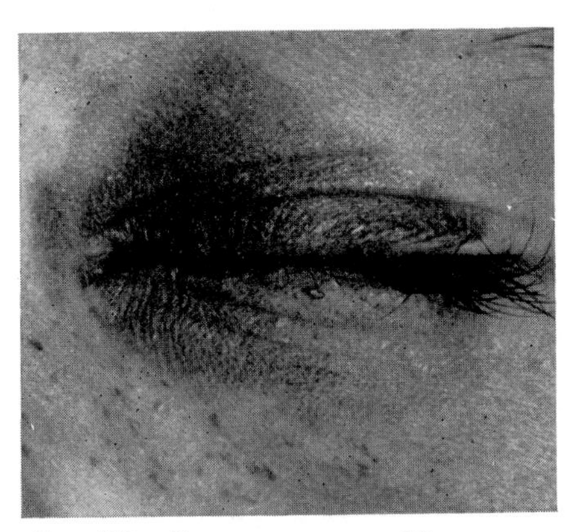

Fig. 580.—Intra-epithelial Melanoma.
Affecting the lids and conjunctiva in a girl aged 4½ (Inst. Ophthal.).

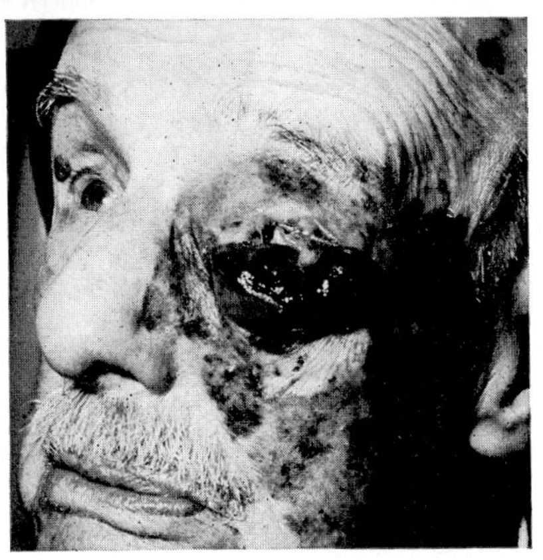

Fig. 581.—Malignant Lentigo of Hutchinson (M. F. Stranc).

The reason for the change of any benign melanoma to malignancy is unknown so that all moles should be regarded as potentially dangerous, but trauma, repeated irritation or ineffectual attempts to remove a simple mole are significant factors appearing as possible ætiological influences in about 30% of cases (Raven, 1950). Malignancy is ushered in by an increase in size and pigmentation in the nævus; the tumour becomes larger, nodular and papillomatous and satellite nodules appear around it. It becomes harder in consistency and fixed; occasionally the melanotic spread is extremely marked and outcrops appear in shoals so that the skin may be peppered with pigment spots as if it were bespattered with powder. Ulceration and hæmorrhage are relatively late symptoms and eventually the lids, conjunctiva, globe and the orbit may all be involved.

In most cases, however, long before this dissemination is advanced but after the intra-epidermal stage has passed, metastases may occur both by the lymphatic channels and the blood-stream. Brownish or black cutaneous lesions varying in size from a pinhead to a hen's egg may be numerous, the pre-auricular and submandibular nodes are heavily implicated and general systemic metastases are advanced. The liver, lungs, heart, central nervous system, bones—every tissue in fact—may be widely affected and a fatal termination is rapid. We have already noted that cases of transplacental metastases from the mother to the fœtus have been recorded (Reynolds, 1955). Frequently, malignant melanoma cells can be demonstrated in the bone marrow or the peripheral blood long before there is any clinical evidence of metastases. Only in the terminal stages does melanin occur in the urine, the discoloration of which, together with the dark mahogany colour of the skin, precedes death by a few days or weeks.

Histopathological Changes. Irrespective of the origin of a malignant melanoma, whether from apparently normal skin, from a benign nævus or from the acquired flat pigmented lesion, the earliest changes within the epidermis are the same. There is an increase in the number of clear cells in the basal layer of the epidermis associated with the migration of pigment towards the surface and also into the dermis where it is phagocytosed by macrophages. There is usually a mild infiltrate with chronic inflammatory cells in the superficial dermis. Features suggestive of malignancy are the diffuse nature of the junctional change and the presence of chronic inflammatory cells in the superficial dermis; as the lesion progresses the intra-epidermal melanocytes form irregular clumps scattered throughout the epidermis and subsequently the constituent cells display pleomorphism. Individual cells are aggregated into large or small clumps irregularly spaced throughout the epidermis, the cells themselves being mostly spaced as though lacking cohesive properties. Finally, malignant cells invade the dermis to produce a frankly malignant melanoma (Figs. 582–3). Cells of a malignant melanoma, derived as they are from the same source as those of nævi, have similar histological appearances, although they show more pronounced pleomorphism. Several cellular types can be recognized.

(1) Epithelioid cells, the most common type, are large polyhedral cells with round or oval nuclei and prominent nucleoli and an abundant cytoplasm usually containing fine melanin granules.

(2) Nævoid cells, smaller than the epithelioid variety but not as small as benign nævus cells. The cytoplasm is more heavily pigmented and the nuclei are large, round and hyperchromatic.

(3) Spindle cells, large or small fusiform cells with oval nuclei and prolonged at the ends to form fibres. The nuclei are oval, vary in size and contain indistinct nucleoli.

(4) Bizarre cells, giant cells with one or more nuclei.

It is interesting that the cellular content has a marked prognostic significance: tumours composed of epithelioid cells or with a high proportion of such cells have the worst prognosis and those composed predominantly of spindle cells the best. The presence and amount of pigment has no prognostic significance.

FIGS. 582 and 583.—MALIGNANT MELANOMA OF THE EYELID (N. Ashton).

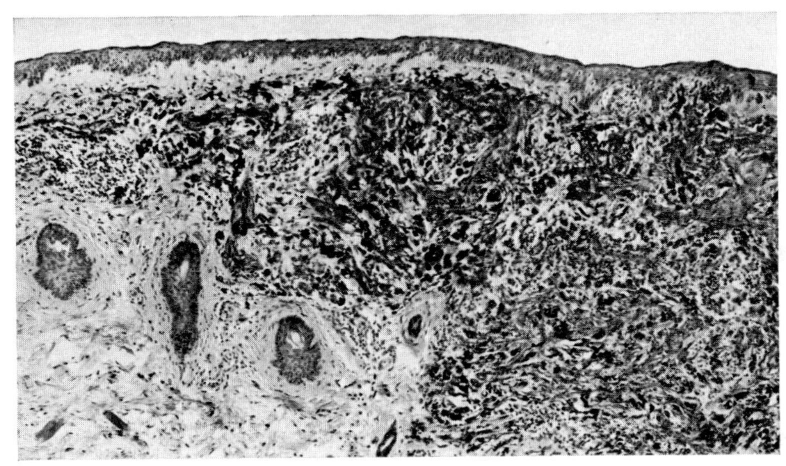

FIG. 582.—A large subepithelial pigmented mass extends down into the dermis (H. & E.; ×60).

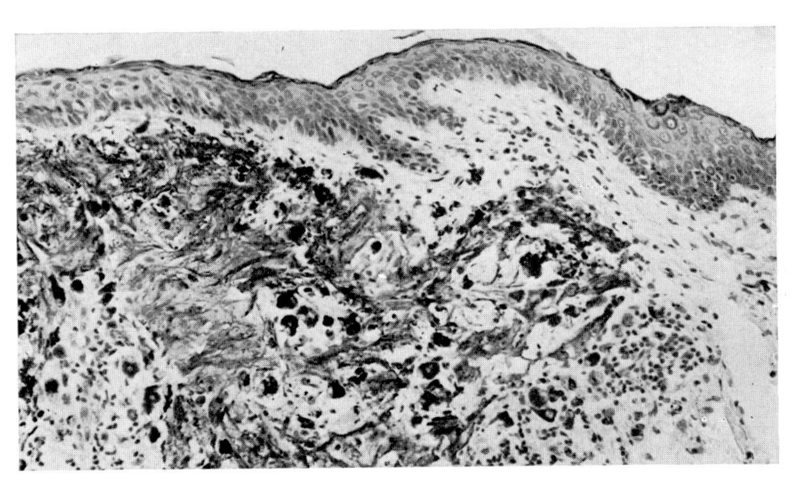

FIG. 583.—The growth is seen to consist of epithelioid and spindle-shaped pigmented cells (H. & E.; ×150).

The clinical *diagnosis* of a malignant melanoma is usually easy, but if there is doubt a biopsy is indicated provided that an adequately wide excision is made and is followed up by appropriate surgical treatment without delay.

It must be said, however, that errors of diagnosis are not infrequent, leading to incorrect treatment or delay in treatment which may prevent any effective therapy; such errors may be made either by the clinician or the pathologist (Bodenham and Lloyd, 1963). When there is little or no pigment or ulceration a malignant melanoma may be diagnosed as a benign lesion; when there is little pigment, rapid growth and the lesion is tense and tender, it may be confused with an acute infection, while an ulcerating lesion may be regarded as a telangiectatic granuloma, carcinoma or hæmangioma. On the other hand, several benign or less malignant conditions may be confused with a melanoma, such as a pigmented basal-cell carcinoma, a thrombosed angioma or a senile papilloma. Whenever there is doubt the clinician should consult with the pathologist and submit to him an adequate biopsy which has not been damaged in the process of excision.

The *treatment* of a malignant melanoma usually involves surgery. The question of prophylactic treatment in relation to the excision of pigmented nævi has already been noted: on the one hand, it is impracticable to excise all nævi; on the other hand, to wait until the pigmented lesion shows evidence of activity is to wait too long and thus to be too late. There is no doubt that when a benign pigmented lesion has been completely removed it does not recur; it is only when a malignant melanoma is mistakenly diagnosed as a benign lesion and inadequately removed that widespread disease is likely to follow.

In the treatment of malignant melanoma a wide surgical excision is the method of choice, involving a resection sufficiently wide in extent and depth to include the surrounding pigmented flare and any satellite nodules, the defect being repaired by some form of plastic procedure. Removal of the regional lymph nodes in continuity when they are involved in the disease or as a prophylactic measure is still a subject of debate. With localized melanomata of the palpebral skin a wide local excision of the full thickness of the lid should be done followed by plastic repair. If the tumour in the skin is extensive and involves the palpebral conjunctiva and if recurrence has appeared after more conservative treatment, exenteration of the orbit is required. When the palpebral conjunctiva is extensively involved but not the skin or the lid-margin, a modified form of exenteration with splitting of the lids and preservation of the skin may be possible, and in localized tumours at either canthus a wide local excision may be possible. It cannot be emphasized too strongly that meddlesome methods of treatment by electrolysis, cauterization, freezing, curettage or the application of caustic and cytotoxic substances are absolutely contra-indicated.

In cases of precancerous melanosis a policy of watchful waiting with repeated photographic observations after a preliminary biopsy should be followed. Only when there is clinical and histological evidence of transition to the cancerous stage, should energetic treatment be instituted without delay; this has usually involved exenteration, but Lederman (1964) found irradiation effective.

It has often been said that malignant melanoma is resistant to ionizing *radiation*, but Lederman (1958–66) has pointed out that this statement does not apply to all sites and there is no doubt that, although the relatively high doses that are required may cause considerable local damage to the tissues, certain melanomata respond to radiation which may sometimes be curative. In the early stages a diffuse precancerous melanosis is radio-sensitive, but in the later stages it is more resistant (Reese, 1943; Braithwaite, 1948); hence the importance of early treatment.

Chemotherapy using cytotoxic drugs administered systemically has practically no place in the management of disseminated disease but there is some evidence that cytotoxic agents may act as potentiators of ionizing radiation. Regional intra-arterial infusion has proved of some value in the control of locally advanced or recurrent lesions in the limbs, either alone or in combination with limited surgical procedures, but is not suitable as a routine technique (Westbury, 1970).

The *prognosis* of malignant melanomata must always be considered as anxious, even after the most effective treatment. Traditionally their development has been considered as a death warrant. This is too drastic a view but the generalization should be regarded as potentially true. In an assessment of these potentialities, however, the long latent period before the "black death"[1] that may occur should be remembered. Although it is frequently difficult to predict the outcome in the individual case, a number of factors should be considered. The age of the patient is relevant; in general the younger the patient the more malignant the tumour. The size and number of the lesions together with their depth of penetration are important; the chances of survival are good if the lesion is small and confined to the dermis. Involvement of the lymph nodes is of serious import: McNeer and Das Gupta (1964–65) found that in malignant melanomata of the skin without their involvement the survival rate after surgery was 62% but with their involvement only 12% survived five years. It is a disease of wide extremes from a most favourable case to the hopeless; the natural history may wax and wane and spontaneous regression occurs in up to 3% of cases (Cade, 1961; Smith and Stehlin, 1965; Bodenham, 1968–69; Lloyd, 1969). Favourable cases which may be cured by local methods and rarely metastasize are those which are slowly growing with a long history, flat lesions without ulceration or bleeding, and those showing histologically lymphocytic infiltration, little invasion of the dermis and no involvement of the lymphatics. At the other extreme the unfavourable cases are those with a short history of rapid growth and with a raised ulcerating and bleeding surface, often with relatively little pigment and showing on histological examination deep invasion involving the lymphatics and with little or no inflammatory cellular reaction (Bodenham, 1968–69).

The presence of a lymphocytic infiltration is of considerable prognostic importance and represents a cellular immunity-reaction of the host to the tumour. When the growth becomes more malignant the cellular response alters and may completely disappear; an intense lymphocytic reaction may be followed by involution of the tumour (Lloyd, 1969). In localized disease auto-antibodies may be demonstrated in the blood (Lewis *et al.*, 1969).

[1] Turner (1939), Tod (1944), Sylven (1949).

Attie and Khafif (1964) presented the five-year survival rates from 33 series of cases and, although there was considerable variation in the results, the majority of five-year survival rates lay between 20–40%. The five years usually quoted in the literature, however, is insufficient; thus among 145 patients with malignant melanomata of the skin followed by Cade (1961) 40% were still living after 5 years, of whom 25% were free of disease, while after ten years the survival rate was only 22%, of whom 14% were free of disease. Petersen and his colleagues (1962) also found a range of 20–58% in the five-year rate of cure in several series.

Malignant melanoma of the eyelids is a rare condition and there is a dearth of statistically significant cases in the literature, but in general the prognosis must be viewed in the light of the facts outlined above. The chances are, however, that a lesion at this site, as elsewhere on the face, is likely to be treated without undue delay. The prognosis for conjunctival melanomata is distinctly better; the five-year cure-rate (survival with no local recurrence) has been quoted as 100% for bulbar tumours, 82% for limbal tumours and 50% for those on the palpebral conjunctiva and the caruncle (Jay, 1965). It must be remembered, however, that recurrences and dissemination may occur up to 10 years and more after the initial treatment.

A flat pigmented lesion in which the melanocytes have not yet invaded the dermis has a favourable prognosis; lymphatic or venous invasion does not occur and there are no metastases. When there is invasion of the dermis but no actual tumour formation, lymphatic invasion is possible but uncommon, so that in uncomplicated cases dissection of the lymph nodes is unnecessary provided adequate primary excision is achieved. When there is a tumour which has invaded the dermis, metastases are common. It is important to examine the surroundings of the primary tumour to see if there is any obvious invasion of the dermal lymphatics since this makes the prognosis worse, indicating that the regional lymph nodes have already been invaded and that lymphatic metastases may arise in the dermis at some point between the lymph nodes and the primary lesion.

SECONDARY MALIGNANT MELANOMATA of the skin may be due to distant metastases or a local recurrence of a tumour. Even when the secondary deposit invades the epidermis it should still be possible to distinguish it from a primary tumour since the metastatic deposits are more sharply delimited from normal epidermis and the lesion is not associated with junctional activity at the dermo-epidermal junction (Allen and Spitz, 1953; Jay, 1964).

INCIDENTALLY PIGMENTED TUMOURS. A few tumours of the skin, either benign or carcinomatous, may show incidental pigmentation and although this may make the clinical diagnosis difficult there is usually no problem with the histological diagnosis.

Ackerman and del Regato. *Cancer: Diagnosis, Prognosis and Treatment*, St. Louis (1970).
Affleck. *Amer. J. Cancer*, **27**, 120 (1936).
Allen. *Cancer*, **2**, 28 (1949).
Arch. Derm., **82**, 325 (1960).
Ann. N.Y. Acad. Sci., **100**, 29 (1963).
Allen and Spitz. *Cancer*, **6**, 1 (1953).
Arch. Derm. Syph. (Chic.), **69**, 150 (1954).
Allington, H. and J. *Arch. Derm.*, **97**, 50 (1968).
Ashton. In *Cancer* (ed. Raven), London, **2**, 608 (1957).

Attie and Khafif. *Melanotic Tumours*, Springfield (1964).
Becker. *Arch. Derm. Syph.* (Chic.), **69**, 11 (1954).
Bodenham. *Ann. roy. Coll. Surg.*, **43**, 218 (1968).
Proc. roy. Soc. Med., **62**, 1090 (1969).
Bodenham and Lloyd. *Postgrad. med. J.*, **39**, 278 (1963).
Boniuk. *Int. Ophthal. Clin.*, **2**, 239 (1962).
Braithwaite. *Brit. J. plast. Surg.*, **1**, 206 (1948).
Cade. *Ann. roy. Coll. Surg.*, **28**, 331 (1961).
Callahan. *Amer. J. Ophthal.*, **29**, 563 (1946).
Chu, Wang, Kung and Hao. *Acta chir. plast.* (Praha), **2**, 314 (1960).
Collenza. *Boll. Oculist.*, **16**, 435 (1937).
Cook. *Trans. ophthal. Soc. U.K.*, **71**, 257 (1951).
Daland. *New Engl. J. Med.*, **260**, 453 (1959).
Dobson. *Amer. J. Surg.*, **89**, 1128 (1955).
Dubreuilh. *Ann. Derm. Syph.* (Paris), **5**, 1092 (1894); **3**, 129, 205 (1912).
Ebert. *Arch. Derm. Syph.* (Chic.), **46**, 604 (1942).
Ehlers. *Arch. Ophthal.*, **73**, 664 (1965).
Acta ophthal. (Kbh.), **47**, 1004 (1969).
Fisher. *Arch. Derm.*, **74**, 227 (1956).
Fuchs, A. *Klin. Mbl. Augenheilk.*, **63**, 678 (1919); **137**, 504 (1960).
Urol. cutan. Rev., **54**, 88 (1950).
Haber. *Trans. St. John's Hosp. derm. Soc.*, **31**, 44 (1952).
Brit. J. Derm., **74**, 224 (1962).
Helwig. *Nat. Cancer Inst.*, Monog. No. 10, 287 (1963).
Hendrix. *Arch. Path.*, **58**, 636 (1954).
Hirshowitz and Mahler. *Brit. J. Ophthal.*, **53**, 343 (1969).
Hogan and Zimmerman. *Ophthalmic Pathology*, 2nd ed., Phila., 207 (1962).
Hutchinson. *Arch. Surg.* (Lond.), **3**, 315 (1891).
Dtsch. med. Wschr., **30**, 1378 (1904).
Jay. *Trans. ophthal. Soc. U.K.*, **84**, 337 (1964).
Naevi and Melanomata of the Conjunctiva (Thesis), Cambridge (1965).
Brit. J. Ophthal., **49**, 169, 359 (1965).
Jones and Dukes. *Amer. J. Ophthal.*, **56**, 816 (1963).
Kaposi. *Pathology and Treatment of Diseases of the Skin* (Transl. Johnston), London (1895).
Kernen and Ackerman. *Cancer*, **13**, 612 (1960).
Kwittken and Negri. *Arch. Derm.*, **94**, 64 (1966).
Lederman. *Trans. ophthal. Soc. U.K.*, **78**, 147 (1958); **84**, 357 (1964).
Brit. J. Radiol., **34**, 21 (1961).
Ocular and Adnexal Tumors (ed. Boniuk), St. Louis (1964).
2nd Cong. europ. ophthal. Soc. (Wien, 1964) (ed. François), Basel, xliii (1966).
Lerman, Murray, O'Hara *et al.* *Cancer*, **25**, 436 (1970).
Levene. *Proc. roy. Soc. Med.*, **65**, 137 (1972).
Lever. *Histopathology of the Skin*, 4th ed., Phila., 700 (1970).
Lewis, Ikonopisov, Nairn *et al.* *Brit. med. J.*, **3**, 547 (1969).
Lloyd. *Proc. roy. Soc. Med.*, **62**, 543 (1969).
Lo. *Chin. med. J.*, **69**, 258 (1951).
Lund and Kraus. *Melanotic Tumors of the Skin*, Washington (1962).
Lund and Stobbe. *Amer. J. Path.*, **25**, 1117 (1949).
McNeer and Das Gupta. *Surgery*, **56**, 512 (1964).
Amer. J. Roentgenol., **93**, 686 (1965).
McWhorter and Woolner. *Cancer*, **7**, 564 (1954).
Montgomery and Kahler. *Amer. J. Cancer*, **36**, 527 (1939).
O'Brien and Braley. *J. Amer. med. Ass.*, **107**, 933 (1936).
Oliver. *Arch. Derm. Syph.* (Chic.), **46**, 605 (1942).
Pack and Davis. *N.Y. St. J. Med.*, **56**, 3498 (1960).
Pack, Lenson and Gerber. *Arch. Surg.*, **65**, 862 (1952).
Paul, Parnell and Fraker. *Int. Ophthal. Clin.*, **2**, 387 (1962).
Petersen, Bodenham and Lloyd. *Brit. J. plast. Surg.*, **15**, 49, 97 (1962).
Rank. *Aust. N.Z. J. Surg.*, **33**, 81 (1963).
Raven. *Ann. roy. Coll. Surg.*, **6**, 28 (1950).
Reese. *Arch. Ophthal.*, **19**, 354 (1938); **29**, 737 (1943).
Tumors of the Eye, 2nd ed., N.Y., 214 (1963).
Amer. J. Ophthal., **61**, 1272 (1966).
Reynolds. *Obstet. Gynec.*, **6**, 205 (1955).
Salamon, Schnyder and Storck. *Dermatologica*, **126**, 65 (1963).
Sanderson. *Textbook of Dermatology* (ed. Rook *et al.*), Oxford, **2**, 1658 (1968).
Skov-Jensen, Hastrup and Lambrethsen. *Cancer*, **19**, 620 (1966).
Smith and Stehlin. *Cancer*, **18**, 1399 (1965).
Spitz. *Amer. J. Path.*, **24**, 591 (1948).
Sylven. *Acta radiol.* (Stockh.), **32**, 33 (1949).
Tod. *Lancet*, **2**, 532 (1944).
Turner. *Trans. St. John's Hosp. derm. Soc.*, 93 (1939).
Welch and Duke. *Amer. J. Ophthal.*, **45**, 415 (1958).
Westbury. *Proc. roy. Soc. Med.*, **63**, 88 (1970).
Wollensak and Meythaler. *Klin. Mbl. Augenheilk.*, **150**, 388 (1967).
Zimmerman. *Arch. Ophthal.*, **76**, 307 (1966).

METASTATIC TUMOURS

The appearance of a metastatic deposit in the eyelid is a rare event; compared to the frequency with which metastases occur in the intra-ocular structures, particularly the uveal tract,[1] and in the orbit,[2] surprisingly few cases have been reported in the lids (15 cases seen at the Mayo Clinic 1922–1969, Riley, 1970). The most common source has been a primary carcinoma of the breast[3] (Figs. 584–5) or the lung and bronchus[4]; other sites of the primary tumour have been the stomach (Cowan, 1952; Riley, 1970), colon

Figs. 584 and 585.—Metastatic Carcinoma of the Eyelid (W. S. Muenzler *et al.*).

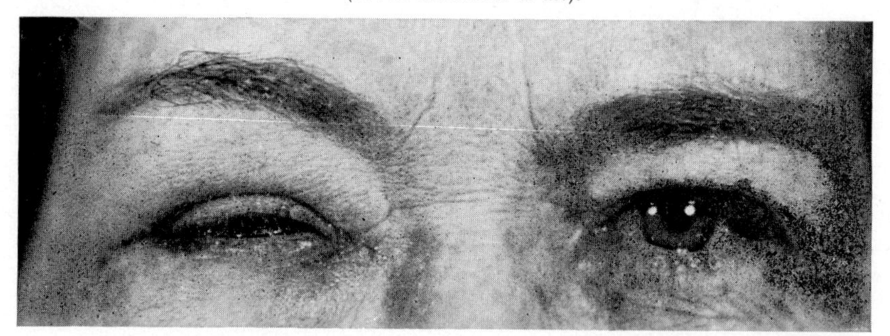

Fig. 584.—In a woman aged 59 who had a scirrhous-type ductal adenocarcinoma removed from the breast.

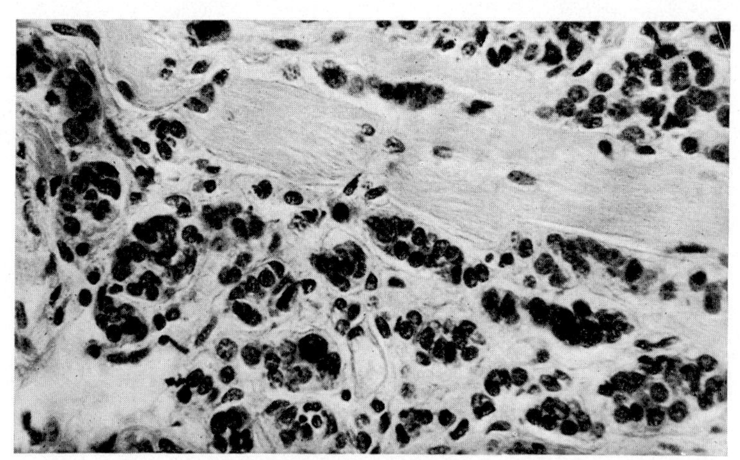

Fig. 585.—Histological section of the tumour seen in Fig. 584, resembling the appearance of the primary tumour in the breast.

[1] Vol. IX, p. 917. [2] p. 1144.

[3] Grönvall (1953), Costner (1960), Wheelock *et al.* (1962), Muenzler *et al.* (1963), Weinstein and Goldman (1963), Riley (1970), Eichholtz (1971), Amoni *et al.* (1973), Antal and Nemeth (1973).

[4] Jaensch (1933), Fleischanderl (1937), Wright and Meger (1962), Aizawa *et al.* (1966), Riley (1970).

(Ostriker, 1957), parotid (Vrabec, 1951; Casanovas, 1966) and the thyroid (Schlagenhauff and Ratzenhofer, 1954; Appalanarsayya and Satyendran, 1964). Rare metastatic tumours include a neuroblastoma (Casanovas, 1966) (Figs. 586–7), renal adenocarcinoma (on the eyebrow, van Arnam and Fine, 1957) (Fig. 588), and an ovarian granular-cell schwannoma (on the upper lid,

FIGS. 586 and 587.—METASTATIC NEUROBLASTOMA OF THE LID.
From a primary mediastinal tumour (J. Casanovas).

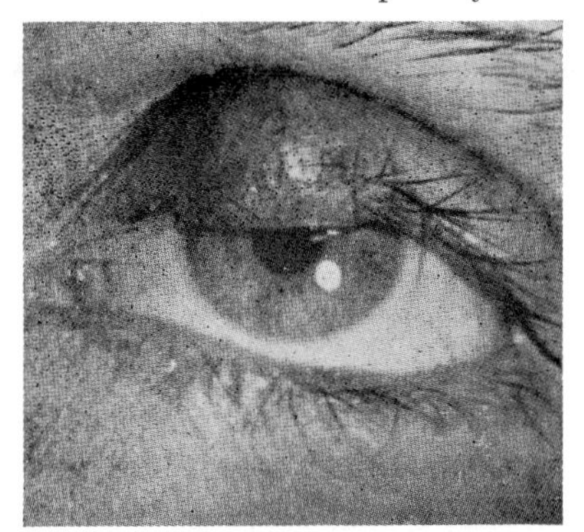

FIG. 586.—Affecting the upper lid in a man aged 21 years.

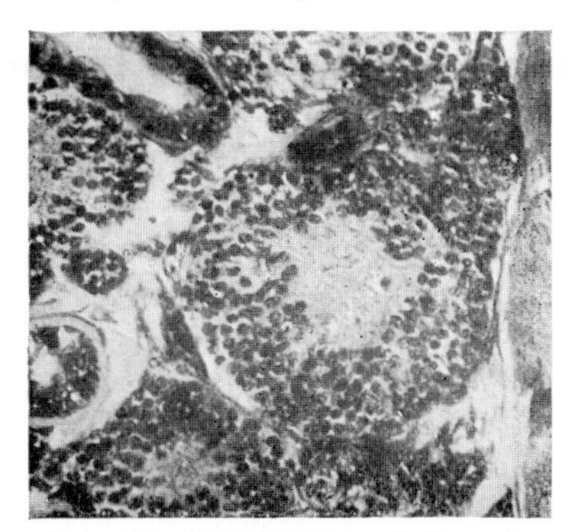

FIG. 587.—The histology indicated a neuroblastoma.

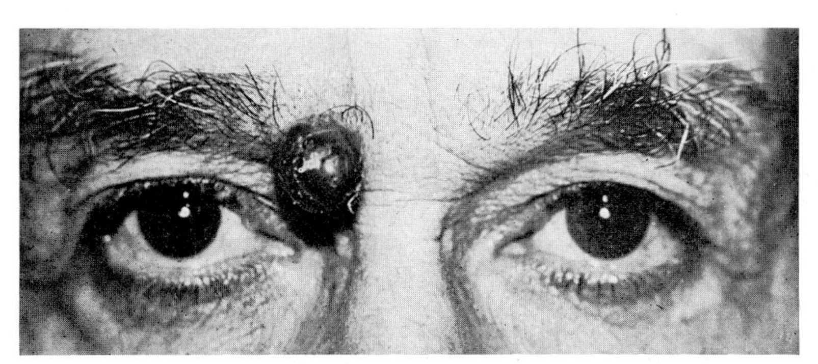

FIG. 588.—METASTATIC CARCINOMA OF THE LID.

In a man aged 72 who had a carcinoma of the renal parenchyma; recovery from the operation on the kidney was uneventful. Seven years later multiple metastases were present in the spine and lung as well as the lesion on the skin of the upper lid and eyebrow (C. E. van Arnam and M. Fine).

Powell, 1946). A primary malignant melanoma of the skin has given rise to metastases in the eyelid (Ascher, 1938; Riley, 1970) as also has a choroidal melanoma (Riley, 1970).

Generally, cutaneous metastases from visceral carcinomata are rare (Lever, 1967) and occur in perhaps 3 to 4% of malignant tumours (Willis, 1960). Arising as a result of hæmatogenous spread they usually indicate widespread dissemination of the disease but may as a rarity be the only detectable metastasis (Lessans, 1973); it is interesting that the lesion in the lid may be the first indication of an occult primary tumour. As a rule the prognosis is gloomy.

Histopathologically metastatic tumours in the lids are recognized by the appearance of abnormal cells which may be sufficiently differentiated to indicate the site of the primary growth, although many are anaplasic. Clinically they may present in three forms—as a solitary subcutaneous painless nodule which may easily be mistaken for a chalazion (Weinstein and Goldman, 1963; Casanovas, 1966) (Fig. 586) and in the case of a malignant melanoma may be pigmented; as a non-tender thickening and induration of the eyelid which may lead to a diffuse swelling of one or more of the lids without discrete nodules, perhaps associated with simultaneous orbital metastases (Fig. 584); or as an ulcerating lesion of the lid which may be confused clinically with a rodent ulcer. Extensive involvement of lymphatics may cause localized œdema while vascular obstruction may lead to necrosis of the tissues; occasionally the vascular reaction may suggest an inflammatory lesion rather than a neoplasm.

Treatment. Excision of a solitary nodular lesion or radiotherapy for the more diffuse type of metastatic disease may be indicated.

Aizawa, Tukahara, Muroya and Takeuchi. *Rinsho Ganka*, **20**, 53 (1966).
Amoni, Rodrigues and Shannon. *Canad. J. Ophthal.*, **8**, 167 (1973).
Antal and Nemeth. *Ann. Ophthal.* (Chic.), **5**, 1213 (1973).
Appalanarsayya and Satyendran. *Orient. Arch. Ophthal.*, **2**, 183 (1964).
van Arnam and Fine. *Arch. Ophthal.*, **57**, 694 (1957).
Ascher. *Klin. Mbl. Augenheilk.*, **101**, 433 (1938).
Casanovas. *Ophthalmologica*, **151**, 272 (1966).
Costner. *J. Tenn. med. Ass.*, **53**, 392 (1960).
Cowan. *Arch. Ophthal.*, **48**, 496 (1952).
Eichholtz. *Klin. Mbl. Augenheilk.*, **158**, 836 (1971).
Fleischanderl. *Z. Augenheilk.*, **93**, 31 (1937).
Grönvall. *Acta ophthal.* (Kbh.), **31**, 153 (1953).
Jaensch. *Klin. Mbl. Augenheilk.*, **90**, 598 (1933).
Lessans. *Amer. J. Ophthal.*, **75**, 458 (1973).
Lever. *Histopathology of the Skin*, 4th ed., Phila., 607 (1967).
Muenzler, Olson and Eubank. *Amer. J. Ophthal.*, **55**, 791 (1963).
Ostriker. *Arch. Ophthal.*, **57**, 279 (1957).
Powell. *Arch. Path.*, **42**, 517 (1946).
Riley. *Amer. J. Ophthal.*, **69**, 259 (1970).
Schlagenhauff and Ratzenhofer. *Wien. med. Wschr.*, **104**, 350 (1954).
Vrabec. *Ophthalmologica*, **122**, 362 (1951).
Weinstein and Goldman. *Amer. J. Ophthal.*, **56**, 960 (1963).
Wheelock, Frable and Urnes. *Amer. J. clin. Path.*, **37**, 475 (1962).
Willis. *Pathology of Tumours*, 3rd ed., London (1960).
Wright and Meger. *Amer. J. Ophthal.*, **54**, 135 (1962).

Developmental Tumours

Dermoids[1] and teratomata[2] have already been discussed.

PHAKOMATOUS CHORISTOMA was the name given to unusual tumours occurring at the medial part of the lower eyelid at birth by Zimmerman (1971). A similar case was described by Filipic and Silva (1972). Histologically they were characterized by dense collagenous connective tissue, rich in mucopolysaccharide with pale epithelial cells arranged in nests, cords and sheets; associated with these were large ovoid cells similar to the bladder cells of Wedl in some cataracts[3] and in all there was thickening

[1] Vol. III, pp. 886 and 956.
[2] Vol. III, p. 967.
[3] Vol. XI, p. 26.

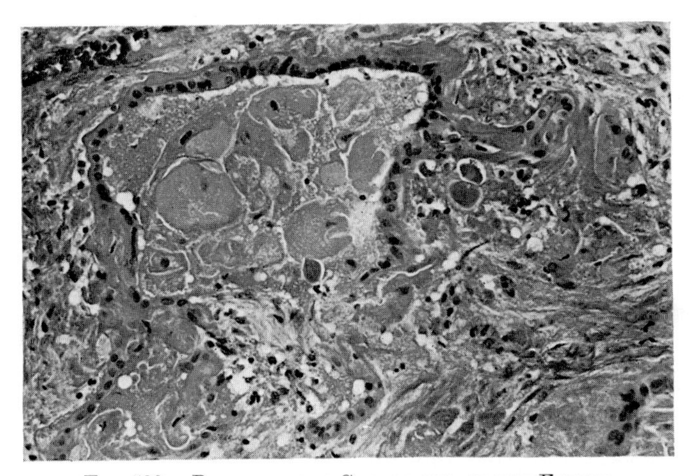

FIG. 589.—PHAKOMATOUS CHORISTOMA OF THE EYELID.

The histology of the tumour shows bladder cells and degenerated material incompletely surrounded by epithelium (M. Filipic and M. Silva).

of the basement membrane (Fig. 589). On the basis of the resemblance to the components of cataractous lenses, Zimmerman regarded them as congenital tumours of lenticular origin, possibly due to burying of the surface ectoderm. Similar tumours have been found in the subconjunctival tissues.[1]

Filipic and Silva. *Arch. Ophthal.*, **88**, 172 (1972).
Zimmerman. *Amer. J. Ophthal.*, **71**, 169 (1971).

[1] Vol. III (2), p. 693.

Fig. 590.—László Blaskovics
[1869–1938]
(Courtesy of Magda Radnót.)

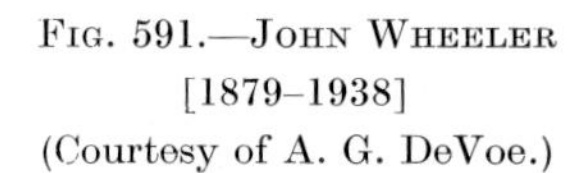

Fig. 591.—John Wheeler
[1879–1938]
(Courtesy of A. G. DeVoe.)

CHAPTER VIII

MOTOR DISORDERS AND DEFORMATIONS OF THE LIDS

DISTURBANCES IN MOTILITY

So many of the disorders of motility and positioning of the lids depend for their amelioration on very specialized plastic surgery that it is suitable to introduce this Chapter with the photographs of two eminent plastic surgeons, one from Europe and one from America.

LÁSZLÓ BLASKOVICS [1869–1938] (Fig. 590), one of the great figures of ophthalmic surgery, studied at Budapest where he spent his whole life; he began his ophthalmological career in the university of that city in 1893. In 1905 he was appointed head of the trachoma ward of St. Stephen's Hospital and in 1907 he became director of the newly established Hungarian State Eye Hospital; twenty years later he occupied the second Chair of Ophthalmology in Budapest, a post which he held until his death. His main interest was surgery, and during his professional life he published accounts of 27 original operations; among these his ingenuity found full scope in the surgery of the eyelids and his technique for the correction of ptosis is still practised with few essential modifications by surgeons all over the world; at his clinic visitors congregated in large numbers. Each plastic operation was planned and demonstrated on a blackboard and then executed with a precision and dexterity that were unique. A quiet, unassuming man, he was immensely popular and was twice President of the Hungarian Ophthalmological Society (1920–25 and 1934–36).

JOHN MARTIN WHEELER [1879–1938] (Fig. 591) was born of English extraction in Burlington, Vermont, and graduated in medicine and science at the university of that city where he subsequently taught anatomy. In 1908 he went to New York, a city where he spent the remainder of his working life. He was appointed professor of ophthalmology at Columbia University in 1928 and three years later when the Eye Institute was built at the Presbyterian Medical Center he was invited to become its director; under his stimulating leadership this institute became one of the foremost ophthalmic centres not only in America but also in the world. Surgery was his great interest and in this his technique could be described without exaggeration as beautiful, being equalled only by his judgement and originality in devising new procedures; here he had no peer. Particularly was this seen in plastic surgery and his contributions to the treatment of disorders of the lids will be appreciated in almost every section of this Chapter. The combination of unusual ability and hard work with his delightful personality made it natural that he played a prominent part in American ophthalmological affairs, culminating in the presidentship of the American Academy of Ophthalmology and Otolaryngology in 1933. The loss of an eye from a malignant melanoma in 1935 did not curtail his surgical dexterity but three years thereafter he died from a cardiac attack.

Of the many motor disorders of the lids, congenital anomalies have been discussed in another Volume of this *System*,[1] while those of neurogenic origin have been dealt with in the Volume on Neurology.[2] These will therefore not be included in this

[1] Vol. III, p. 887. [2] Vol. XII, p. 890.

Chapter which is confined to local disorders arising from causes within the lids themselves.

The lid-reflexes, including Bell's phenomenon and its anomalies, have already been described,[1] so also have those associated with pupillary reactions.[2]

BLEPHAROPTOSIS

PTOSIS (BLEPHAROPTOSIS)—*a drooping of the upper lid below its normal position*—is relatively common. In the usual condition when the levator muscle is at fault, the pendant lid is smooth, inert and unwrinkled, for without the pull of this muscle the tarsal fold is lost. Apart from the cosmetic effect the condition produces no disability until it is so marked that

FIGS. 592 and 593.—BLEPHAROPTOSIS.

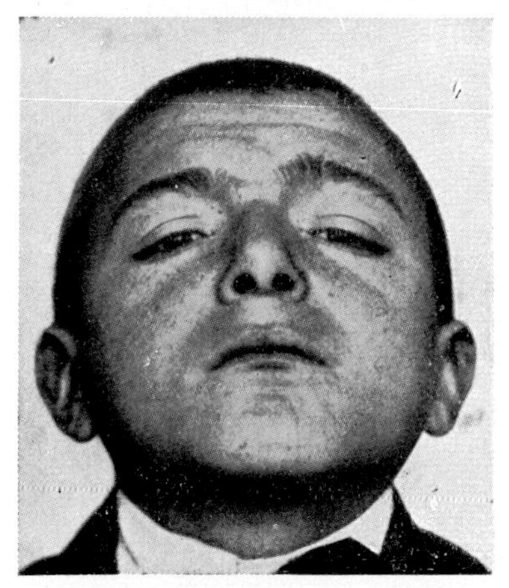

FIG. 592.

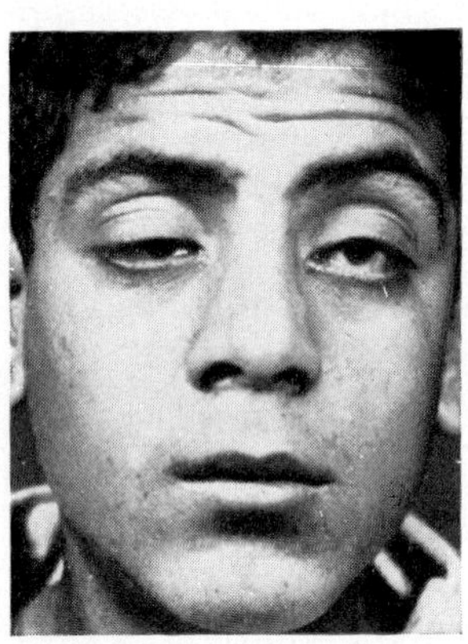

FIG. 593.

FIG. 592.—Ptosis in progressive external ophthalmoplegia, to show the habitual position of the head adopted in compensation, with the chin thrown up and the brows arched (A. Bielschowsky).

FIG. 593.—Simple bilateral asymmetrical congenital ptosis, showing the excessive action of the frontalis muscle (A. Kamel).

the pupillary area is covered and the vision becomes obscured. As this begins to occur, in cases wherein the patient relies on the vision of the eye as in unilateral cases when the only good eye is affected or in bilateral cases of ptosis, compensation is attempted at first by a continued overaction of the frontalis and corrugator muscles which arch up the brows and throw the forehead into horizontal folds laterally and vertical furrows over the root of the nose, and eventually by tilting back the head, producing a characteristic attitude (Figs. 592–3). The contraction of the occipito-frontalis is not entirely voluntary; its mechanism is physiologically comparable to that determining the secondary deviation of a conjugate muscle in paralytic strabismus for, since the levator and the frontalis are normally

[1] Vol. XII, p. 892. [2] Vol. XII, p. 674.

associated muscles, when the former is paralysed, the increased effort expended in attempting to move it involves an increased contraction of the latter. There is, of course, a pseudo-contraction of the upper visual field. If, however, the pupillary aperture is completely covered, the vision in the affected eye is rendered useless, the compensatory effort ceases and the lid, hanging down lax and incapable of active movement, can only be raised mechanically.

The *measurement of ptosis* is sometimes of importance from the prognostic point of view, but no method is entirely reliable. The classical technique is to measure perimetrically the pseudo-contraction of the upper part of the field of fixation as the patient looks upwards, but an associated palsy of the superior rectus may vitiate the readings. Measurement of the vertical width of the palpebral fissure in different directions of gaze is equally unsatisfactory since the result is complicated by movements of the lower lid. Probably the most accurate method is to measure the distance between the ciliary border of the lid and the upper orbital rim ; this is first done when the patient looks down and then as he looks forwards and finally upwards, the surgeon's thumb, supporting the rule, in the meantime pressing firmly on the eyebrow to prevent any activity of the frontalis muscle from affecting the measurement of activity of the levator.

The Treatment of Blepharoptosis

The treatment of ptosis, apart from therapeutic measures directed to the cause, is indicated for cosmetic or visual reasons, and in either case may be palliative or operative. The use of a mechanical crutch may be advisable when the ptosis is paralytic in origin, or forms part of a progressive muscular disease : this applies particularly to the continually changing ptosis of myasthenia gravis. Operation should never be undertaken if the result is to leave the eye permanently open during the waking or sleeping hours lest an exposure keratitis develop, in unilateral cases if elevation of the lid is to lead to the unmasking of a diplopia, or until it has been established that the degree of drooping of the lid has become stabilized, particularly that it will not improve with time. In contradistinction to congenital ptosis, the acquired condition should not be corrected surgically until all other means have been exhausted. It is also to be remembered that in spite of the variety and ingenuity of remedial operations, the results are by no means invariably brilliant. The greatest care is necessary in the selection of cases and the type of procedure adopted must be individually suitable to each case.

Lid crutches to support a drooping lid mechanically were first introduced by Goldzieher (1890) who used as a crutch a shell plate attached to the upper rim of a shell spectacle frame. Several types have since been introduced. Kauffmann (1893) attached an adjustable watch-spring to the spectacle rim, pressing upward and outward near the site of the lacrimal gland, while Meyer (1893) described a wire crutch worn independently of spectacles. The usual type now employed is a wire support in the form of a semilunar loop soldered to the upper part of the rims of a pair of spectacles, or a more flexible single wire soldered at one end only (Dodge, 1935) (Fig. 594):

on adjusting the spectacles the wire hooks up one or both lids and keeps them elevated (Fig. 595). The crutch should project backwards and slightly downwards from the upper part of the frame so that it lies in contact with the curve of the lid, fitting snugly and comfortably under the supra-orbital ridge, and to attain this several adjustments are usually necessary. As a rule such an appliance irritates at first, discomfort being experienced by the forced and permanent opening of the eyes; epiphora and a habit-spasm of the brows and forehead are common initially, but continued use usually brings with it increased acquiescence.

A haptic *contact lens* fitted with a shelf on which the margin of the upper lid rests is another expedient suggested by Dudragne (1946) and employed

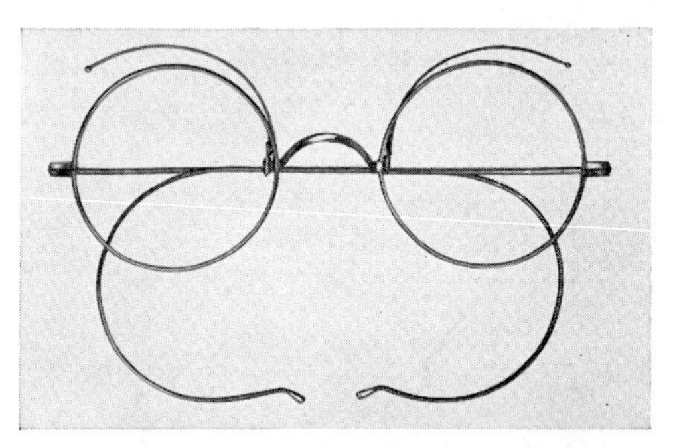

Fig. 594.—Crutch Spectacles for Ptosis.

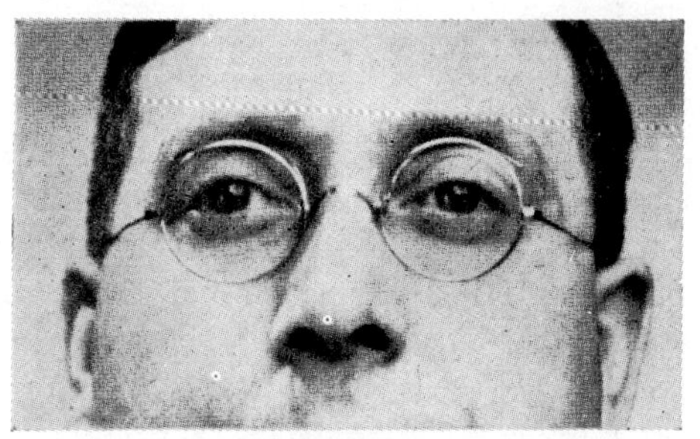

Fig. 595.—The Spectacles in use showing the Invisibility of the Crutches (Hamblin).

by Watillon and Pivont (1957), Cochet and his colleagues (1967) and Davidson (1970) (Figs. 596–9); it is not, however, so generally applicable as the use of crutches on spectacles.

A further suggestion has been the elevation of the lid by *magnetic force* (Conway, 1973). A strip of the highly magnetizable Mu-metal is implanted in the upper lid and a magnet is placed behind the upper rim of the spectacle frame, or more efficiently two small cobalt-platinum magnets are placed one in or on the lid fixed by a skin adhesive and another on the spectacle frame. This procedure avoids the discomfort of the pressure of a crutch spectacle.

SURGICAL TREATMENT is required in most cases to obtain a satisfactory result from the cosmetic and functional points of view. Over 100 modifications of techniques are to be found in the literature, all of which can be resolved into a few general procedures; the more important of these are included in the following summary.

FIGS. 596 to 599.—SCLERAL CONTACT LENSES FOR PTOSIS (M. Ruben).

FIG. 596.

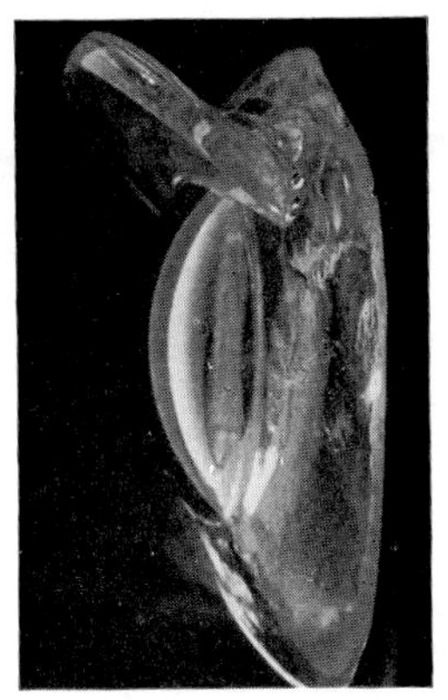

FIG. 597.

FIG. 596.—Ledge-type moulded scleral lens.

FIG. 597.—Solid ledge from thickened optic, the superior haptic portion being cut away.

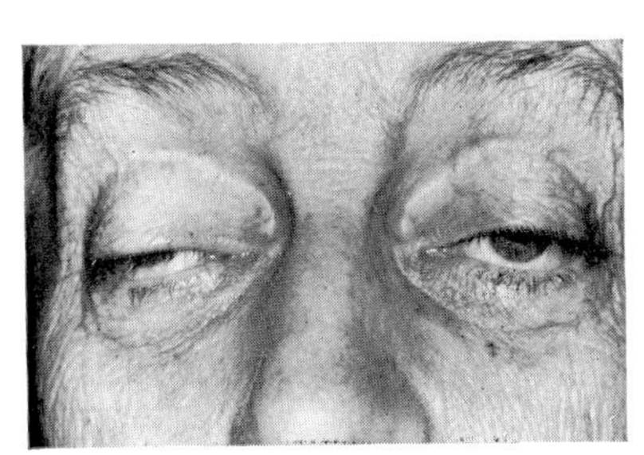

FIG. 598.

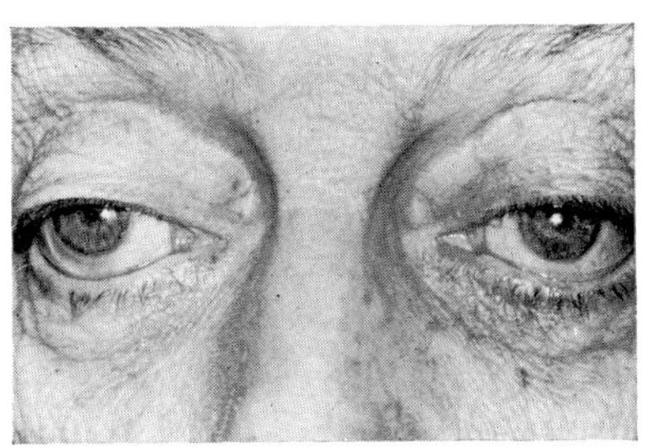

FIG. 599.

FIGS. 598 and 599.—A patient with ocular myopathy before (Fig. 598) and after (Fig. 599) the insertion of the lenses.

(1) In gravitational ptosis *the removal of hyperplasic tissue*, as after chronic inflammations such as trachoma, chalazia or tumours, is a technique originally suggested by Saunders (1811) but obviously requires modifications depending on the lesion involved.

(2) *Shortening of the lid by the removal of an elliptical slip of skin* was an early technique proposed by the Arabian surgeons and adopted and modified by Scarpa (1801), Hunt (1830) and others. It was extended to include an excision of a strip of the orbicularis muscle by von Graefe (1863).

(3) *Reinforcement of the action of the levator palpebræ superioris* in cases in which this muscle has retained some activity.

(a) Resection of the levator by an external approach (through the skin) and if necessary its advancement over the anterior surface of the tarsus was a technique introduced by Eversbusch (1883) and modified by many others.[1]

(b) A resection of the levator and a partial resection of the tarsus (through a conjunctival approach) was a technique pioneered by Bowman (1857), popularized by Blaskovics (1909–23) and modified by many others[2]; among these a very popular procedure for minimal degrees of ptosis is the operation of Fasanella and Servat (1961–73) wherein the upper part of the tarsus with its attached conjunctiva, the palpebral conjunctiva above the tarsus, Müller's muscle and the levator are grasped in the jaws of hæmostats and excised.

(c) A resection of the levator and tarsus from the conjunctival aspect and of a musculo-cutaneous elliptical strip was suggested by Hervouet and Tessier (1956).

(4) *The use of the superior rectus muscle* (provided this muscle is active) to replace the action of the levator muscle is a technique introduced by Motais (1897) who inserted strips of the middle fibres of the tendon of this muscle into the anterior face of the tarsus. Several alternative procedures have been suggested.

(a) In Berke's operation the whole width of the superior rectus divided into three strips is anchored to the tarsus or the margin of the lid and then the belly of the muscle is sewn into its original insertion (Berke and Hackensack, 1949).

(b) Strips of the orbicularis muscle, free at the ends but attached to the tarsal plate in the centre, are sutured to the rectus muscle (Wheeler, 1939).

(c) A loop-suture is passed between the rectus muscle and the upper border of the tarsus (Parinaud, 1897).

(d) Strips of fascia lata sling the anterior surface of the tarsal plate to the rectus muscle (Dickey, 1936; Stallard, 1965).

(5) *The use of the frontalis muscle to elevate the lid* by attaching the upper border of the tarsus subcutaneously to the muscle overlying the eyebrow.

(a) By the use of sutures: catgut (Dransart, 1886); silk (Pagenstecher, 1881; de Wecker, 1882; Hess, 1894; Friedenwald and Guyton, 1948); metals such as gold, silver and platinum (Mules, 1895; Harman, 1903; and others); silicone threads or strips which have the advantage of being elastic (C. and G. Tillett, 1966; Katz and Kuder, 1968). In unilateral cases Callahan (1972) suggested the suspension of both eyelids to their respective frontalis muscles in order to obtain symmetry.

(b) By the use of slips of skin from the upper lid (Panas, 1886; Tansley, 1895; Machek, 1914–15; Fox, 1968); the tendon of the levator muscle cut to the desired length (Angelucci, 1898; Sourdille, 1903; Bietti, 1942; Schimek, 1955; Johnson, 1962); strips of orbicularis muscle (Reese, 1923; Cattaneo, 1934); strips of the frontalis muscle (Fergus, 1901; Roberts, 1916) or the corrugator supercilii (L. Jones and Wilson, 1967).

(c) The use of strips of autogenous fascia lata is the most popular expedient[3]; alternatively, the fascia sterilized by cobalt radiation and stored has been used by Gutman (1965) and Falls and his colleagues (1967), or bovine fascia by Yasuna (1962) and Billet (1968).

(d) Strips of fresh or preserved sclera were employed by Bodian (1968).

(6) *Slinging the tarsus to the orbicularis muscle* by passing a strip of the periorbital

[1] Wolff (1896), de Lapersonne (1903), Elschnig (1903–10), Leahey (1953), Johnson (1954), Berke (1959).

[2] Agatston (1942), Berke (1952), Iliff (1954), Fasanella and Servat (1961), Putterman (1972), Crawford (1973).

[3] Kirschner (1910), Wright (1922), Lexer (1923), Reese (1924), Derby (1928), Stallard (1945), Johnson (1962), Ricci (1966).

part of the muscle through a buttonhole in the upper edge of the tarsal plate was suggested by Sarwar (1952–72).

(7) *The use of a stainless-steel wire spring* incorporated in the tissues of the lid was suggested by Morel-Fatio and Lalardrie (1962–67), a procedure they introduced for the treatment of lagophthalmos in facial palsy. The spring, shaped somewhat after the manner of a safety-pin, is inserted at the outer canthus; one arm is attached to the periosteum of the inferior orbital margin and the other fixed to the tarsus of the upper lid. The two lids are thus separated by the spring which should not be so strong as to prevent their closure. The technique has been modified among others by Grignon and his colleagues (1962), Leopold (1965) and Weinstein and Kaplan (1968).

For the details of this multitude of techniques the original papers or textbooks of plastic and ophthalmic surgery should be consulted.

The aim of any operation should be to lift the ptotic lid above the pupillary aperture when the eyes are in the primary position. Moreover, the height of the two lids whether the ptosis is unilateral or bilateral should be equal; there should also be adequate mobility when blinking, a normal lid-fold and no diplopia. The choice of procedure can be determined only after a careful evaluation of the ætiology of the individual case and an assessment of the remaining functions of the various parts of the ocular motor apparatus. The degree of ptosis should be carefully measured in the primary position and when looking up and down and the actions of the levator palpebræ and of the superior rectus noted. In assessing a weakness of the former, a finger should be pressed firmly on the brow to eliminate the action of the frontalis muscle. In these assessments photographs are valuable and electromyography may assist in determining whether there is activity or not in the levator. The complications which may arise and should be guarded against, apart from an over- or under-correction, are a lid-lag on downward gaze, lagophthalmos with its associated danger to the cornea, the occurrence of entropion or ectropion, the creation of anomalous lid-folds and, as in all surgery, the occurrence of infection.

The multitude of techniques noted above appears very varied and confusing, but they can be reduced to a few general principles (Stallard, 1965; Beard, 1966–69; B. Smith *et al.*, 1969).

Of them all, the excision of an excess of tissue producing a gravitational ptosis is successful in simple cases, such as a tarsectomy for trachoma; but neoplasic diseases provide a very different problem. A neurofibroma, for example, may require extensive surgical procedures which frequently have to be repeated. The excision of skin and subcutaneous tissue has a very limited application; it is usually inadequate and if drastically performed readily leads to lagophthalmos. It may, however, be indicated in cases of senile ptosis, particularly with a heavy fold of redundant skin (Stallard, 1965).

In the usual cases of acquired neurogenic ptosis, the use of foreign materials of whatever kind has been largely abandoned because of the significant incidence of early and late infection and their tendency to become loose or break when subjected to minor trauma (Falls *et al.*, 1967). A considerable amount of mobility of the lid is desirable and for this purpose three sources of power are available: the levator, the frontalis and the superior rectus muscles. If the levator has some activity it is the preferred source of power in most cases. Of the two methods of approach, the external (through the skin) is perhaps preferable particularly in marked cases, that is, a modification of Eversbusch's operation, as by Johnson (1954) and Berke (1959). The internal approach (through the conjunctiva) is a very effective procedure, that is, a modification of Blaskovics's operation as by Fasanella and Servat (1961) for minor degrees or by Agatston (1942), Berke (1952) or Iliff (1954). If the superior rectus is very active it may be employed, but the simple technique of Motais is probably the least satisfactory. The hypotropia which tends to occur is lessened in the procedures devised by Dickey (1936) or Berke and Hackensack (1949), but the post-operative complications of corneal desiccation owing to the impairment of blinking are sometimes unfortunate. The use of the frontalis muscle is probably the most popular technique when the levator muscle is completely ineffective, and of the many expedients available the use of autogenous fascia lata is the safest procedure.

In sympathetic ptosis, as in Horner's syndrome, a small resection of the levator is sufficient.

In myogenic ptosis, as in senile ptosis without redundant skin, a small resection of the levator using either approach is suitable; in progressive external ophthalmoplegia suspension from the brow by fascia lata is usually advisable.

Congenital ptosis often provides the most difficult problem when it is associated with anomalies such as epicanthus or blepharophimosis and other deformities; these require extensive plastic operations such as canthoplasty. So far as the ptosis is concerned the same principles as indicated above are applicable. In the Marcus Gunn jaw-winking phenomenon a resection of the levator is effective in mild cases but a bilateral suspension from the brow by fascia lata is indicated in severe cases.

In traumatic ptosis when the function of the levator muscle is retained its resection is usually sufficient and if the levator muscle itself is lacerated its early re-suture is imperative.

BLEPHAROPTOSIS OF THE LOWER LID is usually due to paralysis, senile degenerative changes or to cicatrization; occasionally, however, an idiopathic sagging of the lower lid occurs spontaneously which is probably caused by a weakness of the capsulo-palpebral fascia or of Müller's muscle in the lower lid. It occurs preferentially in older women and if it constitutes a cosmetic blemish it is most easily rectified by Fox's (1966) modification of the Kuhnt-Szymanowski operation[1] (Fox, 1972).

[1] p. 585.

Agatston. *Arch. Ophthal.*, **27,** 994 (1942).
Angelucci. *Ann. Ottal.*, **27,** 541 (1898).
Beard. *Trans. Amer. ophthal. Soc.*, **64,** 401 (1966).
Ptosis, St. Louis (1969).
Berke. *Arch. Ophthal.*, **48,** 460 (1952); **61,** 177 (1959).
Berke and Hackensack. *Trans. Amer. Acad. Ophthal.*, **53,** 499 (1949).
Bietti. *Boll. Oculist.*, **21,** 721 (1942).
Billet. *Amer. J. Ophthal.*, **65,** 561 (1968).
Blaskovics. *Klin. Mbl. Augenheilk.*, **47** (2), 323 (1909).
Arch. Ophthal., **52,** 563 (1923).
Bodian. *Amer. J. Ophthal.*, **65,** 352 (1968).
Bowman. *Roy. Lond. ophthal. Hosp. Rep.*, **1,** 34 (1857).
Callahan. *Amer. J. Ophthal.*, **74,** 321 (1972).
Cattaneo. *Atti Cong. Soc. oftal. ital.*, **1,** 54 (1934).
Cochet, Marechal-Courtois and Prijot. *Bull. Soc. belge Ophtal.*, No. 145, 1 (1967).
Conway. *Brit. J. Ophthal.*, **57,** 315 (1973).
Crawford. *Canad. J. Ophthal.*, **8,** 19 (1973).
Davidson. *Trans. ophthal. Soc. U.K.*, **90,** 139 (1970).
Derby. *Amer. J. Ophthal.*, **11,** 352 (1928).
Dickey. *Amer. J. Ophthal.*, **19,** 660 (1936).
Dodge. *Arch. Ophthal.*, **14,** 989 (1935).
Dransart. *Ann. Oculist.* (Paris), **84,** 88 (1880).
Dudragne. *Théorie élémentaire et application de l'optique de contact à la vision*, Paris (1946).
Elschnig. *Wien. med. Wschr.*, **53,** 2402 (1903).
Med. Klin., **6,** 771 (1910).
Eversbusch. *Klin. Mbl. Augenheilk.*, **21,** 100 (1883).
Falls, Sloan and Bryson. *Amer. J. Ophthal.*, **64,** 426 (1967).
Fasanella. *Trans. ophthal. Soc. U.K.*, **93,** 425 (1973).
Fasanella and Servat. *Arch. Ophthal.*, **65,** 493 (1961).
Fergus. *Brit. med. J.*, **1,** 762 (1901).
Fox. *Amer. J. Ophthal.*, **65,** 359 (1968); **74,** 330 (1972).
Friedenwald and Guyton. *Amer. J. Ophthal.*, **31,** 411 (1948).
Goldzieher. *Zbl. prakt. Augenheilk.*, **14,** 34 (1890).
von Graefe. *v. Graefes Arch. Ophthal.*, **9** (2), 57 (1863).
Grignon, Chouard and Benoist. *Ann. Otolaryng.* (Paris), **79,** 847 (1962).
Gutman. *Amer. J. Ophthal.*, **59,** 1095 (1965).
Harman. *Brit. med. J.*, **2,** 736 (1903).
Hervouet and Tessier. *Bull. Soc. franç. Ophtal.*, **69,** 239 (1956).
Hess. *Arch. Augenheilk.*, **38,** 22 (1894).
Hunt. *Lond. med. Gaz.*, **7,** 361 (1830).
Iliff. *Amer. J. Ophthal.*, **37,** 529 (1954).
Johnson. *Amer. J. Ophthal.*, **38,** 129 (1954).
Arch. Ophthal., **67,** 18 (1962).
Jones, L., and Wilson. *Trans. Amer. Acad. Ophthal.*, **71,** 889 (1967).
Katz and Kuder. *Canad. J. Ophthal.*, **3,** 353 (1968).
Kauffmann. *Zbl. prakt. Augenheilk.*, **17,** 75 (1893).
Kirschner. *Arch. klin. Chir.*, **92,** 888 (1910).
de Lapersonne. *Arch. Ophtal.*, **23,** 497 (1903).
Leahey. *Arch. Ophthal.*, **50,** 588 (1953).
Leopold. *Arch. Ophtal.*, **25,** 755 (1965).
Lexer. *Klin. Mbl. Augenheilk.*, **70,** 464 (1923).
Machek. *Arch. Augenheilk.*, **76,** 8 (1914).
Arch. Ophthal., **44,** 539 (1915).
Meyer. *Arch. Augenheilk.*, **26,** 153 (1893).
Morel-Fatio and Lalardrie. *Ann. Chir. plast.*, **7,** 275 (1962).
Plastic and Reconstructive Surgery of the Eye and Adnexa (ed. B. Smith *et al.*), St. Louis, 380 (1967).
Motais. *Bull. Soc. franç. Ophtal.*, **15,** 208 (1897).
Ann. Oculist. (Paris), **118,** 5 (1897).
Mules. *Ophthal. Rev.*, **14,** 156 (1895).
Pagenstecher. *Trans. int. med. Cong.*, London, **3,** 108 (1881).
Panas. *Arch. Ophtal.*, **6,** 1 (1886).
Parinaud. *Ann. Oculist.* (Paris), **118,** 13 (1897).
Putterman. *Arch. Ophthal.*, **87,** 655 (1972).
Reese. *Trans. Amer. ophthal. Soc.*, **21,** 71 (1923).
Arch. Ophthal., **53,** 26 (1924).
Ricci. *Ophthalmologica*, **152,** 318 (1966).
Roberts. *Ophthal. Rec.*, **25,** 397 (1916).
Sarwar. *Brit. J. plast. Surg.*, **4,** 293 (1952).
Ann. Ophthal., **4,** 250 (1972).
Saunders. *A Treatise on some Practical Points relating to the Diseases of the Eye*, London (1811).
Scarpa. *Saggio di osservazioni e d'esperienze sulle principali malattie degli occhi*, Pavia (1801).
Schimek. *Arch. Ophthal.*, **54,** 92 (1955).
Smith, B., McCord and Baylis. *Amer. J. Ophthal.*, **68,** 92 (1969).
Sourdille. *Clin. Ophtal.*, **9,** 73 (1903).
Stallard. *Trans. ophthal. Soc. U.K.*, **65,** 68 (1945).
Eye Surgery, 4th ed., Bristol, 164 (1965).
Tansley. *Trans. Amer. ophthal. Soc.*, **7,** 427 (1895).
Tillett, C. and G. *Amer. J. Ophthal.*, **62,** 521 (1966).
Watillon and Pivont. *Bull. Soc. belge Ophtal.*, No. 117, 608 (1957).
de Wecker. *Ann. Oculist.* (Paris), **88,** 29 (1882).
Weinstein and Kaplan. *Ann. Chir. plast.*, **13,** 175 (1968).
Wheeler. *Arch. Ophthal.*, **21,** 1 (1939).
Wolff. *Arch. Augenheilk.*, **33,** 125 (1896).
Wright. *Arch. Ophthal.*, **51,** 99 (1922).
Yasuna. *Amer. J. Ophthal.*, **54,** 1097 (1962).

The Clinical Types of Ptosis

For purposes of description the many types of ptosis which occur may be conveniently classified as follows; it will be seen that most of these conditions are discussed in other Volumes of this *System* so that the only types requiring description here are pseudo-ptosis, mechanical ptosis and myogenic ptosis.

(1) Apparent or pseudo-ptosis.

(2) Congenital ptosis.[1]

(3) Mechanical ptosis due to increased weight of the lid or its involvement in local cicatricial or destructive processes.

(4) Pseudo-paralytic ptosis occurring in conditions of atony of the levator or overaction of the orbicularis, or in senile states; the spasmodic ptosis of hysteria has been discussed[2] as well as the hypertonic type seen in myostatic paralysis.[3]

(5) Myogenic ptosis, due to (i) incidental disease of the levator muscle, (ii) muscular dystrophies, late hereditary ptosis, and (iii) myasthenia gravis.

(6) Paralytic ptosis due to a destructive lesion of the IIIrd cranial nerve in its orbital or intracranial course, the IIIrd nucleus or its supranuclear connections[4]; toxic ptosis.[5]

(7) Sympathetic ptosis due to a lesion of the sympathetic in its peripheral or central course.[6]

(8) Periodic and synkinetic ptosis.[7]

(9) Traumatic ptosis due to palpebral, orbital or intracranial injury.[8]

AN APPARENT OF PSEUDO-PTOSIS appears when the upper lid lacks its normal support, as occurs with an empty socket or an ill-fitting prosthesis, in microphthalmos or with a shrunken eyeball, in enophthalmos or when the orbital fat has been lost as in old age, in facial hemiatrophy, or in downward displacement of the globe after trauma. The intermittent ptosis which occurs in the retraction syndrome on adduction of the affected eye is of this type, for in this movement the globe is retracted several millimetres into the orbit.[9] A RELATIVE PTOSIS may be said to exist in cases of unilateral lid-retraction; not only may the asymmetry suggest ptosis on the sound side rather than retraction on the other, but a compensatory contraction of the palpebral portion of the orbicularis of the retracted lid may involve a bilateral effect actually producing a ptosis (Wolff, 1933).

A MECHANICAL PTOSIS occurs when the lid is weighted downwards as in conditions affecting the conjunctiva (trachoma), or the lid itself (œdema, inflammation, infiltration as in amyloidosis, or a tumour) or has suffered hypertrophy (elephantiasis)[10] or has become weighted with prolapsed fat (*ptosis adiposa*) as in the later stages of blepharochalasis[11] (Fig. 347). Similarly the levator may be incapacitated mechanically or by œdema and infiltration in inflammatory or neoplasic orbital affections so that ptosis is produced (Comberg, 1921); the same result may follow sinus disease (Dejean, 1927; Jain and Agarwal, 1961; Jain and Srivastava, 1966; and others).

[1] Vol. III, p. 887. [2] Vol. XII, p. 897. [3] Vol. XII, p. 854.
[4] Vol. XII, p. 898. [5] Vol. XII, p. 901. [6] Vol. XII, p. 902.
[7] Vol. III, p. 898. [8] Vol. XIV, pp. 217, 283, 424.
[9] Vol. III, p. 991; VI, p. 740. [10] p. 356. [11] p. 350.

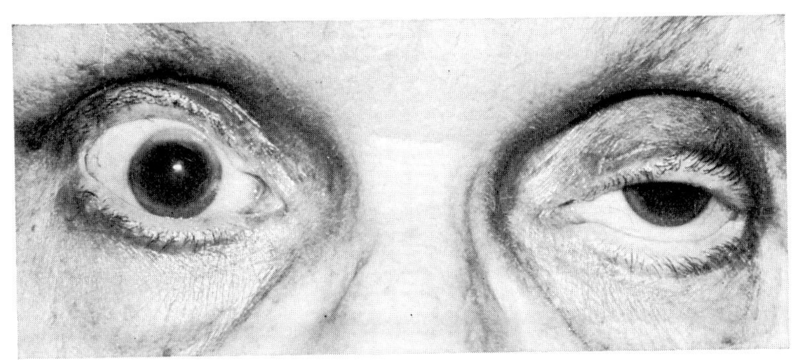

FIG. 600.—SENILE PTOSIS.
Affecting the left eye in a woman aged 88 (Inst. Ophthal.).

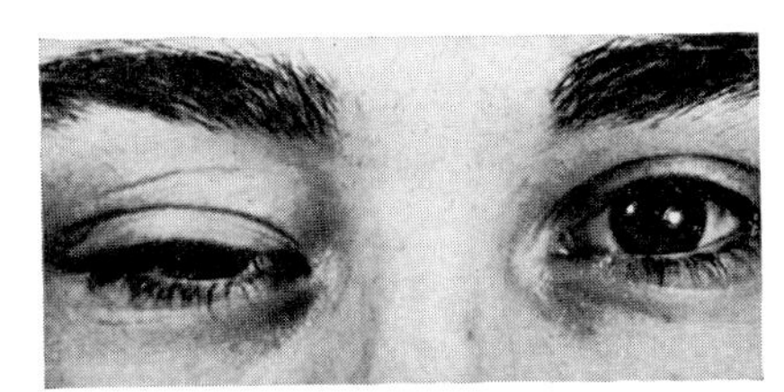

FIG. 601.—TRAUMATIC PTOSIS.
From a laceration of the upper lid severing the levator palpebræ due to the kick of a horse (A. MacIndoe).

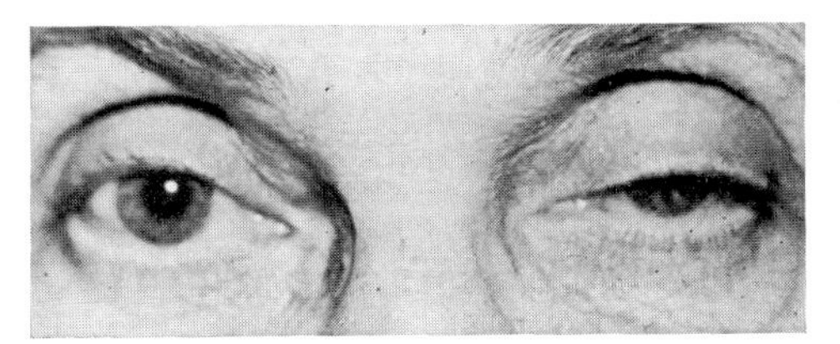

FIG. 602.—UNILATERAL PTOSIS WITH PREGNANCY.
The ptosis developed 29 years previously during an otherwise normal pregnancy (Crowell Beard).

To this category usually belongs the ptosis associated with proptosis in cavernous sinus thrombosis owing to the associated orbital thrombophlebitis, since direct damage to the IIIrd nerve producing complete ophthalmoplegia is rare in this condition. Recurrent œdema and round-celled infiltration together with a general laxity of all the tissues of the lids produce the atonic ptosis of blepharochalasis.[1] A mechanical ptosis also follows a lesion of the upper lid when the tendon or the muscular belly of the levator is cut or rendered ineffective, whether by trauma or after destructive inflammatory processes such as abscess or gangrene of the lid or osteomyelitis of the frontal bone (Fig. 601). The embedding of a contact lens in the tarsal plate was an unusual cause reported by Yassin and his colleagues (1971).

[1] p. 350.

PSEUDO-PARALYTIC PTOSIS of the atonic type occurs when the muscle has lost its tone, such as occurs in senility (Fig. 600), after prolonged bandaging or long predominance of the orbicularis in continued blepharospasm.

A TOXIC PTOSIS may occur in acute infections or metabolic toxæmias such as eclampsia, anæmia and diabetes (Futterweit *et al.*, 1965); it may also develop in pregnancy for unknown reasons (Beard, 1966) (Fig. 602). Ptosis may also complicate the administration of several drugs, such as arsenic, vincristine and the corticosteroids; these are all listed in another Volume.[1]

Beard. *Trans. Amer. ophthal. Soc.*, **64,** 401 (1966).

Comberg. *Z. Augenheilk.*, **46,** 249 (1921).

Dejean. *Arch. Ophtal.*, **44,** 657 (1927).

Futterweit, Schwartz and Mirsky. *J. clin. Endocr.*, **25,** 1280 (1965).

Jain and Agarwal. *J. int. Coll. Surg.*, **35,** 348 (1961).

Jain and Srivastava. *EENT Mthly.*, **45,** 76 (1966).

Wolff. *Trans. ophthal. Soc. U.K.*, **53,** 317 (1933).

Yassin, White and Shannon. *Amer. J. Ophthal.*, **72,** 536 (1971).

MYOGENIC PTOSIS

Disease of the levator produces ptosis of a varying degree, whether it is due to inflammatory, neoplasic or parasitic causes (Spaeth, 1945); a rare example of interest is the involvement of the levator in a general polymyositis (Strümpell, 1891). The muscular dystrophies, ocular myopathy, myotonic dystrophy and myasthenia, however, are deserving of special note. There is considerable doubt regarding the categorization of these conditions and, indeed, it has been suggested that ocular myopathy is a variant of progressive muscular dystrophy with characteristic clinical features (Kiloh and Nevin, 1951; Beckett and Netsky, 1953; Walton, 1964; Magora and Zauberman, 1969). A definite differentiation will probably await a clarification of the underlying biochemical aberrations which presumably determine their incidence. It will be remembered that the clinical aspect of the varying degrees of ophthalmoplegia encountered has been discussed in a previous Volume.[2]

MUSCULAR DYSTROPHIES (MYOPATHIES)

PROGRESSIVE MUSCULAR DYSTROPHY is a hereditary disease of obscure origin in which an increasing wastage and weakness of the muscles occur with no demonstrable involvement of the central nervous system. It usually starts in childhood or adolescence and preferentially attacks different groups of muscles although the general tendency is for all the muscles of the body to be involved eventually in the severe *Duchenne's type* and the patient is incapacitated. The face is usually spared except in the *facio-scapulo-humeral type* of Landouzy and Déjerine (1884) which, in addition, affects the shoulder girdle and upper arms; there is a characteristic facial expression, the eyes cannot be completely closed and the patient cannot whistle. The prognosis in this type is better than in most other forms and the patient often

[1] Vol. XIV, p. 1354.

[2] Vol. VI, p. 758

lives to a normal age although with increasing disability as the dystrophy progresses. In these dystrophies involvement of the ocular muscles is rare but does occur (Walsh and Hoyt, 1969).

CHRONIC PROGRESSIVE EXTERNAL OPHTHALMOPLEGIA (OCULAR MYOPATHY)

This chronic, progressive, bilateral ophthalmoplegia often appearing as a hereditary affection was at one time generally thought to be an internuclear (von Graefe, 1868) or a nuclear palsy (Möbius, 1900), but it has been demonstrated that while some neurogenic cases occur (Daroff *et al.*, 1966; Brion and de Recondo, 1967) most appear to be the result of a primary myopathy; Kiloh and Nevin (1951) therefore suggested the term *ocular myopathy*. An alternative designation was *chronic progressive ocular muscular dystrophy*; Cogan (1956) suggested *abiotrophic ophthalmoplegia*.

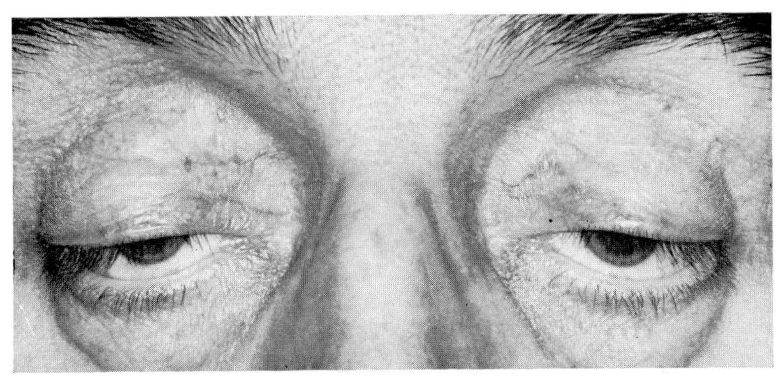

FIG. 603.—PTOSIS IN PROGRESSIVE EXTERNAL OPHTHALMOPLEGIA.
In a man aged 51 (Inst. Ophthal.).

Nevertheless, since the neurological cases are difficult to differentiate from the myopathic, the safest general term may be *progressive external ophthalmoplegia* (Rosenberg *et al.*, 1968).

The condition has long been recognized (Strümpell, 1886; Lawford, 1887; Marina, 1896; Beaumont, 1900; Bradburne, 1912; and others); Wilbrand and Saenger (1900) collected 32 cases from the literature, McMullen and Hine (1921), describing 2 cases, annotated a further 5, and Kiloh and Nevin (1951) reviewed the literature up to that date. Later reports have been numerous.[1]

A rare congenital form of the disease appearing at or shortly after birth may occur, as was first described by von Graefe (1868); in this event the ophthalmoplegia progresses rapidly and is often complete before the age of 5 years.[2] The disorder usually appears during childhood and is

[1] Beckett and Netsky (1953), Schwarz and Liu (1954), Thiel (1954), Henneaux and Hérode (1955), McAuley (1956), Stephens *et al.* (1958), Senita and Fisher (1958), Nicolaissen and Brodal (1959), Magyar *et al.* (1959), Lapresle and Jarlot (1959), Davidson (1960–70), Teasdall and Sears (1960), Andrews (1961), Moses and Heller (1965), Kearns (1965), Drachman (1968), Magora and Zauberman (1969), Daroff (1969), and others.

[2] Vol. III, p. 986.

progressive although its onset may first be noticed between 20 and 30 years of age and has been delayed until the age of 50 (Lawford, 1887). The first symptom is usually ptosis which initially may be unilateral but gradually becomes bilateral and complete, and occasionally soon, but sometimes after an interval of some years, a slowly progressive palsy of all the extra-ocular muscles develops until the eyes are practically immobile (Figs. 592, 603); the internal musculature is, however, unaffected. The clinical nature of the ophthalmoplegia has been discussed in a previous Volume.[1] At any stage

FIGS. 604 and 605.—THE INHERITANCE OF PROGRESSIVE EXTERNAL OPHTHALMOPLEGIA.

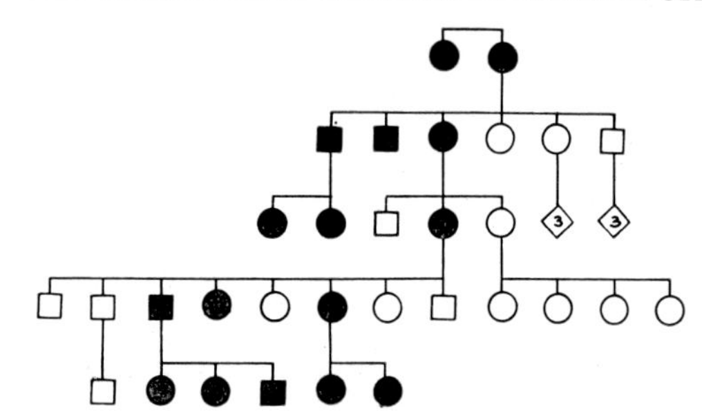

FIG. 604.—Dominant heredity (after A. Bradburne).

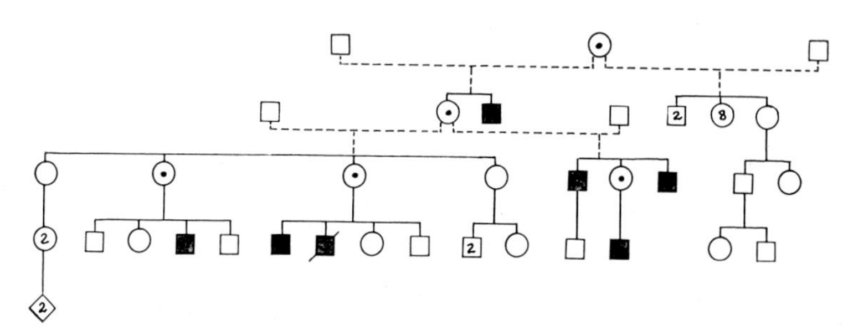

FIG. 605.—Recessive sex-linked inheritance, with myopia and other dysplasias (after A. Salleras and J. C. Ortiz de Zarate).

the disease may come to a standstill, but more often, perhaps with long static periods at any phase in its development, it progresses to complete or nearly complete ophthalmoplegia externa, the progress being extremely slow over a period of 30 to 40 years.

The malady is frequently hereditary in its incidence generally with a dominant inheritance (four generations, Beaumont, 1900; Rath, 1929; three generations, Nyman, 1970; two generations, El-Sheikh, 1958) (Fig. 604), and its familial occurrence is common (Pasetti and Salani, 1907, six cases in one family); a recessive sex-linked transmission has been noted (Salleras and Ortiz de Zarate, 1950; Ortiz de Zarate, 1966) (Fig. 605).

[1] Vol. VI, p. 758.

FIGS. 606 and 607.—PROGRESSIVE EXTERNAL OPHTHALMOPLEGIA.

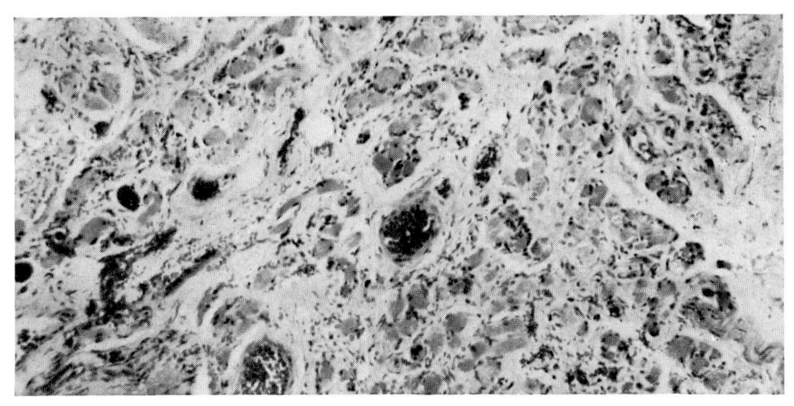

Fig. 606.—The histology of the medial rectus showing atrophy of the muscle fibres with marked proliferation of the sarcolemmal nuclei. There is an increase of the interstitial connective tissue and accumulations of fat between the muscle fibres (H. & E.; ×80) (A. Tarkkanen and V. Tommila).

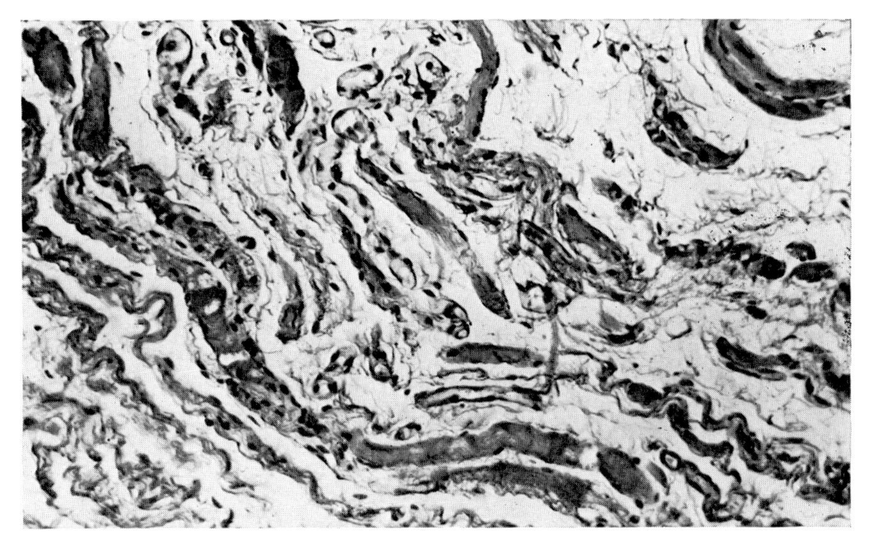

FIG. 607.—The left medial rectus, showing degeneration of the muscle fibres and the loss of cross-striations in many areas (L. Moses and G. L. Heller).

The cause of the disease has given rise to controversy. Based on the findings of Langdon and Cadwalader (1928) the view was generally held that it was due to degenerative changes in the ocular motor nuclei, but later studies at necropsy have failed to confirm this (Beckett and Netsky, 1953; Schwarz and Liu, 1954). On the other hand, the evidence of histological examination in biopsies of the extrinsic ocular muscles have been interpreted as clearly establishing the presence of changes typical of a myopathy in contradistinction to a neurogenic atrophy.[1] The fibres of the same muscle show varying degrees of change, they lose their cross-striations, the nuclei are pyknotic, the cytoplasm is granulated and the connective tissue increased (Figs. 606–7). Electromyographic studies confirm these findings.[2] Although polyphasic potentials are present without fibrillations, positive waves or fasciculations, they are more frequent than in the neuropathies, and the electromyogram shows a higher activation than would be expected from the relative immobility of the muscles (Fig. 608).

Several other conditions may be associated with the palpebral and ocular paralysis. The most frequent is muscular dystrophy. There may be weakness of the orbicularis muscle and the pharynx may be implicated causing dysphagia (*oculopharyngeal dystrophy*).[3] A more widespread involvement of the skeletal muscles may occur particularly of the neck and shoulder

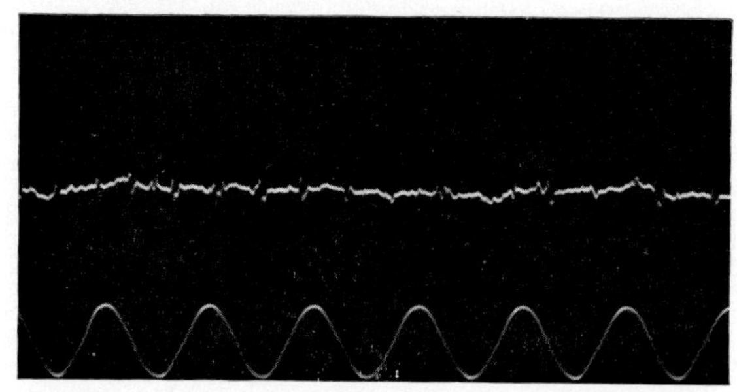

Fig. 608.—Progressive External Ophthalmoplegia.

Electromyograph of the lateral rectus during attempted abduction; the tracing shows low-voltage potentials of short duration. The sine-wave represents time and voltage, the wavelength being equivalent to 10 msec. and the amplitude to 100 V (S. Davidson).

(Fig. 609)[4]; biopsy and electromyographic studies of the non-ocular muscles have been typical of muscular dystrophy. Cardiomyopathy may also occur (Kearns, 1965). A considerable number of authors therefore considers that external ophthalmoplegia is a localized type of muscular dystrophy.[5]

From all this evidence the view that the condition was an ocular myopathy has been generally prevalent for many years, but it has recently been contended that the distinction between myogenic and neurogenic conditions

[1] Fuchs (1890), Sandifer (1946), Kiloh and Nevin (1951), Beckett and Netsky (1953), Schwarz and Liu (1954), McAuley (1956), Thorson and Bell (1959), Mölbert and Doden (1959), Cogan *et al.* (1962), Huber and Wiesendanger (1962), Lind and Prame (1963), Tarkkanen and Tommila (1965), Ellis *et al.* (1966), Zintz and Villiger (1967), Magora and Zauberman (1969), Koerner and Schlote (1972).

[2] Björk and Kugelberg (1953), Breinin (1957–58), Esslen *et al.* (1958), Mölbert and Doden (1959), Busti-Rosner (1960), Davidson (1960), Teasdall and Sears (1960), Huber and Wiesendanger (1962), Huber (1965), Moses and Heller (1965), Barrie and Heathfield (1971).

[3] Amyot (1948), Kiloh and Nevin (1951), Victor *et al.* (1962), Larsen (1963), Walton (1964), Peterman *et al.* (1964), Roberts and Bamforth (1968), Aarli (1969).

[4] Kaesar (1959), Lapresle and Jarlot (1959), Andrews (1961), Schotland and Rowland (1964).

[5] Kiloh and Nevin (1951), Beckett and Netsky (1953), Walton (1964), Moses and Heller (1965), Magora and Zauberman (1969), and others.

is often difficult and sometimes misleading both in biopsies and electromyograms (Daroff *et al.*, 1966; Drachman, 1968; Magora and Zauberman, 1969; Daroff, 1969). Moreover, many of the biopsies have been done when the muscle fibres have been largely replaced by fibrous tissues, while myopathic changes can result from long-standing denervation (Drachman *et al.*, 1967–69; Rosenberg *et al.*, 1968).

The occasional association of a group of neuro-degenerative abiotrophic disorders is of interest in this respect (Cogan, 1956)—atypical pigmentary retinal dystrophy, ataxias including those of Friedreich and Marie, Refsum's syndrome, spongiform encephalopathy, and other conditions (Drachman, 1968; Davidson, 1970). In some instances the disease has therefore been regarded as part of a general degenerative disorder due to some basic defect.

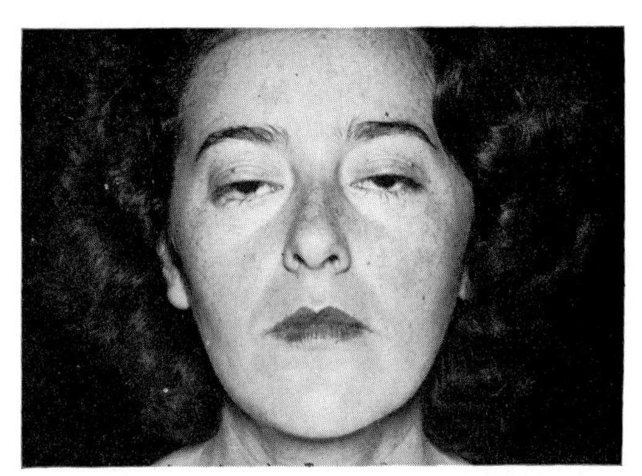

Fig. 609.—Muscular Dystrophy.

In a woman aged 27 with ptosis, paresis of eye movements and weakness and atrophy of the sterno-mastoid and shoulder muscles (Inst. Ophthal.).

It would seem certain that all cases of this condition are not due to the same cause, but our present knowledge does not allow a definite classification.

No *treatment* is effective for the disease and ameliorative measures are generally confined to the ptosis.[1] It is doubtful if its surgical correction is advisable; if this is attempted a sling of fascia lata from the frontalis muscle is usually necessary but, partly owing to the absence of Bell's phenomenon and partly to the weakness of the orbicularis muscle, lagophthalmos and exposure keratitis may result. Fox (1971) suggested that bilateral resection of the levator muscles was a safer procedure. Crutch spectacles are frequently used with success or, alternatively, a haptic contact lens with a supporting shelf (Davidson, 1970) (Figs. 596–9).

Aarli. *Acta neurol. Scand.*, **45**, 484 (1969).
Amyot. *Canad. med. Ass. J.*, **59**, 434 (1948).
Andrews. *Boston med. Quart.*, **12**, 48 (1961).
Barrie and Heathfield. *Brit. J. Ophthal.*, **55**, 633 (1971).
Beaumont. *Trans. ophthal. Soc. U.K.*, **20**, 258 (1900).
Beckett and Netsky. *Arch. Neurol. Psychiat.*, **69**, 64 (1953).

[1] p. 543.

Björk and Kugelberg. *EEG clin. Neurophysiol.*, **5**, 271 (1953).
Bradburne. *Trans. ophthal. Soc. U.K.*, **32**, 142 (1912).
Breinin. *Arch. Ophthal.*, **58**, 375 (1957).
Amer. J. Ophthal., **46** (2), 123 (1958).
Brion and de Recondo. *Rev. neurol.* (Paris), **116**, 383 (1967).
Busti-Rosner. *Ann. Ottal.*, **86**, 157 (1960).
Cogan. *Neurology of the Ocular Muscles*, 2nd ed., Springfield, 41 (1956; 1969).
Cogan, Kuwabara and Richardson. *Bull. Johns Hopk. Hosp.*, **111**, 42 (1962).
Daroff. *Arch. Ophthal.*, **82**, 845 (1969).
Daroff, Solitare, Pincus and Glaser. *Neurology* (Minneap.), **16**, 161 (1966).
Davidson. *Brit. J. Ophthal.*, **44**, 394, 590 (1960).
Trans. ophthal. Soc. U.K., **90**, 139 (1970).
Drachman. *Arch. Neurol.*, **18**, 654 (1968).
Drachman, Murphy, Nigam and Hills. *Arch. Neurol.*, **16**, 14 (1967).
Drachman, Wetzel, Wasserman and Naito. *Arch. Neurol.*, **21**, 170 (1969).
Ellis, Dalessio and Lowe. *Amer. J. Ophthal.*, **61**, 167 (1966).
El-Sheikh. *Acta 1st Afro-Asian Cong. Ophthal.*, Cairo, 697 (1958).
Esslen, Mertens and Papst. *Nervenarzt*, **29**, 10 (1958).
Fox. *Ann. Ophthal.*, **3**, 1033 (1971).
Fuchs. *v. Graefes Arch. Ophthal.*, **36** (1), 234 (1890).
von Graefe. *Berl. klin. Wschr.*, **5**, 125 (1868).
Henneaux and Hérode. *Rev. Oto-neuro-oftal.*, **27**, 257 (1955).
Huber. *Ophthalmologica*, **149**, 359 (1965).
Huber and Wiesendanger. *Ophthalmologica*, **144**, 29 (1962).
Kaeser. *Schweiz. Arch. Neurol. Psychiat.*, **84**, 251 (1959).
Kearns. *Trans. Amer. ophthal. Soc.*, **63**, 559 (1965).
Kiloh and Nevin. *Brain*, **74**, 115 (1951).
Koerner and Schlote. *Arch. Ophthal.*, **88**, 155 (1972).
Landouzy and Déjerine. *C.R. Acad. Sci.* (Paris), **98**, 53 (1884).
Langdon and Cadwalader. *Brain*, **51**, 321 (1928).
Lapresle and Jarlot. *Arch. Ophtal.*, **19**, 384 (1959).
Larsen. *Nord. Med.*, **69**, 731 (1963).
Lawford. *Trans. ophthal. Soc. U.K.*, **7**, 260 (1887).
Lind and Prame. *Acta ophthal.* (Kbh.), **41**, 497 (1963).
McAuley. *Brit. J. Ophthal.*, **40**, 686 (1956).
McMullen and Hine. *Brit. J. Ophthal.*, **5**, 337 (1921).
Magora and Zauberman. *Arch. Neurol.*, **20**, 1 (1969).
Magyar, Grósz and Aszalós. *Ideggyóg. Szle*, **12**, 197 (1959).
Marina. *Ueber multiple Augenmuskellähmungen*, Leipzig (1896).
Möbius. *Dtsch. Z. Nervenheilk.*, **17**, 294, (1900).
Mölbert and Doden. *Ber. dtsch. ophthal. Ges.*, **62**, 392 (1959).
Moses and Heller. *Amer. J. Ophthal.*, **59**, 1051 (1965).
Nicolaissen and Brodal. *Arch. Ophthal.*, **61**, 202 (1959).
Nyman. *Acta psychiat. scand.*, Suppl., **219**, 145 (1970).
Ortiz de Zarate. *Brit. J. Ophthal.*, **50**, 606 (1966).
Pasetti and Salani. *Ann. Ottal.*, **36**, 281 (1907).
Peterman, Lillington and Jamplis. *Arch. Neurol.*, **10**, 38 (1964).
Rath. *Arch. Psychiat. Nervenkr.*, **86**, 360 (1929).
Roberts and Bamforth. *Neurology* (Minneap.), **18**, 645 (1968).
Rosenberg, Schotland, Lovelace and Rowland. *Arch. Neurol.*, **19**, 362 (1968).
Salleras and Ortiz de Zarate. *Brit. J. Ophthal.*, **34**, 662 (1950).
Sandifer. *J. Neurol. Neurosurg. Psychiat.*, **9**, 81 (1946).
Schotland and Rowland. *Arch. Neurol.*, **10**, 433 (1964).
Schwarz and Liu. *Arch. Neurol. Psychiat.*, **71**, 31 (1954).
Senita and Fisher. *Arch. Ophthal.*, **60**, 422 (1958).
Spaeth. *Amer. J. Ophthal.*, **28**, 1073 (1945).
Stephens, Hoover and Denst. *Brain*, **81**, 556 (1958).
Strümpell. *Neurol. Zbl.*, **5**, 25 (1886).
Dtsch. Z. Nervenheilk., **1**, 56 (1891).
Tarkkanen and Tommila. *Brit. J. Ophthal.*, **49**, 102 (1965).
Teasdall and Sears. *Arch. Neurol.*, **2**, 281 (1960).
Thiel. *Folia psychiat.* (Amst.), **57**, 554 (1954).
Thorson and Bell. *Arch. Ophthal.*, **62**, 833 (1959).
Victor, Hayes and Adams. *New Engl. J. Med.*, **267**, 1267 (1962).
Walsh and Hoyt. *Clinical Neuro-ophthalmology*, 3rd ed., Baltimore, 1246 (1969).
Walton. *Brit. med. J.*, **1**, 1271 (1964).
Wilbrand and Saenger. *Die Neurologie d. Auges*, Wiesbaden, **1**, 117 (1900).
Zintz and Villiger. *Ophthalmologica*, **153**, 439 (1967).

MYOTONIC DYSTROPHY (MYOTONIA ATROPHICA)

MYOTONIC DYSTROPHY, a chronic progressive disease wherein there is an excessive contractility combined with difficulty in relaxing the muscles, is associated with several ocular complications, particularly cataract,[1] retinal anomalies[2] and hypotony of the globe[3]; the disease has therefore been described in previous Volumes. It is a hereditary condition being transmitted as an autosomal dominant with varying penetrance, sometimes showing anticipation in succeeding generations (Klein, 1954; Asseman *et al.*, 1965). It affects particularly the face, neck and upper extremities and the drooping of the lids and the expressionless face, smooth forehead and sunken cheeks make a characteristic facies (Figs. 610–11).

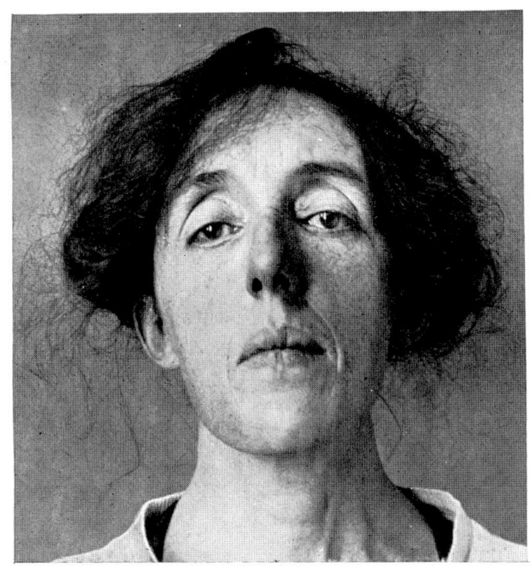

Fig. 610.

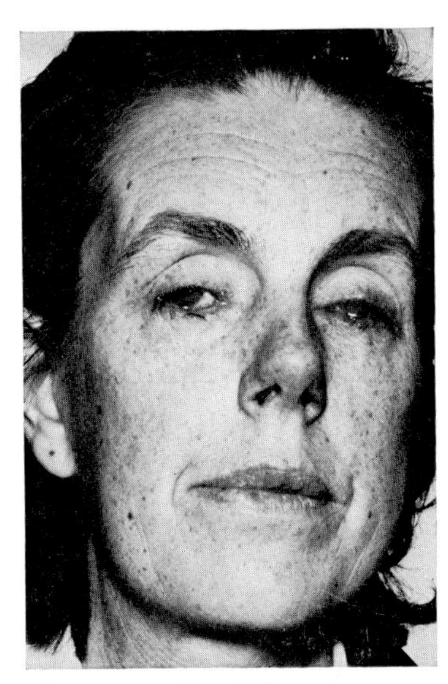

Fig. 611.

Figs. 610 and 611.—Myotonic Dystrophy.

With myopathic facies and wasting of the sterno-mastoid muscles. Fig. 610, Russell Brain; Fig. 611, P. Eustace.

The ptosis is symmetrical and is a constant feature of the disease, as well as a weakness of the orbicularis. The extrinsic ocular muscles may also be affected, giving rise to transient diplopia, defective movements in one or other direction or even ophthalmoplegia.[4] Eventually the levators may become so atrophic that they practically lose their function and become so thinned that the lids are almost transparent.

Occasionally corneal complications not hitherto mentioned in this *System* occur, mainly superficial recurrent inflammatory lesions, perhaps determined by the diminished palpebral function and lagophthalmos due to weakness of the orbicularis muscle (Klein, 1958; Junge, 1966; Burian and Burns, 1967; Eustace, 1969).

The electromyogram shows potentials of short duration, a feature of a myopathic process, trains of high-frequency discharges (the " dive bomber "

[1] Vol. XI, p. 183. [2] Vol. X, p. 615. [3] Vol. XI, p. 732. [4] Vol. VI, p. 753.

effect), a specific finding in myotonic dystrophy, the amplitude of the action potentials varying directly with the degree of ptosis (Davidson, 1961; Eustace, 1969; Lessell *et al.*, 1971) (Figs. 612–13).

Histological examination of the muscles in myotonic dystrophy shows few specific evidences of disease and these become apparent considerably

FIGS. 612 and 613.—THE ELECTROMYOGRAPH IN MYOTONIC DYSTROPHY.

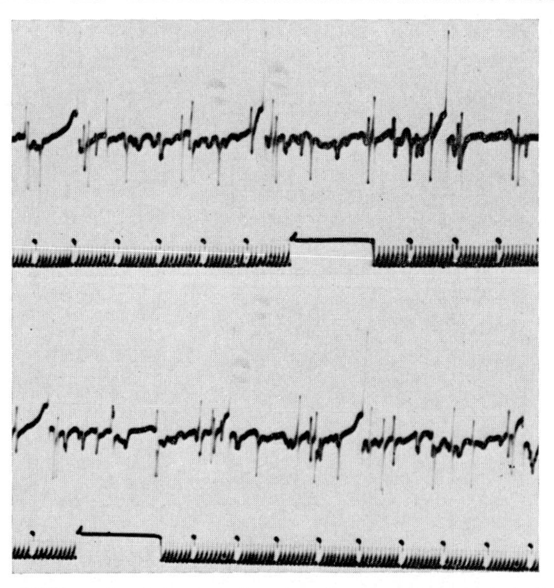

FIG. 612.—EMG of the lateral rectus showing the characteristic high-frequency discharges (S. I. Davidson).

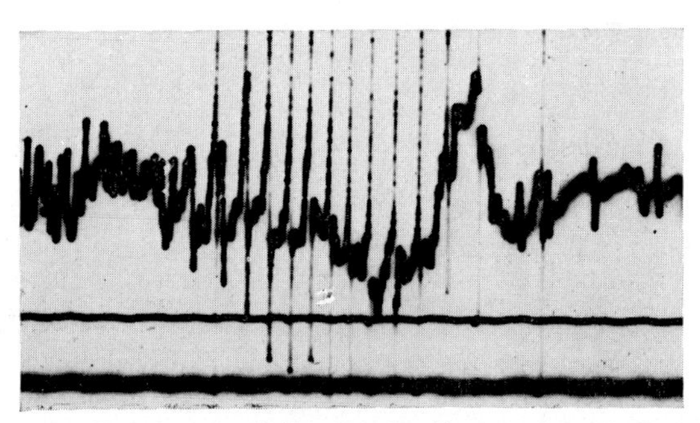

FIG. 613.—Showing the " dive-bomber " potentials (P. Eustace).

after the characteristic electromyographic reactions have developed; the muscle fibres merely show centrally placed lines of nuclei and occasionally hypertrophic fibres with vacuoles may be found as well as degenerative changes (Wohlfart, 1951; Davidson, 1961; Pendefunda *et al.*, 1964; Ketelsen and Schmidt, 1972).

The treatment of the ptosis, if indicated, is as for other types of myopathy.

In the CONGENITAL MYOTONIA OF THOMSEN (1876), a dominantly inherited condition wherein the entire striated musculature of the body is myotonic without atrophy, the palpebral muscles share in the general disease.[1] A lagging of the lids is a prominent feature often combined with a myotonic failure of relaxation of the levator muscle (Sedgwick, 1910; Toomey, 1916; Thomasen, 1948; Davidson, 1961); myotonia may also affect the orbicularis oculi muscle (Klein, 1958) as well as the ocular muscles, giving rise to disturbances of motility (Esslen and Papst, 1961).

In the CONGENITAL PARAMYOTONIA OF EULENBURG (1886) wherein a myotonic phase is followed by a myasthenic phase,[2] the palpebral and extrinsic ocular muscles are affected (de Jong, 1955).

A unique case of difficulty in opening the lids after closure, showing dystrophic changes in the orbicularis was described by Schmidt and Ketelsen (1973).

Asseman, Biervacque and Dufour. *Bull. Soc. Ophtal. Fr.*, **65**, 896 (1965).
Burian and Burns. *Amer. J. Ophthal.*, **63**, 22 (1967).
Davidson. *Brit. J. Ophthal.*, **45**, 183 (1961).
Esslen and Papst. *Bibl. Ophthal.*, **57**, 90 (1961).
Eulenburg. *Neurol. Zbl.*, **5**, 265 (1886).
Eustace. *Brit. J. Ophthal.*, **53**, 633 (1969).
de Jong. *Dystrophia myotonica, Paramyotonia and Myotonia congenita* (Thesis), Utrecht (1955).
Junge. *Docum. ophthal.*, **21**, 1 (1966).
Ketelsen and Schmidt. *v. Graefes Arch. Ophthal.*, **185**, 245 (1972).
Klein. *Confin. neurol.* (Basel), **14**, 169 (1954). *J. Génét. hum.*, Suppl. 7 (1958).
Lessell, Coppeto and Samet. *Amer. J. Ophthal.*, **71**, 1231 (1971).
Pendefunda, Cernea and Dobrescu. *Oftalmologia* (Buc.), **8**, 219 (1964).
Schmidt and Ketelsen. *v. Graefes Arch. Ophthal.*, **186**, 287 (1973).
Sedgwick. *Amer. J. med. Sci.*, **140**, 80 (1910).
Thomasen. *Myotonia. Thomsen's Disease. Paramyotonia. Dystrophia myotonica* (Thesis), Aarhus (1948).
Thomsen. *Arch. Psychiat. Nervenkrankh.*, **6**, 706 (1876).
Toomey. *Amer. J. med. Sci.*, **152**, 738 (1916).
Wohlfart. *J. Neuropath.*, **10**, 109 (1951).

MYASTHENIA GRAVIS

MYASTHENIA GRAVIS is a chronic disease with a tendency for remissions, characterized by abnormal fatiguability of striated muscle eventually leading to permanent wasting and weakening, and may ultimately involve complete and permanent paralysis (Figs. 614–15). It may be confined indefinitely to a single group of muscles; it may become generalized, and in the absence of treatment can be rapidly fatal; or it may progress slowly, involving striking remissions or stationary periods which may last for 20 years or more, until without treatment the patient becomes helpless and emaciated and dies usually from the consequences of respiratory embarrassment. The condition typically becomes evident between the ages of 20 and 50, but some 10% of cases occur in children before puberty, and many familial cases have been reported although no clear pattern of inheritance has yet emerged.[3]

The ætiology of the disease is not yet clarified. There are no gross

[1] Vol. XI, p. 187. [2] Vol. XI, p. 188.
[3] Hart (1927), Rothbart (1937), Riley and Frocht (1943), Walsh (1945), Appelmans and van Vooren (1949), Levin (1949), Walsh and Hoyt (1959), Fournier *et al.* (1967), Herman (1969), Manz and Schmidt (1970), Havard (1973).

pathological evidences in the muscles apart from degenerative and atrophic changes, and the collections of small round cells (" lymphorrhages ") described by Weigert (1901) lying between the muscle fibres and sometimes found in other organs such as the thymus and the adrenal cortex are inconstant and are not now thought to be specific (Genkins *et al.*, 1961; Fenichel and Shy, 1963) although their presence suggests an immune response. Hyperplasia of the thymus and thymic tumours occur in a large percentage of cases (Bell, 1917; Sloan, 1943; Turner, 1959; and many others);

FIGS. 614 and 615.—MYASTHENIA GRAVIS.

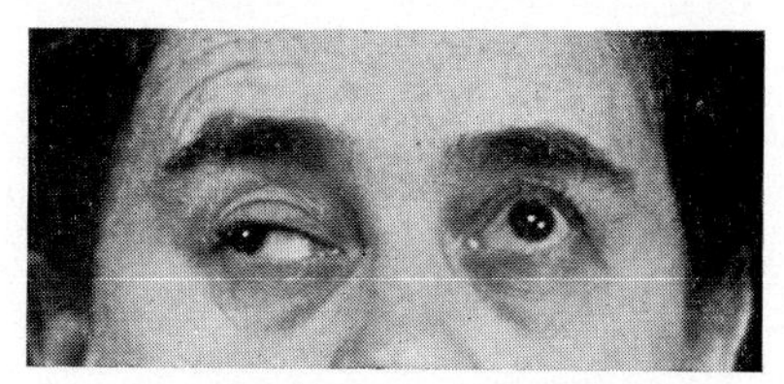

FIG. 614.—An early case showing ptosis on the right side and deficiency of movements (A. Huber).

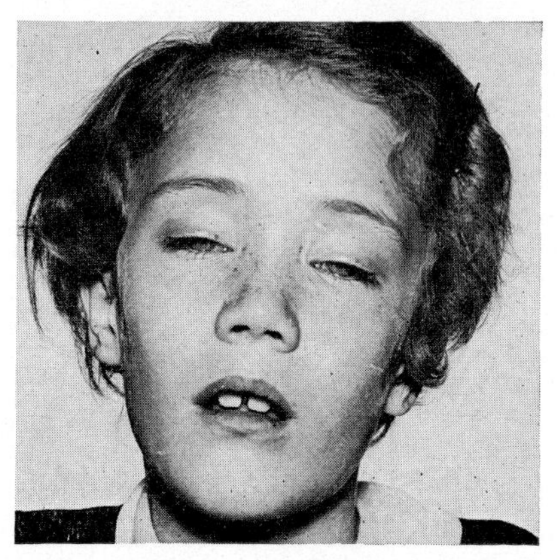

FIG. 615.—A more advanced case in a girl aged 7 with marked ptosis and severe limitation of ocular movements in all directions of gaze in both eyes (Inst. Ophthal.).

but again the association is not invariable. No constant relation has been found with the endocrine system although thyrotoxicosis occurs in some 5% of patients (Silver and Osserman, 1957; Szobor and Környey, 1966). The thymus plays a prominent part in the formation of antibodies or as a source of cells that form these bodies. On the basis of these findings the suggestion has been made that the disease is an auto-immune phenomenon (Simpson, 1960–66; Herman, 1969; Kalden *et al.*, 1969; Matsuda *et al.*, 1970; Frenkel, 1971; Havard, 1973); muscle antibodies, thyroglobulin antibodies and antinuclear antibodies have been shown in the serum of some patients (Kornguth *et al.*, 1970), and that the thymus can produce a substance that induces a neuro-muscular block has been experimentally demonstrated (Goldstein, 1966–68; Goldstein and Whittingham, 1966). The evidence is now incontrovertible that the immediate cause is a defect in the synthesis or storage of acetylcholine which normally initiates muscular contraction at the myoneural junction. At this point no conspicuous anatomical changes have been detected (Coërs and Desmedt, 1958; Bickerstaff *et al.*, 1960). In any event, the muscular weakness is characteristically abolished by the administration of drugs with cholinesterase-inhibiting

FIGS. 616–620.—ANOMALOUS MOVEMENTS OF THE LIDS IN MYASTHENIA GRAVIS.

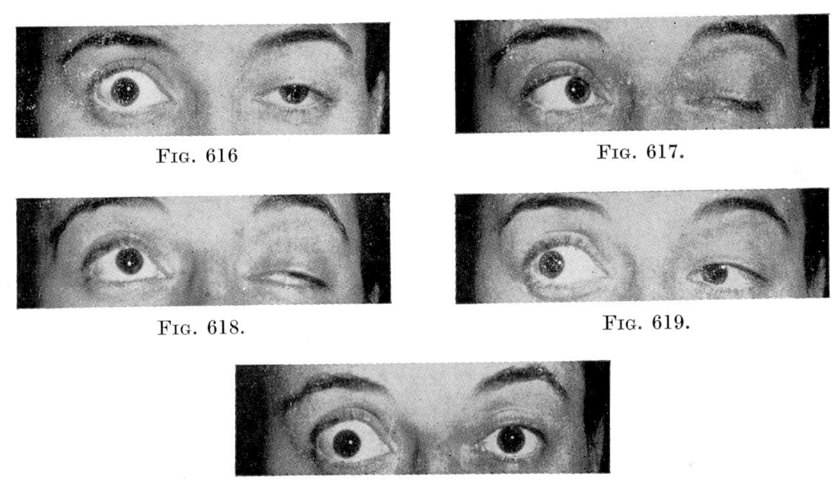

FIG. 616 FIG. 617.

FIG. 618. FIG. 619.

FIG. 620.

Illustrating associated movements of the lids in a 19-year-old girl showing purely ocular signs of the disease (F. B. Walsh).

FIG. 616.—On forward regard there is a compensatory retraction of the right eye and ptosis in the left.

FIG. 617.—On abduction of the left eye the ptosis is increased and is little changed on abduction and elevation in Fig. 618.

FIG. 619.—On adduction there is considerable elevation of the ptotic left lid.

FIG. 620.—The appearance after a diagnostic injection of neostigmine.

properties such as physostigmine and neostigmine (Walker, 1935). It has been shown that the ocular motor group of muscles is peculiarly sensitive to the acetylcholine mechanism,[1] a circumstance which probably explains their early, prominent and almost constant involvement in this disease.

Ptosis of one or both eyes, often fluctuating and variable in its incidence, is the first symptom in a large number of cases (40%, Oppenheim, 1900; 46%, Mattis, 1941), and may for a long time be a single and sometimes a unilateral symptom (Curschmann, 1919) (Fig. 623); it becomes evident eventually in over 95% of all patients (Mattis, 1941). Occasionally an

[1] Vol. VI, p. 48.

added paresis of the orbicularis produces the condition of complete blepharoplegia. It is gradual in its onset: characteristically it appears first in the evening when the patient is tired so that he cannot keep his eyes open, and the palsy disappears in the morning; it is subject to spontaneous remissions, but later it becomes permanent. A compensatory wrinkling of the brows is absent and a myasthenic pupillary reaction and paresis of accommodation may be present. The medial recti may be involved at an early stage resulting in an insufficiency of convergence which may lead to a spasm of accommodation and therefore a pseudo-myopia (Romano and Stark, 1973). In over 80% of cases diplopia with some degree of external ophthalmoplegia follows which is sometimes transient and sometimes permanent.[1]

It is generally admitted that the outlook is most favourable when the ocular muscles are first affected, and it has been reported that 40% of such

Figs. 621 and 622.—Electromyography in Myasthenia Gravis. (A. Huber).

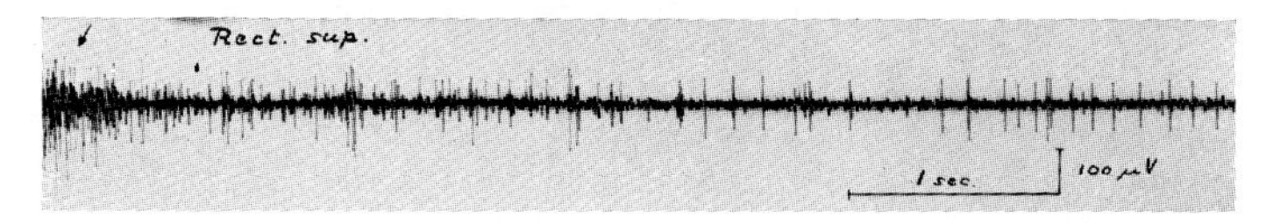

Fig. 621.—The reaction of the superior rectus muscle.

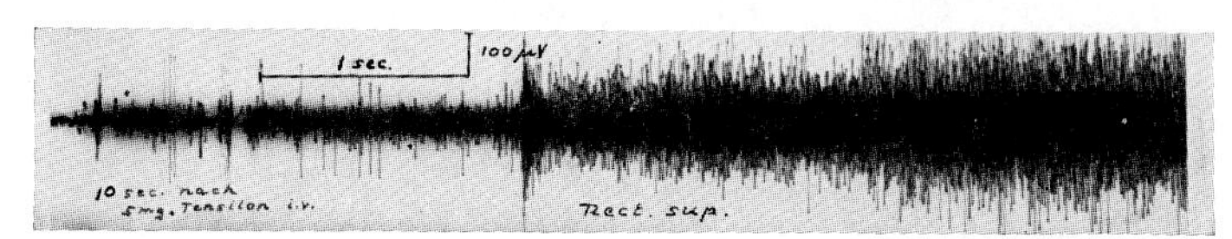

Fig. 622.—The same muscle reacting after an intravenous injection of 5 mg. Tensilon.

patients experience no peripheral signs of myasthenia during the course of their disease (Osserman *et al.*, 1958). Such patients have been observed for periods up to 25 years (Gavey, 1941), and Grob (1953) stated that patients affected by purely ocular symptoms for over 2 years are relatively safe. Occasionally the palsy is unilateral (Moore, 1941) in which case a neurogenic ptosis, particularly neurosyphilitic, may be simulated. Retraction of the lid with some lagging of its movement is a rare and usually transient phenomenon (Buzzard, 1900; Collier, 1927); this has been observed in combination with ptosis of the other lid, and retraction or ptosis may appear as an associated movement when the eyes are moved (Walsh, 1945) (Figs. 616–20). A twitch of the ptotic lid after blinking or on attempted upward gaze was considered by Cogan (1965) to be typical of myasthenia, while a micro-tremor measurable electrically and increased by Tensilon was noted by S. and H. Itoh (1967). The disease is characteristically intermittent;

[1] Vol. VI, p. 755.

although the ptosis is often fluctuating and may disappear for long periods, the ophthalmoplegia, once established, is more constant.

The electromyogram of the muscles is almost silent, showing few changes in potential (Breinin, 1957; Papst and Esslen, 1961; Huber, 1962–65; and many others) but is quickly restored temporarily after the injection of an anticholinesterase drug (Figs. 621–2).

The *diagnosis* of myasthenia gravis is often difficult, for in mild cases the ptosis and occasional diplopia readily excite the suspicion of oculomotor palsy and a severe case closely resembles ocular myopathy. The problem, however, can be resolved pharmacologically by the use of rapidly acting

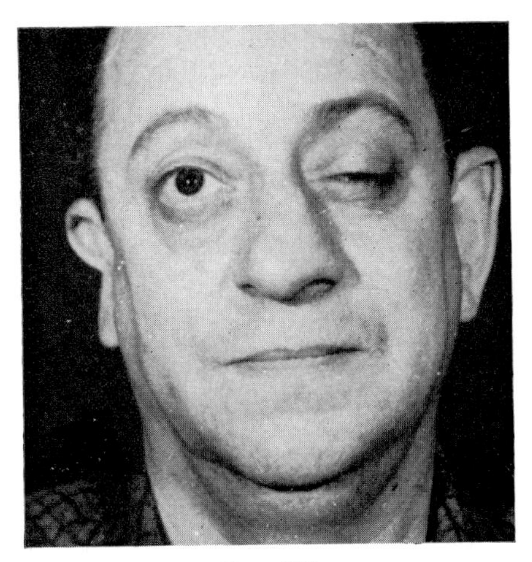

Fig. 623.

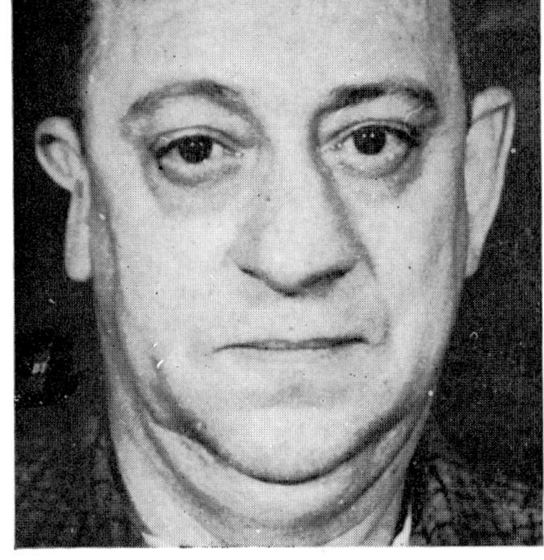

Fig. 624.

Figs. 623 and 624.—Myasthenia Gravis.

Unilateral ptosis before (Fig. 623) and after (Fig. 624) the injection of edrophonium (D. G. Cogan).

anticholinesterase drugs. The drug most commonly employed is edrophonium chloride (Tensilon) (Osserman and Kaplan, 1952–53); 2 mg. (0·2 ml.) is given intravenously, and if no response results a second injection of 8 mg. (0·8 ml.) is given rapidly; a quick relief of the ptosis occurs within 60 to 120 seconds lasting only 1 to 3 minutes (Figs. 623–4). A less dramatic alternative is Prostigmin (neostigmine) (0·5 mg. intravenously or intramuscularly); the ptosis is relieved more slowly (15 minutes) but the effect lasts longer (1 hour). If there is any doubt about the clinical response it can be verified electromyographically (Fig. 622) or, if this somewhat elaborate apparatus is not available, the activity of the muscles before and after the test may be checked on the Hess chart (Parr and McKinnon, 1971) or by assessing tonographically the sudden rise in the ocular tension due to the increase in the tone of the extrinsic muscles (Glaser *et al.*, 1966–67; Campbell *et al.*, 1970). It should be remembered that this test should be used only with great care in patients older than 60 years because of the cholinergic reaction of exciting cardiac arrhythmia; if such a reaction is feared in patients with a labile cardio-vascular system, atropine (0·4 mg.) should be available for intravenous administration as a precaution. This test is specific for myasthenia gravis; other dystrophies and myopathies are unresponsive, but a slight response may occur in the myasthenic syndrome.[1]

The *treatment* of myasthenia gravis has been revolutionized by the pharmacological interpretation of its ætiology. Since the discovery by Mary Walker (1934) that physostigmine salicylate had a transient effect on the muscular weakness and that Prostigmin (neostigmine) was still more efficacious (1935), most cases of the disease can be controlled. The duration of action of neostigmine bromide, however, is relatively short (2 hours) and it has to some extent been replaced by pyridostigmine bromide or ambenonium chloride (Osserman, 1967); the strong, long-lasting anti-cholinesterase drugs such as echothiopate iodide have been employed but they should not be administered unless red-cell esterase activity can be determined. In all cases the avoidance of a cholinergic crisis should be ensured. Overdosage is an easy and may be a fatal mistake since excessive anti-cholinesterase medication produces the same palsies of the skeletal and ocular muscles as myasthenia, and it is tempting to interpret an effect of this type as due to the disease and to increase the dosage of the drug sufficiently to cause respiratory arrest. Indeed, at an international symposium held on this disease[2] it was suggested that more patients die from excessive medication than from myasthenia.

A further procedure is destruction of the thymus by x-rays or thymectomy[3]; the results are difficult to assess but it may be indicated in relatively young girls who do not respond well to pharmacological therapy and in whom the disease is severe and progressive. In such cases Perlo and his team (1966) and Edwards and Wilson (1972) reported improvement in 80% to 84%, but the response is not good in the purely ocular forms of myasthenia.

Other methods of treatment are unimportant. Potassium salts have been advocated in addition to anticholinesterases, or potassium-sparing drugs such as spironolactone (Devic *et al.*, 1964). In severe cases corticosteroids have been suggested to improve the strength of the muscles (Torda and Wolff, 1949; Mount, 1964; Shapiro *et al.*, 1971; Jenkins, 1972), but their effect is unpredictable; they have, however, been used with success in patients resistant to anticholinesterase therapy (Gibberd *et al.*, 1971; Namba, 1972).

Surgical attempts to correct the ptosis or diplopia are frequently contra-indicated because of the tendency for transient remissions; if they are considered necessary for the ptosis, good results have been attained in a limited number of cases by Kapetansky (1972). Crutch spectacles or shelved haptic contact lenses are preferable devices unless a unilateral ptosis eliminates diplopia.

[1] p. 567.
[2] See *Arch. Ophthal.*, **74,** 1 (1965).
[3] Blalock *et al.* (1941), Keynes (1946), Turner (1959), Hugonnier *et al.* (1968), Papatestas *et al.* (1971), Edwards and Wilson (1972).

THE MYASTHENIC SYNDROME (of *Eaton-Lambert*, 1957)

A weakness of the peripheral muscles without constant pathological changes either in the muscles themselves or in the central nervous system occurs in more than one disorder of systemic character. Such a condition may involve the levator and cause ptosis. This is seen occasionally in Addison's disease, in Graves's disease and in other intoxications. The most common association, however, is with *bronchogenic carcinoma*.[1] The appearance and symptoms may simulate myasthenia gravis and although the muscles supplied by the cranial nerves are not usually affected, this may occur late in the disease to a mild degree and may involve ptosis and diplopia. Electromyography may show the fatigue phenomena typical of myasthenia gravis but the Tensilon test shows little response. This association should ensure radiography of the chest and a search for a neoplasm in suspicious cases. Treatment by guanidine hydrochloride (not anticholinesterases) may improve the fatiguability and weakness (Herrmann, 1970).

Anderson, Churchill-Davidson and Richardson. *Lancet*, **2**, 1291 (1953).
Appelmans and van Vooren. *Ophthalmologica*, **117**, 8 (1949).
Bell. *J. nerv. ment. Dis.*, **45**, 130 (1917).
Bickerstaff, Evans and Woolf. *Brain*, **83**, 638 (1960).
Blalock, Harvey, Ford and Lilienthal. *J. Amer. med. Ass.*, **117**, 1259 (1941).
Breinin. *Arch. Ophthal.*, **57**, 161 (1957).
Buzzard. *Brit. med. J.*, **1**, 493 (1900).
Campbell, Simpson, Crombie and Walton. *J. Neurol. Neurosurg. Psychiat.*, **33**, 639 (1970).
Coërs and Desmedt. *Lancet*, **2**, 1124 (1958).
Cogan. *Arch. Ophthal.*, **74**, 217 (1965).
Collier. *Brain*, **50**, 488 (1927).
Croft. *Brit. med. J.*, **1**, 181 (1958).
Curschmann. *Z. ges. Neurol. Psychiat.*, **50**, 131 (1919).
Devic, Thivolet, Michet *et al.* *Rev. Oto-neuro-ophtal.*, **36**, 169 (1964).
Eaton and Lambert. *J. Amer. med. Ass.*, **163**, 1117 (1957).
Edwards and Wilson. *Thorax*, **27**, 513 (1972).
Fenichel and Shy. *Arch. Neurol.*, **9**, 237 (1963).
Fournier, Pauli, Cousin *et al.* *J. Sci. méd. Lille*, **85**, 311 (1967).
Frenkel. *Surv. Ophthal.*, **16**, 43 (1971).
Gavey. *Proc. roy. Soc. Med.*, **35**, 14 (1941).
Genkins, Mendelow, Sobel and Osserman. *Myasthenia Gravis* (ed. Viets), Springfield, 519 (1961).
Gibberd, Navab and Smith. *J. Neurol. Neurosurg. Psychiat.*, **34**, 11 (1971).
Glaser. *Invest. Ophthal.*, **6**, 135 (1967).
Glaser, Miller and Gass. *Arch. Ophthal.*, **76**, 368 (1966).
Goldstein. *Lancet*, **2**, 1164 (1966); **2**, 119 (1968).
Clin. exp. Immunol., **2**, 103 (1967).
Goldstein and Whittingham. *Lancet*, **2**, 315 (1966).
Grob. *J. Amer. med. Ass.*, **153**, 529 (1953).
Hart. *Arch. Neurol. Psychiat.* (Chic.), **18**, 439 (1927).
Havard. *Brit. med. J.*, **3**, 437 (1973).
Heathfield and Williams. *Brain*, **77**, 122 (1954).
Hedges. *Arch. Ophthal.*, **70**, 333 (1963).
Henson, Russell and Wilkinson. *Brain*, **77**, 82 (1954).
Herman, M. *Arch. Neurol.*, **20**, 140 (1969).
Herrmann, C. *Calif. Med.*, **113** (3), 27 (1970).
Huber. *Trans. ophthal. Soc. U.K.*, **82**, 455 (1962).
Ber. dtsch. ophthal. Ges., **67**, 26 (1965).
Hugonnier, Salle, Maret *et al.* *Bull. Soc. Ophtal. Fr.*, **68**, 44 (1968).
Itoh, S. and H. *Acta Soc. ophthal. jap.*, **71**, 15 (1967).
Jenkins. *Lancet*, **1**, 765 (1972).
Kalden, Williamson, Johnston and Irvine. *Clin. exp. Immunol.*, **5**, 319 (1969).
Kapetansky. *Amer. J. Ophthal.*, **74**, 818 (1972).
Keynes. *Lancet*, **1**, 739, 746 (1946).
Kornguth, Hanson and Chun. *Neurology* (Minneap.), **20**, 749 (1970).
Lambert, Eaton and Rooke. *Amer. J. Physiol.*, **187**, 612 (1956).
Levin. *Arch. Neurol. Psychiat.* (Chic.), **62**, 745 (1949).
Mackenzie. *Lancet*, **1**, 108 (1954).
Manz and Schmidt. *Klin. Mbl. Augenheilk.*, **157**, 173 (1970).
Matsuda, Ito, Azumi and Nakamura. *Rinsho Ganka*, **24**, 533 (1970).
Mattis. *Arch. Ophthal.*, **26**, 969 (1941).
Moore. *Arch. Ophthal.*, **26**, 619 (1941).
Mount. *Arch. Neurol.*, **11**, 114 (1964).
Namba. *Arch. Neurol.*, **26**, 144 (1972).
Oppenheim. *Die myasthenische Paralyse*, Berlin (1900).
Osserman. *Invest. Ophthal.*, **6**, 277 (1967).

[1] Anderson *et al.* (1953), Heathfield and Williams (1954), Mackenzie (1954), Henson *et al.* (1954), Lambert *et al.* (1956), Eaton and Lambert (1957), Croft (1958), Wise and MacDermot (1962), Hedges (1963), Herrmann (1970).

Osserman and Kaplan. *J. Amer. med. Ass.* **150,** 265 (1952).
Arch. Neurol. Psychiat. (Chic.), **70,** 385 (1953).
Osserman, Kornfeld, Cohen *et al.* *Arch. intern. Med.*, **102,** 72 (1958).
Papatestas, Alpert, Osserman *et al.* *Amer. J. Med.*, **50,** 465 (1971).
Papst and Esslen. *Klin. Mbl. Augenheilk.*, **139,** 354 (1961).
Ophthalmologica, **141,** 191 (1961).
Parr and McKinnon. *Trans. ophthal. Soc. N.Z.*, **23,** 73 (1971).
Perlo, Poskanzer, Schwab *et al.* *Neurology*, **16,** 431 (1966).
Riley and Frocht. *Arch. Neurol. Psychiat.* (Chic.), **49,** 904 (1943).
Romano and Stark. *Amer. J. Ophthal.*, **75,** 872 (1973).
Rothbart. *J. Amer. med. Ass.*, **108,** 715 (1937).
Shapiro, Namba and Grob. *Arch. Neurol.*, **24,** 65 (1971).
Silver and Osserman. *J. Mt. Sinai Hosp.* (N.Y.), **24,** 1214 (1957).
Simpson. *Scot. med. J.*, **5,** 419 (1960).
Ann. N.Y. Acad. Sci., **135,** 506 (1966).
Sloan. *Surgery* (St. Louis), **13,** 154 (1943).
Szobor and Környey. *Nervenarzt*, **37,** 337 (1966).
Torda and Wolff. *J. clin. Invest.*, **28,** 1228 (1949).
Turner. *Brit. med. J.*, **1,** 778 (1959).
Walker. *Lancet*, **1,** 1200 (1934).
Proc. roy. Soc. Med., **28,** 759 (1935).
Walsh. *Amer. J. Ophthal.*, **28,** 13 (1945).
Walsh and Hoyt. *Amer. J. Ophthal.*, **47** (2), 28 (1959).
Weigert. *Neurol. Zbl.*, **20,** 597 (1901).
Wise and MacDermot. *J. Neurol. Neurosurg. Psychiat.*, **25,** 31 (1962).

LID-RETRACTION AND LID-LAG

LID-RETRACTION

Although in the newborn the margin of the upper lid lies well above the cornea, it descends in youth until in adolescence it normally cuts the limbus 1 to 3 mm. below its highest point. If the globe is normal in size and position, and if the margin of the upper lid rests at or above the limbus so that a band of white sclera is seen above the iris when the eyes are directed horizontally forward without staring, lid-retraction exists. When the eyes are in this position the phenomenon produces an expression of surprise or fear, and gives the appearance of protrusion of the globe (Fig. 625); when the patient looks up the retraction is usually exaggerated and on looking down the lagging of the lid after the eye in its movement also increases the effect.

Lid-retraction may be *physiological* or purely *mechanical*; it may be determined by disease of the levator muscle (*myogenic retraction*) or the nerves supplying the musculature of the lids (neurogenic retraction), either the oculomotor (*spastic retraction*) or the sympathetic (*sympathetic retraction*); finally, it may be spasmodic in nature and associated with other movements (*paradoxical retraction*).

A PHYSIOLOGICAL SPASMODIC RETRACTION of the upper lid occurs as a transient voluntary accompaniment of the normal act of staring—an exaggeration of the raised lid of attention in contradistinction to the drooping lid of inattention and the sleepy state. The same physiological tendency is also frequently seen intensified and perpetuated into a constant tonic retraction in those who are slowly going or have gone blind from such diseases as optic atrophy or chronic glaucoma and have tried hard, long and subconsciously to see (Fig. 625).

A MECHANICAL LID-RETRACTION is also relatively common when the upper lid cannot descend sufficiently for purely physical reasons. Post-traumatic or inflammatory scarring in the upper lid will obviously produce such a result, but more common is the *pseudo-retraction* which must be expected if the eye is too large or protrudes unduly forwards, as occurs in high myopia or buphthalmos, or in cases of exophthalmos and proptosis.

All types of NEUROGENIC LID-RETRACTION, whether due to lesions of the oculomotor or the sympathetic nerve or the paradoxical type occurring synergically with movements of the eye or jaw, have been discussed in a previous Volume on neurology.[1] An interesting case was recorded by Parker and Roy (1972) wherein a retraction of the lid was associated with a partial paresis of the superior rectus muscle, an effect presumably due to excessive innervational impulses. There remain only the myogenic types of retraction to be noted here.

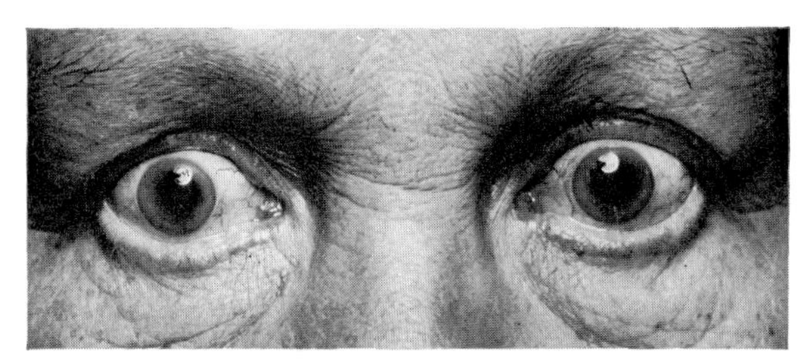

FIG. 625.—SPASMODIC LID-RETRACTION.

In a woman aged 75, almost blind from bilateral optic atrophy of vascular origin (Inst. Ophthal.).

MYOGENIC LID-RETRACTION

Retraction of the upper lid of a minor degree occurs in facial palsies when the normal tone of the levator is unopposed by the orbicularis; this, as well as von Graefe's sign, was noted in congenital unilateral facial myopathy by Strebel (1929). Of more interest, however, is the retraction associated with thyrotoxicosis and endocrine exophthalmos; it may be either bilateral or unilateral (Pochin, 1938; François, 1951; McLean and Norton, 1958) (Figs. 626–7). Thyrotoxicosis, indeed, is the commonest cause of retraction of the upper lid and is associated with three classical symptoms particularly characteristic of this disorder—a staring appearance of the eyes (*Dalrymple's sign*, 1852), infrequent blinking (*Stellwag's sign*, 1869) and, when the globe is moved downwards, an immobility or a lagging of the lid in its descent which is jerky and uneven so that its movements do not correspond to the movements of the eye (*von Graefe's sign*, 1864). Associated with these is *Gifford's sign* (1906), a difficulty in eversion of the upper lid. The clinical occurrence and significance of these signs will be discussed when dealing with Graves's disease.[2]

[1] Vol. XII, p. 905.
[2] p. 940.

The cause of lid-retraction in thyrotoxicosis used generally to be assumed to depend on sympathetic stimulation of Müller's muscle determined by thyroid overactivity. It would seem that spasm of Müller's muscle may account for the phenomenon in thyrotoxicosis in its early stages in which the degree of retraction varies, sometimes rapidly and dramatically, with the emotional state (Mulvany, 1943–44). A similar phenomenon may, indeed, occur after a shock in the absence of established thyrotoxicosis (Savin, 1943). Its relation to thyroid activity is shown by the fact that it may be induced by the administration of thyroid extract (Brain, 1936) but

FIGS. 626 and 627.—LID-RETRACTION IN THYROTOXICOSIS (Inst. Ophthal.).

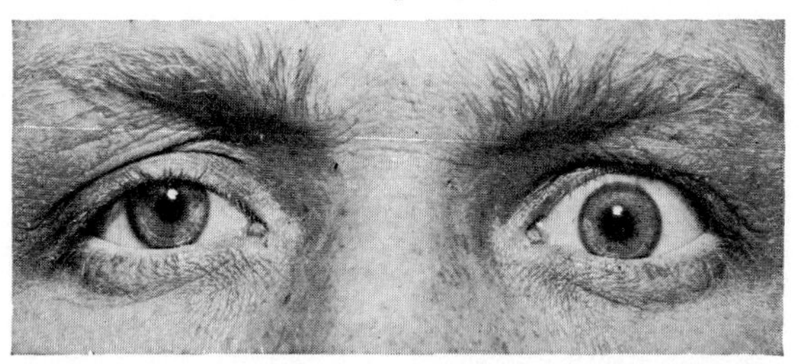

FIG. 626.—In a man aged 24 with unilateral lid-retraction.

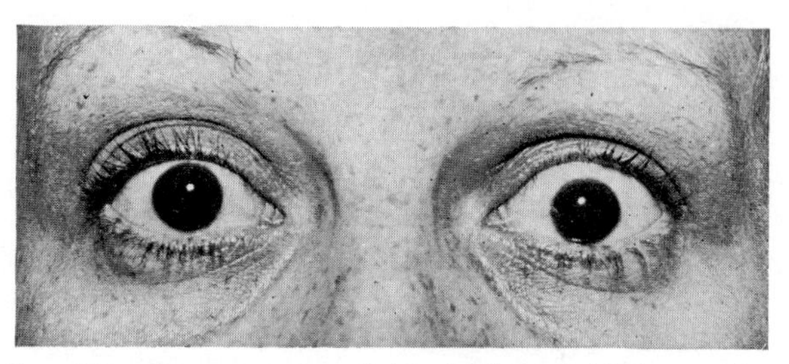

FIG. 627.—In a woman aged 32 with bilateral lid-retraction.

clinically it bears no relation to the presence or degree of exophthalmos or the severity of the thyrotoxicosis (Eden and Trotter, 1942). It is interesting to note that in the treatment of unilateral lid-retraction in thyrotoxicosis by guanethidine eyedrops, ptosis may develop in the other eye owing to systemic absorption of the drug.

The cause of the retraction of the lid and its inability to relax in those cases of endocrine exophthalmos not associated with thyrotoxicosis is not known with certainty but it probably lies in the levator muscle and is not innervational (McLean and Norton, 1958–59)—the original hypothesis advanced by Dalrymple[1] to explain the phenomenon in Graves's disease. It

[1] See White Cooper, *Lancet*, 26th May (1849). It is to be noted that Müller's muscle was unknown at this time; it was first described in 1884.

is almost certainly due to the infiltration and fibrosis which lowers the contractility of its fibres.[1]

A retraction of the upper lid is also seen in the MYOSTATIC PARESIS of parkinsonism[2]; this is also due to muscular rigidity, determined in this case, however, by hypertonia of central origin, a phenomenon confirmed by electromyography (Loeffler *et al.*, 1966) (Fig. 628). The phenomenon is also said to occur in cases of hepatic cirrhosis (Summerskill and Molnar, 1962).

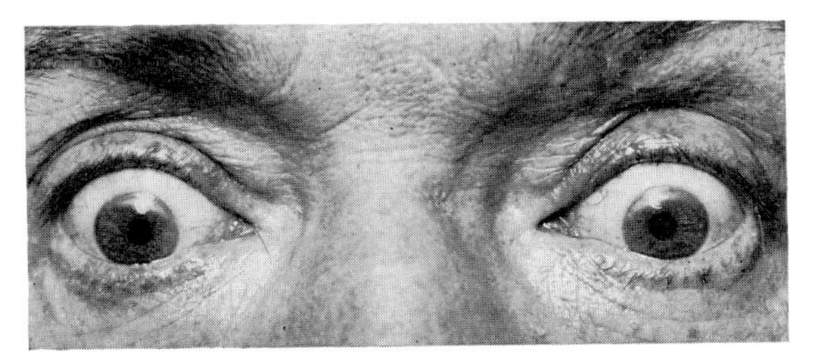

FIG. 628.—BILATERAL LID-RETRACTION.
In post-encephalitic parkinsonism (S. P. Meadows).

FIGS. 629 and 630.—LID-RETRACTION SECONDARY TO CONTRALATERAL PTOSIS (J. S. Gupta *et al.*).

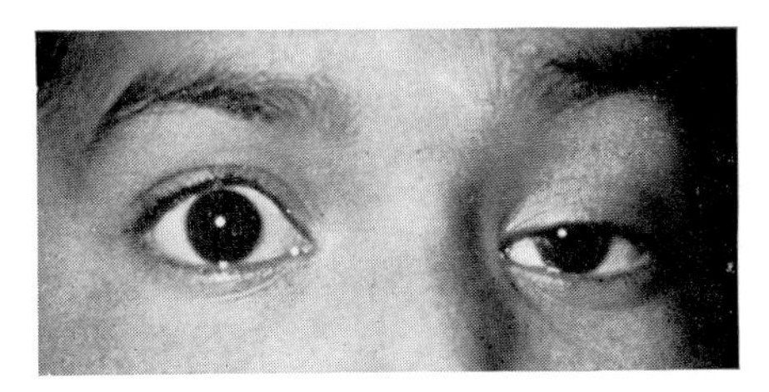

FIG. 629.—Retraction of the right upper lid and left ptosis.

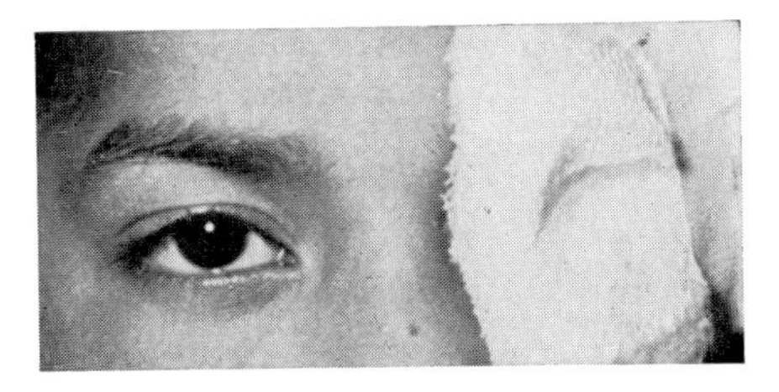

FIG. 630.—When the left eye is covered the right upper lid is normal.

An interesting phenomenon may be noted in passing wherein lid-retraction may occur in cases of unilateral ptosis of the other eye. This has been reported in cases of congenital ptosis (Gupta *et al.*, 1964) and myasthenia gravis (Figs. 616–20) (Gay *et al.*, 1967). The latter authors also reported a case in which ptosis was present on the side opposite to an eye with primary lid-retraction. In all cases, covering the eye with primary ptosis or retraction led to rectification of the deformity in the uncovered eye (Figs. 629–30).

Treatment of lid-retraction is rarely required apart from that proper to any systemic disease causing the condition. If it is marked, however, so that corneal damage may result from the lagophthalmos produced and if it is temporary or variable in nature, a tarsorrhaphy may be indicated. If it is dangerous or unsightly and thought to be permanent and stabilized, the

[1] p. 950. [2] Vol. XII, p. 906.

deformity may be ameliorated by severing the attachment of Müller's muscle to the tarsus and, if necessary, recessing the levator muscle also and fixing it to the conjunctival fornix or the skin underneath the brow (Blaskovics, 1933; Goldstein, 1934; Henderson, 1965). An alternative procedure is an excision of Müller's muscle and a partial tenotomy of the aponeurosis of the levator (Putterman and Urist, 1972). It may be noted that in the early stages of thyrotoxicosis the retraction may be abolished by the instillation of an adrenergic blocking agent such as guanethidine or bethanidine[1]; in established cases this is ineffectual owing to organic changes in the levator muscle. Crombie and Lawson (1967) found that guanethidine sometimes reduced the lid-retraction in apparently idiopathic cases as well as those with thyrotoxicosis.

LID-LAG

A lagging of the upper lid in its descent when the eye is moved downwards so that it does not follow the globe but pauses and then follows it incompletely occurs in several conditions. The most common is thyrotoxicosis when the phenomenon constitutes von Graefe's sign (1864) (Fig. 946); a lagging of the lower lid on looking upwards has been described by Moore (1938). The phenomenon of lid-lag has also been observed in association with encephalitis and parkinsonism, in spastic diplegia, myasthenia gravis, muscular dystrophy and periodic paralysis.

(FAMILIAL) PERIODIC PARALYSIS

(Familial) periodic paralysis is a condition wherein attacks of periodic flaccid paralysis occur typically on awaking from sleep, usually in the first three decades of life, often with a familial incidence. The attacks may last some hours or some days; as a rule they decrease in frequency and intensity as the years pass and the ultimate prognosis is usually good (Talbot, 1941; Gamstorp, 1956; McArdle, 1956–63). Electron-microscopic studies have shown that the condition is characterized by vacuolation and microfibrillar degeneration of the muscle fibres (Engel, 1966; Macdonald *et al.*, 1969). Depending on the concentration of potassium in the blood, the condition has been divided into three forms—hypokalæmic, normokalæmic and hyperkalæmic; of these the first is the most common. In all forms the myopathy may be widespread or localized but one of the most consistent clinical features is a lid-lag which may, indeed, be the only evidence of muscular involvement in the disease (Gamstorp, 1956; van der Meulen *et al.*, 1961; McArdle, 1962; van't Hoff, 1962; Resnick and Engel, 1967). The treatment of the hypokalæmic type is the administration of potassium which rapidly reverses any attack, but since normokalæmic and hyperkalæmic patients are intolerant to potassium which precipitates or aggravates an attack, care must be taken to differentiate the type of the disease.

Berkman and Joseph. *Arch. Ophtal.*, **31,** 617 (1971).
Blaskovics. *Z. Augenheilk.*, **81,** 13 (1933).
Bowden and Rose. *Brit. J. Ophthal.*, **53,** 246 (1969).
Brain. *Lancet*, **1,** 182 (1936).
Cant, Lewis and Harrison. *Brit. J. Ophthal.*, **53,** 233 (1969).
Crombie and Lawson. *Brit. med. J.*, **4,** 592 (1967).

[1] Gay and Wolkstein (1966), Gay *et al.* (1967), Millar (1968), Cant *et al.* (1969), Bowden and Rose (1969), Skinner and Miller (1969), Berkman and Joseph (1971).

Dalrymple. *Pathology of the Human Eye*, London (1852).
Eden and Trotter. *Lancet*, **2**, 385 (1942).
Engel. *Proc. Mayo Clin.*, **41**, 797 (1966).
François. *Bull. Soc. belge Ophtal.*, No. 97, 138 (1951).
Gamstorp. *Acta paediat.* (Stockh.), **45**, Suppl. 108, 1 (1956).
Gay, Salmon and Windsor. *Arch. Ophthal.*, **77**, 157 (1967).
Gay, Salmon and Wolkstein. *Arch. Ophthal.*, **77**, 341 (1967).
Gay and Wolkstein. *Arch. Ophthal.*, **76**, 364 (1966).
Gifford. *Klin. Mbl. Augenheilk.*, **44** (2), 201 (1906).
Goldstein. *Arch. Ophthal.*, **11**, 389 (1934).
von Graefe. *Klin. Mbl. Augenheilk.*, **2**, 183 (1864).
Gupta, Jain and Kumar. *Brit. J. Ophthal.*, **48**, 626 (1964).
Henderson. *Arch. Ophthal.*, **74**, 205 (1965).
van't Hoff. *Amer. J. Med.*, **31**, 385 (1962).
Loeffler, Slatt and Hoyt. *Arch. Ophthal.*, **76**, 178 (1966).
McArdle. *Brit. med. Bull.*, **12**, 226 (1956).
Brain, **85**, 121 (1962).
Amer. J. Med., **35**, 661 (1963).
Macdonald, Rewcastle and Humphrey. *Arch. Neurol.*, **20**, 565 (1969).
McLean and Norton. *Trans. Amer. ophthal. Soc.*, **56**, 327 (1958).
Arch. Ophthal., **61**, 681 (1959).
van der Meulen, Gilbert and Kane. *New Engl. J. Med.*, **264**, 1 (1961).
Millar. *Trans. ophthal. Soc. U.K.*, **88**, 677 (1968).
Moore. *Trans. ophthal. Soc. U.K.*, **58**, 3 (1938).
Mulvany. *Trans. ophthal. Soc. U.K.*, **63**, 22 (1943).
Amer. J. Ophthal., **27**, 589, 693, 820 (1944).
Parker and Roy. *J. pediat. Ophthal.*, **9**, 183 (1972).
Pochin. *Clin. Sci.*, **3**, 197 (1938).
Putterman and Urist. *Arch. Ophthal.*, **87**, 401 (1972).
Resnick and Engel. *J. Neurol. Neurosurg. Psychiat.*, **30**, 47 (1967).
Savin. *Trans. ophthal. Soc. U.K.*, **63**, 9 (1943).
Skinner and Miller. *Amer. J. Ophthal.*, **67**, 764 (1969).
Stellwag von Carion. *Med. Jb.* (Wien), 25 (1869).
Strebel. *Schweiz. med. Wschr.*, **59**, 906 (1929).
Summerskill and Molnar. *New Engl. J. Med.*, **226**, 1244 (1962).
Talbot. *Medicine* (Balt.), **20**, 85 (1941).

DISORDERS IN THE MOTILITY OF THE ORBICULARIS MUSCLE, myokymia, blepharoclonus, facial spasms and tics, blepharospasm, paralyses and paradoxical movements, have been described in the Volume on neurology[1]; so also have the disturbances of the lid reflexes and the sensory disorders, hyperæsthesia and the paræsthesias, neuralgia and anæsthesia.[2]

ABNORMALITIES OF THE PALPEBRAL APERTURE

Several deformations of the palpebral aperture occur as the result of diseases of the lids which it is convenient to group together here; and since most of the causal conditions themselves have been described, their effects on the lids can be rapidly summarized. It will be remembered that many such deformities may be congenital: these we have already discussed.[3] Acquired conditions, however, may determine abnormalities in the configuration of the lid-margin or in the size or shape of the palpebral aperture.

ENTROPION

ENTROPION (ἐν, in; τρέπειν, to turn), *an in-turning of the margin of the lid*, is a common condition usually affecting the lower lid, the seriousness of which lies in the irritation frequently caused to the cornea by the in-turned lashes.

The condition is to be distinguished from trichiasis, a turning inwards of the lashes due to a distortion of the lid-border only or a misalignment of

[1] Vol. XII, p. 929 *et seq.* [2] Vol. XII, p. 945. [3] Vol. III, pp. 840, 862.

the lashes themselves so that they point inwards. In entropion the ciliary border itself is normal but is rolled inwards as a whole so that sometimes it is hidden from view; if the skin can be pulled away towards the orbital margin the lid is unrolled and the lid-margin and lashes occupy their normal position. The two conditions, of course, may be combined, and both have the same distressing effects—chronic and constant corneal irritation, ulceration and eventual opacity.

The mechanism of the development of entropion or ectropion, as was originally pointed out by Stellwag (1886), and amplified by Fuchs (1890) and Czermak (1904), depends on three factors: a muscular factor determining the occurrence of the turning of the lid—the contraction of the fibres of the orbicularis oculi; the element of relative give-and-take between the skin and conjunctiva; and the support provided by the subjacent tissues, both of the latter determining which direction, outwards or inwards, the turning will take. The element of muscular contraction is the most important of the three. It will be remembered that the fibres of the palpebral portion of the orbicularis describe two curves: one in a vertical plane as they encircle the palpebral fissure, the concavity being directed downwards in the upper lid and upwards in the lower; the other in a horizontal plane as they embrace the globe, the concavity in both lids facing backwards. When the fibres contract, the concavity of both arcs tends to shorten to that of the chord with the result that each of the two components of force tends to flatten out one curve, the first approximating the lid-margins, the other pressing the lid against the surface of the eye. As the two lid-margins are forced together in a vertical direction any force acting horizontally will readily deflect them: this may be provided by an inequality between the elasticity of the skin on the one hand and the conjunctiva on the other, so that the lid turns either in or out or, alternatively, lack of support or abnormal pressure from behind may determine a similar direction of turning. In the second place the pressure of the lids upon the globe will ensure perfect contact only so long as they lie flat upon a uniform bed. If, however, the tarsus is already tilted as by swelling of the lid or if it is not provided with a uniform bed to rest upon so that either its free or attached border receives unequal or insufficient support, the tendency will be for it to tilt in one or other direction favouring the development of an entropion or an ectropion, a tendency aggravated if the tarsal fascia is lax (Jones, 1960).

Goldzieher (1908) and Dimmer (1911) raised the question of the relative power of the marginal fibres of the orbicularis compared with that of the palpebral portion of the muscle generally: if the first were more powerfully contracted an entropion would result; if the latter, an ectropion (Bardanzellu, 1933; and others). A more potent factor, however, particularly in senile entropion, is that laxity of the connective tissue of the lid allows the preseptal arcades of the orbicularis muscle to override the tarsus and turn it inwards (Kettesy, 1948; Jones *et al.*, 1963; Dalgleish and Smith, 1966).

Depending on which of these factors is predominant entropion may be of four kinds—spastic, mechanical, senile and cicatricial. We have already seen that it may also occur as a congenital deformity.[1] It is obvious for anatomical reasons that in all types except the cicatricial variety it will affect preferentially the lower lid, largely because of the size and rigidity of the upper tarsus, and partly because the greater development of Müller's muscle in the upper lid tends to keep the tarsus in place, while the cutaneous

[1] Vol. III. p. 863.

attachment of the levator tends to prevent the skin of the upper lid from being drawn downwards to allow an entropion to develop.

Voluntary entropion is a rare accomplishment, but it has been reported as being brought about in the lower lids by forced voluntary contraction (Holth, 1937).

SPASTIC ENTROPION develops typically in conditions of blepharospasm and may be seen equally in children and adults when the spasm has persisted for a long time as in chronic irritative corneal conditions. Its occurrence is facilitated by a loose skin and a swollen and turgid conjunctiva. Bandaging frequently accentuates the tendency; and in all cases the irritation from the in-turned lower lashes augments the spasm and perpetuates the condition. As a rule, however, unless predisposing structural changes are also present, the entropion tends to disappear with the cessation of spasm. This type of entropion is seen essentially in the lower lid (Fig. 631); its occurrence in the upper lid is a great rarity and is usually due to some structural predisposition such as a skin so lax that it overhangs the palpebral aperture (Dimmer, 1911; Elschnig, 1912; Schorr, 1926).

MECHANICAL ENTROPION (ORGANIC ENTROPION of Stellwag, 1886) is usually due to lack of support to the lids—a small or shrunken eye, enophthalmos, or lack of orbital fat. It is frequently seen after removal of the eye when a prosthesis is lacking or too small. Alternatively a swelling of the proximal part of the lid, such as by a chalazion or hyperplasia of the conjunctiva near the fornices, may determine the in-turning of its margin. Blepharophimosis also favours the development of the condition. A rare cause is lack of support owing to removal of the tarsal plate by complete tarsectomy for trachoma.

SENILE ENTROPION is the commonest type, usually seen between the ages of 70 and 75 years (Figs. 632–3). Here accessory factors such as lack of support owing to disappearance of orbital fat, the abundance of loose skin, and frequently a shrunken conjunctiva owing to long-standing and chronic but mild catarrhal inflammation are sufficient to determine its onset with the aid of the normal tone of the orbicularis. The same phenomenon of an in-turning may result from the development of a large fatty hernia into the superficial tissues of the lid (Mawas, 1966–72; Worst, 1972). To these factors may be added a lack of tone of the tarso-orbital fascia (Blaskovics, 1922), a laxity of the tarsus and palpebral ligaments (Butler, 1948; Kirby, 1952) and of the palpebral connective tissue (Kettesy, 1948; McFarlane, 1956); this allows arcades of the orbicularis muscle, including its preseptal portions, to slide upwards over the tarsus thus causing a forward rotation of the lower edge of the tarsal plate along its horizontal axis, a movement aided if the tarso-orbital fascia is lax (Dalgleish and Smith, 1966). Indeed, the lower lid is frequently so lax and flabby that it has to sag in one or other direction: if the orbicularis retains a reasonable amount of tone it will tend to sag inwards, more particularly if the skin is redundant and the eye has become deeply placed. Bandaging the eyes of such a person almost invariably induces an entropion, a complication which requires watching after operations on old people.

L. T. Jones and his colleagues (1972) suggested that senile entropion may be due to a failure of the retractor mechanism of the lower eyelid (the

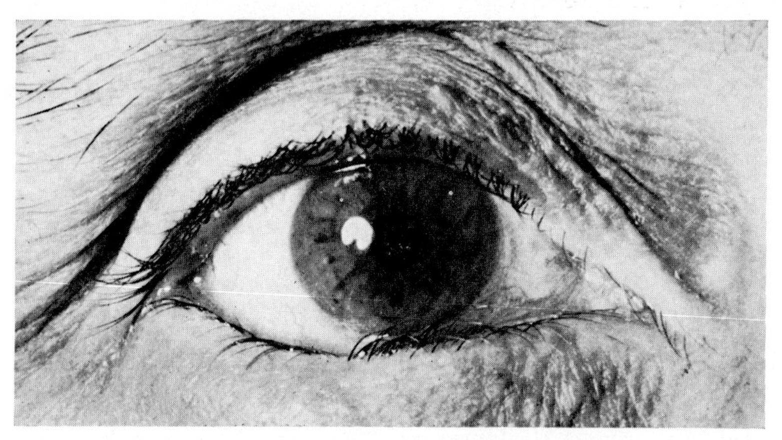

FIG. 631.—SPASTIC ENTROPION OF THE LOWER LID.
In a woman aged 63 (Inst. Ophthal.).

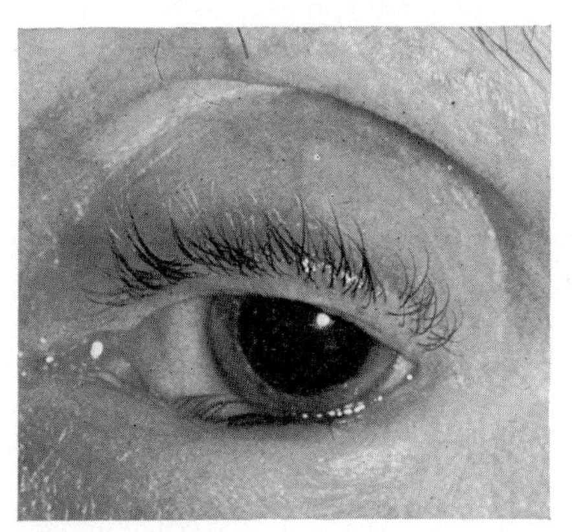

FIG. 632.—MARKED SENILE ENTROPION OF THE LOWER LID WITH INTURNING OF THE LASHES (A. Kornzweig).

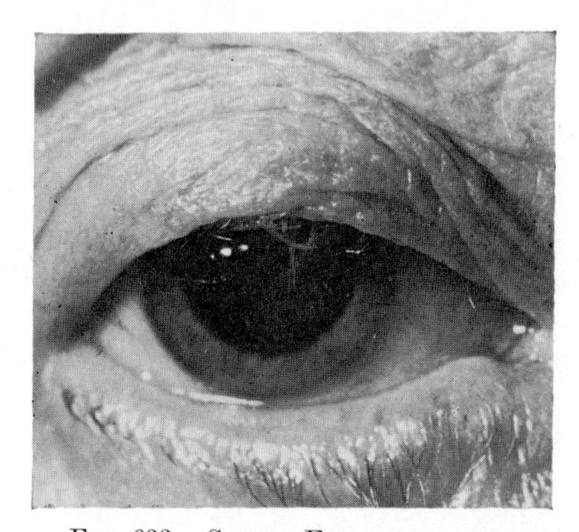

FIG. 633.—SENILE ENTROPION AND ECTROPION.
Entropion of the upper lid with inturning of the lashes, and ectropion of the lower lid (P. MacFaul).

capsulo-palpebral aponeurosis of the inferior rectus inserted into the lower margin of the tarsus), a concept which makes entropion an analogue of blepharoptosis of the upper eyelid.

CICATRICIAL ENTROPION is essentially determined by conjunctival scarring pulling in the lid-margins; it is noteworthy that only in this type is the upper lid frequently affected. It occurs after trauma, burns and severe conjunctival inflammations, particularly purulent and membranous in nature, and trachoma. The last is the most common cause, and here the in-turning of the lid-margins is probably initiated by the contraction of fibrous tissue in the deeper layers of the conjunctiva tending to turn the ciliary margin inwards (Sarkies, 1965) and completed by softening of the tarsus and its consequent deformation as a result of its sharing in the disease itself.

Cicatricial processes in the skin above either canthus may pull the skin of the lower lid upwards and cause entropion. A somewhat similar mechanism accounts for the entropion accompanying epicanthus lateralis.[1]

The *treatment* of entropion depends on whether a temporary expedient is sufficient or a permanent correction is required, and in the latter case whether spasm or laxity is involved in its causation or cicatricial changes require operative correction. In the meantime, if blepharospasm is present it should be controlled so far as that is possible.

Several *temporary expedients* may be adopted which are particularly applicable if it is anticipated that the condition will be short-lived, as in spastic cases. If bandaging has caused the deformation, the eyes should be uncovered or protected only by a shield. The lower lids may be turned out temporarily by pulling them down with narrow strips of adhesive plaster attached near the lid-margin at one end and to the drawn-up skin of the cheek at the other (Fig. 634). A narrow strip of plaster drawn horizontally along the lower margin of the orbit, or a roll of gauze pressed by plaster into the lower lid at the level of the orbital margin, or fixation of the everted lid with collodion may sometimes keep the lid from turning in; but these measures frequently fail particularly when persistent weeping wets such dressings. If they do not produce a satisfactory result, temporary relief can be obtained by rucking the skin in a fold by Michel's clips (Fig. 635) or sutures; the latter may embrace the subcutaneous tissues alone (Gaillard, 1844–47), the lid-margin (Hartleib, 1950), or take in the whole thickness of the lid by threading between the outer lid-margin and the lower part of the conjunctival cul-de-sac (Snellen, 1862). Finally, the wearing of a haptic contact lens may tide over a difficult period. A crutch spectacle was suggested by Oppenheimer (1901) and Salomonsohn (1901) wherein a loop of wire soldered to the lower rim of the frame pushed into the lower lid just above the lower orbital margin; this may succeed in everting the tarsus in mild cases and may sometimes be successful (Wiener, 1943).

When the essential element is a true blepharospasm this may be eliminated by weakening the orbicularis fibres with alcohol: 0·2 to 0·3 ml. of 90% alcohol is injected (after a topical anæsthetic) into the outer fibres of the muscle near the outer canthus, the injection extending into the lower lid

[1] p. 590.

towards its margin.[1] There is little pain and reaction and the injection can, if necessary, be repeated. Alcohol injection into the terminal branches of the facial nerve or into its trunk behind the mandible which may relieve the spasm for several months gives equally good results (Schloesser, 1903; Benedict, 1941). The effects of these injections, however, are often unpredictable and may occasionally cause permanent paralysis.

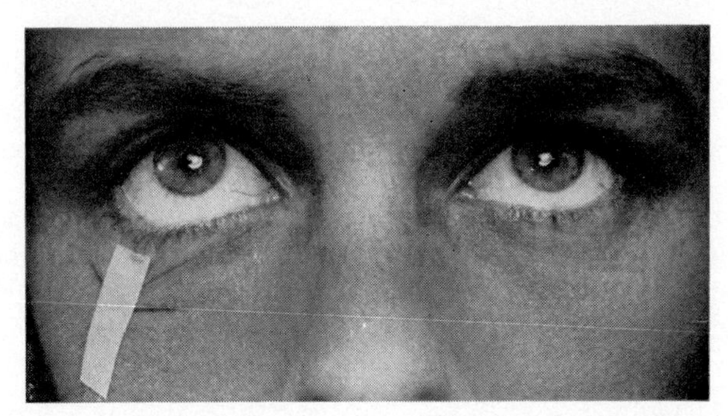

Fig. 634.—Entropion.

Temporary correction by sticking plaster. Note that the plaster is directed downwards and outwards and holds a fold in the skin so that the skin of the lower lid and of the cheek are both taut.

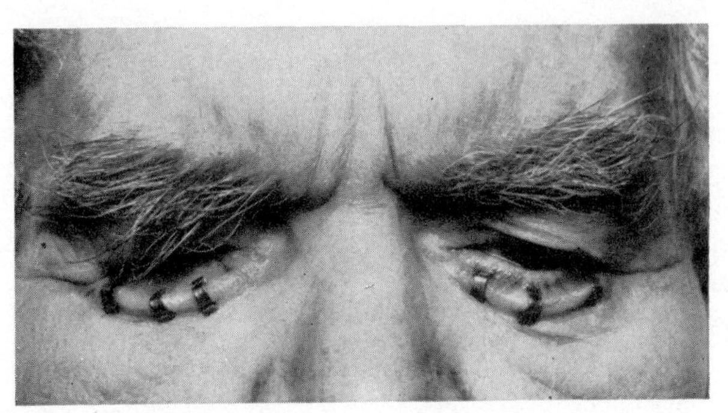

Fig. 635.—Spastic Entropion.

Michel's clips used as a temporary measure until the condition causing the spasm is relieved (MacDonald).

Surgical treatment is necessary if the entropion is permanent and this has excited an unusual interest; more than 90 different techniques are to be found in the literature dating from the drastic procedures used by the ancient Chinese and Hindus and the classical surgeons typified by Celsus and Paulus of Ægina adopted for the relief of the accompanying trichiasis.[2] The more important principles followed are outlined here; for the surgical

[1] Elschnig (1922), Dupuy-Dutemps (1926), Terson (1929), Hughes (1931), Hubbard and Kanski (1973).

[2] p. 383.

techniques the original papers or textbooks of plastic surgery should be consulted. Most procedures were designed for the lower lid.

(1) The excision of skin and a portion of the orbicularis muscle to evert the lid-margin dates back to the procedures practised by Celsus and Paulus of Ægina. The most common technique is the excision of a horizontal ellipse of skin just beneath the lid-margin together with the underlying portion of the muscle in order to eliminate its sphincteric action. This " skin-and-muscle " operation was introduced by Philip Crampton (1805) but is satisfactory only for mild cases. Hotz (1879–82) enhanced the effect by suturing the upper edge of the skin incision to the lower border of the tarsus to turn the lid-margin outwards. von Graefe (1864) excised a triangular area of skin in the middle of the lid, and Goldzieher (1908) at the lower and outer angle of the orbit. Busacca (1936) removed the marginal bundle of fibres, and Kettesy (1948) the entire muscle. If the condition is due to a fatty hernia this should be excised (Worst, 1972).

(2) The eversion of the tarsus was attempted by cutting a horizontal groove in its substance (Streatfeild, 1858) or by cauterization along its length (Terrien, 1902) or by a row of cautery-punctures (Ziegler, 1909). This technique is now rarely employed.

(3) A resection of part of the tissues of the lid to brace it against the globe if it is too long and lax was aimed at by Butler (1948) who excised a wedge of the tarso-conjunctival layer, a procedure elaborated by Fox (1951–72) who in addition excised a triangle of skin and muscle at the outer canthus (1959); a modification of this was suggested by Foulds (1961). A simple whole-thickness wedge was excised by Bick (1966) at the outer canthus.

(4) The transference of a strip of the tarsal portion of the orbicularis muscle downwards to the orbital septum to turn out the upper edge of the tarsus was suggested by Wheeler (1938–39). In addition, Jones (1960) recommended a shortening of the orbital fascia and the technique was modified by Hill and Feldman (1967).

(5) The provision of a septum of scar tissue to prevent the arcades of the orbicularis muscle from riding up to the lid-margin was suggested by Wies (1954) who made a horizontal incision through all the tissues of the lid just below the lid-margin and sutured the conjunctiva at the lower end of the wound to the skin of the upper edge of the wound. This technique was modified by Ffooks (1961), by Jones and his colleagues (1963) who, in addition, excised the tarsal fascia, and also by Beyer and Carroll (1973).

(6) Support of the lower lid by slings has had few advocates. Meek (1940) employed a sling of the orbicularis muscle anchored to the orbital rim to turn the lid outwards. Lebensohn (1953) used strips of skin; Schimek and Newsom (1967) buried strips of collagen tape.

(7) Viewing senile entropion as a failure of retraction of the lower eyelid on looking downwards, a shortening of the retractors of the lower lid to restore its downward movement in senile entropion was suggested by L. T. Jones and his colleagues (1972) through a longitudinal incision at the lower margin of the tarsus, separating the orbital septum from its attachment thereto. The aponeurosis of the capsulo-palpebral head of the inferior rectus muscle is then tucked or resected sufficiently to produce about 4 mm. of retraction of the lower lid when the patient looks downwards.

(8) Upwardly rotated strips of the preseptal bundle of the orbicularis muscle in which the nerve and blood supplies have been retained are sutured to the upper anterior surface of the inferior tarsus thus pulling it outwards when the muscle contracts, a procedure advocated by Sisler (1973) for senile entropion.

Of all the many suggestions those most commonly employed in the treatment of senile entropion are a skin-and-muscle operation in mild cases, perhaps with Hotz's modification, a resection of the tissues of the lower lid by Fox's procedure and support for the lower edge of the tarsus by Wheeler's technique. In severe cases some of these procedures may be combined (Hill and Witzell, 1956).

In cicatricial cases more drastic steps must usually be taken. One of three lines of approach is open, which have already been discussed when we were dealing with trichiasis.[1] In the event of the conjunctiva alone being at fault a mucous membrane graft from the lip usually meets the case. If the tarsus has become deformed its lower part should be everted or a complete tarsectomy performed; alternatively, relief may most easily be sought by the transposition of a conjunctivo-tarsal wedge. The occurrence of entropion of the upper lid following total tarsectomy may be obviated by allowing the tarsal conjunctiva to retract and replacing it by a mucous membrane graft (Kahán and Vén, 1957) or by restoring the integrity of the lid by a graft of auricular cartilage (Goldfeder, 1929; Shimkin, 1938; Christov, 1967) or even an acrylic plate (Chandra and Taneja, 1964) or collagen film (Dortzbach and Callahan, 1971).

Bardanzellu. *Lettura oftal.*, **10,** 365 (1933).
Benedict. *Trans. Amer. ophthal. Soc.*, **39,** 227 (1941).
Beyer and Carroll. *Arch. Ophthal.*, **89,** 33 (1973).
Bick. *Arch. Ophthal.*, **75,** 386 (1966).
Blaskovics. *Klin. Mbl. Augenheilk.*, **69,** 136 (1922).
Busacca. *Z. Augenheilk.*, **88,** 100 (1936).
Arch. Ophthal., **16,** 822 (1936).
Butler. *Arch. Ophthal.*, **40,** 665 (1948).
Chandra and Taneja. *J. All-India ophthal. Soc.*, **12,** 107 (1964).
Christov. *Oftalmologia* (Buc.), **3,** 34 (1967).
Crampton. *An Essay on the Entropion or Inversion of the Eyelids*, London (1805).
Czermak. *Die Augenärztlichen Operationen*, Wien (1904).
Dalgleish and Smith. *Brit. J. Ophthal.*, **50,** 79 (1966).
Dimmer. *Klin. Mbl. Augenheilk.*, **49** (2), 337 (1911).
Dortzbach and Callahan. *Arch. Ophthal.*, **85,** 82 (1971).
Dupuys-Dutemps. *Sem. Hôp.* (Paris), **3,** 64 (1926).
Elschnig. *Klin. Mbl. Augenheilk.*, **50** (1), 17, 335 (1912).
Med. Klin., **18,** 1641 (1922).
Ffooks. *Brit. J. Ophthal.*, **45,** 130 (1961).
Foulds. *Brit. J. Ophthal.*, **45,** 678 (1961).
Fox. *Arch. Ophthal.*, **46,** 424 (1951); **48,** 624 (1952).
Amer. J. Ophthal., **48,** 607 (1959).
Ann. Ophthal. (Chic.), **4,** 217 (1972).
Fuchs. *Lhb. d. Augenheilkunde*, Wien (1890).
Wien. klin. Wschr., **3,** 2 (1890).
Gaillard. *Bull. Soc. Méd. Poitiers* (1844).
Ann. Oculist. (Paris), **18,** 241 (1847).
Goldfeder. *Klin. Mbl. Augenheilk.*, **82,** 809 (1929).
Goldzieher. *Klin. Mbl. Augenheilk.*, **46** (2), 426 (1908).
von Graefe. *v. Graefes Arch. Ophthal.*, **10** (2), 221 (1864).
Hartleib. *Klin. Mbl. Augenheilk.*, **116,** 311 (1950).
Hill and Feldman. *Arch. Ophthal.*, **78,** 621 (1967).
Hill and Witzell. *Trans. Canad. ophthal. Soc.*, **8,** 69 (1956).
Holth. *Acta ophthal.* (Kbh.), **15,** 370 (1937).
Hotz. *v. Graefes Arch. Ophthal.*, **8,** 249 (1879); **11,** 442 (1882).
Hubbard and Kanski. *Proc. roy. Soc. Med.*, **66,** 173 (1973).
Hughes. *Amer. J. Ophthal.*, **14,** 34 (1931).
Jones, L. T. *Amer. J. Ophthal.*, **49,** 29 (1960).
Jones, L. T., Reeh and Tsujimura. *Amer. J. Ophthal.*, **55,** 463 (1963).
Jones, L. T., Reeh and Wobig. *Amer. J. Ophthal.*, **74,** 327 (1972).
Kahán and Vén. *Brit. J. Ophthal.*, **41,** 562 (1957).
Kettesy. *Brit. J. Ophthal.*, **32,** 311 (1948).
Kirby. *Trans. Amer. ophthal. Soc.*, **50,** 359 (1952).
Lebensohn. *Amer. J. Ophthal.*, **36,** 504 (1953).
McFarlane. *Amer. J. Ophthal.*, **41,** 657 (1956).
Mawas. *Travaux d'ophtalmologie moderne*, Paris, 261 (1966).
Bull. Soc. franç. Ophtal., **85,** 420 (1972).
Meek. *Arch. Ophthal.*, **24,** 547 (1940).

[1] p. 384.

Oppenheimer. *Dtsch. med. Wschr.*, **27**, 827 (1901).
Salomonsohn. *Dtsch. med. Wschr.*, **27**, 883 (1901).
Sarkies. *Brit. J. Ophthal.*, **49**, 538 (1965).
Schimek and Newsom. *Arch. Ophthal.*, **77**, 672 (1967).
Schorr. *Klin. Mbl. Augenheilk.*, **76**, 373 (1926).
Schloesser. *Ber. dtsch. ophthal. Ges.*, **31**, 84 (1903).
Shimkin. *Brit. J. Ophthal.*, **22**, 282 (1938). *Rev. int. Trachome*, **15**, 15 (1938).
Sisler. *Ann. Ophthal.* (Chic.), **5**, 483 (1973).
Snellen. *II int. Cong. Ophthal.*, Paris, 236 (1862).
Stellwag von Carion. *Neue Abhdl. a. d. Gebiete d. praktischen Augenheilkunde*, Wien (1886).
Streatfeild. *Roy. Lond. ophthal. Hosp. Rep.*, **1**, 121 (1858).
Terrien. *Chirurgie de l'oeil et de ses annexes*, Paris (1902).
Terson. *Ann. Oculist.* (Paris), **166**, 951 (1929).
Wheeler. *Amer. J. Surg.*, **42**, 7 (1938). *Amer. J. Ophthal.*, **22**, 477 (1939).
Wiener. *Arch. Ophthal.*, **29**, 634 (1943).
Wies. *J. int. Coll. Surg.*, **21**, 758 (1954).
Worst. *Bull. Soc. franç. Ophtal.*, **85**, 418 (1972).
Ziegler. *J. Amer. med. Ass.*, **53**, 183 (1909).

ECTROPION

ECTROPION (ἐκ, out; τρέπειν, to turn), *an out-turning of the lid-margin*, is a very common condition particularly in old people. It occurs in different degrees. The mildest stage is an eversion of the border of the lid wherein the internal ciliary margin just falls away from contact with the globe and begins to curl outwards. A further stage is characterized by a complete eversion of the border exposing the conjunctiva, and the last stage is reached in a complete eversion of the entire lid. This mechanical alteration in the lid entails three unfortunate sequelæ. Owing to eversion of the lacrimal punctum epiphora results, and this tends to intensify the condition by causing eczematous changes in the skin which may pull the lid still further down, a process frequently aided mechanically by the patient constantly wiping his eye. In the second place the conjunctiva suffers from exposure: it either becomes dry, xerotic and epidermal in nature or, becoming swollen and hyperæmic, it may eventually hypertrophy with the production of exuberant granulation-like masses (*ectropion luxurians* or *sarcomatosum*). Here again the result is an accentuation of the ectropion by the conjunctival overgrowth. Finally, in the more severe degrees, closure of the eyes may become impossible so that lagophthalmos results with all the evils of constant exposure of the cornea.

Depending on the mechanism of its causation ectropion may be classified into six types—spastic, senile, marginal, paralytic, mechanical and cicatricial. Its congenital occurrence has already been considered.[1]

SPASTIC ECTROPION. The mechanism of spastic ectropion is the opposite of that causing entropion. This we have already studied[2]; it will be remembered that on strong contraction of the orbicularis muscle the lid-margins will tend to be forced outwards if the skin is taut, if the conjunctiva is resilient or thickened and hypertrophied, and if the eye is prominent or proptosed. It is obvious that all these conditions, particularly the tension of the skin, are found more commonly in young than in old people.

[1] Vol. III, p. 864. [2] p. 574.

The effect of the element of skin traction is seen most dramatically in the total eversion of both lids which so frequently occurs on an attempt to open the eyes of an infant by pulling the lids apart when the orbicularis is in spasm. If reposition is not made within a reasonable time the spasmodic contraction of the peripheral bundles of the palpebral portion of the muscle maintains the lids in their faulty position and produces a congestion and swelling by strangulation so that a permanent spastic ectropion may result requiring an anæsthetic for its reposition (PARAPHIMOSIS PALPEBRÆ). Such an accident has occurred during birth.

SENILE ECTROPION in contradistinction to the spastic type, is seen only in the lower lid. It is essentially the result of relaxation of the tissues of the lids, particularly of the muscle and the lateral palpebral ligament, the eversion being frequently favoured mechanically by the chronic hypertrophic

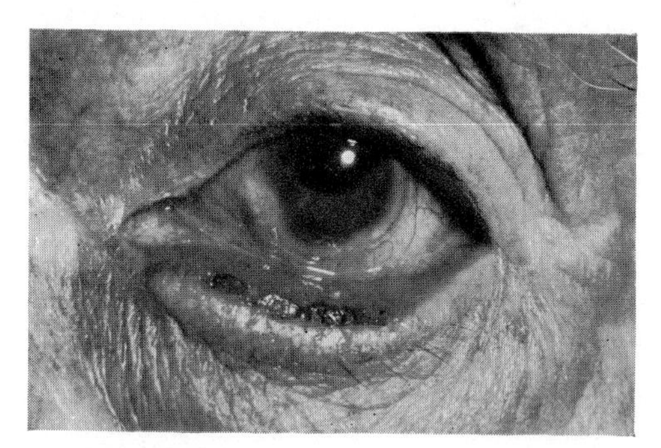

FIG. 636.—SENILE ECTROPION (A. Urrets-Zavalia).

FIGS. 637 and 638.—PARALYTIC ECTROPION.
Due to left-sided facial palsy (T. Pomfret Kilner).

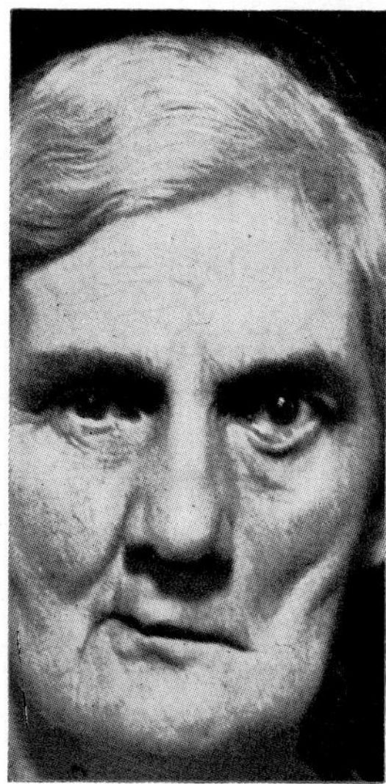

FIG. 637.—The face relaxed and the eyes open.

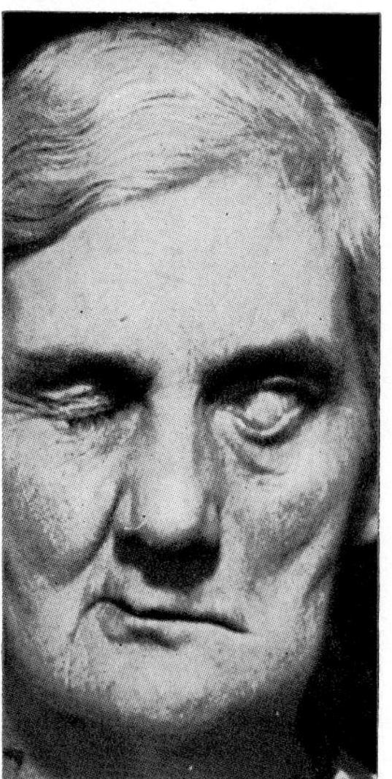

FIG. 638.—Attempted closure of the lids.

senile conjunctivitis so common in old age. If the fibres of the orbicularis retain some tone an eversion of the marginal region only may occur; but if the muscle is atonic the entire lid may sag away from the globe particularly in its inner third (Fig. 636).

MARGINAL (TARSAL) ECTROPION is a condition described by Fox (1960) seen only in middle and old age, which may be unilateral or bilateral. It is characterized by an eversion of the lower lid limited to the tarsal portion; the lid below the tarsus remains in its normal position and shows no looseness and lack of tonicity as is seen in senile ectropion, and it is due to a relaxation of the tarsus and the two canthal ligaments while the skin and muscle retain their normal tonus.

PARALYTIC ECTROPION, occurring in facial palsy or after injury or destruction of the orbicularis fibres, is caused by a similar mechanism of complete relaxation; here the sagging of the lid is so marked that lagophthalmos is the rule (Figs. 637–8). Again, the lower lid only is affected.

MECHANICAL ECTROPION occurs when the outward turning of the lid is determined essentially by a pushing from behind without the occurrence of marked blepharospasm. This occurs when the conjunctiva is markedly thickened or hypertrophic as in gonococcal conjunctivitis or trachoma, in swellings or tumours of the tarsus and when the eye is enlarged (buphthalmos) or proptosed.

CICATRICIAL ECTROPION produces the greatest deformities of the lids and brings the most acute dangers from lagophthalmos; it may affect either lid (Figs. 153, 639–41). It may be seen in contractions of the skin, as in eczema, in which connection the epiphora which accompanies all types of ectropion may accentuate the effect by the eczematous maceration it produces. More pronounced deformities follow destructive lesions, as gangrene, anthrax, syphilis, oriental sore, variola, vaccinia, fungus infections (blastomycosis, coccidioidomycosis, etc.), malignant tumours or bony disease (particularly tuberculosis) of the orbital margin. Chronic skin diseases of the sclerosing type act similarly (lupus vulgaris, pityriasis rubra pilaris, xeroderma pigmentosum, scleroderma, ichthyosiform erythroderma and epidermolysis bullosa). A rare example affecting the upper lid is in association with a fistula of the frontal sinus (McLeod and Lux, 1936). The most pronounced degrees of ectropion commonly occur in epidermolysis bullosa and ichthyosiform erythroderma and after injuries, particularly extensive burns of the face, whereafter cicatrization may lead to a drawing of the upper lid to the brows and the lower over the cheek so that loss of vision owing to corneal damage from exposure is inevitable and perforation of the globe a possibility unless remedial measures are timely and efficient (see Figs. 84–5, 244, 322).

Treatment of a non-operative nature may be effective in slight and temporary cases. Spastic cases may be relieved by a well-fitting pressure bandage, and when the eyes are wiped, care should be taken to apply the handkerchief from below upwards rather than in the usual direction from above downwards. This with the relief of the blepharospasm[1] may suffice. In severe or chronic cases, however, some operative procedure is necessary.

Atonic ectropion (senile and paralytic) always requires operative adjustment. The milder degrees may be corrected by the technique employed by the ancient Greeks and Arabs of cauterizing the conjunctiva to induce cicatrization; the electro-cautery may be used (Ziegler, 1909), and Terson (1931) obtained comparable results by the subconjunctival injection of

FIGS. 639 to 641.—CICATRICIAL ECTROPION.

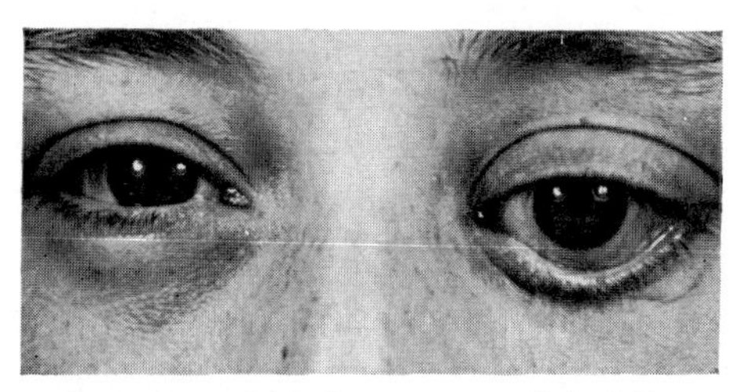

FIG. 639.—Of inflammatory origin following tuberculosis of the malar bone in a girl aged 23 (A. MacIndoe).

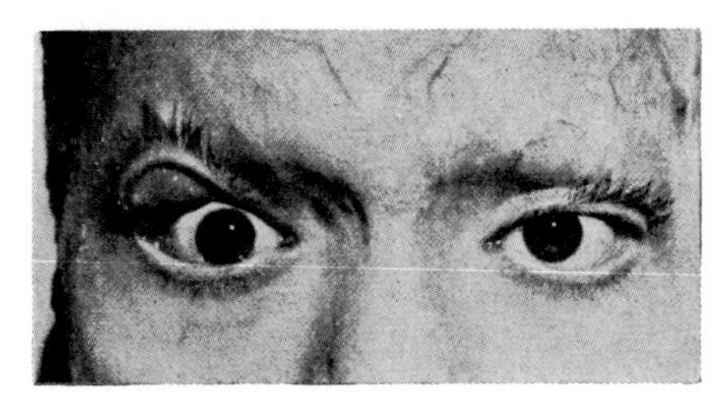

FIG. 640.—Ectropion of the upper lid resulting from a burn (D. B. Kirby).

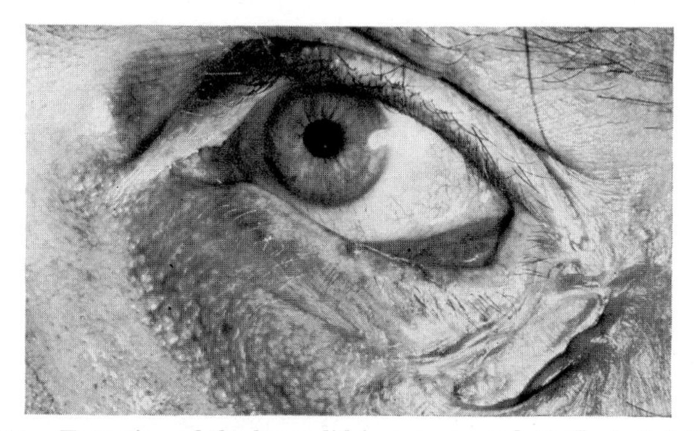

FIG. 641.—Ectropion of the lower lid in a man aged 43 (Inst. Ophthal.).

sclerosing solutions of quinine and urea. A more effective result may be obtained by sutures, either threaded through from the skin surface into the conjunctival cul-de-sac to pull it downwards (Snellen, 1862), or passing horizontally across the lower lid near its margin to give it support. The effect of all these procedures, however, is transitory. A more permanent correction may be obtained by one of three expedients—shortening the palpebral aperture, bracing the lower lid, or supporting it. The first is attained by a lateral tarsorrhaphy, a procedure particularly suitable for paralytic cases. For the rest, many surgical procedures are available. In

[1] Vol. XII, p. 939.

all cases wherein the deformity is accentuated by a hypertrophic chronic conjunctivitis the diseased area should be excised and the lower border of the incision undermined and sutured to the lid-margin.

(1) In mild cases of senile ectropion bracing the lower lid and raising its margin may be accomplished by the simple method of converting a V-shaped incision in the skin into a Y, a technique introduced by Wharton Jones (1847). More effective techniques are the Z-plasty of Elschnig (1912) and the transference of a tongue-shaped skin-flap from the lower border of the orbit into the lower lid (Imre, 1928) while a flap-graft of skin from the upper lid to the lower is a very satisfactory procedure (Wheeler, 1921).

(2) An alternative means of bracing the lower lid is a through-and-through wedge-shaped resection (Adams, 1812), a procedure followed by von Ammon (1831) who excised a triangular piece of the lid near the outer canthus, and used successfully by Molnár (1962); the remainder of the lid was suspended to the lateral rim of the orbit by Bick (1966) and Leone (1970).

(3) In Wheeler's (1920) " halving " technique the lid is split vertically at the grey line and the resections are made in such a way that the superficial flap of skin and orbicularis is on one side and the deep tarso-conjunctival flap is on the other so that on approximation the two are dove-tailed.

(4) A very effective technique is the excision of a triangle of the tarso-conjunctival layer of the lid with its base at the lid-margin. This was proposed by Kuhnt (1883) while the rucks of skin thus caused were eliminated by Müller (1893) by splitting the lid vertically. The best results, however, are obtained by combining such a partial resection of the inner layers of the lid with a similar triangular resection of skin at the outer angle, to replace which the skin of the lower lid is slid outwards so that all the structures of the lid are shortened (Szymanowski, 1870). This technique was modified and improved by Blaskovics (1921) and Fox (1962–72) while Urrets-Zavalia (1959) reinforced the lid by a Z-shaped incision in the skin. In spastic cases Callahan (1954) resected the lower part of the orbicularis muscle.

(5) As an extension of this a technique was suggested by Fox (1960) as being suitable for marginal (tarsal) ectropion whereby the lid is split into two laminæ for about half its medial length extending to the lower edge of the tarsus. The posterior (tarso-conjunctival) lamina is drawn nasally until inversion of the lid is obtained, the excess being resected. The anterior (skin-muscle) flap is drawn temporally and again the excess is resected, thus shortening the lid. An alternative procedure was suggested by White (1971) whereby the lower edge of the tarsus is exposed through a horizontal skin incision and sutured to the skin below the incision so that the tarsus is inverted by the downward pull of the skin.

(6) Attempts have been made to support the lower lid by a sling of fascia lata anchored at either canthus (Wiener, 1928) or to the frontalis and temporalis muscles (Brown *et al.*, 1948).

(7) In cicatricial cases some form of blepharoplasty must be undertaken, if necessary after a tarsorrhaphy in which the adhesion should remain for some time. The milder degrees of deformity may be corrected by local procedures, such as converting a V-shaped incision into a Y; but more extensive contractures necessitate the excision of scar tissue and its replacement by grafting, using either the pedicle or free methods (Hill and Rodrigue, 1971). Wheeler's (1921) technique of using a flap of skin from the upper to the lower lid is very satisfactory if the skin is available; alternatively free grafts from behind the ear (Shindle and Leone, 1973) or the infra-clavicular fossa can be used.

In most of these operations, unless perfect alignment has been attained, it is usually necessary to remove the posterior wall of the canaliculus to cure the epiphora[1]; while in every case wherein closure of the eye is impossible, care should be taken to protect the cornea from exposure until the deformity is corrected.

Adams. *Practical Observations on Ectropion*, London (1812).
von Ammon. *Z. Ophthal.*, **1**, 36 (1831).
Bick. *Arch. Ophthal.*, **75**, 386 (1966).
Blaskovics. *Z. Augenheilk.*, **45**, 1 (1921).
Brown, McDowell and Freyer. *Ann. Surg.*, **127**, 888 (1948).
Callahan. *Amer. J. Ophthal.*, **38**, 787 (1954).
Elschnig. *Klin. Mbl. Augenheilk.*, **50** (1), 599 (1912).
Fox. *Arch. Ophthal.*, **63**, 660 (1960).
Trans. Amer. Acad. Ophthal., **66**, 582 (1962).
Ann. Ophthal. (Chic.), **4**, 217 (1972).
Hill and Rodrigue. *Canad. J. Ophthal.*, **6**, 89 (1971).
Imre. *Lidplastik*, Budapest (1928).
Jones, Wharton. *A Manual of the Principles and Practice of Ophthalmic Medicine and Surgery*, London (1847).
Kuhnt. *Beiträge z. operativen Augenheilkunde*, Jena (1883).
Leone. *Amer. J. Ophthal.*, **70**, 233 (1970).
McLeod and Lux. *Arch. Ophthal.*, **15**, 994 (1936).
Molnár. *Klin. Mbl. Augenheilk.*, **140**, 708 (1962).
Müller. *Klin. Mbl. Augenheilk.*, **31**, 113 (1893).
Shindle and Leone. *Arch. Ophthal.*, **89**, 62 (1973).
Snellen. *II int. Cong. Ophthal.*, Paris, 236 (1862).
Szymanowski. *Hb. d. operativen Chirurgie*, Braunschweig, 243 (1870).
Terson. *Ann. Oculist.* (Paris), **168**, 138 (1931).
Urrets-Zavalia. *Brit. J. Ophthal.*, **43**, 521 (1959).
Wheeler. *Arch. Ophthal.*, **49**, 35 (1920).
Trans. Sect. Ophthal., Amer. med. Ass., 256 (1921).
White. *Amer. J. Ophthal.*, **72**, 615 (1971).
Wiener. *Arch. Ophthal.*, **57**, 597 (1928).
Ziegler. *J. Amer. med. Ass.*, **53**, 183 (1909).

LAGOPHTHALMOS

LAGOPHTHALMOS (λαγώς, a hare, an animal believed to sleep with its eyes open) is a *condition wherein complete closure of the lids over the eyeball is difficult or impossible.* In milder degrees the patient may be able to close the lids by voluntary effort while they may remain open in sleep, but in the more pronounced degrees even forcible squeezing becomes ineffective and the eyes remain permanently unclosed. The seriousness of the condition lies in the evil effects it produces upon the cornea partly from exposure and partly from desiccation, involving either a xerosis or a keratitis e lagophthalmo,[2] both of which leave visually destructive opacities and may terminate in corneal ulceration, perforation and loss of the eye. Fortunately the upturning of the eye in sleep mitigates the effects of exposure, since in mild degrees the conjunctiva below the cornea may alone be exposed; in more pronounced degrees, however, the lower part of the cornea is involved, although rarely the whole of it. A minor complication is epiphora owing to inability of the lower lid to maintain is position in contact with the globe.

The causes of lagophthalmos are varied.

(*a*) A PHYSIOLOGICAL LAGOPHTHALMOS may occur during sleep, a phenomenon met with particularly among Chinese (A. Fuchs and Wu, 1948; A. Fuchs, 1960); a somewhat similar condition has been observed in members of the Amharic race in Ethiopia (Mueller, 1967) and also sometimes in Caucasians (Howitt and Goldstein, 1969).

[1] p. 689.
[2] Vol. VIII, p. 802.

(*b*) MECHANICAL LAGOPHTHALMOS may be due to:

(i) Enlargement or protrusion of the eye so that the lids, although normal, cannot close over it, as in buphthalmos, staphyloma and high axial myopia, or in proptosis and exophthalmos.

(ii) Structural loss or mechanical impairment of the normal movements of the lids, as after trauma, burns, destructive inflammations such as lupus vulgaris, tuberculosis or syphilitic ulcers, yaws or leprosy, atrophic shrinkages such as scleroderma, carcinomata or gangrene (Fig. 642). The same condition may arise in post-inflammatory blepharophimosis.

(iii) Marked ectropion.

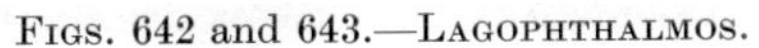

FIGS. 642 and 643.—LAGOPHTHALMOS.

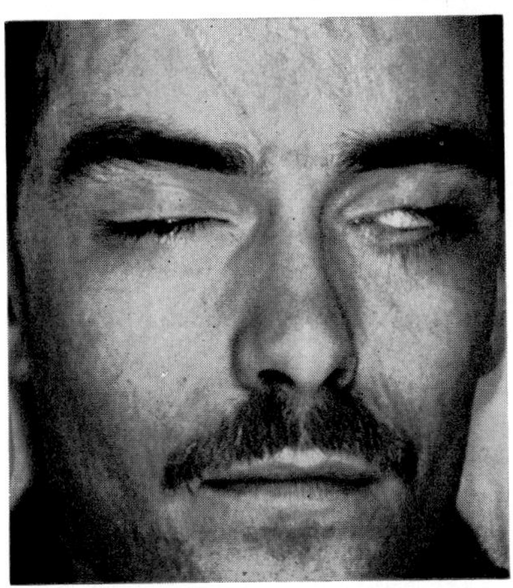

FIG. 642.—Cicatricial lagophthalmos following petrol burns 2 years previously. The patient is trying to close both eyes (A. MacIndoe).

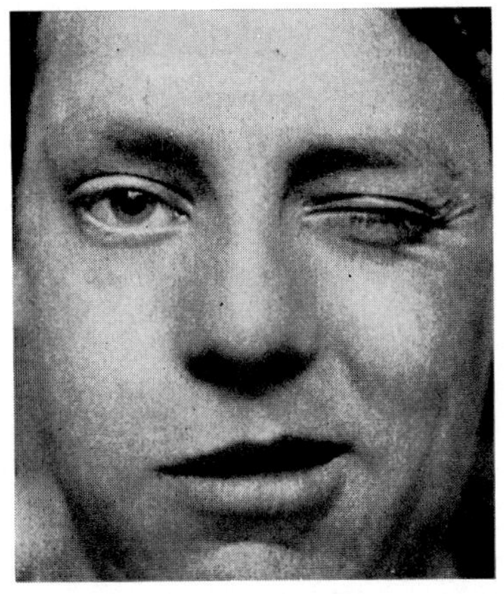

FIG. 643.—Paralytic lagophthalmos following paralysis of the facial nerve. The patient is attempting to close both eyes (Maitland Ramsay).

(*c*) SPASTIC LAGOPHTHALMOS, due to muscular rigidity forbidding descent of the upper lid, as in thyrotoxicosis or endocrine exophthalmos, or spastic paralysis involving lid-retraction.

(*d*) PARALYTIC LAGOPHTHALMOS, due to paralysis of the orbicularis muscle (Figs. 93, 643).

(*e*) FUNCTIONAL LAGOPHTHALMOS. Diminution of corneal sensitivity may involve exposure by the abolition of the corneal reflex, while in extreme debility or unconsciousness the consequences of exposure may result from a failure to initiate the movements for closure of the eyes. Loss of corneal sensation, of course, as in neuropathic keratitis or nerve leprosy, increases the danger to the cornea.

The *treatment* of lagophthalmos should first of all aim at removing the cause, if this is possible. If this is impracticable or until it is accomplished every effort should be made to afford protection to the cornea. As a temporary measure protection may be obtained if the lids can be brought together by keeping them in apposition with narrow strips of surgical plaster after the instillation of yellow paraffin ointment into the conjunctival sac, the whole being covered with a protective pad and bandage. Alternatively, the lashes of both lids may be stuck together with collodion (Gill, 1947), or a skin stitch in the upper lid may be affixed with plaster to the cheek or may be threaded through the skin of the lower lid. When the lids cannot be brought together protection may be obtained by a haptic contact lens. In milder cases it may be sufficient to adopt these measures at night only, and to instruct the patient to lower the upper lid with the finger at frequent intervals throughout the day and to wear protective goggles when exposed to wind and dust.

When it is difficult to close the eye or when the cure of the causal condition takes time, surgical treatment must be considered. A tarsorrhaphy may meet the case, and as a rule a partial junction of the lids suffices, preferably at the outer canthus (external canthorrhaphy). Axenfeld's suture (1922) forms an alternative method: it is a running silk suture threaded round both lid-margins, the ends being tied taut at the outer angle of the orbit after taking a secure bite of the periosteum; Wilczek (1959) tied silk sutures to the periosteum at both canthi. A further suggestion that has given good results has been the threading of a band of silicone rubber subcutaneously encircling both eyelids near the ciliary margin and fixed at each canthus (Arion, 1969; Gotlib and Mawas, 1972; Banuelos, 1973); the elastic element thus introduced gives the eyelids mobility, acting as an artificial orbicularis.

An alternative is to support the lower lid by a sling. This may be done by a strip of fascia lata as has been described in the treatment of paralytic ectropion.[1] The same object has been secured by reflecting slips of the temporal muscle through the upper and lower lid-margins and, after crossing them at the inner canthus, fixing them there, a technique suggested by Lexer (1931) and Rosenthal (1955) and successfully employed by Velhagen (1957) and Huffstadt (1967).

A more ambitious technique has been introduced by Thompson (1971) for cases of facial paralysis, using autogenous free grafts of previously denervated muscle, either the palmaris longus from the forearm or the extensor digitorum brevis from the foot. The muscles transplanted must be complete and are first denervated so that the subsequent change in their metabolism with increased vascularization allows them to survive the initial period of ischæmia after the transplantation. The muscle bellies are sutured to the temporal muscle from which they derive reinnervation, while the muscular tendons in continuity are passed subcutaneously along the margins of the upper and lower lids to be anchored to the medial palpebral ligament. This is followed by reinnervation of the muscular graft (checked histologically and electromyographically) allowing the patient to close the eyes on clenching the teeth.

If there is a mechanical loss of the tissues of the lids, plastic operations may be required to repair the defect, and any ectropion should be corrected.

[1] p. 585.

If the fault lies in retraction of the upper lid (thyrotoxicosis, central lesions causing retraction) relief may be gained by resecting the levator from its insertion and fixing its tendon into the skin under the brow or into the conjunctival fornix, thus causing an artificial ptosis (Blaskovics, 1923–33; Goldstein, 1934); a temporary effect lasting a month or so was obtained by Salvati (1928) by injecting alcohol into the muscle itself. On the other hand, if intra-orbital pressure is causing a proptosis which cannot be controlled and the eye is in danger, in the event of it being impossible to remove the cause, orbital decompression may be necessary.[1]

Finally, mechanical devices have been adopted to correct lagophthalmos. Morel-Fatio and Lalardrie (1962–67) introduced the use of a metal spring shaped like a safety-pin fixed above to the periosteum of the upper orbital rim and below to the perichondrium of the tarsal plate, an idea employed by others[2]; the strength of the spring should be such that the eye is covered by lowering the upper lid but at the same time allows it to be raised by the levator muscle. A weighting of the upper lid by steel (Ambos, 1957) or gold insets (Illig, 1958) is another suggestion. These mechanical devices, however, are not preferable to plastic procedures.

BLEPHAROPHIMOSIS

BLEPHAROPHIMOSIS (βλέφαρον, lid; φιμός, a muzzle; φίμωσις, contraction) may be defined as a condition wherein *the palpebral aperture is decreased without fusion of the lid-margins.* It may occur in several pathological conditions:

(1) SENILE BLEPHAROPHIMOSIS involves mainly a shortening of the palpebral aperture at the outer canthus which may reach or even cover the outer part of the cornea: the deformity can be corrected by pulling the skin of the temple horizontally outwards. It is due to the tone of the orbicularis overcoming the support given to the canthus by the tarso-orbital fascia which has become lax with age (Elschnig, 1912).

(2) SPASTIC BLEPHAROPHIMOSIS has a similar ætiology, the ligamentous supports of the canthi becoming stretched by long-continued and severe blepharospasm; the emphasis, however, lies on the muscular contraction, not on the atonic fascia. It may involve the inner canthus or the outer, and the condition may develop acutely (Dimmer, 1911; von Csapody, 1931).

(3) CICATRICIAL BLEPHAROPHIMOSIS, due to contraction following some destructive lesion, produces the typical appearance of " small eyes " following long-continued disease. It may be a sequel to ulcerative blepharitis or pemphigus, but is seen most typically in the cicatricial stage of trachoma when the tarsus is deformed and shrunken as well as the lid-margins and the

[1] p. 776.
[2] Wexler and Neuman (1967), Guy and Ransohoff (1968), F. and K. English (1972), Levine *et al.* (1972).

conjunctiva. This type of blepharophimosis is frequently accompanied by a lateral epicanthus and entropion.

The treatment of blepharophimosis is surgical; depending on the cause, anchoring the affected canthus to the orbital margin (canthi-fixation) (Elschnig, 1912), or performing a canthoplasty will usually meet the case.

The simple canthoplasty of von Ammon (1839) who slit the lateral canthus and ensured the permanency of the division by suturing the conjunctiva to the skin over it is frequently effective; Johnson (1956) supplemented it by the temporary insertion of thin rubber tissue into the wound. von Ammon's useful method has had many modifications. The most notable of these are the techniques of Blaskovics (1904) who pulled the canthus temporally by excising a triangular fold of skin lateral to it, of Imre (1928) who attained this end by a Z-shaped incision and excising the two triangular flaps, and of Kuhnt (1906) who passed a flap of skin across the cut canthus to lengthen permanently the palpebral fissure. Neméth (1960) excised an oval area of skin and, after fixing the lateral canthal ligament to the outer orbital rim, sutured the skin at the outer canthus to the temporal fascia. An alternative suggested by Wheeler (1938) was to suture strips of the orbicularis muscle to the periosteum of the outer orbital margin. These expedients and many more have been tried; for technical details the original papers or textbooks of plastic surgery should be consulted.

EPICANTHUS LATERALIS

EPICANTHUS LATERALIS (Elschnig, 1912), a condition called *blepharophimosis* by E. Fuchs (1890), is *characterized by the presence of a vertical fold of skin running down at the external angle of the eye and covering it like a sliding screen.* If the skin is drawn backwards over the temple the fold becomes obliterated and the outer canthus appears normal: the blepharophimosis is therefore only apparent. Owing to the fact that the skin over the nose is more securely tied down, an acquired epicanthus very rarely occurs at the inner canthus and that only in old people with a very lax integument. The deformity is due to contraction of the skin near the outer angle after it has been constantly macerated with tears and secretion in chronic conjunctival disease, the pulling of the fold over the eye being encouraged by the associated blepharospasm. In the young in whom the skin is elastic it tends to disappear spontaneously with the conjunctival disease, but in the aged it tends to persist until it is corrected by canthoplasty.

The condition of eczematous dermatitis, as we have seen, produces ectropion in the lower lid. Sometimes the two deformities are found together, but as a rule the pull of the skin-fold in epicanthus lateralis has the effect of producing entropion in the lower lid, which is abolished by eliminating the fold at the outer canthus.

ANKYLOBLEPHARON

ANKYLOBLEPHARON[1] (ἀγκύλη, a thong), indicates *an adhesion between the upper and lower lids along the palpebral margin* so that the aperture is narrowed. It occurs rarely, appearing during the healing process of some destructive lesion involving the ciliary margin, such as trauma, burns and

[1] For congenital ankyloblepharon, see Vol. III, p. 867.

caustic injuries, lupus vulgaris of the skin of the lids, diphtheritic conjunctivitis, trachoma and ulcerative blepharitis: it is artifically induced by the operation of tarsorrhaphy. One or more bands of scar tissue may bridge over the palpebral aperture (*ankyloblepharon filiforme*), but more commonly the margins at one or other angle, particularly the outer, may become united thus shortening the palpebral aperture. In some cases, indeed, this may occur to such an extent that efficient opening of the eye may become difficult. The most marked stage of complete ankyloblepharon is usually accompanied by symblepharon.

Treatment consists of operative separation of the lids, after which a marginal graft may be necessary; if the canthi are involved a canthoplasty, wherein the margins are lined by conjunctiva, is necessary to prevent recurrence,[1] and in cases complicated by symblepharon the latter condition takes operative priority, if, indeed, it is treatable.

SYMBLEPHARON

SYMBLEPHARON, *a cicatricial adhesion between the palpebral conjunctiva and the bulbar conjunctiva or cornea* which is liable to occur when the opposing surfaces have lost their epithelium and when the subepithelial tissues cicatrize has been described in other Volumes of this *System*.[2]

Ambos. *Wien. klin. Wschr.*, **69,** 866 (1957).
von Ammon. *Z. Augenheilk. Chir.*, **2,** 140 (1839).
Arion. *Ann. Chir.*, **23,** 15 (1969).
Axenfeld. *Hb. d. ärztlichen Erfahrungen im Weltkriege 1914/1918*, Leipzig, **5,** 497 (1922).
Banuelos. *Arch. Ophthal.*, **89,** 329 (1973).
Blaskovics. *Z. Augenheilk.*, **12,** 418 (1904); **49,** 30, 94 (1923); **81,** 13 (1933).
Ann. Ottal., **61,** 763 (1933).
von Csapody. *Klin. Mbl. Augenheilk.*, **86,** 789 (1931).
Dimmer. *Klin. Mbl. Augenheilk.*, **49** (2), 337 (1911).
Elschnig. *Klin. Mbl. Augenheilk.*, **50** (1), 17, 335 (1912).
English, F. and K. *Med. J. Aust.*, **1,** 223 (1972).
Fuchs, A. *Klin. Mbl. Augenheilk.*, **137,** 354 (1960).
Fuchs, A., and Wu. *Amer. J. Ophthal.*, **31,** 717 (1948).
Fuchs, E. *Lhb. d. Augenheilkunde*, Wien (1890).
Wien. klin. Wschr., **3,** 2 (1890).
Gill. *Arch. Ophthal.*, **37,** 82 (1947).
Goldstein. *Arch. Ophthal.*, **11,** 389 (1934).
Gotlib and Mawas. *Bull. Soc. franç. Ophtal.*, **85,** 407 (1972).
Guy and Ransohoff. *J. Neurosurg.*, **29,** 431 (1968).
Howitt and Goldstein. *Amer. J. Ophthal.*, **68,** 355 (1969).
Huffstadt. *Arch. chir. neerl.*, **19,** 63 (1967).
Illig. *Klin. Mbl. Augenheilk.*, **132,** 410 (1958).
Imre. *Lidplastik*, Budapest (1928).
Johnson. *Amer. J. Ophthal.*, **41,** 71 (1956).
Kuhnt. *Z. Augenheilk.*, **15,** 238 (1906).
Levine, House and Hitselberger. *Amer. J. Ophthal.*, **73,** 219 (1972).
Lexer. *Die gesamte Wiederherstellungschirurgie*, Leipzig, **1,** 759 (1931).
Morel-Fatio and Lalardrie. *Ann. Chir. plast.*, **7,** 275 (1962).
Plastic and Reconstructive Surgery of the Eye (ed. B. Smith *et al.*), St. Louis, 380 (1967).
Mueller. *Brit. J. Ophthal.*, **51,** 246 (1967).
Neméth. *Amer. J. Ophthal.*, **49,** 1357 (1960).
Rosenthal. *Dtsch. Stomatologie*, **5,** 611 (1955).
Salvati. *Ann. Oculist.* (Paris), **165,** 203 (1928).
Thompson. *Plast. reconstr. Surg.*, **48,** 11 (1971).
Transplantation, **12,** 353 (1971).
Vth Wld. Cong. plast. reconstr. Surg., London, 66 (1971).
Velhagen. *Klin. Mbl. Augenheilk.*, **130,** 1 (1957).
Wexler and Neuman. *Harefuah*, **73,** 166 (1967).
Wheeler. *Amer. J. Surg.*, **42,** 7 (1938).
Wilczek. *Klin. Mbl. Augenheilk.*, **135,** 545 (1959).

[1] p. 590.
[2] Vol. VIII, p. 6; Vol. XIV, pp. 430, 763, 773, 1032, 1048.

Diseases of the Caruncle

In so far as the caruncle shares generally in diseases of the conjunctiva, its involvement in inflammatory and other disease has been discussed in the Volume dealing with this tissue. The developmental anomalies have also been considered,[1] as have dermoid cysts.[2] Since, however, the caruncle is developmentally a part of the lower lid cut off by the growth of the lower canaliculus, the tumours affecting it are more akin to those of the lid than of the conjunctiva, and it may be of value to summarize them here.

TUMOURS OF THE CARUNCLE are relatively rare but have excited considerable attention: they were noted by Galen under the name *encanthis*, and the first modern critical paper came from von Graefe (1854). Subsequent important studies and bibliographies are those of Parsons (1904), Lagrange (1906), Beauvieux (1913), and Morax (1926). Serra (1928) collected 136 cases from the literature, while Evans (1940) gave an analysis of 200 cases; Vail (1933) annotated the mixed (teratoid) tumours and Wetzel (1937) the malignant melanomata, but Wilson (1958) found in Egypt that among 104 biopsies most pigmented tumours were benign. Nævi of the caruncle are relatively common. Daicker (1966), who referred to 22 cases of caruncular nævi, found that they were invariably accompanied by epithelial proliferation and metaplasia of the hair follicles resulting in the formation of strands, cysts and glandular formations.

Cysts are usually sebaceous in nature.[3]

Tumours include papillomata,[4] adenomata,[5] epitheliomata, usually sebaceous carcinomata,[6] fibromata,[7] reticulosarcomata,[8] mycosis fungoides,[9] nævi[10] and malignant melanomata[11]. An ophthalmic curiosity is a psammoma reported by Parzani (1933).

Rare tumours of interest derived from accessory lacrimal or secretory glands are papillary cystadenomata[12] and oncocytomata.[13]

The pathology and literature of all these neoplasms have been cited in previous sections. It is interesting, however, that partly because of their relative isolation and the consequent ease of removal, the prognosis of malignant tumours with reasonably early excision is generally more favourable than when they occur on the conjunctiva or the lid.

Beauvieux. *Arch. Ophtal.*, **33,** 216 (1913).
Daicker. *v. Graefes Arch. Ophthal.*, **170,** 156 (1966).
Evans. *Arch. Ophthal.*, **24,** 83 (1940).
von Graefe. *v. Graefes Arch. Ophthal.*, **1,** 289 (1854).
Lagrange. *Encycl. franç. Ophtal.*, Paris, **5,** 601 (1906).
Morax. *Cancer de l'appareil visuel*, Paris, 192 (1926).
Parsons. *Pathology of the Eye*, London, **1,** 146 (1904).
Parzani. *Policlinico*, **40,** 1492 (1933).
Serra. *Boll. Oculist.*, **7,** 783, 805 (1928).
Vail. *Arch. Ophthal.*, **10,** 593 (1933).
Wetzel. *Amer. J. Ophthal.*, **20,** 675 (1937).
Wilson. *Trans. ophthal. Soc. N.Z.*, **11,** 23 (1958).

[1] Vol. III, p. 860.
[2] Vol. III, p. 823.
[3] Vol. VIII, p. 1140.
[4] p. 404; Vol. VIII, p. 1159.
[5] p. 450; Vol. VIII, p. 1175.
[6] p. 453; Vol. VIII, p. 765.
[7] Vol. VIII, p. 1185.
[8] p. 480.
[9] p. 484.
[10] Vol. VIII, p. 1213.
[11] p. 524.
[12] p. 465.
[13] p. 466.

INDEX

C

E

F

G

H

K

L

M

N

S

U

V

W

X

Y

Z

PRINTED IN GREAT BRITAIN BY THE WHITEFRIARS PRESS LTD.
LONDON AND TONBRIDGE